Millimeter Wave and Optical Dielectric Integrated Guides and Circuits

WILEY SERIES IN MICROWAVE AND OPTICAL ENGINEERING

KAI CHANG, Editor
Texas A & M University

FIBER-OPTIC COMMUNICATION SYSTEMS • *Govind P. Agrawal*

COHERENT OPTICAL COMMUNICATIONS SYSTEMS • *Silvello Betti, Giancarlo De Marchis, and Eugenio Iannone*

HIGH-FREQUENCY ELECTROMAGNETIC TECHNIQUES: RECENT ADVANCES AND APPLICATIONS • *Asoke K. Bhattacharyya*

COMPUTATIONAL METHODS FOR ELECTROMAGNETICS AND MICROWAVES • *Richard C. Booton, Jr.*

MICROWAVE RING CIRCUITS AND ANTENNAS • *Kai Chang*

MICROWAVE SOLID-STATE CIRCUITS AND APPLICATIONS • *Kai Chang*

DIODE LASERS AND PHOTONIC INTEGRATED CIRCUITS • *Larry A. Coldren and Scott W. Corzine*

MULTICONDUCTOR TRANSMISSION-LINE STRUCTURES: MODAL ANALYSIS TECHNIQUES • *J. A. Brandão Faria*

FUNDAMENTALS OF MICROWAVE TRANSMISSION LINES • *Jon C. Freeman*

MICROSTRIP CIRCUITS • *Fred Gardiol*

HIGH-SPEED VLSI INTERCONNECTIONS: MODELING, ANALYSIS AND SIMULATION • *A. K. Goel*

HIGH-FREQUENCY ANALOG INTEGRATED-CIRCUIT DESIGN • *Ravender Goyal, Editor*

OPTICAL COMPUTING: AN INTRODUCTION • *M. A. Karim and A. A. S. Awwal*

MICROWAVE DEVICES, CIRCUITS, AND THEIR INTERACTION • *Charles A. Lee and G. Conrad Dalman*

ANTENNAS FOR RADAR AND COMMUNICATIONS: A POLARIMETRIC APPROACH • *Harold Mott*

FREQUENCY CONTROL OF SEMICONDUCTOR LASERS • *M. Ohtsu, Editor*

SOLAR CELLS AND THEIR APPLICATIONS • *Larry D. Partain, Editor*

ANALYSIS OF MULTICONDUCTOR TRANSMISSION LINES • *Clayton R. Paul*

INTRODUCTION TO ELECTROMAGNETIC COMPATIBILITY • *Clayton R. Paul*

INTRODUCTION TO HIGH SPEED ELECTRONICS AND OPTOELECTRONICS • *Leonard M. Riaziat*

NEW FRONTIERS IN MEDICAL DEVICE TECHNOLOGY • *Arye Rosen and Harel Rosen, Editors*

FREQUENCY SELECTIVE SURFACE AND GRID ARRAY • *T. K. Wu, Editor*

OPTICAL SIGNAL PROCESSING, COMPUTING, AND NEURAL NETWORKS • *Francis T. S. Yu and Suganda Jutamulia*

PHASED ARRAY-BASED SYSTEMS AND APPLICATIONS • *Nick Fourikis*

FINITE ELEMENT SOFTWARE FOR MICROWAVE ENGINEERING • *Tatsuo Itoh*

INTEGRATED ACTIVE ANTENNAS AND SPATIAL POWER COMBINING • *Julio J. Navarro*

NONLINEAR OPTICS • *E. G. Sauter*

MILLIMETER WAVE AND OPTICAL DIELECTRIC INTEGRATED GUIDES AND CIRCUITS • *Shiban K. Koul*

ACTIVE AND QUASI-OPTICAL ARRAYS FOR SOLID-STATE POWER COMBINING • *Robert A. York and Zoya B. Popovic, Editors*

Millimeter Wave and Optical Dielectric Integrated Guides and Circuits

SHIBAN K. KOUL
Indian Institute of Technology, Delhi

A WILEY-INTERSCIENCE PUBLICATION
JOHN WILEY & SONS, INC.
NEW YORK / CHICHESTER / WEINHEIM / BRISBANE / SINGAPORE / TORONTO

This text is printed on acid-free paper.

Lobrary of Congress Cataloging in Publication Data:

Koul, Shiban K.
Millimeter wave and optical dielectric integrated guides and circuits / Shiban K. Koul.
p. cm.
"A Wiley-Interscience publication."
Includes bibliographical references and index.
ISBN 0-471-16841-6 (cloth : alk. paper)
1. Dielectric wave guides. 2. Millimeter wave devices.
3. Integrated optics. I. Title.
TK7871.65.K57 1997
621.381′331—dc20 96-34530

Printed in the United States of America

10 9 8 7 6 5 4 3 2 1

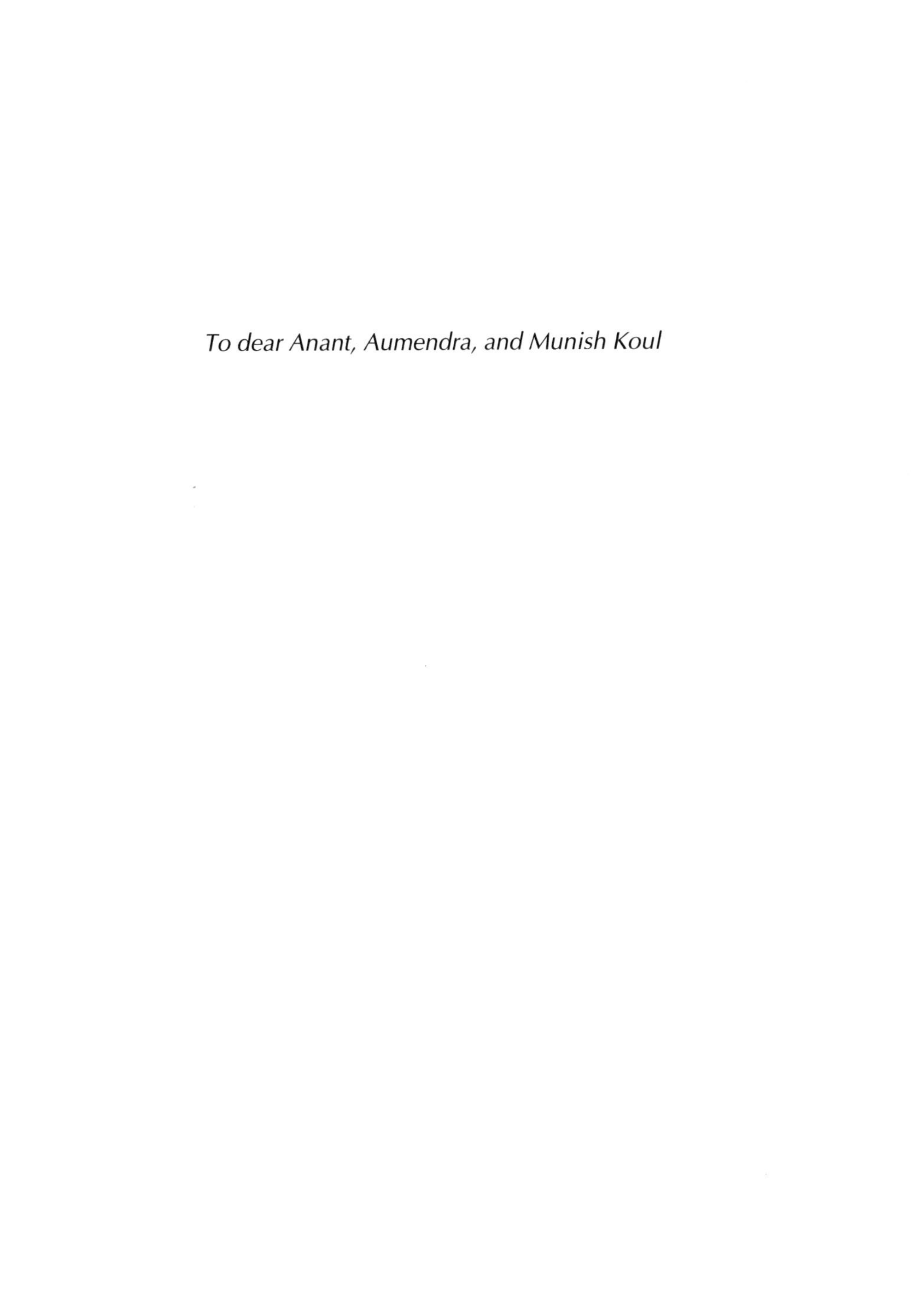

To dear Anant, Aumendra, and Munish Koul

Contents

Foreword xv

Preface xvii

1 Introduction to Dielectric Guides 1

1.1 Historical Perspective 1
1.2 Waveguiding Media for Millimeter Waves 3
1.3 Wave Guidance in Open Homogeneous Dielectric Guides 6
1.3.1 Slab Dielectric Guide 6
1.3.2 Rectangular Dielectric Guide 11
1.3.3 Circular Dielectric Guide 14
1.4 Wave Guidance in Open Composite Planar Dielectric Guides 16
1.4.1 Millimeter Wave Planar Dielectric Guides 16
1.4.2 Optical Planar Dielectric Guides 19
1.4.3 Comparison 22
1.5 Materials and Fabrication Techniques 23
1.6 Integration of Active Devices 29
1.7 Basics of Dielectric Resonators 32
1.8 Application Potential 38
Problems 39
References 41

2 Analysis of Dielectric Integrated Guides: Approximate Methods 46

2.1 Introduction 46
2.2 Electromagnetic Fields in Terms of Potential Functions 47
2.2.1 Vector Potential Functions 47
2.2.2 Field Solutions in Terms of Scalar Potentials 48
2.3 Principle of Effective Dielectric Constant Method 51

2.4 Analysis of Dielectric Integrated Guides: Effective Dielectric Constant Method 54
2.4.1 Illustration of EDC Method Using Field Matching 56
2.4.2 Illustration of EDC Method Using Transverse Transmission Line Technique 66
2.4.3 Analysis of Horizontal Slab Guide Models 70
2.4.4 Analysis of Vertical Layered Guide Models 79
2.4.5 Application to Dielectric Integrated Guides 83
2.5 Network Analysis Method 85
2.5.1 Basic Approach 85
2.5.2 Analysis of a Dielectric Step Junction 86
2.5.3 Application to Image Guide 90
2.6 Conductor and Dielectric Losses 93
2.6.1 Conductor Loss 93
2.6.2 Dielectric Loss 94
Problems 98
References 102
Appendix 2A Characteristic Equations for Horizontal Layered Guide Models Corresponding to E^y_{mn} Modes of Structures in Figures 2.2 to 2.8 104
Appendix 2B Characteristic Equations for Vertical Layered Guide Models Corresponding to E^y_{mn} Modes of Structures in Figures 2.2 to 2.8 107

3 Analysis of Dielectric Integrated Guides: Rigorous Methods 111

3.1 Introduction 111
3.2 Analysis of General Dielectric Guide: Mode-Matching Method 111
3.2.1 Basic Approach 111
3.2.2 Guide Structures and Potential Functions 113
3.2.3 Analysis for TM-to-y Modes 115
3.2.4 Analysis for TE-to-y Modes 122
3.2.5 Practical Guides as Special Cases 127
3.3 Wave Impedance and Attenuation 131
3.3.1 Wave Impedances 131
3.3.2 Attenuation Due to Conductor Loss 132
3.4 Numerical Methods 136
3.4.1 Telegraphist's Equations Method 137
3.4.2 Finite-Element Method 140
Problems 143
References 146

4 Image, Insular Image, Trapped Image, and Other Variant Guides 148

4.1 Introduction 148
4.2 Image Guide 149
4.2.1 Field and Power Distribution 149
4.2.2 Dispersion and Bandwidth 153
4.2.3 Attenuation Characteristics 157
4.3 Insular Image Guide 161
4.3.1 Fields and Dispersion 161
4.3.2 Attenuation Characteristics 165
4.4 Trapped Image Guide 166
4.4.1 Dispersion and Wave Impedance 166
4.4.2 Effect of Insular Layer 169
4.5 Coupled Guide Characteristics 170
4.5.1 Field Distribution 170
4.5.2 Dispersion and Wave Impedance 173
4.6 Strip and Inverted Strip Dielectric Guides 176
4.6.1 Strip Dielectric Guide 178
4.6.2 Inverted Strip Guide 181
4.7 General Cladded Image Guide and Special Structures 182
4.7.1 Cladded Image Guide 183
4.7.2 Hollow Image Guide 185
4.7.3 T-Guide 186
4.8 U-Guide 186
Problems 187
References 191

5 Semiconductor and Planar Optical Guides 193

5.1 Introduction 193
5.2 Principle of Optical Control in Semiconductor Guides 194
5.3 Optically Controlled Rectangular Semiconductor Guide 195
5.4 Optically Controlled Silicon Image, Ridge, and H-Guides 198
5.4.1 Silicon Image Guide 198
5.4.2 Silicon Ridge Guide 201
5.4.3 Silicon H-Guide 201
5.5 Planar Optical Guides 206
5.5.1 Optical Strip Dielectric Guide 207
5.5.2 Optical Rib Guide 209
5.5.3 Embedded Inverted Strip Guide 211
5.5.4 Multilayer Thin-Film Optical Rib Guide 212
Problems 216
References 219

6 Nonradiative Dielectric Guide **222**

6.1 Introduction 222
6.2 Analysis of Dielectric Strip Loaded Parallel-Plate Guide 222
6.2.1 Field Expressions for TM_{mn}^{y} Modes ($H_y = 0$) 223
6.2.2 Field Expressions for TE_{mn}^{y} Modes ($E_y = 0$) 226
6.3 Characteristics of NRD Guide 229
6.3.1 Nonradiative Modes and Fields 229
6.3.2 Dispersion and Bandwidth Characteristics 232
6.3.3 Loss Characteristics 234
6.4 Coupled NRD Guide 236
6.5 Groove NRD Guide 241
Problems 243
References 245

7 Nonplanar Dielectric Guides **247**

7.1 Introduction 247
7.2 Semicircular Dielectric Image Guide 247
7.2.1 Dielectric Attenuation Constant α_d 248
7.2.2 Conductor Attenuation Constant α_c 250
7.3 Semielliptical Dielectric Image Guide 253
7.3.1 Fields and Dispersion 255
7.3.2 Attenuation Characteristics 258
7.4 Y-Type and Triangular Dielectric Guides 258
7.4.1 Y-Dielectric Guide 258
7.4.2 Triangular Dielectric Guide 262
7.5 Tube Contacted Slab Guide 263
Problems 265
References 267

8 H-Guides and Groove Guides **268**

8.1 Introduction 268
8.2 H-Guide 270
8.2.1 Hybrid Modes and Their Characteristics 270
8.2.2 Nonhybrid TE_{0n} Modes 275
8.3 Double-Strip H-Guide 283
8.4 Groove Guide 287
8.4.1 Field Analysis 287
8.4.2 Propagation Characteristics 294
8.5 Double-Groove Guide 298
Problems 303
References 306

9 Dielectric Resonators **308**

9.1 Introduction 308
9.2 Electromagnetic Fields in Cylindrical Coordinates 309
9.3 Isolated Cylindrical Resonator 310
9.3.1 Analysis for $TE_{01\delta}$ Mode 312
9.3.2 Derivation for Radiation Q-Factor 315
9.3.3 Closed Form Expressions 319
9.3.4 $TE_{01\delta}$ Mode Characteristics 321
9.3.5 Higher Order Modes 323
9.4 Cylindrical Resonator in Planar Dielectric Slab Guide Environment 329
9.4.1 EDC Method Combined with Transverse Transmission Line Technique 329
9.4.2 Insular Image Guide Cylindrical Resonator 331
9.4.3 Cylindrical Resonator on a Suspended Substrate 335
9.4.4 Cylindrical Image Resonator with a Top Dielectric Layer 337
9.5 Ring Resonator Structures 339
9.5.1 Isolated Ring Resonator 339
9.5.2 Ring Resonator in Planar Dielectric Slab Guide Environment 345
9.6 Rectangular Resonators 346
9.6.1 EDC Technique of Analysis 346
9.6.2 Isolated Rectangular Resonator 346
9.6.3 Insular Image Rectangular Resonator 350
9.7 Other Resonators 354
9.7.1 Dielectric Ring-Gap Resonator 354
9.7.2 Optically Controlled Dielectric Resonator 355
9.7.3 Hybrid Dielectric/HTS Resonator 357
Problems 359
References 364

Appendix 9A Bessel Function Formulas Used in the Analysis of Cylindrical Resonators 368

10 Discontinuities, Transitions, and Measurement Techniques **369**

10.1 Introduction 369
10.2 Step Discontinuity and Bends 371
10.2.1 Mode Matching and Generalized Scattering Matrix 371
10.2.2 General Analysis of Discontinuities in Shielded Dielectric Guides 374
10.2.3 Example of an E-Plane Dielectric Slab Discontinuity in Rectangular Waveguide 376

10.3 Bends and Junctions 384
10.3.1 Bends 384
10.3.2 Y- and T-Junctions 387
10.4 Transitions 392
10.4.1 Transitions to Image Guide 392
10.4.2 Transitions to NRD Guide 394
10.5 Measurement Techniques 397
10.5.1 Electric Field Probe 397
10.5.2 Measurement of Attenuation Constant and Guide Wavelength 398
10.5.3 Measurement of Radiation Loss at Bends 339
Problems 399
References 402

11 Passive Components 405

11.1 Introduction 405
11.2 Power Transfer in Parallel-Coupled Guides 405
11.2.1 Symmetric Parallel-Coupled Guide 405
11.2.2 Nonsymmetric Parallel-Coupled Guide 408
11.3 Parallel Guide Directional Couplers 412
11.3.1 Symmetric Couplers 413
11.3.2 Nonsymmetric Couplers 417
11.4 Other Directional Couplers 421
11.4.1 Beam-Splitter Type Coupler 421
11.4.2 Image Guide–Microstrip Coupler 428
11.4.3 Branch Guide Coupler 433
11.4.4 Leaky-Wave NRD Guide Coupler 435
11.5 Ring Resonator Filters 435
11.5.1 Basic Ring Circuit 436
11.5.2 Single-Ring Filter 438
11.5.3 Two-Ring Filter 442
11.6 Grating and Other Types of Filters 443
11.6.1 Single Guide Grating Filter 443
11.6.2 Coupled Grating Line Filter 449
11.6.3 Gap-Coupled Dielectric Guide Filter 454
11.6.4 Rectangular Resonator Coupled Filter 456
11.6.5 Periodic Branching Filter 460
11.7 Other Components 464
11.7.1 Power Divider 464
11.7.2 Variable Attenuator and Phase Shifter 466
11.7.3 Reflectometers 468

11.7.4 Isolator 474
11.7.5 Circulator 474
Problems 477
References 480

12 Components Using Semiconductor Devices and Optical Control 484

12.1 Introduction 484
12.2 Image Guide Detector Circuits 484
12.3 Oscillators 488
12.3.1 Image Guide Oscillators 488
12.3.2 NRD Guide Oscillator 495
12.4 Electronic Phase Shifters 495
12.4.1 Phase Shifter Using Contact Injection Technique 496
12.4.2 Phase Shifter Using Optical Injection Technique 497
12.5 Balanced Mixer 497
12.6 Amplifier 500
12.7 Circuit Modules 501
12.7.1 Oscillator-cum-Power Combiner 501
12.7.2 Transceiver Circuit 503
Problems 505
References 507

13 Antennas 509

13.1 Introduction 509
13.2 Dielectric Rod Antennas 510
13.3 Image Guide Antennas 513
13.3.1 Theory of Leaky-Wave Grating Antenna 513
13.3.2 Antennas with Grooves in Dielectric 518
13.3.3 Antennas with Periodic Conducting Patches 519
13.3.4 Antennas with Grooves in Ground Plane 521
13.4 Image and Insular Image Guide-Fed Antennas 527
13.5 Antennas Based on NRD and Groove Guides 532
13.5.1 NRD Guide-Based Antennas 532
13.5.2 Groove Guide Antennas 538
13.5.3 NRD Guide-Fed Planar Antennas 540
13.6 Semiconductor Guide Antennas 544
13.6.1 Silicon Guide Antennas 547
13.6.2 Optically Controlled Silicon Guide Antenna 549
Problems 552
References 553

Index 557

Foreword

After sailing "up and down" through a period of nearly two decades, millimeter wave technology has finally caught on, rather very well in recent years. Not only have millimeter wave systems become a reality, there is now a growing interest in developing cost effective systems for a plethora of radar and wireless communication applications. Presently, most of the practical millimeter wave components and systems are based on rectangular waveguides and planar and quasiplanar transmission lines. The analysis and design of these transmission lines and also millimeter wave circuit techniques and technology employing them are well documented in several books. While these transmission media are known to offer best performance at lower millimeter wave frequencies, there is also available a whole array of alternative guide structures that have the capability to extend the range of operation to the higher end of the millimeter wave band. This includes a class of dielectric integrated guides, which have the potential for developing into a low-cost technology with the attractive possibility of integrating high-performance dielectric antennas to RF front ends, and a class of low-loss H-guides and groove guides, which are particularly suitable for use at higher millimeter wave frequencies. Literature abounds with research articles on these alternative guide structures for millimeter wave applications; however, much work needs to be done to develop circuit techniques and technology, particularly for active device components.

This book arrives at a time when the potentials of a large variety of dielectric integrated guides including the H- and groove guides are well understood, analysis formulas are known, but challenges are open for innovation in circuit techniques and realization into a viable technology. Optically controlled semiconductor guides have opened up new vistas for realizing active millimeter wave components such as phase shifters and switches. Research scientists and engineers interested in entering this challenging area and those who are in pursuit of exploring new techniques and developing components at higher millimeter wave frequencies should find this book a valuable reference.

This book, organized into 13 chapters, covers in a systematic manner virtually all aspects of dielectric integrated guides and their applications thus far reported in the open literature. The chapters on analysis of dielectric guides cover both approximate and rigorous methods and are written in a lucid manner so as to be of direct utility for classroom teaching. A unified analytical treatment based on the well-known effective dielectric constant technique is presented for solving a class of dielectric integrated guides, planar optical guides, and also dielectric resonators placed in a planar slab guide environment. Characteristic equations derived using this technique are listed for a number of practical guides, which could readily be programmed for data generation. The propagation characteristics of image, insular image, and trapped image guides and a number of their variants are covered. Separate chapters are devoted to nonradiative guides, optically controlled semiconductor and planar optical guides, and H- and groove guides. Dielectric resonators, which play a key role in the design of stabilized oscillators and filters, are considered in detail. This book also includes a state-of-the-art review on discontinuities and transitions in dielectric guides; passive components; components using semiconductor devices and optical control; and dielectric and semiconductor guide integrated antennas and arrays. With such a broad and up-to-date coverage of topics in the area of dielectric integrated guides, this book can serve as a text as well as a ready reference for the millimeter wave community.

BHARATHI BHAT

Professor
Centre for Applied Research in Electronics
Indian Institute of Technology
Delhi, India

Preface

The properties of electromagnetic wave guidance along dielectric guides were established as early as the mid-1930s. Since then, surface wave guidance and radiation leakage phenomena have been the subject of extensive research. While these studies have led to the development of a variety of radiative structures, their utility as transmission lines was sidelined by the far more promising metal waveguide. In fact, the metal waveguide that emerged in the 1940s and the class of planar transmission lines that were evolved during the 1960s formed the foundation on which the entire microwave technology fast flourished and matured. The interest in dielectric guides was revived in the 1970s, when there came a strong-felt need to utilize the millimeter wave region, owing to the congestion of the microwave spectrum. Over the last decade, in particular, there has been an upsurge of interest in the millimeter wave band for applications in radar, communications, instrumentation, and so on. Among the various transmission media investigated, the metal waveguide, the planar transmission lines, and the quasiplanar fin-lines are known to offer the best potential at lower millimeter wave frequencies from 30 to about 100 GHz. The techniques and technologies of these guides have matured and are already described in several books. The dielectric integrated guides that form the subject of this book belong to a different class. Over the past two decades, a wide variety of dielectric guides in the form of planar integrated guides, nonradiative guides, dielectric based H-guides, and groove guides have evolved, which together offer the potential for meeting the requirements in the entire millimeter wave band. Dielectric guide-based antennas have also received considerable attention, since they are amenable for direct integration with the rest of the millimeter wave circuitry. Of more recent interest has been the area of optically controlled semiconductor guides for realizing active circuits and scannable antennas.

Dielectric guides have also been of interest to optical scientists. With the emergence of the technology of "optical fibers" and "integrated optics" in the 1970s, dielectric guides having circular cross-section and planar geometry suitable for operation at optical frequencies have been investigated extensively.

A class of planar dielectric guides, in which the guiding mechanism is that of multiple total internal reflections from interfaces parallel to the optical axis with properties similar to the planar guides at millimeter wave frequencies, have evolved thereby bridging the gap between microwaves and optics.

While there exists a vast body of literature on the analysis and design of various millimeter wave dielectric integrated guides and circuit applications, their technology has not matured to the level equivalent to that of the rectangular waveguide or MIC. There appears to be considerable scope for further work, particularly in terms of practical implementation with active circuits and integration of components into subsystems and systems. The objective of this book is to consolidate the information available in the form of a text, which can serve as a single reference for further research and development in this area. The emphasis is on dielectric guide techniques primarily applicable to the millimeter wave band, and planar optical dielectric guides having characteristics similar to those of millimeter wave guides are given peripheral treatment.

The book is divided into 13 chapters. Since the study of dielectric guides dates back to 1910, the introductory chapter (Chapter 1) begins with a historical background and then provides an overview of dielectric guide techniques so as to give the reader a quick guided tour of what is to follow in detail in subsequent chapters. Chapter 2 presents the analysis of a class of planar dielectric integrated guides and NRD guides (excluding H- and groove guides) using approximate methods. Among the various methods, the effective dielectric constant (EDC) technique, which is known for its simplicity and generality, is described in detail. Using this method, a general set of characteristic equations are derived, which cover almost all the dielectric integrated guides hitherto reported. For a more accurate description of the guides, which includes field, wave impedances, and losses, rigorous methods of analysis are presented in Chapter 3. Among these, the mode matching method is systematically applied to a generalized coupled dielectric guide, and formulas are derived that can be applied to a variety of single and coupled guides as special cases. The formulas derived in Chapters 2 and 3 can readily be programmed to generate design data for a wide variety of dielectric guide structures.

Chapter 4 describes the fields and propagation characteristics of a class of single and coupled planar dielectric integrated guides for millimeter wave applications. The structures covered are the image, insular image, trapped image, and a variety of variant guides. The attenuation, impedance, and bandwidth characteristics of some of the important structures are highlighted. Planar semiconductor guides have the advantage that the propagation characteristics can be controlled electronically through optical illumination. Chapter 5 describes the principles of optical control and the characteristics of optically controlled semiconductor guides. The characteristics of some typical planar optical guides, which can be considered as counterpart of some of the millimeter waveguides, are also included in this chapter.

Among the various dielectric guides, the NRD guide has an important status because of its nonradiative property. All aspects of this guide, both in single and coupled versions, are consolidated and presented in Chapter 6. Besides the commonly employed planar and NRD guides, there are certain nonplanar structures—such as the semicircular and semielliptical image guides, Y-guides, and triangular guides—which offer special features. The characteristics of such nonplanar guides are covered in Chapter 7. The H- and groove guide are suited for operation in the upper frequency range of the millimeter wave band. The hybrid mode analysis of these guides, fields, and propagation characteristics are presented in Chapter 8.

With the availability of very-low-loss dielectric materials, high-Q dielectric resonators have emerged as important elements of many millimeter wave circuits and antennas. Chapter 9 presents in detail the analysis of dielectric resonators in isolated form and also in various dielectric guide environment. The EDC method of analysis, modified to suit resonator structures, is systematically developed and closed form expressions are derived for evaluating the resonator characteristics. Discontinuities and transitions, which form integral parts of millimeter wave components, are described in Chapter 10. Techniques for measuring fields, attenuation constant, and guide wavelength in dielectric guides are also included in this chapter.

Chapter 11 deals exclusively with dielectric guide passive components. The analysis and design aspects of directional couplers and filters are covered in detail. The literature on other components—namely, power divider, attenuator, phase shifter, reflectometer, isolator, and circulator—which is rather limited, is also reviewed. Incorporation of active devices in dielectric guides in much more difficult than in other transmission media. However, the use of high-resistivity semiconductor devices lends itself to monolithic integration of active devices and also permits control of characteristics through optical illumination. Chapter 12 is devoted to the description of components using semiconductor devices and optical control. The final chapter reviews the various types of dielectric integrated guide antennas and also the optically controlled semiconductor guide antennas thus far reported for use in the millimeter wave band.

In the 13 chapters, I have tried to include comprehensively almost all aspects of dielectric guides and their applications reported thus far for use in the millimeter wave band. Starting from the fundamental field relations, the book builds up the analytical solutions of various dielectric guides so as to be of direct utility to classroom teaching at the graduate level. The design formulas developed for various guides and resonators and the characteristics provided should be of primary use as design aids to scientists and engineers involved in design and development. The later chapters on passive and active components and antennas, which provide an up-to-date account of the latest developments in the area, should serve as ready reference for microwave engineers in R&D establishments and industry attempting to carry forward the dielectric integrated guide technology toward the much awaited millimeter wave systems in this media.

This book is an outcome of my involvement in teaching a specialized graduate level course in the area of "millimeter wave integrated circuits" and also in research and development in this area over the past several years. I wish to thank the Indian Institute of Technology, New Delhi, and the Centre for Applied Research in Electronics, in particular, for providing the opportunity and academic freedom to write this book.

I am grateful to Professor Bharathi Bhat, at the Centre for Applied Research in Electronics, who spent a considerable amount of her valuable time in assisting me during the course of writing this book.

I am grateful to Professor Tsukasa Yoneyama for providing me the opportunity to work in his laboratory on NRD guide technology during my short visits to Tohoku University, Japan, in 1994 and 1995. I also thank his laboratory staff, in particular, Dr. Futoshi Kuroki, for their cooperation and help.

I thank Mr. George Telecki, Ms. Rose Leo Kish, and Ms. Rosalyn Farakas, of John Wiley & Sons, for their active support, which led to the successful completion of this project. I am also grateful to the anonymous reviewer whose suggestions improved the content of this book.

Finally, I thank my wife, Veena, my parents, Jaya and Bansi Lal, and my sons, Aumendra and Anant, for their moral support.

SHIBAN K. KOUL

Professor & Head
Centre for Applied Research in Electronics
Indian Institute of Technology
Delhi, India

Millimeter Wave and Optical Dielectric Integrated Guides and Circuits

CHAPTER ONE

Introduction to Dielectric Guides

1.1 HISTORICAL PERSPECTIVE

Historically, the study of dielectric guides had its beginning as early as 1910 when Hondros and Debye [1] analyzed the propagation of electromagnetic waves along cylindrical dielectric guides. This was followed by further investigations, both theoretical and experimental, by other investigators, and by about mid-1930, the waveguiding and attenuation characteristics of dielectric guides were quite well established [2, 3]. With the understanding of waveguiding properties of uniform dielectric rods emerged the idea of radiation from rods of finite length. Early investigations on the dielectric rod antennas, carried out mostly in the 1940s, are comprehensively treated in a monograph published by Kiely [4] in 1953. Significant developments have taken place since then, leading to a wide variety of dielectric and dielectric-loaded antennas having desirable properties at microwave and millimeter wave frequencies. A comprehensive account of the developments on this class of antennas as of 1985 is available in a research monograph by Chatterjee [5].

With the rapid developments on metal waveguides during the 1940s, the importance of dielectric guides for microwave applications receded into the background mainly because the metal waveguide can contain the fields within it whereas in a dielectric guide the fields exist outside the structure as well. In the millimeter wave through optical frequency range, however, dielectric guides can be designed such that almost all the electromagnetic energy is concentrated within the dielectric guide. The advent of lasers in the early 1960s and the possibility of using coherent light for long-distance communications created the need for suitable optical guides. Studies on glass fiber guides were reported as early as 1961 by Snitzer and Osterberg [6, 7] and later by Kao and Hockham in 1966 [8]. A typical fiber waveguide consists of a thin central glass core surrounded by a glass cladding of slightly lower refractive index. The glass-cladded glass fiber was the first dielectric guide to be studied for discrete mode propagation at optical frequencies [9]. Since this publication by Kapany [9] in 1967, there has been a virtual explosion of literature on various aspects of optical fiber waveguides. Review papers by Miller et al. [10] and Clarricoats [11], and books by Marcuse [12, 13], Kapany and Burke [14], Arnaud [15], Sodha and Ghatak [16], Midwinter [17], and Owyang [18] together give a complete exposure to the theory of fiber optic guides.

Concurrent with the progress in optical fibers, for long-distance communication, there emerged a need to build optical components and integrate them into optical circuits. With this, the concept of "integrated optics" came into existence sometime in the late 1960s [19]. Investigations on planar optical dielectric guides, which would permit integration of planar optical circuits, gained importance. Significant advances have taken place since 1970 on almost all aspects pertaining to the analysis, design, and fabrication technology of planar optical dielectric guides and optical integrated circuits. The theory and technology relevant to "integrated optics" are well demonstrated in several books [20–24]. An excellent exposition of the theory of planar optical waveguides and fibers is available in a book by Unger [25].

The idea of using dielectric guides for microwave and millimeter wave integrated circuits received attention in the 1950s. A planar dielectric integrated guide consisting of a dielectric strip on metal ground plane, popularly referred to as the "image guide," was first proposed in 1952 by King [26]. Further studies on the dielectric image guide were later reported by King [27, 28], Weiss and Gyorgy [29], and Schlesinger and King [30, 31]. The investigation of dielectric image lines at millimeter wave frequencies was first conducted by Wiltse in 1959 [32] and later by Sobel et al. in 1961 [33] and Knox and Toulios in 1970 [34]. Since then, a wide variety of dielectric guide configurations for millimeter wave integrated circuit applications have emerged. Unlike the planar optical dielectric guides, which are generally devoid of any metallic conductors, the dielectric guides for millimeter wave integrated circuits are invariably backed by metallic ground planes. The ground plane, besides acting as a mechanical support, provides for integration of several components and serves also as a heat sink and dc bias return for active devices. Besides these planar dielectric integrated guides, a class of dielectric surface waveguides in the form of H-guides has been proposed, particularly for use in the higher frequency range of the millimeter wave band.

The need for utilizing the millimeter wave spectrum was felt about two decades ago because of the increasing congestion of the microwave spectrum. Over the years, several types of transmission media have been applied for their potential utility in the millimeter wave band. It is now well established that, in addition to the conventional hollow metal waveguides, the planar and quasiplanar transmission lines offer convenient transmission media at lower millimeter wave frequencies. The techniques and technology of these transmission structures have matured and the vast body of literature generated has already been documented in several books [35–38]. The dielectric integrated guides form a different class of guides and together with H- and groove guides constitute a range of transmission structures that can be used for realizing practical circuits and antennas over the entire millimeter wave band.

1.2 WAVEGUIDING MEDIA FOR MILLIMETER WAVES

Millimeter wave frequencies span typically from 30 to 300 GHz. Several waveguiding media are now available for guided wave transmission in this frequency range. These guiding media may be classified under the following five broad categories:

1. Hollow metal waveguides.
2. Planar transmission lines.
3. Quasiplanar transmission lines.
4. Dielectric integrated guides.
5. H- and groove-guide structures.

The metal waveguide and planar transmission line techniques are basically extensions of the techniques developed originally for applications in the microwave region. The last three categories of guides listed above emerged later, in the 1970s; the quasiplanar and dielectric integrated guides for use in the frequency range from 30 to about 120 GHz, and the H- and groove-guide structures for higher frequencies from about 100 to 300 GHz.

Figure 1.1 shows some of the important representative guides belonging to each of the five categories. Among the hollow metallic waveguides (Fig. 1.1(a), (b)), the TE_{10} mode rectangular waveguide (Fig. 1.1(a)) is generally used for building high-power transmitting systems at lower millimeter wave frequencies up to about 100 GHz. Beyond this frequency, the guide assumes minute dimensions and the cost increases considerably because of the need to maintain increasingly stringent dimensional tolerances and surface finish. The guide suffers higher losses and its power handling capability also becomes limited. In comparison, the TE_{01} mode circular waveguide offers larger dimensions and lower losses. While this guide has received attention for application in long-distance communication, it is not a practical structure for realizing millimeter wave components. This is due to the fact that the TE_{01} mode is not a dominant mode and furthermore the field configuration of this mode is not convenient for component design.

For low- and medium-power circuit applications, the metal waveguide technology has been almost completely replaced by the microwave integrated circuit (MIC) technology. MICs are based on the use of planar transmission lines as the propagating media. Some of the useful planar transimission lines are the microstrip line, slotline, suspended stripline, inverted microstrip, and coplanar line. Of these, the microstrip line (Fig. 1.1(c)) is the most extensively used at microwave frequencies. It offers a simple geometry, easy incorporation of active devices, and

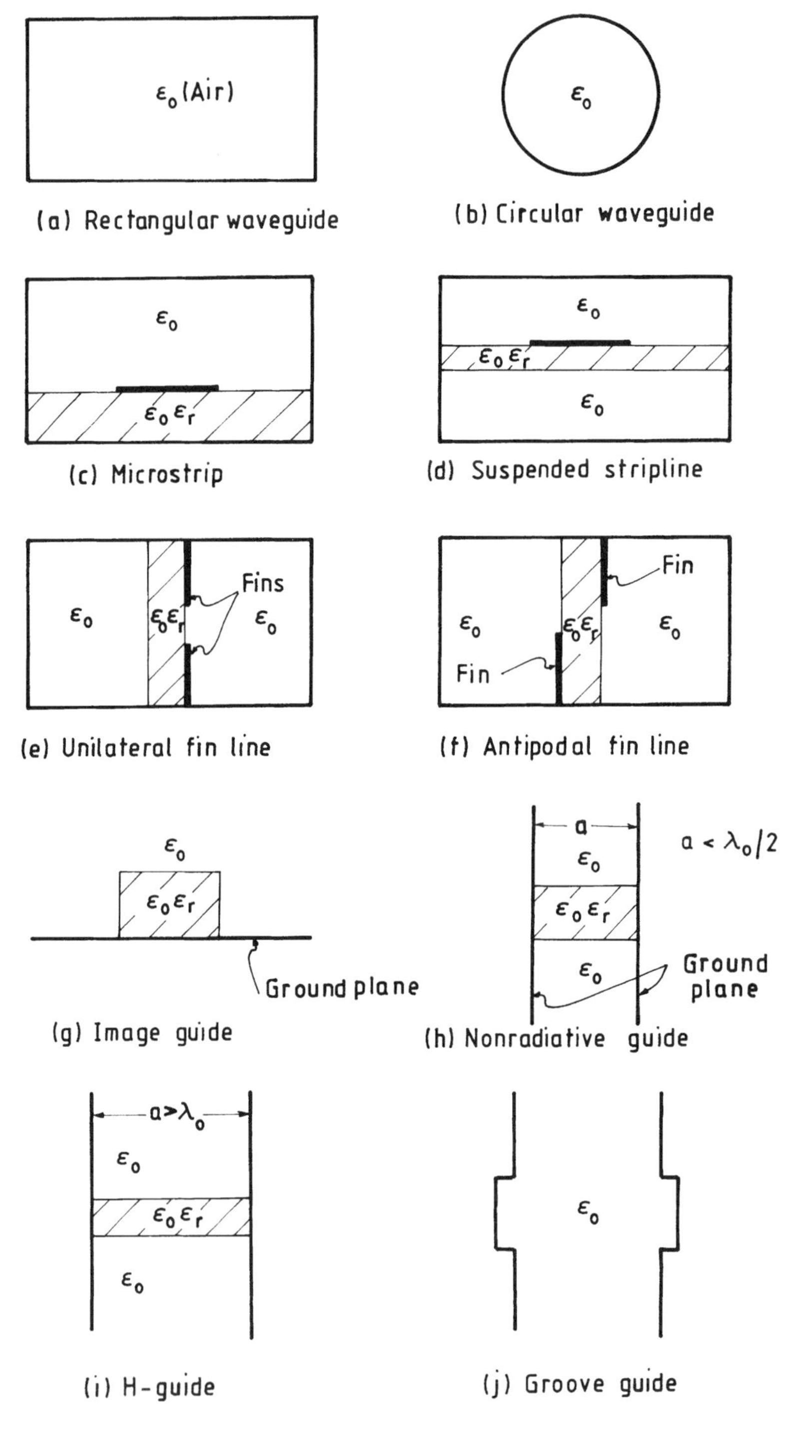

FIGURE 1.1 Examples of millimeter wave transmission structures.

integration of multifunction circuits. The extension of microstrip techniques to millimeter wave frequencies requires the use of progressively thinner substrates having lower dielectric constant in order to reduce spurious coupling and ensure only the dominant mode propagation. With the use of thin low-loss substrates made of materials such as quartz and RT-duroid, microstrip techniques have been extended to lower millimeter wave frequencies up to about 40 GHz. Beyond this frequency, the microstrip poses problems of increased conductor loss and critical dimensional tolerances. Although the utility has been further extended up to about 100 GHz with more sophisticated photolithographic techniques, a preferred planar transmission line for operation at millimeter wave frequencies is the suspended stripline (Fig. 1.1(d)). In view of the airgap introduced between the substrate and the ground plane, this line offers much lower loss and requires less stringent dimensional tolerances as compared with the microstrip. The suspended stripline can therefore be used much more conveniently at millimeter wave frequencies up to about 100 GHz and with more careful fabrication up to about 140 GHz.

Another versatile transmission line that offers low loss and good integration capability in the frequency range 30 to about 120 GHz is the fin line. Fin line is a quasiplanar transmission line and is formed by mounting a dielectric substrate with printed fins on it in the *E*-plane of a standard rectangular waveguide. As examples, Figure 1.1(e) shows the cross-section of an unilateral fin line and Figure 1.1(f) shows that of an antipodal fin line. The fin line overcomes the disadvantages of having to maintain tight dimensional tolerances on the inner walls of the rectangular shield, as in a waveguide, and incorporates the advantageous features of planar technology, including easy mounting of active devices across the slot.

One major problem encountered in the above three classes of transmission media at higher millimeter wave frequencies is the increasing conductor loss. It is well known that the attenuation constant due to the skin effect increases proportional to the square root of the frequency but that due to the dielectric material can be kept relatively low by choosing a dielectric having a low dielectric constant and as low a dielectric loss tangent as possible. From the point of view of achieving low transmission loss, bare dielectric guides are an ideal solution at millimeter wave frequencies. However, for integrated circuit applications, dielectric guides backed by metallic ground planes are more practical.

The simplest form of dielectric guide is the image guide, which consists of a dielectric strip in intimate contact with a ground plane as illustrated in Figure 1.1(g). Another structure that has received considerable attention is the nonradiative dielectric guide shown in Figure 1.1(h). The undesirable radiation at the bends and other discontinuities normally encountered in image guide and other open dielectric guides are suppressed in the nonradiative guide. In addition to these two, a variety of other guide configurations have been reported for use at millimeter wave frequencies. The dielectric integrated guides, in general, offer the combined advantages of low loss, light weight, and ease of fabrication at frequencies in the range of 30 to about 120 GHz. Another important feature of these guides is the possibility of integration of high-performance dielectric antennas with the RF front end.

For use at the higher end of the millimeter wave band, a class of surface waveguiding structures known as the H- and groove guides is found suitable. The configuration of the basic H-guide (Fig. 1.1(i)) resembles that of the nonradiative guide except that the plate separation is greater than a wavelength. It makes use of surface wave guidance at the dielectric interface in one transverse direction and field confinement by parallel plates in the other. The guide supports a hybrid mode, with both *E*- and *H*-lines having a component in the direction of propagation. One interesting feature of this guide is that there is no longitudinal current flow on the metal walls because of the absence of a vertical component of the magnetic field at the walls. The electric field lines are essentially parallel to the conducting walls and the magnetic field lines are parallel to the dielectric surface. Therefore this mode offers low propagation loss, which decreases with an increase in plate separation. The H-guide is reported to have potential over the frequency range from 100 to about 200 GHz. Its operation beyond about 200 GHz is limited because of multimode propagation. This problem is overcome in the groove guide (Fig. 1.1(j)). In this guide, the groove region creates a surface wave effect and supports slow wave propagation. The guide dimensions are convenient to handle at operating frequencies from 100 to 300 GHz. Furthermore, the guide can offer single mode operation with low propagation loss.

A broad comparison of the different categories of transmission structures for millimeter wave integrated circuit applications is provided above in order to indicate the relative utility of the dielectric integrated guides with reference to other guides. A more comprehensive comparison of the characteristics of the various guide structures is available in a review paper by Benson and Tischer [39] and also in a book by Bhat and Koul [38].

1.3 WAVE GUIDANCE IN OPEN HOMOGENEOUS DIELECTRIC GUIDES

In this section, we review the principle of wave guidance in bare dielectric guides. Three basic guide structures are considered: slab dielectric guide (Fig. 1.2(a)), rectangular dielectric guide (Fig. 1.2(b)), and circular dielectric guide (Fig. 1.2(c)). The guide material is assumed to have a relative dielectric constant ε_r that is greater than unity and the surrounding medium is air. Because of the absence of conducting boundaries, the electromagnetic fields in these guides exist both inside and outside the dielectric. The relative amount of energy propagating inside the dielectric increases with an increase in the dielectric constant of the guide, for a given cross-sectional area.

1.3.1 Slab Dielectric Guide

The slab dielectric guide shown in Figure 1.2(a) offers a simple model for understanding the mechanism of wave guidance in dielectric guide structures.

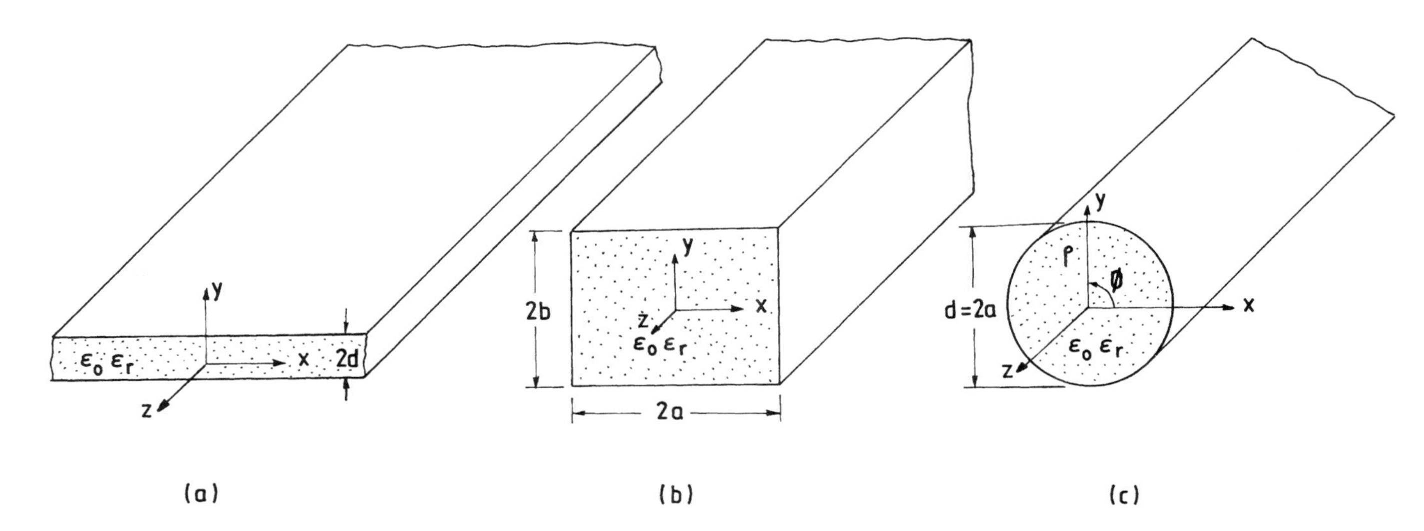

FIGURE 1.2 Basic dielectric guides: (a) slab dielectric guide, (b) rectangular dielectric guide, and (c) circular dielectric guide.

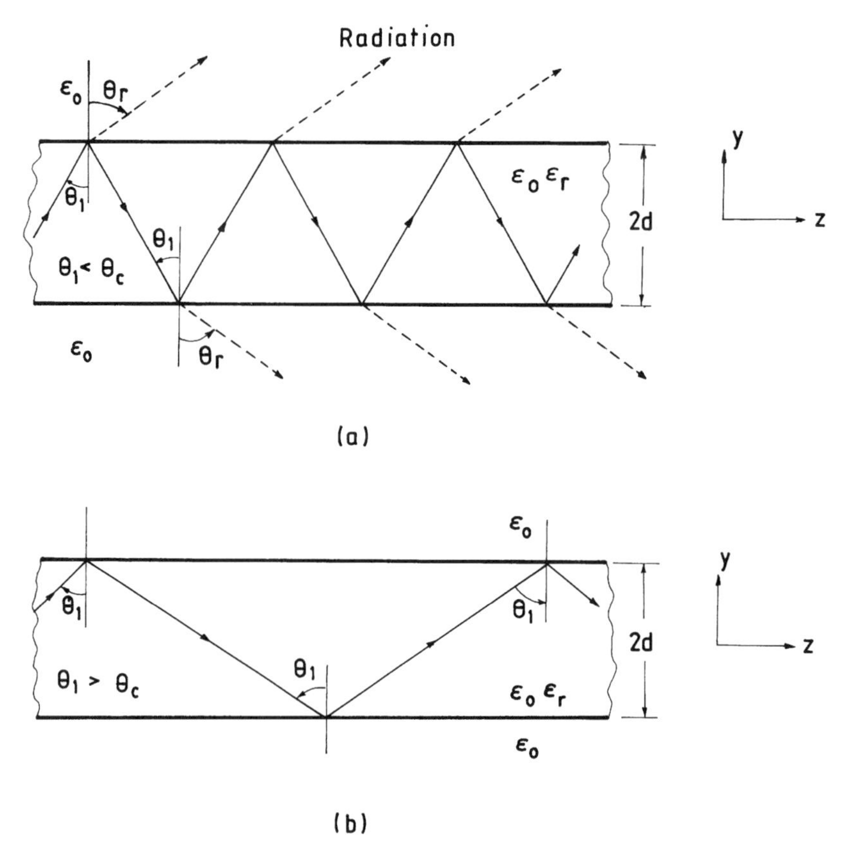

FIGURE 1.3 Optical ray patterns in dielectric slab for (a) radiation mode ($\theta_1 < \theta_c$) and (b) guided mode ($\theta_1 > \theta_c$).

The slab is assumed to be of thickness $2d$ in the y-direction and of infinite extent in the x- and z-directions. We now consider a uniform plane electromagnetic wave incident on the top dielectric–air interface of the slab from the dielectric side. Let the wave normal make an angle θ_1 with respect to the xy-plane as shown in Figure 1.3(a). In terms of geometric optics, this plane wave may be represented by a ray traveling in the direction of the wave normal. It is well known that if the incident angle θ_1 is less than the critical angle θ_c, where θ_c is given by the relation

$$\sqrt{\varepsilon_r}\sin\theta_c = 1 \tag{1.1}$$

the ray, on striking the dielectric–air interface, will be partially reflected back into the dielectric and partially into the air region above. The refracted ray radiates into the space above at an angle θ_r to the normal such that

$$\sin\theta_r = \sqrt{\varepsilon_r}\sin\theta_1 \tag{1.2}$$

The reflected ray strikes the lower boundary at an angle θ_1 with respect to the normal and again undergoes reflection and refraction. The refracted ray now radiates into the space below the slab at angle θ_r with respect to the normal. This process of partial reflection and refraction at the upper and lower dielectric–air interfaces repeats successively along the z-direction. As a consequence, an increasing amount of power from the uniform plane wave represented by the ray leaks into the space. Because of this continuous leakage as radiation, the waves represented by these rays are called radiation modes. Solution of Maxwell's equations for the slab guide, made up of plane waves corresponding to the angles $\theta_c > |\theta_1| > 0$ are known as "leaky waves."

If the angle of incidence θ_1 of the propagating ray is greater than θ_c, then the ray undergoes total internal reflection at both the upper and lower boundaries (Fig. 1.3(b)). The power carried by such rays remains confined to the slab and gets guided by it while associated fields extending outside the slab boundaries decay exponentially in the transverse direction. Thus the condition $\theta_1 > \theta_c$ gives rise to slab guided modes. These modes are referred to as "surface wave" modes.

Surface waves of both TM and TE type may propagate along a dielectric slab. For the TM modes, also known as E-modes, the electric vector is parallel to the plane of incidence (yz-plane) with a component along the direction of propagation (z-direction) whereas the magnetic field has only a transverse x-component. Thus the field components for the TM modes are H_x, E_y, and E_z. For TE modes or H-modes, we interchange the roles of the electric and magnetic fields of TM modes. Thus the field components of TE modes are E_x, H_y, and H_z.

From the point of view of application to millimeter wave integrated circuits, the surface wave mode is of relevance. The surface wave solutions of dielectric slab are well documented in the literature [40]. In the following, some salient features of this guide are summarized. For both types of mode, we assume wave propagation in the z-direction according to $e^{-j\beta z}$ and that away from the slab surface the fields decay according to $e^{-\alpha(|y|-d/2)}$ (refer Fig. 1.2(a)). Since the fields satisfy the wave equation, the propagation constant β(real) and the attenuation constant α are related by

$$\beta^2 = k_0^2 + \alpha^2 \tag{1.3}$$

where k_0 is the free-space propagation constant. The guide wavelength $\lambda\,(=2\pi/\beta)$ and the phase velocity $v_{\mathrm{ph}}\,(=\omega/\beta)$ of the slab guide are less than the corresponding quantities for plane waves in free space.

TM Modes The TM mode solutions of the slab guide may be considered in two parts. One is the symmetric mode for which the tangential electric field vanishes at the plane of symmetry $y=0$ ($E_z=0$ or $\partial H_x/\partial y=0$ at $y=0$). The second is the antisymmetric mode for which the tangential magnetic field H_x vanishes at the plane $y=0$. For symmetric and antisymmetric modes, the single magnetic field component H_x will have the following form [40]: for symmetric modes,

$$H_x = \begin{cases} A\sec(pd)\cos(py), & |y| \leq d \\ Ae^{-\alpha(|y|-d)}, & |y| \geq d \end{cases} \tag{1.4}$$

and for antisymmetric modes,

$$H_x = \begin{cases} Ae^{-\alpha(y-d)}, & y \geq d \\ A\ \operatorname{cosec}(pd)\sin(py), & -d \leq y \leq d \\ -Ae^{\alpha(y+d)}, & y \leq -d \end{cases} \tag{1.5}$$

In (1.4) and (1.5), the propagation factor $e^{-j\beta z}$ is suppressed and p is the y-directed wavenumber inside the slab. The eigenvalue equation for each mode is obtained by matching the tangential fields at $y = \pm d$. They are given by

$$\varepsilon_r \alpha d = pd\tan(pd), \qquad \text{symmetric modes} \tag{1.6a}$$

$$\varepsilon_r \alpha d = -pd\cot(pd), \quad \text{antisymmetric modes} \tag{1.6b}$$

where

$$(\alpha d)^2 + (pd)^2 = (\varepsilon_r - 1)(k_0 d)^2 \tag{1.7}$$

$$\beta^2 = k_0^2 + \alpha^2 = k_0^2 \varepsilon_r - p^2 \tag{1.8}$$

TE Modes In the case of TE modes, the single electric field component E_x may be expressed in the following form: for symmetric modes,

$$E_x = \begin{cases} A\,e^{-\alpha(|y|-d)}, & |y| \geq d \\ A\sec(pd)\cos(py), & |y| \leq d \end{cases} \tag{1.9}$$

and for antisymmetric modes,

$$E_x = \begin{cases} A\,e^{-\alpha(y-d)}, & y \geq d \\ A\operatorname{cosec}(pd)\sin(py), & |y| \leq d \\ -A\,e^{\alpha(y+d)}, & y \leq -d \end{cases} \tag{1.10}$$

The eigenvalue equations are given by

$$\alpha d = pd\tan(pd), \qquad \text{symmetric modes} \tag{1.11a}$$

$$\alpha d = -pd\cot(pd), \quad \text{antisymmetric modes} \tag{1.11b}$$

where α, p, and β are related by Eqs. (1.7) and (1.8).

The solutions of transcendental equations (1.6) and (1.11) yield discrete values for β. These discrete values correspond to the various modes designated as TM_m

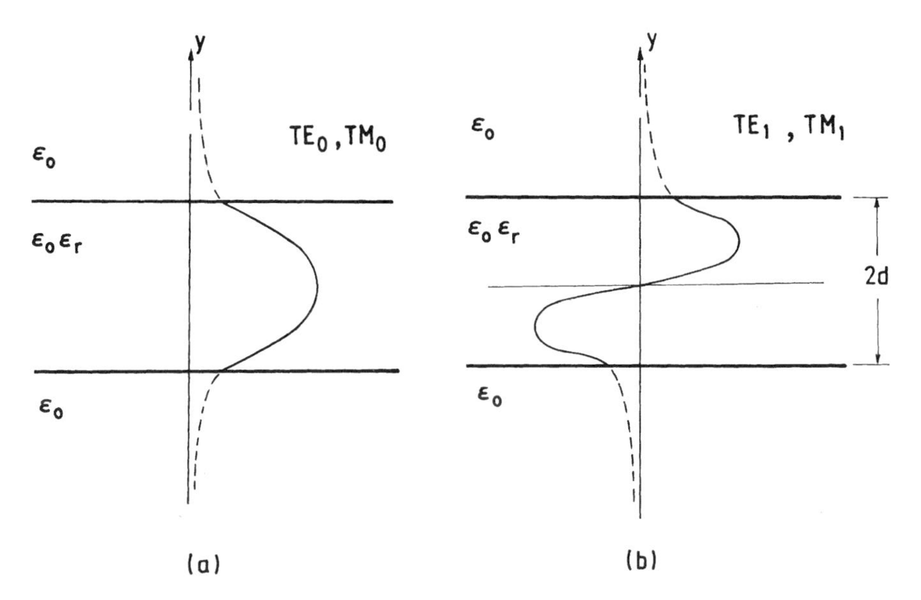

FIGURE 1.4 Transverse distribution of E_x component for TE modes and H_x component for TM modes: (a) TE_0 and TM_0 modes and (b) TE_1 and TM_1 modes.

and TE_m, where $m = 0, 1, 2, \ldots$. The number of modes that can be supported depends on the operating frequency, slab thickness, and slab dielectric constant. The first symmetric modes—namely, TM_0 and TE_0—have no cutoff. All higher order modes have low-frequency cutoff. The cutoff condition is the same for TE and TM modes. The value of $2d/\lambda_0$ at cutoff is given by

$$\frac{2d}{\lambda_0} = \frac{m}{2(\varepsilon_r - 1)^{1/2}}, \quad m = 0, 1, 2, \ldots \tag{1.12}$$

where $m = 0, 2, 4, \ldots$ refer to symmetric modes and $m = 1, 3, 5, \ldots$ refer to antisymmetric modes. Figure 1.4 illustrates typical field intensity distribution of E_x component for TE_0 and TE_1 modes and that of H_x component for TM_0 and TM_1 modes.

1.3.2 Rectangular Dielectric Guide

The rectangular dielectric guide shown in Figure 1.2(b) provides confinement of fields in two dimensions as compared with the slab guide, which can confine the fields only in one dimension. The two-dimensional confinement is necessary not only to guide electromagnetic energy from one point to another but to interconnect circuit elements. The guide supports hybrid modes and wave guidance take place because of total internal reflection at the side walls. At millimeter wave frequencies, these guides are designed with low dielectric constant materials ($\varepsilon_r \approx 2$–10) with the result that total internal reflection occurs when the plane wavelets make a mode impinge at a relatively small angle with respect to the dielectric interface. The modes are therefore of the TE and TM kind with the largest field components being transverse to the axis of the guide. These modes are generally classified as E^y_{mn} (TM-to-y) and E^x_{mn} (TE-to-y) modes. The main transverse field components of E^y_{mn} modes are E_y and H_x, and those of the E^x_{mn} modes are E_x and H_y. The subscripts m and n indicate the number of extrema of electric or magnetic fields in the x- and y-directions, respectively. The dominant modes correspond to $m = n = 1$.

Marcatili [41] has derived an approximate solution to the rectangular dielectric guide problem by assuming guided mode propagation (well above cutoff) along the dielectric and exponential decay of fields transverse to the dielectric surface. Referring to Figure 1.2(b), and assuming propagation in the z-direction according to $e^{-j\beta z}$, the field component E_y corresponding to E^y_{mn} excitation can be written as

$$E_y = \begin{cases} A\cos(k_x x)\cos(k_y y), & |x| \le a, |y| \le b \\ A\cos(k_x a)\cos(k_y y)e^{-\alpha_{x0}(|x|-a)}, & |x| \ge a, |y| \le b \\ A\cos(k_x x)\cos(k_y b)\,e^{-\alpha_{y0}(|y|-b)}, & |x| \le a, |y| \ge b \end{cases} \tag{1.13}$$

In the shadow region $|x| > a, |y| > b$, the field E_y may be approximated to zero. In (1.13), the parameters k_x and k_y are the transverse propagation constants inside the dielectric guide, and α_{x0} and α_{y0} are the attenuation constants outside the guide. k_x and k_y are solutions of the transcendental equations [41]:

$$k_y = \frac{n\pi}{2b} - \frac{1}{b}\tan^{-1}\left[\frac{k_y}{\varepsilon_r\{k_0^2(\varepsilon_r - 1) - k_y^2\}^{1/2}}\right] \tag{1.14a}$$

$$k_x = \frac{m\pi}{2a} - \frac{1}{a}\tan^{-1}\left[\frac{k_x}{\{k_0^2(\varepsilon_r - 1) - k_x^2\}^{1/2}}\right] \tag{1.14b}$$

and

$$\beta^2 = k_0^2\varepsilon_r - k_x^2 - k_y^2 \tag{1.14c}$$

$$\alpha_{x0}^2 = k_0^2(\varepsilon_r - 1) - k_x^2 \tag{1.14d}$$

$$\alpha_{y0}^2 = k_0^2(\varepsilon_r - 1) - k_y^2 \tag{1.14e}$$

Figure 1.5 illustrates typical electric and magnetic field patterns for the dominant E^y_{11} mode. The field intensity distributions of E_y and H_x components are also shown. The field pattern for the E^x_{11} mode is obtained by interchanging the electric and magnetic field lines of the E^y_{11} mode.

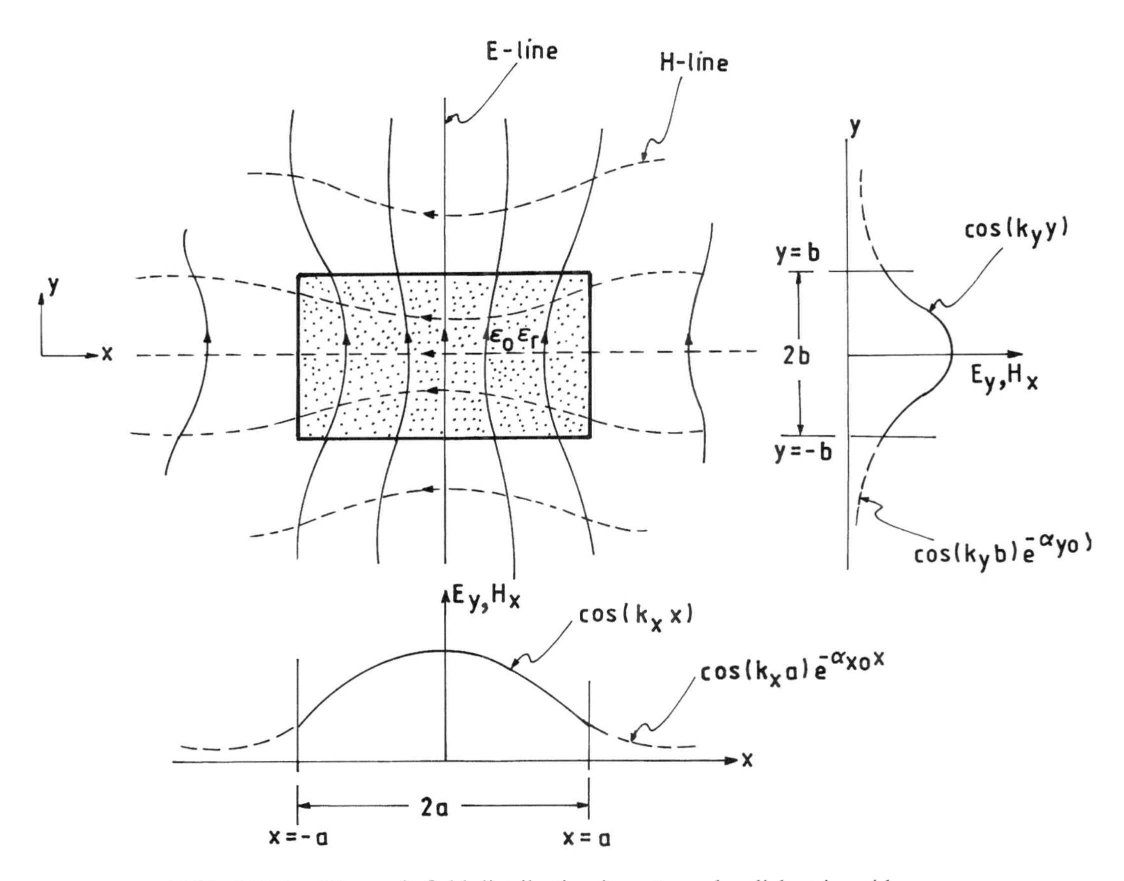

FIGURE 1.5 E_{11}^y mode field distribution in rectangular dielectric guide.

1.3.3 Circular Dielectric Guide

Propagation in the circular dielectric guide (Fig. 1.2(c)) has been treated extensively in the literature [5, 42, 43]. As in the case of the rectangular dielectric guide, mode propagation in the circular guide also takes place because of total internal reflection. The field outside the rod decays approximately exponentially with increasing radial distance. The propagating modes are hybrid in nature and are designated as EH_{mn} and HE_{mn} modes. For $m = 0$, the EH_{mn} modes reduce to TE_{0n} modes and the HE_{mn} modes reduce to TM_{0n} modes. For a wave propagating along a rod of diameter $2a$, the expression for the longitudinal components E_z and H_z of the HE_{mn} modes are given by the following [5]. Inside the rod ($\rho \leq a$), we have

$$E_z = A\,(k_1^2/j\omega\varepsilon_0\varepsilon_r)\,J_m(k_1\rho)\sin(m\phi) \tag{1.15a}$$

$$H_z = -\,B\,(k_1^2/j\omega\mu_0)\,J_m(k_1\rho)\cos(m\phi) \tag{1.15b}$$

Outside the rod ($\rho > a$), we have

$$E_z = C\,(k_2^2/j\omega\varepsilon_0)\,H_m^{(2)}(k_2\rho)\sin(m\phi) \tag{1.16a}$$

$$H_z = -\,D\,(k_2^2/j\omega\mu_0)\,H_m^{(2)}(k_2\rho)\cos(m\phi) \tag{1.16b}$$

where

$$k_1^2 = k_0^2\varepsilon_r - \beta^2 \tag{1.17a}$$

$$k_2^2 = k_0^2 - \beta^2 \tag{1.17b}$$

$$k_0^2 = \omega^2\mu_0\varepsilon_0 \tag{1.17c}$$

In (1.15) to (1.17), β is the propagation constant in the z-direction, and $J_m(x)$ and $H_m^{(2)}(x)$ are a Bessel function of the first kind and a Hankel function of the second kind, respectively. The ρ and ϕ components of the electric and magnetic fields can be obtained by substituting the expressions for E_z and H_z in Maxwell's equations. Then, by matching the tangential E and H components at the rod surface $\rho = a$, the following characteristic equation can be derived for the HE_{mn} modes:

$$\left[\frac{1}{u}\frac{J_m'(u)}{J_m(u)} - \frac{1}{v}\frac{H_m^{(2)\prime}(v)}{H_m^{(2)}(v)}\right]\left[\frac{\varepsilon_r}{u}\frac{J_m'(u)}{J_m(u)} - \frac{1}{v}\frac{H_m^{(2)\prime}(v)}{H_m^{(2)}(v)}\right] = m^2\left[\frac{1}{v^2} - \frac{1}{u^2}\right]\left[\frac{1}{v^2} - \frac{\varepsilon_r}{u^2}\right] \tag{1.18}$$

where

$$u = k_1 a \tag{1.19a}$$

$$v = k_2 a \tag{1.19b}$$

For $m = 0$, the right-hand side of (1.18) vanishes and the two factors on the left-hand side give the characteristic equations for the axially symmetric TE and

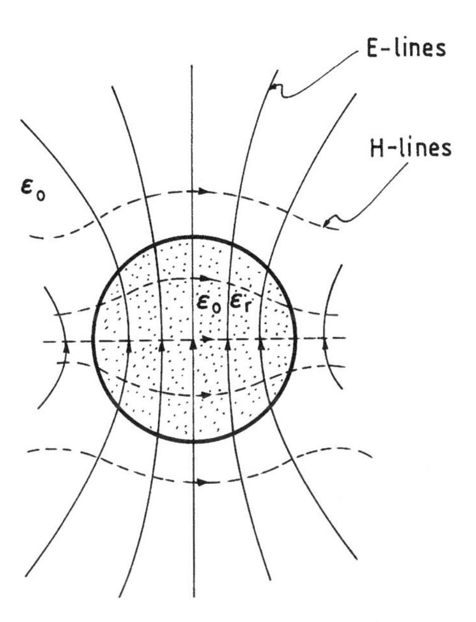

FIGURE 1.6 Dominant HE_{11} mode field distribution in cylindrical dielectric guide.

TM modes. For TE_{0n} modes,

$$u\frac{J_0(u)}{J_0'(u)} = v\frac{H_0^{(2)}(v)}{H_0^{(2)\prime}(v)} \tag{1.20}$$

For TM_{0n} modes,

$$u\frac{J_0(u)}{J_0'(u)} = \varepsilon_r v\frac{H_0^{(2)}(v)}{H_0^{(2)\prime}(v)} \tag{1.21}$$

All the modes exhibit cutoff except the HE_{11} mode. This mode is degenerate with the EH_{11} mode and is the lowest order hybrid mode. Figure 1.6 shows a typical field pattern for the HE_{11} mode.

Both rectangular and circular dielectric guides permit guided mode propagation with two-dimensional field confinement in the transverse plane. The cross-sectional dimensions and the dielectric constant of the guides, however, must be chosen such that almost all of the electromagnetic energy is confined to the inside of the dielectric guide while the loss is kept to a minimum. The dielectric loss can be reduced by reducing the cross-sectional area, but this results in an increased percentage of energy outside the guide. In the case of a rectangular guide, the minimum cross-sectional dimension is governed also by the cutoff frequency of the dominant mode. On the other hand, for the circular guide, the rod diameter can be reduced sufficiently without affecting the dominant mode propagation.

The minimum size of the rod is then governed by mechanical considerations in addition to the need to contain sufficient energy within the guide.

1.4 WAVE GUIDANCE IN OPEN COMPOSITE PLANAR DIELECTRIC GUIDES

For circuit applications, planar dielectric guides are commonly preferred over guides of other geometrical shapes because of the convenience that the planar surface offers for fabrication and integration. The slab guide and the rectangular dielectric guide (see Fig. 1.2(a), (b)) discussed in Section 1.3 are two basic examples of planar guides. Since practical guide structures must provide for lateral confinement of electromagnetic fields, the rectangular dielectric that meets this requirement can be used either singly or in conjunction with a uniform dielectric layered structure. The latter forms a composite planar structure in which the rectangular strip plays the role of ensuring transverse confinement of fields. In the following, we discuss the phenomenon of wave guidance in a class of open planar dielectric guides meant for millimeter wave integrated circuits and compare their features with those of a class of planar guides used in optical integrated circuits.

1.4.1 Millimeter Wave Planar Dielectric Guides

Planar dielectric guides for millimeter wave integrated circuits are invariably backed by metallic ground planes, which serve as mechanical supports and facilitate integration of several components. The ground plane serves also as heat sink and dc bias return for active devices. Figure 1.7 shows some examples of open planar guides for millimeter wave integrated circuits.

The simplest millimeter wave dielectric integrated guide structure is the image guide, the cross-section of which is shown in Figure 1.7(a). It consists of a rectangular dielectric slab of relative permittivity ε_r backed by a perfectly conducting ground plane. It represents the top half of a rectangular dielectric guide of twice the height with an image ground plane at the plane of symmetry.

As in a rectangular dielectric guide, the modes of propagation in an image guide are E^y_{mn} and E^x_{mn} modes. Figure 1.8 shows the field intensity distribution in an image guide and a corresponding rectangular dielectric guide for the lowest order E^y_{11} and E^x_{11} modes. For the E^y_{11} mode, the field intensity distribution in the image guide remains the same as in the top half of the rectangular dielectric guide of twice the height. On the other hand, for the E^x_{11} mode, the metal plane of the image guide shorts the E_x component, thereby essentially suppressing that mode. The image guide therefore permits single mode operation over considerable frequency bandwidth. This is a distinct advantage over the bare rectangular dielectric guide.

The insular image guide and the inverted strip guide shown in Figure 1.7(b) and (c), respectively, are low-loss variants of the image guide. The insular image

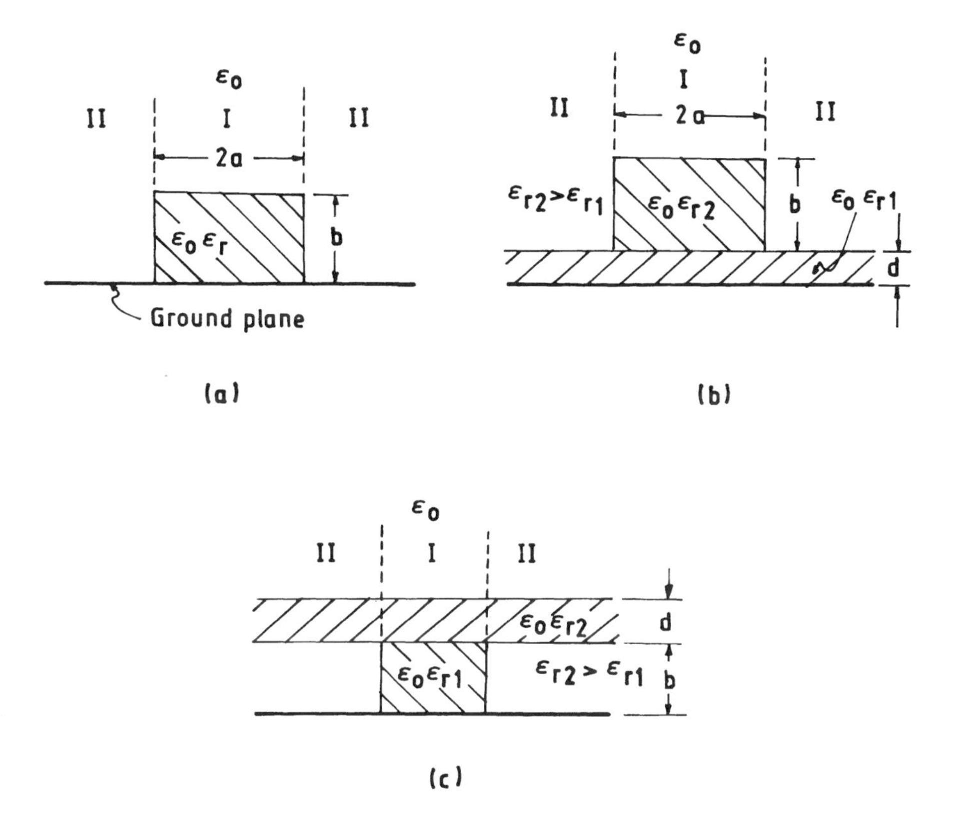

FIGURE 1.7 Examples of planar dielectric guides for millimeter wave integrated circuits: (a) image guide, (b) insular image guide, and (c) inverted strip guide.

guide makes use of an insular layer of lower dielectric constant ($\varepsilon_{r1} < \varepsilon_{r2}$) between the rectangular guiding strip and the ground plane, and the inverted strip guide has a guiding layer of higher dielectric constant (ε_{r2}) separated from the ground plane by a strip of lower dielectric constant ε_{r1}. With $\varepsilon_{r2} > \varepsilon_{r1}$ in both cases, most of the field gets concentrated in the top dielectric strip (Fig. 1.7(b)) or layer (Fig. 1.7(c)), and less near the ground plane. Consequently, the current concentration in the ground plane is reduced, thereby resulting in reduced conductor loss.

It is important to note that the introduction of a large planar dielectric layer in the basic image guide gives rise to new physical effects, which can result in leakage of energy [44–46]. This leakage is caused by the coupling between the TE and TM modes that occurs at the sides of the open dielectric strip. Peng and Oliner [44] were the first to predict this leakage phenomenon and leakage-related resonance effects. A detailed discussion on these physical effects is available in Peng and Oliner [45] and Oliner et al. [46]. In order to provide a qualitative understanding of this special leakage effect, we refer to the open planar guides shown in Figure 1.7 and consider the nature of wave propagation in the two constituent regions marked I and II. Region I refers to the central core region,

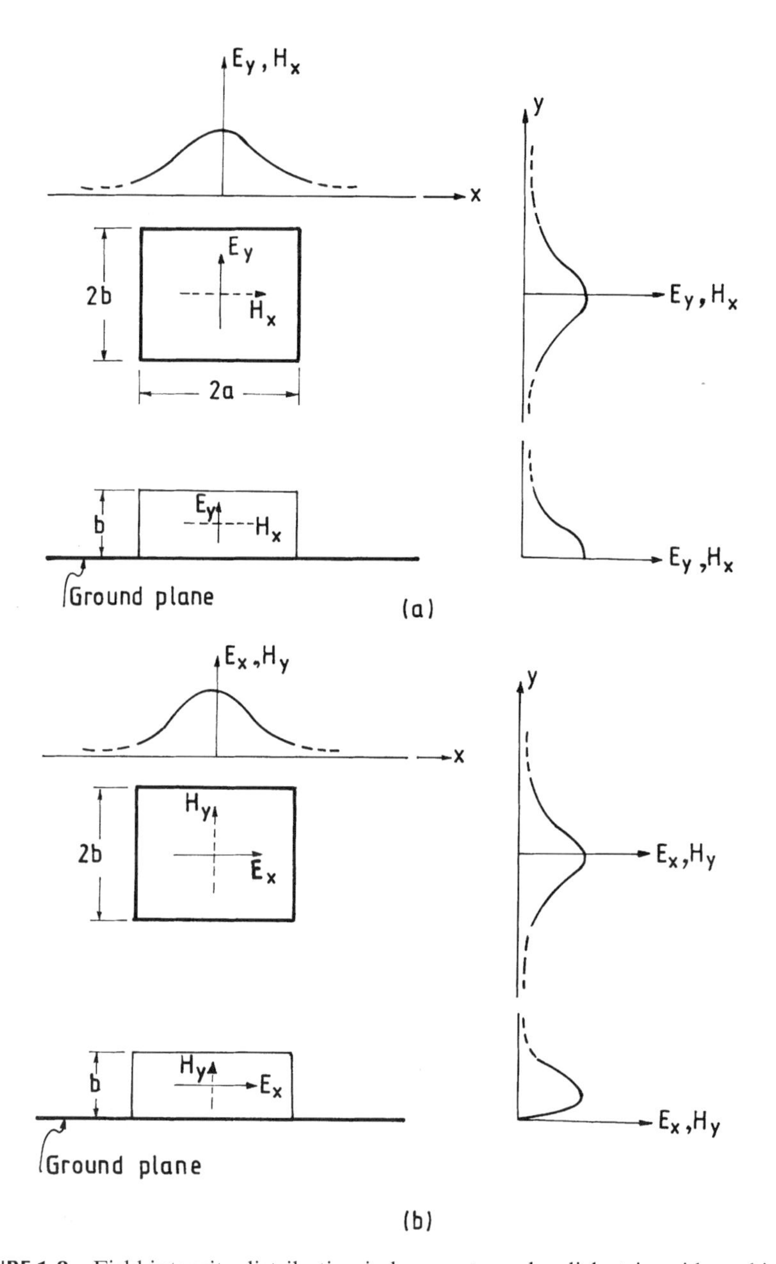

FIGURE 1.8 Field intensity distribution in bare rectangular dielectric guide and image guide for (a) E_{11}^{y} mode and (b) E_{11}^{x} mode.

which includes the dielectric strip, and region II refers to the portion outside. Now consider a TE surface wave traveling at a small angle with respect to the z-direction. While the main components of this wave are H_y, H_z, and E_x, it will also possess small H_x and E_z components. Similarly, a TM surface wave traveling at a small angle with respect to the z-direction will not only have E_y, E_z, and H_x as its main components but will possess additional small E_x and H_z components. Thus, when a TE wave undergoes total internal reflection at the two sides of the strip, the small H_x and E_z components excite a TM surface wave. Similary, when a TM surface wave is incident at a small angle with respect to the z-direction and undergoes total internal reflection, the small E_x and H_z components excite a TE surface wave at the strip sides. Thus TE–TM surface wave mode coupling occurs at the side walls of the rectangular strip.

Because of the TE–TM surface wave mode coupling at the strip edges, the propagating modes in planar guides using rectangular strips are always hybrid modes possessing all six components. If the incident wave is a TE surface wave (having a large vertical magnetic field component), the guided hybrid mode is predominantly TE-like. Similarly, if the incident wave is a TM surface wave (having a large vertical electric field component), the guided hybrid mode is predominantly TM-like.

Oliner et al. [46] have pointed out that because of the TE–TM surface wave mode coupling, there can be leakage of energy in the form of surface waves in certain types of guides. The guides that can leak are those that have a large planar dielectric layer, which can guide a surface wave in the absence of the rectangular strip. Clearly, the image guide cannot leak. All the guided modes in the image guide are therefore purely bound. On the other hand, the insular image guide (Fig. 1.7(b)) and the inverted strip guide (Fig. 1.7(c)) can leak. For example, consider a TE surface wave incident at the strip edge of one of these guides. This would excite a TM surface wave inside and a TM surface wave outside the strip edge. If the transverse wavenumber of the TM surface wave in the outside region (region II in Fig. 1.7(b) or (c)), denoted as k_{x0}^{TM}, satisfies the relation

$$(k_{x0}^{\mathrm{TM}})^2 > 1 \tag{1.22}$$

then this TM surface wave is above cutoff and hence propagates away from the strip guide. Thus the hybrid TE-like guided mode becomes leaky and leakage occurs in the form of TM surface wave energy. The mode-converted TM surface wave, which is inside the strip, is also generally above cutoff and is guided along the strip through total internal reflection. If k_x^{TM} is the transverse wavenumber of this TM surface wave inside the strip and $2a$ is the strip width, then complete cancellation of leakage or resonance occurs when the condition

$$(k_x^{\mathrm{TM}})2a = 2n\pi, \quad n = 1, 2, \ldots \tag{1.23}$$

is satisfied.

1.4.2 Optical Planar Dielectric Guides

In principle, the slab and rectangular dielectric guides presented in Section 1.3 offer possible transmission media for optical integrated circuits also. Practical implementation, however, is prohibitively difficult because of the impractically small physical dimensions that these bare guides assume at optical frequencies. Whereas the cross-sectional dimensions are of the order of a few millimeters at millimeter wave frequencies, the dimensions reduce to submicrometers at optical frequencies. Furthermore, the use of a metal plane as a mechanical support is not desirable because of the high conductor loss that it would introduce at such high frequencies. Planar waveguides for optical integrated circuits make use of thick dielectric substrates as supporting base [19,25]. The rectangular strip (film), which is essentially the guiding vehicle, is either deposited on top of the substrate or diffused into it.

Figure 1.9 illustrates examples of planar dielectric waveguides for optical integrated circuits. The symbols n_1, n_2, and n_0 denote the refractive indices of the various regions. In all these guides, the effective refractive index of the strip-loaded central region is higher than that of the surrounding regions so that there is transverse confinement of fields in the strip region. Wave guidance takes place through total internal reflection at the inner side walls of the strip. For well-guided modes, the fields must decay exponentially perpendicular to the strip both in the air and substrate regions. The raised strip (Fig. 1.9(a)), by virtue of the higher step change in the refractive index at the strip edge, offers better concentration of fields inside it than the embedded strip with its weakly guiding walls (see Fig. 1.9(b), $n_2/n_0 > n_2/n_1$). Because of this feature, the raised strip is preferred over the embedded strip. However, the side walls of the raised strip must be very smooth in order to ensure low scattering loss. The embedded strip, on the other hand, can afford more side wall imperfections without offering excessive scattering loss. Because the step change in refractive index at the strip edges is small, the embedded strip permits larger mode size than the strip guide with raised strip of the same dimension. That is, the evanescent fields outside the side walls of the embedded strip are stronger and extend to a larger distance sideways. This is an advantage in the design of coupled guide components such as couplers.

Further reduction in scattering at the edges of the guiding strip and enhancement in mode size are possible by keeping an even smaller difference in the effective refractive index across the side walls. The strip-loaded film guide shown in Figure 1.9(c) offers this feature. The geometry makes use of a thin layer of dielectric film between the rectangular strip and the substrate. The refractive index n_2 of the film is higher than that of the strip. While the strip serves to provide transverse confinement of fields, most of the field would be concentrated in the portion of the layered film beneath the strip. Because the field concentration is less in the strip region, the dimensional tolerance on the strip edges can be relaxed without suffering undue scattering loss.

The rib guide shown in Figure 1.9(d) is a special case of the strip-loaded film guide, where the refractive index of the film layer and that of the strip are made equal. The structure shown in Figure 1.9(e) is an inverted rib waveguide with embedded rib.

The phenomena of wave guidance and leakage in the class of planar optical waveguides shown in Figure 1.9 are essentially the same as that described in

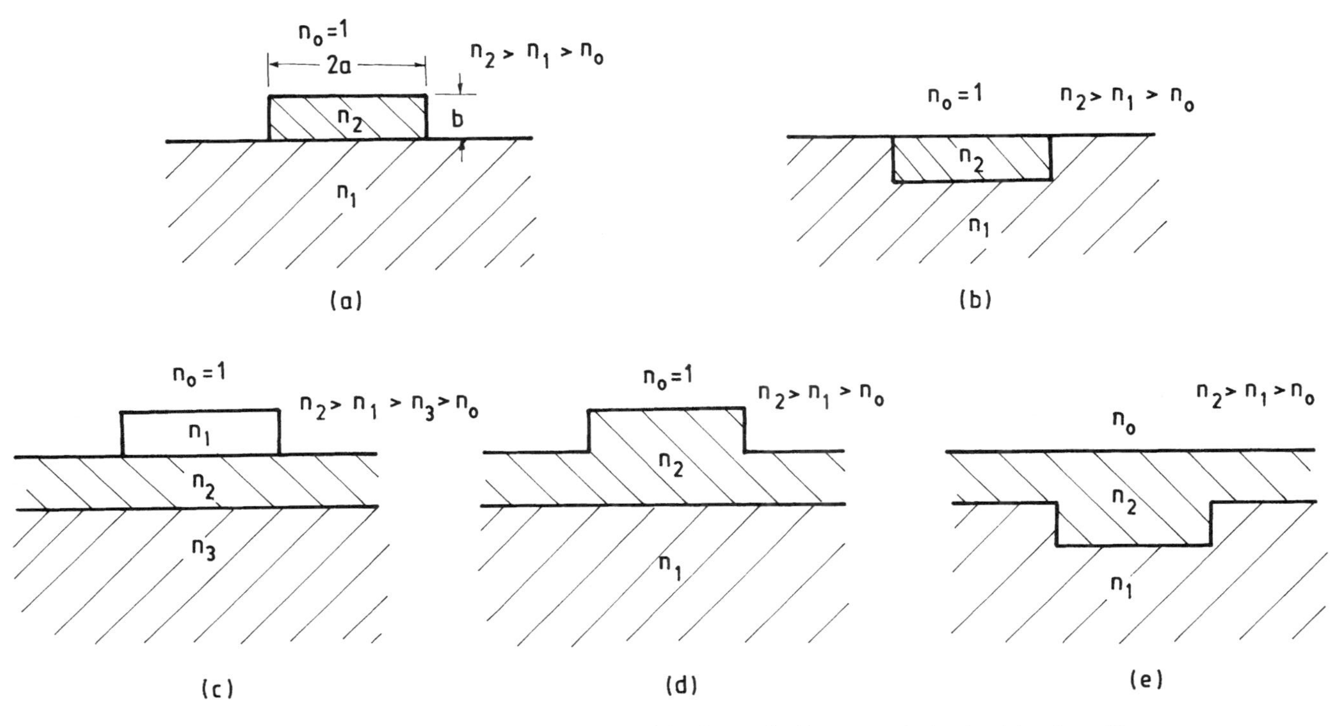

FIGURE 1.9 Examples of planar dielectric guides for optical integrated circuits: (a) strip guide, (b) embedded strip guide, (c) strip loaded film guide, (d) rib guide, and (e) inverted rib guide.

Section 1.4.1 for open planar millimeter wave dielectric guides (Fig. 1.7). Among the various optical guides shown in Figure 1.9, the strip guide (Fig. 1.9(a)) and the embedded strip guide (Fig. 1.9(b)) resemble the image guide in that they do not leak. All the modes in these guides are bound modes. This is because, in the absence of the strip, the dielectric half-space cannot sustain any wave. The substrate serves only to modify the modal behavior and the field distribution. The other guide structures shown in Figure 1.9(c)–(e) make use of a planar dielectric film layer. Even in the absence of the strip, the structure is a planar waveguide that can sustain wave propagation. The guide structures shown in Figure 1.9(c)–(e) can therefore leak energy in the form of leaky surface waves.

1.4.3 Comparison

Table 1.1 provides a comparison of the typical properties of millimeter wave and optical planar dielectric guides. For ease of comparison, we choose the insular image guide (Fig. 1.7(b)) and the optical strip guide (Fig. 1.9(a)) as examples. As discussed in Sections 1.4.1 and 1.4.2, the basic principles of wave guidance in the millimeter wave and optical guide structures are the same. The dominant mode of propagation in both classes of planar open guides with rectangular dielectric strips is the E^y_{11} mode. In the case of millimetric waveguides, the ratio of the refractive index of the core dielectric to that of the surrounding medium is much

TABLE 1.1 Comparison of Typical Millimeter Wave and Optical Dielectric Integrated Guides

Factors	Millimeter Wave Insular Image Guide (Fig. 1.7(b))	Optical Strip Guide (Fig. 1.9(a))
Dielectric material (ε_{r2})	Ceramic dielectric ($\varepsilon_{r2} \approx 4$–$10$) or semiconductor ($\varepsilon_{r2} \approx 12$)	Glass ($\varepsilon_{r2} \approx 2$–$4$) or semiconductor ($\varepsilon_{r2} \approx 12$)
Dielectric material (ε_{r1})	Polymer ($\varepsilon_{r1} \approx 2$–$2.5$)	Glass ($\varepsilon_{r1} \approx 2$–$4$) or semiconductor ($\varepsilon_{r1} \approx 12$)
Refractive index ratio (n_2/n_1 or $\sqrt{\varepsilon_{r2}/\varepsilon_{r1}}$)	~ 2	1.1–1.01
Waveguide strip width ($2a/\lambda$, λ is guide wavelength)	0.5	2–10
Waveguide strip thickness (b/λ)	≤ 0.5	<1
Dominant mode	E^y_{11}	E^y_{11}
Suppression of degenerate E^x_{11} mode	Yes	No
Radius of curvature (in λ)	~ 5	30–1000

higher than that in an optical guide. The use of higher ratio of refractive index (approximately 2) would require a guide width equal to about half-wavelength for supporting the dominant E^y_{11} mode. At millimeter wavelengths, the guide widths assume reasonable dimensions for economical manufacture. For example, typical dimensions of an insular image guide (see Fig. 1.7(b)) having a ratio $\sqrt{\varepsilon_{r2}/\varepsilon_{r1}} = 2$ at a frequency of 60 GHz are $2a = b = 1.34$ mm, $d = 0.13$ mm. The degenerate E^x_{11} mode is suppressed because of the presence of the ground plane. At optical frequencies it is impractical to fabricate guides with width equal to half-wavelength since the wavelengths are in the submicrometer range. It is therefore important to keep n_2 incrementally lower than n_1 (see Fig. 1.9(a)) so that single mode operation in E^y_{11} mode is possible. The use of higher refractive index ratio ($\sqrt{\varepsilon_{r2}/\varepsilon_{r1}}$) in the case of a millimeter waveguide enables better confinement of fields to the core dielectric. As a consequence, in component layout, a bend radius of the order of about five times the guide wavelength is adequate to keep the radiation loss to a minimal level. On the other hand, in the case of optical planar guides, the minimum bend radius required for nearly eliminating the radiation loss is approximately 30 times the guide wavelength.

Besides the above-mentioned differences, the millimeter wave and optical guides differ in terms of materials and fabrication techniques. Substrate materials used for optical integrated guides are of two types: passive and active. Passive materials are incapable of light generation. These include quartz, silicon, lithium niobate, and lithium tantalate. Active materials that are capable of light generation include gallium arsenide (GaAs), gallium aluminium arsenide (GaAlAs), gallium indium arsenide (GaInAs), and other III–V and II–VI direct bandgap semiconductors. Both hybrid and monolithic techniques are used in the fabrication of optical integrated circuits. In the hybrid approach, two or more substrate materials are bonded together to form the guide. In the monolithic approach, a single optically active substrate is used for all the devices. Fabrication of optical integrated guides use techniques such as sputter deposition, diffusion, ion implantation, epitaxial growth, chemical etching, and ion beam etching.

Millimeter wave dielectric guides are usually made of low-loss dielectrics or high-resistivity semiconductor materials. Fabrication techniques include machining, laser cutting, injection molding, green tape cutting and firing, powder sintering, and thick-film printing. A detailed discussion on materials and fabrication techniques used for millimeter wave integrated guides is given in the next section.

1.5 MATERIALS AND FABRICATION TECHNIQUES

Dielectric guides for millimeter wave integrated circuit applications commonly use materials having relative dielectric constants in the range 2–10 and loss tangents less than 10^{-3}. Table 1.2 lists several types of commercially available dielectric materials and semiconductor dielectrics. The table also lists typical

TABLE 1.2 Materials for Millimeter Wave Integrated Guides

Material	ε_r (approximate)	Loss Tangent $\tan\delta$	Thermal Conductivity K (W/cm/°C)	Remarks
		Ceramic Dielectrics [47–51]		
Alumina 99.5% ($f = 10\,\text{GHz}$)	9.6–9.9	1×10^{-4}	0.37	Most commonly used for millimeter wave guides; guides can be formed from "green" Al_2O_3 tapes to suit component layout; low cost of manufacture as compared with sapphire or quartz.
Sapphire ($f = 100\,\text{GHz}$)	9.3–11.7	4×10^{-4}	0.4	Basically anisotropic, not commonly used
Fused quartz ($f = 100\,\text{GHz}$)	3.8	0.8×10^{-4}	0.01	Provide excellent dimensional stability, low thermal coefficient of expansion (5.7×10^{-6} in./in./°C); difficult to fabricate guide section
Z-cut quartz ($f = 100\,\text{GHz}$)	4.4	0.5×10^{-4}	0.01	
		Polymer Dielectrics [49–53]		
RT-Duroid 5880 ($f = 100\,\text{GHz}$)	2.2	9×10^{-4}	0.0026	Easy to machine and shape; Cu-clad substrate can be used as insular layer-cum-ground plane in insular image guide
Rexolite ($f = 10\,\text{GHz}$)	2.55	1×10^{-3}	—	

Teflon (PTFE)($f = 100\,GHz$)	2.07	2×10^{-4}	—	Flexible nonpolar polymers, low-loss materials; ideal for insular layers at lower millimeter wave frequencies; can be used as guide materials at higher millimeter wave frequencies up to about 200 GHz
Polyethylene ($f = 100\,GHz$)	2.3	3×10^{-4}	—	
TPX (poly-4 methylpentent-1) ($f = 100\,GHz$)	2.07	6×10^{-4}	—	Flexible nonpolar polymers, more lossy than Teflon and polyethylene
Polypropylene ($f = 100\,GHz$)	2.26	7×10^{-4}	—	
		Castable Dielectrics [50]		
Paraffin wax	2.22	6×10^{-4}	—	Thermoplastic, low melting point (50°C), low mechanical strength, easy to handle, useful for laboratory experimentation
Stycast resin 35 DA	3.4	2.7×10^{-3}	—	Ceramic-filled resin, stronger than paraffin wax
Custom High K-707	3–25	—	—	Better properties as compared with stycast resin

TABLE 1.2 *(Continued)*

Material	ε_r (approximate)	Loss Tangent $\tan\delta$	Thermal Conductivity K (W/cm/°C)	Remarks
		Semiconductor Materials [51, 56]		
Semi-insulating Si				
$\rho = 2 \times 10^3$–10^5 Ω-cm, $f = 10$ GHz [52, 56]	12	1×10^{-3}	0.9	High-resistivity property is used for low loss propagation in the guide; semiconductor active devices can be directly integrated; useful for monolithic circuits
$\rho = 8 \times 10^3$ Ω-cm, $f = 140$ GHz [51]	11.7	1.3×10^{-3}	—	
Semi-insulating GaAs				
$\rho = 10^7$–10^9 Ω-cm, $f = 10$ GHz [52, 56]	16	1.6×10^{-3}	0.3	
$\rho = 7.8 \times 10^7$ Ω-cm, $f = 140$ GHz [51]	12.9	0.5×10^{-3}	—	

		Other Dielectrics/Ferrites [57, 58]		
Boron nitride dielectric ($f = 60$ GHz)	3.97–4.1	—	—	Useful for passive and active circuit integration above 60 GHz [57]
Magnesium titanate ($f = 94$ GHz)	13, 16	—	—	Ceramic dielectric, guide material compatible for use with ferrite guide elements for the realization of ferrite control elements [58]
Lithium–zinc ferrite $4\pi M_s \approx 4100$ G ($f = 94$ GHz)	12.5	20	0.03	Ferrite guide sections in combination with dielectric guides form ferrite–dielectric composite guides; applications in phase shifters and circulators [58]
Nickel–zinc ferrite $4\pi M_s \approx 5000$ G ($f = 94$ GHz)	14.8	10	0.03	

millimeter wave ferrites, which are used in dielectric–ferrite composite guides for nonreciprocal components.

Ceramic Dielectrics Dielectric materials can be broadly classified into two categories: ceramic dielectrics [47–51] and polymer dielectrics [49–53]. Guides using ceramic dielectrics are normally formed using various machining methods. Alumina guides, however, can be fabricated in the green state prior to firing into final form, thereby enabling economical manufacture. Because of this advantage, alumina is the popularly used ceramic material for millimeter wave dielectric guides. Sapphire is basically anisotropic and hence is used only for special applications. Quartz, with its lower dielectric constant, offers more convenient sizes to fabricate than alumina. This is an advantage particularly at higher millimeter wave frequencies. Another important feature of quartz is its excellent dimensional stability.

Polymer Dielectrics Unlike ceramics, polymer dielectric can easily be cut, sheared, or machined to shape. Several types of low-loss polymers having low dielectric constants in the range 2–2.5 are now commercially available. RT-duroid 5880 (product of Roger Corporation), which is a popularly used substrate material in microstrip and fin line circuits, is suitable for dielectric guides also. The material is composed of inert glass microfiber or inert ceramic filler embedded in a matrix of polytetrafluoroethylene (PTFE). Thin sheets with copper cladding on one side can be used directly as insular layer-cum-ground plane in guides such as the insular image guide. Unclad sheets can easily be machined to form guide strips. Among the other polymer materials listed in Table 1.2, Teflon (PTFE) and polyethylene exhibit the lowest loss. These are ideally suited for use at higher frequencies, that is, typically above 60 GHz.

Castable Dielectrics Dielectrics that can be cast in place rather than requiring machining techniques to achieve the desired shape and size offer considerable fabricational convenience. The fabrication technique is similar to the well-known die casting process. Some of the useful castable dielectrics are paraffin wax ($\varepsilon_r = 2.22$), stycast resin ($\varepsilon_r = 3.4$), and custom high K-707 ($\varepsilon_r = 3$–25) [50]. Paraffin wax is a thermoplastic material. It has a low melting point (50°C) and small mechanical strength. However, since the material is easy to handle, it is convenient for laboratory experimentation. Stycast 35DA is a ceramic filled resin (product of Emerson and Cuming, USA) having greater mechanical strength than paraffin wax. Another material that has even better properties than stycast resin are the castable custom High-K 707 dielectrics (product of 3M Company, Singapore).

Dielectric Pastes for Thick-Film Process An important low-cost manufacturing process that is particularly suitable for frequencies above 60 GHz involves the use of low-loss dielectric paste and thick-film printing technique [54, 55]. The dielectric paste is normally composed of ceramic powder, binder material, and suitable chemical vehicles. In order to achieve large height to width ratios as required in a dielectric guide and to prevent spreading of paste during the setting and drying stages of the deposition process, the paste must possess high initial viscosity. A typical thick-film process for dielectric guide fabrication is as follows [54].

A high-quality gold film is first printed onto an alumina substrate. The dielectric guide pattern is etched onto a screen using photolithographic techniques. Screens with thick emulsions (typically 750 μm thick) are used to delineate the print pattern. The screen is then used to print the paste onto the gold-coated surface of the substrate using a thick-film printer. The waveguide pattern is then subjected to a drying cycle to drive off the organic solvents. This is followed by a firing cycle during which the retained organic materials are burnt off and the film is sintered so as to compact and adhere the guide firmly to the substrate.

Semiconductor Dielectrics For direct integration of active devices into a transmission guide structure, semiconductor guides offer the best potential. High-resistivity silicon (Si), which offers a resistivity in the range 2×10^3–10^5 Ω-cm and semi-insulating gallium arsenide (GaAs) possessing even higher resistivities in the range 10^7–10^9 Ω-cm have been used for millimeter waveguides [52, 56]. The high-resistivity property of these materials is important for achieving low-loss propagation in the guide. Like quartz, these materials are crystalline and can be ground or laser cut before being soldered to the ground plane.

Other Dielectrics/Ferrites Besides the aforementioned general types of dielectrics, there are other dielectrics with desirable properties for specific applications. Table 1.2 lists two such materials—boron nitride ($\varepsilon_r \approx 4$) and magnesium titanate ($\varepsilon_r = 13, 16$). Paul and Chang [57] have reported the potential use of boron nitride image guides for realizing integrated multifunctional modules including passive and active devices. The speciality of magnesium titanate dielectric is that its dielectric constant, either 13 or 16 ($\varepsilon_r = 13$ for Trans Tech type D-13 and $\varepsilon_r = 16$ for D-16) is compatible with ferrites for the realization of ferrite control devices using dielectric waveguide technique. For example, spinel ferrite—namely, nickel–zinc ferrite—has a dielectric constant of 12.5 and lithium–zinc ferrite has a dielectric constant of 14.8. These materials have high saturation magnetization ($4\pi M_s$) and low loss tangent (see Table 1.2) and have been used for realizing a 94-GHz circulator and phase shifter on a magnesium titanate dielectric waveguide [58].

1.6 INTEGRATION OF ACTIVE DEVICES

There are several approaches to active device implementation in dielectric integrated guides. These include (i) mounting of packaged semiconductor devices into the guide, (ii) attaching beam lead diodes using metal contacts, (iii) attaching chip diodes directly on top of a semiconductor dielectric guide, and (iv) monolithic implementation, that is, direct in situ integration of the active device into a semiconductor dielectric guide.

Most of the active device components in dielectric waveguide technology adopt the first two approaches. Mounting of packaged devices in open dielectric guides has an inherent problem in that any discontinuity arising out of mounting the device causes radiation. It is therefore important to introduce some type of

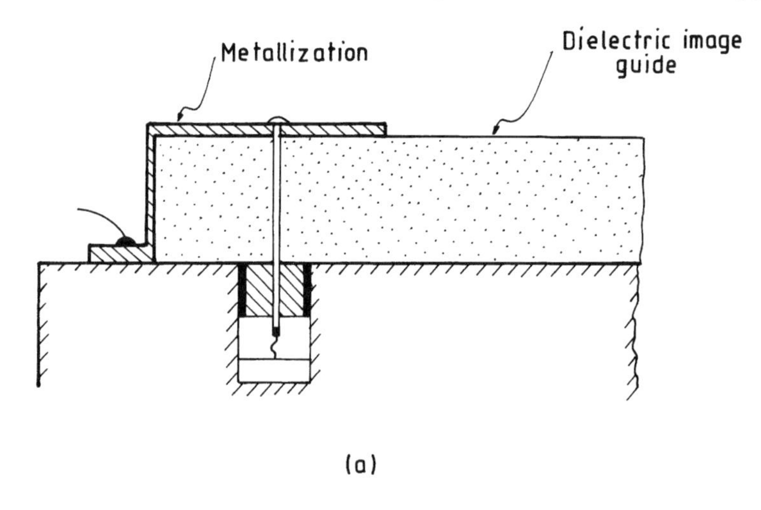

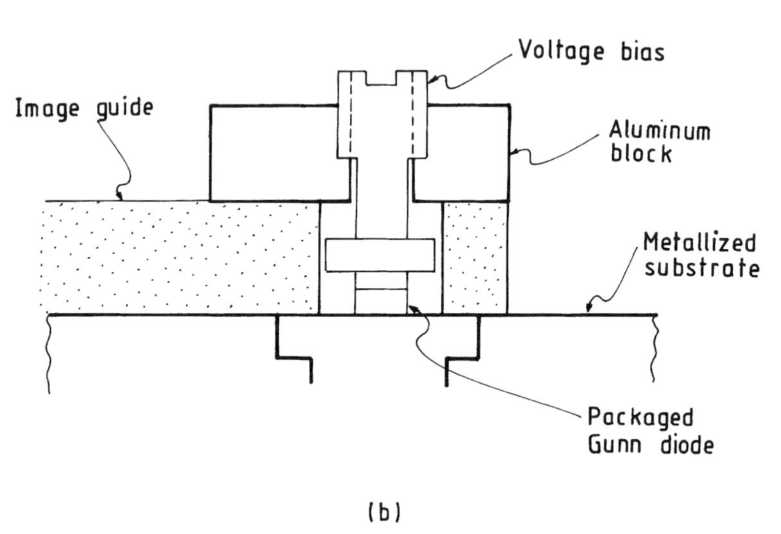

FIGURE 1.10 Techniques for mounting semiconductor devices in dielectric guides: (a) coaxial diode mounted in image guide (after Alywood and Williams [59]); (b) packaged Gunn diode mounted in image guide (after Chang [60]); (c) beam lead diode mounted in image guide (after Mittra et al. [61]); and (d) beam lead diode across slot in ground plane (after Solbach [62]).

metal shielding so as to prevent radiation and also to facilitate higher order mode suppression. One technique is to mount the packaged device in a hollow rectangular metal waveguide section and incorporate a mode launcher to convert the TE_{10} mode of the rectangular waveguide to the appropriate dominant mode of the dielectric guide. While this approach uses the well-proven device mounting technique, the introduction of waveguide section disturbs the planar form of the circuit.

Figure 1.10 illustrates some of the other diode mounting schemes that retain the planar form of the dielectric guide. Figure 1.10(a) shows a typical coaxial detector diode mounted in the ground plane with its center conductor protruding through the dielectric to serve as a probe coupler to the dielectric image guide

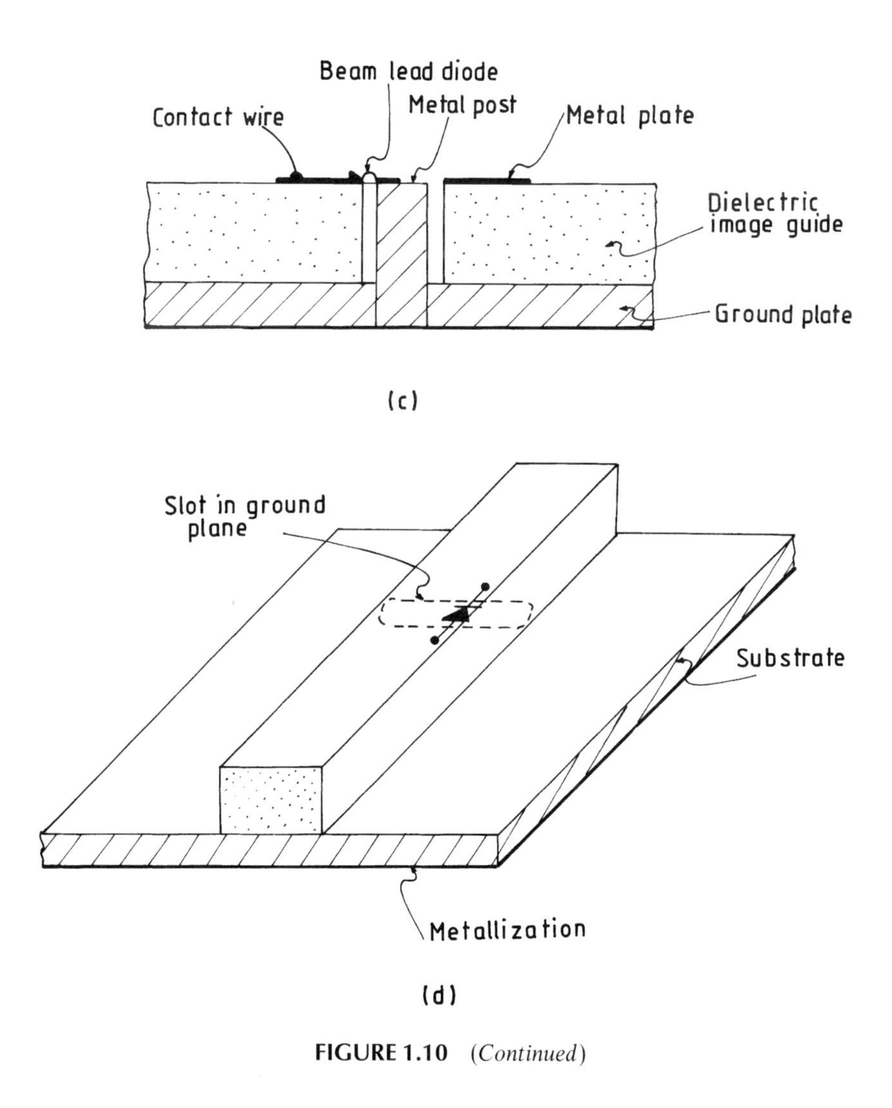

FIGURE 1.10 *(Continued)*

[59]. The probe tip is connected to a metallized layer on top of the guide as shown in the figure. Figure 1.10(b) shows a Gunn diode mounted in a hole within a quartz image guide [60]. The image guide width near the diode is increased and the outer surfaces of the dielectric are metallized to form a cavity around the diode. The Gunn diode is screwed into the ground plane for proper thermal contact.

Schemes for attaching beam lead diodes to dielectric guides are shown in Figure 1.10(c) and (d) [61, 62]. In the arrangement shown in Figure 1.10(c), one end of the beam lead diode is connected to the ground plane through a metal post piercing through the dielectric. The other terminal of the diode is connected to a metallized pattern on top of the guide and surrounding the diode [61]. In the second arrangement shown in Figure 1.10(d), a slot is etched on the ground plane of an insular image guide [62]. The diode is then mounted across the slot.

Semiconductor dielectric guides offer the potential for monolithic integration of both active and passive devices. Monolithic realization of millimeter wave dielectric guide circuits, however, appears to be a technology of the future as there is hardly any literature on this subject at present. One unique device that has been reported on semiconductor dielectric guide is a 70-GHz electronically variable p-i-n diode phase shifter [56]. The structure consists of a triangular shaped p-i-n diode located on top of a semiconductor dielectric guide. When a forward bias is applied to the diode, the p-i-n diode exhibits high conductivity starting from the topmost edge of the triangular structure. As the current is increased, the conducting region extends downward. This causes a reduction in the effective height of the guide, resulting in a phase change.

The above discussion is intended to provide an exposure to the basic features of active device implementation in dielectric guides. A detailed description of the various schemes is included in Chapter 12 on semiconductor components.

1.7 BASICS OF DIELECTRIC RESONATORS

Finite length sections of dielectric guides act as resonating elements. High-Q dielectric resonators find practical applications in millimeter wave filters and oscillators [51, 63–68]. Several types of temperature stable, low-loss ceramics having ε_r in the range 15–100 are now available for use as resonator materials [63–69]. In order to realize resonators that are small but convenient to handle, materials having higher values of ε_r in the range 30–50 are required at microwave frequencies, whereas at millimeter wave frequencies, lower values of ε_r in the range 10–30 are suitable. Table 1.3 lists some of the useful dielectric materials for millimeter wave applications. The properties that are of vital interest for the choice of resonator are listed. They are ε_r, $\tan\delta$ (or product of Q-factor and frequency), and temperature coefficient of resonant frequency (τ_f). The parameter τ_f is defined as $\tau_f = (1/f)(df/dT)$, where f is the resonant frequency and T is the temperature.

TABLE 1.3 Properties of Materials for Millimeter Wave Dielectric Resonators

Material	ε_r	Qf (THz) or $\tan\delta$	Temperature Coefficient of Resonant Frequency, τ_f (ppm/°C)	Reference
$MgTiO_3$	16	$\tan\delta = 2 \times 10^{-4}$	100	[68]
$(Ca, Sr)(Ba, Zr)O_3$	29–32	Qf (THz) = 33	− 50 to 50	[64]
$CaZr_{0.985}Ti_{0.015}O_3$	29	Qf (THz) = 13.2	2	[68]
$CaZrO_3$	30	$\tan\delta = 3 \times 10^{-4}$	40	[68]
$Ba(Zn_{1/3}Nb_{2/3})O_3$ $Ba(Zn_{1/3}Ta_{2/3})O_3$	31.7	Qf (THz) = 90	4	[65]
$Ba_3MgTa_2O_9$	24	$\tan\delta = (3.8 \pm 0.4) \times 10^{-5}$	3.9 ± 1.6	[69]
$Ba(SnMgTa)O_3$	24.4	$\tan\delta = (3.7 \pm 0.4) \times 10^{-5}$	0.1 ± 0.8	[69]
$Ba(MgZrTaNb)O_3$	27	$\tan\delta = (3.6 \pm 0.4) \times 10^{-5}$	0.4 ± 0.7	[69]

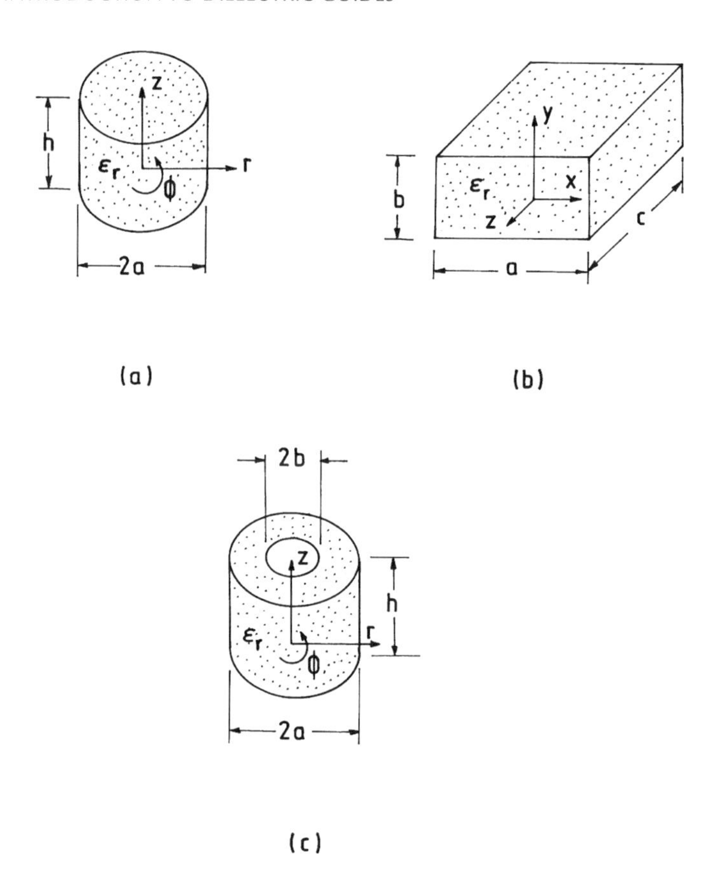

FIGURE 1.11 Isolated dielectric resonators: (a) cylindrical, (b) rectangular, and (c) ring.

Isolated Resonator Figure 1.11 shows three useful shapes of dielectric resonators: cylindrical, rectangular, and ring. The cylindrical and rectangular resonators are the most basic forms and the ring resonator is a variant of the cylindrical resonator. The most commonly used at microwave frequencies is a cylindrical disk in which the height h is smaller than the diameter $2a$. The rectangular resonator, which assumes slightly larger size than the cylindrical resonator for the same dielectric, is preferred at millimeter wave frequencies.

An isolated dielectric resonator in free space can resonate in various modes depending on the shape, dimensions, and the value of ε_r. Each mode represents a particular resonant frequency. For any given mode, the higher the value of ε_r, the larger will be the concentration of fields within the dielectric.

For an isolated cylindrical disk resonator, the dominant mode is designated as the $TE_{01\delta}$ mode with $\delta < 1$. The distribution $\delta < 1$ denotes that the field variation is less than half a cycle in the z-direction. For this mode, only three components—namely, E_ϕ, H_r, and H_z—exist. Inside the resonator, these fields assume the forms

$$E_{\phi} = AJ_1(ur)\cos(\beta z) \tag{1.24a}$$

$$H_z = BJ_0(ur)\cos(\beta z) \tag{1.24b}$$

$$H_r = CJ_1(ur)\sin(\beta z) \tag{1.24c}$$

where u and β are the radial and axial wavenumbers inside the resonator. They are related by

$$u^2 + \beta^2 = k_0^2 \varepsilon_r \tag{1.25}$$

The electric field lines are circular and the associated magnetic field is concentrated essentially along the z-axis of the resonator as illustrated in Figure 1.12(a). Outside the resonator the fields decay exponentially with distance. A typical plot of the electric field intensity variation in the equatorial plane is shown in Figure 1.12(b).

The other resonant modes of the cylindrical disk resonator are the TM-type and the hybrid EH- and HE-type modes (for example $TM_{01\delta}$, $EH_{11\delta}$, and

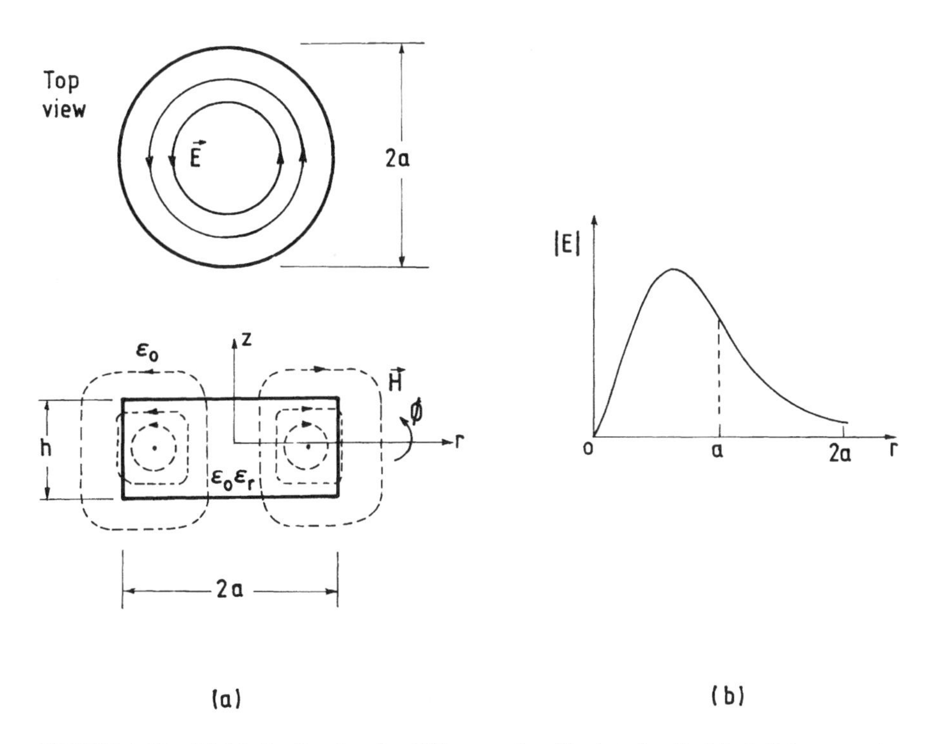

FIGURE 1.12 Field distribution for $TE_{01\delta}$ mode of isolated cylindrical disk resonator: (a) E- and H-lines and (b) typical variation of electric field intensity versus radial distance.

$HE_{11\delta}$). Maximum separation between the resonant frequency of the $TE_{01\delta}$ mode and the next higher order $HE_{11\delta}$ mode occurs when the aspect ratio $h/2a$ of the resonator is around 0.4. For this aspect ratio, the resonant frequency of $HE_{11\delta}$ mode occurs at a frequency that is about 1.3 times the value of the $TE_{01\delta}$ mode. By drilling a small hole through the center of the resonator and forming a ring resonator (Fig. 1.11(c)), better modal separation between the $TE_{01\delta}$ mode and the $HE_{11\delta}$ mode can be achieved [70].

The resonant modes of a rectangular dielectric resonator are normally derived on the basis of the modes of an infinitely long rectangular dielectric waveguide. Thus, corresponding to the E^y_{mn} and E^x_{mn} modes of a rectangular guide, the resonant modes of a rectangular resonator are designated as $E^y_{mn\ell}$ and $E^x_{mn\ell}$ modes, respectively. The subscripts m, n, and ℓ refer to the number of extrema of electric field components inside the resonator in the x-, y-, and z-directions, respectively. The fields inside the resonator assume either sinusoidal or cosinusoidal variation, and outside the resonator they decay exponentially. For most millimeter wave circuit applications, it is practical to use E^y_{111} or the higher order modes because the E^y_{110} mode requires the z-directed dimension c (see Fig. 1.11(b)) to be extremely small [71]. The dominant field components of the E^y_{111} mode are E_y, H_x, and H_z. Figure 1.13 shows the electric field distribution for this mode. The distribution corresponds to a standing wave pattern inside the resonator in all three directions (x, y, and z).

Image and Insular Image Dielectric Resonators Dielectric resonators placed on a ground plane are compatible for use with dielectric image guide and may be termed "dielectric image resonators." Resonators placed on a large dielectric sheet of low dielectric constant and backed by a ground plane are compatible for use with insular image guides and may be termed "insular image resonators." Figure 1.14 shows image and insular image resonator structures. Cylindrical and rectangular resonator elements for use with other dielectric guide structures such

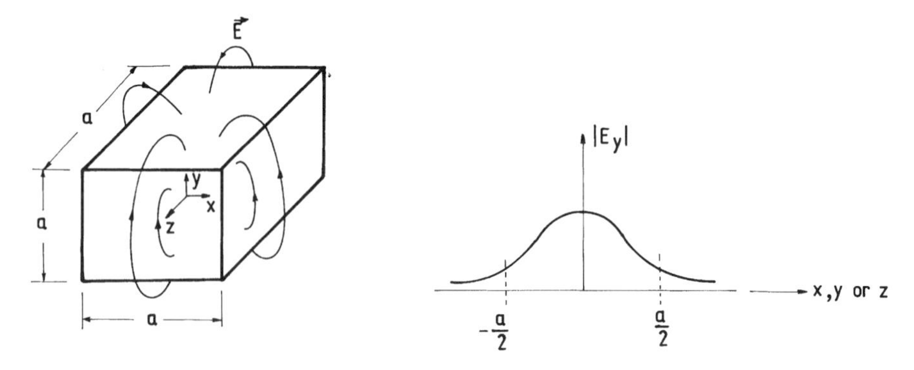

FIGURE 1.13 Electric field distribution for the E^y_{111} mode of a rectangular resonator (cuboid).

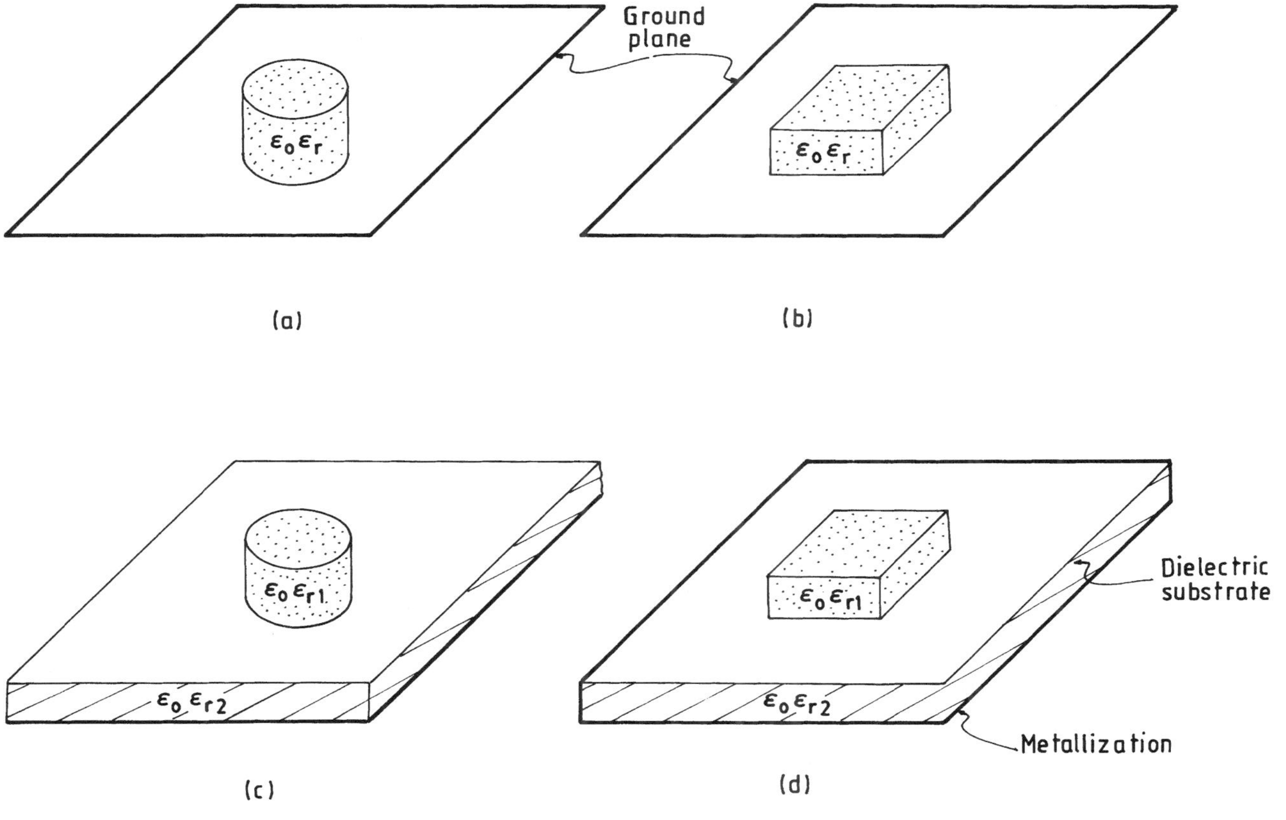

FIGURE 1.14 Examples of dielectric resonator structures for millimeter wave integrated circuits: (a) cylindrical dielectric image resonator, (b) rectangular dielectric image resonator, (c) insular image cylindrical dielectric resonator, and (d) insular image rectangular dielectric resonator.

as the nonradiative guide and the inverted strip guide can be similarly configured. The analysis and resonant characteristics of cylindrical and rectangular resonators in isolated as well as dielectric integrated guide environments are covered in Chapter 9.

1.8 APPLICATION POTENTIAL

Dielectric integrated guide components and subsystems have been under development since the early 1970s. Various dielectric guide structures including the H- and groove guides have been experimented with to realize a range of millimeter wave circuits. The most widely employed guide structures in component development so far have been the image guide (Fig. 1.7(a)) for its simplicity, the insular image guide (Fig. 1.7(b)) for its low-loss property, and the nonradiative guide (Fig. 1.1(h)) for its nonradiative character at bends and discontinuities. The various passive and active components constituting millimeter wave transmitter and receiver systems have been realized in one or the other of the dielectric guide configurations (e.g., see [52], [55–62], [72–75]). The emphasis has been in the frequency range 30 to about 120 GHz.

It is worthwhile to note that for applications in the frequency range 30–120 GHz there are two other technologies—namely, the suspended stripline and fin line. These technologies have already matured, leading to realization of systems. The dielectric guide technology, on the other hand, is yet to mature. The best potential that this technology can offer appears to be at frequencies above 60 GHz and the factors in favor of dielectric guides are low loss, convenient size, low cost, and ease of manufacture. Low loss is possible because of the elimination of current carrying strips/fins and the availability of low-loss dielectric materials ($\tan\delta < 10^{-4}$) to form the guides. Around 6 GHz and up to about 100 GHz, the guide size is of the order of a few millimeters in width and a fraction of a millimeter in thickness and hence is convenient to fabricate using the available machining/casting techniques. Development of low-loss dielectric pastes and the use of thick-film technology to print the planar waveguide circuit appear to be the most viable solutions for realizing dielectric guide components beyond 100 GHz. The upper frequency limit depends on the availability of suitable dielectric paste and the accuracy with which the guides can be printed. Another option for applications beyond 100 GHz is to use the dielectric H-guide and groove-guide structures. The groove guide, in particular, is emerging as an important guide structure for frequencies up to 300 GHz [74–76].

One of the most useful features of dielectric guide techniques is the convenience that it offers for realizing high-performance antennas as an integral part of the circuit. Surface wave antennas in the form of uniform and tapered dielectric rods and leaky wave antennas in the form of periodic structures have been reported [77–79]. Another important application of dielectric guides is in array antennas as feed structures. Such antennas include, for example, an array of slots or grooves cut in the ground plane of an image guide [80] and an array of dielectric radiating elements arranged on the insular layer and adjacent to the guiding strip of an insular image guide [81].

Incorporation of active devices in dielectric guides is more difficult than in suspended striplines or fin lines. The use of hybrid structures that use both dielectric guides and printed transmission lines/slotlines offers a viable alternative. The best potential of dielectric guide techniques can be utilized by using high-resistivity semiconductor dielectric guides and adopting monolithic fabrication to integrate both active and passive devices. Another attractive property of semiconductor guides is that its conductive and dielectric properties can be controlled optically. This feature offers the possibility for realizing a class of dynamically controlled devices such as switches, phase shifters, and attenuators.

PROBLEMS

1.1 **(a)** What are the dominant modes of propagation in (i) rectangular metal waveguide, (ii) microstrip line, (iii) suspended stripline, and (iv) unilateral fin line?

(b) Draw typical electric and magnetic field lines in the cross-sectional plane of each of the above transmission structures for the dominant propagating modes.

1.2 **(a)** What are the limitations of rectangular metal waveguides at millimeter wave frequencies and how are they overcome in the fin line?

(b) What is the practical frequency range of operation of fin lines and what are the limitations beyond this range?

1.3 **(a)** What are the limitations of microstrip at millimeter wave frequencies and how are they overcome in the suspended stripline?

(b) What is the practical frequency range of operation for suspended striplines and what are the limitations beyond this range?

1.4 Consider a homogeneous dielectric slab of thickness $2d$ and relative dielectric constant ε_r as shown in Figure 1.15. The slab extends to infinity in the x- and z-directions.

(a) Discuss the principle of wave guidance in the slab guide.

(b) Distinguish clearly between the slab guided mode and radiation mode.

1.5 **(a)** For the dielectric slab guide shown in Figure 1.15, obtain complete field expressions for TM modes. Assume propagation in the x-direction. Consider solution in terms of symmetric and antisymmetric modes.

(b) Obtain characteristic equations for the above symmetric and antisymmetric cases.

[*Note*: Compare with Eq. (1.6) obtained by assuming propagation in the z-direction.]

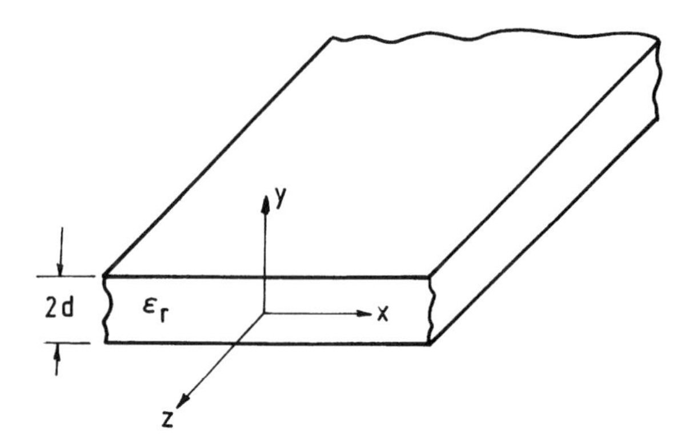

FIGURE 1.15 Dielectric slab guide.

1.6 Repeat problem 1.5 for TE-symmetric and TE-antisymmetric modes.

1.7 Discuss the phenomena of cutoff in a slab guide and compare it with that of a parallel-plate guide.

1.8 Consider the TM-mode solution of the slab guide obtained in Problem 1.5(b). If in Figure 1.15, $\varepsilon_r = 2.56$ and $2d = 4$ mm, what slab guide modes will propagate unattenuated at an operating frequency of 30 GHz. Calculate the cutoff frequencies of these TM modes. Determine the propagation constants of the propagating modes by numerically solving the characteristic equation.

1.9 For the dielectric slab shown in Figure 1.15, what should be the range of values of (relative) dielectric constant ε_r so that waves incident from any oblique angle within the slab propagate as slab guided modes. Do these values depend on thickness $2d$ of the dielectric? Justify your answer.

1.10 Consider a rectangular dielectric guide as shown in Figure 1.16.

(a) Draw typical electric and magnetic field lines for the E_{11}^x mode.

(b) If a ground plane is introduced in the xz-plane at $y = 0$, how will the field lines get modified in the resulting image guide? Draw the field lines.

1.11 **(a)** Discuss the principle of wave guidance in the circular dielectric guide shown in Figure 1.17.

(b) For the dominant (HE_{11}) mode, of the above guide, show that for small radius a, the characteristic Eq. (1.18) reduces to

$$k_1 a = \sqrt{\frac{(\varepsilon_r + 1)}{K_0(k_2 a)}}$$

where $K_0(x)$ is the modified Bessel function of the second kind.

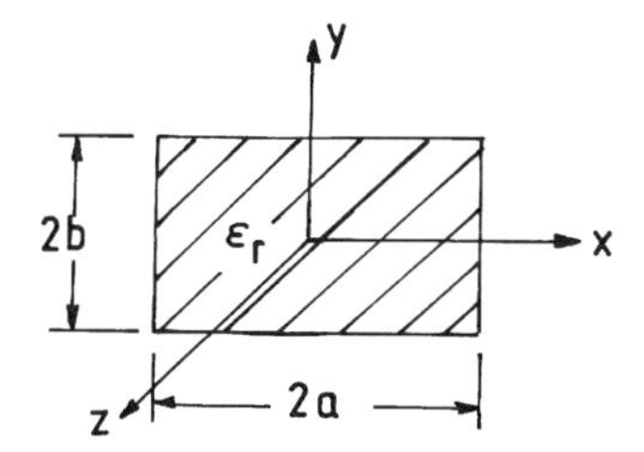

FIGURE 1.16 Rectangular dielectric guide.

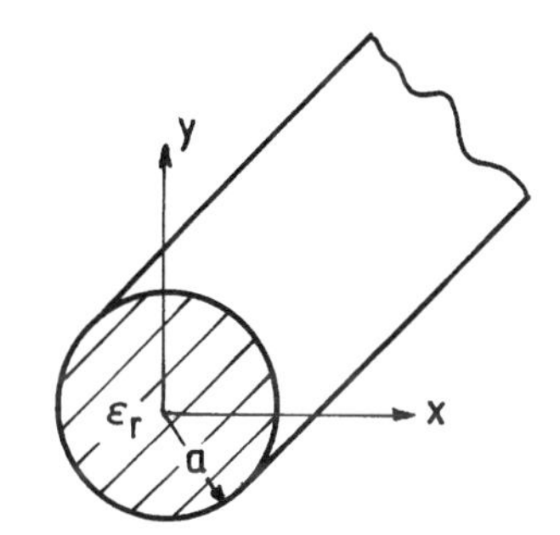

FIGURE 1.17 Circular dielectric guide.

1.12 What are the essential features of dielectric integrated guides that make them suitable for millimeter wave applications? Why are bare dielectric guides not suitable?

1.13 Enumerate the differences between dielectric integrated guides for millimeter wave integrated circuits and planar dielectric guides for optical integrated circuits.

REFERENCES

1. D. Hondros and P. Debye, Electromagnetische Wellen an Dielecktrischen Drahten. *Ann. Phys.*, **32(8)**, 465–476, 1910.
2. G. C. Southworth, Hyperfrequency waveguides—general considerations and experimental results. *Bell. Syst. Tech. J.*, **15**, 284–309, 1936.
3. J. R. Carson, S. P. Mead, and S. A. Schelkunoff, Hyperfrequency waveguides—mathematical theory. *Bell. Syst. Tech. J.*, **15**, 310–333, 1936.
4. D. G. Kiely, *Dielectric Aerials*, Methuen and Co., London, 1953.
5. R. Chatterjee, *Dielectric and Dielectric-Loaded Antennas*, Research Studies Press, Letchworth, UK, 1985.
6. E. Snitzer, Cylindrical dielectric waveguide modes. *J. Opt. Soc. Am.*, **51(5)**, 491–498, May 1961.

7. E. Snitzer and H. Osterberg, Observed dielectric waveguide modes in the visible spectrum. *J. Opt. Soc. Am.*, **51(5)**, 499–505, May 1961.
8. K. C. Kao and G. A. Hockham, Dielectric fibre surface waveguides for optical frequencies. *Proc. IEE*, **113**, 1151–1158, July 1966.
9. N. S. Kapany, *Fibre Optics—Principles and Applications*, Academic Press, New York, 1967.
10. S. E. Miller, E. A. J. Marcatili, and T. Li, Research towards optical fibre transmission system. *Proc. IEEE*, **61**, 1703–1751, 1973.
11. P. J. B. Clarricoats, Theory of optical fibre waveguides—a review. *Progress in Optics*, Vol. 14, E. Wolf, Ed., North-Holland, Amsterdam, 1976.
12. D. Marcuse, *Light Transmission Optics*, Van Nostrand Reinhold, Princeton, NJ, 1972.
13. D. Marcuse, *Theory of Dielectric Optical Waveguides*, Academic Press, New York, 1974.
14. N. S. Kapany and J. J. Burke, *Optical Waveguides*, Academic Press, New York, 1972.
15. J. A. Arnaud, *Beam and Fibre Optics 1*, Academic Press, New York, 1976.
16. M. S. Sodha and A. K. Ghatak, *Inhomogeneous Optical Waveguides*, Plenum Press, New York, 1977.
17. J. E. Midwinter, *Optical Fibres for Transmission*, Wiley, New York, 1979.
18. G. H. Qwyang, *Foundations of Optical Waveguides*, Edward Arnold, London, 1981.
19. S. E. Miller, Integrated optics: an introduction, *Bell Syst. Tech. J.*, **48**, 2059–2069, 1969.
20. T. Tamir, *Integrated Optics*, Springer Verlag, Berlin, 1975.
21. L. D. Hutcheson (Ed.), *Integrated Optical Circuits and Components*, Marcel Dekker, New York, 1987.
22. C. Lin (Ed.), *Optoelectronic Technology and Lightwave Communication Systems*, Van Nostrand Reinhold, New York, 1989.
23. A. K. Ghatak and K. Thyagarajan, *Optical Electronics*, Cambridge University Press, Cambridge, UK, 1989.
24. T. Tamir (Ed.), *Guided Wave Optoelectronics*, Springer Verlag, Berlin, 1988.
25. H. G. Unger, *Planar Optical Waveguides and Fibres*, Clarendon Press, Oxford, 1977.
26. D. D. King, Dielectric image lines. *J. Appl. Phys.*, **23**, 699–700, June 1952.
27. D. D. King, Properties of dielectric image lines. *IRE Trans. Microwave Theory Tech.*, **MTT-3**, 75–81, March 1955.
28. D. D. King, Circuit components in dielectric image lines. *IRE Trans. Microwave Theory Tech.*, **MTT-3**, 35–39, Dec. 1955.
29. M. T. Weiss and E. M. Gyorgy, Low loss dielectric waveguides. *IRE Trans. Microwave Theory Tech.*, **MTT-2**, 38–44, Sept. 1954.
30. S. P. Schlesinger and D. D. King, Dielectric image lines. *IRE Trans. Microwave Theory Tech.*, **MTT-6**, 291–299, July 1958.
31. D. D. King and S. P. Schlesinger, Losses in dielectric image lines. *IRE Trans. Microwave Theory Tech.*, **MTT-5**, 31–35, Jan. 1957.
32. J. C. Wiltse, Some characteristics of dielectric image lines at millimeter wavelengths. *IRE Trans. Microwave Theory Tech.*, **MTT-7**, 65–70, Jan. 1959.

33. F. Sobel, F. L. Wentworth, and J. C. Wiltse, Quasi-optical surface waveguide and other components for the 100–300 Gc Region. *IRE Trans. Microwave Theory Tech.*, **MTT-9**, 512–518, Nov. 1961.
34. R. M. Knox and P. P. Toulios, Integrated circuits for millimetre through optical frequency range. Presented at Symposium on Submillimeter Waves, New York, March 1970.
35. K. C. Gupta, R. Garg, and I. J. Bahl, *Microstrip Lines and Slot Lines*, Artech House, Norwood, MA, 1979.
36. T. C. Edwards, *Foundations for Microstrip Circuit Design*, Wiley, New York, 1981.
37. B. Bhat and S. K. Koul, *Stripline-like Transmission Lines for Microwave Integrated Circuits*, Wiley Eastern, New Delhi, 1989.
38. B. Bhat and S. K. Koul, *Analysis, Design and Applications of Fin Lines*, Artech House, Norwood, MA, 1987.
39. F. A. Benson and F. J. Tischer, Some guiding structures for millimetre waves (IEE Review). *IEE Proc.*, **131(A7)**, 429–449, Sept. 1984.
40. R. E. Collin, *Field Theory of Guided Waves*, McGraw-Hill, New York, 1960.
41. E. A. J. Marcatili, Dielectric rectangular waveguide and directional coupler for integrated optics. *Bell Syst. Tech. J.*, **48**, 2071–2132, Sept. 1969.
42. P. J. B. Clarricoats, Propagation along unbounded and bounded dielectric rods, Part I—Propagation along an unbounded dielectric rod. *Proc. IEE.*, **108(C)**, 170–176, 1961.
43. S. K. Chatterjee and V. Subramanyam, Circular cylindrical rod waveguide. *J. Ind. Inst. Sci.*, **50**, 258–493, 1968.
44. S. T. Peng and A. A. Oliner, Leakage and resonance effects on strip waveguides for integrated optics. *Trans. Electron. Commun. Eng. (Japan)* (Special Issue on Optics and Optical Fibre Communications), **E61**, 151–154, Mar. 1978.
45. S. T. Peng and A. A. Oliner, Guidance and leakage properties of a class of open dielectric waveguides: Part I—Mathematical formulations. *IEEE Trans. Microwave Theory Tech.*, **MTT-29**, 843–854, Sept. 1981.
46. A. A. Oliner et al., Guidance and leakage properties of a class of open dielectric waveguides: Part II—New physical effects. *IEEE Trans. Microwave Theory Tech.*, **MTT-29**, 855–869, Sept. 1981.
47. R. M. Knox, Dielectric waveguide microwave integrated circuits—an overview. *IEEE Trans. Microwave Theory Tech.*, **MTT-24**, 806–814, Nov. 1976.
48. J. R. Birch and T. J. Parker, Dispersive fourier transform spectrometry, *Infrared and Millimetre Waves*, Vol. 2, K. J. Button (Ed.), Academic Press, New York, 1979.
49. C. Yeh, F. I. Shimabukuro, and J. Chu, Dielectric ribbon waveguide: an optimum configuration for ultra-low-loss millimetre/submillimetre dielectric waveguide. *IEEE Trans. Microwave Theory Tech.*, **MTT-38**, 691–701, June 1990.
50. K. Solbach, The fabrication of dielectric image lines using casting resins and the properties of the lines in the millimetre-wave range. *IEEE Trans. Microwave Theory Tech.*, **MTT-24**, 879–881, Nov. 1976.
51. G. B. Morgan, Dielectric resonators for circuits at short millimetre wavelengths. *Microwave J.*, **29**, 107–115, July 1986.
52. M. M. Chrepta and H. Jacobs, Millimetre-wave integrated circuits cost less using dielectric waveguides. *Microwave J.*, **17**, 45–47, Nov. 1974.

53. M. N. Afsar, Precision dielectric measurements of nonpolar polymers in the millimetre wavelength range. *IEEE Trans. Microwave Theory Tech.*, **MTT-33**, 1410–1415, Dec. 1985.

54. M. R. Inggs and N. Williams, Thick-film fabrication techniques for millimetre-wave dielectric waveguide integrated circuits. *Electron. Lett.*, **16(7)**, 245–247, Mar. 1980.

55. R. V. Gelsthorpe, N. Williams, and N. M. Davey, Dielectric waveguide; a low cost technology for millimetre wave integrated circuits. *Radio Electron. Eng.*, **52(11/12)**, 522–528, Nov./Dec. 1982.

56. H. Jacobs and M. M. Chrepta, Electronic phase shifter for millimetre-wave semiconductor dielectric integrated circuits. *IEEE Trans. Microwave Theory Tech.*, **MTT-22**, 411–417, Apr. 1974.

57. J. A. Paul and Y. W. Chang, Millimetre wave image guide integrated passive devices. *IEEE Trans. Microwave Theory Tech.*, **MTT-26**, 751–754, Oct. 1978.

58. R. W. Babbitt and R. A. Stern, Millimetre wave ferrite devices. *IEEE Trans. Magnet.*, **MAG-18(6)**, 1592–1594, Nov. 1982.

59. M. J. Aylward and N. Williams, Feasibility studies of insular guide millimetre wave integrated circuits. AGARD Conference on Millimetre and Submillimetre Wave Propagation and Circuits, Munchen, Sept. 1978, Conf. Reprint No. 245, pp. 30.1–30.11.

60. Y. W. Chang, Millimetre-wave (W-band) quartz image guide Gunn oscillator. *IEEE Trans. Microwave Theory Tech*, **MTT-31**, 194–199, Feb. 1983.

61. R. Mittra, N. Deo, and B. Kirkwood, Active integrated devices on dielectric substrate for millimetre-wave applications. *IEEE MTT-S Int. Microwave Symp. Digest*, 220–221, May 1979.

62. K. Solbach, Millimetre wave dielectric image line detector circuit employing etched slot structure. *IEEE Trans. Microwave Theory Tech.*, **MTT-29**, 953–957, Sept. 1981.

63. N. A. McDonald and M. L. Majewski, Dielectric resonators and their applications. *J. Electrical Electr. Eng. Australia-IE Aust. & IREE Aust.*, **1(1)**, 54–60, Mar. 1981.

64. J. K. Plourde and C. L. Ren, Application of dielectric resonators in microwave components. *IEEE Trans. Microwave Theory Tech.*, **MTT-29**, 754–770, Aug. 1981.

65. Y. Tokumitsu, et al., A 50 GHz MIC transmitter/receiver using a dielectric resonator oscillator. *IEEE MTT-S Int. Microwave Symp. Digest*, 228–230, July 1982.

66. D. Kajfez and P. Guillon (Eds.), *Dielectric Resonators*, Artech House, Norwood, MA, 1986.

67. M. Dydyk, Apply high *Q*-resonators to mm-wave microstrip. *Microwaves*, **19**, 62–63, Dec. 1980.

68. G. B. Morgan, Temperature compensated, high permittivity dielectric resonators for millimetre wave systems. *Int. J. Infrared Millimetre Waves*, **5**, 1–11, Jan. 1984.

69. Y. Kobayashi and S. Nakayama, Design charts for dielectric rods and ring resonators. *IEEE MTT-S Int. Microwave Symp. Digest*, 241–244, 1986.

70. Y. Kobayashi and M. Miura, Optimum design of shielded dielectric rod and ring resonators for obtaining the best mode separation. *IEEE MTT-S Int. Microwave Symp. Digest*, 184–186, June 1984.

71. C. Chang and T. Itoh, Resonant characteristics of dielectric resonators for millimetre wave integrated circuits, *Arch. Elek. Ubertragung.*, **33**, 141–144, Apr. 1979.

72. Y. Shen, D. M. Xu, and C. Ling, The design of an ultra-broad-band 3 dB coupler in dielectric waveguide. *IEEE Trans. Microwave Theory Tech.*, **MTT-38**, 785–787, June 1990.

73. T. Itoh, Open guiding structures for MMW integrated circuits. *Microwave J.*, **25**, 113–125, Sept. 1982.

74. J. Meissner, Groove-guide directional couplers with improved bandwidth. *Electron. Lett.*, **20(17)**, 701–703, Aug. 1984.

75. R. Vahldieck and J. Ruxton, A broadband groove guide coupler for millimetre-wave applications. *IEEE MTT-S Int. Microwave Symp. Digest*, 349–352, 1987.

76. P. Lampariello and A. A. Oliner, Theory and design considerations for millimetre-wave leaky groove guide antenna. *Electron. Lett.*, **19(1)**, 18–20, Jan. 1983.

77. Y. Shiau, Dielectric rod antenna for millimetre-wave integrated circuits. *IEEE Trans. Microwave Theory Tech.*, **MTT-24**, 869–872, Nov. 1976.

78. T. Itoh, Application of gratings in a dielectric waveguide for leaky-wave antenna and band-reject filters. *IEEE Trans. Microwave Theory Tech.*, **MTT-25**, 1134–1138, Dec. 1977.

79. S. Kobayashi, et al., Dielectric rod leaky-wave antenna for millimetre-wave applications. *IEEE Trans. Antennas* Propag., **AP-29**, 822–824, Sept. 1981.

80. T. Hori and T. Itanami, Circularly polarized linear array antenna using a dielectric image line. *IEEE Trans. Microwave Theory Tech.*, **MTT-29**, 967–970, Sept. 1981.

81. M. T. Birand and R. V. Gelsthorpe, Experimental millimetric array using dielectric radiators fed by means of dielectric waveguide. *Electron. Lett.*, **17**, 633–635, Sept. 1981.

CHAPTER TWO

Analysis of Dielectric Integrated Guides: Approximate Methods

2.1 INTRODUCTION

Several analytical techniques, both approximate and rigorous, are available for studying the propagation characteristics, field distribution, loss, and so on of dielectric integrated guides. This chapter deals with the analysis of guides using two important approximate techniques: (i) the effective dielectric constant (EDC) method [1–4] and (ii) the microwave network analysis in conjunction with the transverse resonance method [3–6]. Accurate analyses of guides using a rigorous mode-matching technique and numerical methods are covered in Chapter 3.

Both the EDC and the microwave network analysis techniques are used primarily for determining the dispersion characteristics of dielectric integrated guides. They yield values of propagation constants that are reasonably accurate and adequate for most practical purposes. Between these two techniques, the EDC method is simpler, both analytically and computationally, and has been applied to a larger number of guide structures, whereas the network analysis technique is known to be more accurate particularly at lower frequencies. In the following sections, we first describe the EDC method of analysis and derive a general set of characteristic equations for multilayer slab guide models. Characteristic equations for a number of horizontal slab guides and vertical layered guides are listed in Appendices 2A and 2B, respectively, so that they can readily be used for a large number of practical guides. The microwave network approach is described next. This method involves the use of equivalent network representation of the guide structure and application of transverse resonance conditions. As an illustration, the method is applied to an image guide and its dispersion relation is derived.

2.2 ELECTROMAGNETIC FIELDS IN TERMS OF POTENTIAL FUNCTIONS

2.2.1 Vector Potential Functions

In this section, we derive general expressions for the electromagnetic fields in terms of vector potential functions [7]. We begin with Maxwell's equations for a homogeneous source-free region:

$$\nabla \times \mathbf{E} = -j\omega\mu\mathbf{H} \tag{2.1}$$

$$\nabla \times \mathbf{H} = j\omega\varepsilon\mathbf{E} \tag{2.2}$$

Since $\nabla \cdot \mathbf{E} = 0$ and $\nabla \cdot \mathbf{H} = 0$, we may express $\mathbf{E}$ and $\mathbf{H}$ in the form

$$\mathbf{E} = -\nabla \times \mathbf{F} \tag{2.3}$$

$$\mathbf{H} = \nabla \times \mathbf{A} \tag{2.4}$$

where $\mathbf{F}$ and $\mathbf{A}$ are called the electric vector potential and magnetic vector potential, respectively. Considering (2.1) in conjunction with (2.4), we can write

$$\nabla \times (\mathbf{E} + j\omega\mu\mathbf{A}) = 0 \tag{2.5}$$

or

$$\mathbf{E} + j\omega\mu\mathbf{A} = -\nabla\psi^a \tag{2.6}$$

where ψ^a is an arbitrary electric scalar potential. Substituting for $\mathbf{E}$ and $\mathbf{H}$ from (2.6) and (2.4) in (2.2), we obtain

$$\nabla \times \nabla \times \mathbf{A} = j\omega\varepsilon(-j\omega\mu\mathbf{A} - \nabla\psi^a) \tag{2.7}$$

or

$$\nabla(\nabla \cdot \mathbf{A}) - \nabla^2\mathbf{A} = k^2\mathbf{A} - j\omega\varepsilon\nabla\psi^a \tag{2.8}$$

where

$$k^2 = \omega^2\mu\varepsilon \tag{2.9}$$

Since ψ^a is arbitrary, we may choose

$$\nabla \cdot \mathbf{A} = -j\omega\varepsilon\psi^a \tag{2.10}$$

With this condition in (2.6), the expression for $\mathbf{E}$ can be written as

$$\mathbf{E} = -j\omega\mu\mathbf{A} + \frac{1}{j\omega\varepsilon}\nabla(\nabla \cdot \mathbf{A}) \tag{2.11}$$

and from (2.8), we note that $\mathbf{A}$ satisfies the Helmholtz equation

$$(\nabla^2 + k^2)\mathbf{A} = 0 \tag{2.12}$$

Similarly, considering (2.2) in conjunction with (2.3), we can write

$$\nabla \times (\mathbf{H} + j\omega\varepsilon\mathbf{F}) = 0 \tag{2.13}$$

or

$$\mathbf{H} + j\omega\varepsilon\mathbf{F} = -\nabla\psi^f \tag{2.14}$$

where ψ^f is an arbitrary magnetic scalar potential. Substituting for **E** and **H** from (2.3) and (2.14) in (2.1), we get

$$\nabla \times \nabla \times \mathbf{F} = j\omega\mu(-j\omega\varepsilon\mathbf{F} - \nabla\psi^f) \tag{2.15}$$

or

$$\nabla(\nabla\cdot\mathbf{F}) - \nabla^2\mathbf{F} = k^2\mathbf{F} - j\omega\mu\nabla\psi^f \tag{2.16}$$

Choosing

$$\nabla\cdot\mathbf{F} = -j\omega\mu\psi^f \tag{2.17}$$

and substituting in (2.14) yields

$$\mathbf{H} = -j\omega\varepsilon\mathbf{F} + \frac{1}{j\omega\mu}\nabla(\nabla\cdot\mathbf{F}) \tag{2.18}$$

From (2.16), we note that **F** satisfies the Helmholtz equation

$$(\nabla^2 + k^2)\mathbf{F} = 0 \tag{2.19}$$

From (2.3), (2.4), (2.11), and (2.18), the total **E** and **H** fields can be expressed in the forms

$$\mathbf{E} = -\nabla \times \mathbf{F} - j\omega\mu\mathbf{A} + \frac{1}{j\omega\varepsilon}\nabla(\nabla\cdot\mathbf{A}) \tag{2.20}$$

$$\mathbf{H} = -\nabla \times \mathbf{A} - j\omega\varepsilon\mathbf{F} + \frac{1}{j\omega\mu}\nabla(\nabla\cdot\mathbf{F}) \tag{2.21}$$

2.2.2 Field Solutions in Terms of Scalar Potentials

We now derive general field expressions in terms of certain scalar potential functions in a rectangular coordinate system. The medium chosen is a homogeneous lossless dielectric having parameters $\varepsilon = \varepsilon_0\varepsilon_r$ and $\mu = \mu_0$, where μ_0 and ε_0 are the permeability and permittivity, respectively, of free space; and ε_r is the relative dielectric constant of the dielectric medium.

TM-to-y and TE-to-y Solutions For the TM-to-y solution we consider $\mathbf{F} = 0$ and $\mathbf{A} = \hat{\mathbf{y}} j\omega\varepsilon_0\phi^e$, where ϕ^e is a scalar potential function. Equations (2.20) and

(2.21) then reduce to

$$\mathbf{E} = -j\omega\mu_0\mathbf{A} + \frac{1}{j\omega\varepsilon}\nabla(\nabla\cdot\mathbf{A}) \tag{2.22}$$

$$\mathbf{H} = \nabla \times \mathbf{A} \tag{2.23}$$

Expanding in terms of ϕ^e, we obtain the various field components as

$$E_x = \frac{1}{\varepsilon_r}\frac{\partial^2\phi^e}{\partial x\,\partial y} \tag{2.24a}$$

$$E_y = k_0^2\phi^e + \frac{1}{\varepsilon_r}\frac{\partial^2\phi^e}{\partial y^2} \tag{2.24b}$$

$$E_z = \frac{1}{\varepsilon_r}\frac{\partial^2\phi^e}{\partial y\,\partial z} \tag{2.24c}$$

$$H_x = -j\omega\varepsilon_0\frac{\partial\phi^e}{\partial z} \tag{2.24d}$$

$$H_y = 0 \tag{2.24e}$$

$$H_z = j\omega\varepsilon_0\frac{\partial\phi^e}{\partial x} \tag{2.24f}$$

where $k_0 = \omega\sqrt{\mu_0\varepsilon_0}$ is the free-space propagation constant.

For the TE-to-y solution, we consider $\mathbf{A} = 0$ and $\mathbf{F} = \hat{\mathbf{y}}j\omega\mu_0\phi^h$. From (2.20) and (2.21), we have

$$\mathbf{E} = -\nabla \times \mathbf{F} \tag{2.25}$$

$$\mathbf{H} = -j\omega\varepsilon\mathbf{F} + \frac{1}{j\omega\mu_0}\nabla(\nabla\cdot\mathbf{F}) \tag{2.26}$$

Expanding in terms of the scalar potential function ϕ^h, we can write

$$E_x = j\omega\mu_0\frac{\partial\phi^h}{\partial z} \tag{2.27a}$$

$$E_y = 0 \tag{2.27b}$$

$$E_z = -j\omega\mu_0\frac{\partial\phi^h}{\partial x} \tag{2.27c}$$

$$H_x = \frac{\partial^2\phi^h}{\partial x\,\partial y} \tag{2.27d}$$

$$H_y = k_0^2\varepsilon_r\phi^h + \frac{\partial^2\phi^h}{\partial y^2} \tag{2.27e}$$

$$H_z = \frac{\partial^2\phi^h}{\partial y\,\partial z} \tag{2.27f}$$

The scalar potential functions ϕ^e and ϕ^h satisfy the Helmholtz equation

$$(\nabla^2 + k^2)\begin{Bmatrix}\phi^e \\ \phi^h\end{Bmatrix} = 0 \tag{2.28}$$

Superposing the TE-to-y and TM-to-y solutions and, further, assuming propagation in the z-direction according to $\exp(-j\beta_z z)$, we obtain

$$E_x = \frac{1}{\varepsilon_r}\frac{\partial^2 \phi^e}{\partial x\,\partial y} + \omega\mu_0\beta_z\phi^h \tag{2.29a}$$

$$E_y = \frac{1}{\varepsilon_r}\left(\beta_z^2 - \frac{\partial^2}{\partial x^2}\right)\phi^e \tag{2.29b}$$

$$E_z = -\frac{j\beta_z}{\varepsilon_r}\frac{\partial \phi^e}{\partial y} - j\omega\mu_0\frac{\partial \phi^h}{\partial x} \tag{2.29c}$$

$$H_x = -\omega\varepsilon_0\beta_z\phi^e + \frac{\partial^2 \phi^h}{\partial x\,\partial y} \tag{2.29d}$$

$$H_y = \left(\beta_z^2 - \frac{\partial^2}{\partial x^2}\right)\phi^h \tag{2.29e}$$

$$H_z = j\omega\varepsilon_0\frac{\partial \phi^e}{\partial x} - j\beta_z\frac{\partial \phi^h}{\partial y} \tag{2.29f}$$

We note that for the TE-to-y solution, we set $\phi^e = 0$; and for the TM-to-y solution, we set $\phi^h = 0$ in (2.29).

TM-to-x and TE-to-x Solutions For the TM-to-x solution, we use $\mathbf{F} = 0$ and $\mathbf{A} = \hat{\mathbf{x}}j\omega\varepsilon_0\phi^e$ in (2.20) and (2.21). The various field components are obtained as

$$E_x = \frac{1}{\varepsilon_r}\left(\frac{\partial^2}{\partial x^2} + k^2\right)\phi^e \tag{2.30a}$$

$$E_y = \frac{1}{\varepsilon_r}\frac{\partial^2 \phi^e}{\partial x\,\partial y} \tag{2.30b}$$

$$E_z = \frac{1}{\varepsilon_r}\frac{\partial^2 \phi^e}{\partial x\,\partial y} \tag{2.30c}$$

$$H_x = 0 \tag{2.30d}$$

$$H_y = j\omega\varepsilon_0\frac{\partial \phi^e}{\partial z} \tag{2.30e}$$

$$H_z = -j\omega\varepsilon_0\frac{\partial \phi^e}{\partial y} \tag{2.30f}$$

Similarly, for the TE-to-x solutions, we use $\mathbf{A} = 0$ and $\mathbf{F} = \hat{\mathbf{x}} j\omega\mu_0 \phi^h$ in (2.20) and (2.21). The field components are given by

$$E_x = 0 \tag{2.31a}$$

$$E_y = -j\omega\mu_0 \frac{\partial \phi^h}{\partial z} \tag{2.31b}$$

$$E_z = j\omega\mu_0 \frac{\partial \phi^h}{\partial y} \tag{2.31c}$$

$$H_x = \left(k^2 + \frac{\partial^2}{\partial x^2} \right) \phi^h \tag{2.31d}$$

$$H_y = \frac{\partial^2 \phi^h}{\partial x\, \partial y} \tag{2.31e}$$

$$H_z = \frac{\partial^2 \phi^h}{\partial x\, \partial z} \tag{2.31f}$$

Considering propagation in the z-direction according to $\exp(-j\beta_z z)$, and superposing the TM-to-x and TE-to-x solutions given by (2.30) and (2.31), we obtain

$$E_x = \frac{1}{\varepsilon_r} \left(\beta_z^2 - \frac{\partial^2}{\partial y^2} \right) \phi^e \tag{2.32a}$$

$$E_y = \frac{1}{\varepsilon_r} \frac{\partial^2 \phi^e}{\partial x\, \partial y} - \omega\mu_0 \beta_z \phi^h \tag{2.32b}$$

$$E_z = -\frac{j\beta_z}{\varepsilon_r} \frac{\partial \phi^e}{\partial x} + j\omega\mu_0 \frac{\partial \phi^h}{\partial y} \tag{2.32c}$$

$$H_x = \left(\beta_z^2 - \frac{\partial^2}{\partial y^2} \right) \phi^h \tag{2.32d}$$

$$H_y = \omega\varepsilon_0 \beta_z \phi^e + \frac{\partial^2 \phi^h}{\partial x\, \partial y} \tag{2.32e}$$

$$H_z = -j\omega\varepsilon_0 \frac{\partial \phi^e}{\partial y} - j\beta_z \frac{\partial \phi^h}{\partial x} \tag{2.32f}$$

2.3 PRINCIPLE OF EFFECTIVE DIELECTRIC CONSTANT METHOD

The effective dielectric constant (EDC) method was first developed by Knox and Toulios [1, 8] for the image guide. It is an extension of the simple approximate approach used by Marcatili [9] for analyzing a dielectric rectangular waveguide. The EDC method has subsequently been applied to other guide structures such as the insular image and strip dielectric guides [2], inverted strip guide [10], trapped image guide [11–13], hollow and overlayed image guide [14], and the broadside coupled dielectric guide [15]. In the following, we present the basic principles of the EDC technique.

The modes of propagation in dielectric integrated guides are hybrid in nature. We assume propagation in the z-direction according to $\exp(-j\beta_z z)$ and classify the fields as E^y_{mn} modes (corresponding to TM-to-y fields), and E^x_{mn} modes (corresponding to TE-to-y) fields, where m and n indicate the number of extrema of the dominant electric field in the x- and y-directions, respectively. For E^y_{mn} modes, the principal field components are E_y and H_x, and ϕ^e has the dominant contribution to the modal fields. Similarly, for the E^x_{mn} modes, the principal field components are E_x and H_y, and ϕ^h has the dominant contribution to the modal fields. Hence the solutions for E^y_{mn} and E^x_{mn} modes can be obtained separately from (2.29a) through (2.29f) by setting $\phi^h = 0$ and $\phi^e = 0$, respectively. Since these two sets of modes are similar, we will outline the procedure for only the E^y_{mn} modes. Setting $\phi^h = 0$ in (2.29), the field components of the E^y_{mn} modes are obtained as

$$E_x = \frac{1}{\varepsilon_r}\frac{\partial^2 \phi^e}{\partial x\,\partial y} \tag{2.33a}$$

$$E_y = \frac{1}{\varepsilon_r}\left(\beta_z^2 - \frac{\partial^2}{\partial x^2}\right)\phi^e \tag{2.33b}$$

$$E_z = -\frac{j\beta_z}{\varepsilon_r}\frac{\partial \phi^e}{\partial y} \tag{2.33c}$$

$$H_x = -\,\omega\varepsilon_0\beta_z\phi^e \tag{2.33d}$$

$$H_y = 0 \tag{2.33e}$$

$$H_z = j\omega\varepsilon_0\frac{\partial \phi^e}{\partial x} \tag{2.33f}$$

where ϕ^e satisfies the Helmholtz equation (2.28), and ε_r is the relative dielectric constant of the dielectric medium. We now outline the steps involved in the EDC method of analysis by referring to the image guide shown in Figure 2.1(a):

1. The guide cross-section is divided into constituent regions: region I corresponding to $0 \le x \le a$ and region II corresponding to $x \le 0$, $x \ge a$. The constituent region containing the dielectric (Reg I) is assumed to extend to infinity in the x-direction, resulting in a planar slab guide as shown in Figure 2.1(b). From (2.33), we note that for E^y_{mn} modes,

$$E_z \approx \frac{1}{\varepsilon_r}\frac{\partial \phi^e(y)}{\partial y} \tag{2.34a}$$

$$H_x \approx \phi^e(y) \tag{2.34b}$$

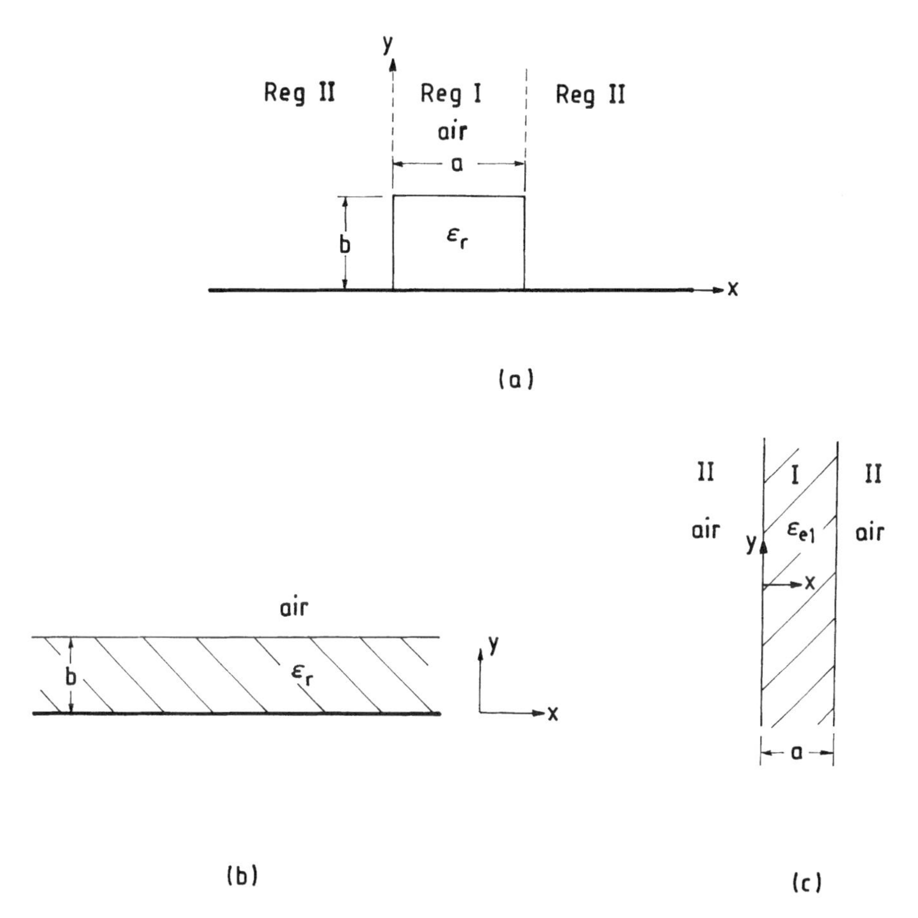

FIGURE 2.1 Image guide and analytical models for applying the EDC technique: (a) cross-section of image guide showing constituent regions, (b) horizontal slab guide representation for region I for determining effective dielectric constant ε_{e1}, and (c) vertical slab model for determining k_z.

By applying the boundary and interface conditions in terms of E_z and H_x, an eigenvalue equation is derived, which describes the wave propagation in the horizontal slab model. The solution of this eigenvalue equation yields the z-directed propagation constant β_{zs} of the slab guide.

2. The effective dielectric constant of region I, denoted as ε_{e1}, is given by

$$\varepsilon_{e1} = (\beta_{zs}/k_0)^2 \tag{2.35}$$

where k_0 is the propagation constant in free space. This effective dielectric constant is interpreted as that of a hypothetical medium in which the propagation constant is the same as that of the slab guide shown in Figure 2.1(b).

3. The original structure is next replaced by a vertical layered model wherein each of the constituent regions is replaced by a uniform vertical layer having the same thickness and a relative dielectric constant equal to the effective dielectric constant of the corresponding horizontal slab guide model. The vertical layered model for Figure 2.1(a) is shown in Figure 2.1(c) wherein region I is represented by a uniform dielectric region of relative dielectric constant ε_{e1} and region II remains as air ($\varepsilon_{e2} = 1$). For the E^y_{mn} modes, we note from (2.33) that

$$E_y \approx \phi^e(x) \tag{2.36a}$$

$$H_z \approx \frac{\partial \phi^e(x)}{\partial x} \tag{2.36b}$$

By matching these tangential field components at the interfaces, we can derive a characteristic equation that describes the propagation in this vertical layered model. The solution of the characteristic equation yields the z-directed propagation constant β_z, which is taken to be the same as that of the original structure.

It may be noted that the EDC method is based on the following two assumptions [4]: (i) each constituent region supports a single surface wave and (ii) the geometrical discontinuities at the strip edges are negligible. Equivalently, in the case of the image guide, the aspect ratio b/a should be much smaller than unity and the relative dielectric constant ε_r of the guide material should be close to that of the adjacent regions.

2.4 ANALYSIS OF DIELECTRIC INTEGRATED GUIDES: EFFECTIVE DIELECTRIC CONSTANT METHOD

In this section, we apply the effective dielectric constant technique to different types of dielectric integrated guides. The guides that are known to be of practical utility and several others that are considered to have potential for millimeter wave integrated circuits are considered. They are classified under the following categories: (i) open dielectric guides with bottom ground plane (Figs. 2.2 and 2.3), (ii) dielectric guides with rectangular metal shield (Fig. 2.4), (iii) trapped dielectric guides (Fig. 2.5), (iv) nonradiative dielectric guides (Fig. 2.6), and (v) asymmetrically coupled dielectric guides (Fig. 2.7). In addition, a class of planar bare dielectric guides without any metallic boundary (Fig. 2.8), which are useful for optical integrated circuits, is included.

In the following, we first illustrate the application of the EDC technique by carrying out in detail the analysis of insular image guide as an example. Similar analysis applies to other guide structures. In order to facilitate ready application of the technique to a wide variety of structures, we derive characteristic equations for several useful horizontal slab guide models as well as vertical layered

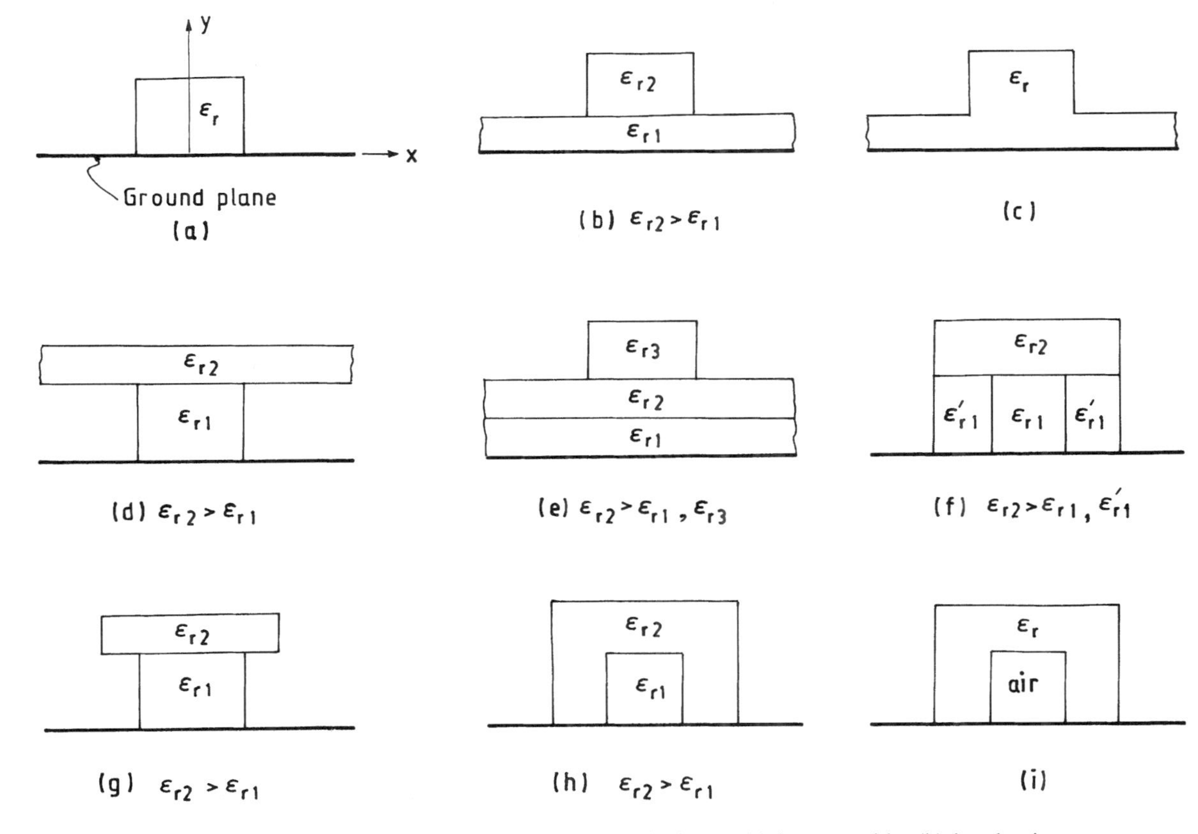

FIGURE 2.2 Open dielectric guides with ground planes: (a) image guide, (b) insular image guide, (c) rib guide, (d) inverted strip guide, (e) strip dielectric guide, (f) general cladded dielectric guide, (g) T-guide, (h) cladded image guide, and (i) hollow image guide.

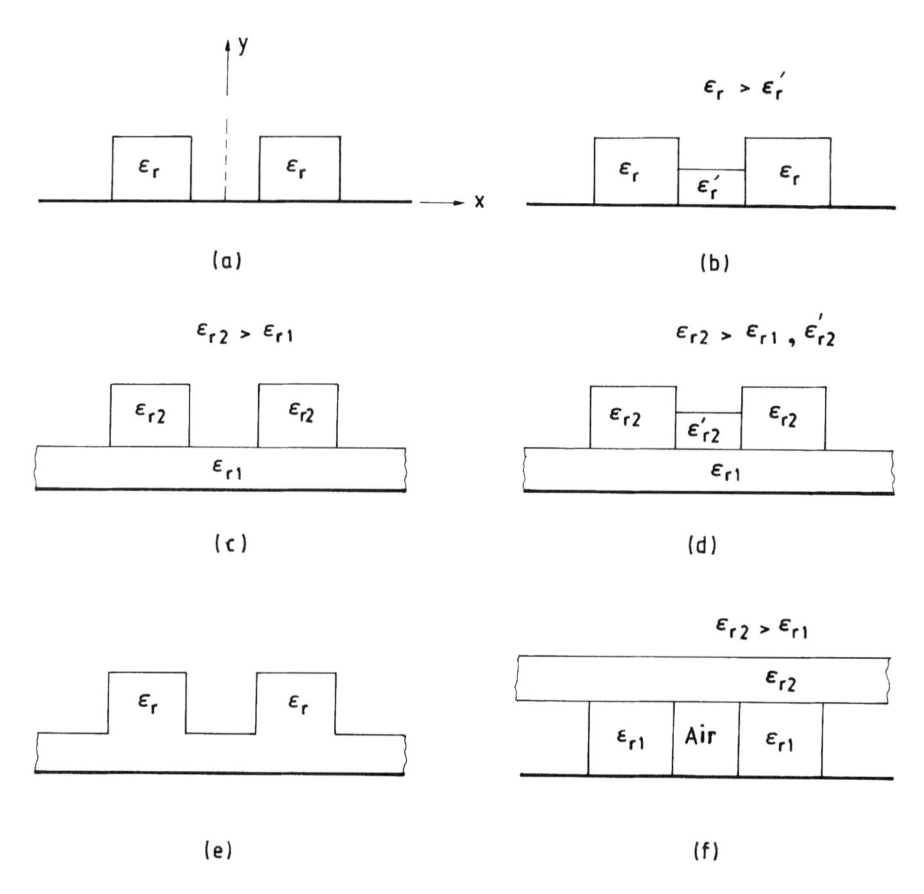

FIGURE 2.3 Open coupled dielectric guides with ground plane: (a) coupled image guide, (b) coupled image guide with dielectric spacer, (c) coupled insular image guide, (d) coupled insular image guide with dielectric spacer, (e) coupled rib guide, and (f) coupled inverted strip guide.

structures. The characteristic equations for horizontal slab guides having up to three dielectric layers are listed in Appendix 2A and for vertical layered guides having up to five layers are listed in Appendix 2B.

2.4.1 Illustration of EDC Method Using Field Matching

We illustrate the EDC method of analysis, by considering the example of an insular image guide. The structure, shown in Figure 2.9(a), is first divided into two constituent regions (marked I and II) and their horizontal slab guide models are used to derive the characteristic equations for determining the y-directed propagation constant and their effective dielectric constants. The slab model corresponding to region I is shown in Figure 2.9(b) and that for region II is shown in

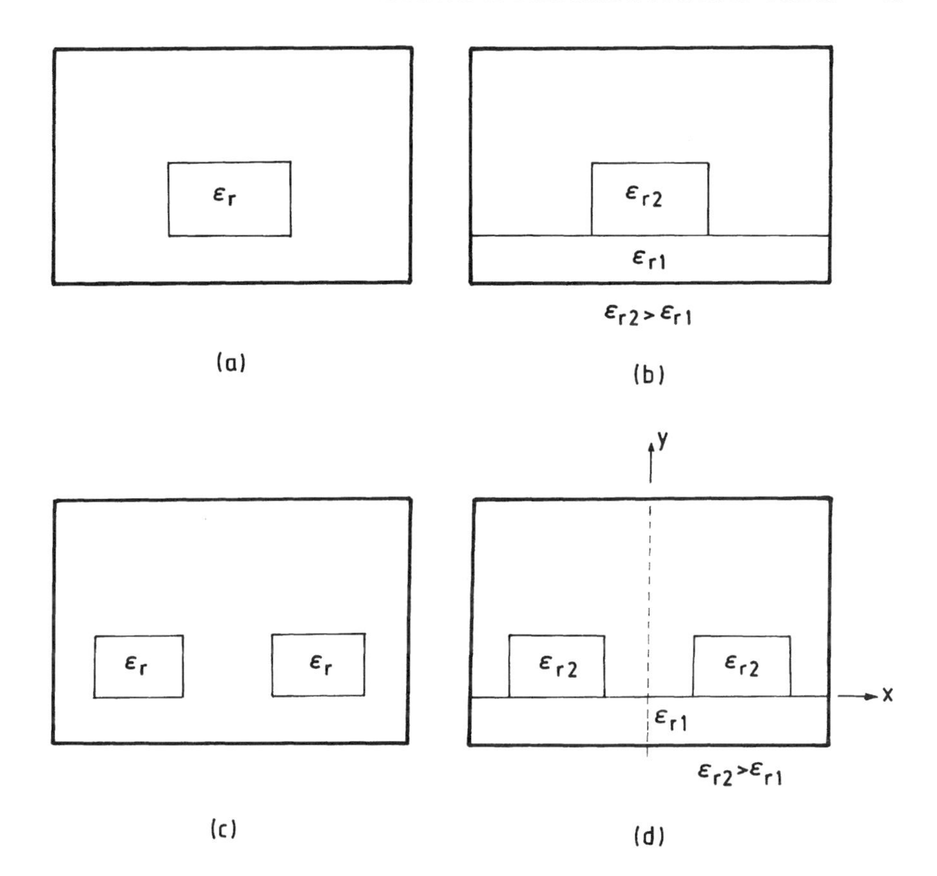

FIGURE 2.4 Single and coupled dielectric guides with rectangular metal shield: (a) shielded rectangular dielectric guide, (b) shielded insular image guide, (c) shielded coupled rectangular dielectric guide, and (d) shielded coupled insular image guide.

Figure 2.9(c). The vertical three-layered equivalent model for determining the z-directed propagation constant of the original structure is shown in Figure 2.9(d). In the following, we carry out the analysis of the structure for the E^y_{mn} (or TM^y_{mn}) modes and also E^x_{mn} (or TE^y_{mn}) modes.

The characteristic equations for the three models can be derived either by using field matching at the various boundaries and interfaces or by applying the transverse transmission line technique. The approach using field matching is presented below and the second approach is illustrated in Section 2.4.2.

Analysis of Insular Image Guide for E^y_{mn} (TM^y_{mn}) Modes (Fig. 2.9)

Horizontal Two-Slab Guide Model (Fig. 2.9(b)) Consider the horizontal two-slab guide model of region I. We assume propagation in the z-direction according

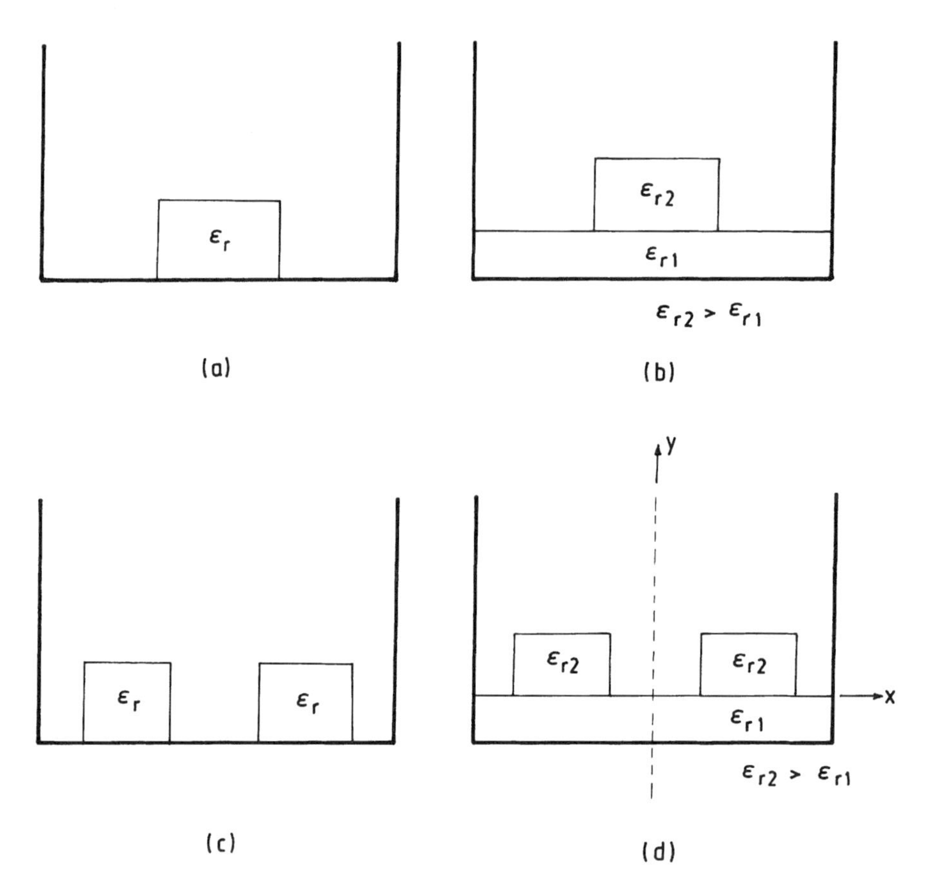

FIGURE 2.5 Single and coupled trapped dielectric guides: (a) trapped image guide, (b) trapped insular image guide, (c) trapped coupled image guide, and (d) trapped coupled insular image guide.

to $\exp(-j\beta_{zs}z)$. Since the structure is of infinite extent in the x-direction, there is no x-variation of fields. From the general field expressions for TM^y_{mn} modes (Eq. (2.33)), we have $H_x \approx \phi^e$ and $E_z \approx (1/\varepsilon_r)\partial\phi^e/\partial y$. The solution of the Helmholtz equation for $\phi^e(y)$ in each slab can now be written in terms of the y-directed propagation constant k_y. Depending on whether the solution corresponds to a traveling wave or a decaying type of wave, k_y may be real or imaginary. In a dielectric guide, the bulk of the energy travels in the region of higher dielectric constant. In the case of the insular image guide (Fig. 2.9(a)), since ε_{r2} is greater than ε_{r1}, we choose the traveling wave solution in slab 2 (see Fig. 2.9(b)). The solution is expressed in terms of a sum of sinusoidal functions with $k_{y2} = \beta_y$, where β_y is real. In slab 1, the solution is expressed in terms of a hyperbolic function with $k_{y1} = -j\zeta_1$, where ζ_1 may be real or imaginary. ζ_1 real and positive refers to decaying waves and ζ_1 imaginary refers to traveling waves

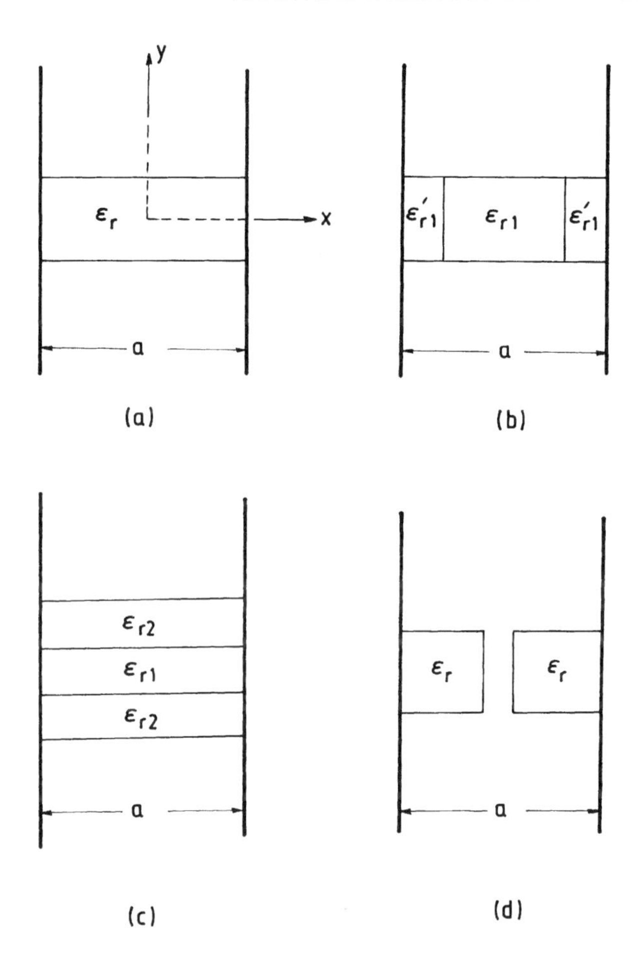

FIGURE 2.6 Single and coupled nonradiative dielectric guides ($a < \lambda/2$): (a) nonradiative dielectric guide, (b) nonradiative dielectric guide with insular layer, (c) edge-coupled nonradiative dielectric guide, and (d) broadside-coupled nonradiative dielectric guide.

in slab 1. In the air region ($y > b$), the fields must vanish as $y \to \infty$ and hence we set $k_{y3} = -j\zeta_0$, where ζ_0 is positive and real.

The solution for $\phi^e(y)$ in three regions of the slab guide (Fig. 2.9(b)) can be written in the form

$$\phi^e(y) = \begin{cases} A\cosh[\zeta_1(y+d)], & -d \le y \le 0 \\ B_1\cos(\beta_y y) + B_2\sin(\beta_y y), & 0 \le y \le b \\ Ce^{-\zeta_0(y-b)}, & y > b \end{cases} \tag{2.37}$$

where

$$\beta_{zs}^2 = k_0^2\varepsilon_{r1} + \zeta_1^2 = k_0^2\varepsilon_{r2} - \beta_y^2 = k_0^2 + \zeta_0^2 \tag{2.38}$$

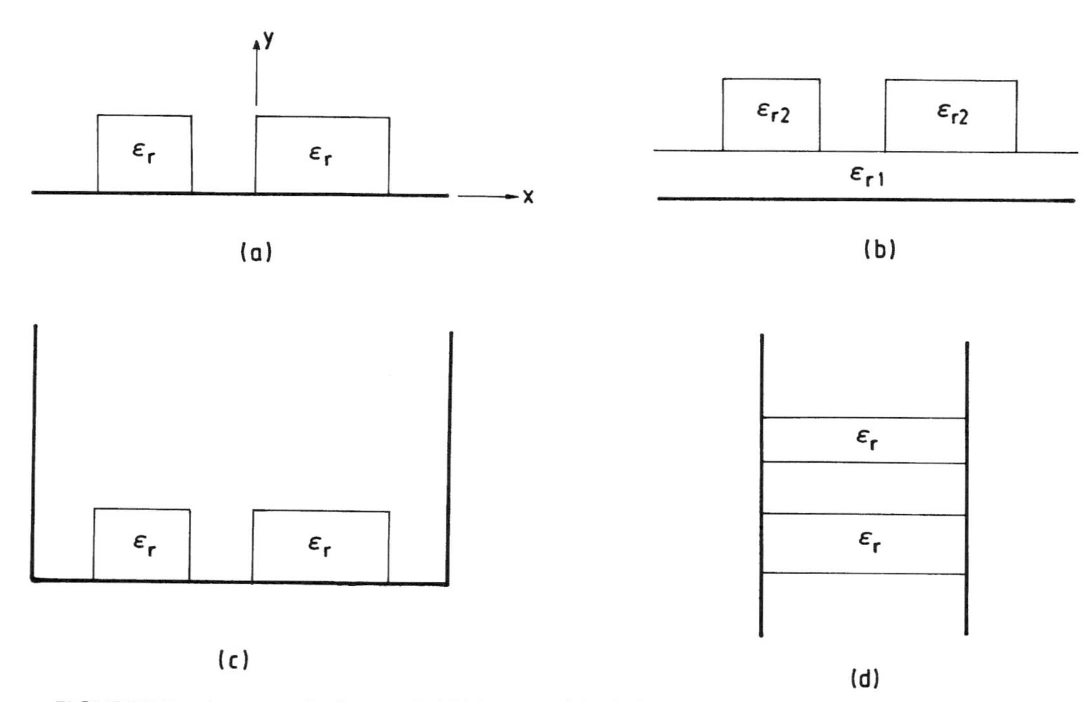

FIGURE 2.7 Asymmetrically coupled (a) image guide, (b) insular image guide, (c) trapped image guide, and (d) edge-coupled nonradiative guide.

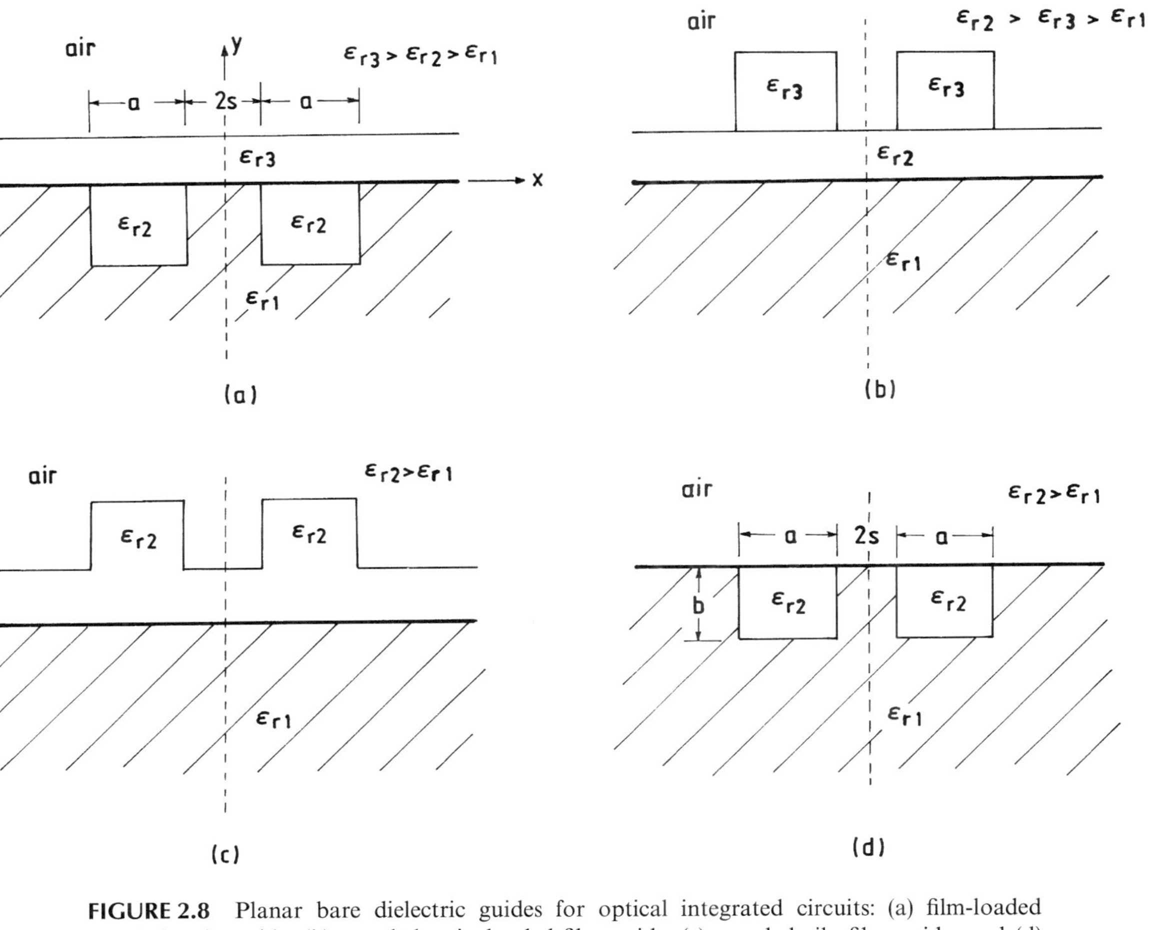

FIGURE 2.8 Planar bare dielectric guides for optical integrated circuits: (a) film-loaded coupled strip guide, (b) coupled strip loaded-film guide, (c) coupled rib–film guide, and (d) embedded coupled strip guide.

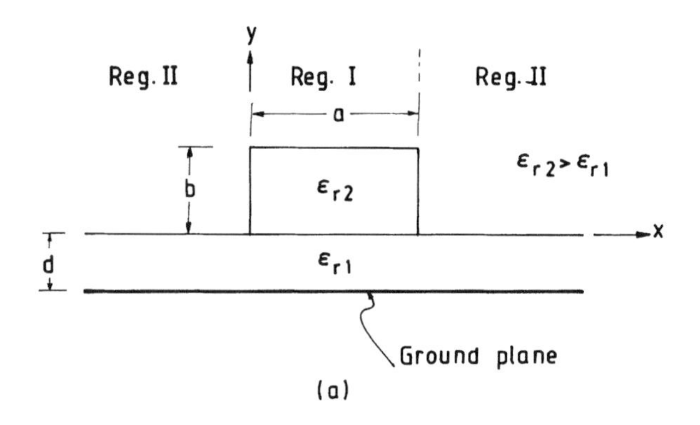

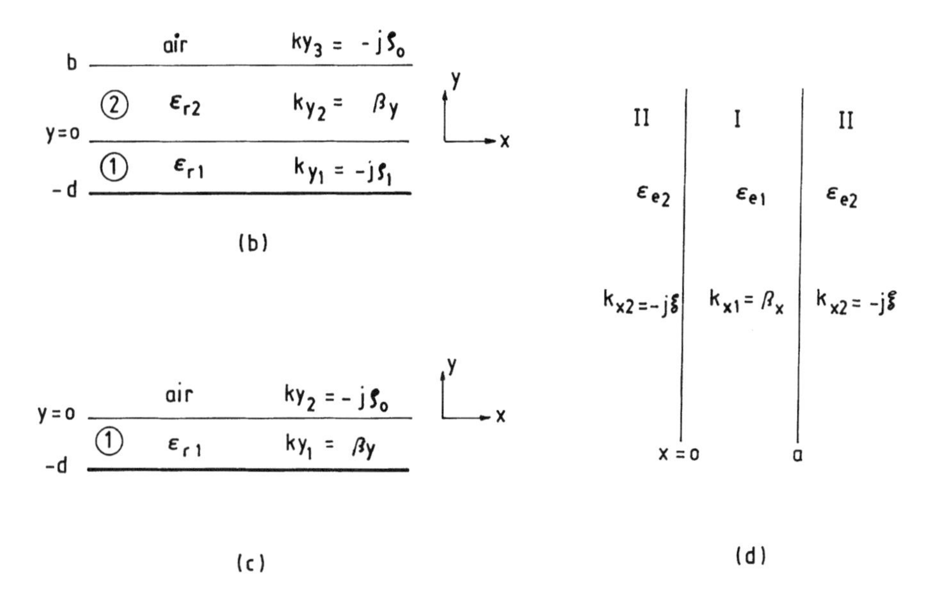

FIGURE 2.9 (a) Insular image guide, (b) horizontal two-slab guide model for region I, (c) horizontal single-slab guide model for region II, and (d) equivalent vertical layered model for determining β_z.

In order to derive the characteristic equation, we match the tangential field components H_x and E_z at the interfaces $y = 0$ and $y = b$. At $y = 0$, we have

$$A \cosh(\zeta_1 d) = B_1 \tag{2.39a}$$

$$A \sinh(\zeta_1 d) = B_2(\beta_y/\zeta_1)(\varepsilon_{r1}/\varepsilon_{r2}) \tag{2.39b}$$

At $y = b$, we have

$$B_1 \cos(\beta_y b) + B_2 \sin(\beta_y b) = C \tag{2.40a}$$

$$-B_1 \sin(\beta_y b) + B_2 \cos(\beta_y b) = -C\zeta_0 \varepsilon_{r2}/\beta_y \tag{2.40b}$$

Eliminating the unknown constants from (2.39) and (2.40), we obtain the following characteristic equation:

$$\left[1 + \left(\frac{\zeta_1}{\beta_y}\right)\left(\frac{\varepsilon_{r2}}{\varepsilon_{r1}}\right) \tanh(\zeta_1 d) \tan(\beta_y b)\right] + \left(\frac{\beta_y}{\zeta_0 \varepsilon_{r2}}\right)\left[-\tan(\beta_y b) + \left(\frac{\zeta_1}{\beta_y}\right)\left(\frac{\varepsilon_{r2}}{\varepsilon_{r1}}\right) \tanh(\zeta_1 d)\right] = 0 \tag{2.41}$$

where

$$\zeta_1 = [k_0^2(\varepsilon_{r2} - \varepsilon_{r1}) - \beta_y^2]^{1/2} \tag{2.42a}$$

$$\zeta_0 = [k_0^2(\varepsilon_{r2} - 1) - \beta_y^2]^{1/2} \tag{2.42b}$$

Using the expressions for ζ_1 and ζ_0 in terms of β_y, we can solve (2.41) for the propagation constant β_y.

With β_y known, the effective dielectric constant for the two-slab guide model corresponding to region I can be calculated using (2.38).

$$\varepsilon_{e1} = (\beta_{zs}/k_0)^2 = \varepsilon_{r2} - (\beta_y/k_0)^2 \tag{2.43}$$

Horizontal Single-Slab Guide Model (Fig. 2.9(c)) In the case of a single-slab guide shown in Figure 2.9(c), we assume fields to be sinusoidally varying in the slab region and exponentially decaying with increasing y above the slab. Referring to Figure 2.9(c), we use $k_{y1} = \beta_y$ and $k_{y2} = -j\zeta_0$. The solution for $\phi^e(y)$ in the two regions can be written as

$$\phi^e(y) = \begin{cases} A\cos[\beta_y(y+d)], & -d \le y \le 0 \\ A\cos(\beta_y d)e^{-\zeta_0 y}, & y \ge 0 \end{cases} \tag{2.44}$$

where

$$\beta_{zs}^2 = k_0^2 \varepsilon_{r1} - \beta_y^2 = k_0^2 + \zeta_0^2 \tag{2.45}$$

Matching E_z at the interface $y = 0$, we obtain

$$1 - (\beta_y/\zeta_0 \varepsilon_{r1}) \tan(\beta_y d) = 0 \tag{2.46}$$

where

$$\zeta_0 = [k_0^2(\varepsilon_{r1} - 1) - \beta_y^2]^{1/2} \tag{2.47}$$

The solution of (2.47) yields β_y for the single-slab guide. The effective dielectric constant for region II is obtained from

$$\varepsilon_{e2} = \varepsilon_{r1} - (\beta_y/k_0)^2 \tag{2.48}$$

Equivalent Vertical Layered Guide Model (Fig. 2.9(d)) We now consider the equivalent vertical layered model shown in Figure 2.9(d), where ε_{e1} and ε_{e2} are the relative dielectric constants of the vertical layers representing regions I and II, respectively. Since these layers are of infinite extent in the y-direction, there is no y-variation in the fields. We assume wave propagation in the z-direction according to $\exp(-j\beta_z z)$, where the propagation constant β_z is assumed to be the same as that of the original structure. From (2.33), we note that the tangential fields satisfy the relations $E_y \approx \phi^e(x)$ and $H_z \approx \partial\phi^e(x)/\partial x$. The solution of the Helmholtz equation for $\phi^e(x)$ can now be expressed in terms of x-directed propagation constants. We assume fields in the form of sinusoidal functions in the slab region $(0 < x < a)$ and allow for exponentially decaying fields outside. The solution for $\phi^e(x)$ can be expressed as

$$\phi^e(x) = \begin{cases} Ae^{\xi x}, & x \le 0 \\ B_1 \cos(\beta_x x) + B_2 \sin(\beta_x x), & 0 \le x \le a \\ Ce^{-\xi(x-a)}, & x \ge a \end{cases} \tag{2.49}$$

where β_x and ξ are real and are related by

$$\beta_z^2 = k_0^2\varepsilon_{e1} - \beta_x^2 = k_0^2\varepsilon_{e2} + \xi^2 \tag{2.50}$$

We now match the tangential field components E_y and H_z at $x = 0$ and $x = a$. At $x = 0$, we have

$$A = B_1 \tag{2.51a}$$

$$A = B_2\beta_x/\xi \tag{2.51b}$$

At $x = a$, we have

$$B_1 \cos(\beta_x a) + B_2 \sin(\beta_x a) = C \tag{2.52a}$$

$$-B_1 \sin(\beta_x a) + B_2 \cos(\beta_x a) = -C\xi/\beta_x \tag{2.52b}$$

Eliminating the unknown constants from (2.51) and (2.52), we obtain the following characteristic equations:

$$(\xi/\beta_x)[1 - (\beta_x/\xi)^2] \tan(\beta_x a) + 2 = 0 \tag{2.53}$$

$$\xi = [k_0^2(\varepsilon_{e1} - \varepsilon_{e2}) - \beta_x^2]^{1/2} \tag{2.54}$$

The solution of (2.53) yields β_x. The propagation constant β_z of the insular image guide is now obtained from

$$\beta_z = [k_0^2 \varepsilon_{e1} - \beta_x^2]^{1/2} \tag{2.55}$$

It may be noted that the transcendental equations (2.41), (2.46), and (2.53) have many roots. For calculating β_z of E_{mn}^y modes, we use the value of β_x corresponding to the mth root of (2.53) and values of β_y corresponding to the nth root of (2.41) and (2.46). Thus, for the dominant E_{11}^y mode ($m=1, n=1$), we make use of the first root of (2.41) in (2.43) to determine ε_{e1}; the first root of (2.46) in (2.49) to determine ε_{e2}; and then use the first root of (2.53) along with the value of ε_{e1} to calculate β_z from (2.55).

Analysis of Insular Image Guide for $E_{mn}^x(TE_{mn}^y)$ Modes (Fig. 2.9)

For E_{mn}^x modes, we make use of the TE-to-y field solutions given by (2.27). For the horizontal slab models ($\partial/\partial x = 0$), the tangential fields E_x and H_z satisfy the relations $E_x \approx \phi^h$ and $H_z \approx \partial\phi^h/\partial y$; and for the vertical layered model ($\partial/\partial y = 0$), the tangential fields satisfy the relations $H_y \approx \varepsilon_r \phi^h$ and $E_z \approx \partial\phi^h/\partial x$.

Horizontal Two-Slab Guide Model (Fig. 2.9(b)) The solution for $\phi^h(y)$ in the three regions of the horizontal slab guide shown in Figure 2.9(b) is given by

$$\phi^h(y) = \begin{cases} A \sinh[\zeta_1(y+d)], & -d \le y \le 0 \\ B_1 \cos(\beta_y y) + B_2 \sin(\beta_y y), & 0 \le y \le b \\ Ce^{-\zeta_0(y-b)}, & y \ge b \end{cases} \tag{2.56}$$

As discussed earlier in the case of E_{mn}^y modes, β_y and ζ_0 are positive and real, whereas ζ_1 may be real or imaginary.

Matching E_x and H_z components at the interfaces $y=0$ and $y=b$, we obtain the following characteristic equation for determining β_y:

$$[1 + (\zeta_0/\beta_y)\tan(\beta_y b)] + (\beta_y/\zeta_1)\tanh(\zeta_1 d)\,[(\zeta_0/\beta_y) - \tan(\beta_y b)] = 0 \tag{2.57}$$

The expressions for ζ_1, ζ_0, and ε_{e1} are the same as given in (2.42a), (2.42b), and (2.43), respectively.

Horizontal Single-Slab Guide Model (Fig. 2.9(c)) Referring to Figure 2.9(c), the solution for $\phi^h(y)$ in the two regions can be written as

$$\phi^h(y) = \begin{cases} A \sinh[\beta_y(y+d)], & -d \le y \le 0 \\ Be^{-\zeta_0 y}, & y \ge 0 \end{cases} \tag{2.58}$$

Matching E_x and H_z components at $y=0$, we obtain

$$1 + (\zeta_0/\beta_y)\tan(\beta_y d) = 0 \tag{2.59}$$

where ζ_0 is related to β_y through (2.47) and the expression for ε_{e2} of the slab guide is given by (2.48).

Equivalent Vertical Layered Guide Model (Fig. 2.9(d)) Referring to the vertical layered guide model shown in Figure 2.9(d), the solution for $\phi^h(x)$ can be written as

$$\phi^h(x) = \begin{cases} Ae^{\xi x}, & x \leq 0 \\ B_1 \cos(\beta_x x) + B_2 \sin(\beta_x x), & 0 \leq x \leq a \\ Ce^{-\xi(x-a)}, & x \geq a \end{cases} \tag{2.60}$$

where β_x and ξ are both real. Matching H_y and E_z components at $x = 0$ and a, we obtain the following characteristic equation for determining the propagation constant β_x:

$$2 + \left(\frac{\varepsilon_{e1}}{\varepsilon_{e2}}\right)\left(\frac{\xi}{\beta_x}\right)\tan(\beta_x a)\left\{1 - \left[\left(\frac{\varepsilon_{e2}}{\varepsilon_{e1}}\right)\left(\frac{\beta_x}{\xi}\right)\right]^2\right\} = 0 \tag{2.61}$$

where ξ is related to β_x through Eq. (2.54). The expression for evaluating β_z is the same as that given by (2.55).

2.4.2 Illustration of EDC Method Using Transverse Transmission Line Technique

The transverse transmission line technique enables the derivation of characteristic equations of slab guides by modeling them as equivalent two-wire transmission line networks and directly applying the standard impedance formulas. This technique eliminates the need to solve a set of simultaneous equations as in the field-matching technique and hence is simpler to apply particularly when the number of layers is large.

Analysis for E^y_{mn}(TM^y_{mn}) Modes

Transmission Line Equivalent of Horizontal Two-Slab Model (see Figs. 2.9(b) and 2.10(a)) The two-wire transmission line representation of the two-slab model considered in Figure 2.9(b) is shown in Figure 2.10(a). The transmission line is short circuited at one end corresponding to the ground plane of the slab guide and extends along the y-direction. Each medium of the slab guide is represented by a section of transmission line having a length equal to the height of the medium. The propagation constant of the transmission line is k_{yi}, and its characteristic impedance Z_{0i} is given by

$$Z_{0i} = k_{yi}/(\omega\varepsilon_0\varepsilon_{ri}) \tag{2.62}$$

where ε_{ri} and k_{yi} are the relative dielectric constant and y-directed propagation

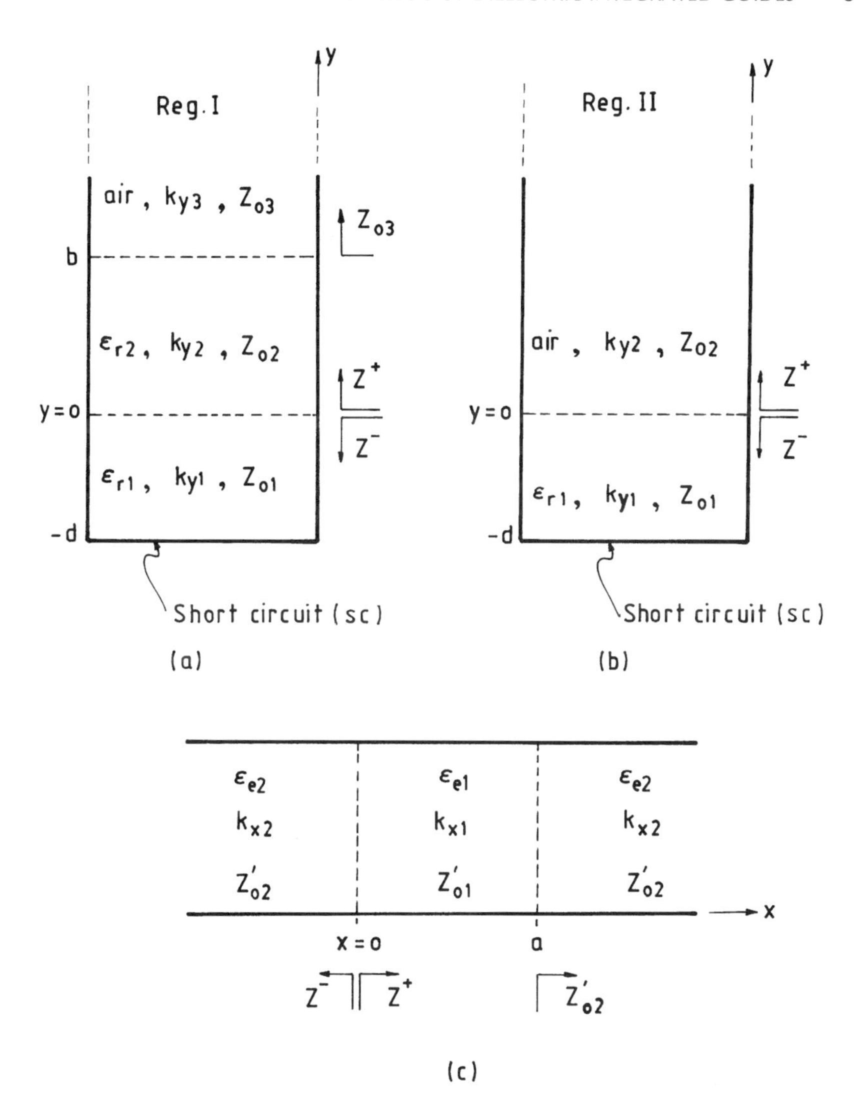

FIGURE 2.10 Equivalent transmission line models for (a) horizontal two-slab guide shown in Figure 2.9(b), (b) horizontal single-slab guide shown in Figure 2.9(c), and (c) vertical layered structure shown in Figure 2.9(d).

constant of the *i*th medium in the slab guide. We note that the characteristic impedance defined in (2.62) corresponds to that of the TM mode in a rectangular waveguide.

Choosing $y = 0$ as a convenient reference plane, the characteristic equation for the two-slab model can be obtained from the following resonance condition:

$$Z^+ + Z^- = 0 \tag{2.63}$$

where Z^+ and Z^- are the input impedances of the transmission line at $y = 0$ when looking toward $+y$ and $-y$ directions, respectively. The standard formula for the input impedance of a section of transmission line having a length l, propagation constant k, characteristic impedance Z_0, and terminated in a load Z_l is given by

$$Z_{\text{in}} = Z_0 \left[\frac{Z_l + jZ_0 \tan(kl)}{Z_0 + jZ_l \tan(kl)} \right] \tag{2.64}$$

Applying this formula to the transmission line shown in Figure 2.10(a), we can write

$$Z^- = jZ_{01} \tan(k_{y1} d) \tag{2.65}$$

$$Z^+ = Z_{02} \left[\frac{Z_{03} + jZ_{02} \tan(k_{y2} b)}{Z_{02} + jZ_{03} \tan(k_{y2} b)} \right] \tag{2.66}$$

where Z_{0i} $(i = 1, 2, 3)$ is given by (2.62). As discussed in Section 2.4.1, with the condition $\varepsilon_{r2} > \varepsilon_{r1}$ for the insular image guide, we may set

$$k_{y1} = -j\zeta_1, \quad k_{y2} = \beta_y, \quad k_{y3} = -j\zeta_0 \tag{2.67}$$

so that

$$Z_{01} = \frac{-j\zeta_1}{\omega\varepsilon_0\varepsilon_{r1}}, \quad Z_{02} = \frac{\beta_y}{\omega\varepsilon_0\varepsilon_{r2}}, \quad Z_{03} = \frac{-j\zeta_0}{\omega\varepsilon_0} \tag{2.68}$$

Substituting for k_{yi} and Z_{0i} $(i = 1, 2, 3)$ in (2.65) and (2.66) and using the resonance condition, we obtain the same characteristic equation as given in (2.41).

Transmission Line Equivalent of Horizontal Single-Slab Model (see Figs. 2.9(c) and 2.10(b)) The transmission line representation of the horizontal single-slab model (Fig. 2.9(c)) is shown in Figure 2.10(b). The expression for the characteristic impedance is the same as that given by (2.62). Referring to Figure 2.10(b), the impedances Z^- and Z^+ are given by

$$Z^- = jZ_{01} \tan(k_{y1} d) \tag{2.69}$$

$$Z^+ = Z_{02} \tag{2.70}$$

where

$$k_{y1} = \beta_y, \quad k_{y2} = -j\zeta_0 \tag{2.71}$$

and

$$Z_{01} = \frac{\beta_y}{\omega\varepsilon_0\varepsilon_{r1}}, \quad Z_{02} = \frac{-j\zeta_0}{\omega\varepsilon_0} \tag{2.72}$$

Using (2.69)–(2.72) in (2.63) yields the same characteristic equation as given by (2.46).

Transmission Line Equivalent of Vertical Layered Model (see Figs. 2.9(d) and 2.10(c)) The two-wire transmission line corresponding to the vertical layered model (Fig. 2.9(d)) extends in the x-direction as shown in Figure 2.10(c). Since there is no variation with respect to y, we note from (2.33) that the propagation in the layered guide corresponds to TE modes. Therefore the characteristic impedance Z'_{0j} of the transmission line is defined in terms of the TE mode as

$$Z'_{0j} = \omega\mu_0/k_{xj} \tag{2.73}$$

where k_{xj} is the x-directed propagation constant of the jth medium in the layered guide. The resonance condition is the same as that given by (2.63), where Z^+ and Z^- are the input impedances of the transmission line at $x = 0$ looking toward $+x$- and $-x$-directions, respectively. Referring to Figure 2.10(c), we can write

$$Z^- = Z'_{02} \tag{2.74}$$

$$Z^+ = Z'_{01}\left[\frac{Z'_{02} + jZ'_{01}\tan(k_{x1}a)}{Z'_{01} + jZ'_{02}\tan(k_{x1}a)}\right] \tag{2.75}$$

As discussed in Section 2.4.1, we consider

$$k_{x1} = \beta_x, \quad k_{x2} = -j\xi \tag{2.76}$$

Using (2.76) in (2.73), we have

$$Z'_{01} = \omega\mu_0/\beta_x, \quad Z'_{02} = \omega\mu_0/(-j\xi) \tag{2.77}$$

Substituting for k_{x1}, k_{x2}, Z'_{01}, and Z'_{02} in (2.74) and (2.75) and using the resonance condition (2.63), we obtain the same characteristic equation as given in (2.53).

The formulas and steps for determining ε_{e1}, ε_{e2}, and β_z are the same as those presented in Section 2.4.1.

Analysis for $E^x_{mn}(TE^y_{mn})$ Modes The analysis procedure is the same as that described above for the E^y_{mn} modes except that the expressions for Z_{0i} and Z'_{0j} correspond to those of the TE and TM modes, respectively, of the waveguide. Thus, while considering the transmission line equivalent for horizontal slab guide models, the characteristic impedance of the ith section is defined as

$$Z_{0i} = \omega\mu_0/k_{yi} \tag{2.78}$$

and for the vertical layered model, the characteristic impedance of the jth layer is defined as

$$Z'_{0j} = k_{xj}/(\omega\varepsilon_0\varepsilon_{ej}) \tag{2.79}$$

2.4.3 Analysis of Horizontal Slab Guide Models

Figures 2.2–2.8 illustrate a wide variety of dielectric integrated guides that can be analyzed using the EDC technique. As described in Sections 2.4.1 and 2.4.2, the analysis of each involves the derivation of characteristic equations for one or more horizontal slab guide models for determining the effective dielectric constants and then a vertical layered model for determining the propagation constant of the original guide. In the following, we derive characteristic equations for horizontal slab guides having up to four dielectric layers so that the expressions could be applied directly to the guides considered in Figures 2.2–2.8. The vertical layered models are considered in Section 2.4.4.

The derivation provided below applies to E^y_{mn} modes of the dielectric integrated guides since practical guides commonly employ this type of excitation. Similar analysis, however, applies to E^x_{mn} modes. In all the horizontal slab guides considered here, wave propagation is assumed to be in the z-direction according to $\exp(-j\beta_{zs}z)$. The effective dielectric constant of the slab guide is defined as $\varepsilon_e = (\beta_{zs}/k_0)^2$.

Three-Slab Guide with Bottom Ground Plane (Fig. 2.11(a)) The slab guide structures backed by ground plane (short circuit, sc) as shown in Figure 2.11(a)–(c) are useful for the analysis of dielectric integrated guides with ground plane (Figs. 2.2, 2.3, and 2.7(a)–(c)) and trapped dielectric guides (Fig. 2.5). We first consider the three-slab guide shown in Figure 2.11(a). Let ε_{si} and k_{yi} denote the relative dielectric constant and the y-directed propagation constant of the ith layer. In the air region ($i = 4$), we have $\varepsilon_{s4} = 1$ and since the fields must vanish as $y \to \infty$, we set $k_{y4} = -j\zeta_0$, where ζ_0 is real and positive. The solution for $\phi^e(y)$ in the four regions of the guide can be written as

$$\phi^e(y) = \begin{cases} A\cos\{k_{y1}(y + y_1)\}, & -y_1 \le y \le 0 \\ B_1\cos(k_{y2}y) + B_2\sin(k_{y2}y), & 0 \le y \le y_2 \\ C_1\cos(k_{y3}y) + C_2\sin(k_{y3}y), & y_2 \le y \le y_2 + y_3 \\ De^{-\zeta_0(y - y_2 - y_3)}, & y \ge y_2 + y_3 \end{cases} \tag{2.80}$$

where

$$\beta_{zs}^2 = k_0^2 + \zeta_0^2 = k_0^2\varepsilon_{si} - k_{yi}^2, \quad i = 1, 2, 3 \tag{2.81}$$

and

$$k_0^2 = \omega^2\mu_0\varepsilon_0 \tag{2.82}$$

The propagation constant k_{yi} ($i = 1, 2, 3$) may be real or imaginary depending on

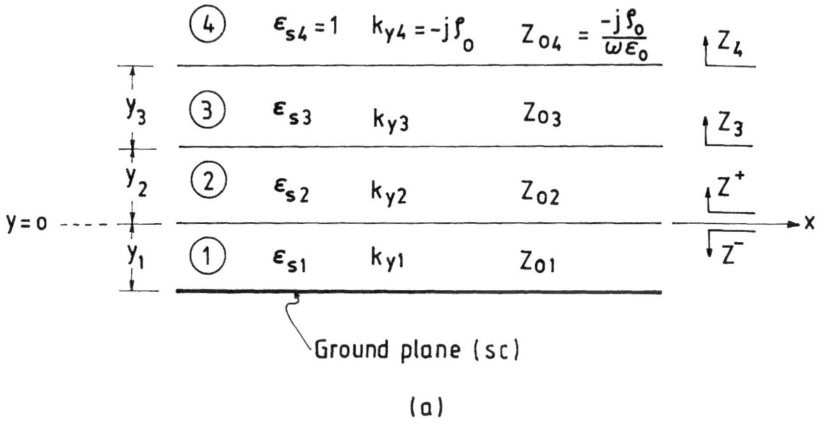

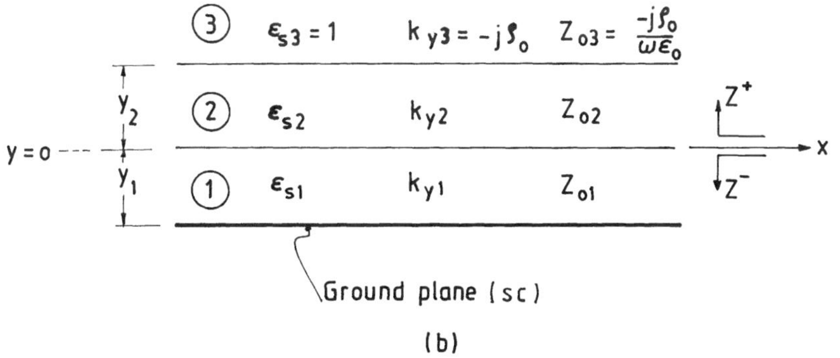

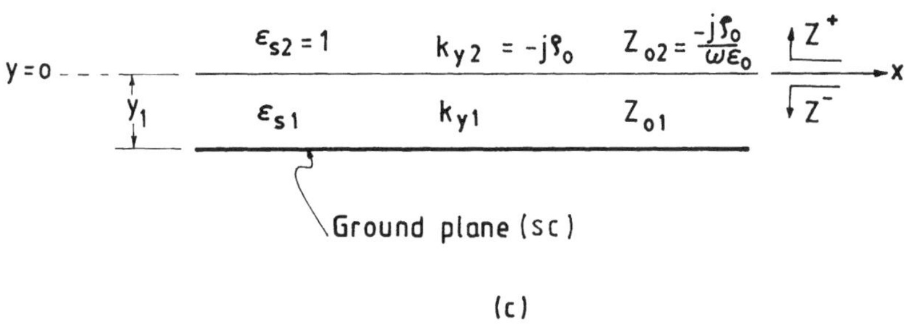

FIGURE 2.11 Horizontal slab guide models and their transmission line equivalents suitable for dielectric integrated guides shown in Figures 2.2–2.8: (a)–(c) layered slab guides with bottom ground plane, (d) layered slab guide with top and bottom ground planes, and (e)–(h) layered slab guides with no ground plane.

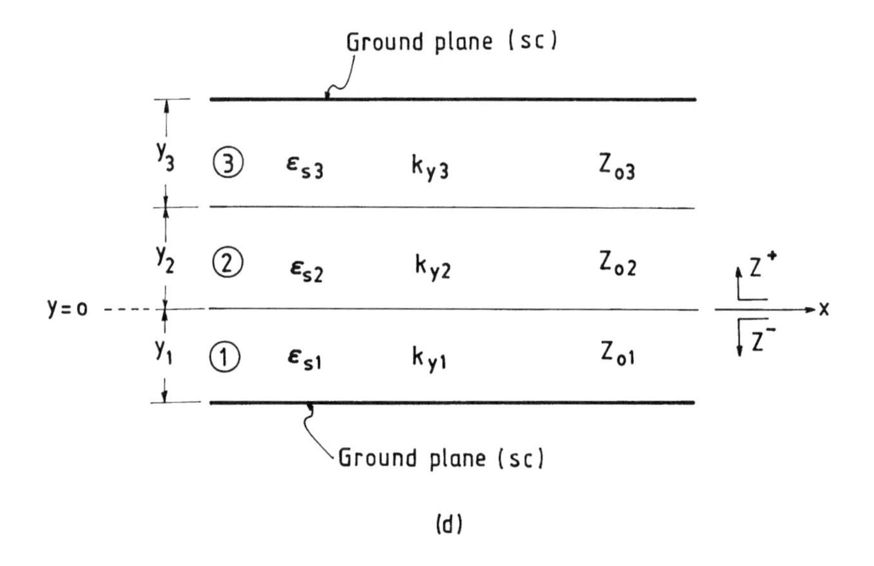

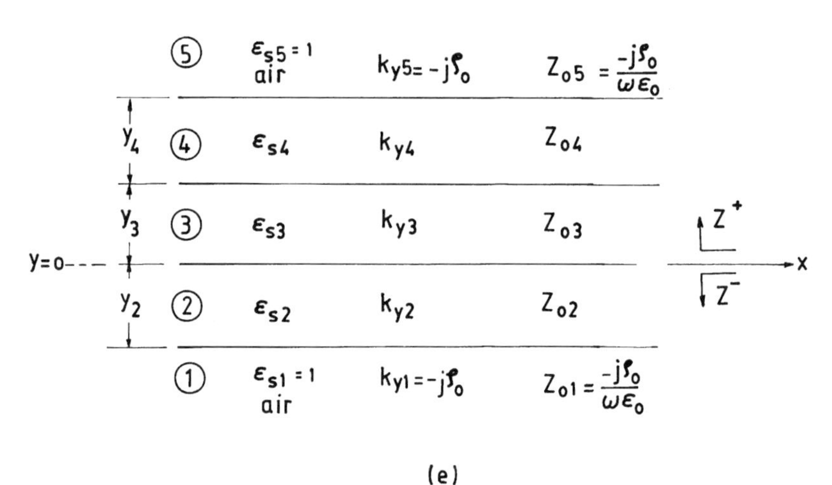

FIGURE 2.11 (*Continued*)

whether the solution corresponds to a traveling wave or a decaying wave. Considering the transmission line equivalent, we can write down the expressions for the various impedances:

$$Z^- = jZ_{01}\tan(k_{y1}y_1) \tag{2.83}$$

$$Z^+ = Z_{02}\left[\frac{Z_3 + jZ_{02}\tan(k_{y2}y_2)}{Z_{02} + jZ_3\tan(k_{y2}y_2)}\right] \tag{2.84}$$

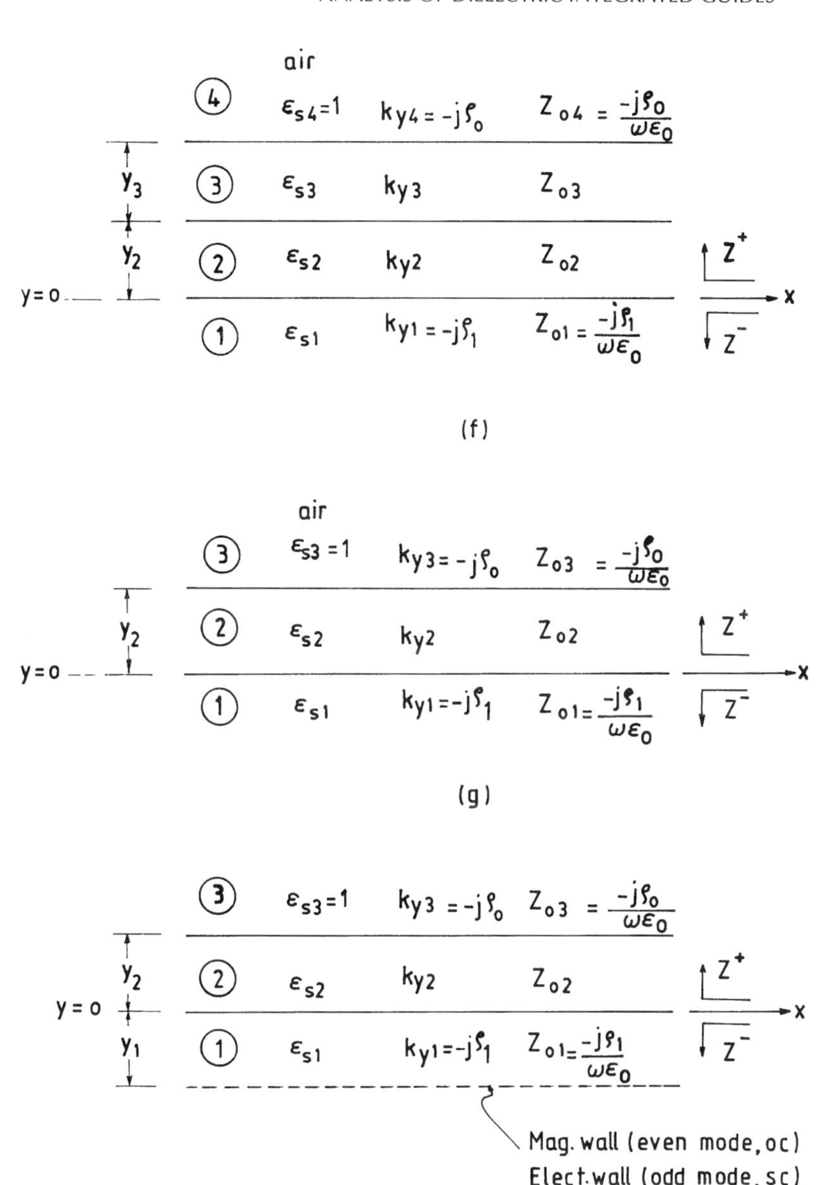

FIGURE 2.11 *(Continued)*

where

$$Z_3 = Z_{03}\left[\frac{Z_{04} + jZ_{03}\tan(k_{y3}y_3)}{Z_{03} + jZ_{04}\tan(k_{y3}y_3)}\right] \tag{2.85a}$$

$$Z_{04} = -j\zeta_0/(\omega\varepsilon_0) \tag{2.85b}$$

$$Z_{0i} = k_{yi}/(\omega\varepsilon_0\varepsilon_{si}), \quad i = 1, 2, 3 \tag{2.85c}$$

Using (2.83), (2.84), and (2.85) in the resonance condition (2.63), we obtain

$$[k_{y3}\tan(k_{y3}y_3) - \zeta_0\varepsilon_{s3}]\left[1 - \left(\frac{\varepsilon_{s2}}{\varepsilon_{s1}}\right)k_{y1}\tan(k_{y1}y_1)\left(\frac{\tan(k_{y2}y_2)}{k_{y2}}\right)\right]$$
$$+\left(\frac{\varepsilon_{s3}}{\varepsilon_{s2}}\right)\left[1 + \zeta_0\varepsilon_{s3}\left(\frac{\tan(k_{y3}y_3)}{k_{y3}}\right)\right]\left[k_{y2}\tan(k_{y2}y_2) + \left(\frac{\varepsilon_{s2}}{\varepsilon_{s1}}\right)k_{y1}\tan(k_{y1}y_1)\right] = 0 \tag{2.86}$$

where ζ_0 and k_{yi} $(i = 1, 2, 3)$ can be expressed in terms of β_{zs} using (2.81). The solution of (2.86) then yields the value of β_{zs}.

We now consider three different cases depending on the relative values of ε_{s1}, ε_{s2}, and ε_{s3}. We choose the traveling wave solution in the slab having the highest dielectric constant while the remaining two slabs may support either traveling waves or exponentially decaying waves.

CASE 1: $\varepsilon_{s1} > \varepsilon_{s2}, \varepsilon_{s3}$ Since slab 1 has the highest dielectric constant, we set $k_{y1} = \beta_y$ (real). Using (2.81), the other parameters can be expressed in terms of β_y as

$$\zeta_0 = [k_0^2(\varepsilon_{s1} - 1) - \beta_y^2]^{1/2} \tag{2.87a}$$

$$k_{y2} = [\beta_y^2 - k_0^2(\varepsilon_{s1} - \varepsilon_{s2})]^{1/2} \tag{2.87b}$$

$$k_{y3} = [\beta_y^2 - k_0^2(\varepsilon_{s1} - \varepsilon_{s3})]^{1/2} \tag{2.87c}$$

Substituting for ζ_0, k_{y2}, and k_{y3} from (2.87) in (2.86), we can solve the characteristic equation for β_y. For determining this root, β_y is varied from 0 to $k_0\sqrt{\varepsilon_{s1} - 1}$. It may be noted that over this range of β_y, β_{zs} varies from $k_0\sqrt{\varepsilon_{s1}}$ to k_0, k_{y2} is real for $\beta_y > k_0\sqrt{\varepsilon_{s1} - \varepsilon_{s2}}$ and imaginary for $0 < \beta_y < k_0\sqrt{\varepsilon_{s1} - \varepsilon_{s2}}$, k_{y3} is real for $\beta_y > k_0\sqrt{\varepsilon_{s1} - \varepsilon_{s3}}$ and imaginary for $0 < \beta_y < k_0\sqrt{\varepsilon_{s1} - \varepsilon_{s3}}$.

CASE 2: $\varepsilon_{s2} > \varepsilon_{s1}, \varepsilon_{s3}$ We set $k_{y2} = \beta_y$ (real) and express ζ_0, k_{y1}, and k_{y3} in terms of β_z. Numerical solution of the characteristic equation (2.86) is found by varying β_y from 0 to $k_0\sqrt{\varepsilon_{s2} - 1}$. Over this range of β_y, k_{y1} and k_{y3} may assume real or imaginary values.

CASE 3: $\varepsilon_{s3} > \varepsilon_{s1}, \varepsilon_{s2}$ In this case, we set $k_{y3} = \beta_y$ (real) and express ζ_0, k_{y1}, and k_{y2} in terms of β_y. Numerical solution of (2.86) is obtained by varying β_y from 0 to $k_0\sqrt{\varepsilon_{s3} - 1}$. Over this range of β_y, k_{y1} and k_{y2} may be real or imaginary.

Two-Slab Guide with Bottom Ground Plane (Fig. 2.11(b)) The characteristic equation for the two-slab guide shown in Figure 2.11(b) can be obtained as a special case of the three-slab guide (Fig. 2.11(a)) by substituting $\varepsilon_{s3} = 1$, $k_{y3} = -j\zeta_0$ (with ζ_0 real), and $y_3 = \infty$.

$$\left[1 - \left(\frac{\varepsilon_{s2}}{\varepsilon_{s1}}\right) k_{y1} \tan(k_{y1} y_1) \left(\frac{\tan(k_{y2} y_2)}{k_{y2}}\right)\right] - \left(\frac{1}{\zeta_0 \varepsilon_{s2}}\right)\left[k_{y2} \tan(k_{y2} y_2) + \left(\frac{\varepsilon_{s2}}{\varepsilon_{s1}}\right) k_{y1} \tan(k_{y1} y_1)\right] = 0 \tag{2.88}$$

where ζ_0 and k_{yi} (for $i = 1$ and 2) are related through Eq. (2.81). In the case of a slab guide having $\varepsilon_{s1} > \varepsilon_{s2}$, we set $k_{y1} = \beta_y$ (real) and express ζ_0 and k_{y2} in terms of β_y. The characteristic equation (2.88) is then solved for β_y by varying its value from 0 to $k_0\sqrt{\varepsilon_{s1} - 1}$. If on the other hand, we set $k_{y2} = \beta_y$ (real) and express ζ_0 and k_{y1} in terms of β_y, the solution of (2.88) is then obtained by varying β_y from 0 to $k_0\sqrt{\varepsilon_{s2} - 1}$.

Single-Slab Guide with Bottom Ground Plane (Fig. 2.11(c)) We consider traveling wave propagation in the slab region. The characteristic equation for the guide is obtained by setting $\varepsilon_{s2} = 1$, $k_{y2} = -j\zeta_0$ (with ζ_0 real), and $y_2 = \infty$ in (2.88). It is given by

$$[1 - (\beta_y/\zeta_0 \varepsilon_{s1}) \tan(\beta_y y_1)] = 0 \tag{2.89}$$

where

$$\zeta_0 = [k_0^2(\varepsilon_{s1} - 1) - \beta_y^2]^{1/2} \tag{2.90}$$

Three-Slab Guide with Top and Bottom Ground Plane (Fig. 2.11(d)) Referring to Figure 2.11(d), the general solution for $\phi^e(y)$ in the three-slab regions can be written as

$$\phi^e(y) = \begin{cases} A\cos[k_{y1}(y + y_1)], & -y_1 \le y \le 0 \\ B_1\cos(k_{y2} y) + B_2 \sin(k_{y2} y), & 0 \le y \le y_2 \\ C\cos[k_{y3}(y - y_2 - y_3)], & y_2 \le y \le y_2 + y_3 \end{cases} \tag{2.91}$$

where

$$\beta_{zs}^2 = k_0^2 \varepsilon_{si} - k_{yi}^2, \quad i = 1, 2, 3 \tag{2.92}$$

From the transmission line model of the slab guide, the expressions for the input impedances Z^- and Z^+ at $y = 0$ can be written as

$$Z^- = jZ_{01}\tan(k_{y1}y_1) \tag{2.93}$$

$$Z^+ = jZ_{02}\left[\frac{Z_{03}\tan(k_{y3}y_3) + Z_{02}\tan(k_{y2}y_2)}{Z_{02} - Z_{03}\tan(k_{y3}y_3)\tan(k_{y2}y_2)}\right] \tag{2.94}$$

where

$$Z_{0i} = k_{yi}/(\omega\varepsilon_0\varepsilon_{si}), \quad i = 1, 2, 3 \tag{2.95}$$

Using Eqs. (2.93)–(2.95) in the resonance condition (2.63) and simplifying, we obtain

$$\begin{aligned} k_{y3}\tan(k_{y3}y_3)&\left[1 - \left(\frac{\varepsilon_{s2}}{\varepsilon_{s1}}\right)k_{y1}\tan(k_{y1}y_1)\left(\frac{\tan(k_{y2}y_2)}{k_{y2}}\right)\right] \\ &+ \left(\frac{\varepsilon_{s3}}{\varepsilon_{s2}}\right)\left[k_{y2}\tan(k_{y2}y_2) + \left(\frac{\varepsilon_{s2}}{\varepsilon_{s1}}\right)k_{y1}\tan(k_{y1}y_1)\right] = 0 \end{aligned} \tag{2.96}$$

This horizontal guide model is useful for analyzing the shielded dielectric guides shown in Figure 2.4. The following special case is of practical interest:

SPECIAL CASE: $\varepsilon_{s2} > \varepsilon_{s1}$ AND $\varepsilon_{s3} = 1$ For this case, we substitute $k_{y3} = -j\zeta_0$, $k_{y2} = \beta_y$, and $\varepsilon_{s3} = 1$ in (2.96) to obtain

$$\begin{aligned} \zeta_0\tanh(\zeta_0 y_3)&\left[1 - \left(\frac{\varepsilon_{s2}}{\varepsilon_{s1}}\right)k_{y1}\tan(k_{y1}y_1)\left(\frac{\tan(\beta_y y_2)}{\beta_y}\right)\right] \\ &- \left(\frac{1}{\varepsilon_{s2}}\right)\left[\beta_y\tan(\beta_y y_2) + \left(\frac{\varepsilon_{s2}}{\varepsilon_{s1}}\right)k_{y1}\tan(k_{y1}y_1)\right] = 0 \end{aligned} \tag{2.97}$$

where

$$\zeta_0 = [k_0^2(\varepsilon_{s2} - 1) - \beta_y^2]^{1/2} \tag{2.98a}$$

$$k_{y1} = [\beta_y^2 - k_0^2(\varepsilon_{s2} - \varepsilon_{s1})]^{1/2} \tag{2.98b}$$

Three-Slab Dielectric Guide with No Ground Plane (Fig. 2.11(e)) The dielectric slab guides without any metallic planes considered in Figure 2.11(e)–(g) are useful for analysis of NRD guides (Figs. 2.6 and 2.7(d)) as well as optical integrated guides (Fig. 2.8).

We first consider the three-slab structure shown in Figure 2.11(e). Since the fields must decay exponentially away from the guide in the two air regions, we can assume $k_{y1} = k_{y5} = -j\zeta_0$, where ζ_0 is real and positive. The solution for $\phi^e(y)$ in the various regions can be written as

$$\phi^e(y) = \begin{cases} Ae^{\zeta_0(y+y_2)}, & y \leq -y_2 \\ B_1 \cos(k_{y2}y) + B_2 \sin(k_{y2}y), & -y_2 \leq y \leq 0 \\ C_1 \cos(k_{y3}y) + C_2 \sin(k_{y3}y), & 0 \leq y \leq y_3 \\ D_1 \cos(k_{y4}y) + D_2 \sin(k_{y4}y), & y_3 \leq y \leq y_3 + y_4 \\ Ee^{-\zeta_0(y-y_3-y_4)}, & y \geq y_3 + y_4 \end{cases} \tag{2.99}$$

where

$$\beta_{zs}^2 = k_0^2 + \zeta_0^2 = k_0^2 \varepsilon_{si} - k_{yi}^2, \quad i = 2, 3, 4 \tag{2.100}$$

Considering the transmission line equivalent of the guide, we can write $Z_{01} = Z_{05} = -j\zeta_0/\omega\varepsilon_0$. The impedances Z^- and Z^+ (at $y = 0$ in Fig. 2.11(e)) are given by

$$Z^- = jZ_{02} \left[\frac{-(\zeta_0/\omega\varepsilon_0) + Z_{02} \tan(k_{y2}y_2)}{Z_{02} + (\zeta_0/\omega\varepsilon_0) \tan(k_{y2}y_2)} \right] \tag{2.101}$$

$$Z^+ = Z_{03} \left[\frac{Z_4 + jZ_{03} \tan(k_{y3}y_3)}{Z_{03} + jZ_4 \tan(k_{y3}y_3)} \right] \tag{2.102}$$

where

$$Z_4 = jZ_{04} \left[\frac{-(\zeta_0/\omega\varepsilon_0) + Z_{04} \tan(k_{y4}y_4)}{Z_{04} + (\zeta_0/\omega\varepsilon_0) \tan(k_{y4}y_4)} \right] \tag{2.103a}$$

$$Z_{0i} = k_{yi}/(\omega\varepsilon_0\varepsilon_{si}), \quad i = 2, 3, 4 \tag{2.103b}$$

Substituting (2.103) in (2.101) and (2.102), and using the resonance condition (2.63), we get

$$\left(\frac{\varepsilon_{s3}}{\varepsilon_{s2}}\right) \left[\frac{-\zeta_0\varepsilon_{s2} + k_{y2} \tan(k_{y2}y_2)}{1 + \zeta_0\varepsilon_{s2} [\tan(k_{y2}y_2)/k_{y2}]} \right]$$
$$+ \left[\frac{[-\zeta_0\varepsilon_{s4} + k_{y4} \tan(k_{y4}y_4)] + (\varepsilon_{s4}/\varepsilon_{s3}) k_{y3} \tan(k_{y3}y_3)[1 + \zeta_0\varepsilon_{s4} \tan(k_{y4}y_4)/k_{y4}]}{(\varepsilon_{s4}/\varepsilon_{s3})\{1 + \zeta_0\varepsilon_{s4}[\tan(k_{y4}y_4)/k_{y4}]\} - [\tan(k_{y3}y_3)/k_{y3}][-\zeta_0\varepsilon_{s4} + k_{y4} \tan(k_{y4}y_4)]} \right] = 0 \tag{2.104}$$

Two-Slab Dielectric Guide with No Ground Plane (Fig. 2.11(f)) The two-slab dielectric guide considered in Figure 2.11(f) is a special case of the guide shown in Figure 2.11(a). The characteristic equation is obtained by substituting $k_{y1} = -j\zeta_1$ and $y_1 = \infty$ in (2.86). The resulting expression is given by

$$[k_{y3}\tan(k_{y3}y_3) - \zeta_0\varepsilon_{s3}]\left[1 + \zeta_1\left(\frac{\varepsilon_{s2}}{\varepsilon_{s1}}\right)\left(\frac{\tan(k_{y2}y_2)}{k_{y2}}\right)\right] + \left(\frac{\varepsilon_{s3}}{\varepsilon_{s2}}\right)\left[1 + \zeta_0\varepsilon_{s3}\left(\frac{\tan(k_{y3}y_3)}{k_{y3}}\right)\right]\left[k_{y2}\tan(k_{y2}y_2) - \left(\frac{\zeta_1\varepsilon_{s2}}{\varepsilon_{s1}}\right)\right] = 0 \tag{2.105}$$

where

$$\beta_{zs}^2 = k_0^2 + \zeta_0^2 = k_0^2\varepsilon_{s1} + \zeta_1^2 = k_0^2\varepsilon_{s2} - k_{y2}^2 = k_0^2\varepsilon_{s3} - k_{y3}^2 \tag{2.106}$$

Single-Slab Dielectric Guide with No Ground Plane (Fig. 2.11(g)) The single-slab guide considered in Figure 2.11(g) is a special case of the guide shown in Figure 2.11(b). The characteristic equation is obtained from (2.88) by substituting $k_{y1} = -j\zeta_1$, $y_1 = \infty$, and $k_{y2} = \beta_y$ (real).

$$[1 + (\zeta_1/\beta_y)(\varepsilon_{s2}/\varepsilon_{s1})\tan(\beta_y y_2)] - (1/\zeta_0\varepsilon_{s2})[\beta_y\tan(\beta_y y_2) - (\zeta_1\varepsilon_{s2}/\varepsilon_{s1})] = 0 \tag{2.107}$$

where

$$\beta_{zs}^2 = k_0^2 + \zeta_0^2 = k_0^2\varepsilon_{s1} + \zeta_1^2 = k_0^2\varepsilon_{s2} - \beta_y^2 \tag{2.108}$$

Symmetric Three-Slab Guide (Fig. 2.11(h)) Figure 2.11(h) shows the top symmetric half of a three-slab guide with no metallic walls. This model is useful for analysis of the edge-coupled nonradiative guide (Fig. 2.6(c)). Using the transmission line equivalent, the expressions for Z^+ and Z^- are obtained as

$$Z^+ = \left(\frac{j}{\omega\varepsilon_0\varepsilon_{s2}}\right)\left[\frac{-1 + (k_{y2}/\zeta_0\varepsilon_{s2})\tan(k_{y2}y_2)}{(1/\zeta_0\varepsilon_{s2}) + [\tan(k_{y2}y_2)/k_{y2}]}\right] \tag{2.109}$$

$$Z^- = \begin{cases} -j(k_{y1}/\omega\varepsilon_0\varepsilon_{s1})\cot(k_{y1}y_1), & \text{for even mode} \quad (2.110\text{a}) \\ j(k_{y1}/\omega\varepsilon_0\varepsilon_{s1})\tan(k_{y1}y_1), & \text{for odd mode} \quad (2.110\text{b}) \end{cases}$$

Using (2.109) and (2.110) in the resonance condition (2.63), we obtain

$$T\left(\frac{\varepsilon_{s1}}{\varepsilon_{s2}}\right)\left[1 - \left(\frac{k_{y2}}{\varepsilon_0\varepsilon_{s2}}\right)\tan(k_{y2}y_2)\right] + \left[\left(\frac{1}{\zeta_0\varepsilon_{s2}}\right) + \left(\frac{\tan(k_{y2}y_2)}{k_{y2}}\right)\right] = 0$$

$$T = \begin{cases} \tan(k_{y1}y_1)/k_{y1}, & \text{even mode} \\ -\cot(k_{y1}y_1)/k_{y1}, & \text{odd mode} \end{cases} \tag{2.111}$$

2.4.4 Analysis of Vertical Layered Guide Models

The structures shown in Figure 2.12 and their special cases serve as vertical layered guide models for all the dielectric integrated guides presented in Figures 2.2–2.8. It may be noted that the dielectric guides with no side walls are covered by letting either $x_3 \to \infty$ in the five-layered models shown in Figure 2.12(a) and (b), or $x_2 \to \infty$ in the three-layered models shown in Figure 2.12(c) and (d). Asymmetrically coupled dielectric guides have applications in the design of directional couplers and filters wherein it is more convenient to use coupled strips of unequal widths rather than unequal dielectric constants. Therefore, in the asymmetric vertical layered models (Fig. 2.12(a), (c)), the two vertical coupled layers marked 2 are chosen to have the same dielectric constant but different widths.

In the vertical layered guides (Fig. 2.12), the dielectric layers are of infinite extent in the y-direction. We consider wave propagation in the z-direction according to $\exp(-j\beta_z z)$. The parameters ε_{ej} and k_{xj} denote the relative dielectric constant and x-directed propagation constant of the layer numbered as j. The characteristic impedance Z'_{0j} of the corresponding transmission line section is defined as in (2.73).

Asymmetric Five-Layered Guide with Metallic Side Walls (Fig. 2.12(a)) The general solution for $\phi^e(y)$ in the various regions of the five-layered guide shown in Figure 2.12(a) can be written as

$$\phi^e(x) = \begin{cases} A\sin[k_{x3}(x + 2x_1 + x'_2 + x_3)], & -(2x_1 + x'_2 + x_3) \le x \le 2x_1 + x'_2 \\ B_1\cos(k_{x2}x) + B_2\sin(k_{x2}x), & -(2x_1 + x'_2) \le x \le 2x_1 \\ C_1\cos(k_{x1}x) + C_2\sin(k_{x1}x), & -2x_1 \le x \le 0 \\ D_1\cos(k_{x2}x) + D_2\sin(k_{x2}x), & 0 \le x \le x_2 \\ E\sin[k_{x3}(x - x_2 - x_3)], & x_2 \le x \le x_2 + x_3 \end{cases} \tag{2.112}$$

where

$$\beta_z^2 = k_0^2\varepsilon_{ej} - k_{xj}^2, \quad j = 1, 2, 3 \tag{2.113}$$

Since this structure is used for modeling coupled dielectric guides, we consider layers marked 2 to represent the strip regions carrying most of the energy of the coupled guide. We therefore consider traveling waves in layers marked 2 and set $k_{x2} = \beta_x$ (real). In layers 1 and 3, the propagation may correspond to traveling waves or decaying waves so that k_{x1} and k_{x3} can be real or imaginary. If we consider an equivalent model of the guide, then the wave impedances are given by (refer Fig. 2.12(a))

$$Z'_{0j} = \omega\mu_0/k_{xj}, \quad j = 1, 2, 3 \tag{2.114}$$

The input impedances Z^+ and Z^- looking toward $+x$ and $-x$ directions,

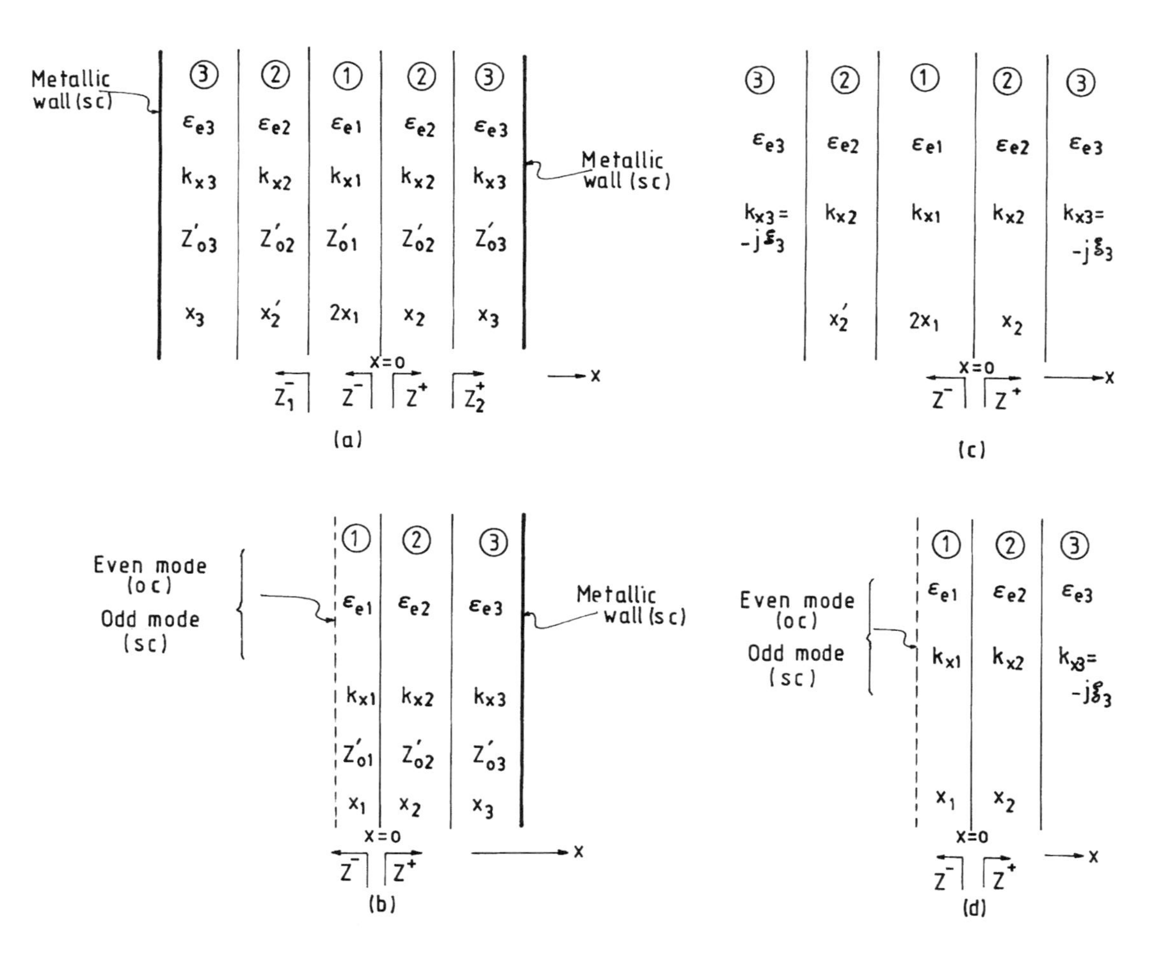

Metallic wall (sc)
Even mode (oc)
Odd mode (sc)
x=0
(a)
(b)
(c)
(d)

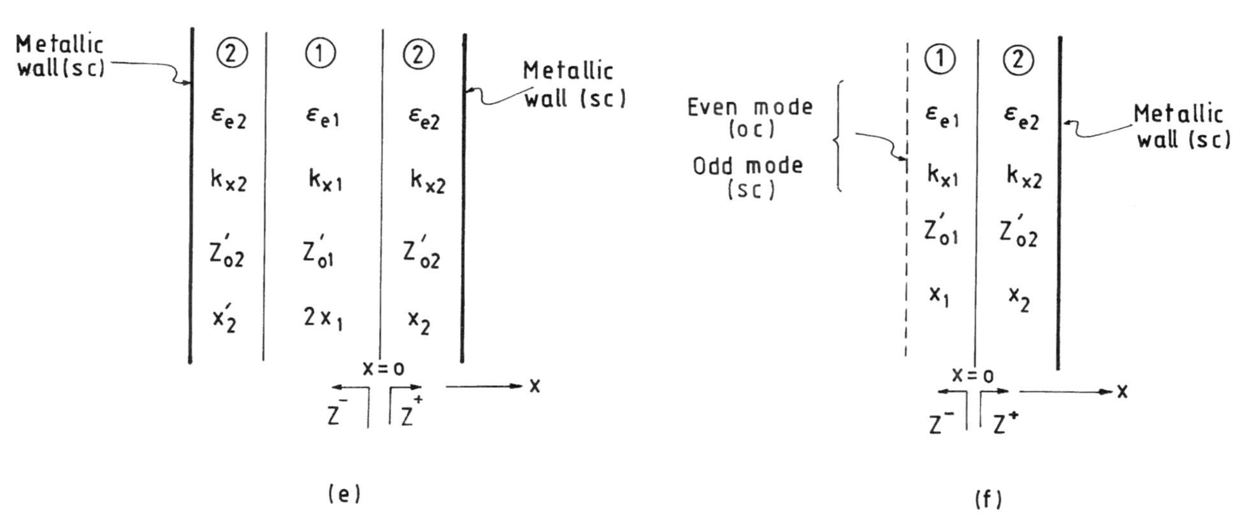

FIGURE 2.12 Vertical layered guide models suitable for dielectric integrated guides shown in Figures 2.2–2.8. Five layered: (a) asymmetric and (b) symmetric guides with side metal walls. Three layered: (c) asymmetric and (d) symmetric guides with no side walls. Three layered: (e) asymmetric and (f) symmetric guides with metal side walls.

respectively, at $x = 0$ are given by

$$Z^+ = j\omega\mu_0 \left[\frac{[\tan(k_{x3}x_3)/k_{x3}] + [\tan(k_{x2}x_2)/k_{x2}]}{1 - k_{x2}\tan(k_{x2}x_2)[\tan(k_{x3}x_3)/k_{x3}]} \right] \tag{2.115}$$

$$Z^- = \left(\frac{\omega\mu_0}{k_{x1}} \right) \left[\frac{Z_1^- + j(\omega\mu_0/k_{x1})\tan(2k_{x1}x_1)}{(\omega\mu_0/k_{x1}) + jZ_1^- \tan(2k_{x1}x_1)} \right] \tag{2.116}$$

where

$$Z_1^- = j\omega\mu_0 \left[\frac{[\tan(k_{x3}x_3)/k_{x3}] + [\tan(k_{x2}x_2')/k_{x2}]}{1 - k_{x2}\tan(k_{x2}x_2')[\tan(k_{x3}x_3)/k_{x3}]} \right] \tag{2.117}$$

Using (2.115)–(2.117) and applying the resonance condition (2.63), we obtain the characteristic equation as

$$\left[\frac{\{[\tan(k_{x3}x_3)/k_{x3}] + [\tan(k_{x2}x_2')/k_{x2}]\} + [\tan(2k_{x1}x_1)/k_{x1}]\{1 - [\tan(k_{x3}x_3)/k_{x3}]k_{x2}\tan(k_{x2}x_2')\}}{\{1 - [\tan(k_{x3}x_3)/k_{x3}]k_{x2}\tan(k_{x2}x_2')\} - k_{x1}\tan(2k_{x1}x_1)\{[\tan(k_{x3}x_3)/k_{x3}] + [\tan(k_{x2}x_2')/k_{x2}]\}} \right]$$

$$+ \left[\frac{[\tan(k_{x3}x_3)/k_{x3}] + [\tan(k_{x2}x_2)/k_{x2}]}{1 - k_{x2}\tan(k_{x2}x_2)[\tan(k_{x3}x_3)/k_{x3}]} \right] = 0 \tag{2.118}$$

Symmetric Five-Layered Guide with Metallic Side Walls (Fig. 2.12(b)) The symmetric five layered guide shown in Figure 2.12(b) is a special case of the asymmetric structure considered in Figure 2.12(a) with $x_2' = x_2$. The characteristic equation is therefore given by (2.118) with x_2' replaced by x_2. The same solution can be obtained by making use of the symmetry of the structure and analyzing it in terms of the even and odd modes. At the symmetry plane $x = -x_1$, the transmission line presents an open circuit for the even mode and a short circuit for the odd mode. Thus the expression for Z^- at $x = 0$ is given by

$$Z^- = -j(\omega\mu_0/k_{x1})\cot(k_{x1}x_1), \quad \text{even mode} \tag{2.119a}$$

$$= j(\omega\mu_0/k_{x1})\tan(k_{x1}x_1), \quad \text{odd mode} \tag{2.119b}$$

Using (2.115) and (2.119) in the resonance relation (2.63), we obtain

$$T\left\{ \left(\frac{\tan(k_{x2}x_2)}{k_{x2}} \right) + \left(\frac{\tan(k_{x3}x_3)}{k_{x3}} \right) \right\} - \left\{ 1 - k_{x2}\tan(k_{x2}x_2) \left(\frac{\tan(k_{x3}x_3)}{k_{x3}} \right) \right\} = 0,$$

$$T = \begin{cases} k_{x1}\tan(k_{x1}x_1), & \text{even mode} \\ -k_{x1}\cot(k_{x1}x_1), & \text{odd mode} \end{cases} \tag{2.120}$$

Asymmetric Three-Layered Guide with No Side Walls (Fig. 2.12(c)) In the asymmetric three-layered guide considered in Figure 2.12(c), region 3 extends to infinity in the x-direction. We therefore consider decaying type waves by choosing $k_{x3} = -j\xi_3$ (with ξ_3 real). The characteristic equation can be obtained from (2.118) by setting $k_{x3} = -j\xi_3$ and $x_3 = \infty$. This is equivalent to replacing $[\tan(k_{x3}x_3)/k_{x3}]$ by $1/\xi_3$.

Symmetric Three-Layered Guide with No Side Walls (Fig. 2.12(d)) The characteristic equation for this guide is obtained as a special case of Figure 2.12(b) with $k_{x3} = -j\xi_3$ and $x_3 = \infty$. It is given by replacing $[\tan(k_{x3}x_3)/k_{x3}]$ by $1/\xi_3$ in (2.120).

Asymmetric Three-Layered Guide with Metal Side Walls (Fig. 2.12(e)) The characteristic equation for this guide is the same as that given by (2.118) for the five-layered guide (Fig. 2.12(a)) with $x_3 = 0$. The expression reduces to

$$\left(\frac{\tan(k_{x2}x_2)}{k_{x2}}\right)\left[1 - k_{x1}\tan(2k_{x1}x_1)\left(\frac{\tan(k_{x2}x_2')}{k_{x2}}\right)\right] + \left[\left(\frac{\tan(k_{x2}x_2')}{k_{x2}}\right) + \left(\frac{\tan(2k_{x1}x_1)}{k_{x1}}\right)\right] = 0 \tag{2.121}$$

Symmetric Three-Layered Guide with Metal Side Walls (Fig. 2.12(f)) The characteristic equation is obtained by setting $x_3 = 0$ in the expression (2.120) derived for Figure 2.12(b). It is given by

$$T\left(\frac{\tan(k_{x2}x_2)}{k_{x2}}\right) - 1 = 0, \quad T = \begin{cases} k_{x1}\tan(k_{x1}x_1), & \text{even mode} \\ -k_{x1}\cot(k_{x1}x_1), & \text{odd mode} \end{cases} \tag{2.122}$$

2.4.5 Application to Dielectric Integrated Guides

Appendices 2A and 2B list the characteristic equations for a range of horizontal and vertical slab guides. These are obtained directly from the expressions derived in Sections 2.4.3 and 2.4.4 for the more generalized structures. The characteristic equations corresponding to the E^y_{mn} modes of any dielectric integrated guide shown in Figures 2.2–2.8 are obtained by identifying the appropriate horizontal slab guide from Appendix 2A and the vertical layered model from Appendix 2B and then substituting the guide parameters in the corresponding expressions. We consider below one example and enumerate the procedure for computing the guide wavelength. A similar procedure applies to all other guides listed in Figures 2.2–2.8.

Example: Cladded Image Guide Figure 2.13 shows the geometry of a cladded image guide indicating the different constitutive regions. Let ε_{e1} and ε_{e2} refer to the effective dielectric constants of the horizontal slab guide models corresponding to regions I and II, respectively. For region III, $\varepsilon_{e3} = 1$. We note that for determining β_y of regions I and II (and hence ε_{e1} and ε_{e2}), the horizontal slab guide models H2 and H1, respectively, of Appendix 2A apply. Next, for determin-

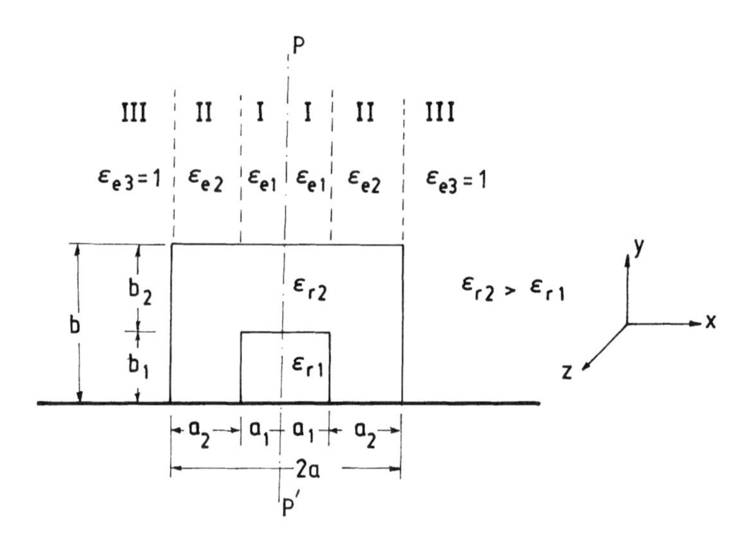

FIGURE 2.13 Cladded image guide showing various constituent regions for application of the EDC formulas.

ing β_z of the original guide, the vertical layered model V6 of Appendix 2B applies with $x_2' = x_2$. Alternatively, we can make use of the symmetry of the guide with respect to the plane PP' (see Fig. 2.13) and use the vertical layered model V2 to solve first for β_x and then determine β_z, for even and odd modes, separately. For a given set of guide parameters (ε_{r1}, ε_{r2}, a_1, a_2, b_1, and b_2) and frequency, the computational procedure for determining the guide wavelength λ is as follows:

- Determine β_y for region I using model H2 of Appendix 2A.
  Substitute $y_1 = b_1$, $y_2 = b_2$, $\varepsilon_{s1} = \varepsilon_{r1}$, and $\varepsilon_{s2} = \varepsilon_{r2}$ in Eq. (2A.2) and solve for β_y at a specified frequency. For this, β_y is varied from 0 to $k_0\sqrt{\varepsilon_{r2} - \varepsilon_{r1}}$ to find the root of the characteristic equation (2A.2a). It may be noted that, over this range of β_y, ζ_0 is real but k_{y1} is imaginary. If we denote $k_{y1} = -j\zeta_1$, then $[\tan(k_{y1}y_1)/k_{y1}]$ in Eq. (2A.2a) gets replaced by $[\tanh(\zeta_1 y_1)/\zeta_1]$, where $\zeta_1 = [k_0^2(\varepsilon_{s2} - \varepsilon_{s1}) - \beta_y^2]^{1/2}$ is real. Let the root of (2A.2) be denoted as β_{y1}.
- Determine ε_{e1} using

$$\varepsilon_{e1} = \varepsilon_{r2} - (\beta_{y1}/k_0)^2 \tag{2.123}$$

- Determine β_y for region II using model H1 of Appendix 2A.
  Substitute $y_1 = b$ and $\varepsilon_{s1} = \varepsilon_{r2}$ in Eq. (2A.1) and solve for β_y by varying β_y from 0 to $k_0\sqrt{\varepsilon_{r2} - 1}$. Let the root be denoted as β_{y2}.
- Determine ε_{e2} using

$$\varepsilon_{e2} = \varepsilon_{r2} - (\beta_{y2}/k_0)^2 \tag{2.124}$$

- Determine β_x for even and odd modes using model V2 of Appendix 2B.

  Substituting $x_1 = a_1$, $x_2 = a_2$, and values of ε_{e1} and ε_{e2} in Eq. (2B.2), we solve for β_x. We note that $\varepsilon_{e3} = 1$ since region III is air, and with $\varepsilon_{r2} > \varepsilon_{r1}$, we have $\varepsilon_{e2} > \varepsilon_{e1}$. The roots β_{xe} for the even mode and β_{xo} for the odd mode are found from Eq. (2B.2a) by varying β_x from 0 to $k_0\sqrt{\varepsilon_{e2} - \varepsilon_{e1}}$. Over this range k_{x1} is imaginary. We may replace $k_{x1}\tan(k_{x1}x_1)$ by $-\zeta_1\tanh(\zeta_1 x_1)$, and $k_{x1}\cot(k_{x1}x_1)$ by $-\zeta_1\coth(\zeta_1 x_1)$ in Eq. (2B.2a), where $\zeta_1 = [k_0^2\sqrt{(\varepsilon_{s2} - \varepsilon_{s1})} - \beta_y^2]^{1/2}$.
- Determine β_z and λ.

  For the even mode

$$\beta_z = \beta_{ze} = (k_0^2\varepsilon_{e2} - \beta_{xe}^2)^{1/2} \tag{2.125a}$$

$$\lambda = \lambda_{oe} = 2\pi/\beta_{ze} \tag{2.125b}$$

For the odd mode

$$\beta_z = \beta_{zo} = (k_0^2\varepsilon_{e2} - \beta_{xo}^2)^{1/2} \tag{2.126a}$$

$$\lambda = \lambda_{oo} = 2\pi/\beta_{zo} \tag{2.126b}$$

2.5 NETWORK ANALYSIS METHOD

2.5.1 Basic Approach

Consider an image guide shown in Figure 2.1. The wave guidance along the guide in Reg I can be considered as surface waves (TE or TM type) bouncing back and forth at an angle to the side surfaces of the dielectric strip. These surface waves are the discrete modes. The coupling of TE and TM modes at the discontinuity interface gives rise to nonsurface waves comprising a continuous spectrum in Reg II. In a general open planar dielectric guide structure having discontinuity in the transverse direction, the complete modal spectrum is comprised of surface waves and nonsurface waves. For mathematical convenience the continuous spectrum can be discretized by considering a perfectly conducting plate above the open planar dielectric guide. The structure then resembles a partially dielectric-filled parallel-plate waveguide, which supports not only a finite number of surface wave modes but also an infinite number of discrete higher order modes, some of which propagate while the others remain cutoff depending on the plate separation.

The EDC technique described in Section 2.4 neglects the effect of discontinuities at the strip edges of the dielectric guide. The "microwave network analysis" method reported by Peng and Oliner [3] and Koshiba et al. [5, 6, 16, 17] overcomes this limitation by taking into account the modal spectrum consisting of surface waves and the nonsurface waves. The nonsurface wave modes form a continuous spectrum. The procedure involves using a network representation for the cross-section of the guide and including the dielectric step or a change in dielectric constant as a transverse discontinuity. Each of the constituent regions in the cross-section of the guide is treated as a uniform transmission line section and the transverse discontinuity is represented as an equivalent lumped network. The determination of the elements of the lumped network requires the use of normalized mode functions of both the surface wave and the nonsurface wave modes in the planar layered structures. The characteristic equation for the guide is obtained by applying the transverse resonance relation to the network; that is, by setting the sum of the admittances looking into two opposite directions at any point in the network system equal to zero.

2.5.2 Analysis of a Dielectric Step Junction

The network analysis technique is described for a step discontinuity between two planar dielectric guides. The generalized analysis due to Koshiba et al. [6], which takes into account the contribution due to the discrete modes as well as the continuous spectrum of modes, is presented.

Consider a planar slab guide structure as shown in Figure 2.14. The structure has a transverse discontinuity at $x = x_0$ in the form of a step change in the dielectric constant. The guide extends to infinity in both the x- and z-directions. We assume wave propagation in the guide along the z-direction according to $\exp[j(\omega t - \beta_z z)]$ and consider the analysis for E^y_{mn} modes ($H_y = 0$) of the guide. Since the equivalent transmission line network extends along the x-direction, we first express the field components in the two constituent regions ($R = \mathrm{I, II}$) in terms of the modal voltage $V^R_n(x)$ and modal current $I^R_n(x)$.

$$E^R_y = \left[\sum_n V^R_n(x) f^R_n(y) + \int_0^\infty V^R(x,\rho) f^R(y,\rho)\, d\rho \right] e^{-j\beta_z z} \tag{2.217}$$

$$H^R_z = \left[\sum_n I^R_n(x) g^R_n(y) + \int_0^\infty I^R(x,\rho) g^R(y,\rho)\, d\rho \right] e^{-j\beta_z z}, \quad R = \mathrm{I, II} \tag{2.218}$$

The mode functions $f^R_n(y)$ and $g^R_n(y)$ for the discrete TM_n modes, and $f^R(y,\rho)$ and $g^R(y,\rho)$ for the continuous TM-mode spectrum are chosen such that they satisfy the following orthonormality condition:

$$\int_0^\infty [f^R_n(y)]^* g^R_l(y)\, dy = \int_0^\infty [g^R_n(y)]^* f^R_l(y)\, dy = \delta_{nl} \tag{2.129a}$$

$$\int_0^\infty [f^R(y,\rho)]^* g^R(y,\rho')\, dy = \int_0^\infty [g^R(y,\rho)]^* f^R(y,\rho')\, dy = \delta(\rho - \rho') \tag{2.129b}$$

where δ_{nl} and $\delta(\rho - \rho')$ denote the Kronecker delta and Dirac delta functions, respectively. The modal voltage $V^R_n(x)$ and the modal current $I^R_n(x)$ satisfy the

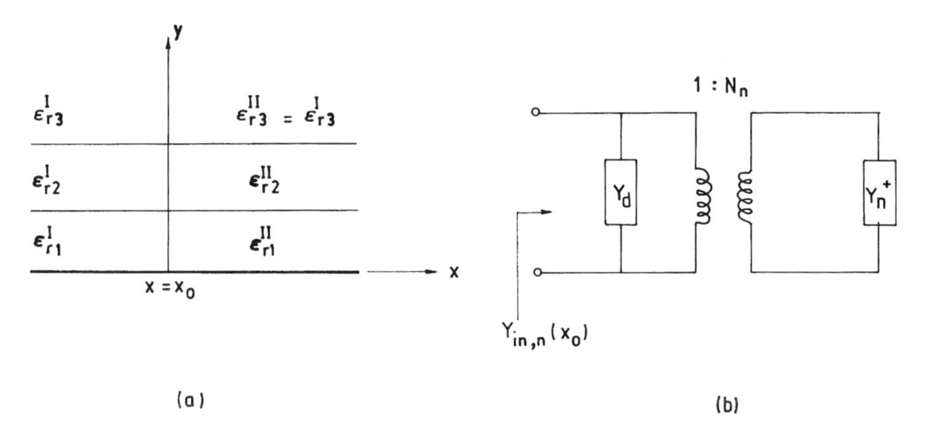

FIGURE 2.14 (a) Transverse step discontinuity in a dielectric guide. (b) Transverse equivalent network.

following transmission line equations:

$$\frac{dV_n^R(x)}{dx} = -jk_{xn}^R Z_n^R I_n^R(x) \tag{2.130a}$$

$$\frac{dI_n^R(x)}{dx} = -jk_{xn}^R Y_n^R V_n^R(x) \tag{2.130b}$$

where

$$\beta_z^2 = k_0^2 \varepsilon_{en}^R - (k_{xn}^R)^2 \tag{2.131a}$$

$$\varepsilon_{en}^R = (\beta_{zs,n}^R / k_0)^2 \tag{2.131b}$$

$$k_0^2 = \omega^2 \mu_0 \varepsilon_0 \tag{2.131c}$$

The parameter ε_{en}^R is the effective dielectric constant of the horizontal slab guide model of region R extending to infinity in the x-direction, and $\beta_{zs,n}^R$ is its z-directed propagation constant. Following Koshiba et al. [6], we define the modal characteristic impedance Z_n^R in the form

$$Z_n^R = \left(\frac{1}{Y_n^R}\right) = \left(\frac{\omega\mu_0\varepsilon_{en}^R}{k_{xn}^R}\right) \tag{2.132}$$

For the continuous spectrum in region R (= I, II), the transmission line equations are given by

$$\frac{dV^R(x,\rho)}{dx} = -jk_x^R(\rho) Z^R(\rho) I^R(x,\rho) \tag{2.133a}$$

$$\frac{dI^R(x,\rho)}{dx} = -jk_x^R(\rho) Y^R(\rho) V^R(x,\rho) \tag{2.133b}$$

where

$$\beta_z^2 = k_0^2 \varepsilon_{r3}^R - [k_x^R(\rho)]^2 - \rho^2 \tag{2.134}$$

The characteristic impedance $Z^R(\rho)$ is given by

$$Z^R(\rho) = \left(\frac{1}{Y^R(\rho)}\right) = \left(\frac{\omega\mu_0 \varepsilon_e^R}{k_x^R(\rho)}\right) \tag{2.135}$$

where

$$k_0^2 \varepsilon_e^R = k_0^2 \varepsilon_{r3}^R - \rho^2 \tag{2.136}$$

At the step discontinuity $x = x_0$, we require E_y and H_z to be continuous. If we assume the unknown E_y field at $x = x_0$ to be $e_y(y)$, we can write

$$\begin{aligned} e_y(y) &= \sum_n V_n^{\mathrm{I}}(x_0) f_n^{\mathrm{I}}(y) + \int_0^\infty V^{\mathrm{I}}(x_0, \rho) f^{\mathrm{I}}(y, \rho)\, d\rho \\ &= \sum_n V_n^{\mathrm{II}}(x_0) f_n^{\mathrm{II}}(y) + \int_0^\infty V^{\mathrm{II}}(x_0) f^{\mathrm{II}}(y, \rho)\, d\rho \end{aligned} \tag{2.137}$$

Similarly, matching the H_z component at $x = x_0$, we can write

$$\begin{aligned} &\sum_n I_n^{\mathrm{I}}(x_0)\, g_n^{\mathrm{I}}(y) + \int_0^\infty I^{\mathrm{I}}(x_0, \rho)\, g^{\mathrm{I}}(y, \rho)\, d\rho \\ &\quad = \sum_n I_n^{\mathrm{II}}(x_0)\, g_n^{\mathrm{II}}(y) + \int_0^\infty I^{\mathrm{II}}(x_0, \rho)\, g^{\mathrm{II}}(y, \rho)\, d\rho \end{aligned} \tag{2.138}$$

Multiplying Eq. (2.138) by $e_y^*(y)$ and integrating from 0 to ∞, we obtain

$$\begin{aligned} &\sum_n I_n^{\mathrm{I}}(x_0) \int_0^\infty e_y^*(y)\, g_n^{\mathrm{I}}(y)\, dy + \int_0^\infty I^{\mathrm{I}}(x_0, \rho)\left[\int_0^\infty e_y^*(y)\, g^{\mathrm{I}}(y, \rho)\, dy\right] d\rho \\ &= \sum_n I_n^{\mathrm{II}}(x_0) \int_0^\infty e_y^*(y)\, g_n^{\mathrm{II}}(y)\, dy + \int_0^\infty I^{\mathrm{II}}(x_0, \rho)\left[\int_0^\infty e_y^*(y)\, g^{\mathrm{II}}(y, \rho)\, dy\right] d\rho \end{aligned} \tag{2.139}$$

Let Y_n^- and $Y^-(\rho)$ denote the admittances of the nth discrete mode and the TM spectrum, respectively, at the junction plane looking in the $-x$-direction; and let Y_n^+ and $Y^+(\rho)$ denote the corresponding quantities looking toward the $+x$-direction. We can then write

$$I_n^{\mathrm{I}}(x_0) = -Y_n^- V_n^{\mathrm{I}}(x_0) = -Y_n^- \int_0^\infty [g_l^{\mathrm{I}}(y')]^* e_y(y')\, dy' \tag{2.140a}$$

$$I_n^{\mathrm{II}}(x_0) = Y_n^+ V_n^{\mathrm{II}}(x_0) = Y_n^+ \int_0^\infty [g_l^{\mathrm{II}}(y')]^* e_y(y')\, dy' \tag{2.140b}$$

$$I^{\mathrm{I}}(x_0, \rho) = -Y^-(\rho)V^{\mathrm{I}}(x_0, \rho) = -Y^-(\rho)\int_0^{\infty} [g^{\mathrm{I}}(y', \rho)]^* e_y(y')\,dy' \quad (2.140c)$$

$$I^{\mathrm{II}}(x_0, \rho) = Y^+(\rho)V^{\mathrm{II}}(x_0, \rho) = Y^+(\rho)\int_0^{\infty} [g^{\mathrm{II}}(y', \rho)]^* e_y(y')\,dy' \quad (2.140d)$$

Using (2.140) in (2.139) and after some algebraic manipulations, we can write

$$\begin{aligned} I_n^{\mathrm{I}}(x_0)\int_0^{\infty} e_y^*(y)\, g_n^{\mathrm{I}}(y)\,dy &= Y_n^+ \left| \int_0^{\infty} [g_n^{\mathrm{II}}(y)]^* e_y(y)\,dy \right|^2 \\ &+ \sum_{R=\mathrm{I,II}} \int_0^{\infty}\int_0^{\infty} e_y^*(y) \left[\sum_{l \neq n} Y_l^{\mp} g_l^R(y)\{g_l^R(y')\}^* \right. \\ &\left. + \int_0^{\infty} Y^{\mp}(\rho)\, g^R(y, \rho)\{g^R(y', \rho)\}^*\, d\rho \right] e_y(y')\,dy\,dy' \end{aligned} \quad (2.141)$$

From (2.140a), we have

$$V_n^{\mathrm{I}}(x_0) = \int_0^{\infty} \{g_n^{\mathrm{I}}(y)\}^* e_y(y)\,dy \quad (2.142)$$

Using the expressions for $I_n^{\mathrm{I}}(x_0)$ and $V_n^{\mathrm{I}}(x_0)$ as given in (2.141) and (2.142), respectively, we can now obtain the input admittance $Y_{\mathrm{in},n}(x_0)$ of the TM_n mode on the left side of the junction $x = x_0$ but looking toward the $+x$-direction.

$$Y_{\mathrm{in},n}(x_0) = \left[\frac{I_n^{\mathrm{I}}(x_0)}{V_n^{\mathrm{I}}(x_0)} \right] = N_n^2 Y_n^+ + Y_d \quad (2.143)$$

where

$$N_n = \frac{\displaystyle\int_0^{\infty} [g_n^{\mathrm{II}}(y)]^* e_y(y)\,dy}{\displaystyle\int_0^{\infty} [g_n^{\mathrm{I}}(y)]^* e_y(y)\,dy} \quad (2.144a)$$

$$Y_d = \frac{\displaystyle\sum_{R=\mathrm{I,II}} \int_0^{\infty}\int_0^{\infty} e_y^*(y) \left[\sum_{l \neq n} Y_l^{\mp} g_l^R(y)[g_l^R(y')]^* + \int_0^{\infty} Y^{\mp}(\rho) g^R(y, \rho)[g^R(y', \rho)]^*\, d\rho \right] e_y(y')\,dy\,dy'}{\left| \displaystyle\int_0^{\infty} [g_n^{\mathrm{I}}(y)]^* e_y(y)\,dy \right|^2} \quad (2.144b)$$

The transverse equivalent network representing the admittance given by (2.143) is shown in Figure 2.14(b).

2.5.3 Application to Image Guide

Figure 2.15 shows the cross-section of an image guide and its transverse equivalent network. Since there are no discrete modes in region II of the image guide (Fig. 2.15(a)), we set $N_n = 0$ in (2.143). If we denote $Y_{in}(a/2)$ as the input admittance at $x = a/2$ looking toward the $+x$-direction from the left side of the step junction, then from (2.143) and (2.144), we can write

$$Y_{in}(a/2) = Y_d = \frac{\displaystyle\int_0^{\infty} Y^{+}(\rho) \left| \int_0^{\infty} e_y(y) [g^{II}(y, \rho)]^* \, dy \right|^2 d\rho}{\displaystyle\left| \int_0^{\infty} [g_n^{I}(y)]^* \, e_y(y) \, dy \right|^2} \tag{2.145}$$

The cross-section of the image guide is therefore represented as a section of transmission line of length a terminated on either side by an admittance Y_d. Considering the right-half of the network (Fig. 2.15(b)) with respect to the plane of symmetry at $x = 0$ and applying the resonance condition, namely, $Y^+ + Y^- = 0$ at $x = a/2$, we can write the characteristic equation as

$$jY_n^{I} T + Y_d = 0, \quad T = \begin{cases} \tan(\beta_{xn}^{I} a/2), & \text{even mode } (n = 1, 3, 5, \ldots) \\ -\cot(\beta_{xn}^{I} a/2), & \text{odd mode } (n = 2, 4, 6, \ldots) \end{cases} \tag{2.146}$$

where

$$Y_n^{I} = \beta_{xn}^{I} / (\omega \mu_0 \varepsilon_{en}^{I}) \tag{2.147a}$$

$$\beta_{xn}^{I} = [(\beta_{zs,n}^{I})^2 - \beta_z^2]^{1/2} \tag{2.147b}$$

$$\varepsilon_{en}^{I} = (\beta_{zs,n}^{I} / k_0)^2 \tag{2.147c}$$

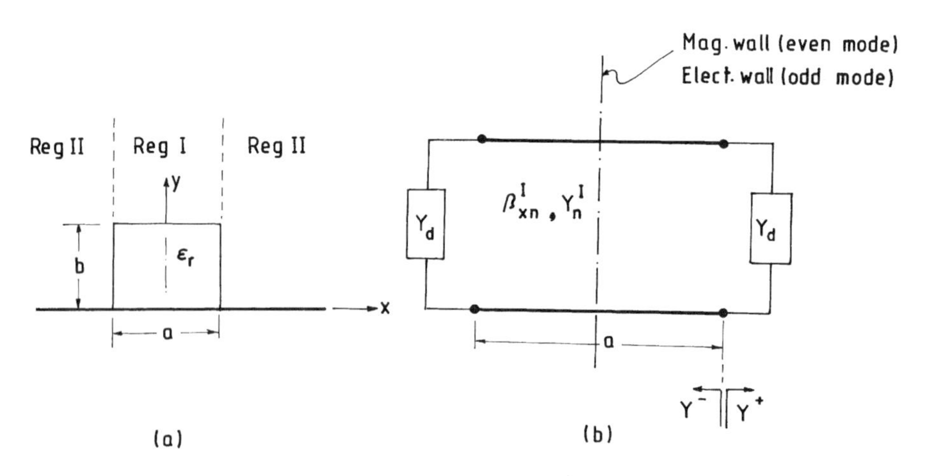

FIGURE 2.15 (a) Cross-section of image guide and (b) its transverse equivalent network.

The parameter $\varepsilon^{\mathrm{I}}_{en}$ is the effective dielectric constant of the horizontal slab guide model of region I and $\beta^{\mathrm{I}}_{zs,n}$ is its z-directed propagation constant. In the expression for Y_d given by (2.145), $Y^+(\rho)$ is the characteristic admittance of region II. It is a function of the wavenumber ρ of the continuous spectrum in the y-direction and the decay coefficient ξ in the x-direction.

$$Y^+(\rho) = -j\xi/(\omega\mu_0\varepsilon^{\mathrm{II}}_e) \tag{2.148}$$

where

$$\xi = (\beta_z^2 + \rho^2 - k_0^2)^{1/2} \tag{2.149a}$$

$$\varepsilon^{\mathrm{II}}_e = (\beta_z/k_0)^2 = (k_0^2 - \rho^2)/k_0^2 \tag{2.149b}$$

Substituting (2.149) in (2.148), we get

$$Y^+(\rho) = \frac{-j\omega\varepsilon_0(\beta_z^2 + \rho^2 - k_0^2)^{1/2}}{k_0^2 - \rho^2} \tag{2.150}$$

In order to evaluate the integral in (2.145), we need to specify the tangential field $e_y(y)$ at the strip edge (at $x = a/2$). Let us assume

$$e_y(y) = f^{\mathrm{I}}_n(y) \tag{2.151}$$

so that

$$\int e_y(y)\,[g^{\mathrm{I}}_n(y)]^*\,dy = \int f^{\mathrm{I}}_n(y)[g^{\mathrm{I}}_n(y)]^*\,dy = 1 \tag{2.152}$$

Using (2.147) and (2.150)–(2.152) in (2.146), we obtain

$$\left(\frac{[(\beta^{\mathrm{I}}_{zs,n})^2 - \beta_z^2]^{1/2}\,T}{(\beta^{\mathrm{I}}_{zs,n})^2}\right) - \int_0^\infty \left(\frac{(\beta_z^2 + \rho^2 - k_0^2)^{1/2}}{(k_0^2 - \rho^2)}\right)\left|\int_0^\infty f^{\mathrm{I}}_n(y)\{g^{\mathrm{II}}(y,\rho)\}^*\,dy\right|^2 d\rho = 0$$

$$T = \begin{cases} \tan\{0.5a[(\beta^{\mathrm{I}}_{zs,n})^2 - \beta_z^2]\}^{1/2}, & \text{even mode } (n = 1, 3, 5, \ldots) \\ -\cot\{0.5a[(\beta^{\mathrm{I}}_{zs,n})^2 - \beta_z^2]\}^{1/2}, & \text{odd mode } (n = 2, 4, 6, \ldots) \end{cases} \tag{2.153}$$

This is the dispersion relation for evaluating the propagation constant β_z of the E^y_{mn} modes of the image guide. The expression is the same as that reported by Koshiba et al. [17]. The propagation constant $\beta^{\mathrm{I}}_{zs,n}$ is determined from the characteristic equation of the horizontal slab guide model of region I as described in Section 2.4. The mode functions may be chosen from the field expressions for the horizontal slab guide model.

For the discrete TM modes, the mode functions are

$$f_n^{\mathrm{I}}(y) = \sqrt{(2/D_n)}(1/\varepsilon_r)\cos(\beta_{yn}^{\mathrm{I}}y) \qquad 0 \leq y \leq b \tag{2.154a}$$

$$= \sqrt{(2/D_n)}\cos(\beta_{yn}^{\mathrm{I}}y)e^{-\zeta(y-b)} \qquad y \geq b \tag{2.154b}$$

$$g_n^{\mathrm{I}}(y) = \sqrt{(2/D_n)}\cos(\beta_{yn}^{\mathrm{I}}y) \qquad 0 \leq y \leq b \tag{2.155a}$$

$$= \sqrt{(2/D_n)}\cos(\beta_{yn}^{\mathrm{I}}y)e^{-\zeta(y-b)} \qquad y \geq b \tag{2.155b}$$

where

$$D_n = \left[\frac{1}{\varepsilon_r}\left(b + \frac{\sin(\beta_{yn}^{\mathrm{I}}b)\cos(\beta_{yn}^{\mathrm{I}}b)}{\beta_{yn}^{\mathrm{I}}}\right) + \left(\frac{\cos^2(\beta_{yn}^{\mathrm{I}}b)}{\zeta}\right)\right] \tag{2.156a}$$

$$\zeta = [k_0^2(\varepsilon_r - 1) - (\beta_{yn}^{\mathrm{I}})^2]^{1/2} \tag{2.156b}$$

and β_{yn}^{I} is the solution of the characteristic equation for the horizontal slab guide model of region I.

For the continuous spectrum, the mode functions are

$$f^{\mathrm{I}}(y, \rho) = [1/\sqrt{D'(\rho)}\varepsilon_r]\cos(\zeta y), \quad 0 \leq y \leq b \tag{2.157a}$$

$$= [1/\sqrt{D'(\rho)}]\{C_1(\rho)\cos[\rho(y-b)]$$

$$+ C_2(\rho)\sin[\rho(y-b)]\}, \quad y \geq b \tag{2.157b}$$

$$g^{\mathrm{I}}(y, \rho) = [1/\sqrt{D'(\rho)}]\cos(\zeta y), \quad 0 \leq y \leq b \tag{2.158a}$$

$$= [1/\sqrt{D'(\rho)}]\{C_1(\rho)\cos[\rho(y-b)]$$

$$+ C_2(\rho)\sin[\rho(y-b)]\}, \quad y \geq b \tag{2.158b}$$

$$f^{\mathrm{II}}(y, \rho) = g^{\mathrm{II}}(y, \rho) = \sqrt{(2/\pi)}\cos(\rho y) \tag{2.159}$$

where

$$\zeta = [k_0^2(\varepsilon_r - 1) + \rho^2]^{1/2} \tag{2.160a}$$

$$D'(\rho) = 0.5\pi[\cos(\zeta b) - (\zeta/\varepsilon_r\rho)\sin(\zeta b)] \tag{2.160b}$$

The procedure outlined above can be applied for deriving the characteristic equations of other guide structures. The expressions include the equivalent lumped circuit due to the edges of the dielectric strips and hence are more accurate than those derived in Section 2.4 using the EDC method. The evaluation of propagation constant, however, requires specification of the mode functions corresponding to the fields at the junction planes.

2.6 CONDUCTOR AND DIELECTRIC LOSSES

The attenuation in a dielectric integrated guide occurs due to conductor and dielectric losses. In open guide structures, the radiation loss forms a third contributing factor to attenuation. Radiation loss results primarily from bends and surface irregularities. In this section, we consider only the conductor and dielectric losses. Radiation loss due to the bends will be treated in Chapter 10.

2.6.1 Conductor Loss

Conductor loss occurs due to the finite resistivity of the metallic plane of the dielectric integrated guide. The attenuation constant α_c due to the conductor loss is normally calculated using the following expression:

$$\alpha_c = \frac{P_l}{2P_T} = \frac{0.5\, R_s \oint |H_{\tan}|^2\, dl}{\sum_n Z_n \int_S \int |H_t|^2\, ds} \tag{2.161}$$

where

P_l = power lost/unit length of the guide
P_T = total power transmitted
R_s = surface resistance of the metal plane
$H_{\tan}$ = tangential magnetic field at the surface of the metal plane
H_t = magnetic field transverse to the direction of propagation
Z_n = wave impedance looking in the direction of propagation

In applying (2.161), the electromagnetic fields for the guide structure are first solved by assuming the metallic planes to be perfect conductors and the magnetic field expressions so derived are substituted. It is therefore implicit that the finite conductivity of the metal planes has negligible effect on the actual field distribution in the guide. This is a reasonable assumption for most metals having high conductivity.

An alternative method, which does not involve the solution of electromagnetic fields, is to derive the characteristic equation taking into account the effect of surface impedance Z_s and determine the complex propagation constant [18].

As an example, we consider the modified equivalent circuit of the horizontal slab guide model of a image guide. The circuit is shown in Figure 2.16 wherein the metal plane of the guide is represented by the surface impedance Z_s. The characteristic equation is given by

$$\varepsilon_r \bar{\zeta} \bar{\beta}_y + j\omega\varepsilon_0\varepsilon_r Z_s [\bar{\zeta}\varepsilon_r \tan(\bar{\beta}_y b) + \bar{\beta}_y] - (\bar{\beta}_y)^2 \tan(\bar{\beta}_y b) = 0 \tag{2.162}$$

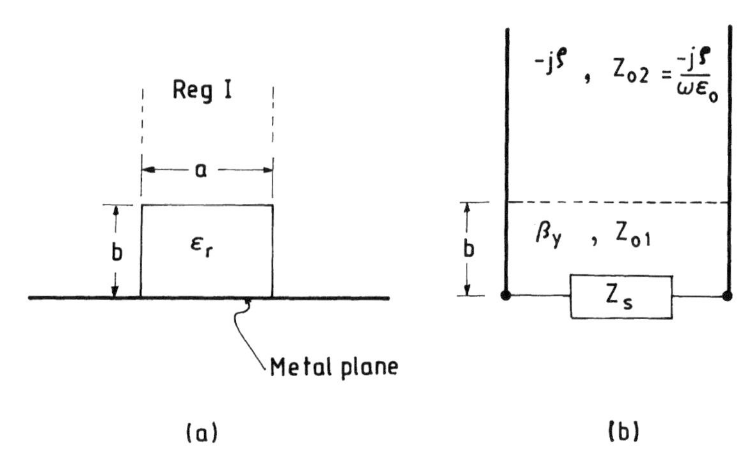

FIGURE 2.16 (a) Image guide with metal plane of finite impedance Z_s. (b) Equivalent network of horizontal slab guide model for region I.

where

$$Z_s = R_s(1+j) \tag{2.163a}$$

$$\bar{\beta}_y = \beta_y - j\alpha_y \tag{2.163b}$$

$$\bar{\zeta} = [k_0^2(\varepsilon_r - 1) - (\bar{\beta}_y)^2]^{1/2} \tag{2.163c}$$

The surface resistance R_s for metals is given by

$$R_s = \sqrt{(\omega\mu_0/2\sigma)} \tag{2.164}$$

where σ is the conductivity of the metal. The solution of (2.162) gives complex values for $\bar{\beta}_y$. The complex effective dielectic constant $\bar{\varepsilon}_e$ is then obtained using

$$\bar{\varepsilon}_e = [\varepsilon_r - (\bar{\beta}_y/k_0)^2] \tag{2.165}$$

This value of $\bar{\varepsilon}_e$ is used in the characteristic equation for the vertical layered model of the image guide to solve for the complex propagation constant $\bar{\beta}_z$. The conductor attenuation constant is given by the imaginary part of $\bar{\beta}_z$.

2.6.2 Dielectric Loss

The attenuation due to dielectric loss is derived by introducing a complex dielectric constant. We give below two different approximate methods of analysis [8, 18].

Method 1 Dydyk [18] has derived a simple expression for the attenuation constant due to the dielectric loss by starting from a complex propagation constant $\bar{\beta}_z$.

$$\bar{\beta}_z = \beta_z - j\alpha_d = [\beta_0^2 \bar{\varepsilon}_r - \beta_t^2]^{1/2} \tag{2.166}$$

where β_t is the transverse propagation constant, α_d is the attenuation constant due to dielectric loss, and $\bar{\varepsilon}_r$ is the complex dielectric constant.

$$\bar{\varepsilon}_r = \varepsilon_r - j\varepsilon_r' = \varepsilon_r(1 - j\tan\delta) \tag{2.167}$$

If we define the unperturbed propagation constant as

$$\beta_{zu} = [\beta_0^2 \varepsilon_r - \beta_t^2]^{1/2} \tag{2.168}$$

then (2.170) can be written as

$$\bar{\beta}_z = \beta_{zu}[1 - j\varepsilon_r'(\beta_0/\beta_{zu})^2]^{1/2} \tag{2.169}$$

Setting

$$\varepsilon_r'(\beta_0/\beta_{zu})^2 = \sinh(\theta) \tag{2.170}$$

and using the relation

$$1 - j\sinh(\theta) = [\cosh(\theta/2) - j\sinh(\theta/2)]^2 \tag{2.171}$$

in (2.168), we obtain the attenuation constant due to dielectric loss α_d as

$$\alpha_d = \beta_{zu}\sinh(\theta/2) = (2\pi/\lambda)\sinh(\theta/2) \tag{2.172}$$

where λ is the guide wavelength. Substituting for θ from (2.170) and setting $\varepsilon_r' = \varepsilon_r \tan\delta$, we obtain the following approximate expression for α_d [18]:

$$\alpha_d = (2\pi/\lambda)\sinh\{0.5\sinh^{-1}[(\lambda/\lambda_0)^2 \varepsilon_r \tan\delta]\} \tag{2.173}$$

Method 2 An approximate expression for the dielectric loss can also be derived by applying a perturbational approach. The characteristic equation is expanded in Taylor series about the unperturbed values corresponding to the lossless case. The first-order terms in the expansion are then used to derive the expression for the attenuation constant α_d. As an example, we present the analysis reported by Toulios and Knox [8] for the even-mode excitation of a image guide.

The characteristic equations for the even-mode excitation of a image guide (Fig. 2.16(a)) are given by

$$1 - (\beta_y/\varepsilon_r\zeta)\tan(\beta_y b) = 0 \tag{2.174}$$

$$1 - (\beta_x/\xi)\tan(\beta_x a/2) = 0 \tag{2.175}$$

where

$$\zeta = [k_0^2(\varepsilon_r - 1) - \beta_y^2]^{1/2} \quad (2.176a)$$

$$\varepsilon_e = \varepsilon_r - (\beta_y/k_0)^2 \quad (2.176b)$$

$$\xi = [k_0^2(\varepsilon_e - 1) - \beta_x^2]^{1/2} \quad (2.176c)$$

$$\beta_z = (2\pi/\lambda) = [k_0^2\varepsilon_e - \beta_x^2]^{1/2} \quad (2.176d)$$

Equations (2.174) and (2.175) can be recast in the form

$$\beta_y b + \tan^{-1}(\beta_y/\varepsilon_r\zeta) - n\pi/2 = 0 \quad (2.177)$$

$$\beta_x a/2 + \tan^{-1}(\beta_x/\xi) - m\pi/2 = 0 \quad (2.178)$$

In order to derive an expression for α_d, we introduce a complex dielectric constant in the form given by (2.167). The real parameters β_y, ζ, β_x, ξ, and ε_e are accordingly replaced by complex values $\bar{\beta}_y$, $\bar{\zeta}$, $\bar{\beta}_x$, $\bar{\xi}$, and $\bar{\varepsilon}_e$, respectively. The characteristic equations for the image guide can be written in the form

$$F(\bar{\beta}_y, \bar{\zeta}, \bar{\varepsilon}_r) = b\bar{\beta}_y + \tan^{-1}(\bar{\beta}_y/\bar{\varepsilon}_r\bar{\zeta}) - n\pi/2 = 0 \quad (2.179)$$

$$G(\bar{\beta}_x, \bar{\xi}) = 0.5a\bar{\beta}_x + \tan^{-1}(\bar{\beta}_x/\bar{\xi}) - m\pi/2 = 0 \quad (2.180)$$

The various complex parameters are given by

$$\bar{\beta}_y = \beta_y - j\alpha_y \quad (2.181a)$$

$$\bar{\beta}_x = \beta_x - j\alpha_x \quad (2.181b)$$

$$\bar{\zeta} = \zeta + j\zeta' = [k_0^2(\bar{\varepsilon}_r - 1) - \bar{\beta}_y^2]^{1/2} \quad (2.181c)$$

$$\bar{\xi} = \xi + j\xi' = [k_0^2(\bar{\varepsilon}_e - 1) - \bar{\beta}_x^2]^{1/2} \quad (2.181d)$$

$$\bar{\varepsilon}_e = \bar{\varepsilon}_r - (\bar{\beta}_y/k_0)^2 \quad (2.181e)$$

Expanding the functions $F(\bar{\beta}_y, \bar{\zeta}, \bar{\varepsilon}_r)$ and $G(\bar{\beta}_x, \bar{\xi})$ in terms of a Taylor series, we can write

$$F(\bar{\beta}_y, \bar{\zeta}, \bar{\varepsilon}_r) = F(\beta_y, \zeta, \varepsilon_r) + (\bar{\beta}_y - \beta_y)\frac{\partial F}{\partial \beta_y} + (\bar{\zeta} - \zeta)\frac{\partial F}{\partial \zeta} + (\bar{\varepsilon}_r - \varepsilon_r)\frac{\partial F}{\partial \varepsilon_r} + \cdots = 0 \quad (2.182)$$

$$G(\bar{\beta}_x, \bar{\xi}) = G(\beta_x, \xi) + (\bar{\beta}_x - \beta_x)\frac{\partial G}{\partial \beta_x} + (\bar{\xi} - \xi)\frac{\partial G}{\partial \xi} + \cdots = 0 \quad (2.183)$$

To a first-order approximation, we may assume

$$\alpha_y \ll \beta_y, \quad \zeta' \ll \zeta, \quad \alpha_x \ll \beta_x, \quad \xi' \ll \xi \quad (2.184)$$

Using these approximations in (2.181d), (2.181e), (2.182), and (2.183), the expressions for α_y, ζ', α_x, and ξ' can be written in the forms:

$$\alpha_y = (\varepsilon_r \tan \delta)\left(\frac{-F_0\beta_y[\zeta^2 + (k_0^2\varepsilon_r/2)]}{b\zeta + F_0\varepsilon_r(\beta_y^2 + \zeta^2)}\right) \tag{2.185a}$$

$$\zeta' = (\varepsilon_r \tan \delta)\left(\frac{(bk_0^2/2) + F_0\zeta[(k_0^2\varepsilon_r/2) - \beta_y^2]}{b\zeta + F_0\varepsilon_r(\beta_y^2 + \zeta^2)}\right) \tag{2.185b}$$

$$\alpha_x = -\left(\frac{\beta_x\{\beta_y\alpha_y + [(k_0^2\varepsilon_r/2)\tan \delta]\}\, G_0}{1 + 0.5a\xi}\right) \tag{2.185c}$$

$$\xi' = \left(\frac{(0.5a + \xi G_0)[\beta_y\alpha_y + (k_0^2\varepsilon_r/2)\tan \delta]}{1 + 0.5a\xi}\right) \tag{2.185d}$$

where

$$F_0 = \frac{1}{\beta_y^2 + \zeta^2\varepsilon_r} \tag{2.186a}$$

$$G_0 = \frac{1}{\beta_x^2 + \xi^2} \tag{2.186b}$$

The complex propagation constant $\bar{\beta}_z$ is given by

$$\bar{\beta}_z = \beta_z - j\alpha_d \tag{2.187}$$

The separation equations involving $\bar{\beta}_z$ and β_z are

$$\beta_x^2 + \beta_y^2 + \beta_z^2 = k_0^2\varepsilon_r \tag{2.188a}$$

$$\bar{\beta}_x^2 + \bar{\beta}_y^2 + \bar{\beta}_z^2 = k_0^2\bar{\varepsilon}_r \tag{2.188b}$$

Using (2.185)–(2.188), the expression for the attenuation constant α_d due to dielectric loss is obtained as [8]

$$\frac{\alpha_d}{\alpha} = \frac{\lambda}{2\pi}\left[\left(\frac{k_0^2\varepsilon_r}{a}\right)\tan \delta + \frac{\beta_x\alpha_x}{\alpha} + \frac{\beta_y\alpha_y}{\alpha}\right] \tag{2.189}$$

where $\lambda\ (=\beta_z/2\pi)$ is the guide wavelength and

$$\alpha = \frac{\pi\sqrt{\varepsilon_r}\tan \delta}{\lambda_0} \tag{2.190}$$

is the attenuation constant of the infinite dielectric medium.

PROBLEMS

2.1 Discuss the principle behind the EDC method. Enumerate its advantages and limitations.

2.2 Consider a dielectric-filled parallel-plate waveguide as shown in Figure 2.17. Assume propagation in the z-direction according to $e^{-j\beta z}$.

(a) Using the transverse resonance technique, derive expressions for the cutoff wavenumbers of TE modes in the guide.

(b) If $\varepsilon_r = 2.56$, $d = 0.5$ cm, and $f = 30$ GHz, calculate the propagation constants of the propagating TE modes.

Answer: $\beta = \sqrt{k^2 - k_c^2}$; $k_c = \dfrac{n\pi}{d}$, $n = 0, 1, 2, \ldots$

$$n = 0; \quad \beta = k = k_0\sqrt{\varepsilon_r} = 2\pi \times 1.6 \text{ rad/cm}$$

$$n = 1; \quad \beta = 2\pi\sqrt{1.56} \text{ rad/cm}$$

2.3 Consider a dielectric slab guide shown in Figure 2.18 having $\varepsilon = 4\varepsilon_0$ and thickness $d = 10$ mm embedded in another dielectric having $\varepsilon = 2.56\varepsilon_0$.

(a) Using the EDC technique, derive the characteristic equation for the TE mode.

(b) Determine the cutoff frequencies for the first three propagating TE modes.

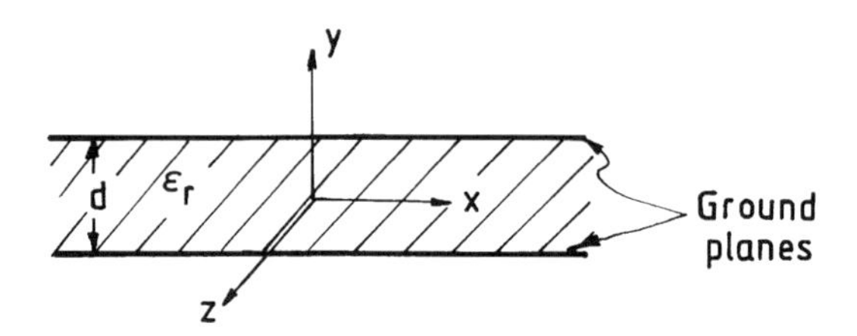

FIGURE 2.17 Dielectric-filled parallel-plate waveguide.

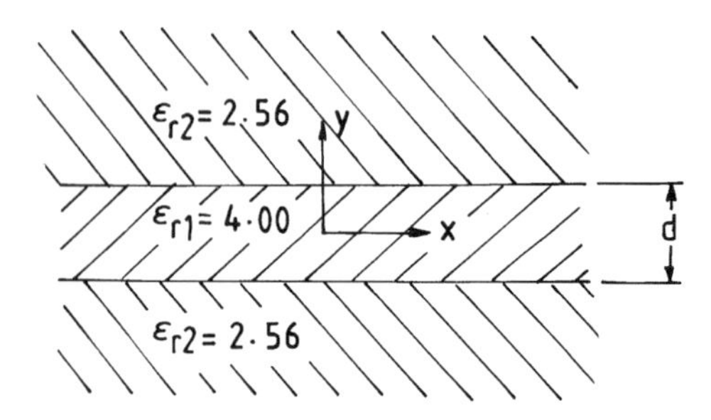

FIGURE 2.18 Dielectric slab guide embedded in another dielectric.

2.4 Consider a rectangular waveguide partially loaded with a horizontal dielectric slab as shown in Figure 2.19.

(a) Draw the transmission line equivalent for TE_{mn} modes and derive the characteristic equation.

(b) Solve the characteristic equation numerically and determine the bandwidth of the dominant mode. Assume $a = 7.11$ mm, $b = 3.56$ mm, $d = 1.5$ mm, $c_1 = c_2$, and $\varepsilon_r = 2.2$.

2.5 Consider a rectangular waveguide partially loaded with a vertical dielectric slab as shown in Figure 2.20.

(a) Draw the transmission line equivalent for TE_{mn} modes and derive the characteristic equation.

(b) Solve the characteristic equation numerically and determine the bandwidth of the dominant mode. Assume $a = 7.11$ mm, $b = 3.56$ mm, $d = 1.5$ mm, $c_1 = c_2$, and $\varepsilon_r = 2.2$. Compare this bandwidth with that obtained in Problem 2.4(b) and for a hollow rectangular waveguide of the same cross-section.

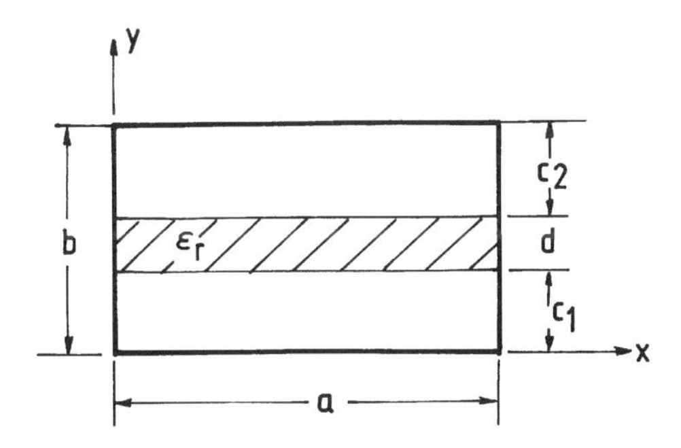

FIGURE 2.19 Rectangular waveguide partially loaded with a horizontal dielectric slab.

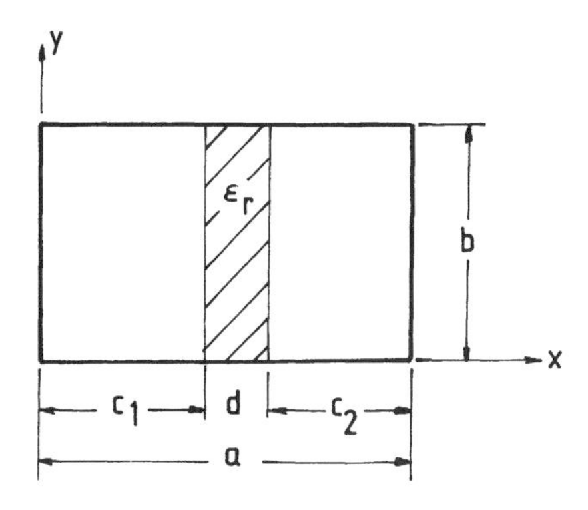

FIGURE 2.20 Rectangular waveguide partially loaded with a vertical dielectric slab.

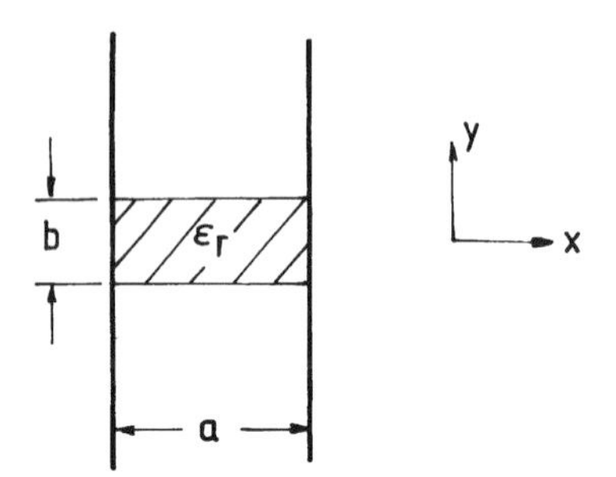

FIGURE 2.21 NRD guide.

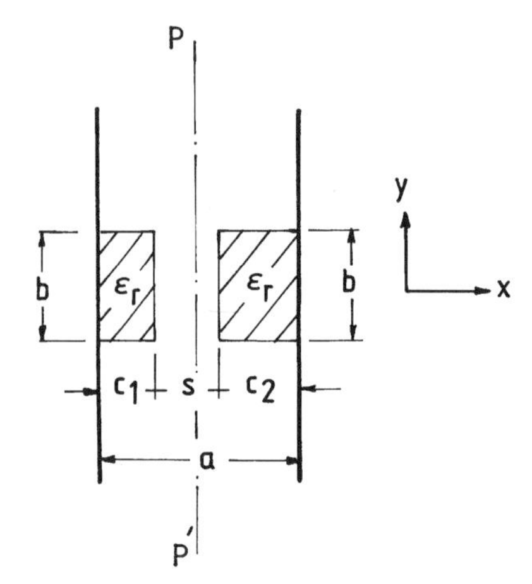

FIGURE 2.22 Broadside-coupled NRD guide.

2.6 **(a)** Using the EDC method, derive the characteristic equation for the E^y_{mn} (TM-to-y) modes of a NRD guide shown in Figure 2.21.

(b) What do m and n stand for? Explain how the propagation constants corresponding to the E^y_{21} and E^y_{12} mode are extracted from the characteristic equation?

2.7 **(a)** Using the EDC method, derive the characteristic equation for the E^y_{mn} (TM-to-y) modes of a broadside coupled NRD guide shown in Figure 2.22.

(b) Rewrite the characteristic equation for the special case $c_1 = c_2$. Verify this solution by rederiving the characteristic equations in terms of even and odd modes with respect to the vertical plane of symmetry PP'.

2.8 Using the EDC method, derive the characteristic equation for E^y_{mn} modes of a shielded insular image guide shown in Figure 2.23. Obtain separate

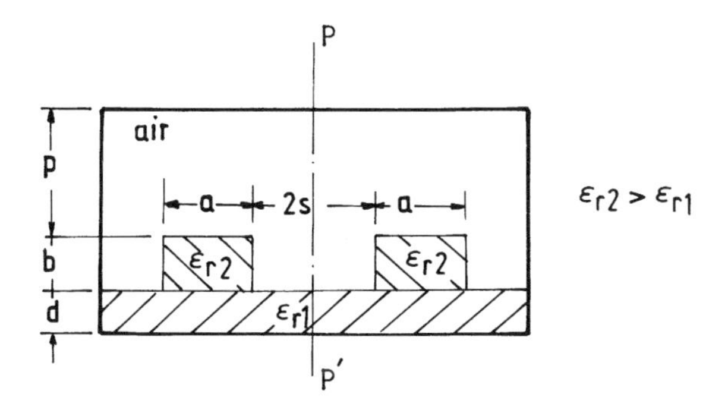

FIGURE 2.23 Shielded insular image guide.

equations for the even and odd modes with respect to the symmetry plane PP'.

2.9 Using the characteristic equations derived in Problem 2.8 to obtain the characteristic equations for the following special cases:

(a) Shielded coupled image guide.

(b) Coupled insular image guide.

(c) Coupled trapped insular image guide.

(d) Single guide versions of (a) to (c).

2.10 Repeat problems 2.6 and 2.7 for E^x_{mn} modes.

2.11 Using the EDC method, derive the characteristic equation for E^y_{mn} modes of a hollow image guide shown in Figure 2.24. Obtain the characteristic equation of an image guide as a special case.

2.12 Apply the network analysis method to a double-layered image guide shown in Figure 2.25 and derive the dispersion relation for evaluating the propagation constant β_z of the E^y_{mn} modes.

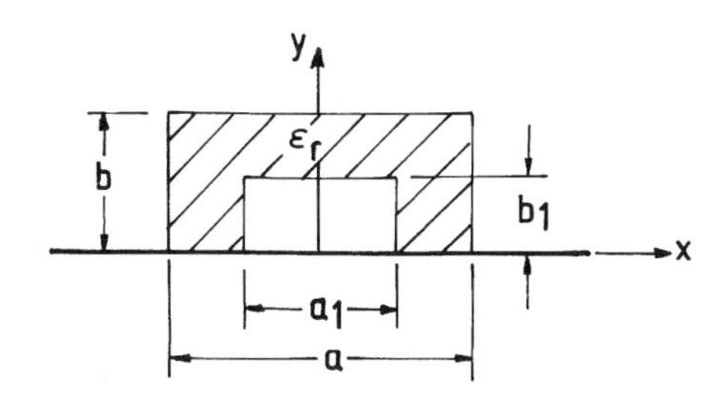

FIGURE 2.24 Hollow image guide.

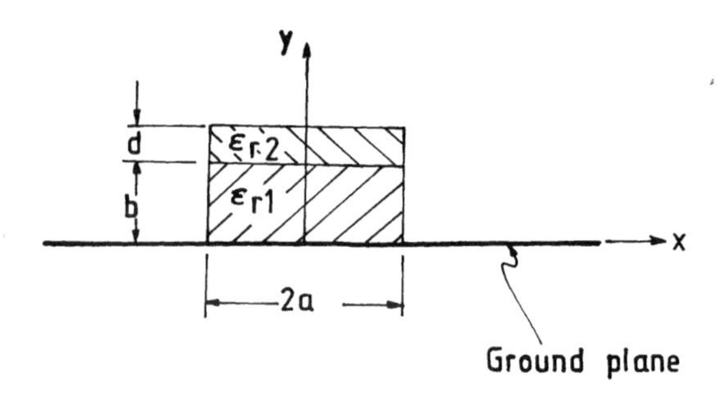

FIGURE 2.25 Double-layered image guide.

REFERENCES

1. R. M. Knox and P. P. Toulios, Integrated circuit for the millimetre through optical frequency range. Proceedings of the Symposium on Submillimetre Waves, New York, Mar./Apr. 1970.
2. W. V. Mclevige, T. Itoh, and R. Mittra, New waveguide structures for millimetre-wave and optical integrated circuits. *IEEE Trans. Microwave Theory Tech.*, **MTT-23**, 788–794, Oct. 1975.
3. S. T. Peng and A. A. Oliner, Guidance and leakage properties of a class of open dielectric waveguides: part I—mathematical formulations. *IEEE Trans. Microwave Theory Tech.*, **MTT-29**, 843–855, Sept. 1981.
4. A. A. Oliner, et al., Guidance and leakage properties of a class of open dielectric waveguides: part II—new physical effects. *IEEE Trans. Microwave Theory Tech.*, **MTT-29**, 855–874, Sept. 1981.
5. M. Koshiba and M. Suzuki, Microwave network analyses of dielectric waveguides for millimetre waves made of dielectric strip and planar dielectric layer. *Trans. Inst. Electron. Commun. Eng. (Japan)*, **E63**, 344–350, May 1980.
6. M. Koshiba, H. Ishii, and M. Suzuki, Improved equivalent network analysis of a dielectric waveguide placed on a ground plane. *Trans. Inst. Electron. Commun. Eng. (Jpn)*, **E65**, 572–578, Oct. 1982.
7. R. F. Harrington, *Time Harmonic Electromagnetic Fields*, McGraw-Hill, New York, 1961.
8. P. P. Toulios and R. M. Knox, Rectangular dielectric image lines for millimetre integrated circuits. Western Electronics Show and Convention, Los Angeles, CA, Digest, pp. 1–10, Aug. 1970.
9. E. A. J. Marcatili, Dielectric rectangular waveguide and directional coupler for integrated optics. *Bell Syst. Tech. J.*, **48(7)**, 2071–2102, Sept. 1969.
10. T. Itoh, Inverted strip dielectric waveguide for millimetre wave integrated circuits. *IEEE Trans. Microwave Theory Tech.*, **MTT-24**, 821–827, Nov. 1976.
11. T. Itoh and B. Adelseck, Trapped image guide for millimetre-wave circuits. *IEEE Trans. Microwave Theory Tech.*, **MTT-28**, 1433–1436, Dec. 1980.

12. W. B. Zhou and T. Itoh, Analysis of trapped image guides using effective dielectric constant and surface impedances. *IEEE Trans. Microwave Theory Tech.*, **MTT-30**, 2163–2166, Dec. 1982.
13. W. B. Zhou and T. Itoh, Fields distribution in the trapped image guide. *Electromagnetics*, **4**, 21–34, 1984.
14. J. F. Miao and T. Itoh, Hollow image guide and overlayed image guide coupler. *IEEE Trans. Microwave Theory Tech.*, **MTT-30**, 1826–1831, Nov. 1982.
15. P. R. Bansal and B. Bhat, Analysis of broadside coupled image guide for millimetre wave applications. *Int. Radar Symp. IRSI Proc.*, 351–356, 1983.
16. M. Koshiba and M. Suzuki, Equivalent network analysis of dielectric thin-film waveguides for optical integrated circuits and its applications. *Radio Sci.*, **17**, 99–107, Jan./Feb. 1982.
17. M. Koshiba, H. Ishii, and M. Suzuki, Simple equivalent network for a rectangular dielectric image guide. *Electron Lett.*, **18(11)**, 473–474, May 1982.
18. M. Dydyk, Image guide: a promising medium for EHF circuits. *Microwaves*, 71–73, Apr. 1981.

APPENDIX 2A CHARACTERISTIC EQUATIONS FOR HORIZONTAL LAYERED GUIDE MODELS CORRESPONDING TO E^y_{mn} MODES OF STRUCTURES IN FIGURES 2.2 TO 2.8

	Structure	Characteristic Equation	
		Note: $\zeta_0, \zeta_1, \zeta_2$, and β_y are real and positive. k_{y1}, k_{y2}, and k_{y3} may be real or imaginary. For $k = -j\zeta$, $[\tan(ky)/k] = [\tanh(\zeta y)/\zeta]$, $k\tan(ky) = -\zeta\tanh(\zeta y)$	
Model H1	$\varepsilon_{s1} > 1$ 1 air $-j\zeta_0$ y_1 ε_{s1} β_y Ground plane (y, x axes)	$[1 - (\beta_y/\zeta_0\varepsilon_{s1})\tan(\beta_y y_1)] = 0$	(2A.1a)
		$\zeta_0 = [k_0^2(\varepsilon_{s1} - 1) - \beta_y^2]^{1/2}$	(2A.1b)
Model H2	$\varepsilon_{s2} > \varepsilon_{s1}$ 1 $-j\zeta_0$ y_2 ε_{s2} β_y y_1 ε_{s1} k_{y1}	$\{1 - (\varepsilon_{s2}/\varepsilon_{s1})[\tan(\beta_y y_2)/\beta_y]k_{y1}\tan(k_{y1}y_1)\}$ $-(1/\zeta_0\varepsilon_{s2})[\beta_y\tan(\beta_y y_2) + (\varepsilon_{s2}/\varepsilon_{s1})k_{y1}\tan(k_{y1}y_1)] = 0$	(2A.2a)
		$\zeta_0 = [k_0^2(\varepsilon_{s2} - 1) - \beta_y^2]^{1/2}$	(2A.2b)
		$k_{y1} = [\beta_y^2 - k_0^2(\varepsilon_{s2} - \varepsilon_{s1})]^{1/2}$	(2A.2c)
Model H3	$\varepsilon_{s2} > \varepsilon_{s1}$ y_3 1 $-j\zeta_0$ y_2 ε_{s2} β_y y_1 ε_{s1} k_{y1}	$\zeta_0\tanh(\zeta_0 y_3)\{1 - (\varepsilon_{s2}/\varepsilon_{s1})[\tan(\beta_y y_2)/\beta_y]k_{y1}\tan(k_{y1}y_1)\}$ $-(1/\varepsilon_{s2})[\beta_y\tan(\beta_y y_2) + (\varepsilon_{s2}/\varepsilon_{s1})k_{y1}\tan(k_{y1}y_1)] = 0$	(2A.3a)
		$\zeta_0 = [k_0^2(\varepsilon_{s2} - 1) - \beta_y^2]^{1/2}$	(2A.3b)
		$k_{y1} = [\beta_y^2 - k_0^2(\varepsilon_{s2} - \varepsilon_{s1})]^{1/2}$	(2A.3c)

Model H4

$\varepsilon_{s2} > \varepsilon_{s1}$

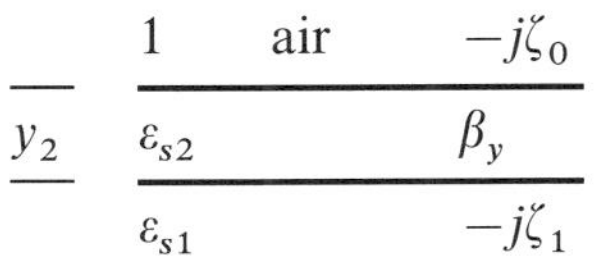

	1 air	$-j\zeta_0$
y_2	ε_{s2}	β_y
	ε_{s1}	$-j\zeta_1$

$$[1+(\zeta_1/\beta_y)(\varepsilon_{s2}/\varepsilon_{s1})\tan(\beta_y y_2)] + (\zeta_1/\zeta_0\varepsilon_{s1})[1-(\varepsilon_{s1}/\varepsilon_{s2})(\beta_y/\zeta_1) \cdot\tan(\beta_y y_2)] = 0 \tag{2A.4a}$$

$$\zeta_0 = [k_0^2(\varepsilon_{s2}-1)-\beta_y^2]^{1/2} \tag{2A.4b}$$

$$\zeta_1 = [k_0^2(\varepsilon_{s2}-\varepsilon_{s1})-\beta_y^2]^{1/2} \tag{2A.4c}$$

Model H5

$\varepsilon_{s1} > \varepsilon_{s2}$

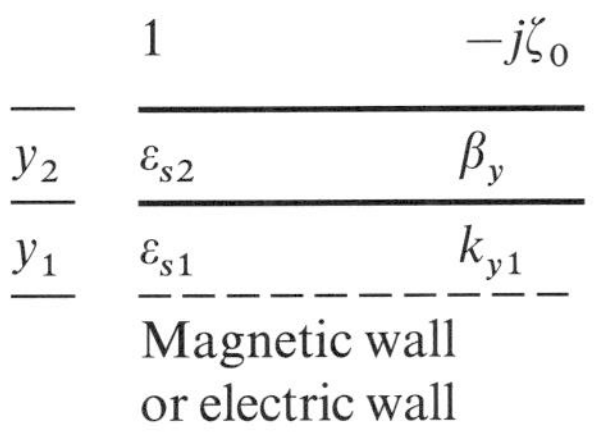

	1	$-j\zeta_0$
y_2	ε_{s2}	β_y
y_1	ε_{s1}	k_{y1}
	Magnetic wall or electric wall	

$$T(\varepsilon_{s1}/\varepsilon_{s2})[1-(\beta_y/\zeta_0\varepsilon_{s2})\tan(\beta_y y_2)] + \{(1/\zeta_0\varepsilon_{s2}) + [\tan(\beta_y y_2)/\beta_y]\} = 0,$$

$$T = \begin{cases} \tan(k_{y1}y_1)/k_{y1}, & \text{for magnetic wall} \\ -\cot(k_{y1}y_1)/k_{y1}, & \text{for electric wall} \end{cases} \tag{2A.5a}$$

$$\zeta_0 = [k_0^2(\varepsilon_{s2}-1)-\beta_y^2]^{1/2} \tag{2A.5b}$$

$$k_{y1} = [\beta_y^2 - k_0^2(\varepsilon_{s2}-\varepsilon_{s1})]^{1/2} \tag{2A.5c}$$

Model H6

$\varepsilon_{s3} > \varepsilon_{s1}, \varepsilon_{s2}$

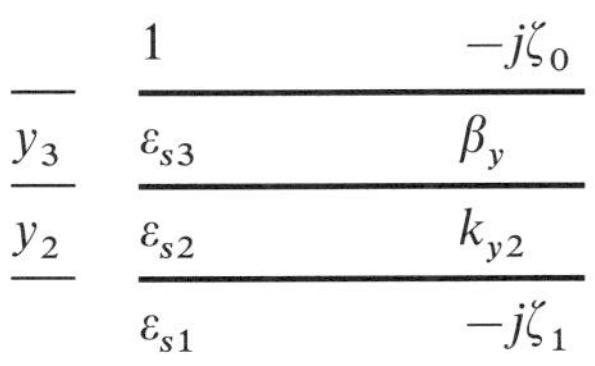

	1	$-j\zeta_0$
y_3	ε_{s3}	β_y
y_2	ε_{s2}	k_{y2}
	ε_{s1}	$-j\zeta_1$

$$[\beta_y\tan(\beta_y y_3) - \zeta_0\varepsilon_{s3}]\{1+(\zeta_1\varepsilon_{s2}/\varepsilon_{s1})[\tan(k_{y2}y_2)/k_{y2}]\} + (\varepsilon_{s3}/\varepsilon_{s2})[1+\zeta_0\varepsilon_{s3}\{\tan(\beta_y y_3)/\beta_y\}][k_{y2}\tan(k_{y2}y_2) - (\zeta_1\varepsilon_{s2}/\varepsilon_{s1})] = 0 \tag{2A.6a}$$

$$\zeta_0 = [k_0^2(\varepsilon_{s3}-1)-\beta_y^2]^{1/2} \tag{2A.6b}$$

$$\zeta_1 = [k_0^2(\varepsilon_{s3}-\varepsilon_{s1})-\beta_y^2]^{1/2} \tag{2A.6c}$$

$$k_{y2} = [\beta_y^2 - k_0^2(\varepsilon_{s3}-\varepsilon_{s2})]^{1/2} \tag{2A.6d}$$

	Structure	Characteristic Equation
Model H7	$\varepsilon_{s2} > \varepsilon_{s1}, \varepsilon_{s3}$ 1 / $-j\zeta_0$ y_3: ε_{s3}, k_{y3} y_2: ε_{s2}, β_y ε_{s1} / $-j\zeta_1$	$[k_{y3}\tan(k_{y3}y_3) - \zeta_0\varepsilon_{s3}]\{1 + (\zeta_1\varepsilon_{s2}/\varepsilon_{s1})[\tan(\beta_y y_2)/\beta_y]\} + (\varepsilon_{s3}/\varepsilon_{s2})\{1 + \zeta_0\varepsilon_{s3}[\tan(k_{y3}y_3)/k_{y3}]\}[\beta_y\tan(\beta_y y_2) - (\zeta_1\varepsilon_{s2}/\varepsilon_{s1})] = 0$ (2A.7a) $\zeta_0 = [k_0^2(\varepsilon_{s2} - 1) - \beta_y^2]^{1/2}$ (2A.7b) $\zeta_1 = [k_0^2(\varepsilon_{s2} - \varepsilon_{s1}) - \beta_y^2]^{1/2}$ (2A.7c) $k_{y3} = [\beta_y^2 - k_0^2(\varepsilon_{s2} - \varepsilon_{s3})]^{1/2}$ (2A.7d)
Model H8	$\varepsilon_{s1} > \varepsilon_{s2}$ 1 / $-j\zeta_0$ y_4: ε_{s2}, β_y y_3: ε_{s3}, k_{y3} y_2: ε_{s2}, β_y 1 / $-j\zeta_0$	$(\{1 + \zeta_0\varepsilon_{s2}(\tan(\beta_y y_4)/\beta_y]\} - (\varepsilon_{s3}/\varepsilon_{s2})[\tan(k_{y3}y_3)/k_{y3})[-\zeta_0\varepsilon_{s2} + \beta_y\tan(\beta_y y_4)]) \cdot[-\zeta_0\varepsilon_{s2} + \beta_y\tan(\beta_y y_2)] + ([-\zeta_0\varepsilon_{s2} + \beta_y\tan(\beta_y y_4)] + (\varepsilon_{s2}/\varepsilon_{s3})k_{y3}\tan(k_{y3}y_3) \cdot\{1 + \zeta_0\varepsilon_{s2}[\tan(\beta_y y_4)/\beta_y]\})\{1 + \zeta_0\varepsilon_{s2}[\tan(\beta_y y_2)/\beta_y]\} = 0$ (2A.8a) $\zeta_0 = [k_0^2(\varepsilon_{s2} - 1) - \beta_y^2]^{1/2}$ (2A.8b) $k_{y3} = [\beta_y^2 - k_0^2(\varepsilon_{s2} - \varepsilon_{s3})]^{1/2}$ (2A.8c)

APPENDIX 2B CHARACTERISTIC EQUATIONS FOR VERTICAL LAYERED GUIDE MODELS CORRESPONDING TO E^y_{mn} MODES OF STRUCTURES IN FIGURES 2.2 TO 2.8

	Structure	Characteristic Equation
		Note: ξ, ξ_1, ξ_2, and β_x are real. k_{x1} and k_{x2} may be real or imaginary.
Model V1	$\varepsilon_{e2} > \varepsilon_{e1}$	
	ε_{e1} | ε_{e2} | ε_{e1} ; $-j\xi$ | β_x | $-j\xi$; width x_2 ; axes y, x	$2+(\xi/\beta_x)[1-(\beta_x/\xi)^2]\tan(\beta_x x_2)=0$ (2B.1a) $\xi=[k_0^2(\varepsilon_{e2}-\varepsilon_{e1})-\beta_x^2]^{1/2}$ (2B.1b)
Model V2	$\varepsilon_{e2} > \varepsilon_{e1}$	$(T/\xi)[(\xi/\beta_x)\tan(\beta_x x_2)+1]+[(\beta_x/\xi)\tan(\beta_x x_2)-1]=0,$
	Magnetic wall (even mode) or Electric wall (odd mode) | ε_{e1} | ε_{e2} | ε_{e3} ; k_{x1} | β_x | $-j\xi$; widths x_1, x_2	$T=\begin{cases} k_{x1}\tan(k_{x1}x_1), & \text{even mode} \\ -k_{x1}\cot(k_{x1}x_1), & \text{odd mode} \end{cases}$ (2B.2a) $\xi=[k_0^2(\varepsilon_{e2}-\varepsilon_{e3})-\beta_x^2]^{1/2}$ (2B.2b) $k_{x1}=[\beta_x^2-k_0^2(\varepsilon_{e2}-\varepsilon_{e1})]^{1/2}$ (2B.2c)

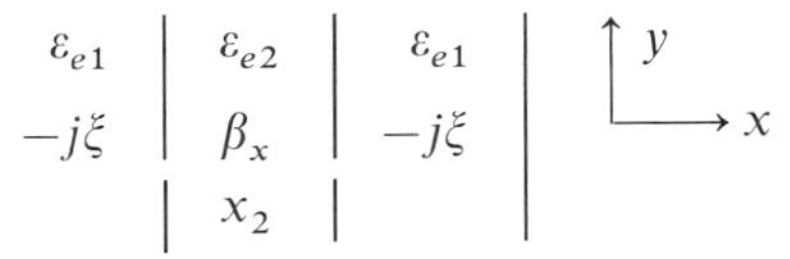

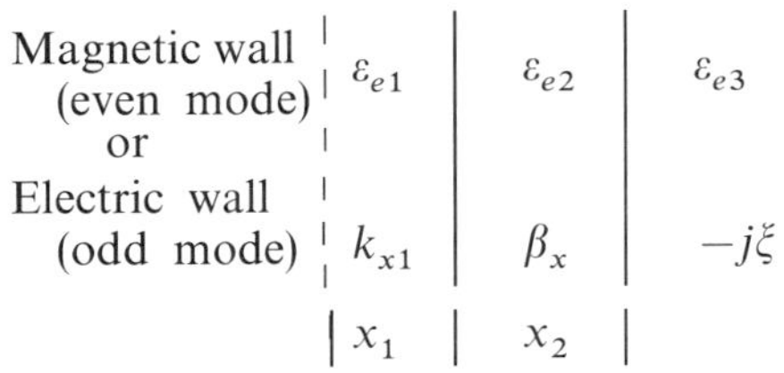

	Structure	Characteristic Equation	
Model V3	$\varepsilon_{e2} > \varepsilon_{e1}, \varepsilon_{e3}$	$T\{[\tan(\beta_x x_2)/\beta_x] + [\tanh(\xi x_3)/\xi]\}$ $-\{1 - [\tanh(\xi x_3)/\xi]\beta_x \tan(\beta_x x_2)\}$,	
	Magnetic wall (even mode) or Electric wall (odd mode) ¦ ε_{e1}, k_{x1}, x_1 | ε_{e2}, β_x, x_2 | ε_{e3}, ξ, x_3 ▌ Metal plane	$T = \begin{cases} k_{x1}\tan(k_{x1}x_1), & \text{even mode} \\ -k_{x1}\cot(k_{x1}x_1), & \text{odd mode} \end{cases}$	(2B.3a)
		$\xi = [k_0^2(\varepsilon_{e2} - \varepsilon_{e3}) - \beta_x^2]^{1/2}$	(2B.3b)
		$k_{x1} = [\beta_x^2 - k_0^2(\varepsilon_{e2} - \varepsilon_{e1})]^{1/2}$	(2B.3c)
Model V4	$\varepsilon_{e2} > \varepsilon_{e1}$	$T\{[\tan(\beta_x x_2)/\beta_x] - 1\} = 0$,	
	Magnetic wall (even mode) or Electric wall (odd mode) ¦ ε_{e1}, k_{x1}, x_1 | ε_{e2}, β_x, x_2 ▌ Metal plane	$T = \begin{cases} k_{x1}\tan(k_{x1}x_1), & \text{even mode} \\ -k_{x1}\cot(k_{x1}x_1), & \text{odd mode} \end{cases}$	(2B.4a)
		$k_{x1} = [\beta_x^2 - k_0^2(\varepsilon_{e2} - \varepsilon_{e1})]^{1/2}$	(2B.4b)

Model V5

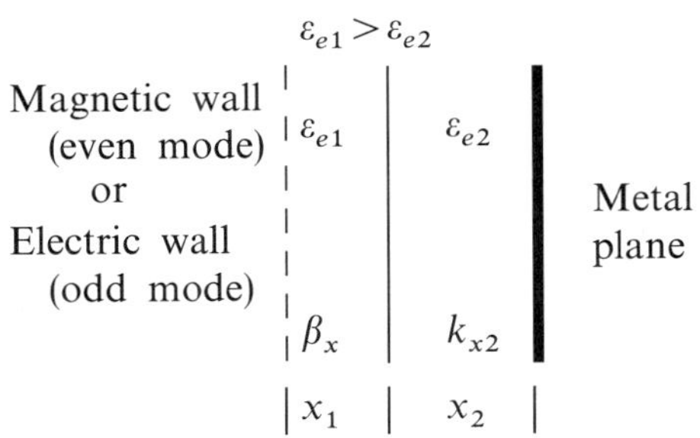

$\varepsilon_{e1} > \varepsilon_{e2}$

Magnetic wall (even mode) or Electric wall (odd mode)	ε_{e1}	ε_{e2}	Metal plane
	β_x	k_{x2}	
	x_1	x_2	

$$T\{[\tan(k_{x2}x_2)/k_{x2}] - 1\} = 0,$$

$$T = \begin{cases} \beta_x \tan(\beta_x x_1), & \text{even mode} \\ -\beta_x \cot(\beta_x x_1), & \text{odd mode} \end{cases} \tag{2B.5a}$$

$$k_{x2} = [\beta_x^2 - k_0^2(\varepsilon_{e1} - \varepsilon_{e2})]^{1/2} \tag{2B.5b}$$

Model V6

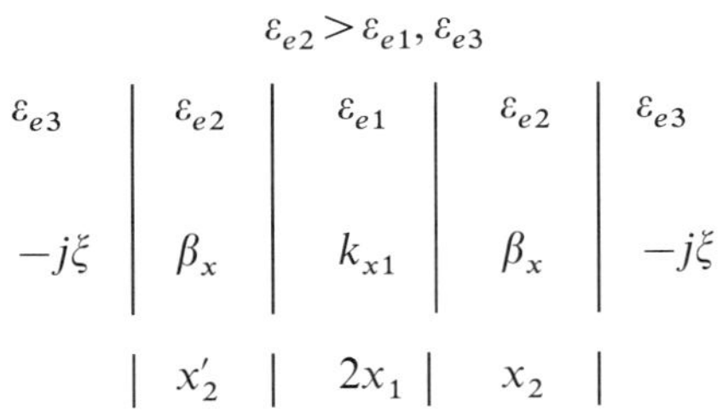

$\varepsilon_{e2} > \varepsilon_{e1}, \varepsilon_{e3}$

ε_{e3}	ε_{e2}	ε_{e1}	ε_{e2}	ε_{e3}
$-j\xi$	β_x	k_{x1}	β_x	$-j\xi$
	x_2'	$2x_1$	x_2	

$$\begin{aligned}\{[1 + (\xi/\beta_x)\tan(\beta_x x_2')] &+ [\tan(2k_{x1}x_1)/k_{x1}] \\ &\cdot[\xi - \beta_x \tan(\beta_x x_2')]\}[\xi - \beta_x \tan(\beta_x x_2)] \\ &+ \{[\xi - \beta_x \tan(\beta_x x_2')] - k_{x1}\tan(2k_{x1}x_1)[1 + (\xi/\beta_x) \\ &\cdot\tan(\beta_x x_2')]\}[1 + (\xi/\beta_x)\tan(\beta_x x_2)] = 0\end{aligned} \tag{2B.6a}$$

$$\xi = [k_0^2(\varepsilon_{e2} - \varepsilon_{e3}) - \beta_x^2]^{1/2} \tag{2B.6b}$$

$$k_{x1} = [\beta_x^2 - k_0^2(\varepsilon_{e2} - \varepsilon_{e1})]^{1/2} \tag{2B.6c}$$

	Structure	Characteristic Equation
Model V7	$\varepsilon_{e2} > \varepsilon_{e1}$ ε_{e1} | ε_{e2} | ε_{e1} | ε_{e2} | ε_{e1} $-j\xi$ | β_x | $-j\xi$ | β_x | $-j\xi$ | x'_2 | $2x_1$ | x_2 |	$\{[1+(\xi/\beta_x)\tan(\beta_x x'_2)]+[\tanh(2\xi x_1)/\xi] \cdot[\xi-\beta_x\tan(\beta_x x'_2)]\}[\xi-\beta_x\tan(\beta_x x_2)] + \{[\xi-\beta_x\tan(\beta_x x'_2)]+\xi\tanh(2\xi x_1) \cdot[1+(\xi/\beta_x)\tan(\beta_x x'_2)]\}[1+(\xi/\beta_x)\tan(\beta_x x_2)]=0$ (2B.7a) $\xi=[k_0^2(\varepsilon_{e2}-\varepsilon_{e1})-\beta_x^2]^{1/2}$ (2B.7b)
Model V8	$\varepsilon_{e2} > \varepsilon_{e1}$ Metal plane | ε_{e1} | ε_{e2} | ε_{e1} | ε_{e2} | ε_{e1} | Metal plane $-j\xi$ | β_x | $-j\xi$ | β_x | $-j\xi$ | x_3 | x'_2 | $2x_1$ | x_2 | x_3 |	$(\{[\tanh(\xi x_3)/\xi]+[\tan(\beta_x x'_2)/\beta_x]\}+[\tan(2\xi x_1)/\xi] \cdot\{1-[\tanh(\xi x_3)/\xi]\beta_x\tan(\beta_x x'_2)\})\cdot\{1-[\tanh(\xi x_3)/\xi] \cdot\beta_x\tan(\beta_x x_2)\}+(\{1-[\tanh(\xi x_3)/\xi] \cdot\beta_x\tan(\beta_x x'_2)\}+\xi\tan(2\xi x_1)\cdot\{[\tanh(\xi x_3)/\xi] +[\tan(\beta_x x'_2)/\beta_x]\})\cdot\{[\tanh(\xi x_3)/\xi] +[\tan(\beta_x x_2)/\beta_x]\}=0$ (2B.8a) $\xi=[k_0^2(\varepsilon_{e2}-\varepsilon_{e1})-\beta_x^2]^{1/2}$ (2B.8b)

CHAPTER THREE

Analysis of Dielectric Integrated Guides: Rigorous Methods

3.1 INTRODUCTION

The effective dielectric constant method and the network analysis method discussed in Chapter 2 are useful primarily for determining the dispersion characteristics of dielectric integrated guides. For obtaining more complete information including accurate field distribution, wave impedance, and losses, it is necessary to apply rigorous methods. The most commonly employed rigorous method for analyzing boundary value problems is the mode-matching technique [1–3]. This technique has been used by several investigators for solving dielectric integrated guides [4–9]. Numerical methods employing telegraphist's equations [10–12] and the finite-element formulation [13–15] and finite-difference method [16, 17] have also been reported. These numerical methods are more versatile and can be applied to dielectric guides of arbitrary cross-sections for which analytical solutions become prohibitively difficult. Since we are concerned with guides of rectangular cross-section, we cover the mode-matching method of analysis in detail. The analysis procedure is presented for a general coupled dielectric guide. The formulas derived can be applied to a number of practical guides as special cases. The telegraphist's equations method and the finite-element method are discussed briefly.

3.2 ANALYSIS OF GENERAL DIELECTRIC GUIDE: MODE-MATCHING METHOD

3.2.1 Basic Approach

The mode-matching method is commonly employed for obtaining solution of boundary value problems. The technique is useful for analyzing structures having a junction between two regions. The formulation involves expansion of unknown fields in the individual regions of the guide in terms of a set of their respective normal modes. In the case of open dielectric waveguides, a complete solution for the fields consists of one or more discrete surface waves having $\beta_z > k_0$, a continuous spectrum of waves with β_z lying in the range $0 < \beta_z < k_0$, and a continuous spectrum of evanescent waves with $\beta_z = -j\alpha_z$ $(0 < \alpha_z < \infty)$; where β_z is the propagation constant of the guide and k_0 is the free-space propagation constant. On the other hand, a guide enclosed completely in a rectangular metal shield supports only discrete modes. Since the formulation involves field expansion only in terms of a discrete set of modes, a closed guide is more convenient to analyze than an open guide structure, which requires inclusion of a continuous spectrum also. The solution of an open guide problem can therefore be obtained more easily by treating it as a limiting case of the solution of a corresponding shielded guide. In a shielded guide since the functional form of normal modes is known, the problem reduces to that of determining the modal coefficients associated with the field expansion in various regions. The modal fields are matched at the various junction interfaces between different regions and by using the orthogonality property of the normal modes, an infinite set of linear simultaneous equations is obtained in terms of the unknown modal coefficients. The solution of this set of equations yields the modal coefficients. Setting the determinant of the coefficient matrix equal to zero and solving the resulting equation yields the propagation constants for various modes. For computational ease, a large but finite number of terms are chosen that can offer reasonably accurate results.

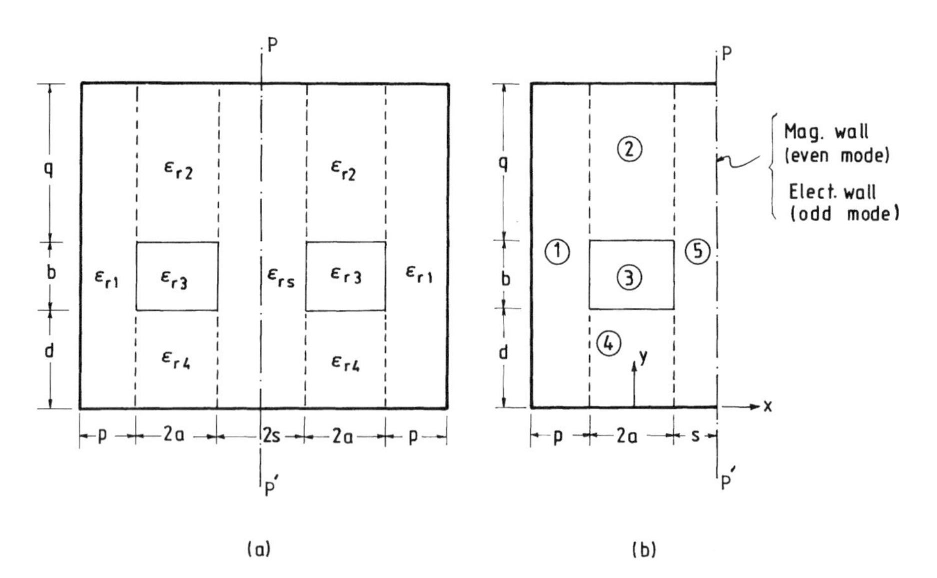

FIGURE 3.1 (a) General coupled dielectric guide. (b) One-half of the structure for the purpose of analysis.

3.2.2 Guide Structures and Potential Functions

Figure 3.1(a) shows the cross-section of a general dielectric guide structure with metallic enclosure. Because of the symmetry of the structure, the analysis can be carried out in terms of the even and odd modes. It therefore suffices to use one symmetric half of the structure as shown in Figure 3.1(b), with the plane PP' at $x = a + s$ considered as a magnetic wall in the case of even mode, and an electric wall in the case of odd mode. The structure supports hybrid modes, which can be classified as TM-to-y (E^y_{mn}) modes and TE-to-y (E^x_{mn}) modes. The analysis presented is general and takes into account the boundary conditions at the dielectric interfaces and also at the metallic walls. We assume guided wave propagation along the z-direction according to $\exp[-j(\omega t - \beta_z z)]$. In order to formulate the boundary value problem, the structure in Figure 3.1(b) is divided into five regions (marked 1 to 5). Since all the fields corresponding to TM-to-y modes can be obtained from a single scalar potential function $\phi^e(x, y)$ as given by (2.24) and those of TE-to-y modes from the potential function $\phi^h(x, y)$ as given by (2.27), we begin with the expressions for $\phi^e(x, y)$ and $\phi^h(x, y)$ in the five regions:

Potential Function $\phi^e(x, y)$ for TM-to-y Modes

Region 1:

$$\phi_1^e(x, y) = \sum_{m=0}^{M} A_m \sin[\beta_{1xm}(x + p + a)] \cos(\beta_{1ym}\, y) \tag{3.1a}$$

Region 2:

$$\phi_2^e(x, y) = \sum_{m=1}^{M} [B_m \cos(\beta_{2xm}\, x) + B'_m \sin(\beta_{2xm}\, x)] \cos\{\beta_{2ym}[y - (d + b + q)]\} \tag{3.1b}$$

Region 3:

$$\phi_3^e(x, y) = \sum_{m=1}^{M} [C_m \cos(\beta_{3xm}\, x) + C'_m \sin(\beta_{3xm}\, x)] \{F_m \cos[\beta_{3ym}(y - d)] + F'_m \sin[\beta_{3ym}(y - d)]\} \tag{3.1c}$$

Region 4:

$$\phi_4^e(x, y) = \sum_{m=1}^{M} [D_m \cos(\beta_{4xm}\, x) + D'_m \sin(\beta_{4xm}\, x)] \cos(\beta_{4ym}\, y) \tag{3.1d}$$

Region 5:

$$\phi_5^e(x, y) = \sum_{m=0}^{M} E_m \cos[\beta_{5xm}(x - s - a)] \cos(\beta_{5ym} y),$$

$$\text{magnetic wall at } x = s + a \tag{3.1e}$$

$$= \sum_{m=0}^{M} E_m \sin[\beta_{5xm}(x - s - a)] \cos(\beta_{5ym} y),$$

$$\text{electric wall at } x = s + a \tag{3.1f}$$

where $\phi_i^e(x, y)$, β_{ixm}, and β_{iym} denote the potential function, x-directed propagation constant, and y-directed propagation constant, respectively, in the ith region. If ε_{ri} is the relative dielectric constant of the ith region, then the propagation constants β_{ixm}, β_{iym}, and β_z are related by

$$\beta_{ixm}^2 + \beta_{iym}^2 + \beta_z^2 = k_0^2 \varepsilon_{ri}, \quad i = 1 \text{ to } 5 \tag{3.2}$$

where k_0 is the propagation constant in free space.

Potential Function $\phi^h(x, y)$ for TE-to-y Modes

Region 1:

$$\phi_1^h(x, y) = \sum_{m=0}^{M} A_m \cos[\beta_{1xm}(x + p + a)] \sin(\beta_{1ym} y) \tag{3.3a}$$

Region 2:

$$\phi_2^h(x, y) = \sum_{m=1}^{M} [B_m \cos(\beta_{2xm} x) + B'_m \sin(\beta_{2xm} x)] \sin\{\beta_{2ym}[y - (d + b + q)]\} \tag{3.3b}$$

Region 3:

$$\phi_3^h(x, y) = \sum_{m=1}^{M} [C_m \cos(\beta_{3xm} x) + C'_m \sin(\beta_{3xm} x)] \{F_m \cos[\beta_{3ym}(y - d)] + F'_m \sin[\beta_{3ym}(y - d)]\} \tag{3.3c}$$

Region 4:

$$\phi_4^h(x, y) = \sum_{m=1}^{M} [D_m \cos(\beta_{4xm} x) + D'_m \sin(\beta_{4xm} x)] \sin(\beta_{4ym} y) \tag{3.3d}$$

Region 5:

$$\phi_5^h(x, y) = \sum_{m=0}^{M} E_m \sin[\beta_{5\,xm}(x - s - a)] \sin(\beta_{5\,ym}\, y),$$

magnetic wall at $x = s + a$ (3.3e)

$$= \sum_{m=0}^{M} E_m \cos[\beta_{5\,xm}(x - s - a)] \sin(\beta_{5\,ym}\, y),$$

electric wall at $x = s + a$ (3.3f)

where β_{ixm}, β_{iym}, and β_z satisfy the relation

$$\beta_{ixm}^2 + \beta_{iym}^2 + \beta_z^2 = k_0^2\, \varepsilon_{ri}, \quad i = 1 \text{ to } 5 \tag{3.4}$$

3.2.3 Analysis for TM-to-*y* Modes

For TM-to-y modes, the expressions for electric and magnetic fields in the five regions of Figure 3.1(b) are obtained by substituting (3.1) in (2.24).

Region 1:

$$E_{1x} = -\frac{1}{\varepsilon_{r1}} \sum_{m=0}^{M} A_m \beta_{1\,xm} \beta_{1\,ym} \cos[\beta_{1\,xm}(x + p + a)] \sin(\beta_{1\,ym}\, y) \tag{3.5a}$$

$$E_{1y} = \frac{1}{\varepsilon_{r1}} \sum_{m=0}^{M} A_m (\beta_{1\,xm}^2 + \beta_z^2) \sin[\beta_{1\,xm}(x + p + a)] \cos(\beta_{1\,ym}\, y) \tag{3.5b}$$

$$E_{1z} = \frac{j}{\varepsilon_{r1}} \sum_{m=0}^{M} A_m \beta_{1\,ym} \beta_z \sin[\beta_{1\,xm}(x + p + a)] \sin(\beta_{1\,ym}\, y) \tag{3.5c}$$

$$H_{1x} = -\,\omega\varepsilon_0 \sum_{m=0}^{M} A_m \beta_z \sin[\beta_{1\,xm}(x + p + a)] \cos(\beta_{1\,ym}\, y) \tag{3.5d}$$

$$H_{1y} = 0 \tag{3.5e}$$

$$H_{1z} = j\omega\varepsilon_0 \sum_{m=0}^{M} A_m \beta_{1\,xm} \cos[\beta_{1\,xm}(x + p + a)] \cos(\beta_{1\,ym}\, y) \tag{3.5f}$$

Region 2:

$$E_{2x} = \frac{1}{\varepsilon_{r2}} \sum_{m=1}^{M'} \beta_{2\,xm} \beta_{2\,ym} [B_m \sin(\beta_{2\,xm}\, x) - B'_m \cos(\beta_{2\,xm}\, x)]$$

$$\cdot \sin\{\beta_{2\,ym}[y - (d + b + q)]\} \tag{3.6a}$$

$$E_{2y} = \frac{1}{\varepsilon_{r2}} \sum_{m=1}^{M'} (\beta_{2xm}^2 + \beta_z^2)[B_m \cos(\beta_{2xm} x) + B'_m \sin(\beta_{2xm} x)]$$
$$\cdot \cos\{\beta_{2ym}[y - (d + b + q)]\} \tag{3.6b}$$

$$E_{2z} = \frac{j}{\varepsilon_{r2}} \sum_{m=1}^{M'} \beta_z \beta_{2ym} [B_m \cos(\beta_{2xm} x) + B'_m \sin(\beta_{2xm} x)]$$
$$\cdot \sin\{\beta_{2ym}[y - (d + b + q)]\} \tag{3.6c}$$

$$H_{2x} = -\omega\varepsilon_0 \sum_{m=1}^{M'} \beta_z [B_m \cos(\beta_{2xm} x) + B'_m \sin(\beta_{2xm} x)]$$
$$\cdot \cos\{\beta_{2ym}[y - (d + b + q)]\} \tag{3.6d}$$

$$H_{2y} = 0 \tag{3.6e}$$

$$H_{2z} = -j\omega\varepsilon_0 \sum_{m=1}^{M'} \beta_{2xm} [B_m \sin(\beta_{2xm} x) - B'_m \cos(\beta_{2xm} x)]$$
$$\cdot \cos\{\beta_{2ym}[y - (d + b + q)]\} \tag{3.6f}$$

Region 3:

$$E_{3x} = \frac{1}{\varepsilon_{r3}} \sum_{m=1}^{N} \beta_{3xm} \beta_{3ym} [C_m \sin(\beta_{3xm} x) - C'_m \cos(\beta_{3xm} x)]$$
$$\cdot \{F_m \sin[\beta_{3ym}(y - d)] - F'_m \cos[\beta_{3ym}(y - d)\} \tag{3.7a}$$

$$E_{3y} = \frac{1}{\varepsilon_{r3}} \sum_{m=1}^{N} (\beta_{3xm}^2 + \beta_z^2)[C_m \cos(\beta_{3xm} x) + C'_m \sin(\beta_{3xm} x)]$$
$$\cdot \{F_m \cos[\beta_{3ym}(y - d)] + F'_m \sin[\beta_{3ym}(y - d)]\} \tag{3.7b}$$

$$E_{3z} = \frac{j}{\varepsilon_{r3}} \sum_{m=1}^{N} \beta_z \beta_{3ym} [C_m \cos(\beta_{3xm} x) + C'_m \sin(\beta_{3xm} x)]$$
$$\cdot \{F_m \sin[(\beta_{3ym}(y - d)] - F'_m \cos[\beta_{3ym}(y - d)]\} \tag{3.7c}$$

$$H_{3x} = -\omega\varepsilon_0 \sum_{m=1}^{N} \beta_z [C_m \cos(\beta_{3xm} x) + C'_m \sin(\beta_{3xm} x)]$$
$$\cdot \{F_m \cos[\beta_{3ym}(y - d)] + F'_m \sin[\beta_{3ym}(y - d)]\} \tag{3.7d}$$

$$H_{3y} = 0 \tag{3.7e}$$

$$H_{3z} = -j\omega\varepsilon_0 \sum_{m=1}^{N} \beta_{3xm} [C_m \sin(\beta_{3xm} x) - C'_m \cos(\beta_{3xm} x)]$$
$$\cdot \{F_m \cos[\beta_{3ym}(y - d)] + F'_m \sin[\beta_{3ym}(y - d)]\} \tag{3.7f}$$

Region 4:

$$E_{4x} = \frac{1}{\varepsilon_{r4}} \sum_{m=1}^{N'} \beta_{4xm}\beta_{4ym} [D_m \sin(\beta_{4xm} x) - D'_m \cos(\beta_{4xm} x)] \cdot \sin(\beta_{4ym} y) \tag{3.8a}$$

$$E_{4y} = \frac{1}{\varepsilon_{r4}} \sum_{m=1}^{N'} (\beta_{4xm}^2 + \beta_z^2) [D_m \cos(\beta_{4xm} x) + D'_m \sin(\beta_{4xm} x)] \cdot \cos(\beta_{4ym} y) \tag{3.8b}$$

$$E_{4z} = \frac{j}{\varepsilon_{r4}} \sum_{m=1}^{N'} \beta_z \beta_{4ym} [D_m \cos(\beta_{4xm} x) + D'_m \sin(\beta_{4xm} x)] \cdot \sin(\beta_{4ym} y) \tag{3.8c}$$

$$H_{4x} = -\omega\varepsilon_0 \sum_{m=1}^{N'} \beta_z [D_m \cos(\beta_{4xm} x) + D'_m \sin(\beta_{4xm} x)] \cdot \cos(\beta_{4ym} y) \tag{3.8d}$$

$$H_{4y} = 0 \tag{3.8e}$$

$$H_{4z} = -j\omega\varepsilon_0 \sum_{m=1}^{N'} \beta_{4xm} [D_m \sin(\beta_{4xm} x) - D'_m \cos(\beta_{4xm} x)] \cdot \cos(\beta_{4ym} y) \tag{3.8f}$$

Region 5:

$$E_{5x} = \frac{1}{\varepsilon_{r5}} \sum_{m=0}^{N''} E_m \beta_{5xm} \beta_{5ym} \sin(\beta_{5ym} y) \cdot \begin{cases} \sin(\beta_{5xm}[x-(s+a)]\}, & \text{even mode} \\ -\cos\{\beta_{5xm}[x-(s+a)]\} & \text{odd mode} \end{cases} \tag{3.9a}$$

$$E_{5y} = \frac{1}{\varepsilon_{r5}} \sum_{m=0}^{N''} E_m (\beta_{5xm}^2 + \beta_z^2) \cos(\beta_{5ym} y) \cdot \begin{cases} \cos\{\beta_{5xm}[x-(s+a)]\}, & \text{even mode} \\ \sin\{\beta_{5xm}[x-(s+a)]\}, & \text{odd mode} \end{cases} \tag{3.9b}$$

$$E_{5z} = \frac{j}{\varepsilon_{r5}} \sum_{m=0}^{N''} E_m \beta_{5ym} \beta_z \sin(\beta_{5ym} y) \cdot \begin{cases} \cos\{\beta_{5xm}[x-(s+a)]\}, & \text{even mode} \\ \sin\{\beta_{5xm}[x-(s+a)]\}, & \text{odd mode} \end{cases} \tag{3.9c}$$

$$H_{5x} = -\omega\varepsilon_0 \sum_{m=0}^{N''} E_m \beta_z \cos(\beta_{5ym} y) \cdot \begin{cases} \cos\{\beta_{5xm}[x-(s+a)]\}, & \text{even mode} \\ \sin\{\beta_{5xm}[x-(s+a)]\}, & \text{odd mode} \end{cases} \tag{3.9d}$$

$$H_{5y} = 0 \tag{3.9e}$$

$$H_{5z} = -j\omega\varepsilon_0 \sum_{m=0}^{N''} E_m \beta_{5xm} \cos(\beta_{5ym} y) \cdot \begin{cases} \sin\{\beta_{5xm}[x-(s+a)]\}, & \text{even mode} \\ -\cos\{\beta_{5xm}[x-(s+a)]\}, & \text{odd mode} \end{cases} \tag{3.9f}$$

The above equations form a complete solution for the fields if the upper limit M, M', N, N', and N'' become infinite. However, for computational feasibility, we may set $M = M' = N = N' = N''$ and choose M sufficiently large so as to achieve desired accuracy in the results. The wavenumbers in the various regions satisfy the relation (3.2):

$$\beta_{ixm}^2 + \beta_{iym}^2 + \beta_z^2 = k_0^2 \varepsilon_{ri}, \quad i = 1 \text{ to } 5 \tag{3.10}$$

Because the tangential components of the electric field are zero at the metallic walls, we can set

$$\beta_{1ym} = \beta_{5ym} = \frac{m\pi}{d+b+q}, \quad m = 0, 1, 2, \ldots \tag{3.11}$$

The boundary conditions at the horizontal planes $y = d$ and $y = d + b$ over $-a \le x \le a$ can be satisfied separately independent of the remaining conditions. In order to maintain continuity of tangential electric and magnetic fields at these interfaces, it is necessary that

$$\beta_{2xm} = \beta_{3xm} = \beta_{4xm} = \beta_{xm} \tag{3.12}$$

We now match the boundary conditions at the various interfaces.

Match E_z and H_z Between Regions 2 and 3 at $y = d + b$ Setting $E_{2z} = E_{3z}$ and $H_{2z} = H_{3z}$ at $y = d + b$ and equating the coefficients of $\cos(\beta_{xm}x)$ and $\sin(\beta_{xm}x)$ independently, we obtain the following results:

$$\frac{B_m}{C_m} = \left(\frac{B'_m}{C'_m}\right) = \left(\frac{\beta_{3ym}\,\varepsilon_{r2}}{\beta_{2ym}\,\varepsilon_{r3}}\right)\left[\frac{-F_m \sin(\beta_{3ym} b) + F'_m \cos(\beta_{3ym} b)}{\sin(\beta_{2ym} q)}\right] = \frac{F_m \cos(\beta_{3ym} b) + F'_m \sin(\beta_{3ym} b)}{\cos(\beta_{2ym} q)} \tag{3.13}$$

From (3.13), we can write

$$\left(\frac{\beta_{2ym}}{\varepsilon_{r2}}\right)\tan(\beta_{2ym}q)=\left(\frac{\beta_{3ym}}{\varepsilon_{r3}}\right)\left[\frac{-(F_m/F'_m)\tan(\beta_{3ym}b)+1}{(F_m/F'_m)+\tan(\beta_{3ym}b)}\right] \tag{3.14}$$

Match E_z and H_z Between Regions 3 and 4 at $y=d$ Setting $E_{3z}=E_{4z}$ and $H_{3z}=H_{4z}$ at $y=d$ and equating the coefficients of $\cos(\beta_{xm}x)$ and $\sin(\beta_{xm}x)$ separately, we get

$$\frac{D_m}{C_m}=\frac{D'_m}{C'_m}=-F'_m\left[\frac{\beta_{3ym}\,\varepsilon_{r4}}{\beta_{4ym}\,\varepsilon_{r3}}\frac{1}{\sin(\beta_{4ym}d)}\right]=\frac{F_m}{\cos(\beta_{4ym}d)} \tag{3.15}$$

From (3.15), we can write

$$\frac{F_m}{F'_m}=-\left(\frac{\beta_{3ym}\,\varepsilon_{r4}}{\beta_{4ym}\,\varepsilon_{r3}}\right)\cot(\beta_{4ym}d) \tag{3.16}$$

Using (3.16) in (3.13), we can eliminate F'_m to yield

$$\frac{B_m}{C_m}=\frac{B'_m}{C'_m}=\frac{F_m\cos(\beta_{3ym}b)\,G(m)}{\beta_{3ym}\,\varepsilon_{r4}\cos(\beta_{2ym}q)} \tag{3.17}$$

where

$$G(m)=\beta_{3ym}\,\varepsilon_{r4}-\beta_{4ym}\,\varepsilon_{r3}\tan(\beta_{3ym}b)\tan(\beta_{4ym}d) \tag{3.18}$$

Substituting (3.16) in (3.14), we obtain

$$\left(\frac{\beta_{2ym}}{\varepsilon_{r2}}\right)\tan(\beta_{2ym}q)+\left(\frac{\beta_{3ym}}{\varepsilon_{r3}}\right)\tan(\beta_{3ym}b)+\left(\frac{\beta_{4ym}}{\varepsilon_{r4}}\right)\tan(\beta_{4ym}d)$$
$$=\beta_{2ym}\left(\frac{\beta_{4ym}}{\beta_{3ym}}\right)\left(\frac{\varepsilon_{r3}}{\varepsilon_{r2}\varepsilon_{r4}}\right)\tan(\beta_{2ym}q)\tan(\beta_{3ym}b)\tan(\beta_{4ym}d) \tag{3.19}$$

From (3.10), we can express β_{2ym} and β_{4ym} in terms of β_{3ym}:

$$\beta_{2ym}=[\beta_{3ym}^2-k_0^2(\varepsilon_{r3}-\varepsilon_{r2})]^{1/2} \tag{3.20a}$$
$$\beta_{4ym}=[\beta_{3ym}^2-k_0^2(\varepsilon_{r3}-\varepsilon_{r4})]^{1/2} \tag{3.20b}$$

The transcendental equation (3.19) can be solved for β_{3ym} and then from (3.20), β_{2ym} and β_{4ym} can be determined.

Match E_y Component at $x=-a$ Using the field expressions (3.5b), (3.6b), (3.7b), and (3.8b) along with the relations (3.15)–(3.18), and then applying the orthogonality property of trigonometric functions, we obtain

$$\delta_n A_n(\beta_{1\,xn}^2 + \beta_z^2)\,[(d+b+q)/2]\sin(\beta_{1\,xn}\,p)$$
$$= \sum_{m=1}^{M} (\beta_{xm}^2 + \beta_z^2)\,[B_m\cos(\beta_{xm}\,a) - B'_m\sin(\beta_{xm}\,a)]\,P_{mn} \tag{3.21}$$

where

$$P_{mn} = \left[\frac{I_1(n)}{\varepsilon_{r2}} + \frac{\cos(\beta_{2\,ym}\,q)}{\varepsilon_{r3}\cos(\beta_{3\,ym}\,b)\,G(m)}[\beta_{3\,ym}\,\varepsilon_{r4}\,I_2(n) - \beta_{4\,ym}\,\varepsilon_{r3}\tan(\beta_{4\,ym}\,d)\,I_3(n)]\right.$$
$$\left. + \frac{\beta_{3\,ym}\cos(\beta_{2\,ym}\,q)\,I_4(n)}{\cos(\beta_{4\,ym}\,d)\cos(\beta_{3\,ym}\,b)\,G(m)}\right], \quad n = 0, 1, 2, \ldots, M-1 \tag{3.22}$$

and

$$I_1(n) = \int_{d+b}^{d+b+q} \cos\{\beta_{2\,ym}[y-(d+b+q)]\}\cos[n\pi y/(d+b+q)]\,dy \tag{3.23a}$$

$$I_2(n) = \int_{d}^{d+b} \cos[\beta_{3\,ym}(y-d)\cos[n\pi y/(d+b+q)]\,dy \tag{3.23b}$$

$$I_3(n) = \int_{d}^{d+b} \sin[\beta_{3\,ym}(y-d)\cos[n\pi y/(d+b+q)]\,dy \tag{3.23c}$$

$$I_4(n) = \int_{0}^{d} \cos(\beta_{4\,ym}\,y)\cos[n\pi y/(d+b+q)]\,dy \tag{3.23d}$$

$$\delta_n = \begin{cases} 2, & n = 0 \\ 1, & n > 0 \end{cases} \tag{3.23e}$$

Match H_z at $x = -a$ Using (3.5f), (3.6f), (3.7f), and (3.8f) along with (3.15)–(3.18), and applying the principle of orthogonality, we obtain

$$\delta_n A_n \beta_{1\,xn}\,[(d+b+q)/2]\cos(\beta_{1\,xn}\,p) = \sum_{m=1}^{M} \beta_{xm}[B_m\sin(\beta_{xm}\,a) + B'_m\cos(\beta_{xm}\,a)]\,Q_{mn} \tag{3.24}$$

where

$$Q_{mn} = \left[I_1(n) + \frac{\cos(\beta_{2\,ym}\,q)}{\cos(\beta_{3\,ym}\,b)\,G(m)}[\beta_{3\,ym}\varepsilon_{r4}I_2(n) - \beta_{4\,ym}\varepsilon_{r3}\tan(\beta_{4\,ym}\,d)\,I_3(n)]\right.$$
$$\left. + \frac{\beta_{3\,ym}\varepsilon_{r4}\cos(\beta_{2\,ym}\,q)\,I_4(n)}{\cos(\beta_{4\,ym}\,d)\cos(\beta_{3\,ym}\,b)\,G(m)}\right], \quad n = 0, 1, 2, \ldots, M-1 \tag{3.25}$$

Match E_y at $x = a$ Using (3.6b), (3.7b), (3.8b), and (3.9b) along with (3.15)–(3.18), and applying the principle of orthogonality, we obtain

$$\delta_n E_n(\beta_{5xn}^2 + \beta_z^2)[(d+b+q)/2]\ T_1(s) = \sum_{m=1}^{M} (\beta_{xm}^2 + \beta_z^2)[B_m \cos(\beta_{xm} a) + B'_m \sin(\beta_{xm} a)] P_{mn},$$

$$T_1(s) = \begin{cases} \cos(\beta_{5xn} s), & \text{even mode} \\ -\sin(\beta_{5xn} s), & \text{odd mode} \end{cases} \quad n = 0, 1, 2, \ldots, M-1 \tag{3.26}$$

Match H_z at $x = a$ Using (3.6f), (3.7f), (3.8f), and (3.9f) along with (3.15)–(3.18), and applying the principle of orthogonality, we get

$$\delta_n E_n \beta_{5xn}[(d+b+q)/2]\ T_2(s) = \sum_{m=1}^{M} \beta_{xm}[-B_m \sin(\beta_{xm} a) + B'_m \cos(\beta_{xm} a)] Q_{mn}$$

$$T_2(s) = \begin{cases} \sin(\beta_{5xn} s), & \text{even mode} \\ -\cos(\beta_{5xn} s), & \text{odd mode} \end{cases} \quad n = 0, 1, 2, \ldots, M-1 \tag{3.27}$$

Dividing (3.21) by (3.24) eliminates the unknown constant A_n. The resulting equation is given by

$$\left[\frac{(\beta_{1xn}^2 + \beta_z^2)\tan(\beta_{1xn} p)}{\beta_{1xn}}\right] \sum_{m=1}^{M} \beta_{xm}[B_m \sin(\beta_{xm} a) + B'_m \cos(\beta_{xm} a)] Q_{mn}$$
$$= \sum_{m=1}^{M} (\beta_{xm}^2 + \beta_z^2)[B_m \cos(\beta_{xm} a) + B'_m \sin(\beta_{xm} a)] P_{mn}, \quad n = 0, 1, 2, \ldots, M-1 \tag{3.28}$$

Similarly, dividing (3.26) by (3.27) eliminates the unknown constant E_n to give

$$\left[\frac{(\beta_{5xn}^2 + \beta_z^2)}{\beta_{5xn}\ T(s)}\right] \sum_{m=1}^{M} \beta_{xm}[-B_m \sin(\beta_{xm} a) + B'_m \cos(\beta_{xm} a)] Q_{mn}$$
$$= \sum_{m=1}^{M} (\beta_{xm}^2 + \beta_z^2)[B_m \cos(\beta_{xm} a) + B'_m \sin(\beta_{xm} a)] P_{mn}$$

$$T(s) = \begin{cases} \tan(\beta_{5xn} s), & \text{even mode} \\ -\cot(\beta_{5xn} s), & \text{odd mode} \end{cases} \quad n = 0, 1, 2, \ldots, M-1 \tag{3.29}$$

Equations (3.28) and (3.29) form the final two sets of equations in terms of the unknown constants B_m and B'_m. For a given set of guide parameters and frequency, we can obtain β_{3ym} by solving (3.19) in conjunction with (3.20). The values of β_{2ym} and β_{4ym} are obtained from (3.20), and those of β_{1ym} and β_{5ym} from (3.11). Using (3.10) and (3.12), β_{xm} and β_{5xm} can be expressed in terms of β_z and substituted in (3.28) and (3.29). For the computation of β_z, we set the determinant of the coefficient matrix of Eqs. (3.28) and (3.29) equal to zero and solve the resulting equation for its roots. Since all the wavenumbers are now known, the constants B_m and B'_m can be determined from (3.28) and (3.29). With the knowledge of B_m and B'_m, all other unknown constants associated with the field expressions can be calculated from (3.15)–(3.18), (3.21), and (3.26). Thus the field distribution in the guide structure can be determined accurately.

3.2.4 Analysis for TE-to-y Modes

Analysis of the guide structures in Figure 3.1 for the TE-to-y modes is similar to the analysis presented in Section 3.2.3 for the TM-to-y modes. We therefore delete the details and provide only the salient expressions.

The electric and magnetic fields in the five regions of Figure 3.1(b) can be obtained by substituting (3.3) in (2.27).

Region 1:

$$E_{1x} = \omega\mu_0\beta_z \sum_{m=0}^{M} A_m \cos[\beta_{1xm}(x+p+a)] \sin(\beta_{1ym} y) \tag{3.30a}$$

$$E_{1y} = 0 \tag{3.30b}$$

$$E_{1z} = j\omega\mu_0 \sum_{m=0}^{M} A_m\beta_{1xm} \sin[\beta_{1xm}(x+p+a)] \sin(\beta_{1ym} y) \tag{3.30c}$$

$$H_{1x} = -\sum_{m=0}^{M} A_m\beta_{1xm}\beta_{1ym} \sin[\beta_{1xm}(x+p+a)] \cos(\beta_{1ym} y) \tag{3.30d}$$

$$H_{1y} = \sum_{m=0}^{M} A_m(\beta_{1xm}^2 + \beta_z^2) \cos[\beta_{1xm}(x+p+a)] \sin(\beta_{1ym} y) \tag{3.30e}$$

$$H_{1z} = -j\beta_z \sum_{m=0}^{M} A_m\beta_{1ym} \cos[\beta_{1xm}(x+p+a)] \cos(\beta_{1ym} y) \tag{3.30f}$$

Region 2:

$$E_{2x} = \omega\mu_0\beta_z \sum_{m=1}^{M} [B_m \cos(\beta_{2xm} x) + B'_m \sin(\beta_{2xm} x)] \cdot\sin\{\beta_{2ym}[y-(d+b+q)]\} \tag{3.31a}$$

$$E_{2y} = 0 \tag{3.31b}$$

$$E_{2z} = j\omega\mu_0 \sum_{m=1}^{M} \beta_{2xm}[B_m \sin(\beta_{2xm} x) - B'_m \cos(\beta_{2xm} x)] \cdot\sin\{\beta_{2ym}[y-(d+b+q)]\} \tag{3.31c}$$

$$H_{2x} = \sum_{m=1}^{M} \beta_{2xm}\beta_{2ym}[-B_m \sin(\beta_{2xm}x) + B'_m \cos(\beta_{2xm}x)]$$
$$\cdot\cos\{\beta_{2ym}[y-(d+b+q)]\} \quad (3.31\text{d})$$

$$H_{2y} = \sum_{m=1}^{M} (\beta_{2xm}^2 + \beta_z^2)[B_m \cos(\beta_{2xm}x) + B'_m \sin(\beta_{2xm}x)]$$
$$\cdot\sin\{\beta_{2ym}[y-(d+b+q)]\} \quad (3.31\text{e})$$

$$H_{2z} = -j\beta_z \sum_{m=1}^{M} \beta_{2ym}[B_m \cos(\beta_{2xm}x) + B'_m \sin(\beta_{2xm}x)]$$
$$\cdot\cos\{\beta_{2ym}[y-(d+b+q)]\} \quad (3.31\text{f})$$

Region 3:

$$E_{3x} = \omega\mu_0\beta_z \sum_{m=1}^{M} [C_m \cos(\beta_{3xm}x) + C'_m \sin(\beta_{3xm}x)]$$
$$\cdot\{F_m \cos[\beta_{3ym}(y-d)] + F'_m \sin[\beta_{3ym}(y-d)]\} \quad (3.32\text{a})$$

$$E_{3y} = 0 \quad (3.32\text{b})$$

$$E_{3z} = j\omega\mu_0 \sum_{m=1}^{M} \beta_{3xm}[C_m \sin(\beta_{3xm}x) - C'_m \cos(\beta_{3xm}x)]$$
$$\cdot\{F_m \cos[\beta_{3ym}(y-d)] + F'_m \sin[\beta_{3ym}(y-d)]\} \quad (3.32\text{c})$$

$$H_{3x} = \sum_{m=1}^{M} \beta_{3xm}\beta_{3ym}[C_m \sin(\beta_{3xm}x) - C'_m \cos(\beta_{3xm}x)]$$
$$\cdot\{F_m \sin[\beta_{3ym}(y-d)] - F'_m \cos[\beta_{3ym}(y-d)]\} \quad (3.32\text{d})$$

$$H_{3y} = \sum_{m=1}^{M} (\beta_{3xm}^2 + \beta_z^2)[C_m \cos(\beta_{3xm}x) + C'_m \sin(\beta_{3xm}x)]$$
$$\cdot\{F_m \cos[\beta_{3ym}(y-d)] + F'_m \sin[\beta_{3ym}(y-d)]\} \quad (3.32\text{e})$$

$$H_{3z} = -j\beta_z \sum_{m=1}^{M} \beta_{3ym}[C_m \cos(\beta_{3xm}x) + C'_m \sin(\beta_{3xm}x)]$$
$$\cdot\{-F_m \sin[\beta_{3ym}(y-d)] + F'_m \cos[\beta_{3ym}(y-d)]\} \quad (3.32\text{f})$$

Region 4:

$$E_{4x} = \omega\mu_0\beta_z \sum_{m=1}^{M} [D_m \cos(\beta_{4xm}x) + D'_m \sin(\beta_{4xm}x)] \sin(\beta_{4ym}y) \quad (3.33\text{a})$$

$$E_{4y} = 0 \quad (3.33\text{b})$$

$$E_{4z} = -j\omega\mu_0 \sum_{m=1}^{M} \beta_{4xm} [-D_m \sin(\beta_{4xm} x) + D'_m \cos(\beta_{4xm} x)] \sin(\beta_{4ym} y) \tag{3.33c}$$

$$H_{4x} = \sum_{m=1}^{M} \beta_{4xm}\beta_{4ym} [-D_m \sin(\beta_{4xm} x) + D'_m \cos(\beta_{4xm} x)] \cos(\beta_{4ym} y) \tag{3.33d}$$

$$H_{4y} = \sum_{m=1}^{M} (\beta_{4xm}^2 + \beta_z^2) [D_m \cos(\beta_{4xm} x) + D'_m \sin(\beta_{4xm} x)] \sin(\beta_{4ym} y) \tag{3.33e}$$

$$H_{4z} = -j\beta_z \sum_{m=1}^{M} \beta_{4ym} [D_m \cos(\beta_{4xm} x) + D'_m \sin(\beta_{4xm} x)] \cos(\beta_{4ym} y) \tag{3.33f}$$

Region 5:

$$E_{5x} = \omega\mu_0\beta_z \sum_{m=0}^{M} E_m \sin(\beta_{5ym} y) \cdot \begin{cases} \sin\{\beta_{5xm}[x-(s+a)]\}, & \text{even mode} \\ \cos\{\beta_{5xm}[x-(s+a)]\}, & \text{odd mode} \end{cases} \tag{3.34a}$$

$$E_{5y} = 0 \tag{3.34b}$$

$$E_{5z} = -j\omega\mu_0 \sum_{m=0}^{M} E_m\beta_{5xm} \sin(\beta_{5ym} y) \cdot \begin{cases} \cos\{\beta_{5xm}[x-(s+a)]\}, & \text{even mode} \\ -\sin\{\beta_{5xm}[x-(s+a)]\}, & \text{odd mode} \end{cases} \tag{3.34c}$$

$$H_{5x} = \sum_{m=0}^{M} E_m\beta_{5xm}\beta_{5ym} \cos(\beta_{5ym} y) \cdot \begin{cases} \cos\{\beta_{5xm}[x-(s+a)]\}, & \text{even mode} \\ -\sin\{\beta_{5xm}[x-(s+a)]\}, & \text{odd mode} \end{cases} \tag{3.34d}$$

$$H_{5y} = \sum_{m=0}^{M} E_m(\beta_{5xm}^2 + \beta_z^2) \sin(\beta_{5ym} y) \cdot \begin{cases} \sin\{\beta_{5xm}[x-(s+a)]\}, & \text{even mode} \\ \cos\{\beta_{5xm}[x-(s+a)]\}, & \text{odd mode} \end{cases} \tag{3.34e}$$

$$H_{5z} = -j\beta_z \sum_{m=0}^{M} E_m\beta_{5ym} \cos(\beta_{5ym} y) \cdot \begin{cases} \sin\{\beta_{5xm}[x-(s+a)]\}, & \text{even mode} \\ \cos\{\beta_{5xm}[x-(s+a)]\}, & \text{odd mode} \end{cases} \tag{3.34f}$$

The wave equation in various regions satisfies the following relations:

$$\beta_{ixm}^2 + \beta_{iym}^2 + \beta_z^2 = k_0^2\,\varepsilon_{ri}, \quad i = 1 \text{ to } 5 \tag{3.35}$$

$$\beta_{1\,ym} = \beta_{5\,ym} = \frac{m\pi}{d+b+q}, \qquad m = 0, 1, 2, \ldots \tag{3.36}$$

$$\beta_{2\,xm} = \beta_{3\,xm} = \beta_{4\,xm} = \beta_{xm} \tag{3.37}$$

Matching the tangential components E_z and H_z at $y = d$ and $y = d + b$, and using (3.37), we obtain the following results:

$$\frac{B_m}{C_m} = \frac{B'_m}{C'_m} = \frac{-F_m\,G(m)}{\beta_{3\,ym}\sin(\beta_{2\,ym}\,q)} \tag{3.38}$$

and

$$G(m) = [\beta_{3ym}\cos(\beta_{3\,ym}\,b) + \beta_{4\,ym}\sin(\beta_{3\,ym}\,b)\cot(\beta_{4\,ym}\,d)] \tag{3.39}$$

$$\frac{D_m}{C_m} = \frac{D'_m}{C'_m} = F'_m\left[\frac{\beta_{3\,ym}}{\beta_{4\,ym}\cos(\beta_{4\,ym}\,d)}\right] = \frac{F_m}{\sin(\beta_{4\,ym}\,d)} \tag{3.40}$$

$$\frac{\tan(\beta_{2\,ym}\,q)}{\beta_{2\,ym}} + \frac{\tan(\beta_{3\,ym}\,b)}{\beta_{3\,ym}} + \frac{\tan(\beta_{4\,ym}\,d)}{\beta_{4\,ym}}$$
$$= \left(\frac{\beta_{3\,ym}}{\beta_{2\,ym}\,\beta_{4\,ym}}\right)\tan(\beta_{2\,ym}\,q)\tan(\beta_{3\,ym}\,b)\tan(\beta_{4\,ym}\,d) \tag{3.41}$$

From (3.36), we can express $\beta_{2\,ym}$ and $\beta_{4\,ym}$ in terms of $\beta_{3\,ym}$:

$$\beta_{2\,ym} = [\beta_{3\,ym}^2 - k_0^2\,(\varepsilon_{r3} - \varepsilon_{r2})]^{1/2} \tag{3.42a}$$

$$\beta_{4\,ym} = [\beta_{3\,ym}^2 - k_0^2\,(\varepsilon_{r3} - \varepsilon_{r4})]^{1/2} \tag{3.42b}$$

Matching the H_y and E_z components at the interface $x = -a$ and using the relations (3.36)–(3.40) as well as the principle of orthogonality of trigonometric functions, we obtain

$$A_n(\beta_{1\,xn}^2 + \beta_z^2)\,[(d+b+q)/2]\cos(\beta_{1\,xn}\,p) = \sum_{m=1}^{M}(\beta_{1\,xn}^2 + \beta_z^2)$$
$$\cdot[B_m\cos(\beta_{xm}\,a) - B'_m\sin(\beta_{xm}\,a)]\,P_{mn}, \quad n = 1, 2, \ldots, M \tag{3.43}$$

$$A_n\beta_{1\,xn}\,[(d+b+q)/2]\sin(\beta_{1\,xn}\,p) = -\sum_{m=1}^{M}\beta_{xm}\,[B_m\sin(\beta_{xm}\,a)$$
$$+ B'_m\cos(\beta_{xm}\,a)]\,P_{mn}, \quad n = 1, 2, \ldots, M \tag{3.44}$$

where

$$P_{mn} = \left[I_1(n) - \left(\frac{\sin(\beta_{2ym} q)}{G(m)} \right) [\beta_{3ym} I_2(n) + \beta_{4ym} \cot(\beta_{4ym} d) I_3(n)] - \frac{\beta_{3ym} \sin(\beta_{2ym} q) I_4(n)}{\sin(\beta_{4ym} d) G(m)} \right] \tag{3.45}$$

$$I_1(n) = \int_{d+b}^{d+b+q} \sin\{\beta_{2ym} [y - (d+b+q)]\} \sin[n\pi y/(d+b+q)] \, dy \tag{3.46a}$$

$$I_2(n) = \int_{d}^{d+b} \cos[\beta_{3ym} (y-d)] \sin[n\pi y/(d+b+q)] \, dy \tag{3.46b}$$

$$I_3(n) = \int_{d}^{d+b} \sin[\beta_{3ym} (y-d)] \sin[n\pi y/(d+b+q)] \, dy \tag{3.46c}$$

$$I_4(n) = \int_{0}^{d} \sin(\beta_{4ym} y) \sin[n\pi y/(d+b+q)] \, dy \tag{3.46d}$$

Matching H_y and E_z components at the interface $x = a$ and using the relations (3.36)–(3.40) as well as the principle of orthogonality, we obtain

$$E_n(\beta_{5xn}^2 + \beta_z^2) [(d+b+q)/2] \, T_1(s) = - \sum_{m=1}^{M} (\beta_{xm}^2 + \beta_z^2) [B_m \cos(\beta_{xm} a) + B'_m \sin(\beta_{xm} a)] \, P_{mn}, \quad T_1(s) = \begin{cases} \sin(\beta_{5xn} s), & \text{even mode} \\ -\cos(\beta_{5xn} s), & \text{odd mode} \end{cases} \tag{3.47}$$

$$E_n \beta_{5xn} [(d+b+q)/2] \, T_2(s) = - \sum_{m=1}^{M} \beta_{xm} [B_m \sin(\beta_{xm} a) - B'_m \cos(\beta_{xm} a)] \, P_{mn}, \quad T_2(s) = \begin{cases} \cos(\beta_{5xn} s), & \text{even mode} \\ \sin(\beta_{5xn} s), & \text{odd mode} \end{cases} \tag{3.48}$$

Eliminating A_n from (3.43) and (3.44), and also eliminating E_n from (3.47) and (3.48), we obtain the following two expressions in terms of the remaining unknown constants B_m and B'_m:

$$\left[\frac{(\beta_{1xn}^2 + \beta_z^2)}{\beta_{1xn} \tan(\beta_{1xn} p)} \right] \sum_{m=1}^{M} \beta_{xm} [B_m \sin(\beta_{xm} a) + B'_m \cos(\beta_{xm} a)] \, P_{mn} = - \sum_{m=1}^{M} (\beta_{1xm}^2 + \beta_z^2) [B_m \cos(\beta_{xm} a) - B'_m \sin(\beta_{xm} a)] \, P_{mn}, \quad n = 1, 2, \ldots, M \tag{3.49}$$

$$\left[\frac{(\beta_{5xn}^2 + \beta_z^2)\,T(s)}{\beta_{5xn}}\right] \sum_{m=1}^{M} \beta_{xm}[B_m \sin(\beta_{xm} a) - B'_m \cos(\beta_{xm} a)]\, P_{mn}$$

$$= \sum_{m=1}^{M} (\beta_{xm}^2 + \beta_z^2)\,[B_m \cos(\beta_{xm} a) + B'_m \sin(\beta_{xm} a)]\, P_{mn}$$

$$T(s) = \begin{cases} \tan(\beta_{5xn} s), & \text{even mode} \\ -\cot(\beta_{5xn} s), & \text{odd mode} \end{cases} \quad n = 1, 2, \ldots, M \tag{3.50}$$

These are the final sets of expressions for the evaluation of β_z of TE-to-y modes. For a given set of guide parameters and frequency, the wavenumbers β_{iym} for $i = 1$ to 5 are evaluated from (3.36), (3.41), and (3.42). Using (3.35) and (3.37), the wavenumbers β_{xm} and β_{5xm} can be expressed in terms of β_z. The determinant of the coefficient matrix of (3.49) and (3.50) then involves only β_z as the unknown parameter. The values of β_z for the dominant and higher order TE-to-y modes are obtained by setting the determinant equal to zero and solving the resulting equation for its roots.

With the evaluation of all the wavenumbers, the constants B_m and B'_m can be determined from (3.49) and (3.50). Using the values of B_m and B'_m, all other unknown constants associated with the field expressions can be evaluated from (3.38)–(3.40), (3.43), and (3.48). Thus the field distribution in the guide for TE-to-y mode can be determined accurately.

3.2.5 Practical Guides as Special Cases

By substituting suitable dimensional parameters and values of dielectric constants, the general guide structure shown in Figure 3.1 reduces to a number of practical guides.

Coupled-Guide Structures (Fig. 3.2)

1. Trapped coupled insular image guide (Fig. 3.2(a)):

$$\varepsilon_{r1} = \varepsilon_{r2} = \varepsilon_{r5} = 1, \quad \varepsilon_{r3} = \varepsilon_r, \quad \varepsilon_{r4} = \varepsilon'_r, \quad \varepsilon'_r < \varepsilon_r, \quad d \ll b, \quad q \to \infty$$

(q is sufficiently large such that effect of the top wall is negligible).

2. Trapped coupled image guide (Fig. 3.2(b)):

$$\varepsilon_{r1} = \varepsilon_{r2} = \varepsilon_{r4} = \varepsilon_{r5} = 1, \quad \varepsilon_{r3} = \varepsilon_r, \quad d = 0, \quad q \to \infty$$

3. Coupled insular image guide (Fig. 3.2(c)):

$$\varepsilon_{r1} = \varepsilon_{r2} = \varepsilon_{r5} = 1, \quad \varepsilon_{r3} = \varepsilon_r, \quad \varepsilon_{r4} = \varepsilon'_r, \quad \varepsilon'_r < \varepsilon_r, \quad p = q \to \infty$$

4. Coupled image guide (Fig. 3.2(d)):

$$\varepsilon_{r1} = \varepsilon_{r2} = \varepsilon_{r4} = \varepsilon_{r5} = 1, \quad \varepsilon_{r3} = \varepsilon_r, \quad d = 0, \quad p = q \to \infty$$

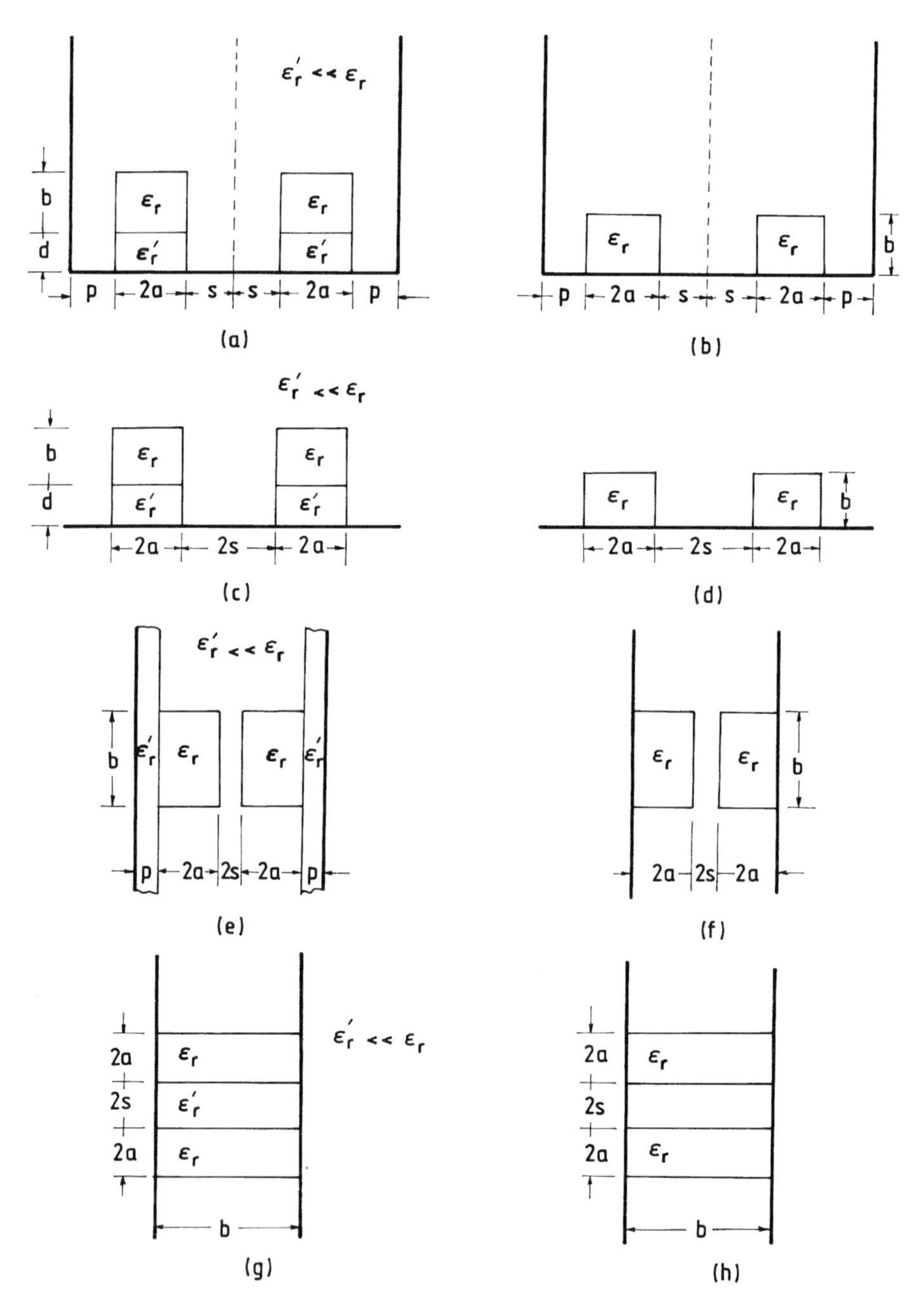

FIGURE 3.2 Examples of coupled-guide structures: (a) trapped coupled insular image guide, (b) trapped coupled image guide, (c) coupled insular image guide, (d) coupled image guide, (e) broadside-coupled insulated nonradiative guide, (f) broadside-coupled nonradiative guide, (g) end-coupled nonradiative guide with dielectric spacer, and (h) end-coupled nonradiative guide.

5. Broadside-coupled insulated nonradiative guide (Fig. 3.2(e)):

$$\varepsilon_{r2} = \varepsilon_{r4} = \varepsilon_{r5} = 1, \quad \varepsilon_{r3} = \varepsilon_r, \quad \varepsilon_{r1} = \varepsilon_r', \quad \varepsilon_r' < \varepsilon_r, \quad p \ll 2a, \quad q = d \to \infty$$

6. Broadside-coupled nonradiative guide (Fig. 3.2(f)):

$$\varepsilon_{r1} = \varepsilon_{r2} = \varepsilon_{r4} = \varepsilon_{r5} = 1, \quad \varepsilon_{r3} = \varepsilon_r, \quad p = 0, \quad q = d \to \infty$$

7. End-coupled nonradiative guide with dielectric spacer (Fig. 3.2(g)):

$$\varepsilon_{r1} = \varepsilon_{r2} = \varepsilon_{r4} = 1, \quad \varepsilon_{r3} = \varepsilon_r, \quad \varepsilon_{r5} = \varepsilon_r', \quad \varepsilon_r' < \varepsilon_r, \quad d = q = 0, \quad p \to \infty$$

8. End-coupled nonradiative guide (Fig. 3.2(h)):

$$\varepsilon_{r1} = \varepsilon_{r2} = \varepsilon_{r4} = \varepsilon_{r5} = 1, \quad \varepsilon_{r3} = \varepsilon_r, \quad d = q = 0, \quad p \to \infty$$

In practice, the structures shown in Figure 3.2(a)–(e) are excited with rectangular waveguides such that E_y and H_x are the dominant field components. Therefore the solution corresponding to TM-to-y modes (E^y_{mn} modes) is applicable. For the end-coupled nonradiative guides shown in Figure 3.2(f) and (g), the TE-to-y solution can be used since the dominant E-field is parallel to the image plane.

Single-Guide Structures (Fig. 3.3) The solution for several single-guide structures can be obtained as a special case of the corresponding coupled-guide structure.

1. Trapped insular image guide (Fig. 3.3(a)): odd-mode solution of Figure 3.2(a) with $p = s$.
2. Trapped image guide (Fig. 3.3(b)): odd-mode solution of Figure 3.2(b) with $p = s$.
3. Insular image guide (Fig. 3.3(c)): solution of Figure 3.2(c) with $s \to \infty$.
4. Image guide (Fig. 3.3(d)): solution of Figure 3.2(d) with $s \to \infty$.
5. Insular nonradiative dielectric guide (Fig. 3.3(e)): odd-mode TM-to-y solution of Figure 3.1(b) with

$$\varepsilon_{r1} = \varepsilon_{r5} = \varepsilon_r', \quad \varepsilon_{r2} = \varepsilon_{r4} = 1, \quad \varepsilon_{r3} = \varepsilon_r, \quad p = s \text{ small}, \quad q = d \to \infty$$

6. Nonradiative dielectric guide (Fig. 3.3(f)): odd-mode TM-to-y solution of Figure 3.1(b) with

$$\varepsilon_{r1} = \varepsilon_{r2} = \varepsilon_{r4} = \varepsilon_{r5} = 1, \quad \varepsilon_{r3} = \varepsilon_r, \quad p = s = 0, \quad q = d \to \infty$$

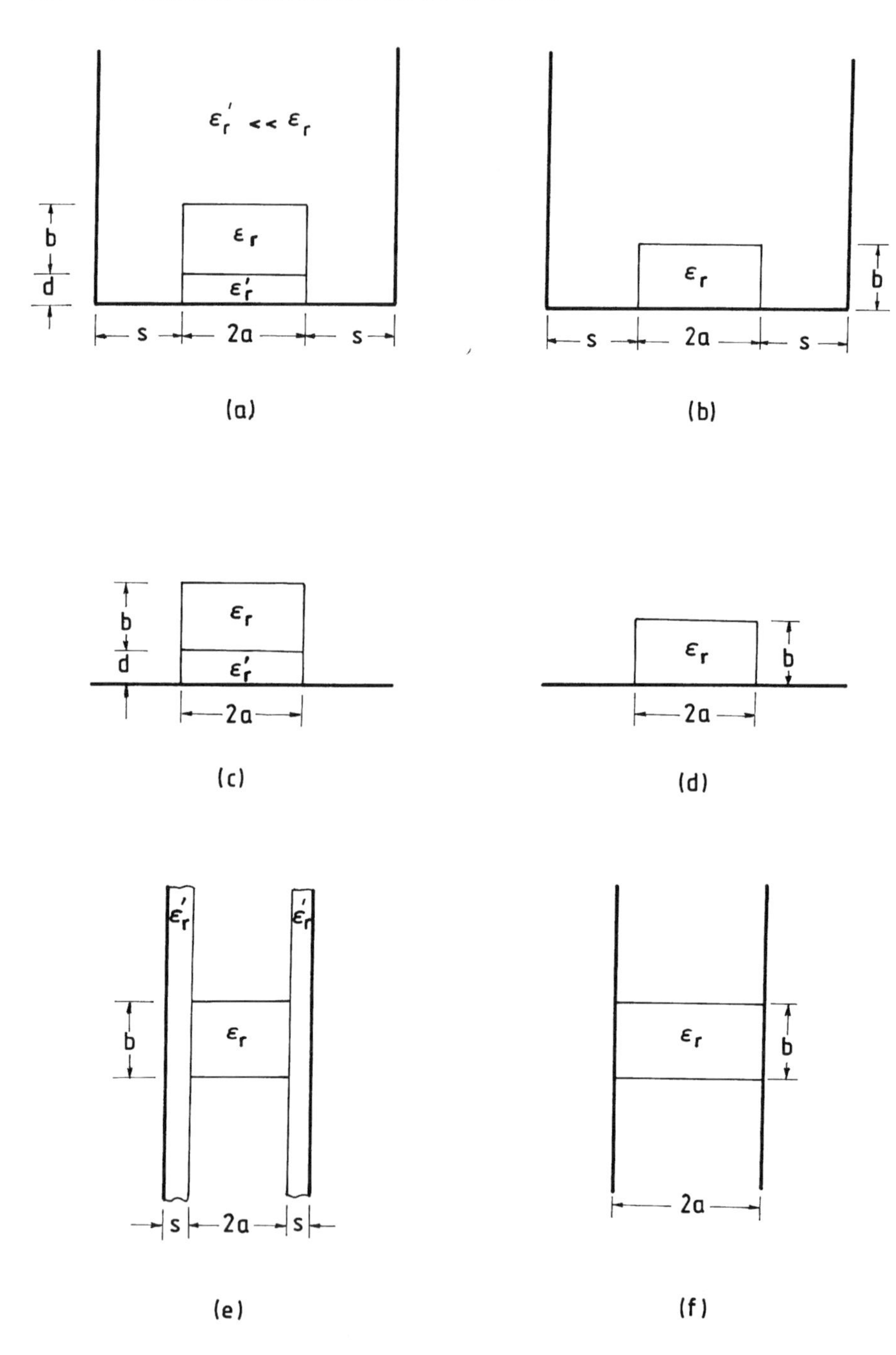

FIGURE 3.3 Examples of single-guide structures: (a) trapped insular image guide, (b) trapped image guide, (c) insular image guide, (d) image guide, (e) insular nonradiative dielectric guide, and (f) nonradiative dielectric guide.

3.3 WAVE IMPEDANCE AND ATTENUATION

The wave impedance of a dielectric guide is obtained as the ratio of transverse electric field to magnetic field in the dielectric region having the highest dielectric constant. Next, an expression is derived for the conductor attenuation coefficient of a fairly general single guide structure. The attenuation constants of most practical dielectric integrated guides can be obtained as special cases of the general expression. With the availability of low-loss dielectric materials, the dielectric loss in dielectric integrated guides is much smaller than the conductor loss. Solbach [18] has shown that the results of the dielectric loss in an image guide calculated using the mode-matching method are fairly close to those evaluated using the approximate theory of Toulios and Knox [19]. The dielectric loss coefficient of the dielectric integrated guides can therefore the determined using the approximate technique presented in Chapter 2.

3.3.1 Wave Impedances

For the TM-to-y mode, the wave impedance can be obtained from

$$Z_w = \frac{(E_x^2 + E_y^2)^{1/2}}{H_x} \tag{3.51}$$

Since the principal field components are E_y and H_x, the magnitude of E_x may be neglected in comparison with E_y. Hence we may approximate (3.51) by

$$Z_w = -\frac{E_y}{H_x} \tag{3.52}$$

For practical calculations, we may choose the field expressions corresponding to the dielectric region having the highest dielectric constant (region carrying most of the guided wave energy). For example, if in Figure 3.2(b), region 3 supports most of the energy, the wave impedance can be obtained by using Eqs. (3.7 b) and (3.7 d) in Eq. (3.52). Under single-mode approximation, the expression is given by

$$Z_w = \frac{\beta_z^2 + \beta_{3x}^2}{\omega \varepsilon_0 \varepsilon_{r3} \beta_z} = \frac{k_0^2 \varepsilon_{r3} - \beta_{3y}^2}{\omega \varepsilon_0 \varepsilon_{r3} \beta_z} \tag{3.53}$$

where β_{3x} and β_{3y} are the x- and y-directed wavenumbers in region 3.

For the TE-to-y mode, E_x and H_y are the principal transverse field components. Neglecting H_x in comparison with H_y, the wave impedance can be approximated by

$$Z_w = \frac{E_x}{H_y} \tag{3.54}$$

Choosing the field expression (3.32a) for E_x and (3.32e) for H_y corresponding to the dielectric core region (region 3 in Fig. 3.2(b)) and further using a single-mode approximation, we can write

$$Z_w = \frac{\omega \mu_0 \beta_z}{\beta_z^2 + \beta_{3x}^2} = \frac{\omega \mu_0 \beta_z}{k_0^2 \varepsilon_{r3} - \beta_{3y}^2} \tag{3.55}$$

3.3.2 Attenuation Due to Conductor Loss

The general expression for determining the attenuation coefficient due to conductor loss is given by

$$\alpha_c = \frac{R_s \oint |H_t|^2 \, dl}{2 \int_S \int \hat{\mathbf{z}} \cdot (\mathbf{E} \times \mathbf{H}^*) \, ds} \quad \text{nepers/unit length} \tag{3.56}$$

where H_t is the tangential magnetic field at the surface of the conductor walls and $R_s = 1/\sigma\delta$ is the surface resistance of the conductor walls (σ is the conductivity and δ is the skin depth of the conductor). The line integral in the numerator is carried out on the conducting wall boundaries and the surface integration in the denominator is taken over the complete cross-section of the guide. Using the field expressions presented in Section 3.2 in (3.56), the attenuation constant α_c for the various dielectric integrated guides with metal walls can be evaluated.

In order to facilitate the computation of conductor loss of different types of useful dielectric guide structures illustrated in Figure 3.3, we derive below the expressions for the conductor loss coefficient of a general guide structure shown in Figure 3.4. The derivation is carried out for the dominant TM-to-y hybrid mode (E_{11}^y mode). Referring to the structure in Figure 3.4 and considering the TM-to-y modes, we can rewrite (3.56) in the form

$$\alpha_c = \frac{R_s \oint |H_t|^2 \, dl}{2 \sum_{i=1}^{5} Z_{wi} \int\int |H_{ix}|^2 \, dx \, dy} \tag{3.57}$$

where Z_{wi} denotes the wave impedance for the ith region. It is given by

$$Z_{wi} = \frac{\beta_z^2 + \beta_{ix}^2}{\omega \varepsilon_0 \varepsilon_{ri} \beta_z} = \frac{k_0^2 \varepsilon_{ri} - \beta_{iy}^2}{\omega \varepsilon_0 \varepsilon_{ri} \beta_z}, \quad i = 1 \text{ to } 5 \tag{3.58}$$

The magnitude of the tangential magnetic field $|H_t|$ is given by

$$|H_t|^2 = |H_x|^2 + |H_z|^2 \tag{3.59}$$

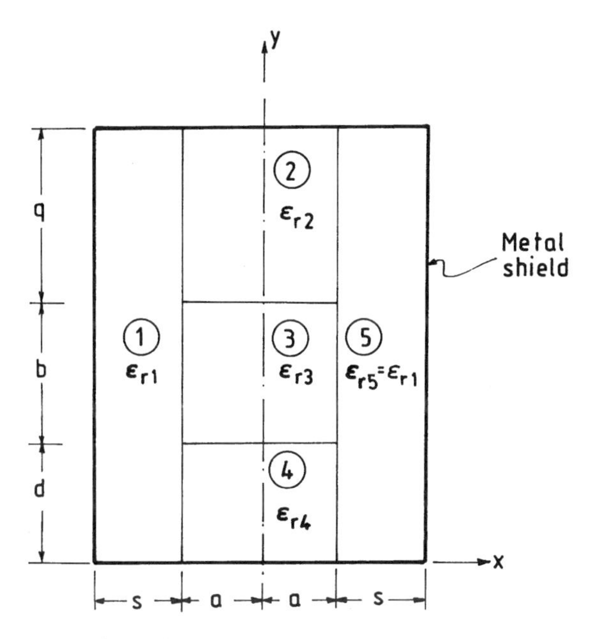

FIGURE 3.4 Dielectric guide structure for derivation of conductor loss.

Using the divergence relation $\nabla\cdot\mathbf{H}$, we can express $|H_z|$ in terms of H_x as

$$|H_z|^2 = \left(\frac{1}{\beta_z^2}\right)\left|\frac{\partial H_x}{\partial x}\right|^2 \tag{3.60}$$

so that

$$|H_t|^2 = |H_x|^2 + \frac{1}{\beta_z^2}\left|\frac{\partial H_x}{\partial x}\right|^2 \tag{3.61}$$

The expressions for H_x in the various regions of the generalized coupled guide structure shown in Figure 3.1 are given in Section 3.2.3. As an example, we derive below an expression for the conductor loss of a single guide structure shown in Figure 3.4. The structure is the same as that shown in Figure 3.1(b) with PP' as the electric wall, $p = s$, and $\varepsilon_{r5} = \varepsilon_{r1}$. We then have

$$\beta_{5x} = \beta_{1x} \tag{3.62a}$$

$$\beta_{5y} = \beta_{1y} \tag{3.62b}$$

$$\beta_{2x} = \beta_{3x} = \beta_{4x} = \beta_x \tag{3.62c}$$

Considering a single term of the series expansion in (3.5d), (3.6d), (3.7d), (3.8d) and the odd-mode expression of (3.9d) (the subscript m associated with the various parameters is dropped), and further using the result $B'/B = C'/C = D'/D$ from (3.15) and (3.17), we can write

$$H_{1x} = -\omega\varepsilon_0\beta_z A \sin[\beta_{1x}(x+s+a)]\cos(\beta_{1y}y) \quad (3.63a)$$

$$H_{2x} = -\omega\varepsilon_0\beta_z B[\cos(\beta_x x) + (B'/B)\sin(\beta_x x)] \cdot\cos[\beta_{2y}(y-d-b-q)] \quad (3.63b)$$

$$H_{3x} = -\omega\varepsilon_0\beta_z CF[\cos(\beta_x x) + (B'/B)\sin(\beta_x x)] \cdot\{\cos[\beta_{3y}(y-d)] + (F'/F)\sin[\beta_{3y}(y-d)]\} \quad (3.63c)$$

$$H_{4x} = -\omega\varepsilon_0\beta_z D[\cos(\beta_x x) + (B'/B)\sin(\beta_x x)]\cos(\beta_{4y}y) \quad (3.63d)$$

$$H_{5x} = -\omega\varepsilon_0\beta_z E \sin[\beta_{1x}(x-s-a)]\cos(\beta_{1y}y) \quad (3.63e)$$

Using (3.16) and (3.21)–(3.26), we can solve for the various constants and express all the H_x components given in (3.63) in terms of a single constant B as follows:

$$H_{1x} = -\omega\varepsilon_0\beta_z B\xi_1 \sin[\beta_{1x}(x+s+a)]\cos(\beta_{1y}y) \quad (3.64a)$$

$$H_{2x} = -\omega\varepsilon_0\beta_z B[\cos(\beta_x x) + \xi_3\sin(\beta_x x)]\cos[\beta_{2y}(y-d-b-q)] \quad (3.64b)$$

$$H_{3x} = -\omega\varepsilon_0\beta_z B\xi_2[\cos(\beta_x x) + \xi_3\sin(\beta_x x)]\{\cos[\beta_{3y}(y-d)] + \xi_4 \cdot\sin[\beta_{3y}(y-d)]\} \quad (3.64c)$$

$$H_{4x} = -\omega\varepsilon_0\beta_z B\xi_2[\cos(\beta_x x) + \xi_3\sin(\beta_x x)]\frac{\cos(\beta_{4y}y)}{\cos(\beta_{4y}d)} \quad (3.64d)$$

$$H_{5x} = -\omega\varepsilon_0\beta_z B\xi_5 \sin[\beta_{1x}(x-s-a)]\cos(\beta_{1y}y) \quad (3.64e)$$

where

$$\xi_1 = \left(\frac{2PQ}{d+b+q}\right)\left[\frac{\beta_x(\beta_x^2+\beta_z^2)}{\cos(\beta_x a)\cos(\beta_{1x}s)[\beta_{1x}(\beta_x^2+\beta_z^2)P\tan(\beta_x a) + \beta_x(\beta_{1x}^2+\beta_z^2)Q\tan(\beta_{1x}s)]}\right] \quad (3.65a)$$

$$\xi_2 = \frac{\varepsilon_{r4}\beta_{4y}\cos(\beta_{2y}q)}{\cos(\beta_{3y}b)[\beta_{3y}\varepsilon_{r4} - \beta_{4y}\varepsilon_{r3}\tan(\beta_{3y}b)\tan(\beta_{4y}d)]} \quad (3.65b)$$

$$\xi_3 = [(d+b+q)/2](\xi_1\beta_{1x}/Q\beta_x)\cos(\beta_{1x}s) - \tan(\beta_x a) \quad (3.65c)$$

$$\xi_4 = -(\varepsilon_{r3}\beta_{4y}/\varepsilon_{r4}\beta_{3y})\tan(\beta_{4y}d) \quad (3.65d)$$

$$\xi_5 = -\left(\frac{2P}{d+b+q}\right)\left(\frac{\beta_x^2+\beta_z^2}{\beta_{1x}^2+\beta_z^2}\right)\left(\frac{\cos(\beta_x a)}{\sin(\beta_{1x}s)}\right)[1+\xi_3\tan(\beta_x a)] \quad (3.65e)$$

$$P = \frac{I_1}{\varepsilon_{r2}} + \left(\frac{\xi_2}{\varepsilon_{r3}}\right)\left[I_2 - \left(\frac{\varepsilon_{r3}}{\varepsilon_{r4}}\right)\left(\frac{\beta_{4y}}{\beta_{3y}}\right)\tan(\beta_{4y} d) I_3 + \left(\frac{\varepsilon_{r3}}{\varepsilon_{r4}}\right)\left(\frac{I_4}{\cos(\beta_{4y} d)}\right)\right] \tag{3.66a}$$

$$Q = I_1 + \xi_2\left[I_2 - \left(\frac{\varepsilon_{r3}}{\varepsilon_{r4}}\right)\left(\frac{\beta_{4y}}{\beta_{3y}}\right)\tan(\beta_{4y} d)\, I_3 + \left(\frac{I_4}{\cos(\beta_{4y} d)}\right)\right] \tag{3.66b}$$

The integrals I_1, I_2, I_3, and I_4 are given by Eq. (3.23) with $n = 1$.

Using the field expressions (3.64) along with (3.60) and (3.61) and (3.57), we can derive the expression for α_c. Practical dielectric integrated guides generally do not use the top conducting wall or it is sufficiently away from the guiding strip (q large in Fig. 3.4) so that its effect on the propagation is negligible. Thus, in the evaluation of the numerator of the expression for α_c (Eq. (3.57)), we may neglect the contribution to power dissipation due to the top wall and carry out the line integration on the bottom wall and the two side walls of the structure. Denoting the numerator as NR, we can write

$$\begin{aligned} NR = R_s \oint |H_t|^2\, dl = 2R_s \Bigg[& \int_0^a (|H_{4x}|^2 + |H_{4z}|^2)|_{y=0}\, dx \\ & + \int_a^{a+s} (|H_{5x}|^2 + |H_{5z}|^2)|_{y=0}\, dx \\ & + \int_0^{d+b+q} (|H_{5z}|^2)|_{x=a+s}\, dy \Bigg] \end{aligned} \tag{3.67}$$

The factor 2 in Eq. (3.67) appears because of the symmetry of the structure (Fig. 3.4) about the vertical axis at $x = 0$. Substituting the necessary field expressions and performing the integration, we obtain

$$NR = R_s |B|^2 (\omega\varepsilon_0)^2 (\beta_z^2 \xi_2^2 U_0 U_1 + \beta_x^2 \xi_2^2 U_0 U_2 + \beta_z^2 \xi_5^2 U_3 + \beta_{1x}^2 \xi_5^2 U_4 + \beta_{1x}^2 \xi_5^2 U_5) \tag{3.68}$$

where

$$U_0 = 1/\cos^2(\beta_{4y} d) \tag{3.69a}$$

$$\begin{aligned} U_1 = {} & \{a + [\sin(2\beta_x a)/2\beta_x]\} + \xi_3^2 \{a - [\sin(2\beta_x a)/2\beta_x]\} \\ & + (\xi_3/\beta_x)[1 - \cos(2\beta_x a)] \end{aligned} \tag{3.69b}$$

$$\begin{aligned} U_2 = {} & \{s - [\sin(2\beta_{1x} s)/2\beta_{1x}]\} + \xi_3^2 \{s + [\sin(2\beta_{1x} s)/2\beta_{1x}]\} \\ & - (\xi_3/\beta_x)\{\cos(2\beta_x a) - \cos[2\beta_x (a + s)]\} \end{aligned} \tag{3.69c}$$

$$U_3 = s - [\sin(2\beta_{1x} s)/2\beta_{1x}] \tag{3.69d}$$

$$U_4 = s + [\sin(2\beta_{1x} s)/2\beta_{1x}] \tag{3.69e}$$

$$U_5 = (d + b + q) + \{\sin[2\beta_{1y}(d + b + q)]/2\beta_{1y}\} \tag{3.69f}$$

The denominator (DR) of the expression (3.57) is given by

$$DR = 2 \sum_{i=1}^{5} Z_{wi} \iint |H_{ix}|^2 \, ds$$

$$= 4 \Bigg[Z_{w2} \int_0^a \int_{d+b}^{d+b+q} |H_{2x}|^2 \, dx \, dy + Z_{w3} \int_0^a \int_d^{d+b} |H_{3x}|^2 \, dx \, dy$$

$$+ Z_{w4} \int_0^a \int_0^d |H_{4x}|^2 \, dx \, dy + Z_{w5} \int_a^{a+s} \int_0^{d+b+q} |H_{5x}|^2 \, dx \, dy \Bigg] \quad (3.70)$$

In writing (3.70), we have set $Z_{w1} = Z_{w5}$ (because $\varepsilon_{r1} = \varepsilon_{r5}$) and used the symmetry of the structure about the y-axis at $x = 0$. Substituting the relevant magnetic field expressions from (3.63) and performing the integration, we obtain

$$DR = |B|^2 (\omega\varepsilon_0)^2 \, \beta_z^2 \, (Z_{w2} U_1 S_1 + Z_{w3} \xi_3^2 U_1 S_2 + Z_{w4} \xi_2^2 U_0 U_1 S_3 + Z_{w5} \xi_5^2 U_3 U_5) \quad (3.71)$$

where

$$S_1 = q + [\sin(2\beta_{2y} q)/2\beta_{2y}] \quad (3.72a)$$

$$S_2 = \{b + [\sin(2\beta_{3y} b)/2\beta_{3y}]\} + \xi_4^2 \{b - [\sin(2\beta_{3y} b)/2\beta_{3y}]\}$$
$$+ \{(\xi_4/\beta_{3y})[1 - \cos(2\beta_{3y} b)]\} \quad (3.72b)$$

$$S_3 = d + [\sin(2\beta_{4y} d)/2\beta_{4y}] \quad (3.72c)$$

The wave impedances Z_{wi} ($i = 2$ to 5) are given by (3.58) and the parameters U_0, U_1, U_3, and U_5 are given by (3.69). Dividing Eq. (3.70) by (3.71), the final expression for the attenuation coefficient α_c due to the conductor loss is obtained as

$$\alpha_c = \frac{R_s \{\xi_2^2 U_0 (\beta_z^2 U_1 + \beta_x^2 U_2) + \xi_5^2 [\beta_z^2 U_3 + \beta_{1x}^2 (U_4 + U_5)]\}}{\beta_z^2 \{U_1 [Z_{w2} S_1 + \xi_2^2 (Z_{w3} S_2 + Z_{w4} U_0 S_3)] + \xi_5^2 Z_{w5} U_3 U_5\}} \quad (3.73)$$

The computation of α_c, however, requires knowledge of the propagation constant β_z, which is to be evaluated first by solving the characteristic equation of the structure.

3.4 NUMERICAL METHODS

Different forms of numerical methods, such as the telegraphist's equations [10–12], finite-element method [13–15], and finite-difference method [16–17], have been used for the analysis of dielectric guides. These numerical methods have the advantage that they can be used for solving complicated geometries for which analytical solutions are difficult to obtain. In the following, we provide a brief outline of the telegraphist's equations method and the finite-element method.

3.4.1 Telegraphist's Equations Method

The telegraphist's method of analysis [10–12] is simpler to apply to shielded structures. Open dielectric guides are therefore analyzed by enclosing them in metallic enclosures. As an example, Figure 3.5 shows an insular image guide with a metal shield. The procedure involves derivation of generalized telegraphist's equations for mode voltages and mode currents using the two-dimensional Green's theorem. Elimination of mode currents from the generalized telegraphist's equations yields the characteristic eigenvalue equation. Once the eigenvalues and their eigenvectors are known, the various field components can be determined.

For an inhomogeneous dielectric guide extending uniformly in the z-direction, Maxwell's equations can be written as

$$\nabla \times \mathbf{E} = -j\omega\mu_0 \mathbf{H} \tag{3.74}$$

$$\nabla \times \mathbf{H} = j\omega\varepsilon_0\varepsilon_r(x, y)\,\mathbf{E} \tag{3.75}$$

Eliminating the longitudinal field components E_z and H_z from (3.74) and (3.75), we can write

$$\frac{\partial \mathbf{E}_t}{\partial t} = \left(\frac{1}{j\omega}\right)\nabla_t\left[\left(\frac{1}{\varepsilon_0\varepsilon_r(x, y)}\right)\nabla_t\cdot(\mathbf{H}_t \times \hat{\mathbf{z}})\right] - j\omega\mu_0(\mathbf{H}_t \times \hat{\mathbf{z}}) \tag{3.76}$$

$$\frac{\partial \mathbf{H}_t}{\partial z} = \left(\frac{1}{j\omega\mu_0}\right)\nabla_t\nabla_t\cdot(\hat{\mathbf{z}} \times \mathbf{E}_t) - j\omega\varepsilon_0\varepsilon_r(x, y)(\hat{\mathbf{z}} \times \mathbf{E}_t) \tag{3.77}$$

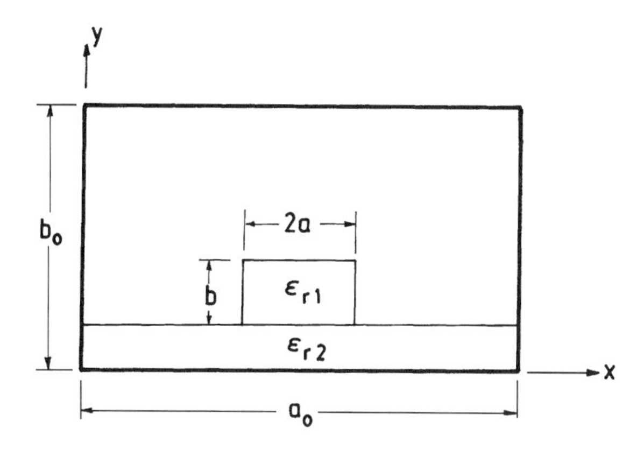

FIGURE 3.5 Shielded insular image guide.

where the subscript t denotes transverse component, and ∇_t is the two-dimensional del operator.

$$\mathbf{E}_t = \hat{\mathbf{x}} E_x + \hat{\mathbf{y}} E_y \tag{3.78a}$$

$$\mathbf{H}_t = \hat{\mathbf{x}} H_x + \hat{\mathbf{y}} H_y \tag{3.78b}$$

$$\nabla_t = \hat{\mathbf{x}} \frac{\partial}{\partial x} + \hat{\mathbf{y}} \frac{\partial}{\partial y} \tag{3.78c}$$

The transverse electric and magnetic fields $\mathbf{E}_t$ and $\mathbf{H}_t$ can be expanded in terms of the orthogonal mode functions $\mathbf{f}(x, y)$ and $\mathbf{g}(x, y)$ of the empty rectangular metal waveguides and also the unknown modal voltages $V(z)$ and modal currents $I(z)$ as follows:

$$\mathbf{E}_t = \sum_p^{\infty} V_p^e(z) \mathbf{f}_p^e(x, y) + \sum_q^{\infty} V_q^h(z) \mathbf{f}_q^h(x, y) \tag{3.79a}$$

$$\mathbf{H}_t = \sum_p^{\infty} I_p^e(z) \mathbf{g}_p^e(x, y) + \sum_q^{\infty} I_q^h(z) \mathbf{g}_q^h(x, y) \tag{3.79b}$$

The subscripts e and h refer to the TE and TM modes, respectively, and p and q denote double index such as mn. The mode functions $\mathbf{f}_p^e$, $\mathbf{g}_p^e$, $\mathbf{f}_q^h$, and $\mathbf{g}_q^h$ are defined in terms of scalar potential functions $\phi_p^e(x, y)$ and $\phi_q^h(x, y)$ as follows: for TE modes, we have

$$\mathbf{f}_p^e(x, y) = \hat{\mathbf{z}} \times \nabla_t \phi_p^e(x, y) \tag{3.80a}$$

$$\mathbf{g}_p^e(x, y) = -\nabla_t \phi_p^e(x, y) \tag{3.80b}$$

and for the TM modes,

$$\mathbf{f}_q^h(x, y) = -\nabla_t \phi_q^h(x, y) \tag{3.81a}$$

$$\mathbf{g}_q^h(x, y) = -\hat{\mathbf{z}} \times \nabla_t \phi_q^h(x, y) \tag{3.81b}$$

The scalar functions $\phi^e(x, y)$ and $\phi^h(x, y)$ satisfy the two-dimensional Helmholtz equation

$$(\nabla_t^2 + k_c^2) \begin{Bmatrix} \phi^e(x, y) \\ \phi^h(x, y) \end{Bmatrix} = 0 \tag{3.82}$$

where k_c is the cut off wavenumber of the mn mode of the rectangular waveguide of dimensions $a \times b$. It is given by

$$k_c^2 = (m\pi/a)^2 + (n\pi/b)^2 \tag{3.83}$$

The expressions for $\phi^e(x, y)$ and $\phi^h(x, y)$ are

$$\phi^e(x, y) = [2/(\sqrt{ab}\, k_c)] \sin(m\pi x/a) \sin(n\pi y/b), \quad m = 1, 2, 3, \ldots \tag{3.84a}$$

$$\phi^h(x, y) = [\sqrt{\varepsilon_m \varepsilon_n}(\sqrt{ab}\, k_c)] \cos(m\pi x/a) \cos(n\pi y/b), \quad m, n = 0, 1, 2, 3, \ldots$$

(mode $m = n = 0$ is excluded);

$$\varepsilon_{m,n} = \begin{cases} 1, & m, n = 0 \\ 2, & m, n \neq 0 \end{cases} \tag{3.84b}$$

In order to derive the generalized telegraphist's equations in terms of the mode voltages and currents, we first substitute (3.79) in (3.76) and (3.77) and take inner products with the known mode functions $\mathbf{f}_p^e(x, y)$, $\mathbf{g}_p^e(x, y)$, $\mathbf{f}_q^h(x, y)$, and $\mathbf{g}_q^h(x, y)$ for different values of p and q. Using the orthogonality property of mode functions and the two-dimensional Green's theorem, the following generalized telegraphist's equations are obtained [12]:

$$\frac{dV_n^h(z)}{dz} = -\left(\frac{1}{j\omega\varepsilon_0}\right) \sum_q^{\infty} Z_1(n, q)\, I_q^h(z) - j\omega\mu_0 I_n^h(z) \tag{3.85a}$$

$$\frac{dV_m^e(z)}{dz} = -j\omega\mu_0 I_m^e(z) \tag{3.85b}$$

$$\frac{dI_n^h(z)}{dz} = -j\omega\varepsilon_0 \left[\sum_q^{\infty} Y_1(n, q)\, V_q^h(z) + \sum_q^{\infty} Y_2(n, p)\, V_p^e(z) \right] \tag{3.85c}$$

$$\frac{dI_m^e(z)}{dz} = -j\omega\varepsilon_0 \left[\sum_q^{\infty} Y_3(m, q)\, V_q^h(z) + \sum_p^{\infty} Y_4(m, p)\, V_p^e(z) \right] - \left[\frac{(k_{cm}^e)^2}{j\omega\mu_0} \right] V_m^e(z) \tag{3.85d}$$

where

$$Z_1(n, q) = (k_{cq}^h)^2 (k_{cn}^h)^2 \iint (1/\varepsilon_r)\, \phi_q^h(x, y)\, \phi_n^h(x, y)\, dx\, dy \tag{3.86a}$$

$$Y_1(n, q) = \iint \varepsilon_r(x, y)\, \mathbf{g}_q^h(x, y)\, \mathbf{g}_n^h(x, y)\, dx\, dy \tag{3.86b}$$

$$Y_2(n, q) = \iint \varepsilon_r(x, y)\, \mathbf{g}_p^e(x, y)\, \mathbf{g}_n^h(x, y)\, dx\, dy \tag{3.86c}$$

$$Y_3(m, q) = \iint \varepsilon_r(x, y)\, \mathbf{g}_q^h(x, y)\, \mathbf{g}_m^e(x, y)\, dx\, dy \tag{3.86d}$$

$$Y_4(m, q) = \iint \varepsilon_r(x, y)\, \mathbf{g}_p^e(x, y)\, \mathbf{g}_m^e(x, y)\, dx\, dy \tag{3.86e}$$

In (3.86), the integration is carried out over the cross-section of the waveguide.

We now assume wave propagation along the z-direction according to $\exp(-j\beta_z z)$ so that

$$V_p^e(z) = V_p^e e^{-j\beta_z z} \tag{3.87a}$$

$$V_q^h(z) = V_q^h e^{-j\beta_z z} \tag{3.87b}$$

Considering M terms for the TE modes and N terms for the TM modes in (3.62), and eliminating the mode currents from (3.85a)–(3.85d), we obtain the following characteristic eigenvalue equation:

$$\begin{bmatrix} C_{11} & C_{12} & \cdots & C_{1\,N+M} \\ C_{21} & C_{22} & \cdots & C_{2\,N+M} \\ \cdot & \cdot & \cdots & \cdot \\ \cdot & \cdot & & \cdot \\ \cdot & \cdot & & \cdot \\ \cdot & \cdot & & \cdot \\ \cdot & \cdot & & \cdot \\ C_{N+M\,1} & \cdot & \cdots & C_{N+M\,N+M} \end{bmatrix} \begin{bmatrix} V_1^h \\ \cdot \\ V_N^h \\ V_1^e \\ \cdot \\ \cdot \\ \cdot \\ V_M^e \end{bmatrix} = \beta_z^2 \begin{bmatrix} V_1^h \\ \cdot \\ V_N^h \\ V_1^e \\ \cdot \\ \cdot \\ \cdot \\ V_M^e \end{bmatrix} \tag{3.88}$$

where the elements C_{pq} $(p, q = 1, 2, \ldots, N + M)$ are functions of the dielectric constants and cross-sectional dimensions of the guide. With the determination of the elements C_{pq}, we can solve (3.88) for the eigenvalues β_z and eigenvectors $V_1^e, \ldots, V_M^e, V_1^h, \ldots, V_N^h$. The mode voltages and currents are related by

$$V_p^e = Z_p^e I_p^e \tag{3.89a}$$

$$V_q^h = Z_q^h I_q^h \tag{3.89b}$$

where

$$Z_p^e = \frac{\omega \mu_0}{\sqrt{k_0^2 - (k_{cp}^e)^2}} \tag{3.90a}$$

$$Z_q^h = \frac{\sqrt{k_0^2 - (k_{cq}^h)^2}}{\omega \varepsilon_0} \tag{3.90b}$$

With the determination of eigenvalues and eigenvectors, all the electric and magnetic field components of the shielded dielectric guide corresponding to E_{mn}^y and E_{mn}^x modes can be obtained.

It may be noted that since the generalized telegraphist's equations are derived through Galerkin's procedure, the resulting expression for β_z is stationary.

3.4.2 Finite-Element Method

The finite-element method is a versatile numerical technique, applicable to both open and shielded dielectric guides [12–15]. In principle, the technique involves the following steps: (i) discretizing the cross-sectional region of the guide into a number of finite elements, commonly triangular elements; (ii) representing the field components in each element in terms of polynomials; and (iii) assembling of all elements in the cross-sectional region and solving the resulting system of equations.

In the case of dielectric guides, the magnetic field components H_x, H_y, and H_z are continuous across the discontinuity boundaries between different dielectric regions. For finite-element formulation, it is therefore more convenient to use the transverse magnetic field components for each triangular element. Such a formulation has been reported by Su [14] and is outlined below.

Starting from Maxwell's curl equations for an inhomogeneous dielectric guide extending uniformly in the z-direction, we can write

$$\nabla \times \left(\frac{\nabla \times \mathbf{H}}{\varepsilon_r(x, y)} \right) - k_0^2 \mathbf{H} = 0 \tag{3.91}$$

where k_0 is the free-space propagation constant. Using the vector identity

$$\nabla \times \phi \mathbf{A} = \phi \nabla \times \mathbf{A} + \nabla \phi \times \mathbf{A} \tag{3.92a}$$

and the relation

$$\nabla \cdot \mathbf{H} = 0 \tag{3.92b}$$

in (3.91), we can write

$$\nabla^2 \mathbf{H} + k_0^2 \varepsilon_r(x, y) \mathbf{H} - (\nabla \times \mathbf{H}) \times \left(\frac{\nabla \varepsilon_r(x, y)}{\varepsilon_r(x, y)} \right) = 0 \tag{3.93}$$

Multiplying (3.93) by some arbitrary vector function $\mathbf{H}^a$, which is independent of $\mathbf{H}$, and integrating over the entire space, we obtain

$$\iint \left\{ k_0^2 \varepsilon_r(x, y) \mathbf{H} \cdot \mathbf{H}^a + (\nabla^2 \mathbf{H}) \cdot \mathbf{H}^a - \left[(\nabla \times \mathbf{H}) \times \left(\frac{\nabla \varepsilon_r(x, y)}{\varepsilon_r(x, y)} \right) \right] \cdot \mathbf{H}^a \right\} dV = 0 \tag{3.94}$$

We assume propagation in the z-direction according to $\exp(-j\beta_z z)$ and substitute $\mathbf{H}^a = \hat{\mathbf{x}} H_x^a \exp(-j\beta_z z)$ and $\hat{\mathbf{y}} H_y^a \exp(-j\beta_z z)$ in (3.94). H_x^a and H_y^a are independent of H_x and H_y and the choice of $\exp(-j\beta_z z)$ makes (3.94) independent of the z-coordinate. The resulting equations are given by [14]

$$\begin{aligned} \iint \Bigg[(k_0^2 \varepsilon_r - \beta_z^2) H_x H_x^a + \left(\frac{\partial H_x}{\partial x} \frac{\partial H_x^a}{\partial x} \right) + \left(\frac{\partial H_x}{\partial y} \frac{\partial H_x^a}{\partial y} \right) \\ + \left(\frac{\partial H_y}{\partial x} - \frac{\partial H_x}{\partial y} \right) \left(\frac{H_x^a}{\varepsilon_r} \right) \left(\frac{\partial \varepsilon_r}{\partial y} \right) \Bigg] dx\, dy = 0 \end{aligned} \tag{3.95a}$$

$$\begin{aligned} \iint \Bigg[(k_0^2 \varepsilon_r - \beta_z^2) H_y H_y^a + \left(\frac{\partial H_y}{\partial x} \frac{\partial H_y^a}{\partial x} \right) + \left(\frac{\partial H_y}{\partial y} \frac{\partial H_y^a}{\partial y} \right) \\ + \left(\frac{\partial H_x}{\partial y} - \frac{\partial H_y}{\partial x} \right) \left(\frac{H_y^a}{\varepsilon_r} \right) \left(\frac{\partial \varepsilon_r}{\partial x} \right) \Bigg] dx\, dy = 0 \end{aligned} \tag{3.95b}$$

It may be noted from (3.95) that H_z has been uncoupled from H_x and H_y.

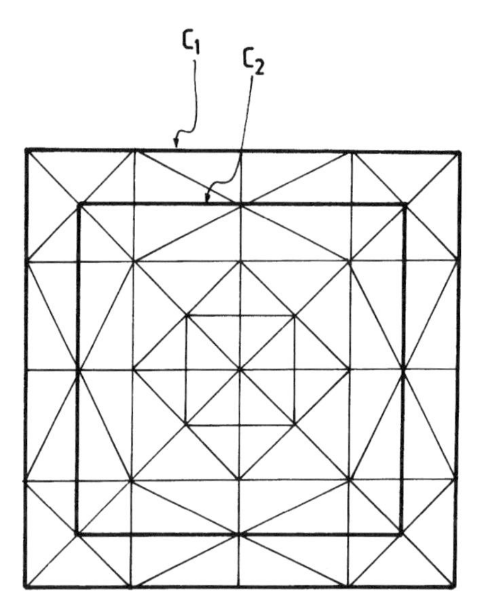

FIGURE 3.6 Segmentation of a region into rectangular elements for finite-element treatment.

In order to apply the numerical procedure, the cross-sectional area of the guide is divided into several triangular elements as illustrated in the example shown in Figure 3.6. Within each element, the associated fields—namely, H_x, H_y, H_x^a, and H_y^a—are expanded in some local basis functions with the expansion coefficients being the field nodal values. If N is the total number of nodes, then by considering variation of (3.95a) and (3.95b) with respect to H_x^a and H_y^a at all the nodes, we obtain $2N$ simultaneous equations in terms of H_x and H_y. If M is the number of nodes in the outermost boundary C_1, there will be $2M$ equations corresponding to these nodes and $2(N-M)$ equations corresponding to the inner nodes. Using these $2(N-M)$ equations, one can solve the matrix form of relations between the fields of the nodes on the outermost boundary C_1 and those of the next inner boundary marked C_2. This yields explicit relations between the associated fields H_x and H_y and their normal derivatives at the boundaries of the guide structure. The next step is to obtain the longitudinal field components E_z and H_z from the transverse H fields and their normal derivatives. Expressions relating these fields can be derived from Maxwell's equations (3.75) and (3.92). They are given by [20]

$$j\omega E_z = \left(\frac{1}{\varepsilon_0 \varepsilon_r}\right)\left(\frac{\partial H_n}{\partial l} - \frac{\partial H_l}{\partial n}\right) \tag{3.96a}$$

$$j\beta_z H_z = \frac{\partial H_n}{\partial n} + \frac{\partial H_l}{\partial l} \tag{3.96b}$$

where H_n denotes the component of H normal to the boundary surface, and H_l denotes the transverse component tangential to the boundary surface. The components H_n and H_l can be expressed in terms of H_x and H_y

$$H_n = H_x \cos\theta + H_y \sin\theta \tag{3.97a}$$

$$H_l = H_x \sin\theta - H_y \cos\theta \tag{3.97b}$$

where θ is the angle from the x-axis to the unit normal $\hat{\mathbf{n}}$ at the boundary surface. Using (3.96) and (3.97), the longitudinal fields in the entire guide structure can be obtained in terms of H_x and H_y and their normal derivatives at all the boundaries. Next, matching the tangential fields E_z and H_z at all the boundaries of the guide, the propagation characteristics are determined.

PROBLEMS

3.1 Discuss the principle of mode-matching method. Enumerate its advantages over the EDC method.

3.2 Consider a grounded dielectric slab of thickness d, relative dielectric constant ε_r, and relative permeability μ_r as shown in Figure 3.7. The guide extends to infinity in the xz-plane. Consider TM (to z) modes and propagation in the z-direction according to $e^{-j\beta z}$.

(a) Starting from Maxwell's equation, derive the modal fields in both the dielectric and air regions.

(b) Derive characteristic equation by applying the boundary conditions and obtain expressions for the cutoff frequency of TM_n modes.

3.3 Repeat Problem 3.2 for TE (to z) modes.

3.4 From the solutions obtained for Problems 3.2 and 3.3 identify the first three propagating surface wave modes. If $\varepsilon_r = 4$, $\mu_r = 1$, and $d = 0.5$ cm, calculate the cutoff frequencies for these modes. Plot the propagation constants as a function of frequency from 0 to 30 GHz.

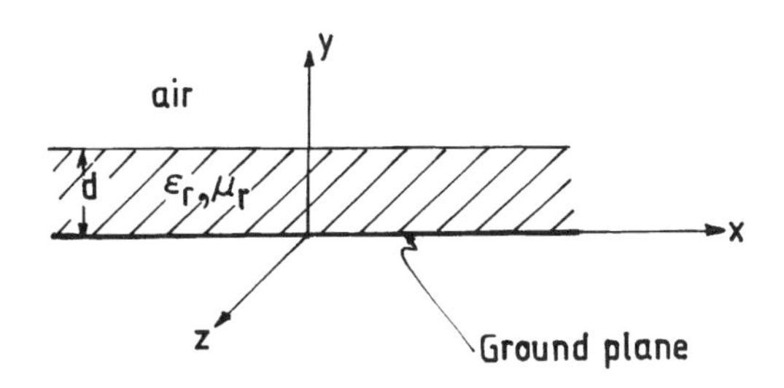

FIGURE 3.7 Grounded dielectric slab.

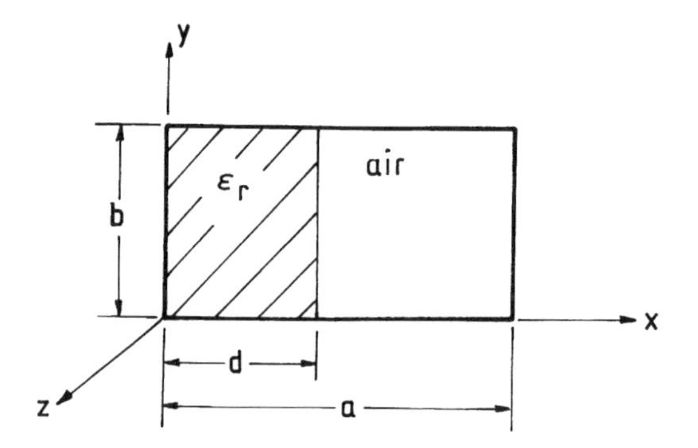

FIGURE 3.8 Partially loaded rectangular waveguide.

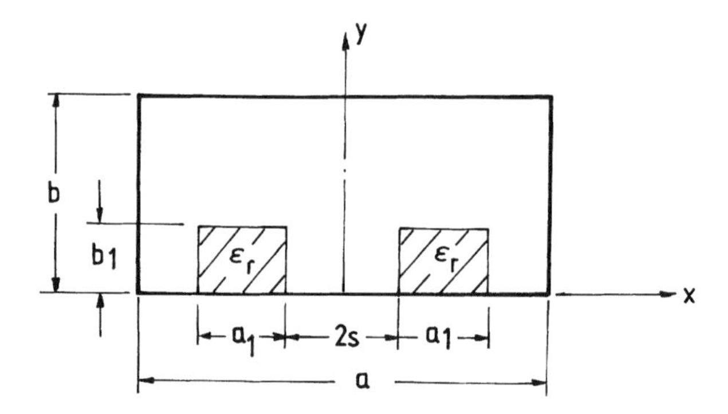

FIGURE 3.9 Shielded coupled image guide.

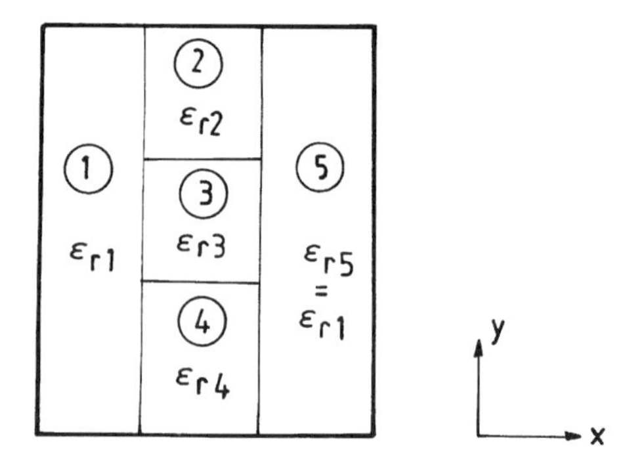

FIGURE 3.10 General dielectric guide structure.

3.5 Consider a partially loaded rectangular waveguide as shown in Figure 3.8. Assume wave propagation in the z-direction according to $e^{-j\beta z}$.

(a) Derive field expressions for TE_{m0} modes (w.r.t. z) and obtain the characteristic equation.

(b) Solve the above characteristic equation to obtain cutoff frequency of the TE_{10} mode. Assume the following guide parameters: $a = 22.86$ mm, $b = d = 10.16$ mm, and $\varepsilon_r = 2.56$.

3.6 For the partially loaded rectangular guide of Problem 3.5, derive expressions for attenuation of the TE_{10} mode due to imperfectly conducting walls.

3.7 Consider a shielded coupled image guide shown in Figure 3.9 fed by a hollow rectangular waveguide having the cross section $a \times b$.

(a) What types of mode will propagate in the structure if it is excited by the TE_{10} mode of rectangular waveguide? Designate the dominant mode and draw the field lines.

(b) Write the modal solution for one symmetric half of the structure for the dominant mode.

(c) Obtain the expression for wave impedance.

(d) Derive expressions for the attenuation of this mode due to imperfectly conducting walls.

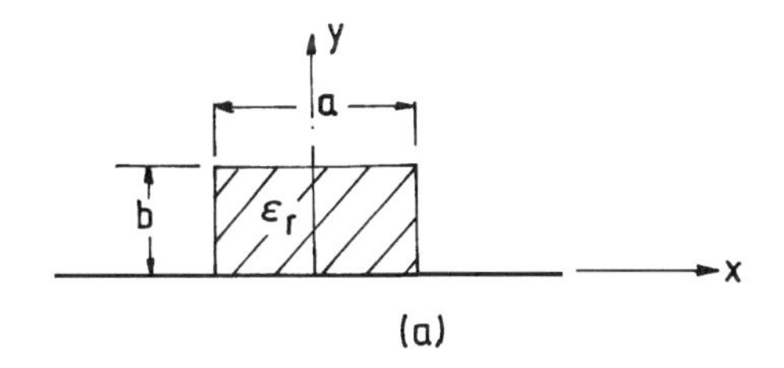

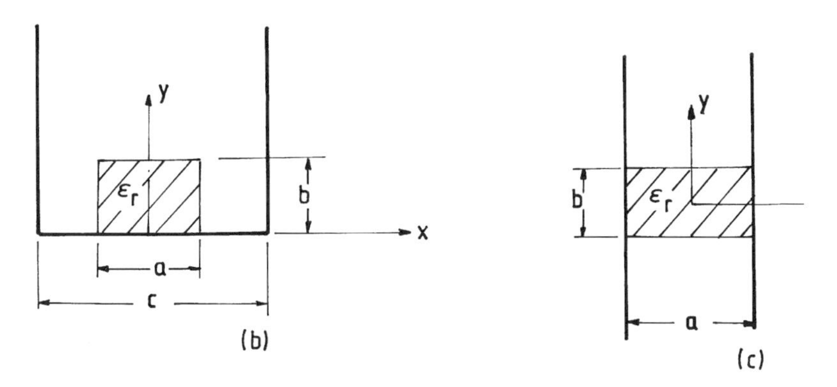

FIGURE 3.11 (a) Image guide. (b) Trapped image guide. (c) NRD guide.

3.8 Repeat Problem 3.7, if the structure shown in Figure 3.9 is excited by the TE_{20} mode of a rectangular waveguide having cross-section $a \times b$.

3.9 For the dielectric guide structure shown in Figure 3.10, derive expressions for the conductor loss coefficient for the dominant TE-to-y (E^x_{11}) mode. Use the field expressions given in the text.

3.10 Derive expressions for the conductor loss coefficients of the dominant TM-to-y (E^y_{11}) and TE-to-y (E^x_{11}) modes in the case of the following guides.

(a) Image guide (Fig. 3.11(a)).

(b) Trapped image guide (Fig. 3.11(b)).

(c) NRD guide (Fig. 3.11(c)).

REFERENCES

1. R. F. Harrington, *Time-Harmonic Electromagnetic Fields*, McGraw-Hill, New York, 1961.
2. T. Itoh (Ed.), *Numerical Techniques for Microwave and Millimeter Wave Passive Structures*, Wiley, New York, 1989.
3. R. Mittra and S. W. Lee, *Analytical Techniques in the Theory of Guided Waves*, MacMillan, New York, 1971.
4. K. Solbach and I. Wolff, Electromagnetic fields and phase constants of dielectric image lines. *IEEE Trans. Microwave Theory Tech.*, **MTT-26**, 266–274, Apr. 1978.
5. R. Mittra, Y. L. Hou, and V. Jamnejad, Analysis of open dielectric waveguides using mode-matching technique and variational methods. *IEEE Trans. Microwave Theory Tech.*, **MTT-28**, 36–43, Jan. 1980.
6. N. Deo and R. Mittra, A new technique for analyzing planar dielectric waveguides for millimetre wave integrated circuits. *Arch. Elek. Ubertragung.*, **37**, 236–244, July/Aug. 1983.
7. A. K. Tiwari and B. Bhat, Analysis of trapped single and coupled image guides using the mode-matching technique. *Arch. Elek. Ubertragung.*, **38**, 181–185, Mar. 1984.
8. B. Bhat and A. K. Tiwari, Analysis of low-loss broadside-coupled dielectric image guide using the mode-matching technique. *IEEE Trans. Microwave Theory Tech.*, **MTT-32**, 711–717, July 1984.
9. A. K. Tiwari, B. Bhat, and R. P. Singh, Generalized coupled dielectric waveguide and its variants for millimetre-wave applications. *IEEE Trans. Microwave Theory Tech.*, **MTT-34**, 869–875, Aug. 1986.
10. S. A. Schelkunoff, Generalised telegraphist's equations for waveguides. *Bell Syst. Tech. J.*, **31**, 784–801, July 1952.
11. K. Ogusu and K. Hongo, Analysis of dielectric waveguides by the generalized telegraphist's equations. *Trans. Inst. Electron. Commun. Eng. Japan*, **J60-B**, 358–359, May 1977.
12. K. Ogusu, Numerical analysis of the rectangular dielectric waveguide and its modifications. *IEEE Trans. Microwave Theory Tech.*, **MTT-25**, 874–885, Nov. 1977.

13. Z. J. Csendes and P. Silvester, Numerical solution of dielectric loaded waveguides: I Finite-element analysis. *IEEE Trans. Microwave Theory Tech.*, **MTT-18**, 1124–1131, Dec. 1970.
14. C. C. Su, A combined method for dielectric waveguides using the finite-element technique and the surface integral equations method. *IEEE Trans. Microwave Theory and Tech.*, **MTT-34**, 1140–1146, Nov. 1986.
15. M. Ikeuchi, H. Sawami, and H. Niki, Analysis of open-type dielectric waveguides by the finite-element method. *IEEE Trans. Microwave Theory Tech.*, **MTT-29**, 234–239, Mar. 1981.
16. E. Schweig and W. B. Bridges, Computer analysis of dielectric waveguides: a finite difference method. *IEEE Trans. Microwave Theory Tech.*, **MTT-32**, 531–541, May 1984.
17. K. Bierwirth, N. Schulz, and F. Arndt, Finite-difference analysis of rectangular dielectric waveguide structures. *IEEE Trans. Microwave Theory Tech.*, **MTT-34**, 1104–1114, Nov. 1986.
18. K. Solbach, Calculation and measurement of the attenuation constants of dielectric image lines of rectangular cross-sections. *Arch. Elek. Ubertragung*, **32**, (8), 321–328, 1978.
19. P. P. Toulios and R. M. Knox, Rectangular dielectric image lines for millimetre integrated circuits. Weston Electronics Show and Convention, CA, Digest, 1–10, Aug. 1970.
20. C. C. Su, A surface integral equations method for homogeneous optical fibres and coupled image lines of arbitrary cross-sections. *IEEE Trans. Microwave Theory and Tech.*, **MTT-33**, 1114–1119, Nov. 1985.

CHAPTER FOUR

Image, Insular Image, Trapped Image, and Other Variant Guides

4.1 INTRODUCTION

The analyses of dielectric integrated guides in general are covered in Chapters 2 and 3. The formulas developed using the EDC technique in Chapter 2 are very easy to program. They are particularly useful for the computation of dispersion characteristics and approximate evaluation of dielectric and conductor losses. The rigorous formulas derived by applying the mode-matching method in Chapter 3 are more complicated but they enable accurate computation of fields, propagation parameters, and loss. This chapter presents the characteristics of the basic image guide, insular image guide, trapped image guide, and several miscellaneous variant guides for millimeter wave applications. The variant structures reported here include the strip dielectric guide, inverted strip guide, ridge guide, cladded image guide, hollow image guide (or π-guide), T-guide, and U-guide.

The propagating modes in the guides referred to above are hybrid in nature, having all the components of the electric and magnetic fields. For describing the modal fields, we retain the designation E^y_{mn}(TM-to-y) and E^x_{mn}(TE-to-y) as used in Chapters 2 and 3. It may be recalled that for the E^y_{mn} modes, E_y, H_x, and E_z components dominate over the other three field components, whereas for the E^x_{mn} modes, E_x, H_y, and H_z components are dominant over the remaining field components. The subscripts m and n refer to the number of extrema of the dominant transverse fields in the dielectric region, with respect to x- and y-directions, respectively. The lowest order mode in the TM-to-y category is the E^y_{11} mode and in the TE-to-y category it is the E^x_{11} mode. For most circuit applications employing these guides, the desirable mode of excitation is the dominant E^y_{11} mode for which the main transverse electric field E_y is perpendicular to the ground plane.

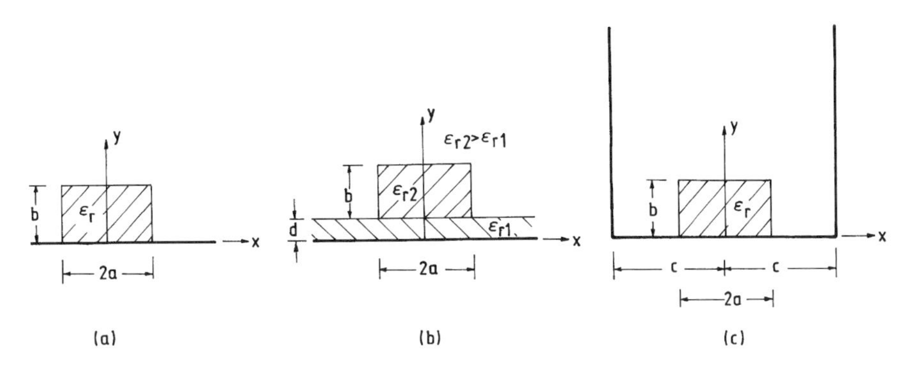

FIGURE 4.1 Cross-sectional views of (a) image guide, (b) insular image guide, and (c) trapped image guide.

Of the various guides, the rectangular image guide (Fig. 4.1(a)) is the simplest to fabricate. The insular image guide (Fig. 4.1(b)) adopts a thin dielectric layer of lower dielectric constant between the dielectric strip and the ground plane, thereby enabling considerable reduction in the conductor loss over that of the image guide. Both image and insular image guides, because of their open nature, give rise to radiation leakage laterally at bends and curved sections. This problem is circumvented in the trapped image guide (Fig. 4.1(c)). The side metallic walls of the guide can be suitably configured at the bends and curved sections so as to redirect the end-radiated energy back to the guiding strip. The other guide structures considered are essentially variants of the image guide and have the advantages of reduced transmission loss or reduced radiation loss or certain other features specific to millimeter wave integrated circuits.

A completely shielded dielectric guide structure may permit higher order waveguide modes. In a practical situation, where the excitation is in the dominant E^y_{11} mode, the top shielding wall is placed at an appropriate height such that higher order modes are not excited while at the same time its effect on the dominant mode is negligibly small. In the following section, the effects of the top shielding wall on the characteristics of dielectric guide structures are presented only for the dominant mode.

Throughout this chapter, the symbols v_0, λ_0, k_0, β, and λ appear in several places. The symbols v_0, λ_0, and k_0 denote the electromagnetic wave velocity, wavelength, and propagation constant, respectively, in free space; β and λ are the propagation constant (phase constant) and wavelength in the guide (single) under consideration.

4.2 IMAGE GUIDE

4.2.1 Field and Power Distribution

Typical variations of the various field components of the E^y_{11} mode in an image guide are shown in Figure 4.2 and that of the E^x_{11} mode in Figure 4.3 [1]. The

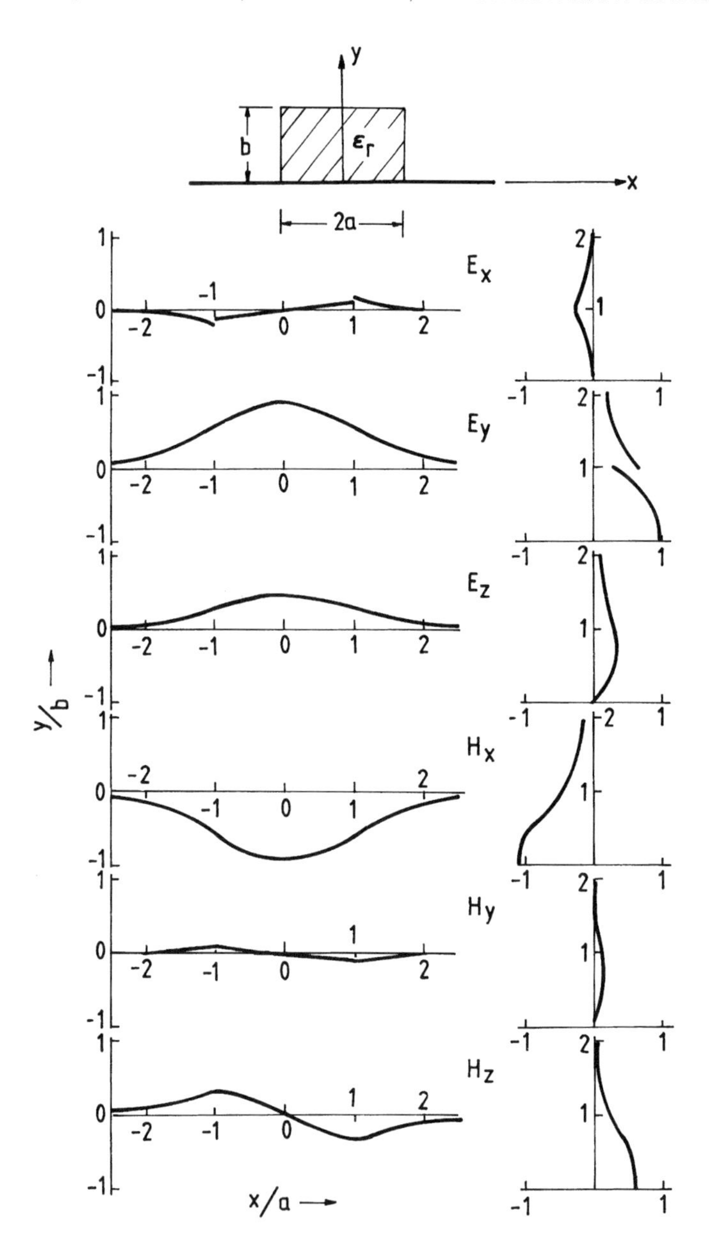

FIGURE 4.2 Normalized calculated field distribution of the E_{11}^y mode of the image guide in a horizontal plane at $y = 0.8b$ and in a vertical plane at $x = 0.9a$: $a/b = 1$, $\varepsilon_r = 2.22$, $\beta/k_0 = 1.1854$, and $B = (4b/\lambda_0)\sqrt{\varepsilon_r - 1} = 2$. (From Solbach and Wolff [1]. Copyright © 1978 IEEE, reprinted with permission.)

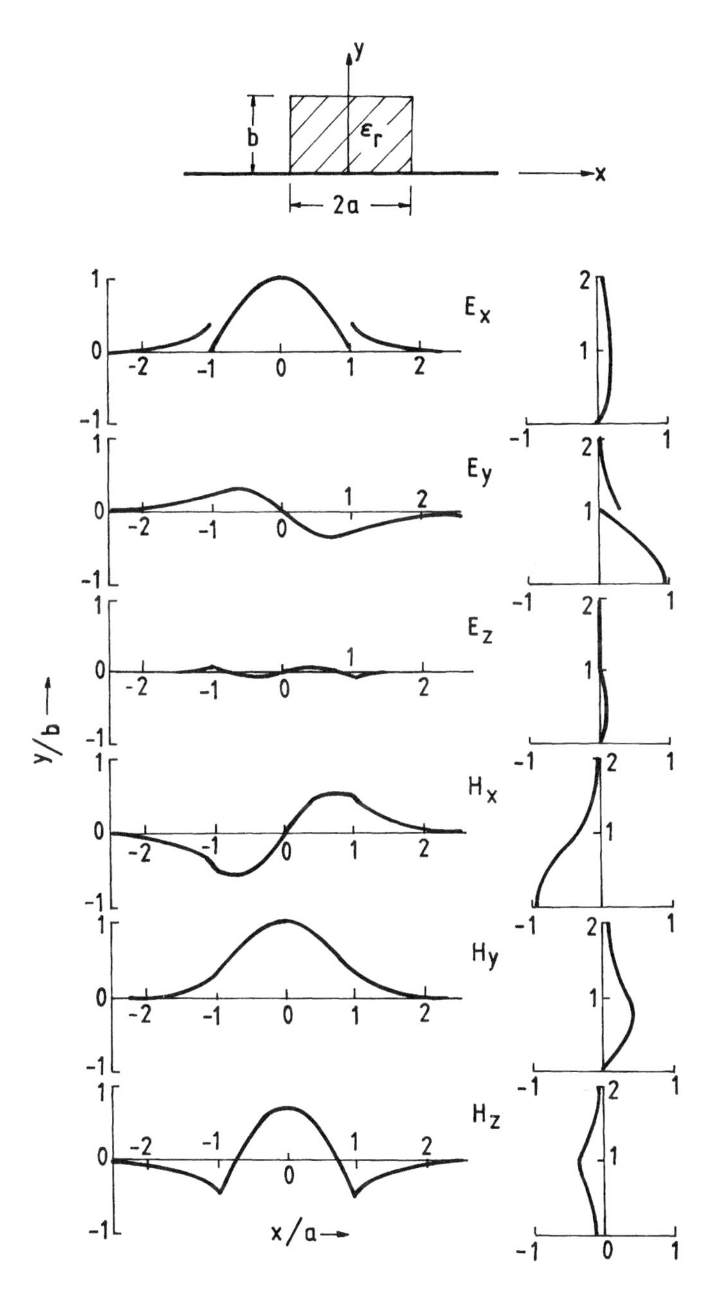

FIGURE 4.3 Normalized calculated field distribution of the E_{11}^x mode of image guide in a horizontal plane at $y = 0.8b$ and in a vertical plane at $x = 0.9a$: $a/b = 1$, $\varepsilon_r = 10$, $\beta/k_0 = 1.9183$, and $B = (4b/\lambda_0)\sqrt{\varepsilon_r - 1} = 2$. (From Solbach and Wolff [1]. Copyright © 1978 IEEE, reprinted with permission.)

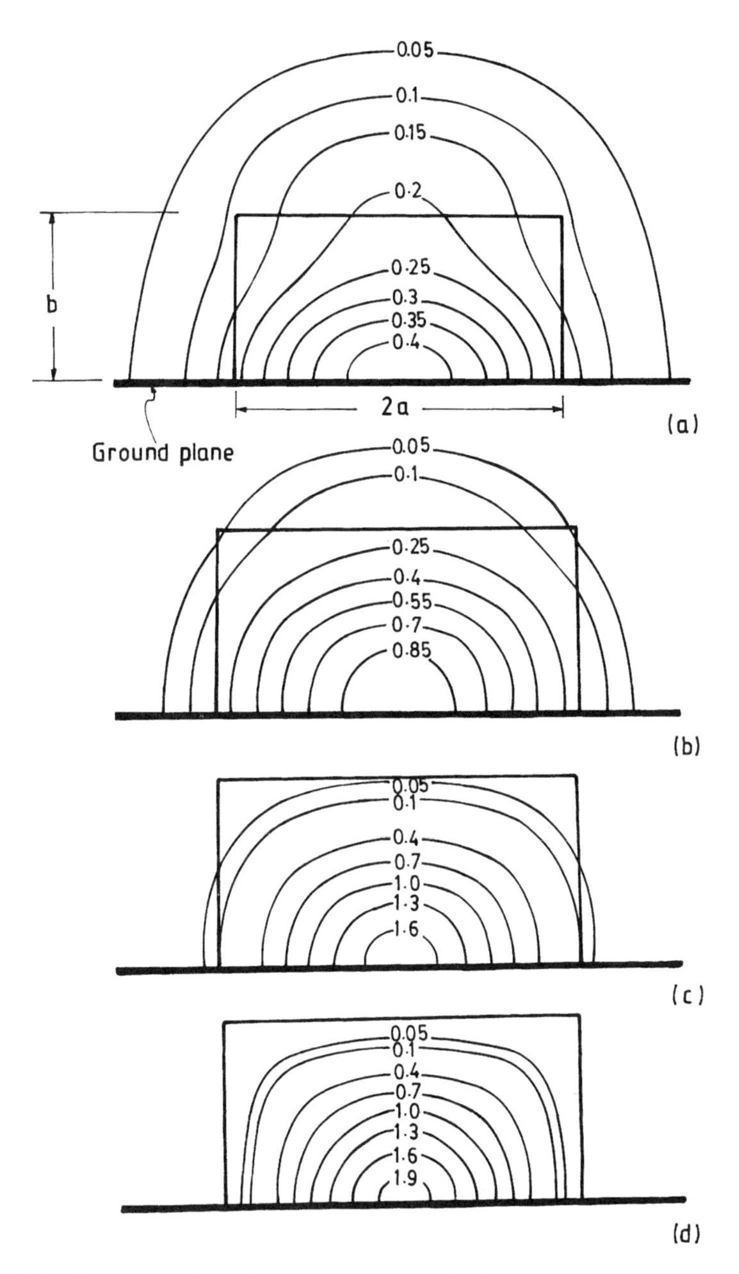

FIGURE 4.4 Frequency dependence of power distribution for E^y_{11} mode in image guide with $a/b = 1$ and $\varepsilon_r = 2.5$: (a) $k_0\, a = 1.5$, (b) $k_0\, a = 2.5$, (c) $k_0\, a = 5$, and (d) $k_0\, a = 20$. (From Ogusu [2]. Copyright © 1977 IEEE, reprinted with permission.)

plots correspond to variation in a horizontal plane at $y = 0.8b$ and in a vertical plane at $x = 0.9a$ for a guide with parameters $a/b = 1, \varepsilon_r = 2.22, \beta/k_0 = 1.1854$, and $B = 2$, where B is the normalized frequency parameter defined as

$$B = (4b/\lambda_0)\sqrt{\varepsilon_r - 1} = (4fb/v_0)\sqrt{\varepsilon_r - 1} \tag{4.1}$$

As per the modal nomenclature used, the main transverse fields E_y and H_x for the E_{11}^y mode show a single maximum in both the x- and y-directions. Similarly, for the E_{11}^x mode, the E_x and H_y components show a single maximum in each of the x- and y-directions. Also, as expected, the normal component of E at $x = \pm a$ and $y = b$ (E_x at $x = \pm a$ and E_y at $y = b$) is discontinuous. The fields vary nearly sinusoidally within the dielectric strip and decay exponentially away from the strip in the air region.

The spatial power distribution of the dominant E_{11}^y mode in the transverse plane and its frequency dependence are illustrated in Figure 4.4 [2]. The solid curved lines represent contours of constant power level in the transverse plane. Plots are shown for a guide with parameters $a/b = 1$ and $\varepsilon_r = 2.5$ for four different values of k_0a. For a fixed $k_0a = 1.5$, Figure 4.4(a) shows that the constant power contours marked 0.2 and below extend outside the dielectric, whereas for a high value of $k_0a = 20$, these contours are confined entirely to the dielectric (refer to Fig. 4.4(d)).

The E_{11}^y and E_{11}^x modes have their main transverse electric fields orthogonal to each other. Each of these modes can be excited by a TE_{10} mode rectangular waveguide with a suitable transition. The rectangular waveguide must be oriented such that its E-vector lies perpendicular to the image guide ground plane for exciting the E_{11}^y mode and parallel to the ground plane for exciting the E_{11}^x mode.

4.2.2 Dispersion and Bandwidth

Dispersion Characteristics In order to illustrate the dispersion characteristics of different modes and to understand the validity of the EDC technique, we reproduce in Figures 4.5–4.7 the graphs reported by Deo and Mittra [3] for three different sets of guide parameters. Figure 4.5 is for an image guide having low permittivity and moderate aspect ratio ($\varepsilon_r = 2.22$, $a/b = 1$), Figure 4.6 is for a guide with low permittivity and high aspect ratio ($\varepsilon_r = 2.22$, $a/b = 5$), and Figure 4.7 is for a guide with high permittivity and moderate aspect ratio ($\varepsilon_r = 12$, $a/b = 1$). The plots show variation of the normalized propagation constant β/k_0 as a function of the normalized frequency parameter B. The parameter B represents the ratio of the operating frequency to the cutoff frequency of the first higher order mode (TE_1) of the dielectric slab having the same height b and relative dielectric constant ε_r as for the image guide under consideration. The graphs show comparison of the results computed using a field-expansion method by Deo and Mittra [3], mode-matching method by Solbach and Wolff [1], the EDC method by Knox and Toulios [4], and the approximate theory by Marcatili [5].

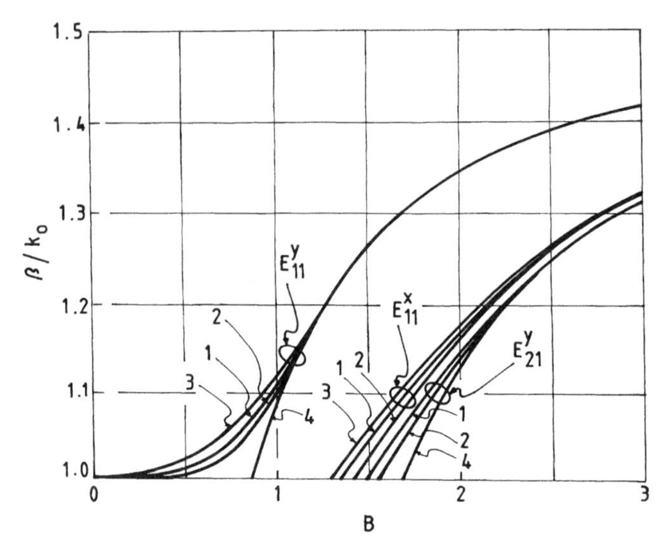

FIGURE 4.5 Normalized propagation constant β/k_0 of low-permittivity image guide modes as a function of normalized frequency parameter B: $\varepsilon_r = 2.22$, $a/b = 1$, and $B = (4b/\lambda_0)\sqrt{\varepsilon_r - 1}$. Curve 1, field-expansion method [3]; curve 2, mode-matching method [1]; curve 3, EDC method [4]; and curve 4, Marcatili's approximation [5]. (From Deo and Mittra [3]. Reprinted from AEU-37, 1983.)

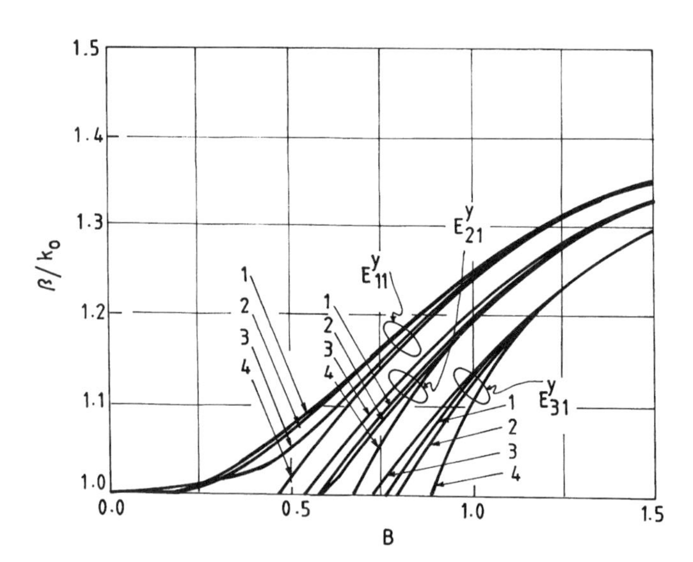

FIGURE 4.6 Normalized propagation constant β/k_0 for flat, low-permittivity image guide as a function of normalized frequency parameter B: $\varepsilon_r = 2.22$, $a/b = 5$, and $B = (4b/\lambda_0)\sqrt{\varepsilon_r - 1}$. Curve 1, field-expansion method [3]; curve 2, mode-matching method [1]; curve 3, EDC method [4]; and curve 4, Marcatili's approximation [5]. (From Deo and Mittra [3]. Reprinted from AEU-37, 1983.)

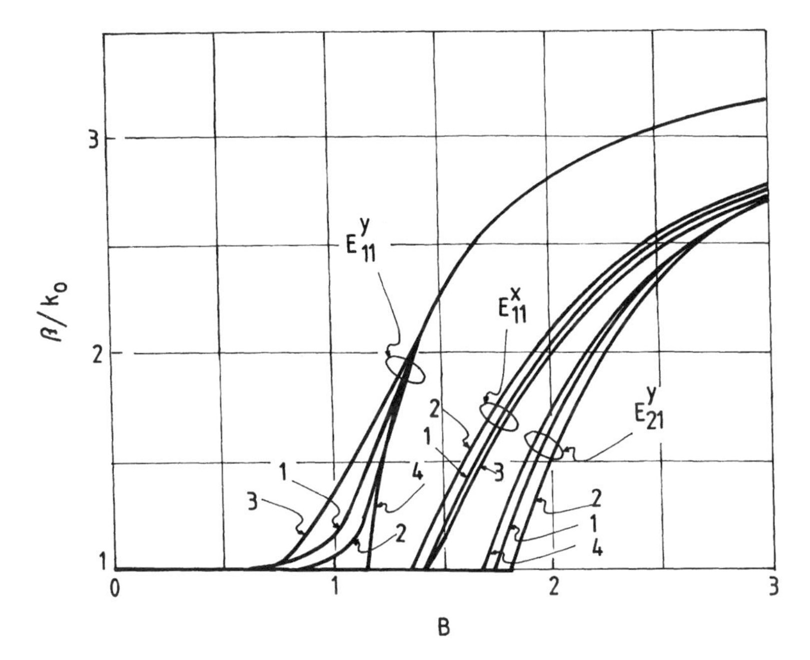

FIGURE 4.7 Normalized propagation constant β/k_0 of high-permittivity guide as a function of normalized frequency parameter B: $\varepsilon_r = 12, a/b = 1$, and $B = (4b/\lambda_0)\sqrt{\varepsilon_r - 1}$. Curve 1, field expansion method [3]; curve 2, mode-matching method [1]; curve 3, EDC method [4]; and curve 4, Marcatili's approximation [5]. (From Deo and Mittra [3]. Reprinted from AEU-37, 1983.)

The computed results due to the accurate theories [1, 3], as shown in Figures 4.5–4.7, indicate that the dominant E_{11}^y mode of the image guide has no low-frequency cutoff. That is, the fundamental mode exists down to zero frequency and for any height of the guide. However, all the higher order modes exhibit a low-frequency cutoff at a distinct value of B. The first set of higher order modes shown in Figure 4.5 are E_{11}^x and E_{21}^y. For the E_{11}^x mode, E_x is even symmetric in x and has one maximum in the y-direction, and for the E_{21}^y mode, E_y is antisymmetric with two maxima in the x-direction and one maximum in the y-direction. It may be noted from Figures 4.5 and 4.6 that for low-permittivity guides, the results of Marcatili's approximate theory [5] deviate significantly from the rigorous methods [1, 3] at and in the vicinity of cutoff. This is because Marcatili's method [5], which assumes the fields to be predominantly transverse to the direction of propagation, is inaccurate for low values of ε_r. On the other hand, the results of the EDC technique match well with the accurate theories in Figure 4.5 and with even improved accuracy in Figure 4.6. When the aspect ratio a/b is large, as considered in Figure 4.6 ($a/b = 5$), the image guide resembles a planar dielectric slab guide. For this case, the fundamental E_{11}^y mode of the guide has negligible contribution from the E_x, H_y, and H_z components, and hence it becomes nearly transverse magnetic in nature. Furthermore, the fields get confined to a greater extent in the x-direction than in a guide with lower aspect ratio. For large width guides, fewer terms in the modal expansion would be adequate to describe the propagation characteristics accurately [3]. Consequently, the EDC technique, which is essentially a single-term approximation of the mode-matching method, matches very well with the accurate methods. With an increase in the dielectric constant of the guide, the accuracy of the EDC technique reduces, whereas that of Marcatili's method improves [1]. This feature may be observed by comparing Figures 4.5 and 4.7. However, for the dominant E^y_{11} mode, the EDC technique still yields accurate values of the propagation constant at higher frequencies (larger values of B).

Bandwidth The image guide is normally excited in the E^y_{11} mode. The E^x_{11} mode, which is orthogonal to the E^y_{11} mode, is not excited. The first higher order mode to appear in such a guide would be the E^y_{21} mode for moderate to large aspect ratio, and the E^y_{12} mode for small aspect ratio. For monomode operation in the E^y_{11} mode, the upper frequency limit is governed by the cutoff frequency of the first higher order mode. It may be noted that the cutoff frequency of the first higher order mode (E^y_{21} or E^y_{12}) increases with a decrease in the aspect ratio a/b as well as ε_r. The lower frequency limit is governed by practical considerations such as dispersion, waveguiding efficiency, and impedance variation [6], although ideally the image guide has no low-frequency cutoff. Collier and Birch [6] have described the image guide bandwidth based on dispersion and wave guidance. Their graphs are reproduced in Figures 4.8 and 4.9. The plots are normalized to a cutoff frequency of 14.53 GHz, which corresponds to a guide with $a/b = 1$ and $\varepsilon_r = 2.5$. For defining the dispersion bandwidth (see Fig. 4.8), the lower usable frequency is chosen as that at which the phase velocity has changed by 20% of the value at the next higher order mode. It can be seen that for a given aspect ratio, the dispersion bandwidth increases with a decrease in ε_r. For defining the waveguiding bandwidth, the lower frequency is chosen as that at which the power in the air region becomes equal to that in the dielectric strip. Below this frequency, energy traveling in the air region would be greater than in the dielectric region. It can be seen from Figure 4.9 that the bandwidth increases with an increase in ε_r and, for a given dielectric, maximum bandwidth is achieved when a/b approaches unity.

A third factor that may be considered for determining the frequency limit is the characteristic impedance of the guide. Figure 4.10 shows the variation in the impedance with frequency for $a/b = 1$ and for different values of ε_r [6]. The formula used for defining the impedance Z is

$$Z = b^2 E^2 / 2P_i \qquad (4.2)$$

where P_i is the total power flow in the dielectric and air regions, and E is the maximum value of the electric field. It can be seen from Figure 4.10 that the larger

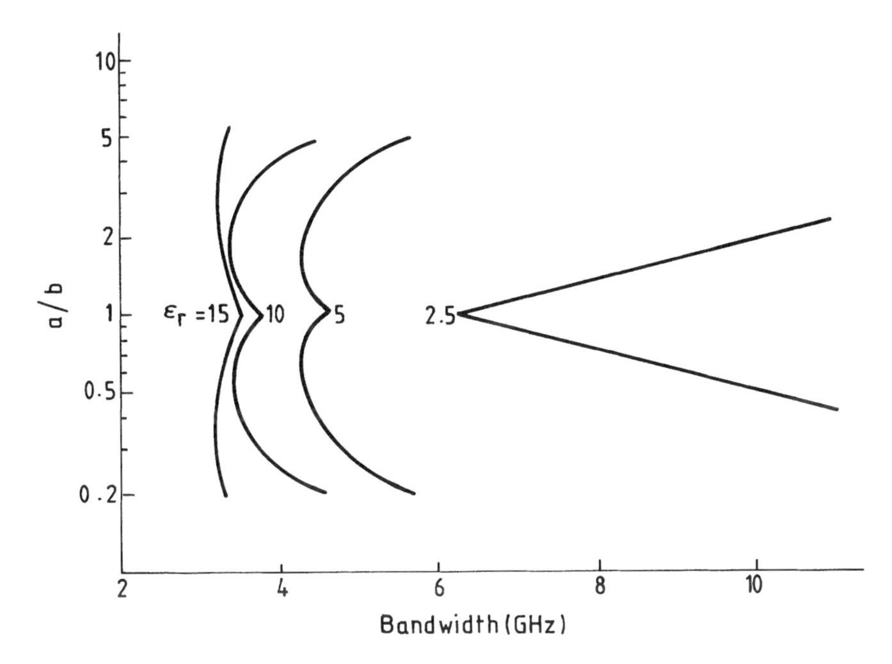

FIGURE 4.8 Dispersion bandwidth of image guide as a function of aspect ratio with ε_r as parameter. (From Collier and Birch [6]. Copyright © 1980 IEEE, reprinted with permission.)

the dielectric constant, the smaller is the variation in impedance as a function of frequency, but the bandwidth in terms of percentage change in Z is independent of the dielectric constant. Another factor that limits the practical bandwidth is radiation from curved bends. A straight image guide has no leaky modes and hence does not radiate, but leakage radiation does take place at the curved sections [7, 8]. Increasing the dielectric constant offers better confinement of energy to the dielectric, thereby reducing leakage radiation from such curved sections. For a given dielectric, the radiation loss can be made negligible by choosing a sufficiently large radius of curvature. Considering various factors that affect the low-frequency performance, Collier and Birch [6] have reported that for a image guide with unity aspect ratio, maximum bandwidth is achieved for $\varepsilon_r = 2.5$.

4.2.3 Attenuation Characteristics

The procedure for calculating the conductor and dielectric attenuation constants is presented in Chapter 2. For the E^y_{11} mode of the image guide, Toulios and Knox [9] have reported theoretical graphs of attenuation as a function of frequency for various values of a/b and ε_r. Figures 4.11 and 4.12 show the variation in the normalized conductor attenuation constant $(\alpha_c \lambda_0 / R_s)$ and the

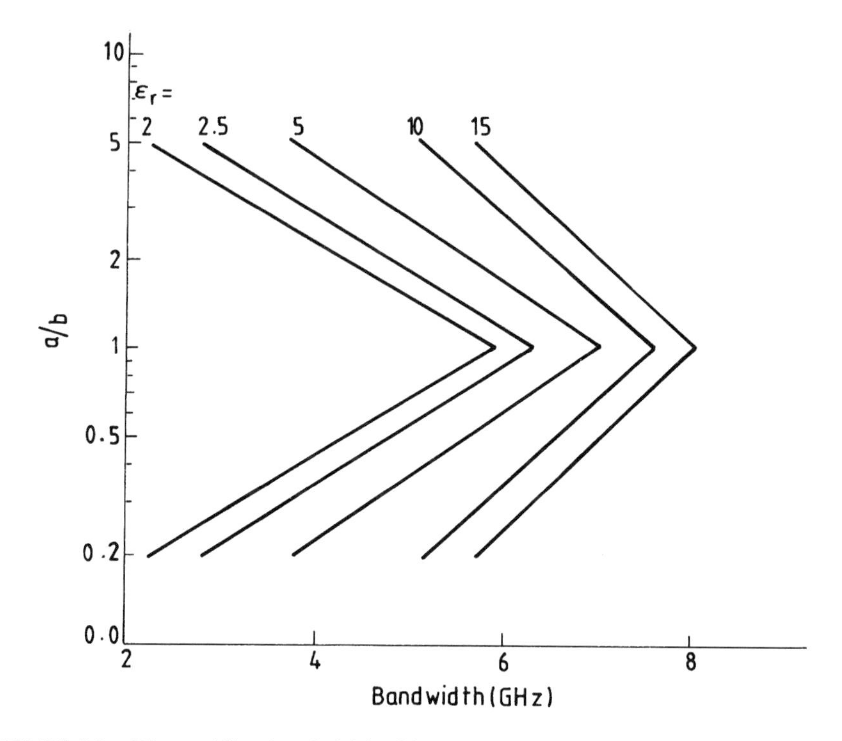

FIGURE 4.9 Waveguiding bandwidth of image guide as a function of aspect ratio a/b with ε_r as parameter. (From Collier and Birch [6]. Copyright © 1980 IEEE, reprinted with permission.)

normalized dielectric attenuation constant α_d/α_0, respectively, as a function of normalized frequency parameter B [9]. The parameter B is defined in (4.1), α_c is the conductor attenuation constant, R_s is the surface resistance of the conductor, α_d is the dielectric attenuation constant, and α_0 is the attenuation constant of the infinite dielectric medium. The expressions for α_0 is

$$\alpha_0 = 8.686(\pi\sqrt{\varepsilon_r}/\lambda_0)\tan\delta \tag{4.3}$$

where $\tan\delta$ is the loss tangent of the dielectric material. It can be seen from Figures 4.11 and 4.12 that for lower values of B (equivalently lower frequency or smaller guide height b), as the aspect ratio a/b is increased, the normalized conductor loss decreases whereas the dielectric loss increases. The same trend remains as B is increased. Beyond a sufficiently large value of B, both the conductor and dielectric losses become nearly independent of the aspect ratio and the ratio α_d/α_0 approaches unity. This is because of greater confinement of energy within the dielectric with an increase in either the frequency or the guide height b.

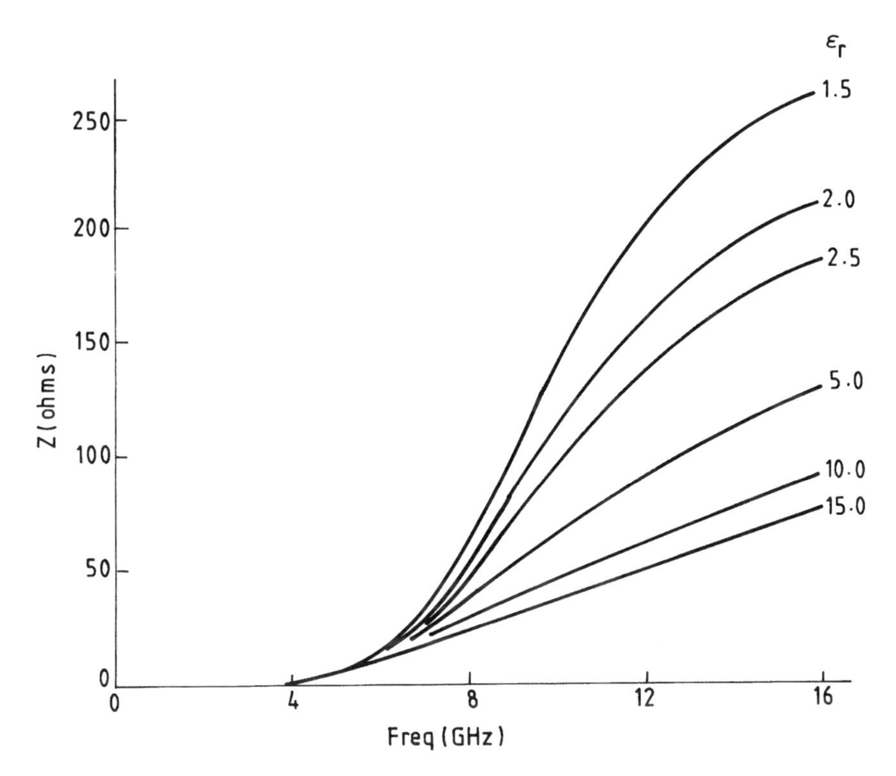

FIGURE 4.10 Variation of impedance Z of image guide as a function of frequency with ε_r as parameter: $a/b = 1$. (From Collier and Birch [6]. Copyright © 1980 IEEE, reprinted with permission.)

The E^x_{11} mode is the lowest of the E^x_{mn} modes and has its dominant transverse electric field E_x parallel to the ground plane. Since the tangential magnetic field at the ground plane is nearly zero, the conductor loss is small. The electric field confinement in the dielectric is less than that for the E^y_{11} mode, resulting also in less dielectric loss. Figure 4.13 shows a comparison of conductor and dielectric attenuation constants of the E^x_{11} mode with those of the E^y_{11} mode [10]. As an example, for an image guide having parameters $\varepsilon_r = 2$, $\tan\delta = 1.5 \times 10^{-4}$, $\sigma = 6.17 \times 10^7$ mhos/m (silver), and operating at 50 GHz, the theoretical total loss (conductor and dielectric) is reported to be 2.26 dB/m for the E^y_{11} mode and 0.99 dB/m for the E^x_{11} mode. The radiation loss at the curved bends, however, is higher for the E^x_{11} mode than for the E^y_{11} mode. For an image guide with parameters $\varepsilon_r = 2$, $a/b = 1$, and bend radius 3 cm, as the frequency is increased from 50 to 54 GHz, the radiation loss for the E^y_{11} mode is reported to decrease from 0.43 to 0.12 dB/rad, whereas that for the E^x_{11} mode decreases from 0.75 to 0.5 dB/rad [10]. It may be noted that the radiation loss of both modes can be reduced by increasing either ε_r or the aspect ratio a/b.

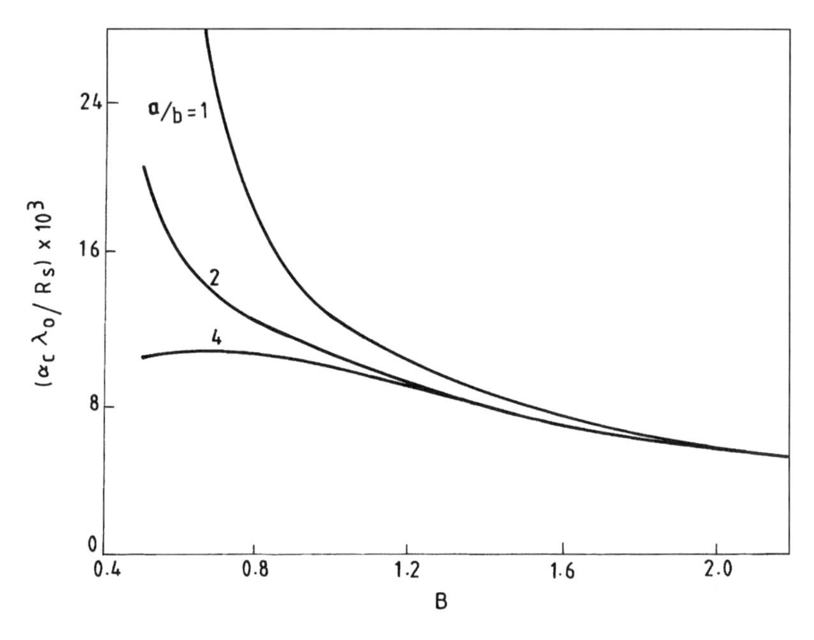

FIGURE 4.11 Normalized conductor attenuation for E_{11}^y mode of image guide as a function of normalized frequency parameter B: $\varepsilon_r = 2.25$ and $B = (4b/\lambda_0)\sqrt{\varepsilon_r - 1}$. (After Toulios and Knox [9].)

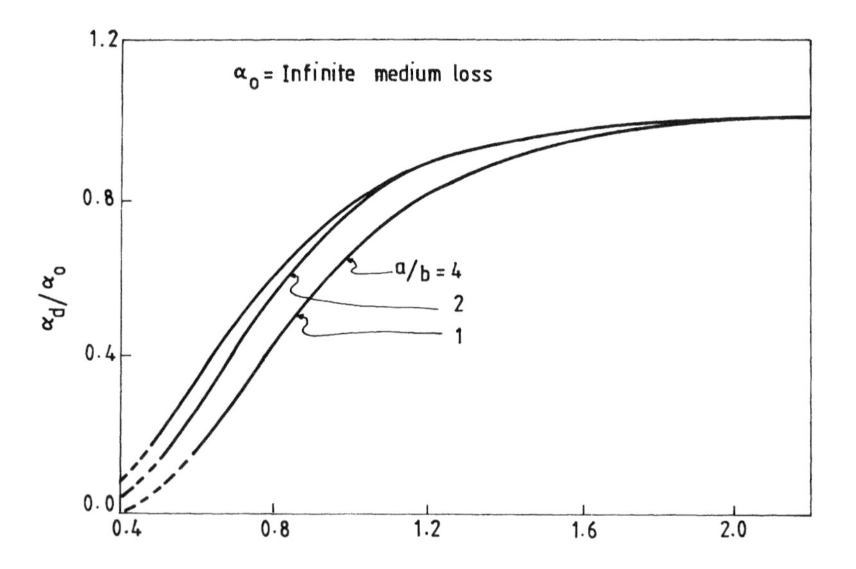

FIGURE 4.12 Normalized dielectric attenuation α_d/α_0 of E_{11}^y mode of image guide as a function of normalized frequency parameter B: $\varepsilon_r = 2.25$ and $B = (4b/\lambda_0)\sqrt{\varepsilon_r - 1}$. (After Toulios and Knox [9].)

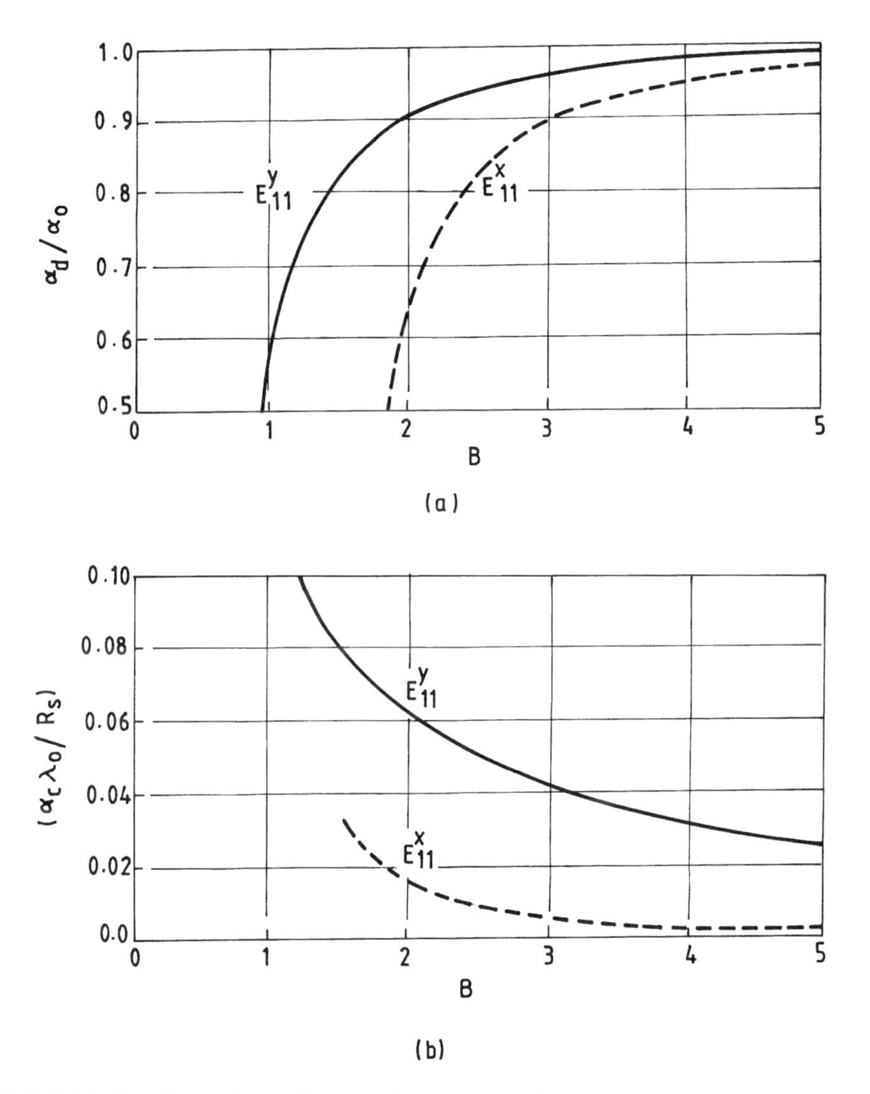

FIGURE 4.13 Comparison of attenuation constants for E^y_{11} and E^x_{11} modes of image guide: (a) normalized dielectric attenuation versus B and (b) normalized conductor attenuation versus B. $B = (4b/\lambda_0)\sqrt{\varepsilon_r - 1}$. (From Shindo and Itanami [10]. Copyright © 1978 IEEE, reprinted with permission.)

4.3 INSULAR IMAGE GUIDE

4.3.1 Fields and Dispersion

The image guide operating in its dominant E^y_{11} mode suffers from high conductor loss because of the large field concentration near the metal ground plane. This problem is circumvented in the insular image guide (IIG) (Fig. 4.1(b)) by

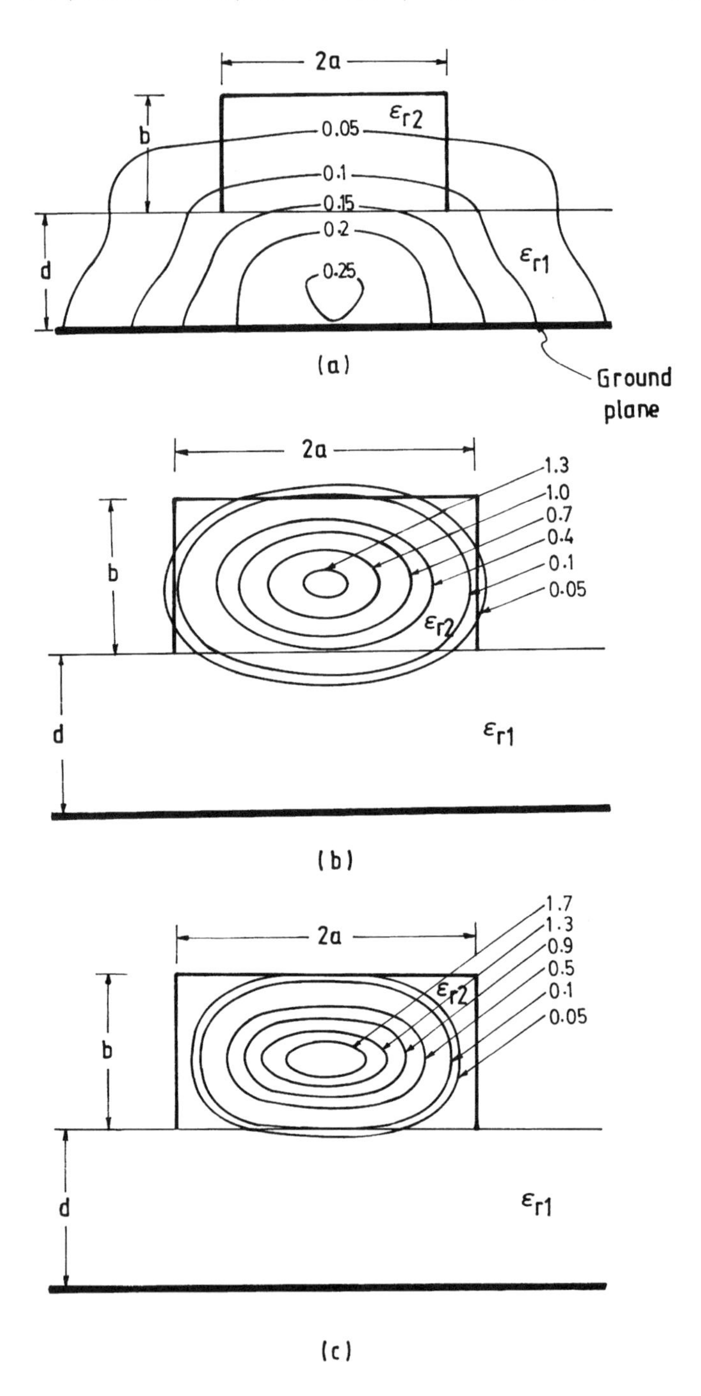

FIGURE 4.14 Frequency dependence of the power distribution for E^y_{11} mode in insular image guide with $d/a = b/a = 1$, $\varepsilon_{r2} = 3$, and $\varepsilon_{r1} = 2$: (a) $k_0 a = 1$, (b) $k_0 a = 5$, and (c) $k_0 a = 20$. (From Ogusu [2]. Copyright © 1977 IEEE, reprinted with permission.)

introducing a thin insular layer of low dielectric constant between the dielectric strip and the ground plane [11]. The fields decay exponentially in the insular layer with most of the energy confined to the strip. Consequently, the conductor loss in the ground plane decreases and also it enables relaxed fabricational tolerances for the guide. The relative field concentration in the dielectric strip with respect to that in the insular layer increases with an increase in the frequency. This feature is illustrated in Figure 4.14, which shows typical frequency dependence of the power distribution for the E^y_{11} mode in an IIG with $\varepsilon_{r1} = 2$ and $\varepsilon_{r2} = 3$ [2]. It can be seen that at high frequencies (see Fig. 4.14(c) for $k_0 a = 20$), the field penetration in the air region and insular layer is negligible. Furthermore, maximum energy is located at the center of the dielectric strip with the field distribution resembling that of a bare rectangular dielectric guide.

The dispersion characteristics of the insular image guide have been reported by several investigators [2, 11–14]. The mode classification for the IIG is identical to that for the image guide. Figure 4.15 shows the effect of varying the insular layer thickness on the dispersion characteristics of the guide [11]. The plots show variation in the normalized propagation constant β/k_0 of the guide as a function of the normalized frequency parameter B defined as

$$B = (4b/\lambda_0)\sqrt{\varepsilon_{r2} - 1} \tag{4.4}$$

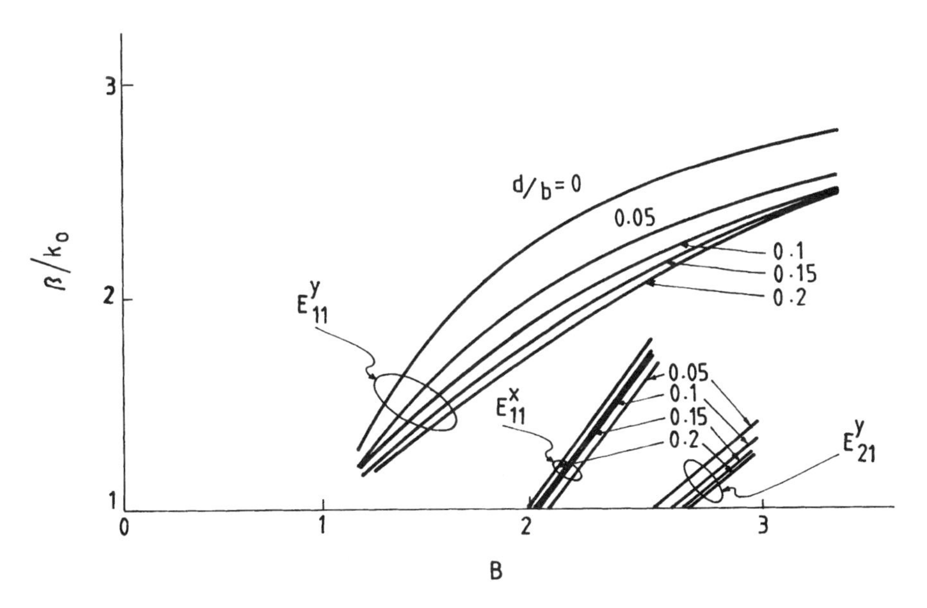

FIGURE 4.15 Normalized propagation constant of insular image guide as a function of normalized frequency parameter B (structure in Fig. 4.1(b)): $a/b = 0.5$, $\varepsilon_{r1} = 2.25$ (polyethylene), $\varepsilon_{r2} = 9.8(Al_2O_3)$, and $B = (4b/\lambda_0)\sqrt{\varepsilon_r - 1}$. (From Knox [11]. Copyright © 1976 IEEE, reprinted with permission.)

For a fixed value of B, as the insular layer thickness d is increased, β/k_0 decreases or equivalently the guide wavelength increases. Furthermore, the rate of decrease of β/k_0 is largest for very small values of d/b. This effect is the result of a decrease in the effective guide height that takes place when an insular layer of low dielectric constant is introduced. The same effect applies to the E^y_{21} mode. The propagation constant of the E^x_{11} mode, however, follows the opposite trend since its dominant E-field (E_x) is parallel to the ground plane. This is a disadvantageous feature because the cutoff of this mode decreases with an increase in d/b, thereby reducing the single-mode bandwidth. A compromise choice reported by Knox [11] is to choose a value of d/b in the range 0.015–0.15 so that maximum bandwidth is achieved (with respect to the E^x_{11} mode cutoff) with minimum dependence of guide wavelength on the insular layer thickness variation. Furthermore, a choice of aspect ratio $a/b = 0.5$ is reported to be useful from the point of view of optimum bandwidth and suppression of higher order modes [11].

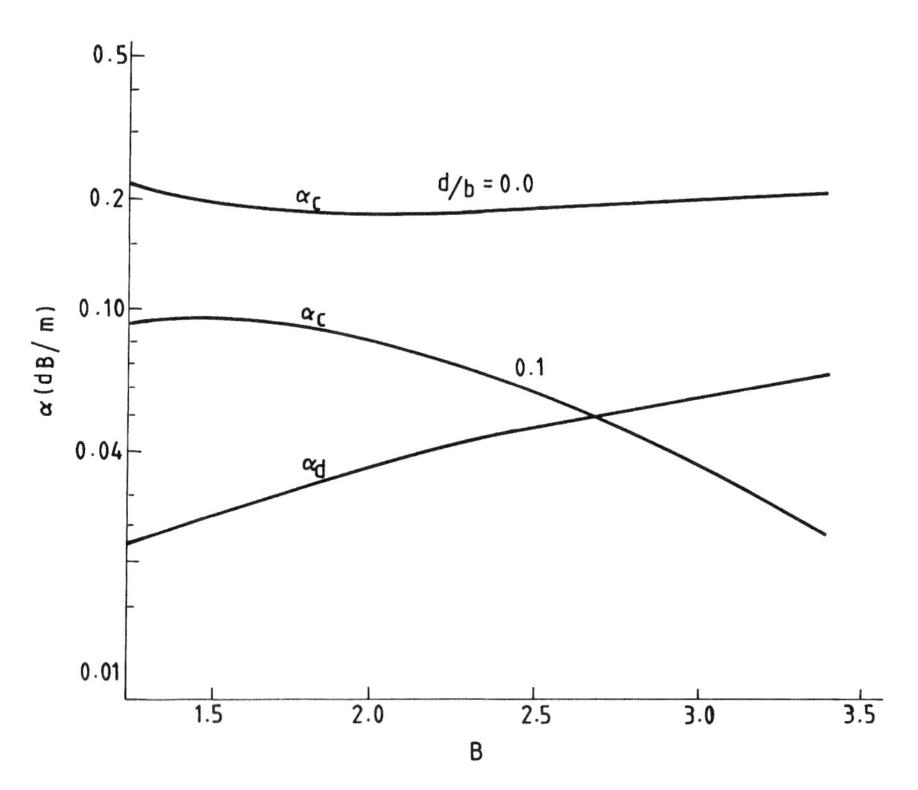

FIGURE 4.16 Attenuation constants due to conductor and dielectric losses in insular image guide as a function of B (structure in Fig. 4.1(b)): $a/b = 0.5$, $\varepsilon_{r1} = 2.25$ (polyethylene), $\varepsilon_{r2} = 9.8(Al_2O_3)$, $\tan\delta = 10^{-4}(Al_2O_3)$, $\sigma = 3.72 \times 10^7$ mhos/m (aluminum), and $B = (4b/\lambda_0)\sqrt{\varepsilon_r - 1}$. (From Knox [11]. Copyright © 1976 IEEE, reprinted with permission.)

4.3.2 Attenuation Characteristics

Figure 4.16 shows typical attenuation characteristics of the insular image guide [11]. Comparison of curves for the image guide ($d/b = 0$) and IIG having $d/b = 0.1$ shows that the conductor loss expressed in dB/m in the IIG is less than half that in the image guide. Furthermore, with an increase in B (equivalently, an increase in frequency or b), the conductor loss in the IIG decreases because of greater field confinement to the high dielectric constant guide. For the same reason, attenuation due to dielectric loss increases.

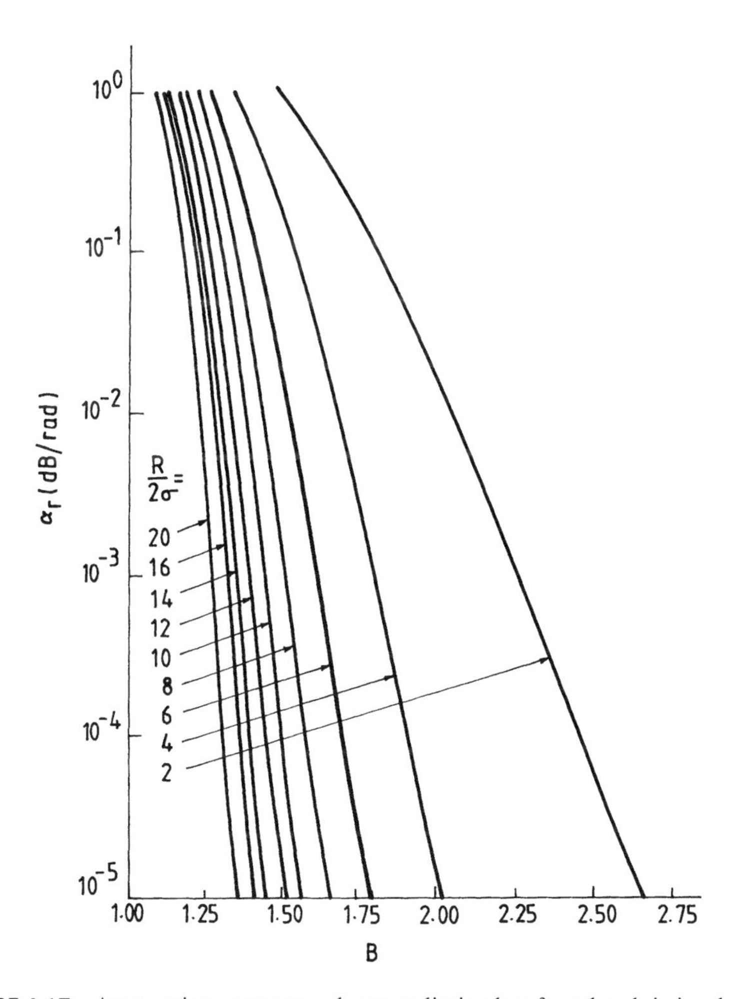

FIGURE 4.17 Attenuation constant α_r due to radiation loss from bends in insular image guide as a function of B with normalized radius of curvature as parameter (structure in Fig. 4.1(b)): $a/b = 0.5$, $\varepsilon_{r1} = 2.25$, $\varepsilon_{r2} = 9.8$, $d/b = 0.1$, and $B = (4b/\lambda_0)\sqrt{\varepsilon_r - 1}$. (From Knox [11]. Copyright © 1976 IEEE, reprinted with permission.)

The radiation loss from bends in IIG depends on a/b as well as the dielectric constant ratio $\sqrt{\varepsilon_{r2}/\varepsilon_{r1}}$. For a guide with parameters $a/b = 0.5$, $d/b = 0.1$, $\varepsilon_{r1} = 2.25$, and $\varepsilon_{r2} = 9.8$, Figure 4.17 shows plots of the attenuation constant due to radiation loss versus B with $R/2a$ as a parameter, where R is the radius of curvature at the bend [11]. For a given guide, the radiation loss can be reduced to an acceptable level by appropriately choosing the operating frequency (value of B) and the normalized bend radius ($R/2a$).

4.4 TRAPPED IMAGE GUIDE

4.4.1 Dispersion and Wave Impedance

The trapped image guide proposed by Itoh and Adelseck [15] is basically an image guide with metallic side walls (Fig. 4.1(c)). Propagation takes place along the dielectric strip with fields decaying exponentially away from the strip in the air region. The main advantage of this guide over the image and insular image guides is the reduction in radiation loss at the curved sections. The metallic side walls enable energy leaking from the bends to be reflected back to the dielectric strip so as to couple again to the guided mode. The height of the metal walls must be sufficiently large so that the fields decay to a negligible level at the opening. The propagation characteristics of the trapped image guide have been reported using the EDC method [15, 16] and also the rigorous mode-matching method [17–19].

Effect of Top Wall The trapped image guide can easily be excited in its dominant E^y_{11} mode from a TE_{10} mode rectangular waveguide by means of a transition. In practice, the guide may be closed by a top metallic plate placed at appropriate height. In order to have an estimate of the minimum height required to allow fields to decay to a negligible level, we consider in Figure 4.18 the typical variation in the normalized propagation constant β/k_0 for the E^y_{11} mode of a shielded image guide as a function of the normalized height h/b [18]. It may be noted from the graph that for guides having $\varepsilon_r \geq 2.5$ and at frequencies above 20 GHz, the top metal plate has negligible effect on the guide wavelength of E^y_{11} mode if the condition $h/b \geq 4$ is satisfied.

Dispersion Figures 4.19 and 4.20 show the dispersion characteristics of the trapped image guide for aspect ratios $a/b = 1$ and 0.5, respectively [19]. The rigorous mode-matching formulas (see Chapter 3) are used to compute the results. From Figure 4.19, we note that as the spacing s between the dielectric strip and the adjacent side wall increases, the cutoff frequencies of the dominant E^y_{11} mode and the higher order E^y_{21} mode decrease and the frequency separation between the two modes also decreases. The decrease in frequency separation, however, is not significant, particularly for smaller values of s, and the modal operational bandwidth remains nearly the same. Comparing the dispersion

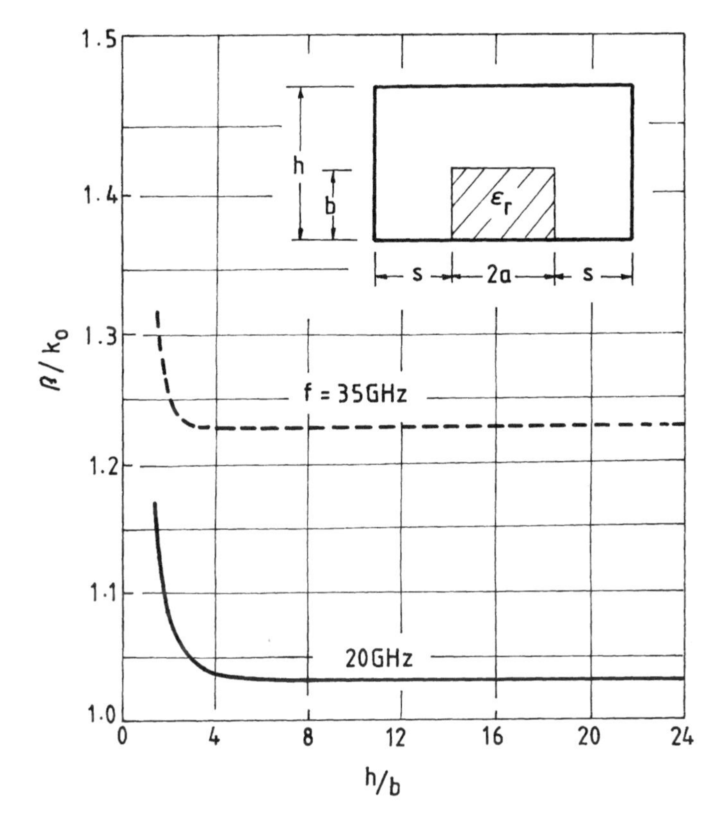

FIGURE 4.18 Normalized propagation constant β/k_0 versus h/b for E^y_{11} mode of trapped image guide: $a/b = 1, b = 2$ mm, $\varepsilon_r = 2.56$, and $s \rightarrow \infty$. (From Tiwari and Bhat [18]. Reprinted from AEU-38, 1984.)

curves for the E^y_{11} mode in Figure 4.19 (for $a/b = 1$) with those in Figure 4.20 (for $a/b = 0.5$) shows that for a fixed spacing s, as a/b is increased, the cutoff frequency decreases but the value of β/k_0 for a given frequency above cutoff increases. A similar trend occurs when ε_r is increased with all other parameters kept fixed [18].

Wave Impedance Figure 4.21 shows the typical variation in the wave impedance Z of a trapped image guide as a function of frequency calculated using the modal fields in the relation $Z = E_y/H_x$ [19]. The wave impedance curves are drawn starting with a frequency corresponding to $\beta/k_0 = 1$. The special case of $s = \infty$ refers to the image guide. The wave impedance of the image guide for the E^y_{11} mode increases with an increase in frequency and nearly saturates at higher frequencies. As the side walls are brought closer to the dielectric slab, the wave impedance increases. For extremely small spacing, the guide offers very high

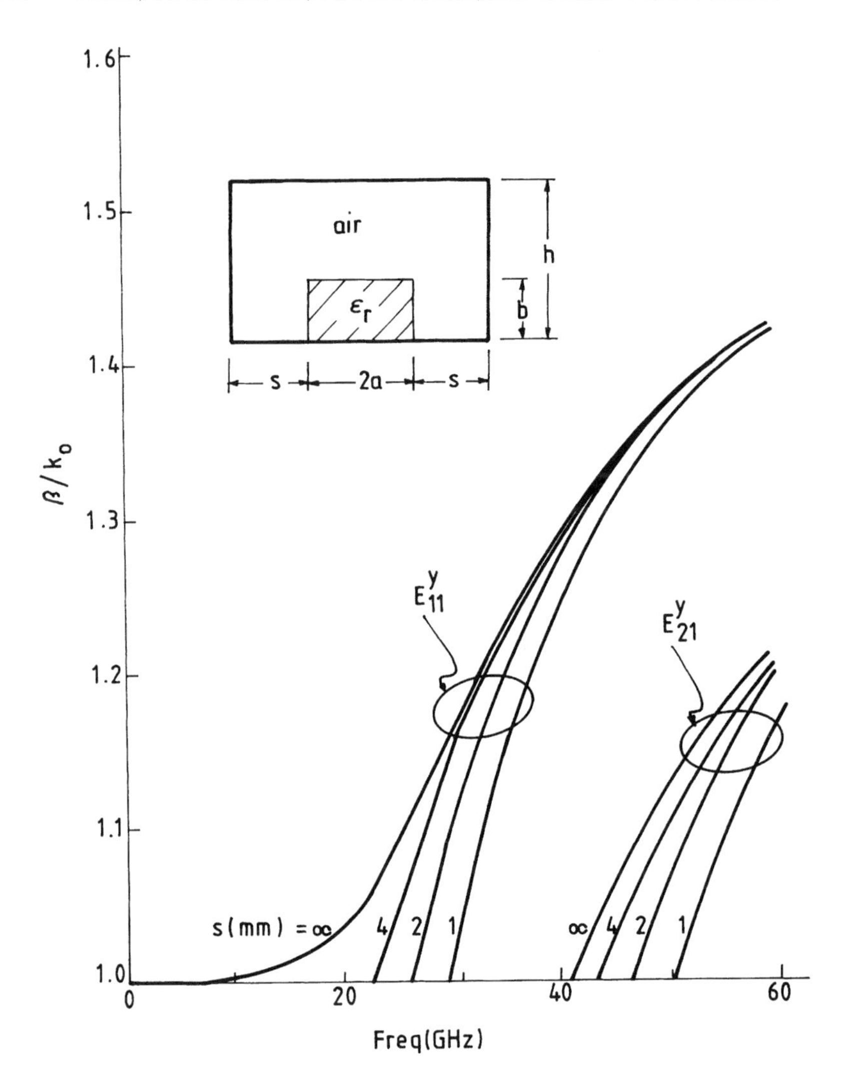

FIGURE 4.19 Variation of β/k_0 of trapped image guide as a function of frequency for $a/b = 1$, $b = 2$ mm, $h = 8$ mm, and $\varepsilon_r = 2.56$. (After Tiwari [19].)

impedance near $\beta/k_0 = 1$, and as the frequency is increased, the impedance decreases. At higher frequencies, the wave impedances for all spacings approach each other and reach nearly a constant value. For the E_{21}^y mode, the wave impedance starts with a high value near $\beta/k_0 = 1$ and then decreases with increasing frequency for all values of s. For a fixed set of parameters b, s, and ε_r, the wave impedance is reported to increase with a decrease in the aspect ratio a/b [19].

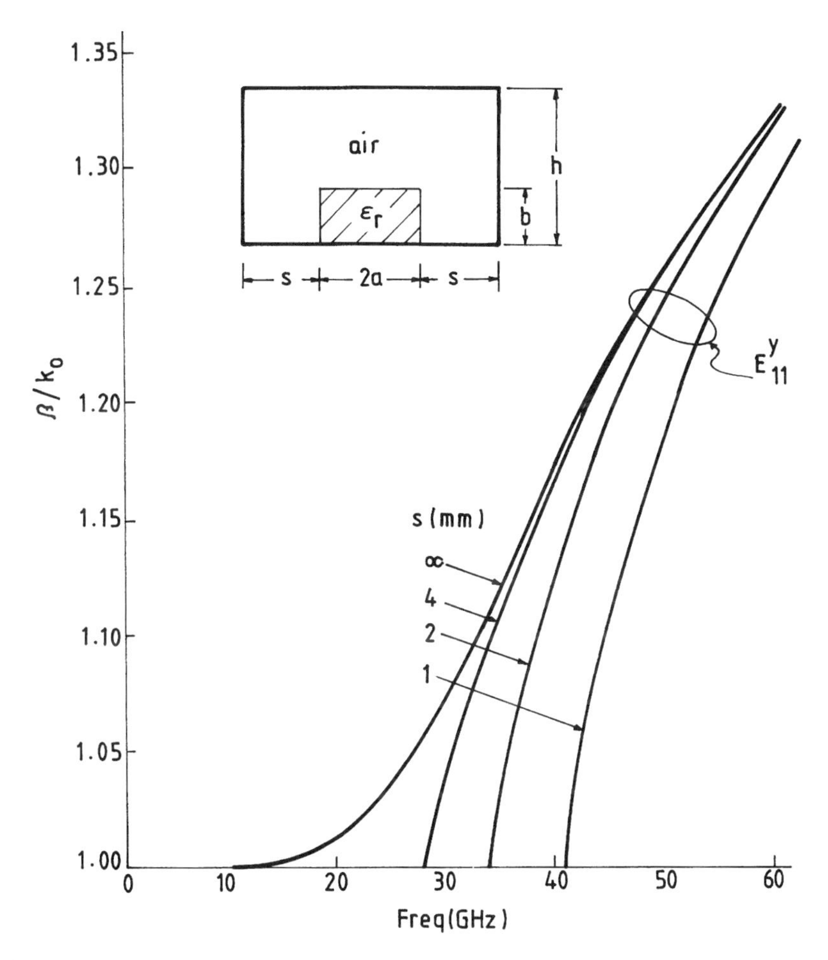

FIGURE 4.20 Variation of β/k_0 for E^y_{11} mode of trapped image guide as a function of frequency for $a/b = 0.5$, $b = 2$ mm, $h = 8$ mm, and $\varepsilon_r = 2.56$. (After Tiwari [19].)

4.4.2 Effect of Insular Layer

The effect of an air insular layer between the dielectric strip and the ground plane on the propagation constant of the E^y_{11} mode of the trapped image guide is illustrated in Figure 4.22 [17]. The top metal plate is kept at an appropriate height above the dielectric ($q = 6$ cm) so that propagation characteristics of the E^y_{11} mode are not significantly altered. The characteristics are plotted for the condition $\beta/k_0 > 1$ (corresponding to image guide modes). It can be seen that for a fixed frequency, β/k_0 decreases with an increase in the insular layer thickness. Furthermore, with all other parameters fixed, the cutoff frequency corresponding to $\beta/k_0 = 1$ increases with an increase in the layer thickness.

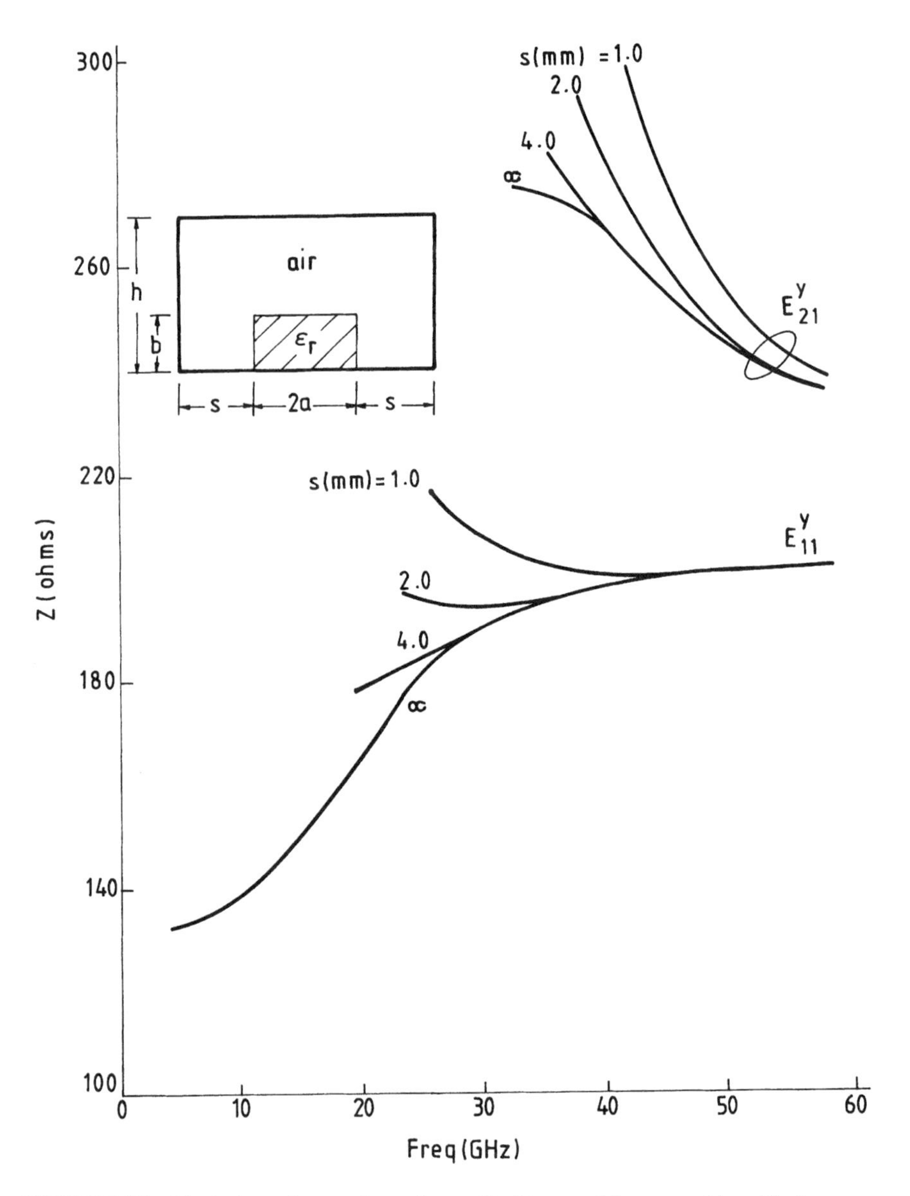

FIGURE 4.21 Variation of wave impedance Z of trapped image guide with frequency for $a/b = 1$, $b = 2\,\text{mm}$, $h = 8\,\text{mm}$, and $\varepsilon_r = 3.4$. (After Tiwari [19].)

4.5 COUPLED GUIDE CHARACTERISTICS

4.5.1 Field Distribution

In order to illustrate the field distribution in coupled guides, we consider a shielded structure shown in Figure 4.23. We denote the dominant E^y_{11} mode for

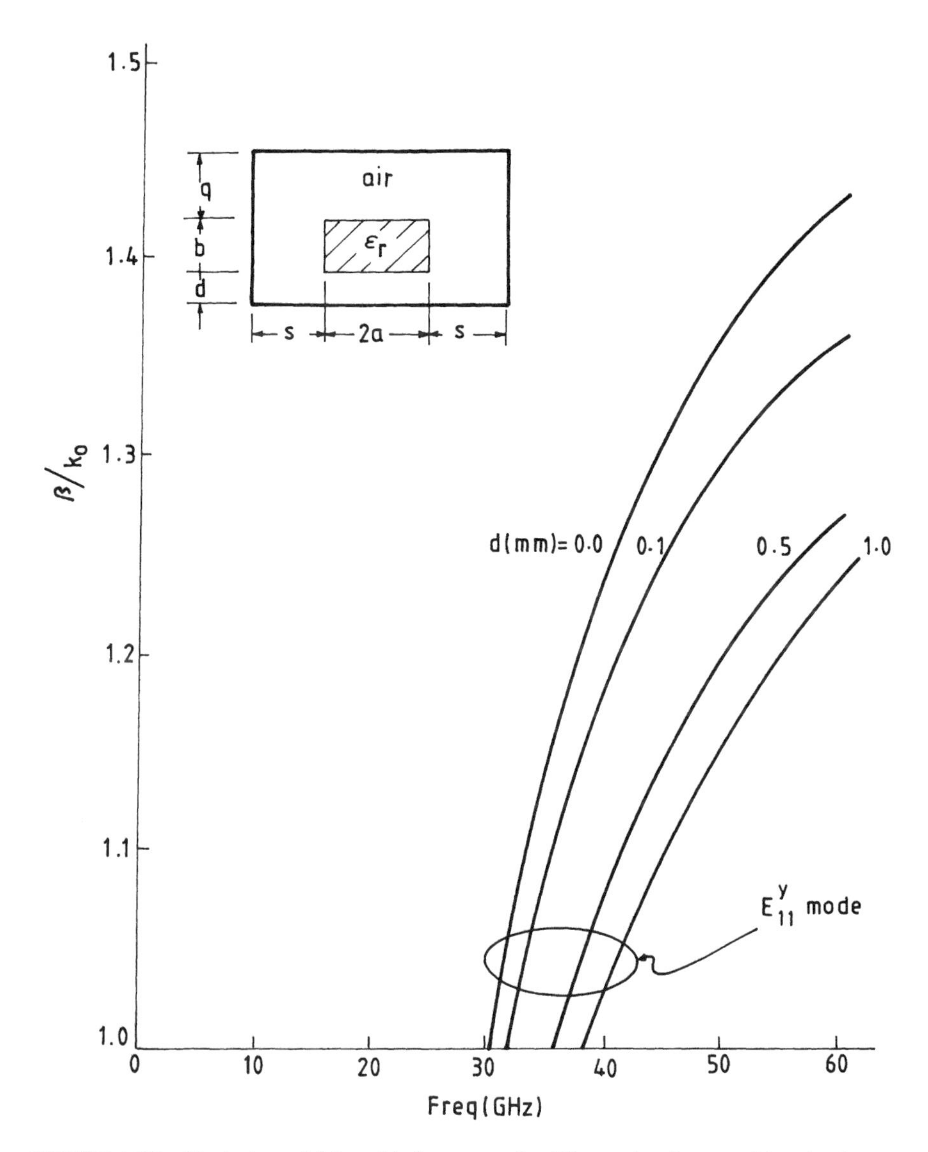

FIGURE 4.22 Variation of β/k_0 with frequency for E^y_{11} mode of trapped insular image guide for various values of insular layer thickness d: $a/b = 1$, $q/b = 3$, $b = 2\,\text{mm}$, $s = 1\,\text{mm}$, and $\varepsilon_r = 2.56$. (From Tiwari et al. [17]. Copyright © 1986 IEEE, reprinted with permission.)

even excitation as E^y_{11e} and for the odd excitation as E^y_{11o}. For the even mode, the plane of symmetry PP' represents a magnetic wall and for the odd mode, it represents an electric wall. For the E^y_{11e} and E^y_{11o} modes of the trapped coupled insular image guide, Figure 4.23 shows plots of field distribution at the horizontal plane $y = d + 0.8b$ for various values of the dielectric spacing [19]. With the

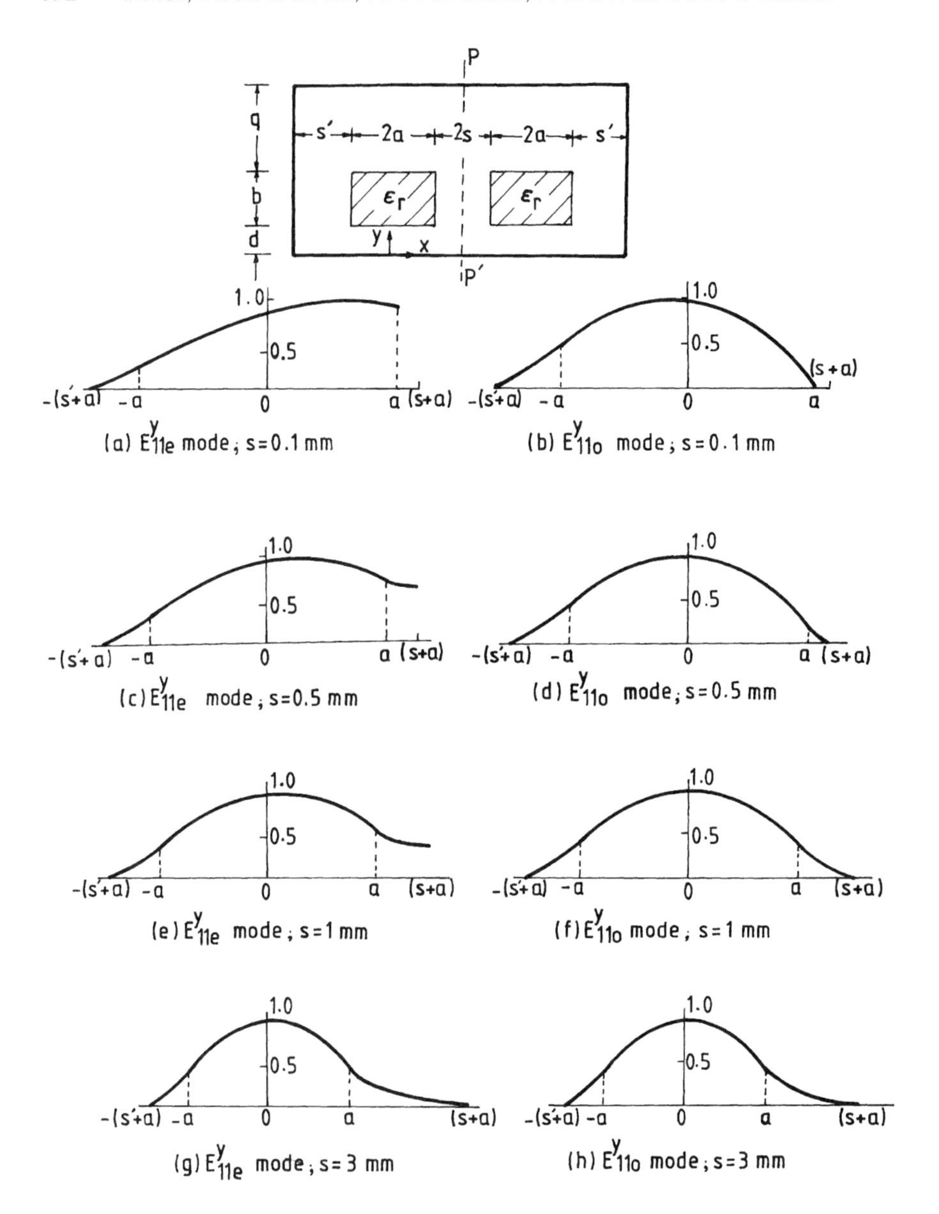

FIGURE 4.23 Normalized calculated field distribution for the E_y component of the E^y_{11e} and E^y_{11o} modes of coupled trapped insular image guide in a horizontal plane $y = d + 0.8\,b$: $a/b = 1$, $q/b = 3$, $b = 2\,\text{mm}$, $d = 0.5\,\text{mm}$, $s' = 1\,\text{mm}$, $\varepsilon_r = 2.56$, and $f = 60\,\text{GHz}$. (After Tiwari [19].)

choice of $q/b = 3$, the top wall is sufficiently away from the dielectric so as to have negligible effect on the propagation of the guide at 60 GHz. The plots are shown for the left-half of the guide cross-section. Since the structure is symmetric, the plot for the right-half is the mirror image of the plot for the left-half for the even mode, whereas for the odd mode it is 180° out of phase. For the E^y_{11e} mode, the E_y field maximum occurs near the right edge of the dielectric strip for the condition $s \ll s'$ (Fig. 4.23(a)), and it shifts toward the center of the strip as s is increased. For the E^y_{11o} mode, the E_y field maximum occurs toward the left side of the dielectric strip for $s \ll s'$ (Fig. 4.23(b)) and shifts toward the center of the strip as s is increased. When $s = s'$ ($= 1$ mm), the E_y maximum for the E^y_{11o} mode occurs at the center of the strip (Fig. 4.23(f)).

4.5.2 Dispersion and Wave Impedance

Coupled Image Guide Figure 4.24 shows the cross-section of a shielded coupled image guide and its dispersion characteristics calculated using the mode-matching method [18]. The characteristics shown in Figure 4.24(a) are for the special case of a coupled image guide obtained by setting $s' = \infty$ and $h/b = 4$. With $h/b = 4$, the effect of the top wall on the dominant mode propagation characteristics is negligible for frequencies above 20 GHz. We note from the graph that the E^y_{11e} mode has no cutoff and its propagation constants are higher than those for the E^y_{11o} mode. The difference between the propagation constants of the even and odd modes decreases as the spacing between the strips increases. Beyond a sufficiently large value of s when the two modes are decoupled, their propagation constants become equal.

Trapped Coupled Image Guide Figure 4.24(b) illustrates typical dispersion characteristics of trapped coupled image guide ($h/b = 4$). The plots show the effect of varying the spacing on the normalized propagation constants of E^y_{11e} and E^y_{11o} modes. The separation between the side walls is kept fixed, $2(s + 2a + s') =$ 12 mm. The effect of the side wall is to introduce a definite cutoff for the even mode. As s is increased from a small value, the cutoff frequency of the even mode increases whereas that of the odd mode first decreases, reaches a minimum for $s = s'$, and then increases again for $s > s'$. However, the difference in the even- and odd-mode cutoff frequencies decreases with an increase in s for a fixed value of side wall spacing, $2(s + 2a + s')$.

Figure 4.24(c) shows the effect of moving the side walls (varying s') on the propagation constants of E^y_{11e} and E^y_{11o} modes of the trapped coupled image guide when all other parameters are held fixed. It may be noted that as s' is increased from a small value, the cutoff frequencies of both the E^y_{11e} and E^y_{11o} modes decrease. In the limit $s' \to \infty$, the propagation constants reduce to those of the coupled image guide [18].

The effect of moving the side walls on the wave impedances of the E^y_{11e} and E^y_{11o} modes of the trapped coupled image guide is illustrated in Figure 4.25 [19].

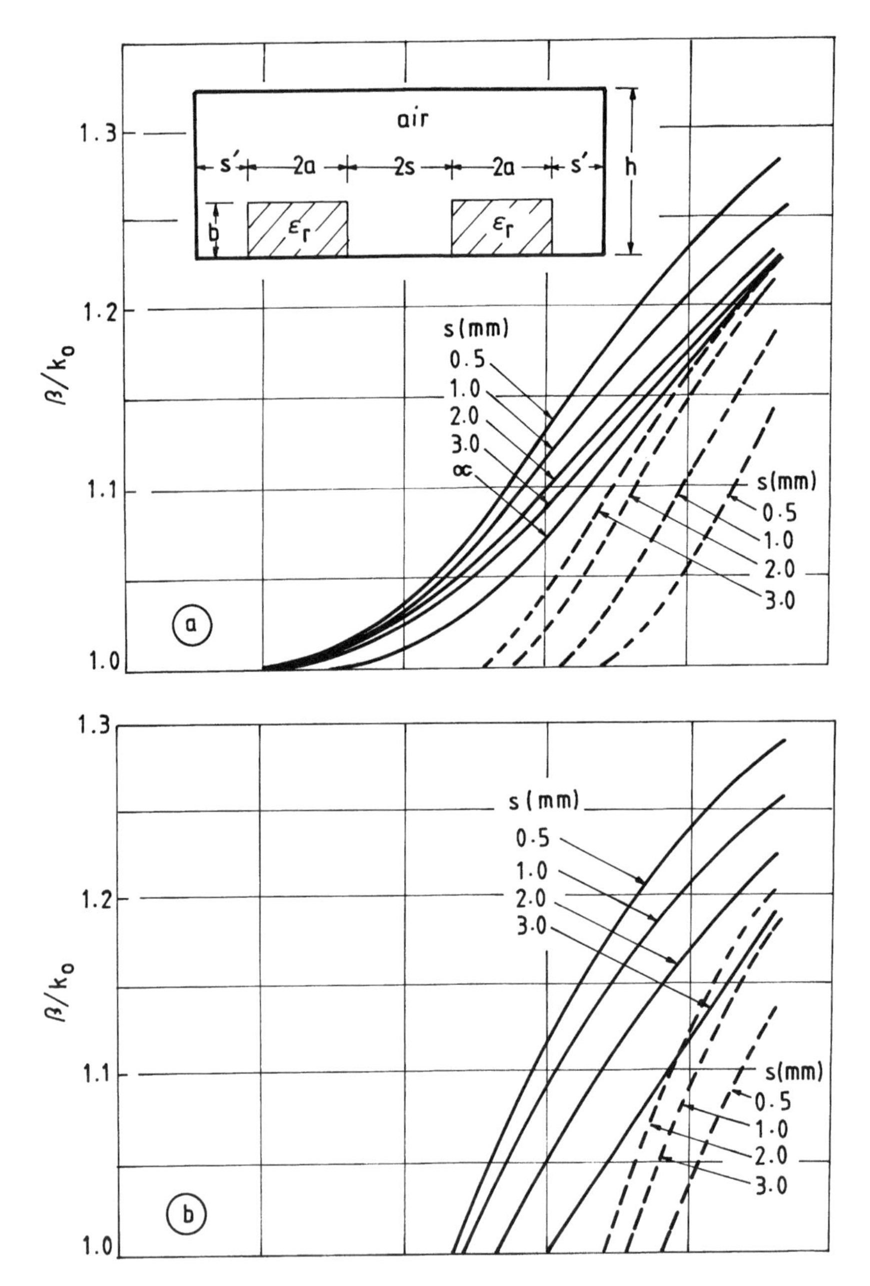

FIGURE 4.24 Normalized propagation constant β/k_0 versus frequency of coupled image and trapped coupled image guides for E^y_{11e} (———) and E^y_{11o} (------) modes: (a) coupled image guide ($s' = \infty$, $h/b = 4$), $a/b = 0.5$, $b = 2\,\text{mm}$, $\varepsilon_r = 2.56$; (b) trapped coupled image guide ($h/b = 4$), $a/b = 0.5$, $(s + 2a + s')/b = 3$, $b = 2\,\text{mm}$, $\varepsilon_r = 2.56$; and (c) trapped coupled image guide ($h/b = 4$), $a/b = 1$, $s/b = 1$, $b = 2\,\text{mm}$, $\varepsilon_r = 2.56$. (From Tiwari and Bhat [18]. Reprinted from AEU-38, 1984.)

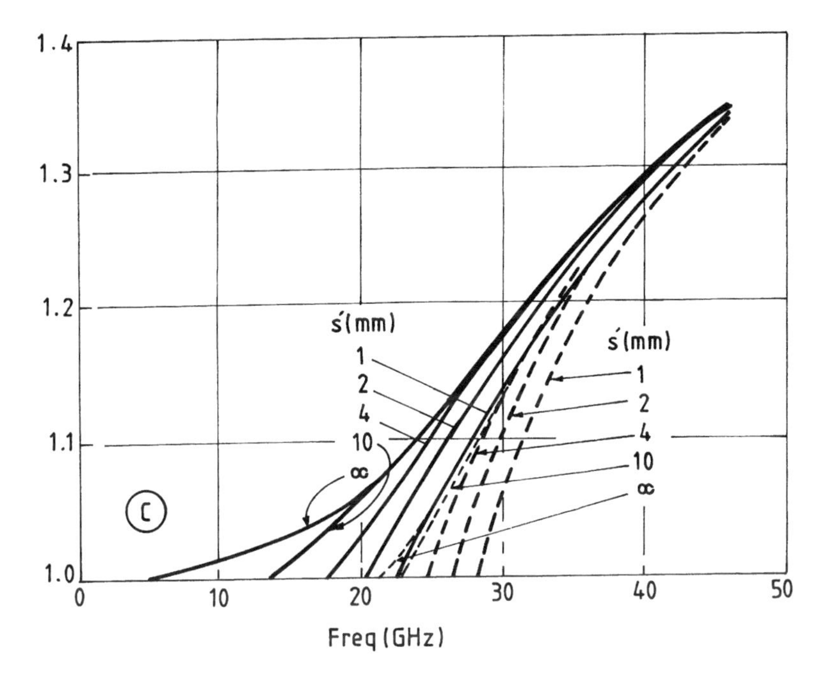

FIGURE 4.24 (*Continued*)

We note that the impedance curves start from a frequency corresponding to $\beta/k_0 = 1$. At a fixed frequency if s' is increased from a low value, both the even- and odd-mode impedances decrease. For a fixed value of s', if the frequency is increased above cutoff, the even mode impedance increases whereas the odd mode impedance decreases and at sufficiently high frequencies the two impedances tend to converge. From Figure 4.26, we observe that for a fixed value of side wall spacing, $2(s + 2a + s')$, as s is increased from a low value, the wave impedance of the E^y_{11e} mode increases, whereas that for the E^y_{11o} mode first decreases, reaches a minimum value at $s = s'$, and then increases again for $s > s'$ [19].

Trapped Coupled Insular Image Guide The effect of introducing an insular layer in the trapped coupled image guide is illustrated in Figure 4.27 [17]. It can be seen that for a given frequency above the cutoff, as the insular layer thickness d is increased, the propagation constants of both the E^y_{11e} and E^y_{11o} modes decrease. The cutoff frequencies of both these modes and also the separation between these two frequencies increase with an increase in d. The reason for this

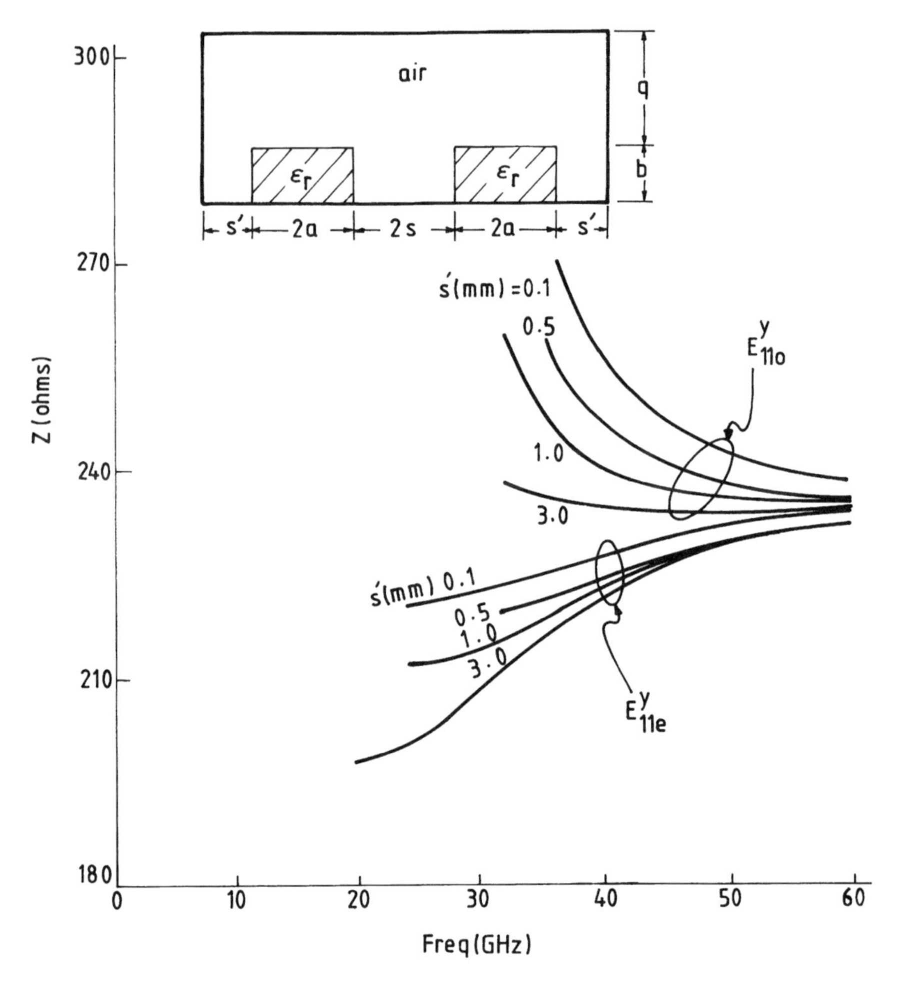

FIGURE 4.25 Variation of wave impedance Z versus frequency for the E^y_{11e} and E^y_{11o} modes of trapped coupled image guide for different values of s': $a/b = 1.0$, $q/b = 3$, $b = 2\,\text{mm}$, and $\varepsilon_r = 2.56$. (After Tiwari [19].)

trend is that with an increase in d, the effective dielectric constant of the structure reduces, resulting in a decrease in the propagation constant.

4.6 STRIP AND INVERTED STRIP DIELECTRIC GUIDES

The strip dielectric gide (SDG) (Fig. 4.28(a)) and the inverted strip dielectric guide (ISG) (Fig. 4.28(b)) have the feature that they adopt a planar guiding layer with a dielectric constant higher than that of the strip for guiding most

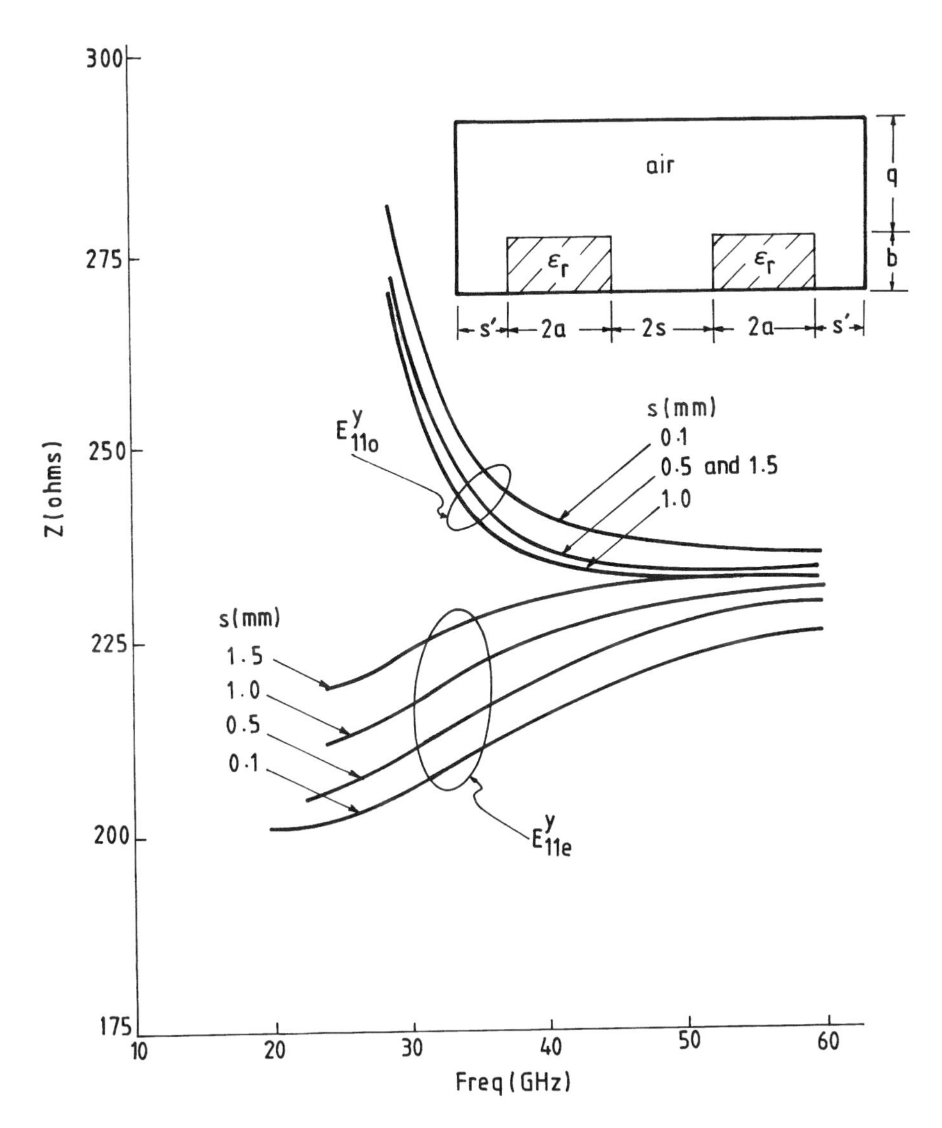

FIGURE 4.26 Variation of wave impedance Z versus frequency for the E^y_{11e} and E^y_{11o} modes of trapped coupled image guide for different values of s: $a/b = 1.0$, $q/b = 3$, $(s + 2a + s')/b = 3$, $b = 2\,\text{mm}$, and $\varepsilon_r = 2.56$. (After Tiwari [19].)

of the electromagnetic energy [20]. In the SDG, the guiding layer having the higher relative dielectric constant ε_{r2} is placed between the substrate and the top dielectric strip ($\varepsilon_{r2} > \varepsilon_{r1}, \varepsilon_{r3}$), and in the ISG, the guiding layer is placed immediately above the dielectric strip ($\varepsilon_{r2} > \varepsilon_{r1}$). In both structures, the ground plane may be used for dc biasing of active devices and also heat sinking.

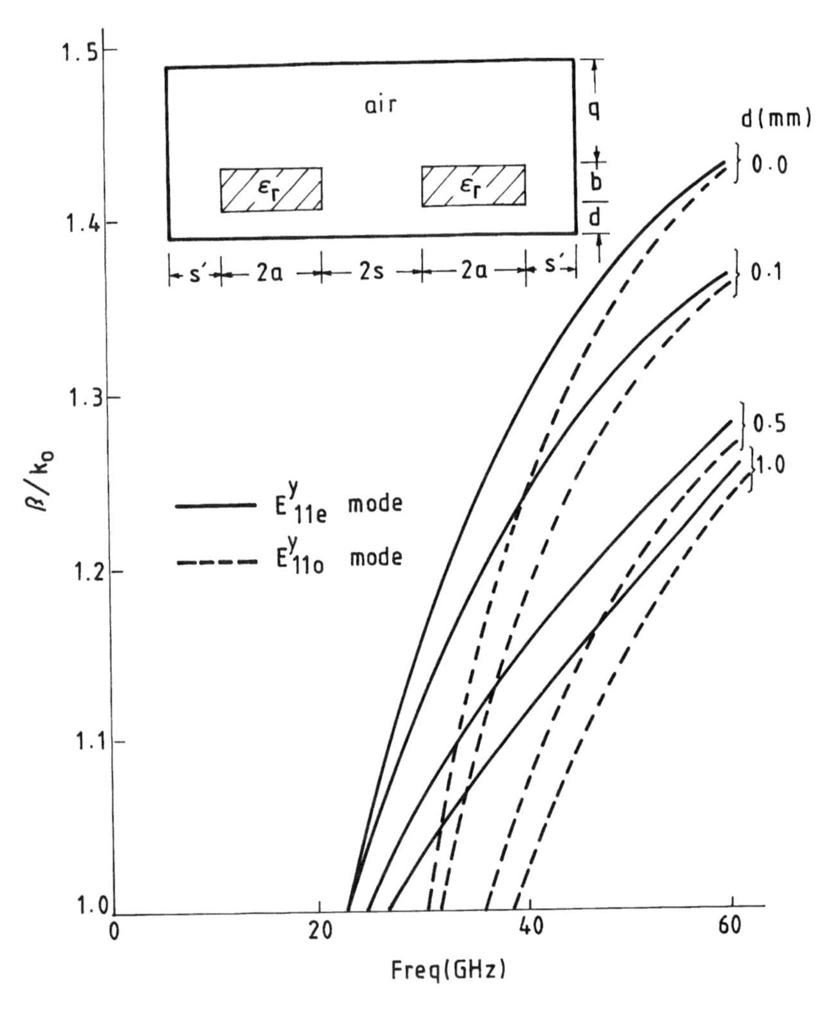

FIGURE 4.27 Variation of β/k_0 versus frequency for E^y_{11e} and E^y_{11o} modes of trapped coupled insular image guide for various values of insular layer thickness d:$a/b = 1$, $s/b = 0.5$, $s'/b = 0.5$, $q/b = 3$, $b = 2$ mm, and $\varepsilon_r = 2.56$. (From Tiwari et al. [17]. Copyright © 1986 IEEE, reprinted with permission.)

4.6.1 Strip Dielectric Guide

The strip dielectric guide resembles the insular image guide except that the confinement of electromagnetic energy is in a planar layer instead of in the top dielectric strip. As discussed in Section 4.3, the IIG (see Fig. 4.1(b)) by virtue of having an insular layer of low dielectric constant next to the ground plane offers considerable reduction in conductor loss over the image guide. The same advantage is retained in the SDG shown in Figure 4.28(a). With $\varepsilon_{r2} \gg \varepsilon_{r1}$, the field

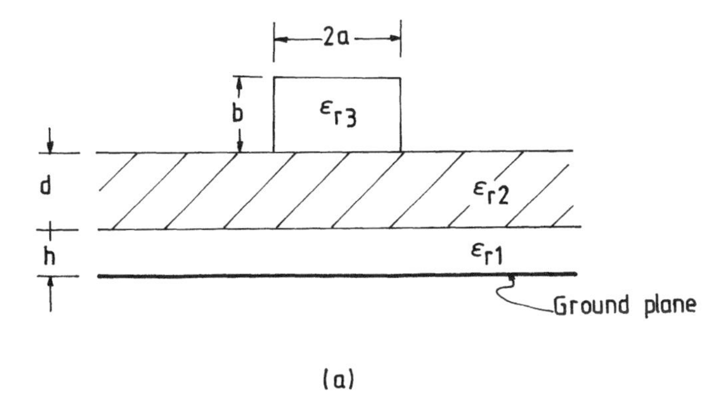

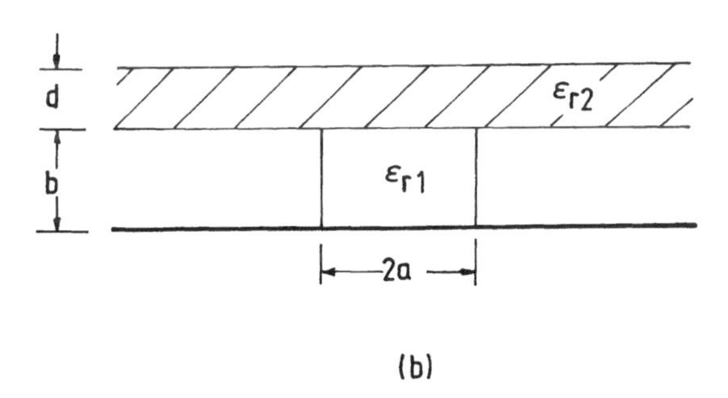

FIGURE 4.28 Cross-sectional views of (a) strip dielectric guide and (b) inverted strip guide (After Itoh [20].)

concentration near the ground plane is reduced, resulting in reduced conductor loss. An additional advantage in the SDG, by virtue of having $\varepsilon_{r2} > \varepsilon_{r3}$, is the reduction in radiation loss that takes place due to surface roughness. This is because the guiding layer, which concentrates most of the energy, is easier to fabricate with extremely smooth surface finish. On the other hand, in the IIG, any surface roughness on the side surfaces of the guiding strip enhances the radiation.

In the special case when the substrate thickness h is set equal to zero, the geometry of the SDG becomes identical to the IIG. However, since maximum power is concentrated in the region of higher dielectric constant, the two guides differ in the mechanism of propagation. Figure 4.29 shows the dispersion characteristics of the E_{11}^{y} mode of a strip dielectric guide with $h = 0$, the insular

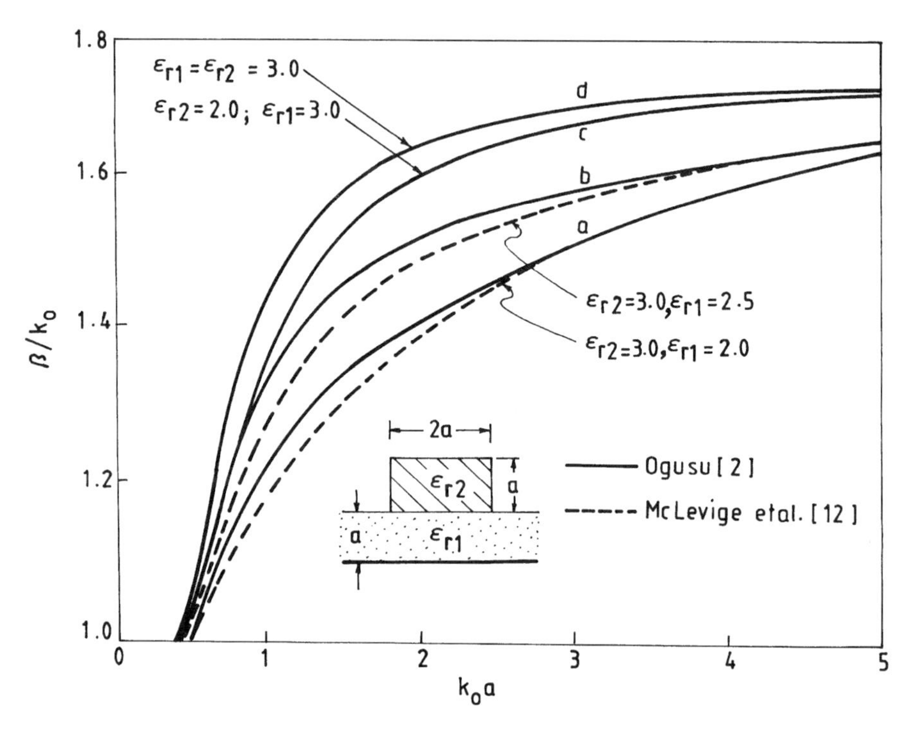

FIGURE 4.29 Dispersion characteristics for the E^y_{11} mode in insulated image guide (curves *a* and *b*), strip dielectric guide (curve *c*, structure in Fig. 4.28(a) with $h = 0$, $\varepsilon_{r2} = \varepsilon_{r1}$, $\varepsilon_{r3} = \varepsilon_{r2}$), and ridge guide (curve *d*). (From Ogusu [2]. Copyright © 1977 IEEE, reprinted with permission.)

image guide, and also a ridge guide [2]. The ridge guide is a special case of the IIG with $\varepsilon_{r1} = \varepsilon_{r2}$. Plots show the normalized propagation constant β/k_0 as a function of normalized frequency k_0a. The solid lines show the results of a rigorous numerical analysis due to Ogusu [2] and the dashed lines indicate the results of the EDC method for IIG due to McLevige et al. [12]. Comparing the curves marked *a* and *c*, we note that for a fixed frequency and guide dimensions, when the values of ε_{r1}(layer) and ε_{r2}(strip) are interchanged, the SDG offers higher propagation constant than the IIG. This is because, when the planar layer having the same thickness as the strip height ($d = b$) assumes higher dielectric constant ($\varepsilon_{r1} > \varepsilon_{r2}$), the effective dielectric constant of the guide increases, thereby increasing the propagation constant. Similarly, the propagation constant of the ridge guide having $\varepsilon_{r1} = \varepsilon_{r2} = 3$ (curve *d*) is higher than that of the SDG having $\varepsilon_{r1} = 3$ and $\varepsilon_{r2} = 2.5$ (curve *c*) because of the increase in the effective dielectric constant.

4.6.2 Inverted Strip Guide

The inverted strip guide operates on the same principle as the strip dielectric guide. With $\varepsilon_{r2} > \varepsilon_{r1}$ in the ISG (Fig. 4.28(b)), most of the energy is guided through the top dielectric layer. Hence it retains the advantages of the SDG—namely, low conductor and radiation losses. The ISG can also offer reduced dielectric loss since it eliminates the use of dielectric substrate. Furthermore, the construction of the ISG is simpler than the SDG since bonding of the guiding layer to the substrate is eliminated.

Figure 4.30 illustrates the variation in the field strength $|E_y|$ at $x = 3a$ and some value of y (equal to y_0) normalized with respect to $|E_y|$ *at* $x = 0$, $y = 0$; and also the field strength $|H_x|$ at $x = 0$, $y = 0$ on the ground plane normalized with respect to the maximum value of H_x on the y-axis ($x = 0$) as a function of the ratio b/d [20]. Plots are shown for two different values of the guiding layer thickness d. The variation in the normalized value of $|E_y|$ shows a minimum for a certain value of b/d, indicating that there is an optimum ratio b/d at which maximum energy is concentrated toward the center of the guide. Thus for good guiding ability, the value of b/d corresponding to the dip in the $|E_y|$ curve is chosen. The variation in the normalized value of $|H_x|$ at the surface of the conductor is a measure of conductor loss. It can be seen from Figure 4.30 that the larger the ratio b/d, the smaller is the conductor loss. Since the bulk of the energy propagates in the top layer, a larger value of b/d is desirable also from the point of view of reducing the

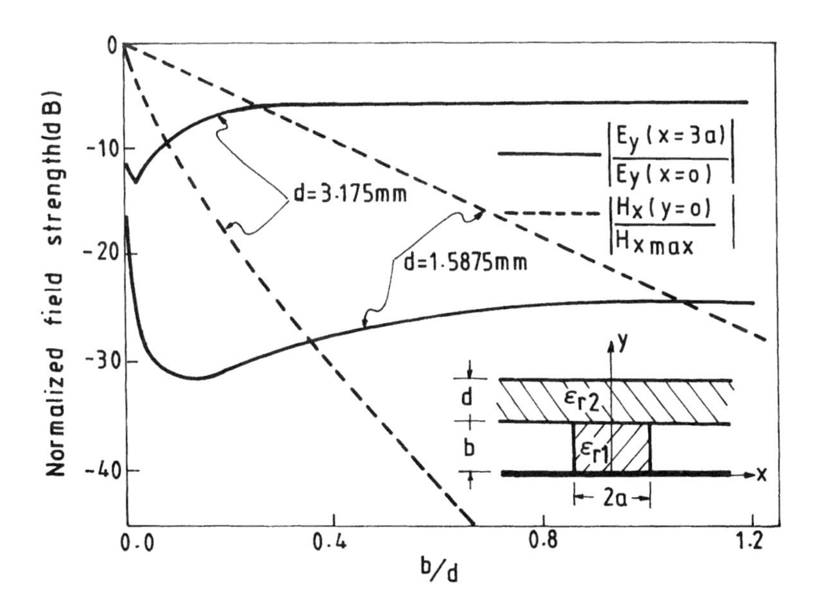

FIGURE 4.30 Computed normalized field strength as a function of b/d for E^y_{11} mode of inverted strip guide: $a = 2$ mm, $\varepsilon_{r1} = 2.1$, $\varepsilon_{r2} = 3.8$, and $f = 81.7$ GHz. (From Itoh [20]. Copyright © 1976 IEEE, reprinted with permission.)

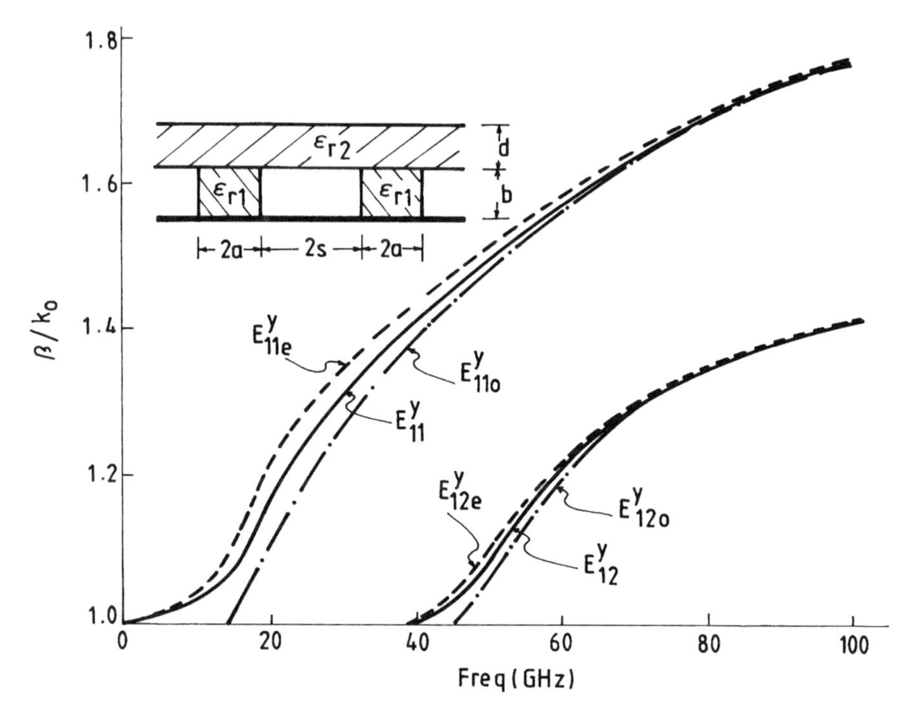

FIGURE 4.31 Dispersion characteristics of coupled and uncoupled versions of inverted strip guide: $a = 2\,\text{mm}$, $b = d = 1.5875\,\text{mm}$, $\varepsilon_{r1} = 2.1$, $\varepsilon_{r2} = 3.8$, and $s = 1\,\text{mm}$ (for coupled guide). (From Itoh [20]. Copyright © 1976 IEEE, reprinted with permission.)

effect of the ground plane on the propagation constant β [20]. This would permit relaxed dimensional tolerances for the strip. Thus the optimum choice of d/b for the design of the ISG depends on the requirements on guiding ability, permissible conductor loss, and dimensional tolerances.

Figure 4.31 shows typical dispersion characteristics of single and coupled ISGs [20]. For the coupled ISG, the plots are shown for a spacing $2s = 2\,\text{mm}$, and the even and odd modes are distinguished by the subscript e and o, respectively. As the spacing $2s$ is increased, the coupling between the adjacent guides decreases, and the even- and odd-mode dispersion curves approach that of the uncoupled guide. For a fixed spacing $2s$, as the operating frequency gets higher, the coupling between the guides decreases, with the even- and odd-mode propagation constants approaching that of the uncoupled guide.

4.7 GENERAL CLADDED IMAGE GUIDE AND SPECIAL STRUCTURES

In the strip dielectric guide and the inverted strip guide, the guiding layer extends to infinity in the lateral direction. The general cladded image guide shown in Figure 4.32(a) is a modified version of the image guide that limits the lateral

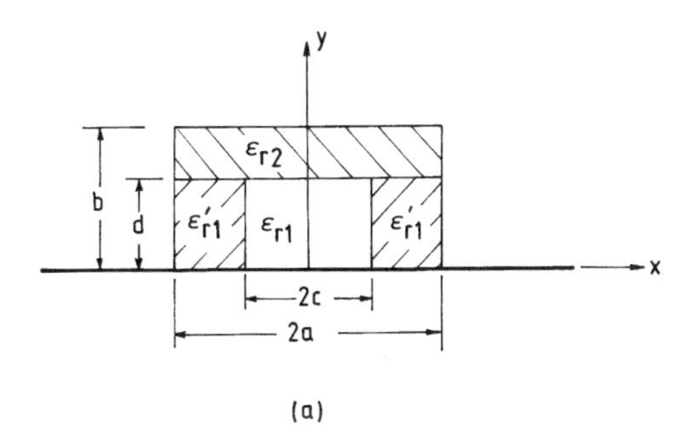

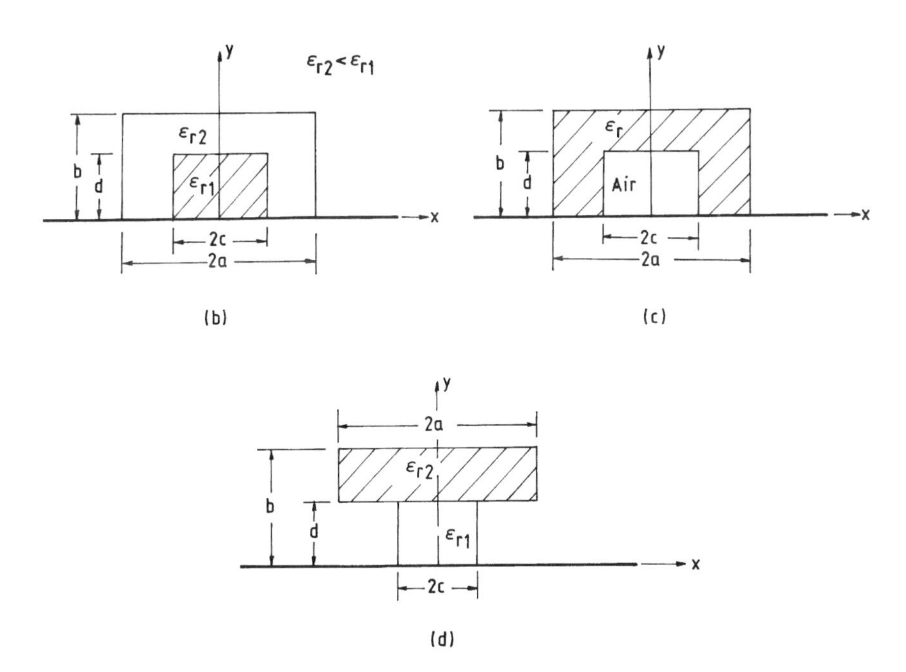

FIGURE 4.32 Cross-sectional views of (a) general cladded image guide, (b) cladded image guide, (c) hollow image guide, and (d) T-guide.

extent of the dielectric to a finite width. In the following, we consider three special cases of this structure—namely, the cladded image guide (Fig. 4.32(b)), the hollow image guide (also called the π-guide) (Fig. 4.32(c)) [2, 21, 22], and the T-guide (Fig. 4.32(d)) [21, 23].

4.7.1 Cladded Image Guide

The cladded image guide (Fig. 4.32(b)) incorporates a cladding of lower dielectric constant than the core ($\varepsilon_{r2} < \varepsilon_{r1}$), so that the field penetration in air is reduced as

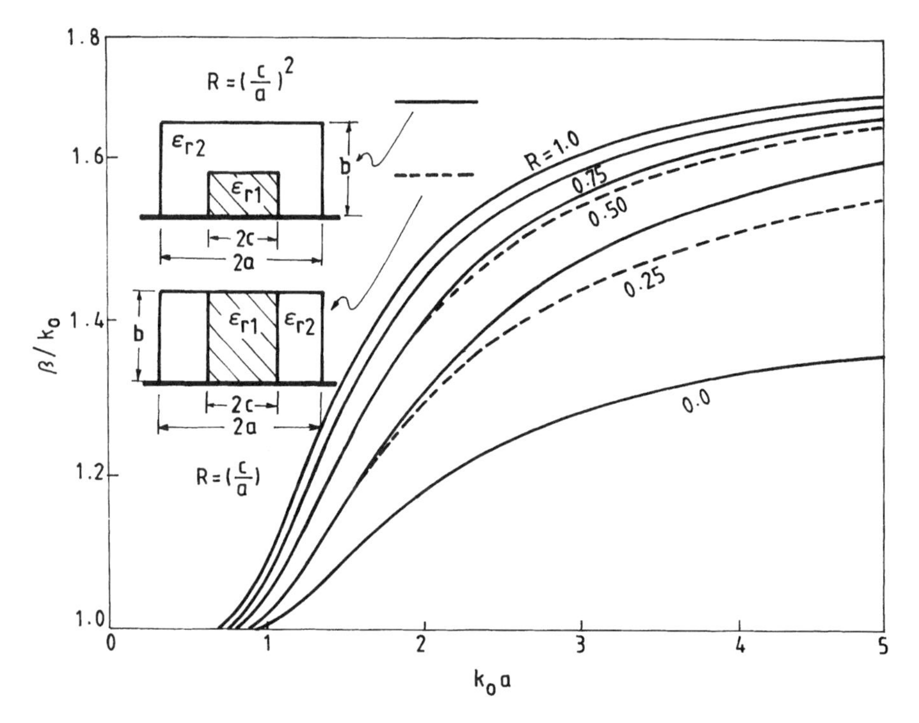

FIGURE 4.33 Dispersion characteristics of the E^y_{11} mode in a cladded image guide: $a/b = 1$, $\varepsilon_{r1} = 3$, and $\varepsilon_{r2} = 2$. (From Ogusu [2]. Copyright © 1977 IEEE, reprinted with permission.)

compared with the image guide. Ogusu [2] has pointed out that with cladding the bending loss of the image guide can be reduced and furthermore the power carried by the guide can be isolated from the environment to some extent.

Figure 4.33 illustrates the effect of cladding on the dispersion characteristics of the E^y_{11} mode in an image guide [2]. The aspect ratio of the complete guide is $a/b = 1$; the relative dielectric constant of the core is $\varepsilon_{r1} = 3$ and that of the cladding is $\varepsilon_{r2} = 2$. The plots show variation in the normalized propagation constant β/k_0 as a function of normalized frequency $k_0 a$ with $R = (c/a)^2$ as a parameter, where R is the ratio of the cross-sectional area of the core to that of the guide. The dashed curves give the dispersion characteristics of the guide for the case with no cladding on the top side of the core. The curves for $R = 0$ and $R = 1$ refer to the limiting cases of image guide with cross-section $2a \times a$ but relative dielectric constant equal to 2 and 3, respectively. The dispersion curves for the cladded guide having the same cross-section $2a \times a$ lie in-between these two limiting cases.

4.7.2 Hollow Image Guide

The hollow image guide shown in Figure 4.32(c) is basically an image guide with a hollow core. The structure may also be viewed as two image guides of height d strongly coupled by means of a dielectric overlay of thickness $b - d$. The structure can be used either as a single guide or a coupled guide. As a coupled guide, the degree of coupling may be varied by varying the dimensions c, b, or d. Other interesting features of the guide are (i) a reduced conductor loss owing to reduced field concentration near the ground plane, (ii) ease of mounting solid-state devices in the core, and (iii) easy control of field distribution and propagation constant without altering the exterior dimensions of the guide [24].

Figure 4.34 compares the field distribution for a hollow image guide with that for an image guide having the same propagation constant [24]. We observe from the graph that the hollow image guide has larger exterior dimensions than the image guide. The peak of the electric field is shifted toward the side walls of the guide. Also, the field strength outside the dielectric is higher than in the image guide. Figure 4.35 compares the dispersion characteristics of the hollow image

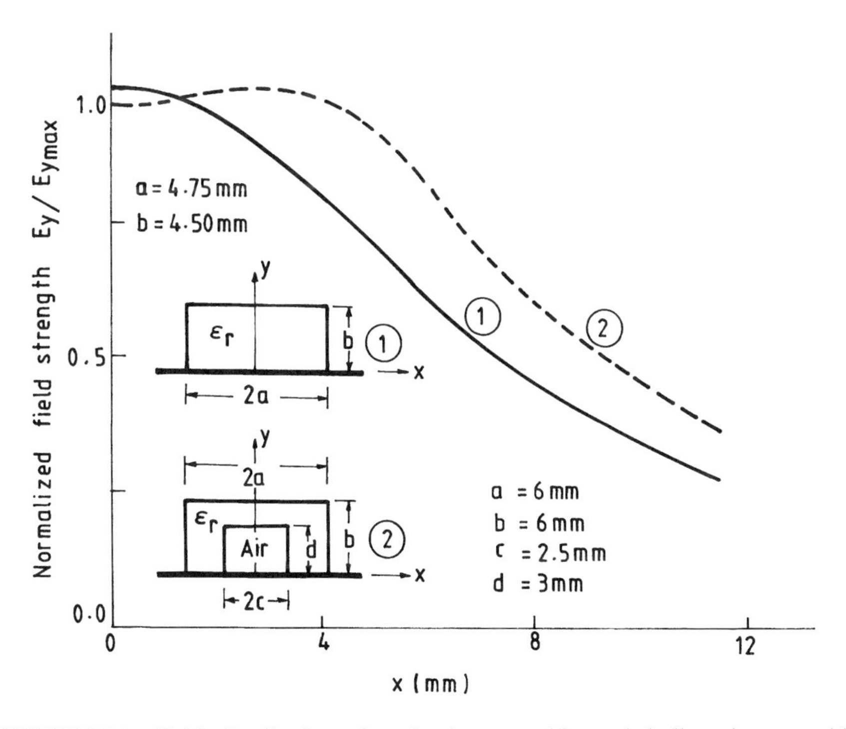

FIGURE 4.34 Field distributions for the image guide and hollow image guide with identical propagation constants: $f = 14\,\text{GHz}$, $\beta/k_0 = 1.117$, $\varepsilon_r = 2.23$, ——— hollow image guide, and ------- image guide. (From Miao and Itoh [24]. Copyright © 1982 IEEE, reprinted with permission.)

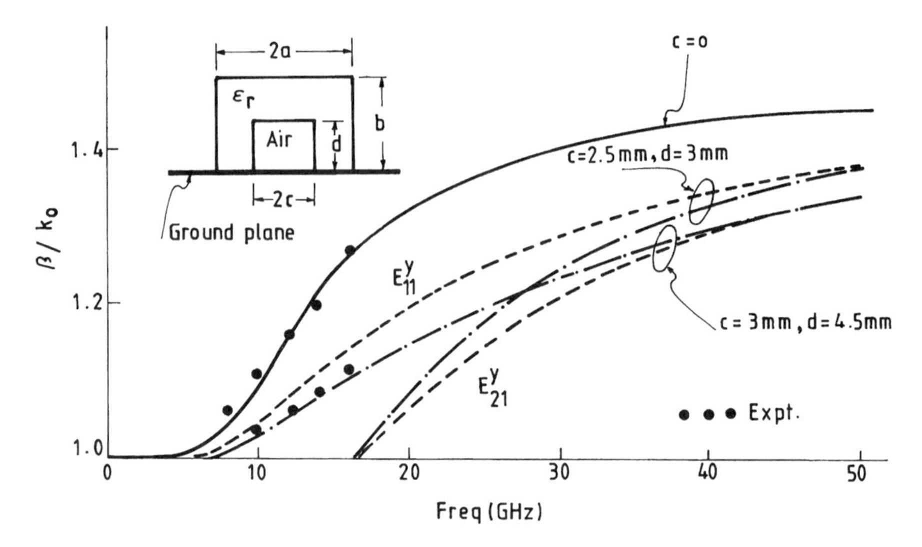

FIGURE 4.35 Normalized propagation constant β/k_0 versus frequency for the E_{11}^y and E_{21}^y modes of hollow image guide, $a = 6\,\text{mm}$, $b = 6\,\text{mm}$, $\varepsilon_r = 2.23$ (From Miao and Itoh [24]. Copyright © 1982 IEEE, reprinted with permission.)

guide for two different core sizes with that of the image guide having the same exterior dimensions [24]. The plots clearly reveal that the hollow image guide offers considerable flexibility for adjusting the propagation constant by simply altering the core size. Rodriguez and Prieto [22] have investigated the coupling characteristics of the two coupled hollow image guides (referred as π-guides) and have shown that by suitably choosing the dimensions and separation distance between the guides, the difference between the even- and odd-mode propagation constants can be made nearly constant with frequency over a wide band. The broadband coupling characteristics of the coupled guide and its applications to directional couplers are discussed in Chapter 11.

4.7.3 T-Guide

The T-guide shown in Figure 4.32(d) can be considered as an inverted strip guide with the top dielectric restricted to a finite width. The conductor loss in the guide is reduced by choosing $\varepsilon_{r2} > \varepsilon_{r1}$. Furthermore, two coupled T-guides offer the flexibility of adjusting the coupling by simply trimming the width of the top dielectric layer.

4.8 U-GUIDE

The U-guide shown in Figure 4.36 consists of two image guides with an intervening dielectric spacer of lower dielectric constant ($\varepsilon_r' < \varepsilon_r$) and lower height ($d < b$)

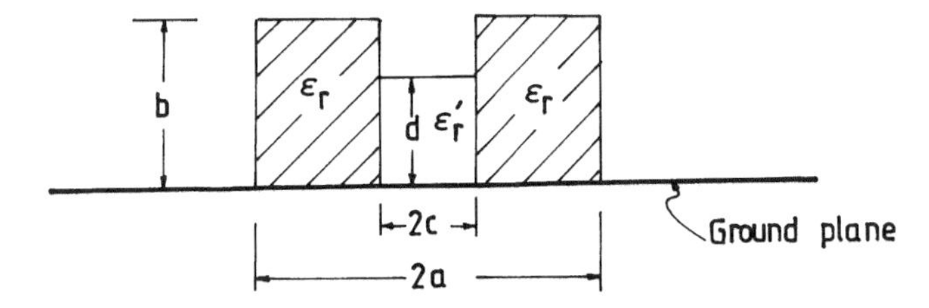

FIGURE 4.36 Cross-sectional view of U-guide.

than the guiding strip. Varying the spacer height offers a convenient technique of controlling the coupling between the two guides. Kim et al. [25] have reported the dispersion and coupling characteristics of the U-guide for the special case $\varepsilon_r = \varepsilon_r'$ (referred to as directly connected image guide in [25]). Figure 4.37 illustrates the effect of introducing a dielectric spacer on the dispersion and coupling characteristics of a coupled image guide [25]. The variation in the normalized propagation constants and the 3-dB coupling length as a function of the frequency for a coupled image guide is shown in Figure 4.37(a) and that for the same guide with a spacer having $\varepsilon_r' = \varepsilon_r$ is shown in Figure 4.37(b). The 3-dB coupling length is given by $\pi/2(\beta_{oe} - \beta_{oo})$, where β_{oe} and β_{oo} are the propagation constants of the dominant even mode (E^y_{11e} mode) and the dominant odd mode (E^y_{11o} mode), respectively. Comparison of Figure 4.37(a) with 4.37(b) shows that the introduction of a dielectric spacer lowers the cutoff frequencies of the dominant as well as the higher order modes of the coupled image guide. Another effect is to make the difference between the even- and odd-mode propagation constants fairly constant over a wider frequency range. Consequently, as shown in Figure 4.37(b), the coupling versus frequency curve is much flatter for the U-guide. Kim et al. [25] have shown that, for single-mode operation, there exists an optimum value of d for which the U-guide offers extremely good broadband coupling performance. This is an important feature for the design of broadband directional couplers.

PROBLEMS

4.1 Explain the following:

(a) For the dominant E^y_{11} mode, an image guide ideally has no low-frequency cutoff, whereas an insular image guide has a low-frequency cutoff.

(b) In an image guide, the conductor loss for the E^x_{11} mode is less than that for the E^y_{11} mode.

(c) The conductor loss in an image guide is higher than that in a corresponding insular image guide having the same aspect ratio a/b and ε_r for the dielectric strip.

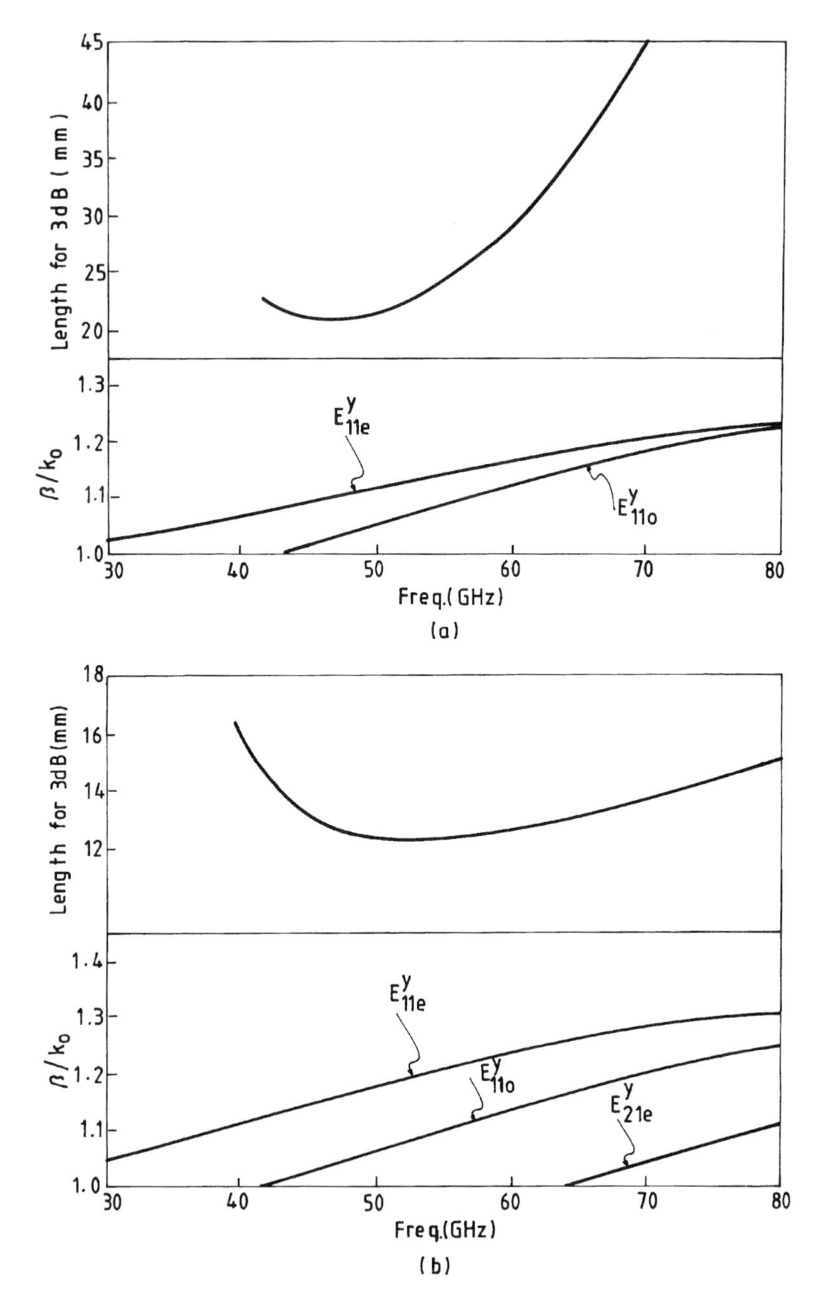

FIGURE 4.37 Dispersion curves and coupling length for 3-dB directional couplers for (a) conventional coupled image guide: $a = 3\,\text{mm}$, $b = 1.5\,\text{mm}$, $c = 1.5\,\text{mm}$, $d = 0$, $\varepsilon_r = 2.1$; and (b) U-guide: $a = 3\,\text{mm}$, $b = 1.5\,\text{mm}$, $c = 1\,\text{mm}$, $d = 1.25\,\text{mm}$, $\varepsilon_r = \varepsilon_r' = 2.1$ (Refer to Fig. 4.36.) (From Kim et al. [25]. Copyright © 1984 IEEE, reprinted with permission.)

4.2 Compare the propagation characteristics of the image guide shown in Figures 4.5–4.7 and explain why the EDC method shows increasingly better agreement with the mode-matching method for larger values of a/b and smaller values of ε_r. What is the approximate range of a/b and ε_r over which the EDC method can be recommended for evaluation of propagation constants for practical applications?

4.3 Distinguish between "modal bandwidth," "disperion bandwidth," and "waveguiding bandwidth" for an image guide. What are the optimum values of aspect ratio a/b and ε_r for achieving maximum practical bandwidth? [*Note*: First determine modal bandwidth, disperion bandwidth, and waveguiding bandwidth. Calculate the practical bandwidth by determining the overlapping frequency range.]

4.4 Derive an expression for the conductor attenuation coefficient α_c in the trapped image guide (Fig. 4.38) for the E^y_{11} mode.

(a) Assume $a = 4\,\text{mm}$, $b = 2\,\text{mm}$, $\varepsilon_r = 2.25$, and $f = 35\,\text{GHz}$. Plot α_c as a function of s.

(b) Assume $a = 4\,\text{mm}$, $b = 2\,\text{mm}$, $\varepsilon_r = 2.25$, and $s = 1\,\text{mm}$. Plot α_c as a function of frequency.

(c) Explain the variation of α_c as a function of s and frequency.

4.5 Repeat Problem 4.4 for the E^x_{11} mode. Compare the attenuation characteristics obtained for the E^x_{11} mode with those of the E^y_{11} mode.

4.6 A trapped image guide approaches an image guide as the side walls are moved away.

(a) How do the propagation constant, cutoff frequency, modal bandwidth, and wave impedance vary as the side walls are moved away with all other parameters held fixed?

(b) What are the advantages of the trapped image guide over the image guide?

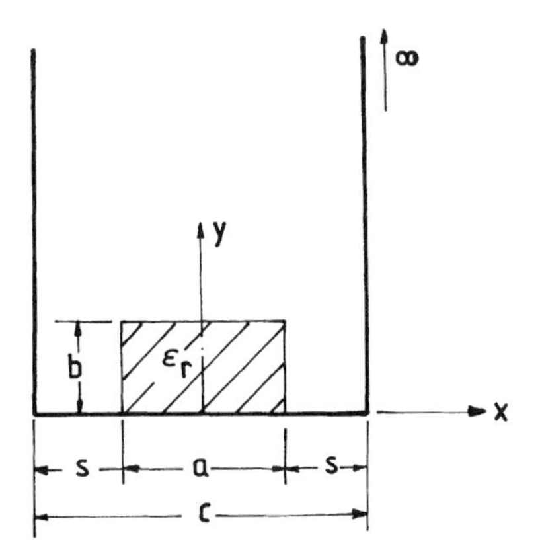

FIGURE 4.38 Trapped image guide.

4.7 Explain the following:

(a) In a coupled image guide, the even mode ideally has no low-frequency cutoff, whereas the odd mode has a low frequency cutoff.

(b) In a trapped coupled image guide, as the spacing s is increased from a small value, the odd mode cutoff frequency decreases first and then increases again, whereas in the case of the coupled image guide, the odd mode cutoff frequency decreases continuously.

(c) Introducing an insular layer of lower dielectric constant in a trapped coupled image guide increases the cutoff frequencies of both the even and odd modes.

4.8 **(a)** What is the mechanism of propagation in a strip dielectric guide (SDG)? How is it different from that in an insular image guide?

(b) How does an inverted strip guide (ISG) differ from a SDG? What are its advantages over the SDG? How do you determine the optimum parameters of an ISG from the point of view of good guidability?

4.9 Enumerate the advantages of (a) the cladded image guide and (b) the hollow image guide over an image guide having the same outer cross-section $a \times b$ for the dielectric.

4.10 Consider a coupled image guide with a dielectric overlay as shown in Figure 4.39.

(a) Using the EDC technique, obtain the even- and odd-mode propagation constants of the dominant E^y_{11} mode (E^y_{11e}, E^y_{11o}) and the first higher order E^y_{21} mode (E^y_{21e}, E^y_{21o}).

(b) Assume $a = 3$ mm, $b = 1.5$ mm, $c = 1$ mm, and $\varepsilon_{r1} = \varepsilon_{r2} = 2.1$. Plot $\beta_{11e,o}$ and $\beta_{21e,o}$ as a function of frequency for $d = 0$, 0.5 mm, and 1.25 mm.

(c) Plot $\pi/2(\beta_{11e} - \beta_{11o})$ as a function of frequency for the above three values of d.

4.11 Compare the characteristics generated in Problem 4.10 for $d = 1.25$ mm with those in Figure 4.37 for the U-guide.

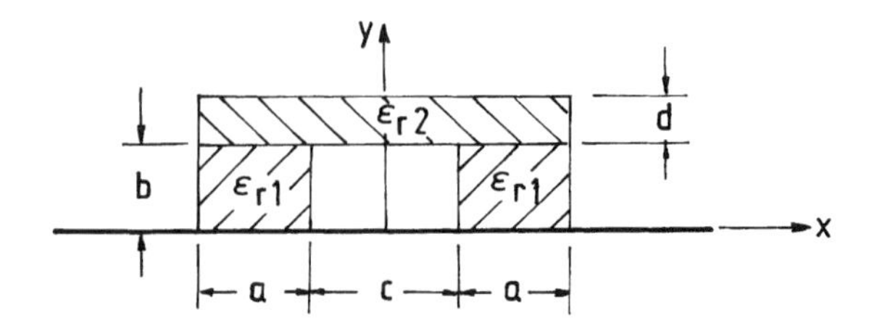

FIGURE 4.39 Coupled image guide with dielectric overlay.

REFERENCES

1. K. Solbach and I. Wolff, The electromagnetic fields and the phase constants of dielectric image line. *IEEE Trans. Microwave Theory Tech.*, **MTT-26**, 266–274, Apr. 1978.
2. K. Ogusu, Numerical analysis of the rectangular dielectric waveguide and its modifications. *IEEE Trans. Microwave Theory Tech.*, **MTT-25**, 874–885, Nov. 1977.
3. N. Deo and R. Mittra, A technique for analyzing planar dielectric waveguides for millimetre wave integrated circuits. *Arch. Elek. Ubertragung*, **AEU-37**, 236–244, July/Aug. 1983.
4. R. M. Knox and P. P. Toulios, Integrated circuits for the millimetre through optical frequency range. In: *Proceedings of the Symposium on Submillimetre Waves*, Polytechnic Press of Polytechnic Institute of Brooklyn, Brooklyn, NY, 1970, pp. 497–516.
5. E. A. J. Marcatili, Dielectric rectangular waveguide and directional coupler for integrated optics. *Bell Syst. Tech. J.*, **48**, 2071–2102, Sept. 1969.
6. R. J. Collier and R. D. Birch, The bandwidth of image guide. *IEEE Trans. Microwave Theory Tech.*, **MTT-28**, 932–935, Aug. 1980.
7. S. T. Peng and A. A. Oliner, Guidance and leakage properties of a class of open dielectric waveguides, Part I–Mathematical formulations. *IEEE Trans. Microwave Theory Tech.*, **MTT-29**, 843–855, Sept. 1981.
8. A. A. Oliner et al., Guidance and leakage properties of a class of open dielectric waveguides, Part II—New physical effects. *IEEE Trans. Microwave Theory Tech.*, **MTT-29**, 855–869, Sept. 1981.
9. P. P. Toulios and R. M. Knox, Rectangular dielectric image lines for millimetre integrated circuits. Weston Electronics Show and Convention, Los Angeles (CA), Aug. 1970, pp. 1–10.
10. S. Shindo and T. Itanami, Low-loss rectangular dielectric image line for millimetre wave integrated circuits. *IEEE Trans. Microwave Theory Tech.*, **MTT-26**, 747–751, Oct. 1978.
11. R. M. Knox, Dielectric waveguide microwave integrated circuits—an overview. *IEEE Trans. Microwave Theory Tech.*, **MTT-24**, 806–814, Nov. 1976.
12. W. V. McLevige, T. Itoh, and R. Mittra, New waveguiding structures for millimetre-wave and optical integrated circuits. *IEEE Trans. Microwave Theory Tech.*, **MTT-23**, 788–794, Oct. 1975.
13. M. Koshiba and M. Suzuki, Microwave network analyses of dielectric waveguides for millimetre waves made of dielectric strip and planar dielectric layer. *Trans. IECE Japan*, **E63**, 344–350, May 1980.
14. M. Koshiba, H. Ishii, and M. Suzuki, Improved equivalent network analysis of a dielectric waveguide placed on a ground plane. *Trans. IECE Japan*, **E65**, 572–578, Oct. 1982.
15. T. Itoh and B. Adelseck, Trapped image guide for millimetre wave circuits. *IEEE Trans. Microwave Theory Tech.*, **MTT-28**, 1433–1436, Dec. 1980.
16. W. B. Zhou and T. Itoh, Analysis of trapped image guides using effective dielectric constant and surface impedances. *IEEE Trans. Microwave Theory Tech.*, **MTT-30**, 2163–2166, Dec. 1982.

17. A. K. Tiwari, B. Bhat, and R. P. Singh, Generalized coupled dielectric waveguide and its variants for millimetre-wave applications. *IEEE Trans. Microwave Theory Tech.*, **MTT-34**, 869–875, Aug. 1986.
18. A. K. Tiwari and B. Bhat, Analysis of trapped single and coupled image guides using the mode-matching technique. *Arch. Elek. Ubertragung.*, **AEU-38**, 181–185, Mar. 1984.
19. A. K. Tiwari, *Investigations on Dielectric Integrated Guides for Millimetre Wave Applications*, Ph.D Thesis, Department of Electrical Engineering, Indian Institute of Technology, New Delhi, 1984.
20. T. Itoh, Inverted strip dielectric waveguide for millimetre-wave integrated circuits. *IEEE Trans. Microwave Theory Tech.*, **MTT-24**, 821–827, Nov. 1976.
21. E. Rubio, J. L. Garcia, and A. Prieto, Estudio de las guias dielectricas T y π. *An. Fis. B*, **78(1)**, 22–26, Jan.–Apr. 1982.
22. J. Rodriguez and A. Prieto, Wide-band directional couplers in dielectric waveguide. *IEEE Trans. Microwave Theory Tech.*, **MTT-35**, 681–687, Aug. 1987.
23. J. Rodriguez, E. Rubio, and A. Prieto, Estudio teorico experimental del factor de acoplo entre diversas guias dielectricas. *An. Fis. B*, **80**, 44–52, 1984.
24. J. F. Miao and T. Itoh, Hollow image guide and overlayed image guide coupler. *IEEE Trans. Microwave Theory Tech.*, **MTT-30**, 1826–1831, Nov. 1982.
25. D. I. Kim et al., Directly connected image guide 3 dB couplers with very flat couplings. *IEEE Trans. Microwave Theory Tech.*, **MTT-32**, 621–627, June 1984.

CHAPTER FIVE

Semiconductor and Planar Optical Guides

5.1 INTRODUCTION

Semiconductor guides made of silicon (Si) and gallium arsenide (GaAs) have received considerable attention for application in millimeter wave integrated circuits [1–4] and antennas [5–9]. The use of semiconductor material introduces two important features. First, the propagation parameters of the guide can be controlled dynamically through optical illumination [10–16]. Second, the guide enables monolithic integration of solid-state devices, which can also be activated through optical illumination. Investigations on direct optical control of a variety of solid-state devices such as IMPATT, MESFET, transferred electron devices, and HEMT have been reported [4, 17–20]. Important advantages of optical control of semiconductor guides and devices are near perfect isolation between the controlling and controlled devices, picosecond precision, and fast response [10]. The advent of picosecond and femtosecond lasers has increased the potential for applying optical control techniques for achieving ultrafast switching capability in switches and phase shifters and phase control in amplifiers and finer frequency control in oscillators.

While a wide variety of semiconductor guide structures similar to the dielectric integrated guides are possible, investigations reported thus far have been confined mostly to the following four geometries: the bare rectangular semiconductor guide [1, 2], the image guide [13], the ridge guide [14], and the H-guide [15, 16]. The effective dielectric constant (EDC) method has been used for analyzing the propagation characteristics of these guides with and without the optically induced plasma [1, 10, 13–16]. In this chapter, we present the salient features of optical control phenomena on millimeter wave propagation in semiconductor guides. The dispersion and attenuation characteristics of the above four guides under optical control are discussed.

Another topic that is covered in this chapter is planar optical guides. These guides are similar in geometry to the planar dielectric integrated guides except that the guide is normally supported on a thick dielectric substrate. The ground plane, which is the cause of excessive loss at optical frequencies, is eliminated. The characteristics of some selected guide structures—namely, the optical strip guide [21], rib guide [21–23], buried strip guide [21], inverted strip guide (called the strip–slab guide [24]), and multilayer thin-film polarization-maintaining guide [25]—are covered.

5.2 PRINCIPLE OF OPTICAL CONTROL IN SEMICONDUCTOR GUIDES

When a semiconductor guide is illuminated by laser radiation having energy greater than the bandgap energy of the semiconductor, photons are absorbed, creating photocarriers (electron–hole pairs) in the material. A thin plasma layer is thus created near the surface of the waveguide. The presence of plasma modifies the dielectric and conductive properties in the plasma-occupied region. The change in the dielectric constant is known to follow the predictions of the Drude–Lorentz theory [10, 20]. The expression for the complex relative dielectric constant is given by [10]

$$\varepsilon_{rp} = \varepsilon_{rs} - \sum_{i=e,h} \left[\frac{\omega_{pi}^2 (1 + j\nu_i/\omega)}{(\omega^2 + \nu_i^2)} \right] = \varepsilon'_{rp} - j\varepsilon''_{rp} \tag{5.1}$$

where ε_{rs} is the relative dielectric constant of the host lattice including the contribution from the bound charges; ν_i is the collision frequency, which is related to the relaxation time τ_i of the charge carriers by $\nu_i = 1/\tau_i$; ω_{pi} is the plasma frequency; and the symbols e and h refer to electrons and holes, respectively. The plasma frequency is given by

$$\omega_{pi}^2 = e^2 n_i / m_i^* \varepsilon_0, \quad i = e, h \tag{5.2}$$

where n_e is the electron density in the conduction band, n_h is the hole density in the valence band, m_i^* is the effective mass of electron/hole, e is the electronic charge, and ε_0 is the free-space permittivity.

Figure 5.1 illustrates typical variation in the complex relative permittivity of a plasma region of silicon as a function of plasma density at millimeter wave frequencies [13]. Typical parameters for silicon are $\varepsilon_r = 11.8$, $m_e^* = 0.259\, m_o$, $m_h^* = 0.38\, m_o$, $\nu_e = 4.52 \times 10^{12}\, \text{sec}^{-1}$, and $\nu_h = 7.71 \times 10^{12}\, \text{sec}^{-1}$. The graph shows that the real part of the permittivity ε'_{rp} remains rather unchanged for a change in the plasma density at low density levels of the order of 10^{15}–$10^{16}\, \text{cm}^{-3}$ and has negligible dependence on frequency. On the other hand, the imaginary part of ε''_{rp} (equivalently, conductivity) rises rapidly with an increase in the plasma density.

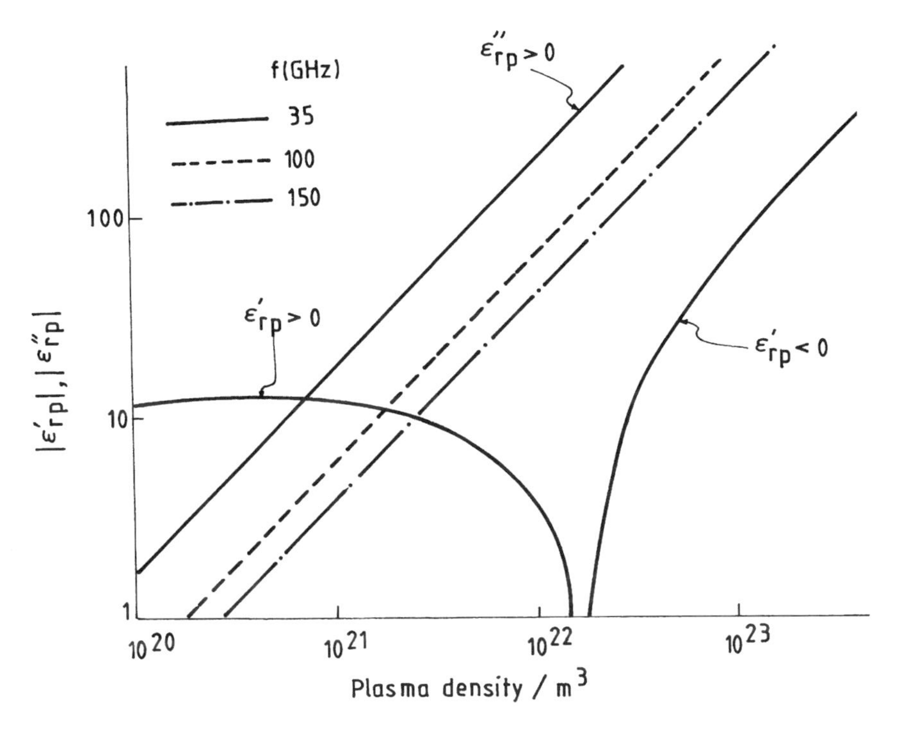

FIGURE 5.1 Variation of real and imaginary parts of complex permittivity ($\varepsilon'_{rp} - j\varepsilon''_{rp}$) of plasma region in silicon as a function of plasma density. (From Alphones [13], reprinted with permission.)

5.3 OPTICALLY CONTROLLED RECTANGULAR SEMICONDUCTOR GUIDE

The schematic of an optically controlled rectangular semiconductor guide is shown in Figure 5.2 [10]. The guide can be made of Si, GaAs, silicon-on-sapphire, or silicon-on-alumina. The cross-sectional dimensions (a and b) are chosen such that the guide supports a E^y_{11} mode. The ends of the guide are tapered so as to provide an efficient transition to empty rectangular metal waveguides at the input and output ports. The guide is illuminated on its broad wall by a laser beam. The absorption of photon energy creates a plasma layer at the air–semiconductor interface as illustrated in the figure. The initial depth of plasma injection is controlled by appropriately choosing the wavelength of optical radiation and the absorption properties of the semiconductor material. For example, the absorption coefficient of GaAs for a photon energy of 1.55 eV is approximately $10^4\ \text{cm}^{-1}$. The plasma layer thickness corresponding to $1/e$ absorption depth for light is less than 1 μm. Varying the optical illumination intensity changes the plasma density, thereby changing the refractive index of the plasma-occupied region.

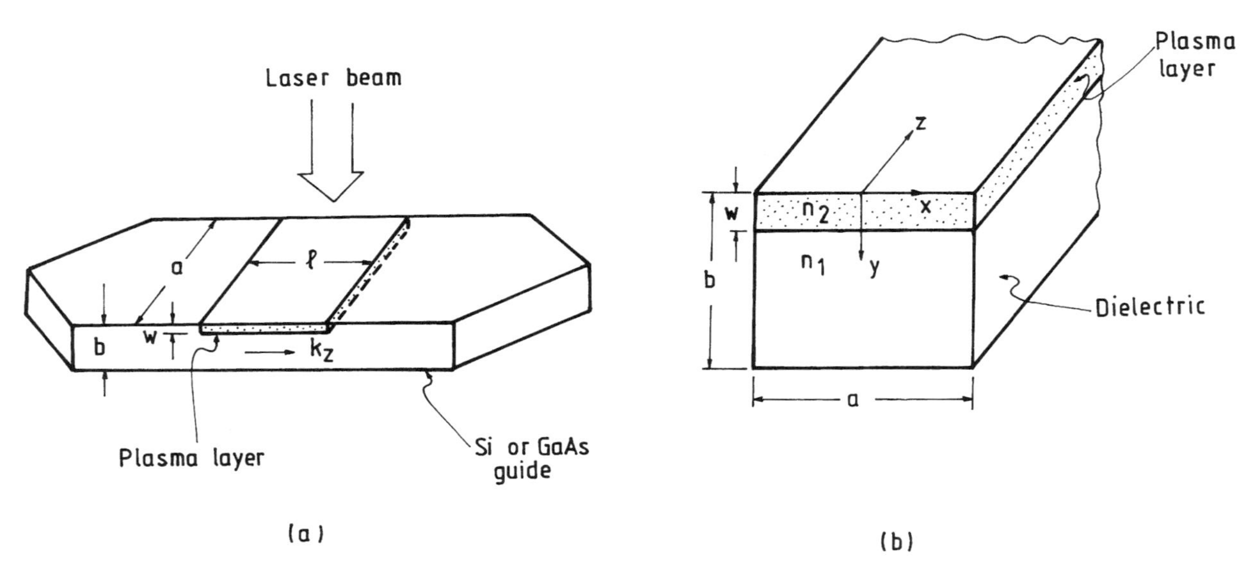

FIGURE 5.2 (a) Schematic of optically controlled rectangular semiconductor guide. (After Lee et al. [10].) (b) Dielectric–plasma waveguide model. (After Vaucher et al. [2].)

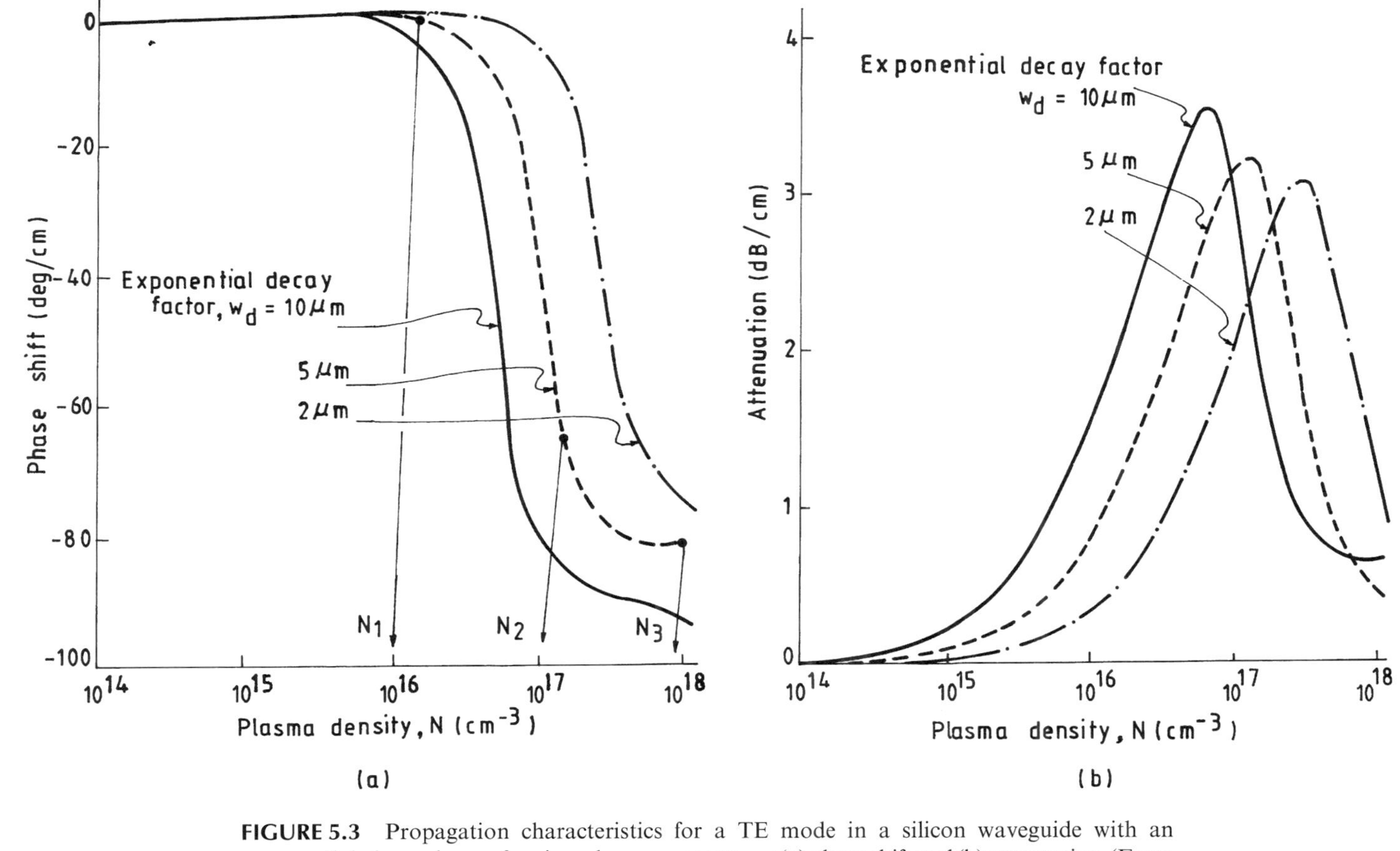

FIGURE 5.3 Propagation characteristics for a TE mode in a silicon waveguide with an exponential plasma layer of various decay constants w_d: (a) phase shift and (b) attenuation. (From Butler et al. [26]. Copyright © 1986 IEEE, reprinted with permission.)

The phase shift and attenuation characteristics of a optically illuminated guide have been studied by modeling the device as a dielectric–plasma waveguide [2, 10, 26]. Both TE- and TM-mode solutions have been reported. The model analyzed by Vaucher et al. [2] is a rectangular dielectric guide with a thin surface plasma layer of uniform carrier concentration as shown in Figure 5.2(b). In the model considered by Butler et al. [26], the guiding structure is assumed to extend to infinity in the lateral direction (x-direction in Fig. 5.2(b)) and the plasma layer is assumed to have a nonuniform carrier concentration corresponding to an exponentially absorbed optical beam. Figure 5.3 shows the theoretical phase shift and attenuation characteristics for a TE mode in a Si guide with an exponential plasma layer [26]. The plasma density profile is represented as $N(y) = N \exp(-y/w_d)$, where N is the plasma density at the surface ($y = 0$). Plots are shown as a function of N for three different values of the decay factor w_d. The frequency is 94 GHz. For low values of plasma density for which the plasma frequency is less than the operating frequency ($\omega_p < \omega$), the plasma layer does not have much effect on the propagation. As the plasma density increases with an increase in the optical beam intensity, the interaction between the wave and the plasma becomes stronger. In the density range where the plasma frequency crosses the operating frequency, the attenuation is large and the phase shift increases rapidly. As the density is increased further, the attenuation decreases, but the phase shift through the device saturates. At sufficiently large densities for which the skin depth of millimeter waves approaches the thickness of the plasma layer, the plasma region behaves like a metallic layer. The dielectric waveguide then behaves essentially like an image guide. It can be seen from Figure 5.3 that the guide can offer large phase shift with low attenuation. The optically induced phase shift $\Delta\phi$ over a section of guide of length ℓ is given by $(\beta' - \beta)\ell$, where β' and β denote the z-directed phase constants in the guide with and without the plasma, respectively. Lee et al. [10] have reported measured phase shifts as high as 300°/cm in a Si waveguide having a cross-section of $2.4 \times 1.0\,\text{mm}^2$ and 1400°/cm in a GaAs waveguide having a cross-section of $2.4 \times 0.5\,\text{mm}^2$ at a frequency of 94 GHz.

5.4 OPTICALLY CONTROLLED SILICON IMAGE, RIDGE, AND H-GUIDES

5.4.1 Silicon Image Guide

Figure 5.4 shows the cross-section of an optically controlled silicon image guide investigated by Alphones and Tsutsumi [14]. The guide has a metallic ground plane of width w equal to the width of the guide on one surface and the opposite surface is optically illuminated by LEDs. The induced plasma is assumed to be uniform in the x-direction and extends up to a depth t_p. The relative permittivity (ε_{rs}) of the plasma-unoccupied region is real, whereas that of the plasma layer (ε_{rp}) is complex as given in Eq. (5.1). The complex dielectric constant of the plasma

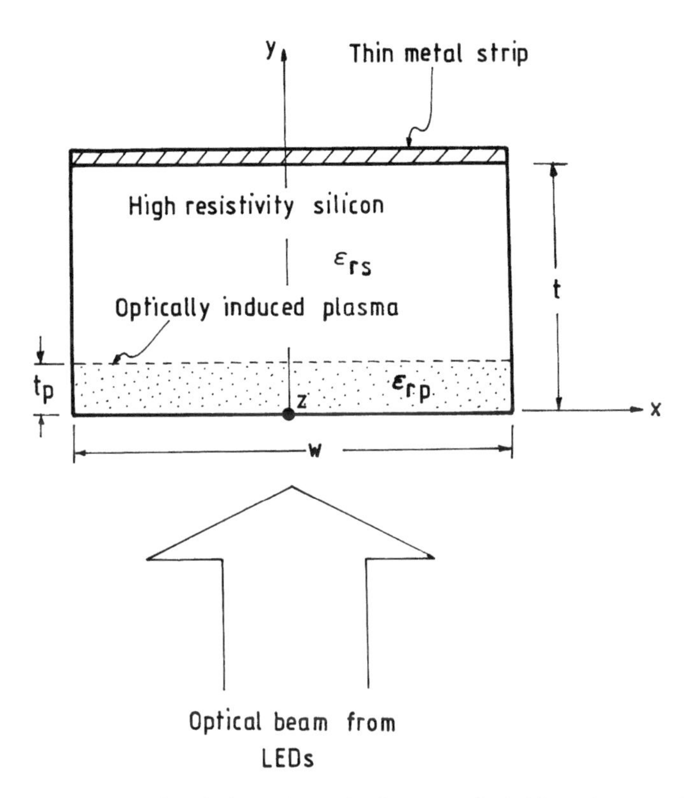

FIGURE 5.4 Cross-sectional view of a optically controlled silicon image guide. (From Alphones [13], reprinted with permission.)

layer results in a complex propagation constant $k'_z = \beta - j\alpha$. Alphones and Tsutsumi [14] have used the EDC method for deriving the dispersion relation and have obtained the complex propagation constant k'_z by applying Muller's algorithm with deflation procedure. The computed results on phase shift and attenuation as reported by Alphones [13] are reproduced in Figures 5.5–5.8. The phase shift per unit length of the guide is given by

$$\Delta\beta = \text{Re}(k'_z - k_z) \tag{5.3}$$

where k'_z and k_z denote the propagation constants of the guide with and without the plasma, respectively. Figure 5.5(a) shows the variation in $\Delta\beta$ as a function of plasma density for different thicknesses of the plasma layer in a silicon guide. The guide parameters are $\varepsilon_{rs} = 11$, $w = 300\,\mu\text{m}$, and $t = 400\,\mu\text{m}$; and the frequency is 100 GHz. For the same set of parameters, Figure 5.5(b) shows the variation in attenuation constant α as a function of plasma density. These plots show that a thinner plasma layer (smaller t_p) yields smaller phase shift as well as attenuation. Wave–plasma interaction is quite strong for values of plasma density around

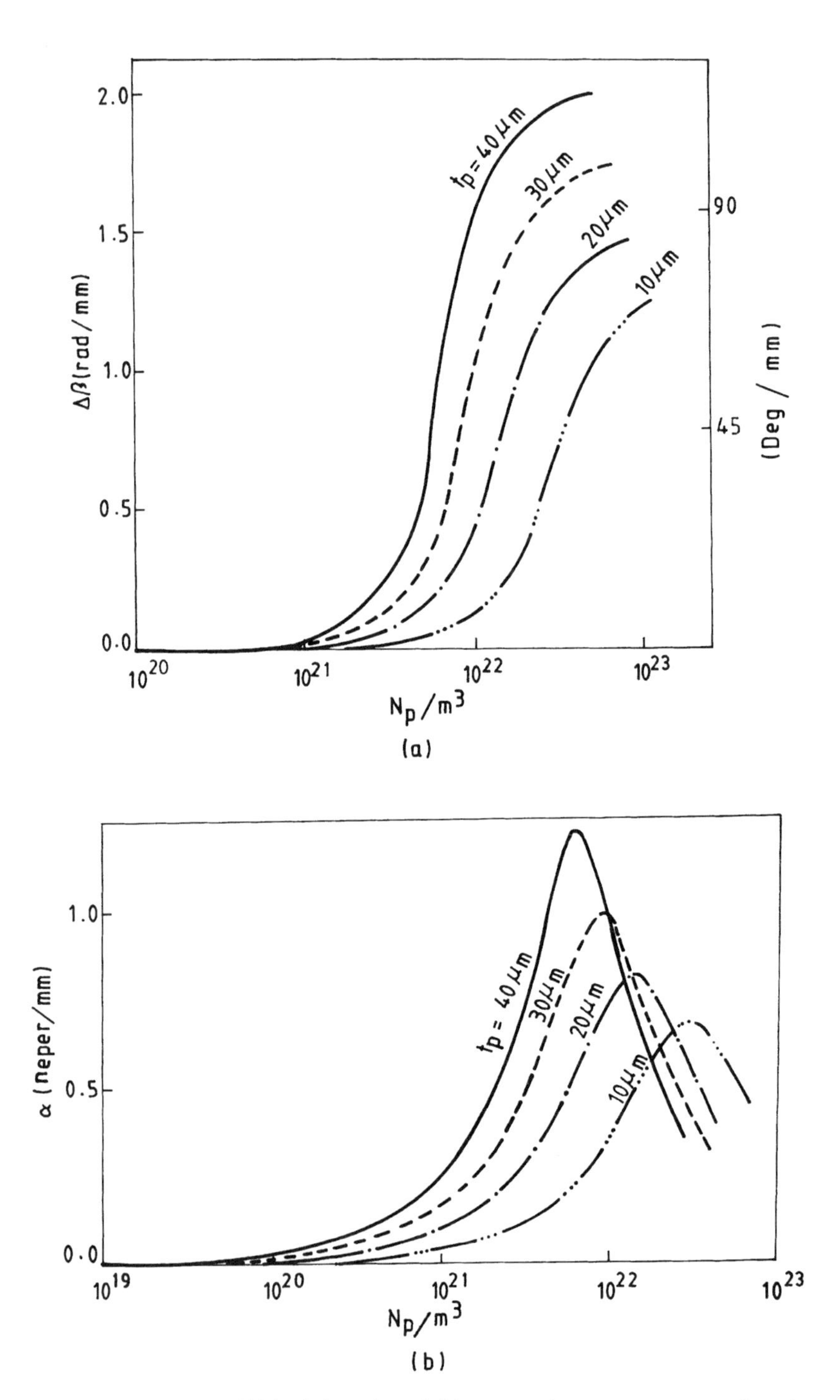

FIGURE 5.5 (a) Phase shift/unit length and (b) attenuation constant α as a function of plasma density in optically controlled silicon image guide (structure in Fig. 5.4). $\varepsilon_{rs} = 11.8$, $w = 300\,\mu m$, $t = 400\,\mu m$, and $f = 100\,GHz$. (From Alphones [13], reprinted with permission.)

$10^{22}\,\mathrm{m}^{-3}$. The phase shift undergoes a rapid change and the attenuation is also large. The attenuation goes through a maximum and then decreases with a further increase in the plasma density. For a fixed plasma density–thickness product ($N_p t_p = 2 \times 10^{17}/\mathrm{m}^2$), Figure 5.6 shows the variation in $\Delta\beta$ and α as a function of width w of the guide at 145 GHz. As expected, with an increase in w, higher order modes—namely E_{21}^y, E_{31}^y, and E_{41}^y—are supported by the guide. We note from Figure 5.6 that the higher order modes offer larger phase shift as well as higher attenuation than the lower order modes [13].

5.4.2 Silicon Ridge Guide

Besides the silicon image guide, another structure that has been investigated extensively for its optically controlled characteristics is the silicon ridge guide [14]. (This is referred to as a rib guide in [14].) The cross-sectional view of the silicon ridge guide with optically induced plasma layer is shown as an inset in Figure 5.7(a). The ridge guide is made of a high-resistivity Si substrate of height h, with a ridge of height t and width w, and is backed by a ground plane. The plasma layer of thickness t_p is assumed to be uniform over the ridge width w.

Figures 5.7 and 5.8 depict the propagation characteristics of the guide computed using the EDC method [14]. The guide parameters are $\varepsilon_{rs} = 11.8$, $w = 2$ mm, and $h = t = 400\,\mu$m. In Figure 5.7(a) is shown the dispersion characteristics of the ridge guide as a function of frequency with the product $N_p t_p$ as parameter for the dominant E_{11}^y mode and the first higher order mode E_{21}^y. The corresponding variation in the attenuation constant α is shown in Figure 5.7(b). The plots show that an increase in the plasma density–thickness product $N_p t_p$, at low values—that is, from $10^{16}\,\mathrm{m}^{-2}$ to $10^{17}\,\mathrm{m}^{-2}$—has a small effect on the phase constant β, whereas the increase in α is considerable. As the value of $N_p t_p$ is increased above $10^{17}\,\mathrm{m}^{-2}$, both $\Delta\beta$ and α show a sharp increase as illustrated in Figure 5.8. As in the case of the silicon image guide, the attenuation reaches a maximum and then decreases with a further increase in the value of $N_p t_p$.

Figure 5.9 shows schematics of test sections of a silicon ridge guide fed by rectangular metal waveguide via a mode launcher [14]. The mode launcher consists of a metal horn loaded partially with a tapered Teflon slab. The arrangement for uniform optical illumination of the guide surface consists of a linear array of LEDs as shown in Figure 5.9(a). An alternate arrangement using a xenon arc lamp is shown in Figure 5.9(b). In this case, the optical radiation is focused by a lens and is guided by an array of optical fibers to the top surface of the ridge guide.

5.4.3 Silicon H-Guide

The basic configuration of the H-guide is shown in Figure 1.1(f). Satomura and Tsutsumi [15, 16] have investigated the propagation characteristics of millimeter waves in the silicon H-guide containing the optically induced plasma region, both theoretically and experimentally. The experimental setup consists of a silicon

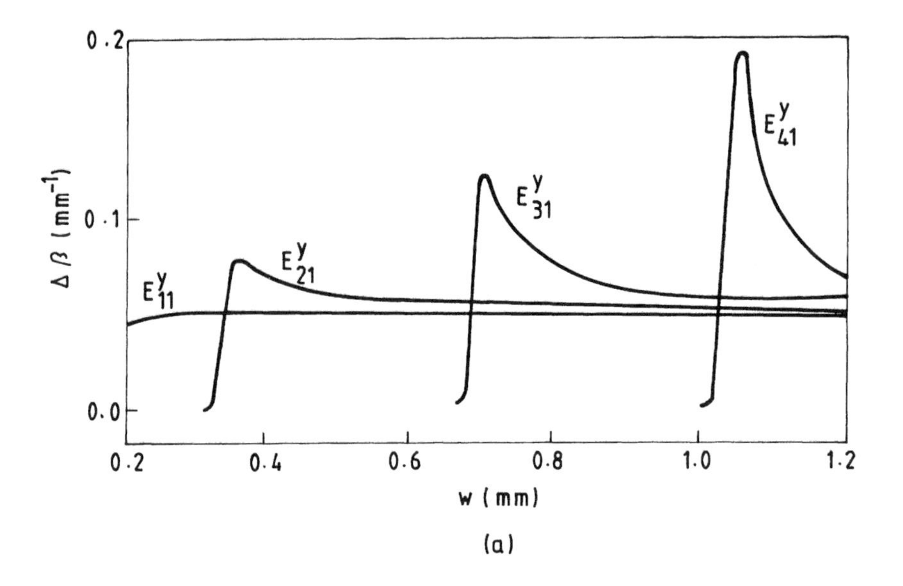

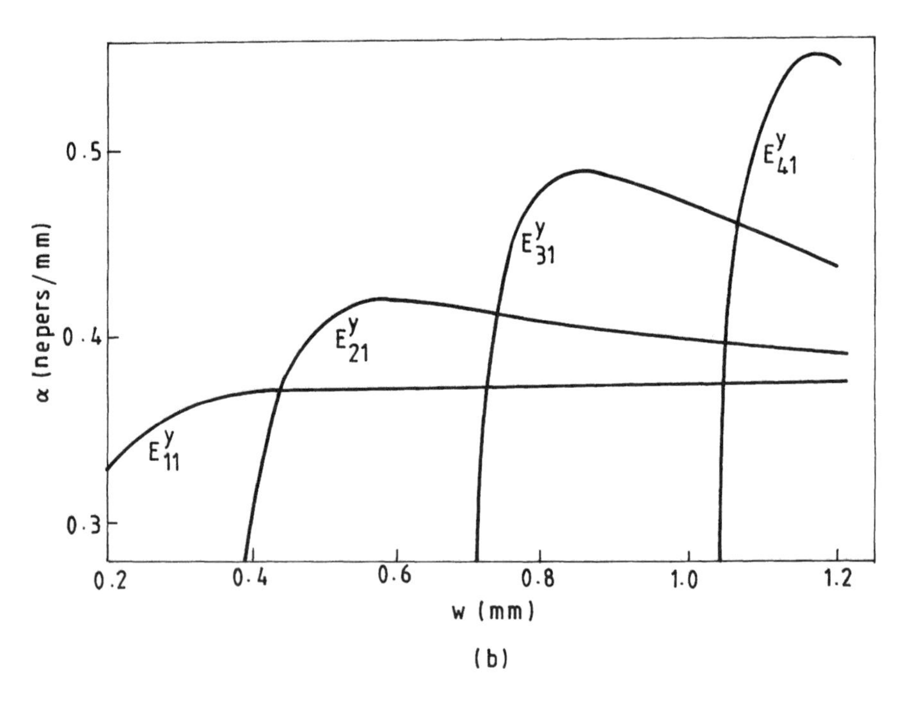

FIGURE 5.6 Variation of (a) phase shift/unit length and (b) attenuation constant α as a function of guide width w in an optically controlled silicon image guide (structure in Fig. 5.4). $\varepsilon_{rs} = 11.8$, $w = 300\,\mu\text{m}$, $t = 400\,\mu\text{m}$, $f = 145\,\text{GHz}$, and $N_p t_p = 2 \times 10^{17}\,\text{m}^{-2}$. (From Alphones [13], reprinted with permission.)

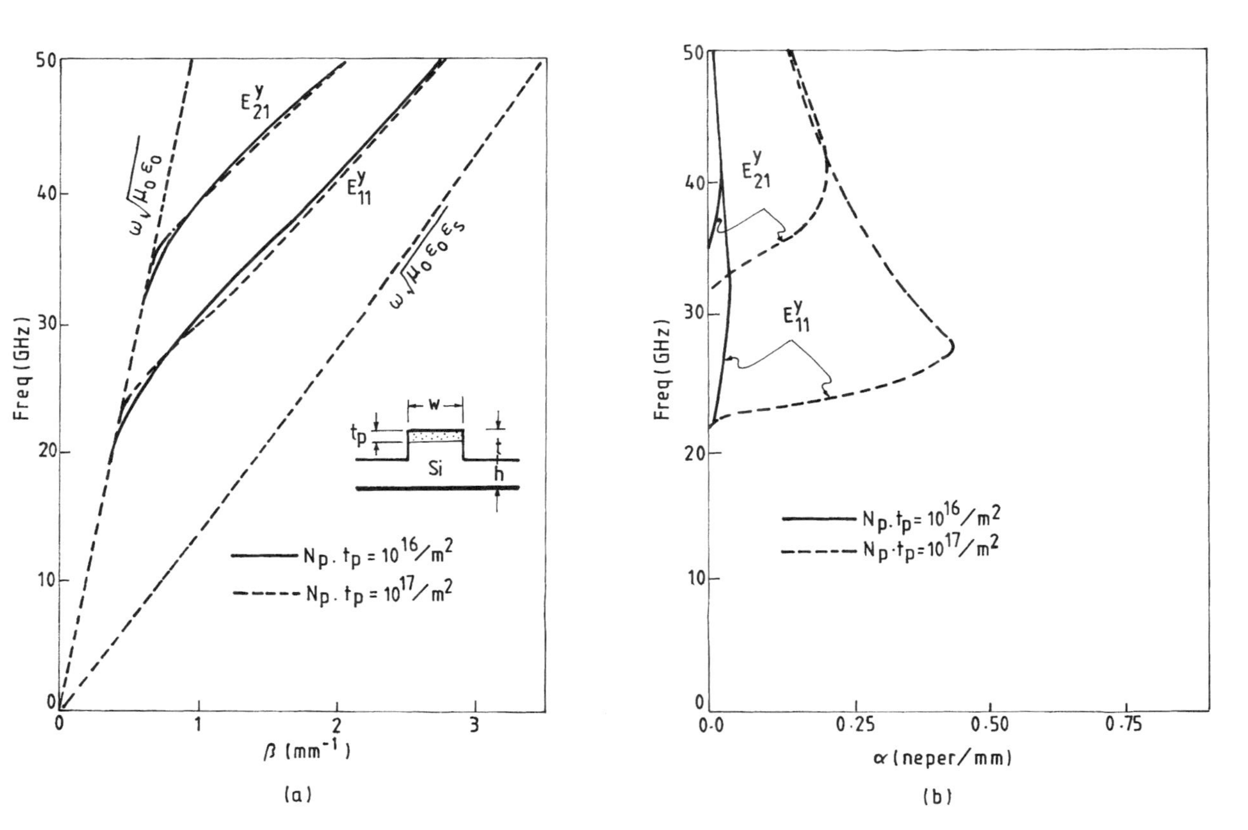

FIGURE 5.7 Variation of (a) phase constant β and (b) attenuation constant α as a function of frequency in a silicon ridge guide with optically induced plasma: $w = 2$ mm, $h = t = 400$ μm, and $t_p = 10$ μm. (From Alphones and Tsutsumi [14], reprinted with permission of Communication Engineers, Denshi Joho Tsushin Gakkai, Tokyo, Japan.)

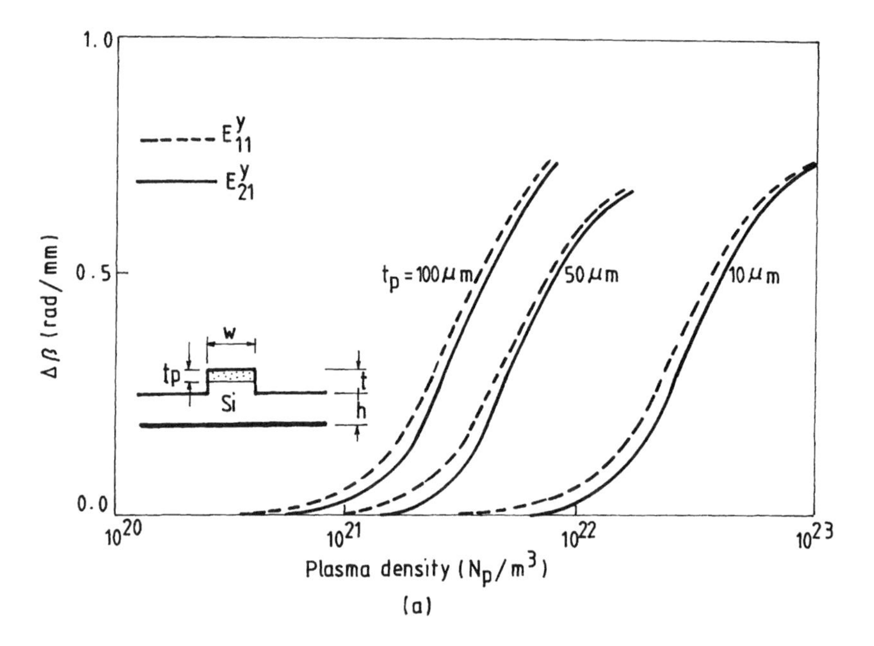

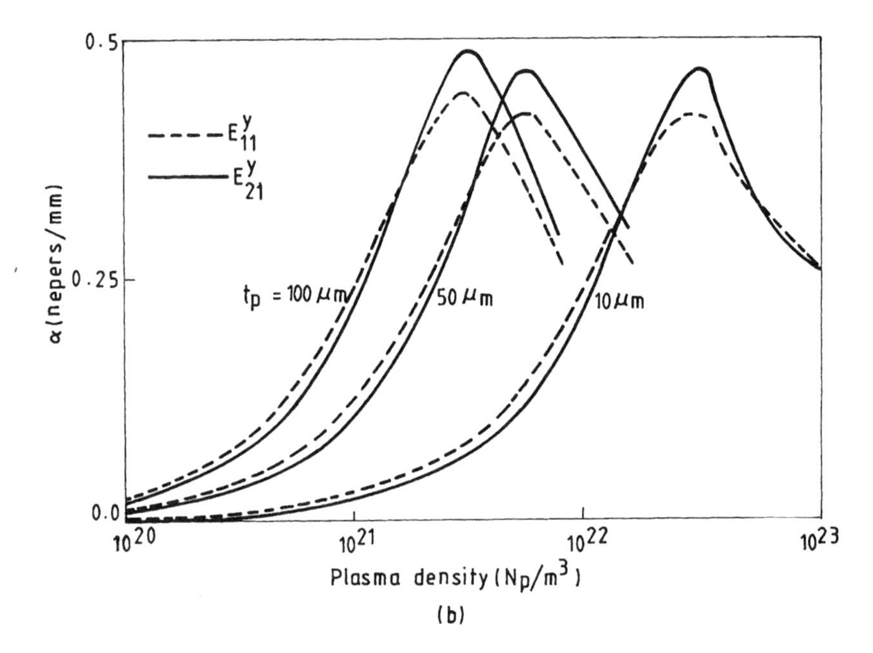

FIGURE 5.8 Variation of (a) phase shift/unit length and (b) attenuation constant α as a function of plasma density in a silicon ridge guide with optically induced plasma: $w = 2\,\text{mm}$, $h = t = 400\,\mu\text{m}$, and $f = 40\,\text{GHz}$. (From Alphones and Tsutsumi [14], reprinted with permission of Communication Engineers, Denshi Joho Tsushin Gakkai, Tokyo, Japan.)

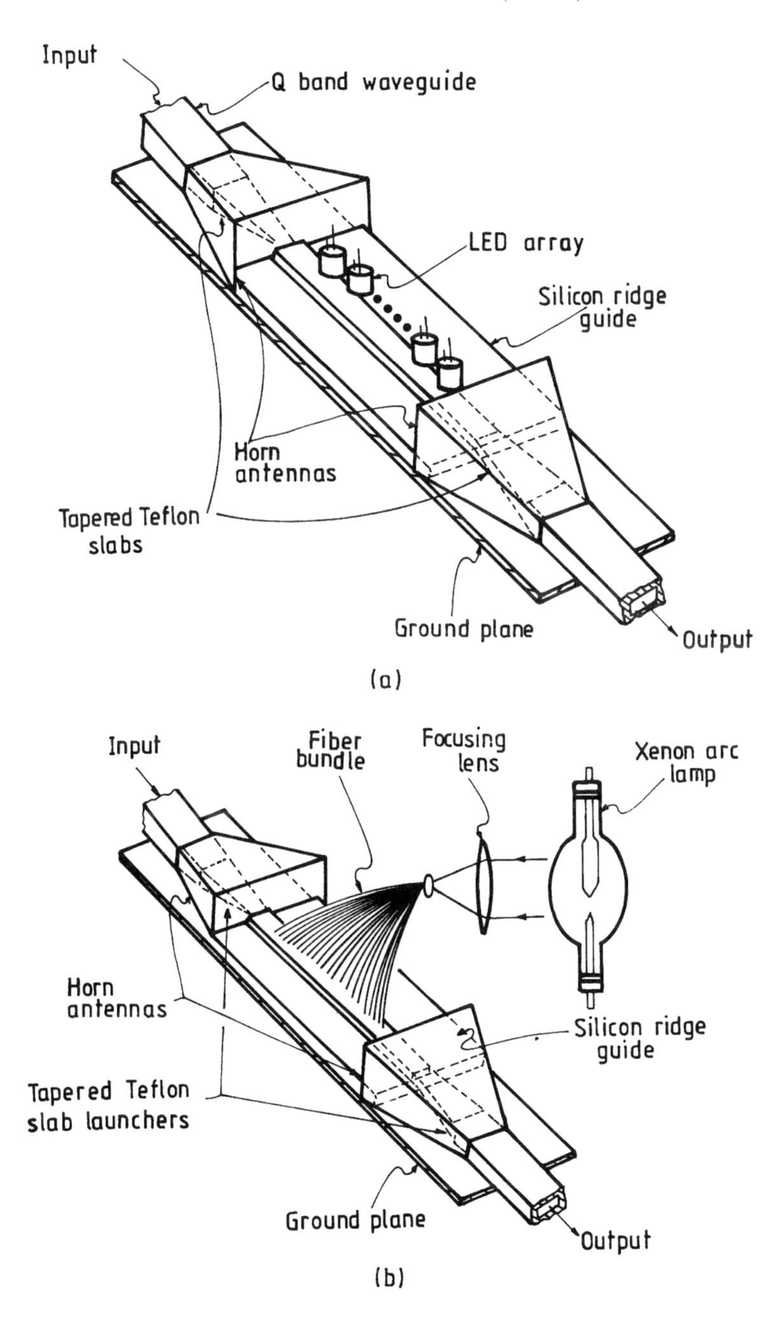

FIGURE 5.9 Schematic diagram of experimental arrangement of silicon ridge guide with (a) LED illumination and (b) xenon lamp illumination. (From Alphones and Tsutsumi [14], reprinted with permission of Communication Engineers, Denshi Joho Tsushin Gakkai, Tokyo, Japan.)

chip of length 35 mm and thickness 400 μm inserted and cascaded between two Teflon slabs of thickness 3.2 mm and width 6.5 mm (same as plate separation a) in the H-guide. A light guide tube connected with the xenon arc lamp is utilized to induce photoconductivity optically in the section of silicon guide, thereby causing change of its dielectric properties. Satomura and Tsutsumi [15, 16] have also reported the dispersion characteristics of the dominant LSE mode for the silicon H-guide having induced plasma region for different plasma density at the Q-band. An important observation reported is the transformation of the incident wave into a leaky wave at the specified frequency for large values of plasma density.

5.5 PLANAR OPTICAL GUIDES

Dielectric guides for optical integrated circuits are commonly supported on thick dielectric substrate [21–24, 27]. Figure 5.10 shows some examples of such structures in the literature [21–24, 27]. The strip or the film or the strip–

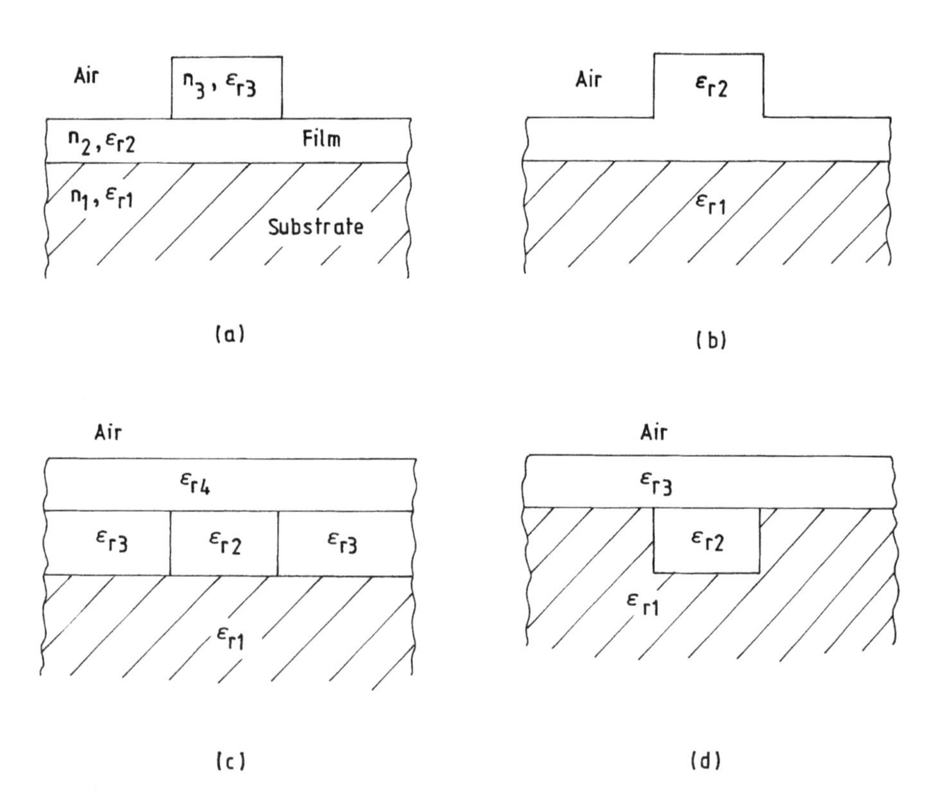

FIGURE 5.10 Examples of planar optical dielectric guides: (a) strip dielectric guide, (b) rib guide, (c) buried strip guide, and (d) embedded inverted strip guide; $n_i = \sqrt{\varepsilon_{ri}}$, $i = 1$ to 4.

film combination serves as the main guiding vehicle and the bottom substrate serves only to modify the modal behavior and the field distribution. As in the case of millimeter wave dielectric integrated guides, the propagating modes are hybrid. Therefore the same mode classification in terms of E^y_{mn} and E^x_{mn} modes can be used to describe the fields. Also, the many methods of analysis used for millimeter wave dielectric integrated guides are applicable for planar optical guides. In particular, the effective dielectric constant (EDC) method [24, 28, 29] has been used for obtaining approximate results; and the mode-matching method [30–32], the equivalent network analysis method [33, 34], and the finite-difference analysis [35] have been applied for obtaining more accurate solutions.

As an illustration of the meaning of the modal indices m and n in the E^y_{mn} and E^x_{mn} modes of planar optical guides, we show in Figure 5.11 the approximate field distribution reported by Marcatili [27] for an ideal slab-coupled guide. It is assumed that the main transverse components vary sinusoidally within the core, decay exponentially away from the core in the slab region, and vanish at the edges of the guide. The main transverse fields of a mode have m field extrema in the x-direction and n field extrema in the y-direction. For a guided mode, the slab thickness d is always smaller than the half-period of the mode in the core along the y-direction [27].

5.5.1 Optical Strip Dielectric Guide

The optical strip dielectric guide shown in Figure 5.10(a) is the optical counterpart of the millimeter wave strip dielectric guide/insular image guide with the ground plane replaced by a thick dielectric substrate. If the refractive index n_3 of the loading strip film is lower than that of the planar film layer ($n_3 < n_2$, $n_2 = \sqrt{\varepsilon_{r2}}$, $n_3 = \sqrt{\varepsilon_{r3}}$) with $n_1 < n_2$, the optical wave would be confined to the film guide just below the loading strip. The thickness of the loading strip serves as a controlling parameter for varying the propagation constant of the guide [29]. The effect of the loading strip can be enhanced by choosing its refractive index higher than that of the lower film layer ($n_3 > n_2$). Figure 5.12 illustrates the effect of varying the film layer thickness d and also the strip thickness b on the normalized propagation constant β/k_0 of the guide [28]. The parameters chosen are $n_1 = 2.14$, $n_2 = 2.142$, and $\lambda_0 = 1.05\,\mu\text{m}$. The solid and dashed curves represent the E^x_{11} and E^x_{21} modes, respectively, for the case of film loading with a higher refractive index ($n_3 = 2.2$, $n_3 > n_2$). The dotted curves show the variation for the case of low-index film loading ($n_3 = 2.14$, $n_3 < n_2$) and strip thickness $b = \infty$. It is observed that the propagation constant is far more sensitive to variation in strip thickness b in the case of high-index film loading than low-index film loading. Furthermore, this variation is shown to be maximum when the change of b is around the cutoff thickness required for supporting the optical wave. For the loading film formed directly on the substrate having a refractive index $n_1 = 2.142$, the cutoff thickness b is reported to be $0.4368\,\mu\text{m}$ [28].

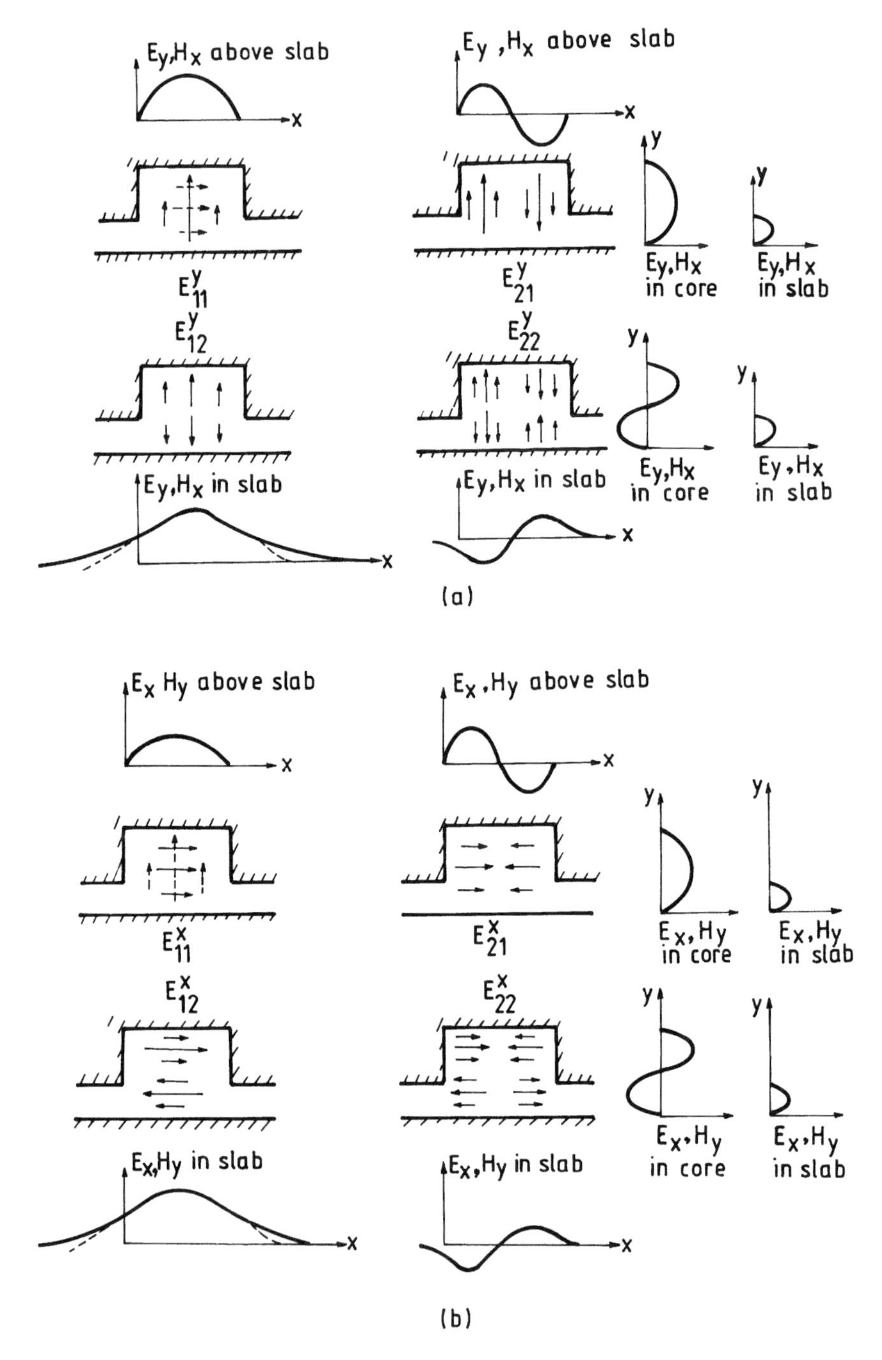

FIGURE 5.11 Typical field distribution in a dielectric rib guide in free space: (a) E^y_{mn} modes and (b) E^x_{mn} modes. (From Marcatili [27]. Copyright © 1974 AT & T. All rights reserved. Reprinted with permission.)

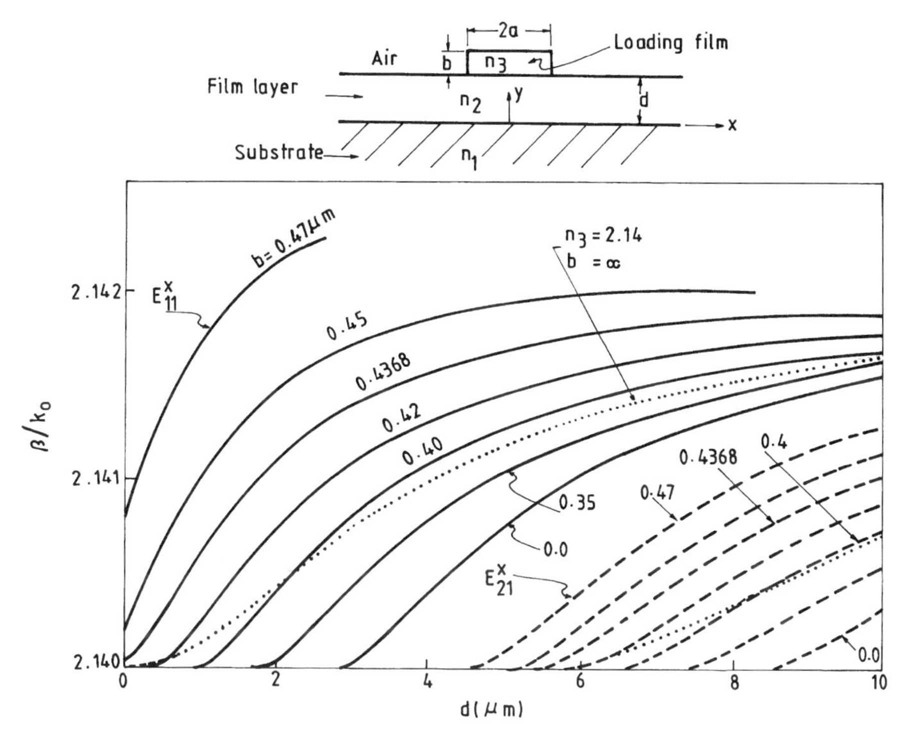

FIGURE 5.12 Normalized propagation constant β/k_0 as a function of film layer thickness for the strip dielectric guide: $\lambda_0 = 1.05\,\mu\text{m}$, $n_1 = \sqrt{\varepsilon_{r1}} = 2.14$, and $n_2 = \sqrt{\varepsilon_{r2}} = 2.142$. Solid ($E^x_{11}$) and broken ($E^x_{21}$) curves are for $n_3 = 2.2$ with b as parameter. Dotted curves represent E^x_{11} and E^x_{21} modes for $n_3 = \sqrt{\varepsilon_{r3}} = 2.14$ and $b = \infty$. (From Uchida [28], *Applied Optics*, Vol. 15, pp. 179–182, 1976, reprinted with permission of Optical Society of America, Inc.)

5.5.2 Optical Rib Guide

Of the various planar optical guides, the rib guide (Fig. 5.10(b)) is the most extensively studied [22, 23, 31–34, 36]. It is a special case of the strip dielectric guide when the loading strip has the same refractive index as the film layer ($n_3 = n_2$, $n_1 < n_2$). The guide has considerably less scattering loss from the wall imperfections and hence has the advantage of permitting relaxed dimensional tolerance on edge smoothness during fabrication.

Figures 5.13 and 5.14 show typical dispersion characteristics of a wide strip rib guide for the E^x_{mn} and E^y_{mn} modes, respectively [34]. The parameters of the rib guide are $n_1 = \sqrt{\varepsilon_{r1}} = 1.69$, $n_2 = \sqrt{\varepsilon_{r2}} = 1.742$, $a = 3d$, and $b = d$. For a rib guide with a large width and small discontinuity height, the first higher order TE-to-y mode would be E^x_{21} and the first higher order TM-to-y mode would be E^y_{21}. For the dimensions considered in Figures 5.13 and 5.14, the first four modes that are

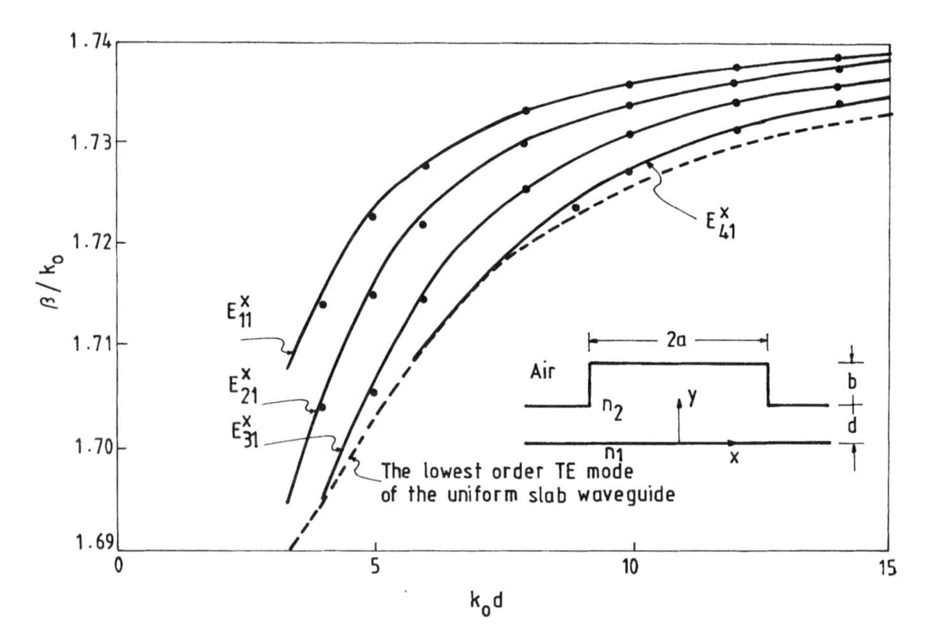

FIGURE 5.13 Dispersion characteristics for the E^x_{mn} modes of a optical rib guide: $n_1 = \sqrt{\varepsilon_{r1}} = 1.69$, $n_2 = \sqrt{\varepsilon_{r2}} = 1.742$, $a = 3d$, $b = d$, ——— equivalent network analysis method [34], and ●——●——● mode-matching method [31]. (From Koshiba and Suzuki [34], *Radio Science*, Vol. 17, pp. 99–107, 1982, Copyright © by the American Geophysical Union.)

supported correspond to $n = 1$ and $m = 1$ to 4. The dotted curve shows the dispersion characteristics for a uniform slab guide of thickness d.

Oliner et al. [22] have discussed the phenomenon of leakage and resonance effects in a rib guide. Figure 5.15 shows curves of attenuation constant α due to the leakage of TE surface waves as a function of the strip width $2a$ for the lowest TM-like guided mode [22]. The effect of the continuous spectrum and the material losses are neglected. It can be seen that the attenuation due to leakage is lower for a smaller rib height b. This trend is expected since a smaller rib results in a smaller discontinuity. The curves also show sharp dips for specific values of the rib width. Oliner et al. [22] have pointed out that this resonance effect occurs because of the mode-converted surface wave bouncing back and forth above cutoff in the inside region. That is, a TM surface wave incident on the strip edge converts in part into a TE surface wave outside and a TE surface wave inside the strip. If the TE surface wave outside the strip is above cutoff, then the guided mode leaks energy. The mode-converted TE surface wave bounces back and forth above cutoff in the inside region. For certain values of $2a$ the leakage is canceled, thereby creating a resonance effect. The condition for resonances is given by [22]

$$(k_x^{\mathrm{TE}})\,2a = 2m\pi, \quad m = 1, 2, \ldots \tag{5.4}$$

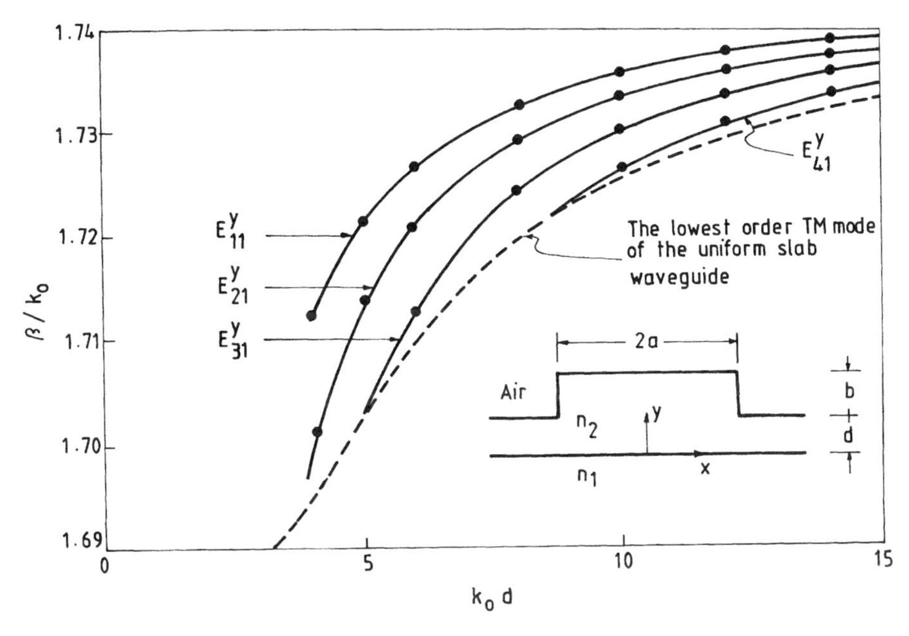

FIGURE 5.14 Dispersion characteristics for the E^{y}_{mn} modes of a optical rib guide: $n_1 = \sqrt{\varepsilon_{r1}} = 1.69$, $n_2 = \sqrt{\varepsilon_{r2}} = 1.742$, $a = 3d$, $b = d$, ——— equivalent network analysis method [34], and ●——●——● mode-matching method [31]. (From Koshiba and Suzuki [34], *Radio Science*, Vol. 17, pp. 99–107, 1982, Copyright © by the American Geophysical Union.)

where k_x^{TE} is the transverse wavenumber of the TE surface wave in the strip region. The resonance peaks decrease with an increase in the strip width. Thus larger strip width is desirable from the point of view of keeping the leakage small. Larger strip width is also reported to reduce the leakage angle (with respect to the slab surface) of the leaking surface wave [22].

5.5.3 Embedded Inverted Strip Guide

The buried strip guide is a general form of optical guide [21, 35] from which several useful planar optical guides can be derived as special cases. The embedded inverted strip guide shown in Figure 5.10(d) is one such structure.The propagation characteristics of this guide in single and coupled form have been reported by McLevige et al. [24] and are reproduced in Figure 5.16. (This guide is referred to as a strip–slab guide in [24].) It may be noted that the cutoff guide wavelength occurs at $\beta = k_0\sqrt{2.5}$, where 2.5 is the relative dielectric constant of the base dielectric substrate and k_0 is the free-space wavenumber. By comparison, in a similar millimeter waveguide, such as an inverted strip guide or an insular image guide with a ground conductor, cutoff occurs at $\beta_z = k_0$.

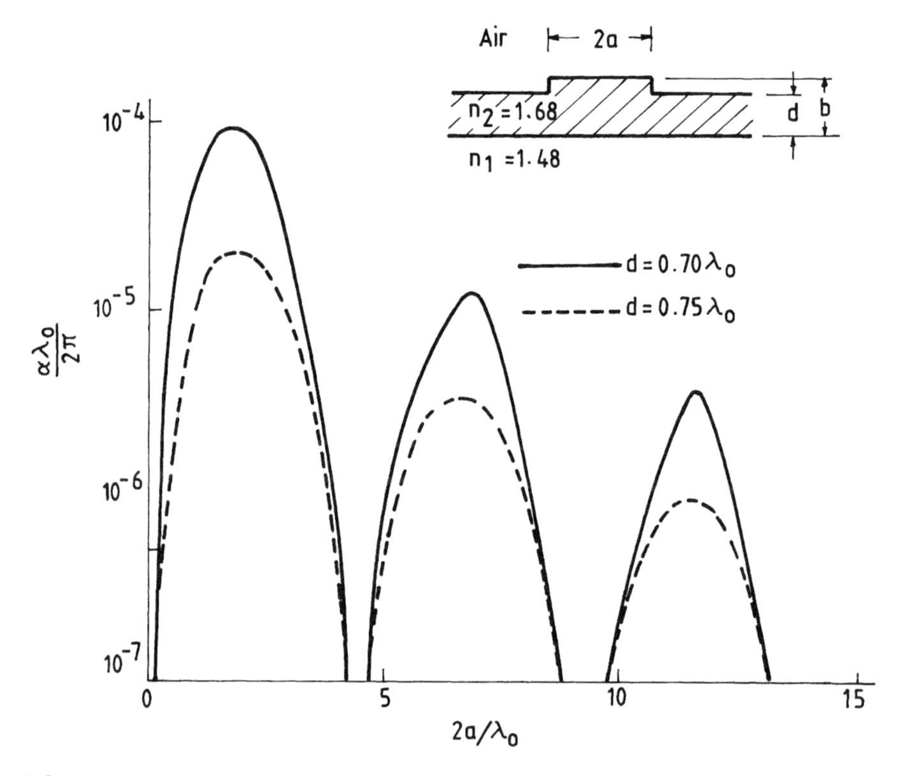

FIGURE 5.15 Curves of attenuation constant α, due to leakage of TE surface wave, versus strip width $2a$ for the lowest TM-like guided mode on a rib guide. $n_1 = 1.48$, $n_2 = 1.68$, and $b = 0.8\,\lambda$. (From Oliner et al. [22]. Copyright © 1981 IEEE, reprinted with permission.)

5.5.4 Multilayer Thin-Film Optical Rib Guide

In planar optical guides employing two or three dielectric thin-film layers, the TM-like (E^y_{mn}) and the TE-like (E^x_{mn}) modes are nearly degenerate, with the result that it is difficult to achieve single-mode operation in either the E^y_{11} or E^x_{11} mode. Single-mode and polarization-maintaining planar optical guides have considerable potential in optical integrated circuits. Pang et al. [25] have proposed a multilayer thin-film rib type optical guide (Fig. 5.17), which can offer larger modal separation between the dominant TM-like and TE-like modes. The operation is based on the property that a multilayer film structure having alternate layers of two optical materials with different refractive indices exhibit artificial birefringence effect [37, 38]. Figure 5.17(a) shows such a multilayered structure consisting of thin-film laminations of TiO_2 and SiO_2 on a substrate and Figure 5.17(b) shows a rib-type guide employing such a multilayered film structure [25]. The lamination provides for modal separation, and the rib-type

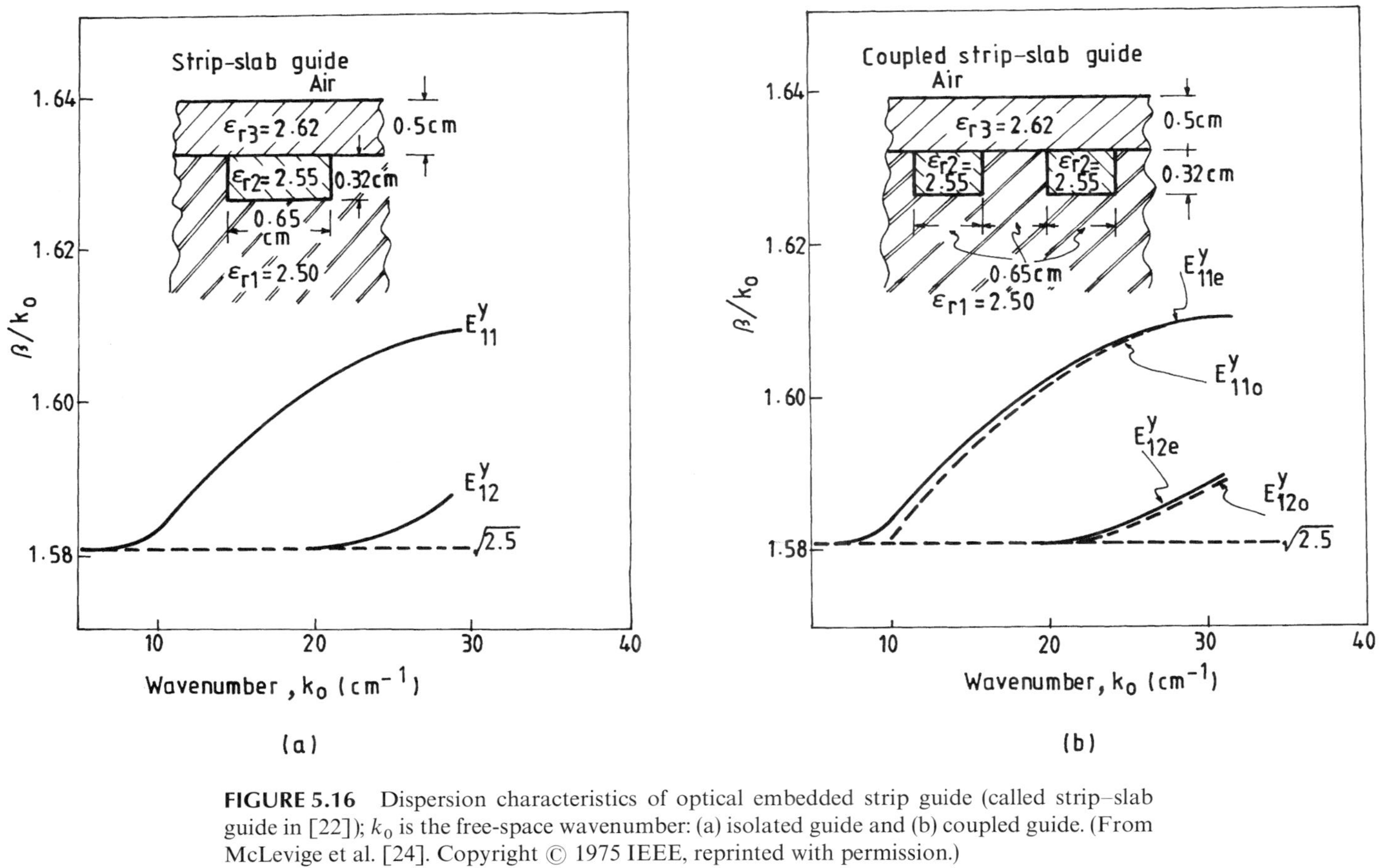

FIGURE 5.16 Dispersion characteristics of optical embedded strip guide (called strip–slab guide in [22]); k_0 is the free-space wavenumber: (a) isolated guide and (b) coupled guide. (From McLevige et al. [24]. Copyright © 1975 IEEE, reprinted with permission.)

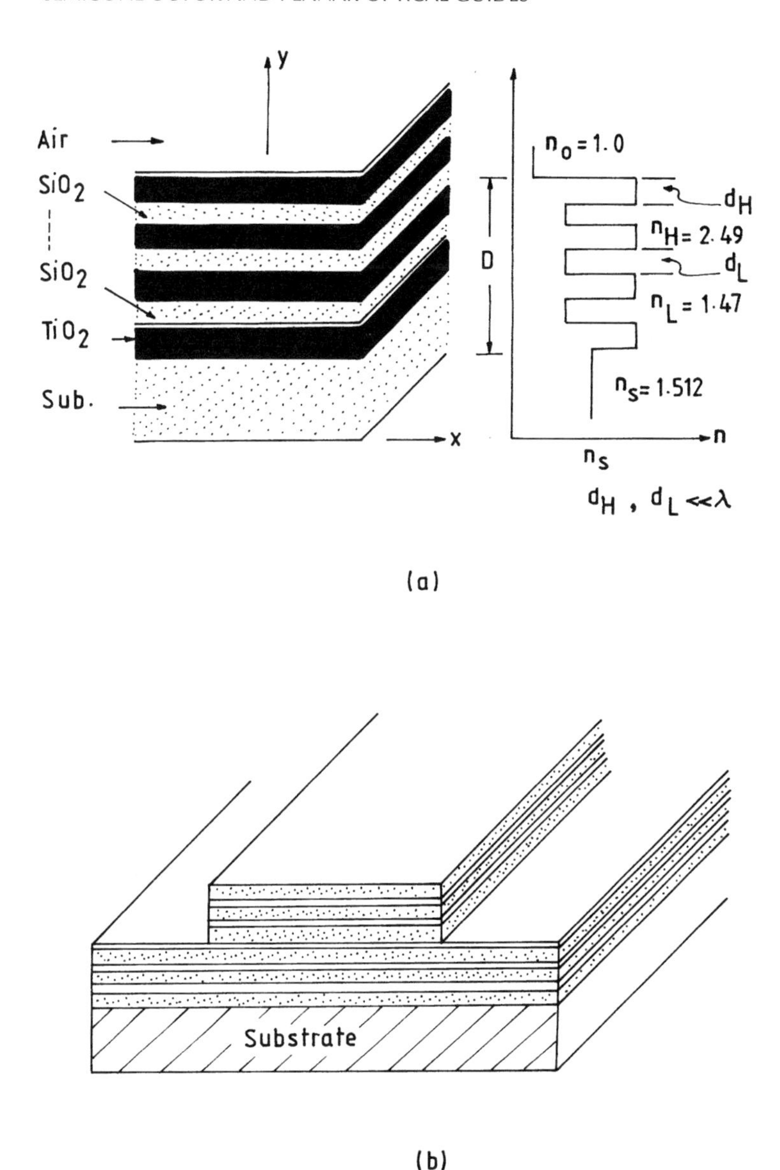

FIGURE 5.17 (a) Multilayered thin-film slab structure and refractive index distribution. (b) Rib-type multilayered thin-film optical guide. (From Jui-Pang et al. [25]. Copyright © 1991 IEEE, reprinted with permission.)

configuration provides for confinement of the optical field in the lateral direction. Pang et al. [25] have computed the characteristics of the guide by applying the equivalent multimode transmission line model and the transverse resonance method. The refractive indices of TiO_2, SiO_2, and the substrate are chosen as

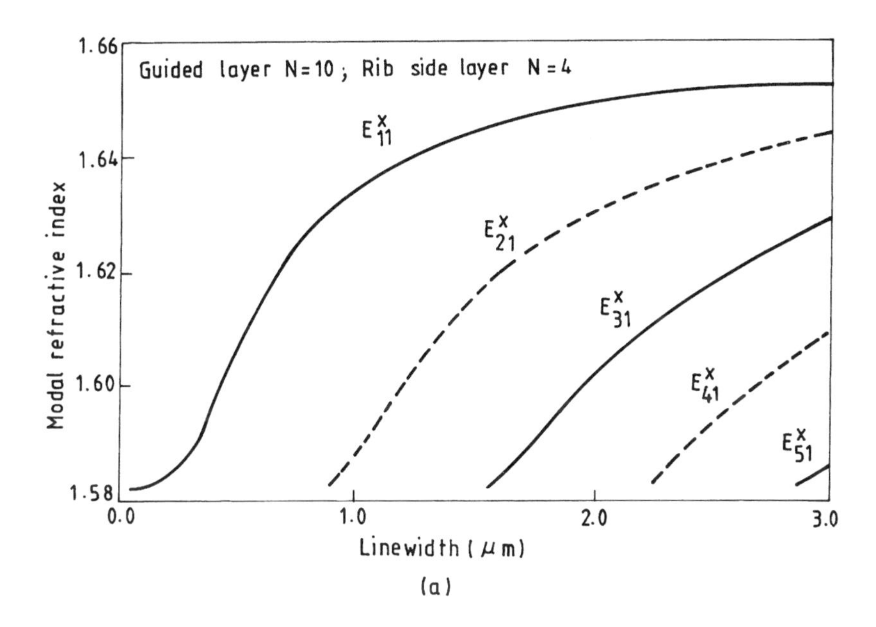

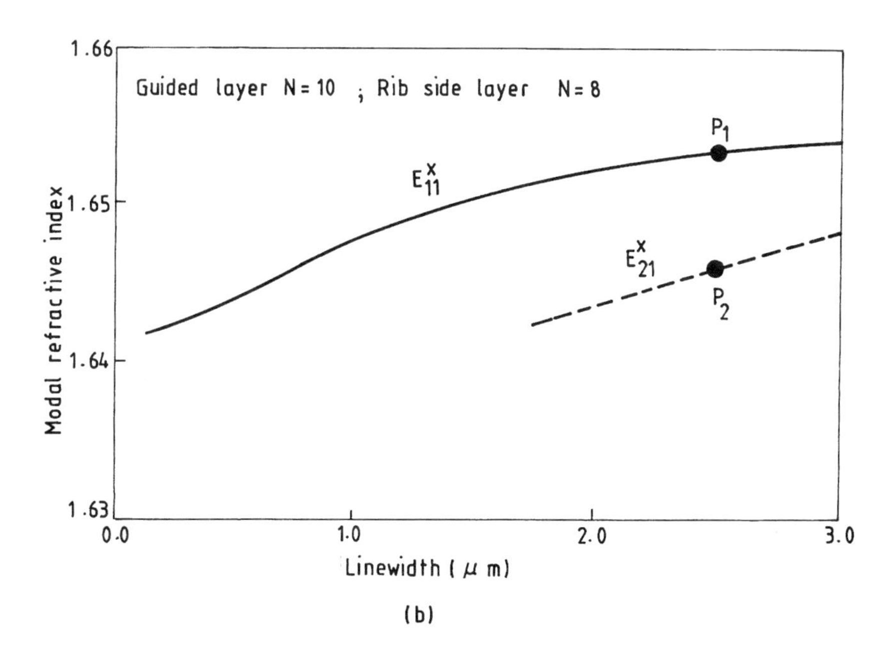

FIGURE 5.18 Modal refractive index as a function of linewidth for multilayered thin-film optical guide. Refractive indices: $n_s = 1.512$ (substrate), $n_H = 2.49$ (TiO_2), $n_L = 1.47$ (SiO_2), $d_H/d_L = 0.2$, $D = 0.6\,\mu m$, and $\lambda = 0.6\,\mu m$. (From Jui-Pang et al. [25]. Copyright © 1991 IEEE, reprinted with permission.)

$n_H = 2.49$, $n_L = 14.7$, and $n_s = 1.512$, respectively. The other parameters are total thickness of film layer $D = 0.6\,\mu\text{m}$, operating wavelength $\lambda = 0.6328\,\mu\text{m}$, and $d_H/d_L = 0.2$, where d_H and d_L are the thicknesses of the TiO_2 and SiO_2 films, respectively. With these parameters, it is shown that about ten film layers in the slab guide structure (Fig. 5.17(a)) would be adequate to achieve large modal separation between the TE-like and TM-like modes, and the refractive index for each mode saturates to a certain value when the number of layers exceeds ten. Figure 5.18 shows variation in the modal refractive index as a function of the linewidth for a rib-type guide (Fig. 5.17(b)) with ten guided layers and for rib-side layers equal to four and eight. It can be seen that with four layers for the rib-side, the guide can be operated in a single TE-polarized mode (E^x_{11} mode) for linewidth up to 1 μm, and with eight layers the range increases to 2 μm [25].

PROBLEMS

5.1 Discuss the main features of planar optical guides and compare with those of planar dielectric integrated guides.

5.2 Using the EDC technique, derive the characteristic equation for the optically illuminated silicon guide shown in Figure 5.19. Assume the plasma layer created by optical illumination to be of thickness t and uniform in the x- and z-directions. The relative dielectric constant of the plasma-unoccupied region of the guide is ε_{rs} (real) and that of the plasma layer is $\varepsilon_{rp} = \varepsilon'_{rp} - j\varepsilon''_{rp}$ (ε'_{rp} and ε''_{rp} are real). Derive expressions for the E^y_{11} and E^x_{11} modes.

5.3 The parameters of an optically illuminated silicon guide are: $\varepsilon_{rs} = 11.8$, $m^*_e = 0.259\,m_o$, $m^*_h = 0.38\,m_o$, $\nu_e = 4.52 \times 10^{12}\,\text{sec}^{-1}$, and $\nu_h = 7.7 \times 10^{12}\,\text{sec}^{-1}$. For $m_o = 5 \times 10^{20}\,\text{m}^{-3}$, calculate using Eq. (5.1) the real and imaginary parts of the complex relative dielectric constant ε_{rp} at $f = 30\,\text{GHz}$ and 40 GHz.

5.4 **(a)** Consider a silicon image guide optically illuminated over a length ℓ as shown in Figure 5.20. The plasma density is assumed to be uniform throughout the length ℓ and depth t of the plasma layer. Using the values of ε_{rs} and ε_{rp} of Problem 5.3 in the characteristic equations derived in Problem 5.2, calculate the differential phase shift in the guide at 30 GHz and 40 GHz. Assume $a = b = 1$ mm, $t = 0.05$ mm, and $\ell = 2$ mm.

(b) If the plasma is assumed to be collision free ($\nu_e = \nu_h = 0$), what is the differential phase shift for the same guide parameters as in (a).

5.5 Draw a typical field distribution in a coupled dielectric rib guide in free space (Fig. 5.21) for (a) E^y_{mn} even and odd modes for $m = 1, 2$; $n = 1, 2$; (b) E^x_{mn} even and odd modes for $m = 1, 2$; $n = 1, 2$. The modal indices m and n refer to the single dielectric rib guide.

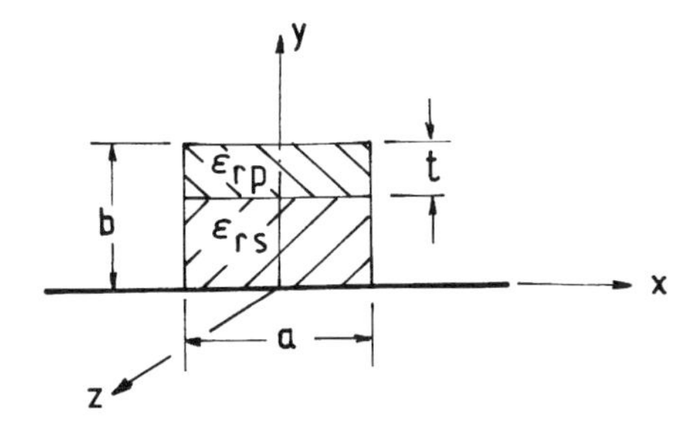

FIGURE 5.19 Optically illuminated silicon guide.

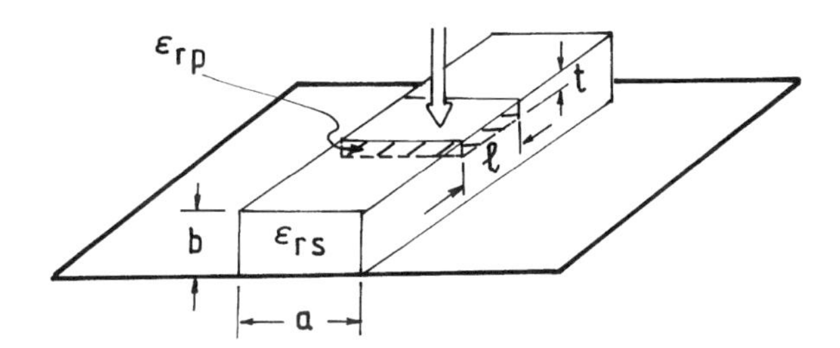

FIGURE 5.20 Silicon image guide optically illuminated over a length ℓ.

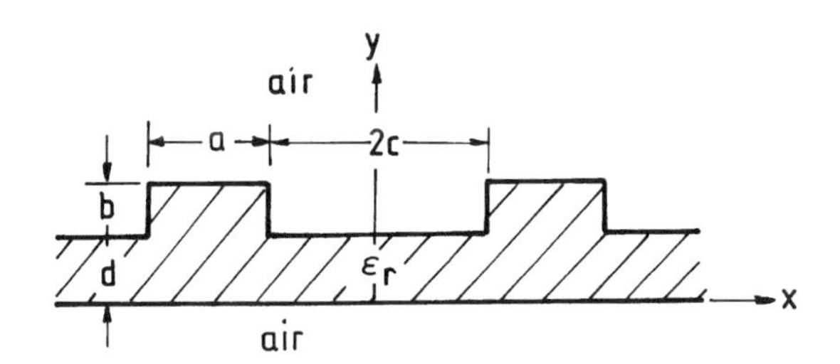

FIGURE 5.21 Coupled dielectric rib guide in free space.

5.6 **(a)** Using the EDC technique, derive characteristic equations for a coupled optical dielectric rib guide shown in Figure 5.22 for the E_{11}^{y} and E_{21}^{y} modes. Obtain separate equations for the even and odd modes in each case.

(b) Assume $a = 3d$, $c = d$, $n_0 = 1$, $n_1 = \sqrt{\varepsilon_{r1}} = 1.68$, and $n_2 = \sqrt{\varepsilon_{r2}} = 1.44$. Plot β/k_0 as a function of k_0 for $b = d$, $d/2$, and $d/4$.

(c) Set $n_2 = 1$ in (b) above and obtain the characteristic equations for the coupled dielectric rib guide embedded in free space. What is the effect of introducing a dielectric ε_{r2} below the rib guide?

5.7 Repeat Problem 5.6 for the E_{11}^{x} and E_{21}^{x} modes. Compare the modal bandwidth with that obtained in Problem 5.6 for $b = d$, $d/2$, and $d/4$.

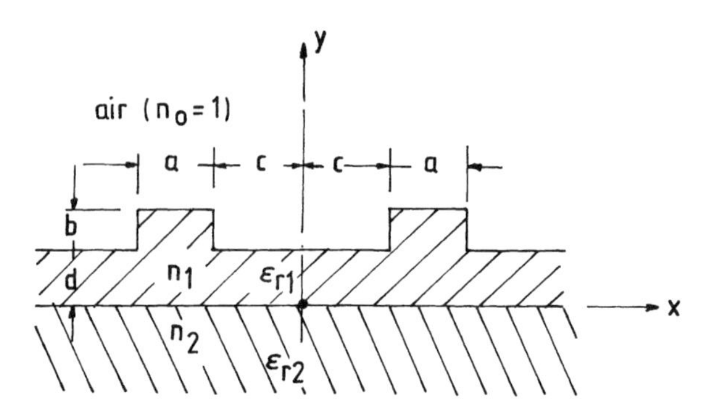

FIGURE 5.22 Coupled optical dielectric rib guide.

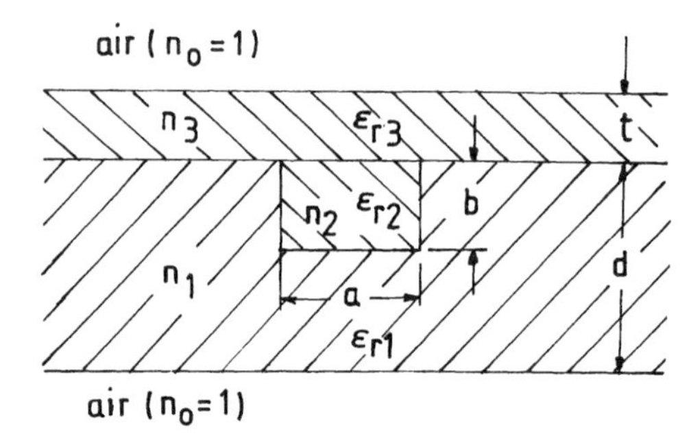

FIGURE 5.23 Embedded inverted strip–slab guide.

5.8 Draw a typical field distribution for an embedded inverted strip–slab guide shown in Figure 5.23 for (a) E^y_{mn} mode, $m = 1$, $n = 1$, 2; and (b) E^x_{mn} mode, $m = 1$, $n = 1, 2$. Assume $t \ll d$, $\varepsilon_{r2} > \varepsilon_{r3} > \varepsilon_{r1}$, $n_1 = \sqrt{\varepsilon_{r1}}$, $n_2 = \sqrt{\varepsilon_{r2}}$, and $n_3 = \sqrt{\varepsilon_{r3}}$.

5.9 **(a)** Using the EDC technique, derive the characteristic equation for an embedded coupled inverted strip–slab guide shown in Figure 5.24 for the following modes: (i) E^y_{mn} even and odd modes and (ii) E^x_{mn} even and odd modes.

(b) Obtain the characteristic equation of the single inverted strip-slab guide shown in Figure 5.23 by setting $c = \infty$.

5.10 Consider the limiting case of the coupled structure in Figure 5.24 for $t = 0$ (or $n_3 = 1$). Assume $n_1 = 1.44$, $n_2 = 1.68$, $a = 3b$, $c = 2b$, and $b/\lambda = 0.5$.

(a) Calculate and plot β_{11e}/k_0 and β_{11o}/k_0 of the E^y_{11} mode as a function of d/λ from 0.5 to 2.0 (*Note*: $k_0 = 2\pi/\lambda$.)

(b) What is the effect of the dielectric slab thickness d on the propagation parameters?

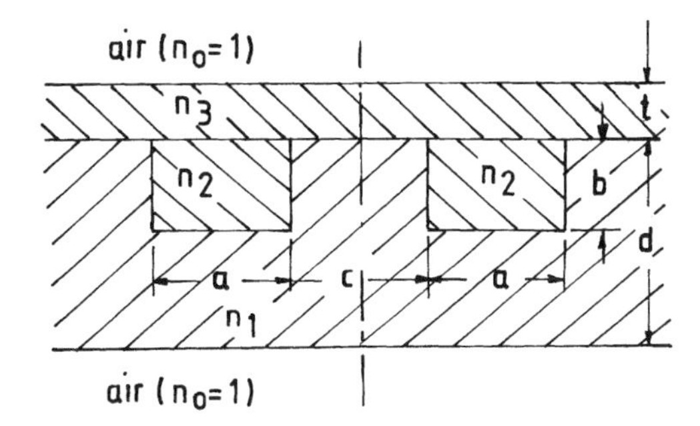

FIGURE 5.24 Embedded coupled inverted strip–slab guide.

REFERENCES

1. H. Jacobs and M. M. Chrepta, Electronic phase shifter for millimeter-wave semiconductor dielectric integrated circuits. *IEEE Trans. Microwave Theory Tech.*, **MTT-22**, 411–417, Apr. 1974.
2. A. M. Vaucher, C. D. Striffler, and C. H. Lee, Theory of optically controlled millimeter-wave phase shifters. *IEEE Trans. Microwave Theory Tech.*, **MTT-31**, 209–216, Feb. 1983.
3. W. Platte, LED-induced distributed Bragg reflection microwave filter with fibre-optically controlled change of centre frequency via photoconductivity gratings. *IEEE Trans. Microwave Theory Tech.*, **MTT-39**, 359–363, 1991.
4. *Special Issue on Applications of Light-Wave Technology to Microwave Devices, Circuits and Systems, IEEE Trans. Microwave Theory Tech.*, **MTT-38**, May 1990.
5. M. Matsumoto, M. Tsutsumi, and N. Kumagai, Radiation of millimeter waves from a leaky dielectric waveguide with a light induced grating layer. *IEEE Trans. Microwave Theory Tech.*, **MTT-35**, 1033–1042, 1987.
6. K. L. Klohn et al., Silicon waveguide frequency scanning linear array antenna. *IEEE Trans. Microwave Theory Tech.*, **MTT-26**, 764–773, Oct. 1978.
7. R. E. Horn et al., Electronic modulated beam steerable silicon waveguide array antenna. *IEEE Trans. Microwave Theory Tech.*, **MTT-28**, 647–653, June 1980.
8. R. E. Horn et al., Single frequency electronic modulated analog line scanning using a dielectric antenna. *IEEE Trans. Microwave Theory Tech.*, **MTT-30**, 816–820, May 1982.
9. C. Yao et al., Monolithic integration of a dielectric millimeter wave antenna and mixer diode: an embrionic millimeter-wave IC. *IEEE Trans. Microwave Theory Tech.*, **MTT-30**, 1241–1247, Aug. 1982.
10. C. H. Lee, P. S. Mak, and A. P. DeFonzo, Optical control of millimeter-wave propagation in dielectric waveguide. *IEEE J. Quantum Electron.*, **QE-16**, 277–288, Mar. 1980.
11. C. H. Lee et al., Optoelectronic techniques for microwave and millimeter-wave applications. *IEEE MTT-S Int. Microwave Symp. Digest*, 178–181, 1985.

12. C. H. Lee, Pico-second optoelectronic switching in GaAs. *Appl. Phys. Lett.*, **30**, 84–86, 1977.
13. A. Alphones, *Studies on Optically Controlled Millimeter Wave Circuits*, D. Eng. Thesis, Department of Electronics and Information Science, Kyoto Institute of Technology, Kyoto, Japan.
14. A. Alphones and M. Tsutsumi, Optical control of millimeter wave in silicon rib guides, Paper of Technical Group on Electronics and Communications. *IEICE Japan*, **MW 90-148**, 19–26, Feb. 1991.
15. Y. Satomura and M. Tsutsumi, Switching characteristics of optically controlled millimeter waves in the silicon H-guide, 19th International Conference on Infrared and Millimetre Waves, *Conf. Digest.*, 512–513, Dec. 1994.
16. M. Tsutsumi and Y. Satomura, Optical control of millimeter waves in the silicon waveguide, 18th International Conference on Infrared and Millimetre Waves, *Conf. Digest.*, 539–540, Dec. 1993.
17. R. N. Simons, *Optical Control of Microwave Devices*, Artech House, Norwood, MA, 1990.
18. H. W. Yen et al., Switching of GaAs IMPATT diode oscillator by optical illumination. *Appl. Phys. Lett.*, **31**, 120–122, 1977.
19. A. A. A. DeSalles, Optical control of GaAs MESFETs, *IEEE Trans. Microwave Theory Tech.*, **MTT-31**, 812–820, 1983.
20. C. H. Lee, *Picosecond Optoelectronic Devices*, Academic Press, Orlando, FL, 1984.
21. H. G. Unger, *Planar Optical Waveguides and Fibres*, Clarendon Press, Oxford, 1977.
22. A. A. Oliner et al., Guidance and leakage properties of a class of open dielectric waveguides: Part II—New physical effects. *IEEE Trans. Microwave Theory Tech.*, **MTT-29**, 855–869, Sept. 1981.
23. J. E. Goell, Rib waveguides for integrated optical circuits. *Appl. Opt.*, **12**, 2797–2798, Dec. 1973.
24. W. V. McLevige, T. Itoh, and R. Mittra, New waveguiding structures for millimetre-wave and optical integrated circuits. *IEEE Trans. Microwave Theory Tech.*, **MTT-23**, 788–794, Oct. 1975.
25. Hsui Jui-Pang, T. Anada, S. Nakamura, and T. Kobayashi, Propagation properties of multi-layer thin film polarization-maintaining optical 3-D waveguide. *IEEE MTT-S Int. Microwave Symp. Digest*, 615–618, 1991.
26. J. K. Butler, T. F. Wu, and M. W. Scott, Non-uniform layer model of a millimetre-wave phase shifter. *IEEE Trans. Microwave Theory Tech.*, **MTT-34**, 147–155, Jan. 1986.
27. E. A. J. Marcatili, Slab-coupled guides. *Bell Syst. Tech. J.*, **53(4)**, 645–674, Apr. 1974.
28. N. Uchida, Optical waveguide loaded with high refractive index strip film. *Appl. Opt.* **15**, 179–182, 1976.
29. V. Ramaswamy, Strip loaded film waveguide. *Bell Syst. Tech. J.*, **53(4)**, 697–704, 1974.
30. K. Ogusu, S. Kawakami, and S. Nishida, Optical strip waveguide: an analysis. *Appl. Opt.* **18**, 908–914, Mar. 1979. [Also correction in *Appl. Opt.*, **18**, 3725, Nov. 1979.]
31. K. Yasuura, K. Shimohara, and T. Miyamoto, Numerical analysis of a thin-film waveguide for mode-matching method. *J. Opt. Soc. Am.*, **70**, 183–191, 1980.

32. S. T. Peng and A. A. Oliner, Leakage and resonance effects on strip waveguides for integrated optics. *Trans. Inst. Electron. Commun. Eng. Japan* (Special issue on Integrated Optics and Optical Fibre Communications), **E61**, 151–154, Mar. 1978.
33. N. Dagli and C. G. Fonstad, Analysis of rib waveguides. *IEEE J. Quantum Electron.*, **QE-21**, 315–321, Apr. 1985.
34. M. Koshiba and M. Suzuki, Equivalent network analysis of dielectric thin-film waveguides for optical integrated circuits and its applications. *Radio Sci.*, **17(1)**, 99–107, Jan./Feb. 1982.
35. K. Bierwirth, N. Schulz, and F. Arndt, Finite-difference analysis of rectangular dielectric waveguide structures. *IEEE Trans. Microwave Theory Tech.*, **MTT-34**, 1104–1114, Nov. 1986.
36. Z. M. Lu, S. L. Wang, and S. T. Peng, Polarization conversions in dielectric strip waveguides, 13th International Conference on Infrared and Millimetre Waves, *Conf. Digest.*, 243–244, Dec. 1988.
37. M. Born and E. Wolff, *Principles of Optics*, Macmillan, New York, 1964.
38. M. Kitagawa and M. Tateda, Form birefringence of SiO_2/TaO_2 periodic multilayers. *Appl. Opt.*, **24(20)**, 3359–3362, Oct. 1985.

CHAPTER SIX

Nonradiative Dielectric Guide

6.1 INTRODUCTION

The nonradiative dielectric guide, first proposed by Yoneyama and Nishida [1] is a low-loss transmission line having a unique feature that it can suppress radiation almost completely. It consists of a parallel-plate guide with a dielectric strip inserted between the plates (Fig. 6.1). Structurally, it is identical to the H-guide except that the plate separation is less than half the free-space wavelength ($\lambda_0/2$).

In a conventional parallel-plate guide with air as the intervening medium, the TE modes, which have the electric fields parallel to the metal plates, form the low-loss modes. For a plate separation less than $\lambda_0/2$, these TE modes are cut off. In the NRD guide, the presence of the dielectric strip enables electromagnetic waves to propagate along the strip whereas the radiated waves, if any, are suppressed because of the cutoff nature of the air-filled region.

The above two important features of the NRD guide—namely, low transmission loss and absence of radiation from bends and discontinuities—make it suitable for realizing compact and high-performance millimeter wave integrated circuits. Considering the physical size of the guide and the convenience of fabrication, the most useful frequency range for practical applications is 30 GHz to about 100 GHz. Because of its high potential in the realization of millimeter wave integrated circuits, the NRD guide has received extensive attention both theoretically and experimentally [1–11].

In this chapter, we first derive general expressions for the hybrid-mode fields of a parallel-plate guide with a dielectric strip. The fields and propagation characteristics of the NRD guide are discussed. The characteristics of the coupled NRD guide and the groove NRD guide are also presented.

6.2 ANALYSIS OF DIELECTRIC STRIP LOADED PARALLEL-PLATE GUIDE

In this section, we derive general expressions for the hybrid-mode fields of a parallel-plate guide loaded with a dielectric strip. The fields of the NRD guide

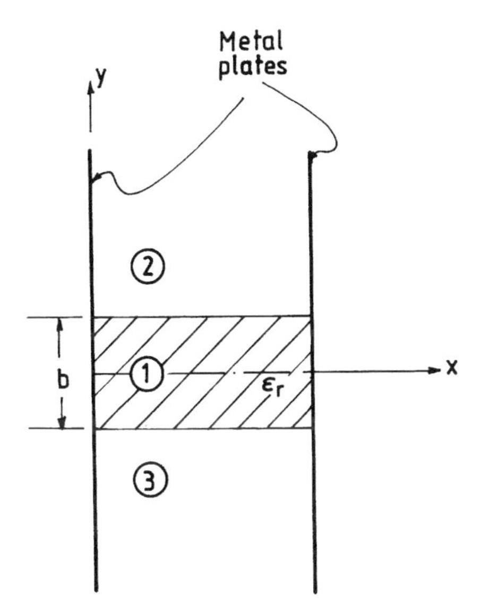

FIGURE 6.1 Cross-section of a parallel-plate waveguide loaded with a dielectric strip.

operating modes are obtained as a special case by assigning suitable modal numbers. Referring to the structure and the coordinate system shown in Figure 6.1, the hybrid modes of the guide are classified into a combination of TM modes to y (TM^y_{mn}) and TE modes to y (TE^y_{mn}). As discussed in Chapters 2 and 3, the TM^y_{mn} modes are also referred to as E^y_{mn} modes to indicate that E_y is the dominant electric field component. For this set of modes, the magnetic field lines lie entirely in planes perpendicular to the y-axis ($H_y = 0$). Similarly, the TE^y_{mn} modes are referred to as E^x_{mn} modes to indicate that E_x is the dominant electric field component. For this set of modes, the electric field lines lie entirely in planes perpendicular to the y-axis ($E_y = 0$). For ease of analysis, we make use of the symmetry of the structure about the $y = 0$ plane and identify TM^y_{mn} modes as symmetric or antisymmetric depending on whether the E_y component is symmetric or antisymmetric about $y = 0$. Similarly, we identify TE^y_{mn} modes as symmetric or antisymmetric depending on whether the E_x component is symmetric or antisymmetric about same plane.

6.2.1 Field Expressions for TM^y_{mn} Modes ($H_y = 0$)

Referring to the coordinate system shown in Figure 6.1, we assume wave propagation along the guide with fields varying according to $e^{j(\omega t - \beta z)}$, where β is the propagation constant for the TM^y modes. The electromagnetic fields of these modes can be expressed in terms of a scalar potential ϕ^e as given in Eq. (2.24) of Chapter 2. We note that E_y is proportional to ϕ^e as given by the relation

$$E_y = \frac{1}{\varepsilon_r}\left(\beta^2 - \frac{\partial^2}{\partial x^2}\right)\phi^e \tag{6.1}$$

Thus the symmetric and antisymmetric TM^y fields can be obtained from the expressions for ϕ^e, which are symmetric and antisymmetric about $y = 0$. For symmetric modes, the plane $y = 0$ represents an electric wall and for antisymmetric modes, it represents a magnetic wall.

TM^y_{mn} Symmetric Modes Referring to Figure 6.1, the expressions for ϕ^e in the dielectric and air regions can be written as

$$\phi^e = A\sin(m\pi x/a)\cos(\beta_y y) \qquad |y| > b/2 \tag{6.2a}$$

$$= A\cos(\beta_y b/2)\sin(m\pi x/a)\, e^{\zeta(b/2 - |y|)} \qquad |y| < b/2 \tag{6.2b}$$

where A is an arbitrary constant, β_y is the y-directed propagation constant in the dielectric region, and ζ is the decay constant in the air region. The common factor $\exp[j(\omega t - \beta z)]$ has been suppressed from (6.2). The parameters β, β_y, and ζ are related by

$$\beta^2 = k_0^2\,\varepsilon_r - (m\pi/a)^2 - \beta_y^2 = k_0^2 - (m\pi/a)^2 + \zeta^2 \tag{6.3}$$

The various field components of the TM^y symmetric modes can now be obtained in terms of the single unknown constant A by substituting for ϕ^e from (6.2) in (2.24).

Region 1 ($|y| < b/2$):

$$E_x = -A(\beta_y/\varepsilon_r)(m\pi/a)\cos(m\pi x/a)\sin(\beta_y y) \tag{6.4a}$$

$$E_y = (A/\varepsilon_r)g^2\sin(m\pi x/a)\cos(\beta_y y) \tag{6.4b}$$

$$E_z = j(A/\varepsilon_r)\beta\,\beta_y\sin(m\pi x/a)\sin(\beta_y y) \tag{6.4c}$$

$$H_x = -A\omega\varepsilon_0\beta\sin(m\pi x/a)\cos(\beta_y y) \tag{6.4d}$$

$$H_y = 0 \tag{6.4e}$$

$$H_z = jA\omega\varepsilon_0(m\pi/a)\cos(m\pi x/a)\cos(\beta_y y) \tag{6.4f}$$

Regions 2 and 3 ($|y| > b/2$):

$$E_x = \mp A\zeta(m\pi/a)\cos(m\pi x/a)\cos(\beta_y b/2)e^{\zeta(b/2 - |y|)} \tag{6.5a}$$

$$E_y = Ag^2\sin(m\pi x/a)\cos(\beta_y b/2)e^{\zeta(b/2 - |y|)} \tag{6.5b}$$

$$E_x = \pm jA\beta\zeta\sin(m\pi x/a)\cos(\beta_y b/2)e^{\zeta(b/2 - |y|)} \tag{6.5c}$$

$$H_x = -A\omega\varepsilon_0\beta\sin(m\pi x/a)\cos(\beta_y b/2)e^{\zeta(b/2 - |y|)} \tag{6.5d}$$

$$H_y = 0 \tag{6.5e}$$

$$H_z = jA\omega\varepsilon_0(m\pi/a)\cos(m\pi x/a)\cos(\beta_y b/2)e^{\zeta(b/2 - |y|)} \tag{6.5f}$$

where the upper and lower signs in (6.5) apply to the air regions 2 and 3, respectively, and g is given by

$$g^2 = \beta^2 + (m\pi/a)^2 = k_0^2\,\varepsilon_r - \beta_y^2 = k_0^2 + \zeta^2 \tag{6.6}$$

Matching E_x at the air–dielectric interfaces yields the characteristic equation

$$\beta_y \tan(\beta_y b/2) = \varepsilon_r \zeta \tag{6.7}$$

where, from (6.6), we have

$$\zeta^2 = k_0^2\,(\varepsilon_r - 1) - \beta_y^2 \tag{6.8}$$

Equation (6.7) is identical to that of the symmetric TM modes of a dielectric slab guide of thickness b and relative dielectric constant ε_r. The same relation can also be obtained by applying the EDC technique described in Chapter 2. The solution of (6.7) gives a system of eigenvalues for β_{yn} and ζ_n with odd values of n ($n = 1, 3, 5, \ldots$). The propagation constants of TM_{mn}^y symmetric modes are then obtained from

$$\beta_{mn}^2 = k_0^2\,\varepsilon_r - (m\pi/a)^2 - \beta_{yn}^2, \quad m = 1, 2, \ldots; \quad n = 1, 3, 5, \ldots \tag{6.9}$$

TM_{mn}^y Antisymmetric Modes Referring to Figure 6.1(a), the expressions for ϕ^e in the dielectric and air regions are given by

$$\phi^e = B\sin(m\pi x/a)\sin(\beta_y y) \qquad |y| < b/2 \tag{6.10a}$$

$$= B\sin(m\pi x/a)\sin(\beta_y b/2)e^{\zeta(b/2 - y)} \qquad y > b/2 \tag{6.10b}$$

$$= -\,B\sin(m\pi x/a)\sin(\beta_y b/2)e^{\zeta(b/2 - y)} \qquad y < -b/2 \tag{6.10c}$$

where the parameters β, β_y, and ζ are related by (6.3). The various field com ponents of the TM^y antisymmetric modes can be obtained in terms of the single unknown constant B by substituting for ϕ^e from (6.10) in (2.27) of Chapter 2.

Region 1 ($|y| < b/2$):

$$E_x = B\,(\beta_y/\varepsilon_r)\,(m\pi/a)\cos(m\pi x/a)\cos(\beta_y y) \tag{6.11a}$$

$$E_y = (B/\varepsilon_r)g^2\sin(m\pi x/a)\sin(\beta_y y) \tag{6.11b}$$

$$E_z = -j(B/\varepsilon_r)\beta\beta_y\sin(m\pi x/a)\cos(\beta_y y) \tag{6.11c}$$

$$H_x = -\,B\omega\varepsilon_0\beta\sin(m\pi x/a)\sin(\beta_y y) \tag{6.11d}$$

$$H_y = 0 \tag{6.11e}$$

$$H_z = jB\omega\varepsilon_0(m\pi/a)\cos(m\pi x/a)\sin(\beta_y y) \tag{6.11f}$$

Regions 2 and 3 ($|y| > b/2$):

$$E_x = -B\zeta(m\pi/a)\cos(m\pi x/a)\sin(\beta_y b/2)e^{\zeta(b/2-|y|)} \tag{6.12a}$$

$$E_y = \pm Bg^2 \sin(m\pi x/a)\sin(\beta_y b/2)e^{\zeta(b/2-|y|)} \tag{6.12b}$$

$$E_z = jB\beta\zeta \sin(m\pi x/a)\sin(\beta_y b/2)e^{\zeta(b/2-|y|)} \tag{6.12c}$$

$$H_x = \mp B\omega\varepsilon_0\beta \sin(m\pi x/a)\sin(\beta_y b/2)e^{\zeta(b/2-|y|)} \tag{6.12d}$$

$$H_y = 0 \tag{6.12e}$$

$$H_z = \pm jB\omega\varepsilon_0(m\pi/a)\cos(m\pi x/a)\sin(\beta_y b/2)e^{\zeta(b/2-|y|)} \tag{6.12f}$$

where the upper and lower signs in (6.12) apply to the air regions 2 and 3, respectively, and g is given by (6.6). Matching E_x at the air–dielectric interfaces yields the relation

$$\beta_y \cot(\beta_y b/2) = -\zeta\varepsilon_r \tag{6.13}$$

where ζ and β_y are related by (6.8). It may be noted that the characteristic equation (6.13) is identical to that of the antisymmetric TM modes of a dielectric slab guide of thickness b and relative dielectric constant ε_r. The solution of (6.13) gives a system of eigenvalues for β_{yn} and ζ_n with n assuming even values ($n = 2, 4, 6, \ldots$). The propagation constants are obtained from (6.9) for $m = 1, 2, 3, \ldots$ and $n = 2, 4, 6, \ldots$.

6.2.2 Field Expressions for TE^y_{mn} Modes ($E_y = 0$)

Consider propagation of TE^y modes along the guide with fields varying according to $e^{j(\omega t - \beta z)}$, where β is the propagation constant in the z-direction. The fields of these modes can be expressed in terms of a scalar potential ϕ^h as in Eq. (2.27) of Chapter 2. We note that E_x is proportional to ϕ^h and is given by

$$E_x = -\omega\mu_0\beta\phi^h \tag{6.14}$$

Thus the symmetric and antisymmetric TE^y fields can be obtained from the expressions for ϕ^h, which are symmetric and antisymmetric about $y = 0$. For symmetric modes, the plane $y = 0$ represents a magnetic wall, and for antisymmetric modes, it represents an electric wall.

TE^y_{mn} Symmetric Modes Referring to Figure 6.1, the expressions for ϕ^h in the dielectric and air regions can be written as

$$\phi^h = C\cos(m\pi x/a)\cos(\beta_y y) \qquad |y| < b/2 \tag{6.15a}$$

$$= C\cos(\beta_y b/2)\cos(m\pi x/a)e^{\zeta(b/2-|y|)} \qquad |y| > b/2 \tag{6.15b}$$

where the parameters β_y and ζ are related to β by the expression

$$\beta^2 = k_0^2 \varepsilon_r - (m\pi/a)^2 - \beta_y^2 = k_0^2 - (m\pi/a)^2 + \zeta^2 \tag{6.16}$$

Substituting for ϕ^h from (6.15) in (2.27) and replacing $\partial/\partial z$ by $-j\beta$, we obtain the various field expressions in terms of a single unknown constant C.

Region 1 ($|y| < b/2$):

$$E_x = C\omega\mu_0\beta\cos(m\pi x/a)\cos(\beta_y y) \tag{6.17a}$$

$$E_y = 0 \tag{6.17b}$$

$$E_z = jC\omega\mu_0(m\pi/a)\sin(m\pi x/a)\cos(\beta_y y) \tag{6.17c}$$

$$H_x = C\beta_y(m\pi/a)\sin(m\pi x/a)\sin(\beta_y y) \tag{6.17d}$$

$$H_y = Cg^2\cos(m\pi x/a)\cos(\beta_y y) \tag{6.17e}$$

$$H_z = jC\beta\beta_y\cos(m\pi x/a)\sin(\beta_y y) \tag{6.17f}$$

Regions 2 and 3 ($|y| > b/2$):

$$E_x = C\omega\mu_0\beta\cos(m\pi x/a)\cos(\beta_y b/2)e^{\zeta(b/2-|y|)} \tag{6.18a}$$

$$E_y = 0 \tag{6.18b}$$

$$E_z = jC\omega\mu_0(m\pi/a)\sin(m\pi x/a)\cos(\beta_y b/2)e^{\zeta(b/2-|y|)} \tag{6.18c}$$

$$H_x = \pm C\zeta(m\pi/a)\sin(m\pi x/a)\cos(\beta_y b/2)e^{\zeta(b/2-|y|)} \tag{6.18d}$$

$$H_y = Cg^2\cos(m\pi x/a)\cos(\beta_y b/2)e^{\zeta(b/2-|y|)} \tag{6.18e}$$

$$H_z = \pm jC\beta\zeta\cos(m\pi x/a)\cos(\beta_y b/2)e^{\zeta(b/2-|y|)} \tag{6.18f}$$

where the upper and lower sign in (6.18) apply to the air regions 2 and 3, respectively, and g is given by

$$g^2 = \beta^2 + (m\pi/a)^2 = k_0^2\varepsilon_r - \beta_y^2 = k_0^2 + \zeta^2 \tag{6.19}$$

Matching H_x at the air–dielectric interfaces yields the characteristic equation

$$\beta_y \tan(\beta_y b/2) = \zeta \tag{6.20}$$

where, from (6.17), we have

$$\zeta^2 = k_0^2(\varepsilon_r - 1) - \beta_y^2 \tag{6.21}$$

Equation (6.20) is the same as the characteristic equation for the symmetric TE modes of a dielectric slab guide of thickness b and relative dielectric constant ε_r.

The solution of (6.21) gives a system of eigenvalues with n assuming odd integers ($n = 1, 3, 5, \ldots$). The propagation constants of symmetric TE^y modes are then obtained from

$$\beta_{mn}^2 = k_0^2 \varepsilon_r - (m\pi/a)^2 - \beta_{yn}^2, \quad m = 0, 1, 2, \ldots; \quad n = 1, 3, 5, \ldots \tag{6.22}$$

***TE_{mn}^y* Antisymmetric Modes** For this set of modes, E_x is antisymmetric about the $y = 0$ plane. The scalar potential ϕ^h takes the form

$$\phi^h = D\cos(m\pi x/a)\sin(\beta_y y) \qquad |y| < b/2 \tag{6.23a}$$

$$= D\cos(m\pi x/a)\sin(\beta_y b/2)e^{\zeta(b/2-y)} \qquad y > b/2 \tag{6.23b}$$

$$= -D\cos(m\pi x/a)\sin(\beta_y b/2)e^{\zeta(b/2+y)} \qquad y < -b/2 \tag{6.23c}$$

where the parameters β_y and ζ are related by (6.16). Substituting for ϕ^h from (6.23) in (2.27) and replacing $\partial/\partial z$ by $-j\beta$, we obtain the following field expressions:

Region 1 ($|y| < b/2$):

$$E_x = D\omega\mu_0\beta\cos(m\pi x/a)\sin(\beta_y y) \tag{6.24a}$$

$$E_y = 0 \tag{6.24b}$$

$$E_z = jD\omega\mu_0(m\pi/a)\sin(m\pi x/a)\sin(\beta_y y) \tag{6.24c}$$

$$H_x = -D\beta_y(m\pi/a)\sin(m\pi x/a)\cos(\beta_y y) \tag{6.24d}$$

$$H_y = Dg^2\cos(m\pi x/a)\sin(\beta_y y) \tag{6.24e}$$

$$H_z = -jD\beta\beta_y\cos(m\pi x/a)\cos(\beta_y y) \tag{6.24f}$$

Regions 2 and 3 ($|y| > b/2$):

$$E_x = \pm D\omega\mu_0\beta\cos(m\pi x/a)\sin(\beta_y b/2)e^{\zeta(b/2-|y|)} \tag{6.25a}$$

$$E_y = 0 \tag{6.25b}$$

$$E_z = \pm jD\omega\mu_0(m\pi/a)\sin(m\pi x/a)\sin(\beta_y b/2)e^{\zeta(b/2-|y|)} \tag{6.25c}$$

$$H_x = D\zeta(m\pi/a)\sin(m\pi x/a)\sin(\beta_y b/2)e^{\zeta(b/2-|y|)} \tag{6.25d}$$

$$H_y = \pm Dg^2\cos(m\pi x/a)\sin(\beta_y b/2)e^{\zeta(b/2-|y|)} \tag{6.25e}$$

$$H_z = jD\beta\zeta\cos(m\pi x/a)\sin(\beta_y b/2)e^{\zeta(b/2-|y|)} \tag{6.25f}$$

Matching H_x at the air–dielectric interfaces yields the following characteristic equation:

$$\beta_y \cot(\beta_y b/2) = -\zeta \tag{6.26}$$

where β_y and ζ are related by (6.21). Equation (6.26) is the same as the characteristic equation for the antisymmetric TE modes of a dielectric slab guide of thickness b and relative dielectric constant ε_r. The solution of (6.26) gives a system of eigenvalues with n assuming even integers ($n = 2, 4, 6, \ldots$). The propagation constants of antisymmetric TE^y modes are obtained from

$$\beta_{mn}^2 = k_0^2 \varepsilon_r - (m\pi/a)^2 - \beta_{yn}^2, \quad m = 1, 2, \ldots; \quad n = 2, 4, 6, \ldots \tag{6.27}$$

From the field expressions derived above, it may be noted that for $m \geq 1$ and $n \geq 1$, the TM_{mn}^y and TE_{mn}^y modes are hybrid modes. The structure cannot support TEM and TM_{0n} modes. The only nonhybrid modes that it can support are TE_{0n}^y modes having E_x, H_y, and H_z as the nonzero field components. The E_x and H_y components are symmetric about the $y = 0$ plane for n odd ($n = 1, 2, \ldots$) and antisymmetric about the same plane for n even ($n = 2, 4, \ldots$).

6.3 CHARACTERISTICS OF NRD GUIDE

6.3.1 Nonradiative Modes and Fields

The structure shown in Figure 6.1 operates as a NRD guide when the spacing between the metal plates is less than half the free-space wavelength ($a < \lambda_0/2$). With this condition, the two fundamental modes, TM_{11}^y and TE_{11}^y are non-radiative in nature. It may be noted that the TM^y and TE^y modes are the LSM (longitudinal-section-magnetic) and LSE (longitudinal-section-electric) modes referred to by Yoneyama [5]. According to the mode nomenclature chosen in the preceding section, the TM_{11}^y and TE_{11}^y modes are identical to the dominant LSM_{11} (referred as LSM_{01} in [5]) and LSE_{11} (referred as LSE_{01} in [5]) modes, respectively. The field expressions for the LSM_{11} mode are given by (6.4) and (6.5) and for the LSE_{11} mode are given by (6.17) and (6.18) with $m = 1$. Figure 6.2 shows typical field lines for these modes.

Fields for LSM_{11} Mode (See Fig. 6.2 (a))

Dielectric Region ($|y| < b/2$):

$$E_x = -A(\beta_y/\varepsilon_r)(\pi/a)\cos(\pi x/a)\sin(\beta_y y) \tag{6.28a}$$

$$E_y = (A/\varepsilon_r)(k_0^2 \varepsilon_r - \beta_y^2)\sin(\pi x/a)\cos(\beta_y y) \tag{6.28b}$$

$$E_z = j(A/\varepsilon_r)\beta\beta_y \sin(\pi x/a)\sin(\beta_y y) \tag{6.28c}$$

$$H_x = -A\omega\varepsilon_0 \beta \sin(\pi x/a)\cos(\beta_y y) \tag{6.28d}$$

$$H_y = 0 \tag{6.28e}$$

$$H_z = jA\omega\varepsilon_0(\pi/a)\cos(\pi x/a)\cos(\beta_y y) \tag{6.28f}$$

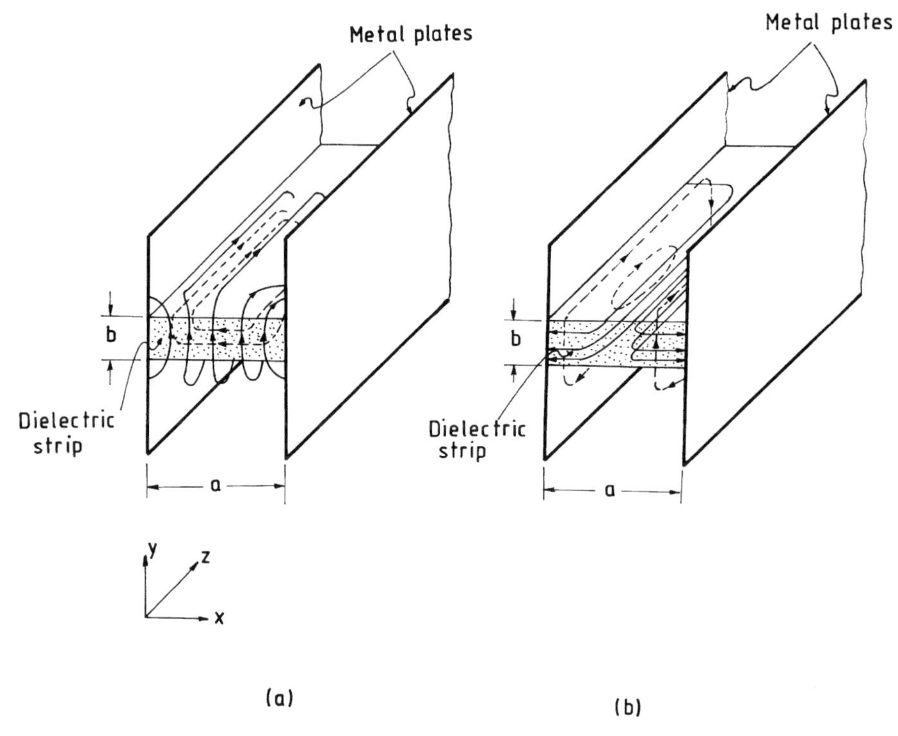

FIGURE 6.2 Typical field lines in NRD guide ($a < \lambda_0/2$): (a) dominant LSM_{11} mode, (b) dominant LSE_{11} mode. ——— E lines, ------- H lines.

Air Region ($|y| > b/2$):

$$E_x = \mp A\zeta(\pi/a)\cos(\pi x/a)\cos(\beta_y b/2)e^{\zeta(b/2-|y|)} \tag{6.29a}$$

$$E_y = A(k_0^2\varepsilon_r - \beta_y^2)\sin(\pi x/a)\cos(\beta_y b/2)e^{\zeta(b/2-|y|)} \tag{6.29b}$$

$$E_z = \pm jA\beta\zeta\sin(\pi x/a)\cos(\beta_y b/2)e^{\zeta(b/2-|y|)} \tag{6.29c}$$

$$H_x = -A\omega\varepsilon_0\beta\sin(\pi x/a)\cos(\beta_y b/2)e^{\zeta(b/2-|y|)} \tag{6.29d}$$

$$H_y = 0 \tag{6.29e}$$

$$H_z = jA\omega\varepsilon_0(\pi/a)\cos(\pi x/a)\cos(\beta_y b/2)e^{\zeta(b/2-|y|)} \tag{6.29f}$$

where β, β_y, and ζ are obtained from the relations

$$\beta^2 = k_0^2\varepsilon_r - (\pi/a)^2 - \beta_y^2 \tag{6.30a}$$

$$\beta_y\tan(\beta_y b/2) = \varepsilon_r\zeta \tag{6.30b}$$

$$\zeta^2 = k_0^2(\varepsilon_r - 1) - \beta_y^2 \tag{6.30c}$$

The value of β_y is obtained as the first root of the characteristic equation (6.30b). Substituting this value of β_y in (6.30a) yields the propagation constant β of the LSM_{11} mode.

Fields for LSE_{11} Mode (See Fig. 6.2(a))

Dielectric Region ($|y| < b/2$):

$$E_x = C\omega\mu_0\beta\cos(\pi x/a)\cos(\beta_y y) \tag{6.31a}$$

$$E_y = 0 \tag{6.31b}$$

$$E_z = jC\omega\mu_0(\pi/a)\sin(\pi x/a)\cos(\beta_y y) \tag{6.31c}$$

$$H_x = C\beta_y(\pi/a)\sin(\pi x/a)\sin(\beta_y y) \tag{6.31d}$$

$$H_y = C(k_0^2\varepsilon_r - \beta_y^2)\cos(\pi x/a)\cos(\beta_y y) \tag{6.31e}$$

$$H_z = jC\beta\beta_y\cos(\pi x/a)\sin(\beta_y y) \tag{6.31f}$$

Air Region ($|y| > b/2$):

$$E_x = C\omega\mu_0\beta\cos(\pi x/a)\cos(\beta_y b/2)e^{\zeta(b/2 - |y|)} \tag{6.32a}$$

$$E_y = 0 \tag{6.32b}$$

$$E_z = jC\omega\mu_0(\pi/a)\sin(\pi x/a)\cos(\beta_y b/2)e^{\zeta(b/2 - |y|)} \tag{6.32c}$$

$$H_x = \pm C\zeta(\pi/a)\sin(\pi x/a)\cos(\beta_y b/2)e^{\zeta(b/2 - |y|)} \tag{6.32d}$$

$$H_y = C(k_0^2\varepsilon_r - \beta_y^2)\cos(\beta_y b/2)\cos(\pi x/a)e^{\zeta(b/2 - |y|)} \tag{6.32e}$$

$$H_z = \pm jC\beta\zeta\cos(\pi x/a)\cos(\beta_y b/2)e^{\zeta(b/2 - |y|)} \tag{6.32f}$$

where β, β_y, and ζ are related by

$$\beta^2 = k_0^2\varepsilon_r - (\pi/a)^2 - \beta_y^2 \tag{6.33a}$$

$$\beta_y\tan(\beta_y b/2) = \zeta \tag{6.33b}$$

$$\zeta^2 = k_0^2(\varepsilon_r - 1) - \beta_y^2 \tag{6.33c}$$

Solving (6.33b) for its first root yields β_y. Substituting this value of β_y in (6.33a) gives the propagation constant β of the LSE_{11} mode. Of the two nonradiative modes, the dominant LSM_{11} mode is the desirable operating mode of the NRD guide and the LSE_{11} mode is considered as parasitic. The LSM_{11} mode has its electric field component predominantly parallel to the plates and hence forms the low-loss propagating mode. This mode can easily be excited by means of a TE_{10} mode rectangular waveguide feed oriented such that its E-field lies parallel to the metal walls of the NRD guide.

6.3.2 Dispersion and Bandwidth Characteristics

The propagation constant of the LSM_{mn} mode of a NRD guide is given by

$$\beta_{mn} = [k_0^2 \varepsilon_r - (m\pi/a)^2 - \beta_{yn}^2]^{1/2}, \quad m = 1, 2, \ldots \tag{6.34}$$

where the values of β_{yn} are the roots of the characteristic equation (6.7) for n odd ($n = 1, 3, 5, \ldots$) and of (6.13) for n even ($n = 2, 4, 6, \ldots$). At cutoff $\beta_{mn} = 0$. The cutoff condition for the dominant LSM_{11} mode and the higher order LSM_{12} (referred as LSM_{02} in [5]) and LSM_{21} (referred as LSM_{11} in [5]) modes are given by

$$LSM_{11} \text{ mode cutoff } (\beta_{11} = 0): \quad a = \lambda_{c11}/2, \quad \lambda_{c11} = 2\pi/\sqrt{k_0^2 \varepsilon_r - \beta_{y1}^2} \tag{6.35a}$$

$$LSM_{12} \text{ mode cutoff } (\beta_{12} = 0): \quad a = \lambda_{c12}/2, \quad \lambda_{c12} = 2\pi/\sqrt{k_0^2 \varepsilon_r - \beta_{y2}^2} \tag{6.35b}$$

$$LSM_{21} \text{ mode cutoff } (\beta_{21} = 0): \quad a = \lambda_{c11} \tag{6.35c}$$

The requirement that the NRD guide, in addition to being nonradiative ($a < \lambda_0/2$), operates only in the dominant LSM_{11} mode is met by the condition

$$\lambda_{c11}/2 < a < \lambda_{c11}, \lambda_{c12}/2 \tag{6.36}$$

Figure 6.3 shows a typical operational diagram of a NRD guide using alumina ($\varepsilon_r = 9.5$) strip wherein the normalized parameter $(b/\lambda_0)\sqrt{\varepsilon_r - 1}$ is plotted as a function of a/λ_0 [5]. The single-mode (LSM_{11}) operating region is bound by the curves $\lambda_{c11} = 2a$, $\lambda_{c11} = a$, and $\lambda_{c12} = 2a$ and the vertical line $\lambda_0 = 2a$. It can be seen that the onset of the LSM_{21} mode (curve c; $\lambda_{c11} = a$) reduces the area of the single-mode operating region. It has been pointed out [5] that the higher order LSM_{21} curve can be eliminated by choosing dielectric materials having ε_r lower than a threshold value of 6.8.

Figure 6.4 shows variation in the relative bandwidth of the NRD guide for the dominant LSM_{11} mode as a function of the parameter $b\sqrt{\varepsilon_r - 1}/a$ for three different dielectrics: polystyrene ($\varepsilon_r = 2.56$), fused quartz ($\varepsilon_r = 3.8$), and alumina ($\varepsilon_r = 9.5$) [5]. The relative bandwidth is defined a

$$BW = \left[\frac{2(f_2 - f_1)}{(f_2 + f_1)}\right] \times 100\% \tag{6.37}$$

where the lower frequency f_1 is the cutoff frequency of the dominant LSM_{11} mode and the upper frequency f_2 is the lowest among the cutoff frequencies of LSM_{21} and LSM_{12} modes and the frequency corresponding to $a = \lambda_0/2$. It can be seen from Figure 6.4 that for guides with low dielectric constant materials ($\varepsilon_r = 2.56, 3.8$), the bandwidth is not affected by the LSM_{21} mode. For the alumina NRD guide, the bandwidth is reduced by the presence of the LSM_{21} mode over

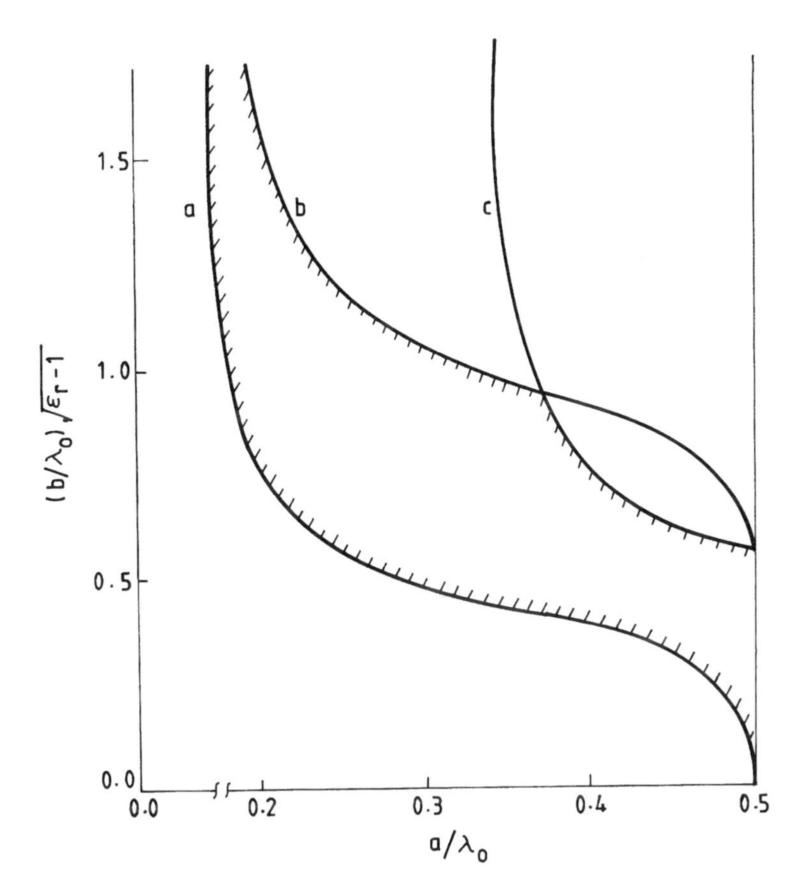

FIGURE 6.3 Operational diagram of a NRD guide: dielectric used, alumina ($\varepsilon_r = 9.5$). Curve a: $\lambda_{c11} = 2a$; curve b: $\lambda_{c12} = 2a$; curve c: $\lambda_{c11} = a$. The extra LSM_{21} mode with the critical curve c acts to reduce the area of the single-mode operating region. (From Yoneyama [5], Copyright © 1984 Academic Press, Inc., Orlando, FL, reprinted with permission.)

a certain range of $b\sqrt{\varepsilon_r - 1}/a$. It may, however, be noted that, in practice, the LSM_{21} mode, which is odd with respect to the symmetry plane between the plates, is not likely to be excited if structural symmetry is maintained. In all three cases, there is a certain value of $b\sqrt{\varepsilon_r - 1}/a$ at which the bandwidth reaches a maximum. It has been shown that the maximum achievable bandwidth increases with an increase in ε_r reaching about 45% when ε_r attains a threshold value of 6.8. For ε_r greater than 6.8, the maximum bandwidth remains nearly constant at 45% if the LSM_{21} mode is taken as the second hybrid mode but continues to increase at a slow rate if, instead of LSM_{21}, the LSM_{12} mode is taken into account. Based on the dispersion and bandwidth characteristics, Yoneyama

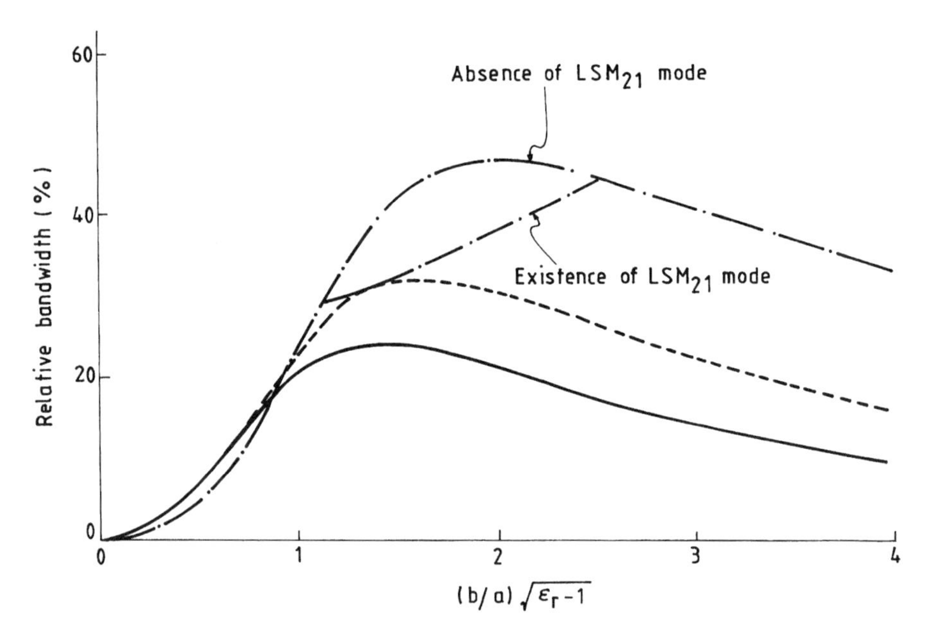

FIGURE 6.4 Relative bandwidth of NRD guide as a function of $(b/a)\sqrt{\varepsilon_r - 1}$ with different dielectrics: —— polystyrene ($\varepsilon_r = 2.56$), -------- fused quartz ($\varepsilon_r = 3.8$), and –·–·– alumina ($\varepsilon_r = 9.5$). Bandwidth is limited by the presence of LSM_{21} mode for certain values of $(b/a)\sqrt{\varepsilon_r - 1}$. (From Yoneyama [5], Copyright © 1984 Academic Press, Inc., Orlando, FL, reprinted with permission.)

[6] has reported the following relations for the design of the NRD guide:

$$a/\lambda_0 \approx 0.45 \tag{6.38a}$$

$$(b/\lambda_0)\sqrt{\varepsilon_r - 1} \approx 0.4\text{–}0.6 \tag{6.38b}$$

6.3.3 Loss Characteristics

The attenuation constant α_t due to transmission loss in the guide is the sum of the attenuation constants α_c due to the conductor loss and α_d due to the dielectric loss:

$$\alpha_t = \alpha_c + \alpha_d \tag{6.39}$$

where

$$\alpha_c = P_c/2P_i \tag{6.40a}$$

$$\alpha_d = P_d/2P_i \tag{6.40b}$$

In (6.40), P_c and P_d denote the power dissipated per unit length of the guide due to the conductor loss and dielectric loss, respectively, and P_i is the power carried by the dominant LSM_{11} mode. In terms of the NRD guide (see Fig. 6.1) field components, we have

$$P_c = 2R_s \int_{-\infty}^{\infty} |H_z(x=0)|^2 \, dy \tag{6.41a}$$

$$P_d = \omega \varepsilon_0 \varepsilon_r (\tan \delta) \int_{-b/2}^{b/2} \int_0^a (|E_x|^2 + |E_y|^2 + |E_z|^2) \, dx \, dy \tag{6.41b}$$

$$P_i = -\int_{-\infty}^{\infty} \int_0^a E_y H_x^* \, dx \, dy \tag{6.41c}$$

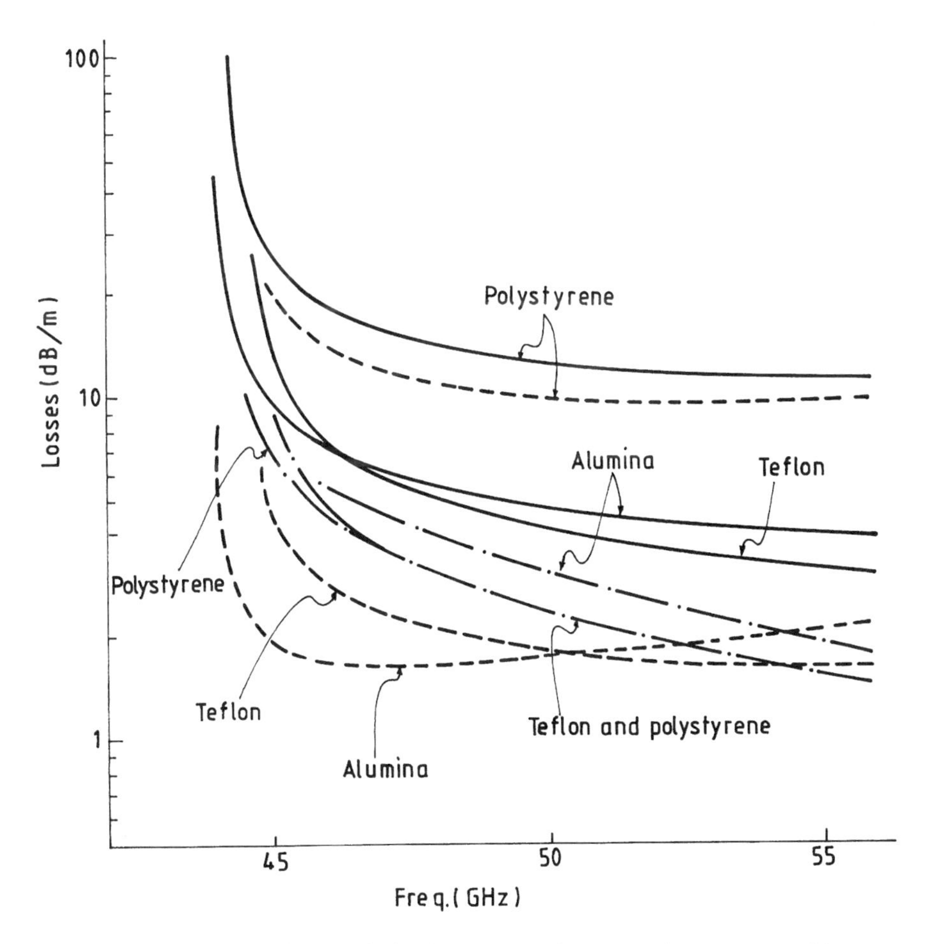

FIGURE 6.5 Theoretical transmission loss curves of NRD guides: —— total loss, ----- dielectric loss, –·–·– conductor loss. Teflon ($\varepsilon_r = 2.04$, $\tan \delta = 1.5 \times 10^{-4}$, $b = 3.5$ mm), polystyrene ($\varepsilon_r = 2.56$, $\tan \delta = 9 \times 10^{-4}$, $b = 2.4$ mm), and alumina ($\varepsilon_r = 9.5$, $\tan \delta = 10^{-4}$, $b = 0.93$ mm) are assumed for strip materials; $\sigma = 5.8 \times 10^7$ mhos/m, $a = 2.7$ mm. (From Yoneyama [5], Copyright © 1984 Academic Press, Inc., Orlando, FL, reprinted with permission.)

where $\tan\delta$ is the loss tangent of the dielectric material, $R_s = \sqrt{\omega\mu_0/2\sigma}$ is the surface resistance of the metal plates, and σ is the conductivity of the metal plates. Substituting for the fields from (6.28) and (6.29) in (6.41) and then using the resulting expressions after simplification in (6.40), we obtain [5]

$$\alpha_c = \left[\frac{2R_s\omega\varepsilon_0\varepsilon_r\pi^2}{(k_0^2\varepsilon_r - \beta_y^2)\beta a^3}\right]\left[\frac{\beta_y b + [1 + (\beta_y^2/\zeta^2\varepsilon_r)]\sin(\beta_y b)}{\beta_y b + [k_0^2(\varepsilon_r - 1)/\zeta^2]\sin(\beta_y b)}\right] \tag{6.42}$$

$$\alpha_d = \left[\frac{k_0^2\varepsilon_r\tan\delta}{2\beta}\right]\left[\frac{\beta_y b + [1 - (2\beta_y^2/k_0^2\varepsilon_r)]\sin(\beta_y b)}{\beta_y b + [k_0^2(\varepsilon_r - 1)/\zeta^2]\sin(\beta_y b)}\right] \tag{6.43}$$

where β, β_y, and ζ are the parameters of the LSM_{11} mode.

Figure 6.5 illustrates typical transmission loss characteristics of the NRD guide for three different dielectric strips: Teflon ($\varepsilon_r = 2.04$, $\tan\delta = 1.5 \times 10^{-4}$, $b = 3.5$ mm), polystyrene ($\varepsilon_r = 2.56$, $\tan\delta = 9 \times 10^{-4}$, $b = 2.4$ mm), and alumina ($\varepsilon_r = 9.5$, $\tan\delta = 10^{-4}$, $b = 0.93$ mm). For all three guides, the conductor material is copper ($\sigma = 5.8 \times 10^7$ mhos/m) and the plate separation is fixed at 2.7 mm. As expected from the field structure for the LSM_{11} mode, the conductor loss for all three guides is small. The dielectric loss for each guide is nearly the same as that for an infinite dielectric medium. This is an expected result because most of the electromagnetic energy is confined within the dielectric strip. Since the dielectric loss is proportional to the loss tangent of the dielectric material, the Teflon and alumina NRD guides show much lower dielectric loss than the polystyrene guide. Correspondingly, the overall transmission loss in Teflon and alumina guides is also much smaller. As can be seen, typical transmission loss for Teflon and alumina guides at 50 GHz is about 4 dB/m and 5 dB/m, respectively. As a comparison, the transmission loss of a microstrip line on alumina substrate at the same frequency is reported to be 57 dB/m [12].

6.4 COUPLED NRD GUIDE

Coupled NRD guides employing edge-coupled dielectric strips are useful in the design of directional couplers and filters. For a symmetrically coupled NRD guide as shown in Figure 6.6, the low-loss LSM fields may be considered as a superposition of even and odd modes propagating longitudinally with different propagation constants β_e and β_o, respectively. The symmetry plane $y = 0$ is assumed to be a magnetic wall ($E_y = 0$ at $y = 0$) for the even mode and an electric wall ($E_x = 0$ at $y = 0$) for the odd mode. For the LSM-type fields, the scalar potentials ϕ^e satisfying the Helmholtz equation (2.28) in the three regions marked in Figure 6.6 can be expressed as

$$\phi^e = A\sin(m\pi x/a)e^{-\zeta y} \qquad y > b + d/2 \tag{6.44a}$$

$$= [B\cos(\beta_y y) + C\sin(\beta_y y)]\sin(m\pi x/a) \qquad d/2 < y < b + d/2 \tag{6.44b}$$

$$= D\sin(m\pi x/a)\begin{cases}\sinh(\zeta y), & \text{even mode}\\ \cosh(\zeta y), & \text{odd mode}\end{cases} \qquad 0 < y < d/2 \tag{6.44c}$$

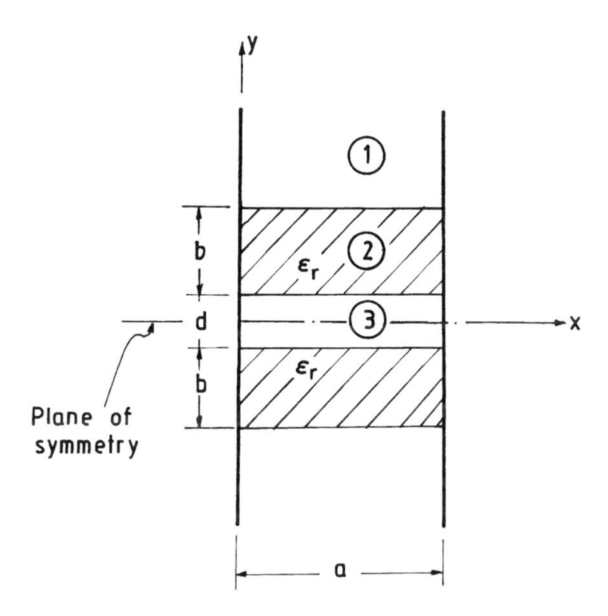

FIGURE 6.6 Parallel-coupled NRD guide.

Substituting for ϕ^e in (2.24) and setting $\partial/\partial z = -j\beta$, we can obtain expressions for all the field components. The E_x and E_y components are given by

$$E_x = -A(m\pi/a)\zeta\cos(m\pi x/a)e^{-\zeta y} \qquad y > b + d/2 \quad (6.45a)$$

$$= (1/\varepsilon_r)(m\pi/a)\beta_y\cos(m\pi x/a)[-B\sin(\beta_y y) + C\cos(\beta_y y)] \qquad d/2 < y < b + d/2 \quad (6.45b)$$

$$= D(m\pi/a)\zeta\cos(m\pi x/a)\begin{cases}\cosh(\zeta y), & \text{even mode}\\ \sinh(\zeta y), & \text{odd mode}\end{cases} \qquad 0 < y < d/2 \quad (6.45c)$$

$$E_y = Ag^2\sin(m\pi x/a)e^{-\zeta y} \qquad y > b + d/2 \quad (6.46a)$$

$$= (g^2/\varepsilon_r)\cos(m\pi x/a)[B\cos(\beta_y y) + C\sin(\beta_y y)] \qquad d/2 < y < b + d/2 \quad (6.46b)$$

$$= Dg^2\sin(m\pi x/a)\begin{cases}\sinh(\zeta y), & \text{even mode}\\ \cosh(\zeta y), & \text{odd mode}\end{cases} \qquad 0 < y < d/2 \quad (6.46c)$$

where

$$g^2 = \beta^2 + (m\pi/a)^2 = k_0^2\varepsilon_r - \beta_y^2 = k_0^2 + \zeta^2 \qquad (6.47)$$

Applying the boundary conditions at the air–dielectric interfaces $y = d/2$ and

$y = b + d/2$, we obtain the following characteristic equation:

$$\frac{\beta_y[\beta_y \tan(\beta_y b) - \varepsilon_r \zeta]}{\varepsilon_r \zeta[\varepsilon_r \zeta \tan(\beta_y b) + \beta_y]} = \begin{cases} \coth(\zeta d/2), & \text{even mode} \\ \tanh(\zeta d/2), & \text{odd mode} \end{cases} \tag{6.48}$$

The solution of (6.48) in conjunction with (6.47) yields the even- and odd-mode propagation constants β_e and β_o of the dominant *LSM* mode.

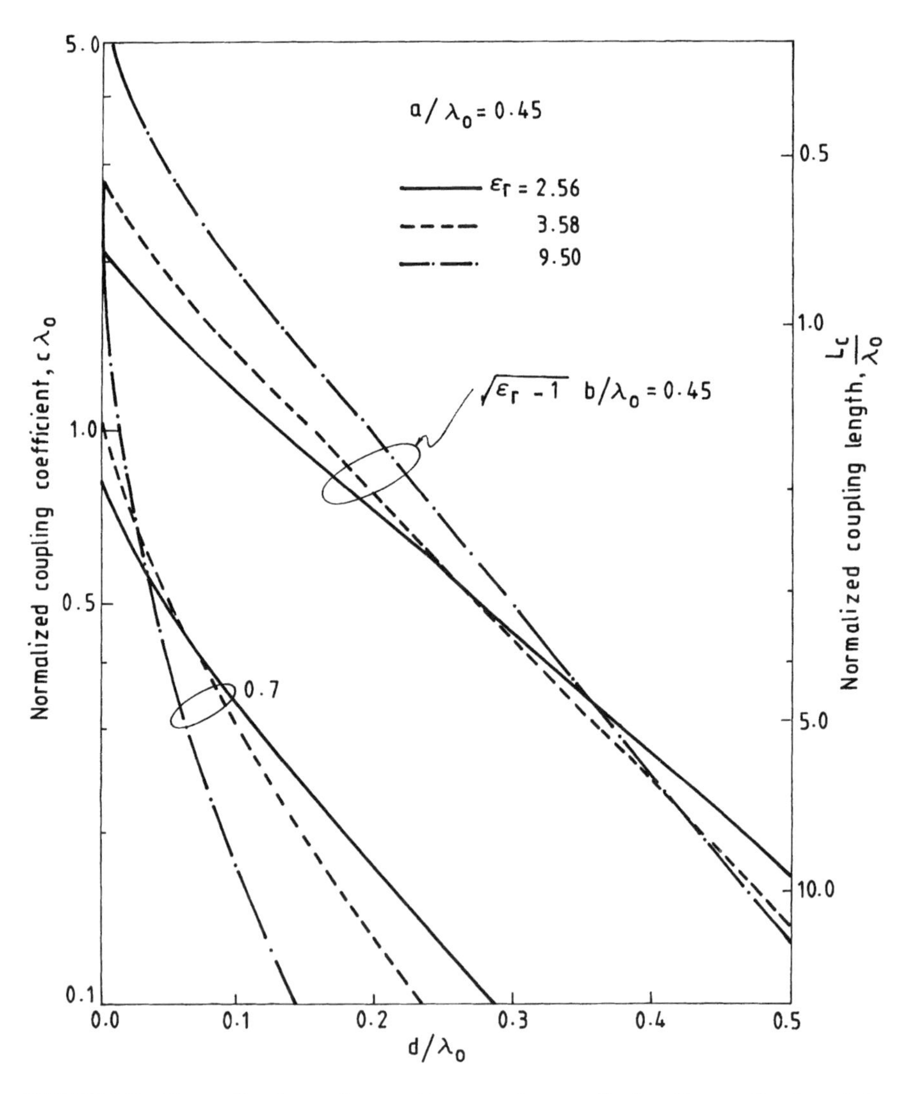

FIGURE 6.7 Normalized coupling coefficient and normalized 0-dB coupling length for coupled strips of alumina ($\varepsilon_r = 9.5$), fused quartz ($\varepsilon_r = 3.8$), and polystyrene ($\varepsilon_r = 2.56$). (From Yoneyama and Nishida [1]. Copyright © 1981 IEEE, reprinted with permission.)

Figure 6.7 shows the coupled guide characteristics in terms of the coupling coefficient and 0-dB coupling length as a function of the strip spacing normalized with respect to the free-space wavelength λ_0 [1]. The coupling coefficient is defined as

$$C = 0.5\,(\beta_e - \beta_o) \tag{6.49}$$

and the 0-dB coupling length L_c required for complete power transfer from one guide to the other is given by [6]

$$L_c = \frac{\pi}{2C} = \frac{\pi}{\beta_e - \beta_o} \tag{6.50}$$

It can be seen from Figure 6.7 that for all three dielectrics chosen ($\varepsilon_r = 2.56$, 3.8, 9.5), a smaller value of $b\sqrt{(\varepsilon_r - 1)}/a$ gives a larger coupling coefficient. This is an expected result since, for strips of smaller width b and lower dielectric constant ε_r, fields extend over larger distances in the transverse direction, thereby enhancing the coupling.

Another coupled NRD guide geometry realized by introducing a vertical slit at the center of the dielectric strip is shown in Figure 6.8. The even- and odd-*LSM* modes are defined with respect to the symmetry plane parallel to the metal plates. The structure is basically a broadside-coupled image guide. The even- and odd-mode characteristics of this guide for excitation in the TM^y (*LSM*) mode are

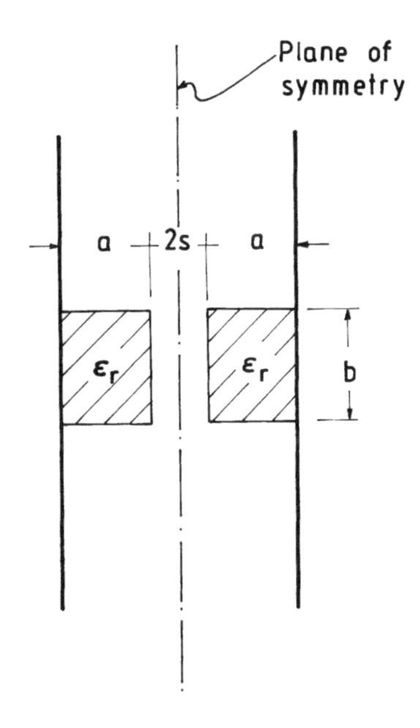

FIGURE 6.8 Broadside-coupled NRD guide.

reported in Bhat and Tiwari [13]. It is shown that the cutoff frequency for the odd mode is higher than that for the even mode. For the structure to be nonradiative, the cutoff frequency of the odd mode must be less than that for an air-filled structure with plate separation $2(a+s)$. This can be achieved by choosing sufficiently large values of $b/2a$ and ε_r. Figure 6.9 shows typical dispersion characteristics of a shielded broadside-coupled NRD guide having $b/2a = 1.5$ and

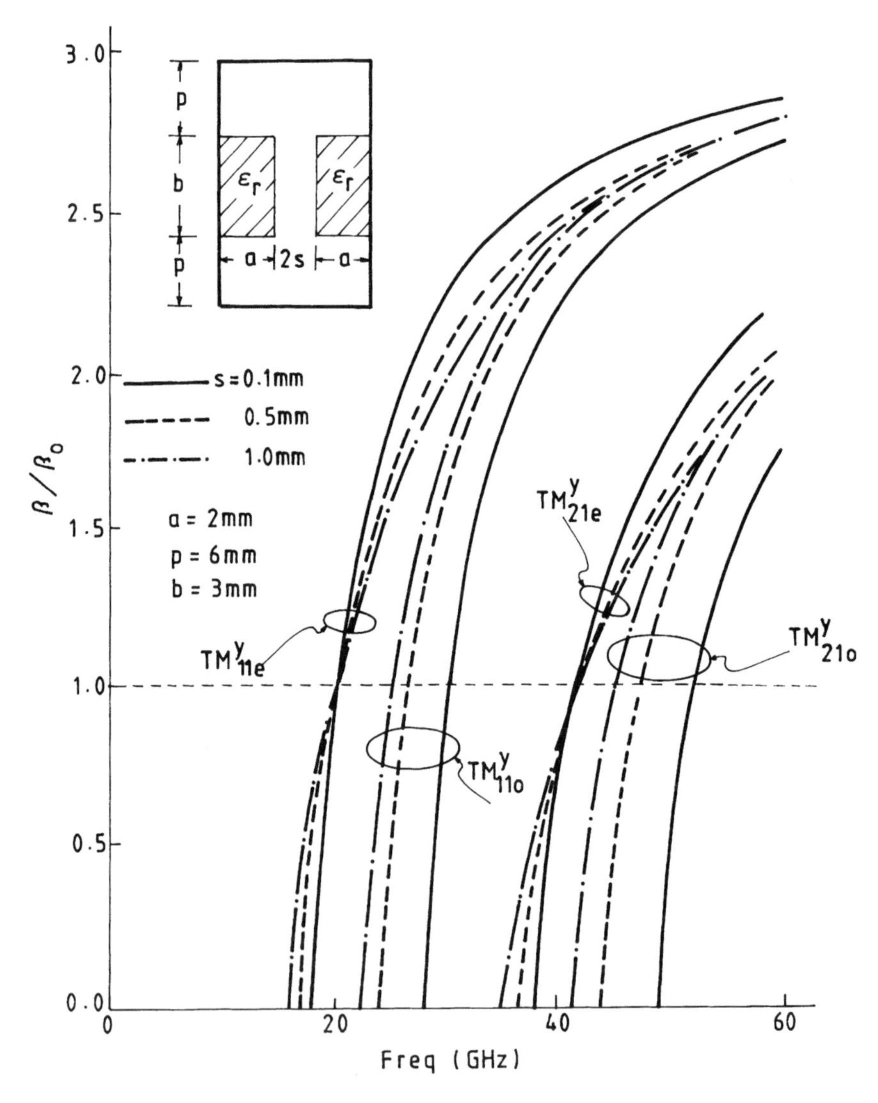

FIGURE 6.9 Normalized propagation constant β/β_0 versus frequency of broadside-coupled NRD guide for TM^y (LSM) modes. (From Bhat and Tiwari [13]. Copyright © 1984 IEEE, reprinted with permission.)

$\varepsilon_r = 9.6$ [13]. The top and bottom metal plates of the structure are chosen sufficiently away ($p = 6$ mm) such that their influence on the propagation characteristics of the guide is negligible. It may be observed from the graph that in the case of the even mode, the dispersion curves for different spacing $2s$ cross each other at $\beta/\beta_0 = 1$, whereas for the odd mode they do not cross. The region $\beta/\beta_0 < 1$ corresponds to the waveguide modes, whereas the region $\beta/\beta_0 \geq 1$ corresponds to the nonradiative guide modes. Furthermore, for $\beta/\beta_0 > 1$, the fields in the space between the dielectric strips are evanescent. The coupling between the strips therefore decreases as s increases and for sufficiently large values of s, the even- and odd-mode propagation constants tend to become equal.

6.5 GROOVE NRD GUIDE

The groove NRD guide, the cross-section of which is shown in Figure 6.10, is a useful variant of the NRD guide [14, 15]. Besides acting as a support for the dielectric strip, the grooves help in enhancing the field concentration in the dielectric strip. With the dielectric strip filling the entire groove region, the nonradiative condition can be written as

$$(a - 2d) < \lambda_0/2 \tag{6.51}$$

where d is the groove depth. Tong and Blundell [15] have studied the dispersion characteristics of the groove NRD assuming square cross-section for the groove region ($a = b$) and small groove depth ($d = 0.1\,a$). Figure 6.11 shows a comparison of the dispersion characteristics of a groove NRD with that of a simple NRD

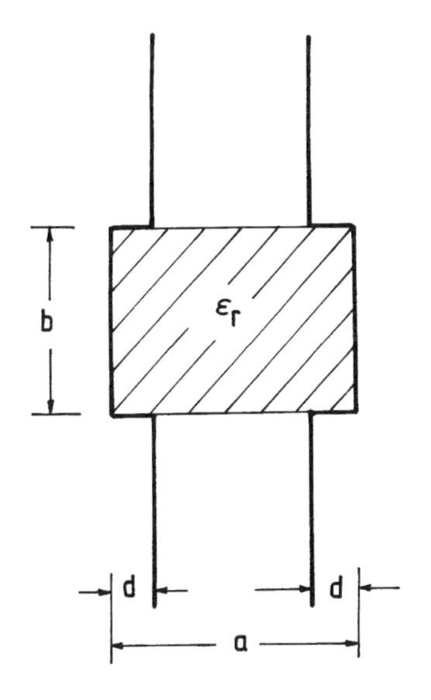

FIGURE 6.10 Groove NRD guide.

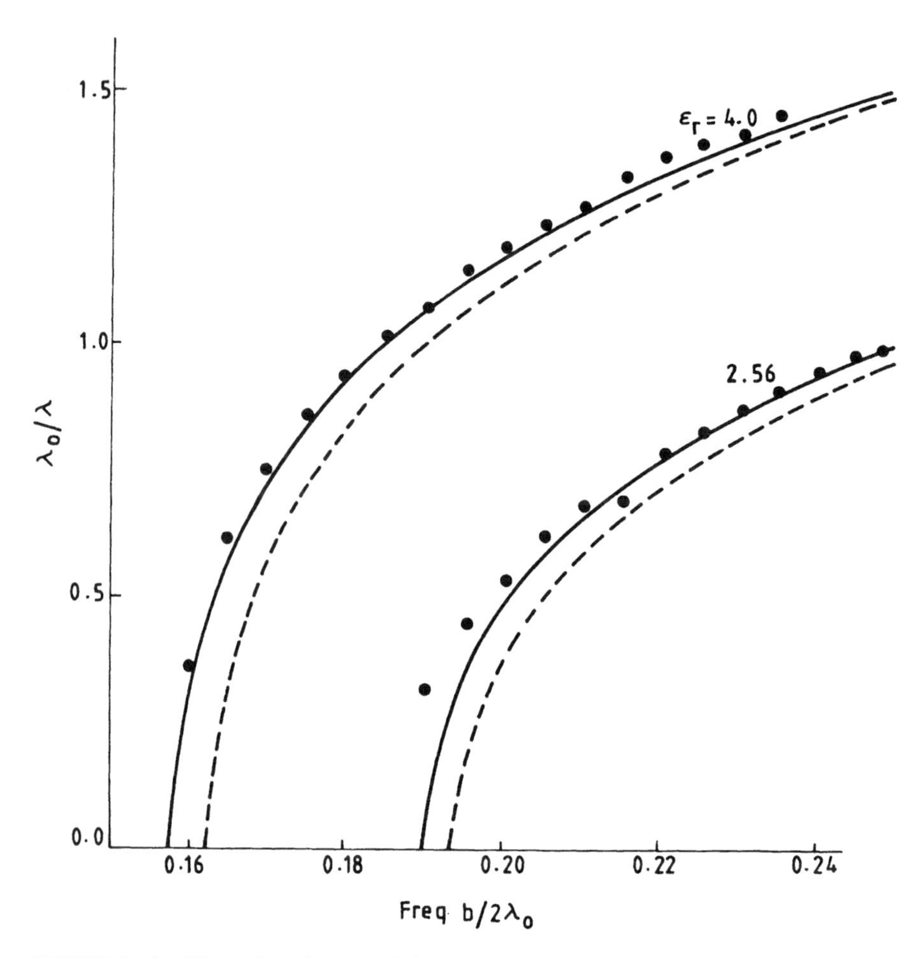

FIGURE 6.11 Dispersion characteristics of groove NRD guide; measured (•) and computed (———) for $a = b$ and $d = 0.1a$. Also shown are the computed characteristics of a simple NRD guide ($d = 0$) (-------) [11]. (From Tong and Blundell [15], reprinted with permission of IEE.)

guide ($d = 0$) for the dominant LSM_{11} mode. The plots show variation in λ_0/λ (where λ is the guide wavelength) as a function of $b/2\lambda_0$ for two different dielectrics, stycast HiK ($\varepsilon_r = 4$) and rexolite ($\varepsilon_r = 2.56$) [15]. The strip cross-section is $a \times b = 12\,\text{mm} \times 12\,\text{mm}$ and the groove depth is $d = 0.1\,a = 1.2\,\text{mm}$. The effect of the grooves is clearly to lower the cutoff frequency of the LSM_{11} mode of the NRD guide but only marginally up to about 4%. It has been shown [15] that for the next higher order LSM_{12} mode, the cutoff frequency increases over that of the simple NRD guide for small values of ε_r but decreases for higher ε_r. Thus, with dielectric strips of lower dielectric constant, the useful bandwidth of a groove NRD is higher than that of a simple NRD. For example, the bandwidth is reported to increase from 24% to 33.4% for $\varepsilon_r = 2.5$, to increase from 33.5% to 36.6% for $\varepsilon_r = 4$, and to decrease from 42.2% to 40.4% for $\varepsilon_r = 10$ [15].

PROBLEMS

6.1 **(a)** Draw the field intensity distribution of the dominant field components of the LSM_{11} and LSE_{11} modes in a NRD guide.

(b) Why are these two modes called nonradiative?

6.2 Calculate and plot cutoff frequency versus a/b for the dominant LSM_{11} mode and the next two higher order LSM modes of a NRD guide. Assume $b = 2$ mm and $\varepsilon_r = 2.56$. Determine the value of a that yields the maximum monomode bandwidth for the dominant mode. What is the corresponding frequency range?

6.3 Design a NRD guide for operation only in the LSM_{11} mode over the frequency 34–36 GHz. Choose polystyrene having $\varepsilon_r = 2.56$.

6.4 Derive the following closed-form expression for the characteristic impedance of the LSM_{11} mode of a NRD guide (Fig. 6.12) based on the power–voltage definition:

$$Z_c = \frac{V^2}{2P}$$

where

$$P = \frac{1}{2}\,\mathrm{Re}\int_S (\mathbf{E}_y \times \mathbf{H}_x^*)\cdot\hat{\mathbf{z}}\,ds$$

$$V = \int_{-\infty}^{\infty} E_y\,dy \quad \text{at } x = \frac{a}{2}$$

Answer:

$$Z_c = \frac{16\sin^2\left(\dfrac{k_1 b}{2}\right)\left[\dfrac{(k^2 - k_1^2)}{k_1} + \dfrac{k_1(k_0^2 + k_2^2)^2}{k_2^2}\right]^2}{\omega\varepsilon_0\beta a\left[\varepsilon_r(k_1 b + \sin k_1 b)\dfrac{(k^2 - k_1^2)}{k_1} + 2k_1^2\dfrac{(k_0^2 + k_2^2)\sin^2(k_1 a)}{k_2^3}\right]}$$

where $k_0^2 = \omega^2\mu_0\varepsilon_0$, $k^2 = k_0^2\varepsilon_r$, k_1 and k_2 are wavenumbers in the y-direction in the dielectric and air regions, respectively, and β is the propagation constant in the z-direction.

6.5 Consider a NRD guide with a thin insular layer of thickness d and having a low dielectric constant ($\varepsilon_{r2} < \varepsilon_{r1}$) as shown in Figure 6.13.

(a) Using the EDC technique, derive the characteristic equation for the LSM_{mn} modes.

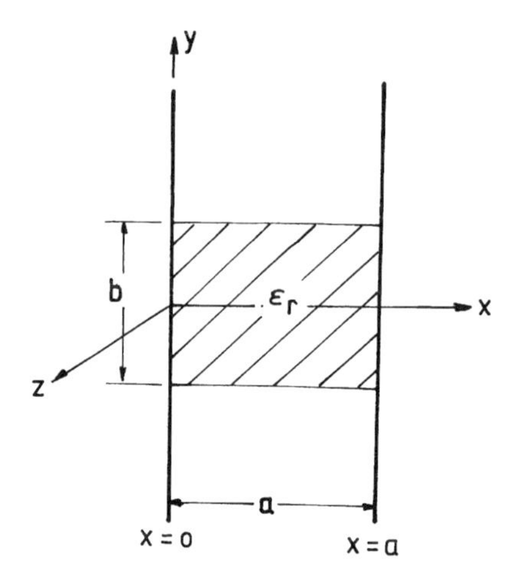

FIGURE 6.12 NRD guide.

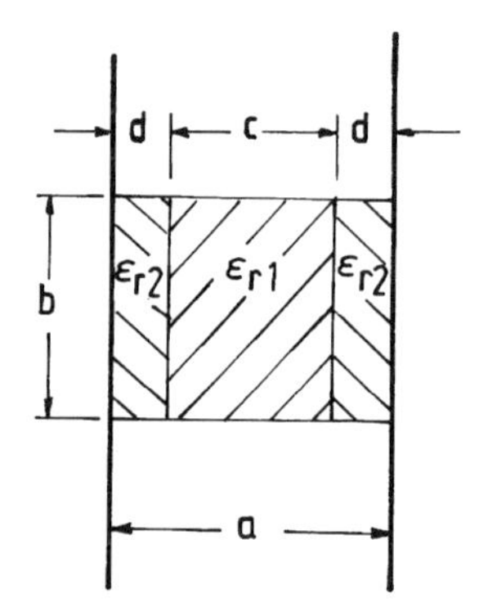

FIGURE 6.13 NRD guide with a thin insular guide.

(b) Assume $b = 2\,\text{mm}$, $c = 3\,\text{mm}$, and $d = 0.5\,\text{mm}$; plot the dispersion characteristic as a function of frequency for the LSM_{11} mode and the first higher order mode for the following two cases. (i) $\varepsilon_{r1} = 2.56$, $\varepsilon_{r2} = 1.0$; and (ii) $\varepsilon_{r1} = \varepsilon_{r2} = 2.56$. Which of the two cases gives a larger bandwidth?

6.6 (a) For the structure shown in Figure 6.13, obtain fields in various regions.

(b) Derive an expression for the conductor attenuation constant α_c.

(c) For the parameters considered in Problem 6.5, calculate α_c in dB/m at $f = 1.2 f_c$, where f_c is the cutoff frequency of the LSM_{11} mode. Discuss the effect of an insular layer on the conductor loss of a NRD guide.

6.7 Draw typical field lines for the LSM_{11} even- and odd-modes in the coupled NRD guide shown in Figure 6.14.

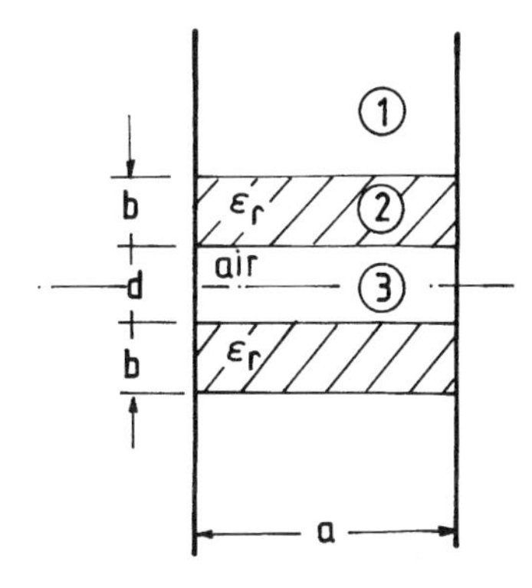

FIGURE 6.14 Coupled NRD guide.

6.8 Consider the LSM_{11} even mode of the coupled NRD guide shown in Figure 6.14. For the parameters $a = 4.5\,\text{mm}$, $b = 3.5\,\text{mm}$, and $\varepsilon_r = 2.56$, compute and plot the propagation constant β_e as a function of frequency for $d = 0$, 0.5 mm, and 2 mm. Discuss the effect of an insular layer d on the propagation constant and cutoff frequency.

6.9 For the structure shown in Figure 6.14, compute the monomode bandwidth (for the LSM_{11} mode). Choose the same parameters as in Problem 6.8. What is the effect of the insular layer d on the bandwidth?

6.10 Derive an expression for the conductor attenuation constant α_c of the LSM_{11} mode for the NRD guide shown in Figure 6.14. Assume $a = 4.5\,\text{mm}$, $b = 3.5\,\text{mm}$, and $\varepsilon_r = 2.56$. Calculate α_c in dB/m at $f = 1.2\,f_c$ (where f_c is the cutoff frequency of the LSM_{11} mode) for $d = 0$, 0.5 mm, and 2 mm. Discuss the effect of an insular layer d on the conductor loss.

6.11 Draw typical field lines for the LSM_{11} mode in a groove NRD guide. Discuss the effect of grooves on the cutoff frequency and bandwidth of the guide.

REFERENCES

1. T. Yoneyama and S. Nishida, Non-radiative dielectric waveguide for millimetre-wave integrated circuits. *IEEE Trans. Microwave Theory Tech.*, **MTT-29**, 1188–1192, Nov. 1981.
2. T. Yoneyama, H. Tamaki, and S. Nishida, Analysis and measurements of non-radiative dielectric waveguide. *IEEE Trans. Microwave Theory Tech.*, **MTT-34**, 876–882, Aug. 1986.
3. T. Yoneyama, N. Tozawa, and S. Nishida, Coupling characteristics of non-radiative dielectric waveguide. *IEEE Trans. Microwave Theory Tech.*, **MTT-31**, 648–654, Aug. 1983.

4. T. Yoneyama, N. Tozawa, and S. Nishida, Loss measurements of non-radiative dielectric waveguide. *IEEE Trans. Microwave Theory Tech.*, **MTT-32**, 943–946, Aug. 1984.
5. T. Yoneyama, Nonradiative dielectric waveguide. *Infrared and Millimeter Waves*, Vol. 11, K. J. Button (Ed.), Chap. 2, Academic Press, Orlando, FL, 1984.
6. T. Yoneyama, Millimetre wave integrated circuits using non-radiative dielectric waveguide, Proceedings of the Yagi Symposium on Advanced Technology Bridging the Gap Between Light and Microwaves, pp. 57–66, Sept. 1990.
7. T. Yoneyama, S. Fujita, and S. Nishida, Insulated non-radiative dielectric waveguide for millimetre-wave integrated circuits. *IEEE Trans. Microwave Theory Tech.*, **MTT-31**, 1002–1008, Dec. 1983.
8. T. Yoneyama, M. Yamaguchi, and S. Nishida, Bends in non-radiative dielectric waveguides. *IEEE Trans. Microwave Theory Tech.*, **MTT-30**, 2146–2150, Dec. 1982.
9. T. Yoneyama and S. Nishida, Non-radiative dielectric waveguide circuit components. *Int. J. Infrared Millimetre Waves*, **4(3)**, 439–449, 1983.
10. T. Yoneyama, F. Kuroki, and S. Nishida, Design of non-radiative dielectric waveguide filters. *IEEE Trans. Microwave Theory Tech.*, **MTT-32**, 1659–1662, Dec. 1984.
11. F. Kuroki and T. Yoneyama, Non-radiative dielectric waveguide circuit components using beam lead diodes. *Trans. IEICE (C-I)*, **J-73-C-I(2)**, 71–76, Feb. 1990.
12. T. Tokumitsu, M. Ishizaki, and T. Saito, 50 GHz IC components using alumina substrates. *IEEE Trans. Microwave Theory Tech.*, **MTT-31**, 121–128, Feb. 1983.
13. B. Bhat and A. K. Tiwari, Analysis of low-loss broadside-coupled dielectric image guide using the mode-matching technique. *IEEE Trans. Microwave Theory Tech.*, **MTT-32**, 711–717, July 1984.
14. W. X. Zhang and L. Zhu, New leaky-wave antenna for millimetre-wave constructed from groove NRD waveguide. *Electron. Lett.*, **23(22)**, 1191–1192, 1987.
15. C. E. Tong and R. Blundell, Study of groove nonradiative dielectric waveguide. *Electron. Lett.*, **25(14)**, 934–936, July 1989.

CHAPTER SEVEN

Nonplanar Dielectric Guides

7.1 INTRODUCTION

Most of the dielectric guides used for millimeter wave circuit application are of the planar type, having rectangular shaped dielectrics. Guides belonging to this class are covered in Chapters 2–6. In this chapter, we present the characteristics of a few special guides, which employ nonrectangular shaped dielectrics. These guides include the semicircular dielectric image guide [1–3], the semielliptical image guide [4, 5], the Y-dielectric guide [6], the triangular dielectric guide [6], and the dielectric tube contacted slab guide [7].

7.2 SEMICIRCULAR DIELECTRIC IMAGE GUIDE

The propagating modes in a circular dielectric rod are hybrid modes, designated as EH_{mn} and HE_{mn} modes. Pure TE and TM modes are possible only for $m = 0$, that is, when the field is independent of the azimuthal angle. For this case, the EH_{mn} modes reduce to the TE_{0n} modes and the HE_{mn} modes reduce to the TM_{0n} modes. The dominant mode in the guide is the EH_{11} mode (same as HE_{11} mode), also known as the dipole mode. Figure 7.1 shows the field configuration for this mode. The dispersion characteristics for the first three modes of the guide are shown in Figure 7.2 [8]. The parameters β and k_0 $(= 2\pi/\lambda_0)$ marked in the figure are the propagation constants in the dielectric guide and free space, respectively. As can be seen, the dominant EH_{11} mode has no low-frequency cutoff, whereas the higher order modes have finite cutoff frequencies.

The semicircular dielectric image guide (shown in Fig. 7.3) makes use of the symmetry of the EH_{11} mode of the circular dielectric guide to replace one-half of the dielectric rod by an image ground plane [1–3]. The ground plane provides a support for the guide and also facilitates the design of bends. This guide can easily be excited by the TE_{11} mode circular metal waveguide through a tapered transition. As in the case of the rectangular image guide, the losses in the

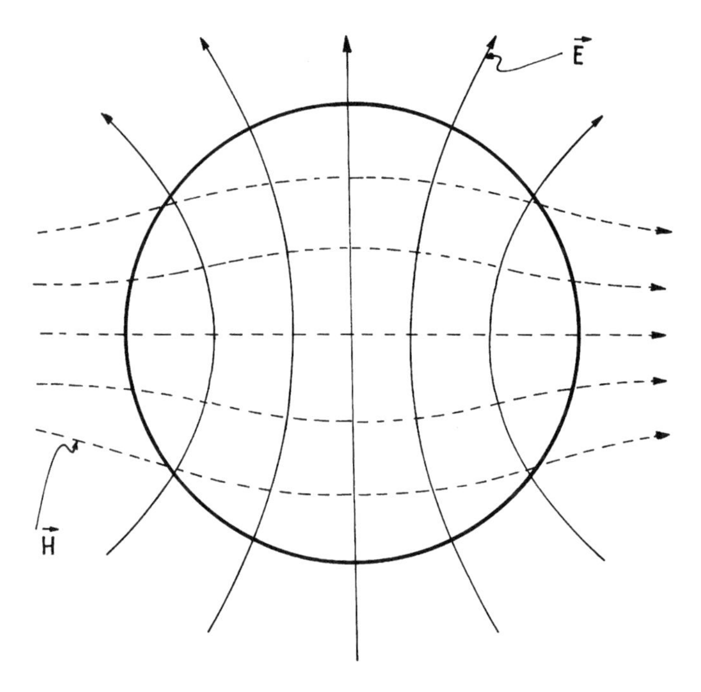

FIGURE 7.1 Field configuration of the EH_{11} mode in a circular dielectric rod.

semicircular image guide are of two types: conductor loss due to the current in the ground plane and dielectric loss due to the finite loss tangent of the dielectric. Explicit expressions have been reported for the conductor attenuation constant α_c [2] and also for the dielectric attenuation constant α_d [9].

7.2.1 Dielectric Attenuation Constant α_d

The dielectric attenuation constant [9] is given by

$$\alpha_d = 27.3[(\varepsilon_r \tan \delta)/\lambda_0]R \tag{7.1}$$

where λ_0 is the free-space wavelength and $\tan \delta$ is the loss tangent of the dielectric. R is a ratio factor given by

$$R = \frac{1}{D}\left[\left(\frac{\varepsilon_r - 1}{q^2}\right)\left(\frac{F^2 + 1/p^2 - 1/p^4}{1/p^2 + 1/q^2}\right) + (U^2 + V^2)X + \frac{4UV}{p^4}\right] \tag{7.2}$$

where

$$D = UX(\varepsilon_r + V^2) + UY(1 + V^2) + (2V/p^4)(\varepsilon_r + U^2) - (2V/q^4)(1 + U^2) \tag{7.3a}$$

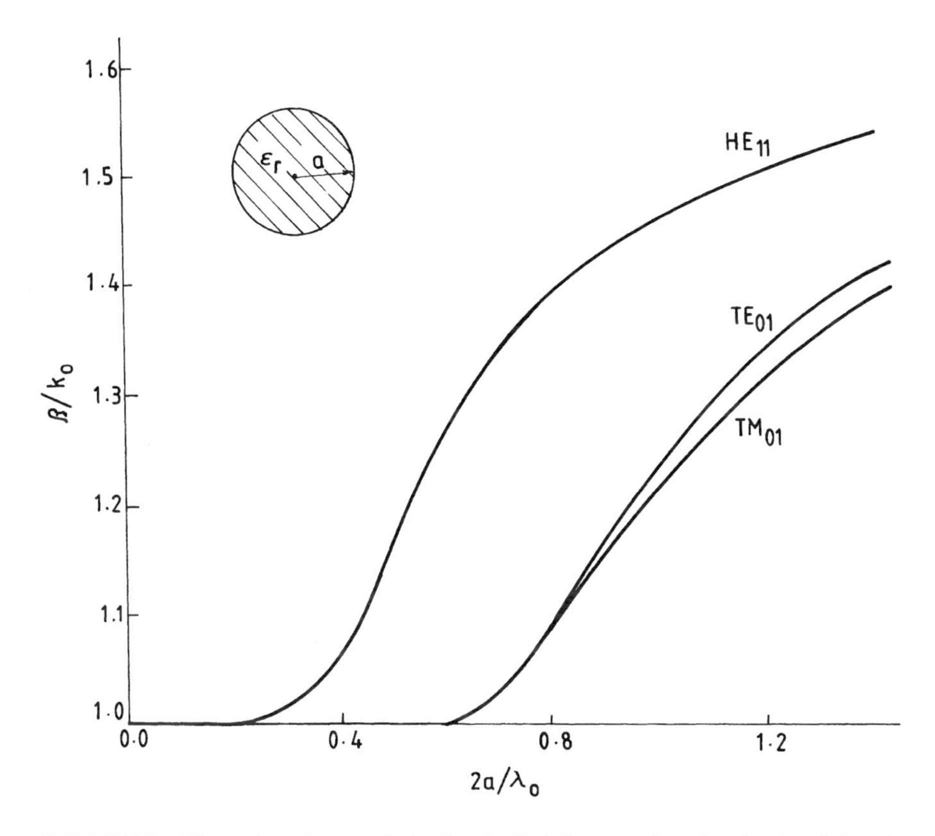

FIGURE 7.2 Dispersion characteristics for the first three modes of a circular dielectric rod: $\varepsilon_r = 2.56$ (polystyrene). (After Chatterjee [8].)

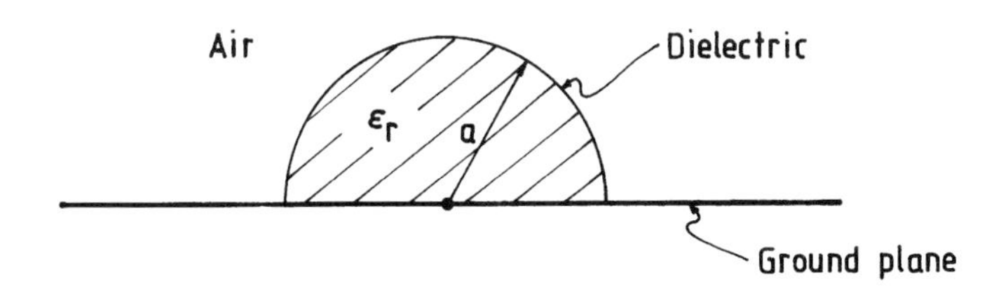

FIGURE 7.3 Cross-sectional view of a semicircular dielectric image guide.

$$U = \left(\frac{\varepsilon_r/p^2 + 1/q^2}{1/p^2 + 1/q^2} \right)^{1/2} \tag{7.3b}$$

$$V = \left(\frac{\varepsilon_r F + G}{F + G} \right)^{1/2} \tag{7.3c}$$

$$X = F^2 + \frac{2F + 1}{p^2} - \frac{1}{p^4} \tag{7.3d}$$

$$Y = -G^2 - \frac{2G-1}{q^2} + \frac{1}{q^4} \tag{7.3e}$$

$$F = \frac{J_1'(p)}{pJ_1(p)} \tag{7.3f}$$

$$G = \frac{K_1'(q)}{qK_1(q)} \tag{7.3g}$$

The parameters p and q are related to the propagation constant β by the relation

$$\beta^2 = k_0^2\varepsilon_r - (p/a)^2 = k_0^2 + (q/a)^2 \tag{7.3h}$$

7.2.2 Conductor Attenuation Constant α_c

The conductor attenuation constant [2] is given by

$$\alpha_c = 69.5(R_s/\eta_0\lambda_0)R' \quad \text{dB/m} \tag{7.4}$$

where R_s is the surface resistivity of the image line and η_0 is the free-space wave impedance. The ratio factor R' is given by

$$R' = (1/\pi D)(\lambda_0/2a)[\{f(I)/J_1^2(p)\} + \{f(H)/K_1^2(q)\}] \tag{7.5a}$$

where D and p are as given in (7.3) and a is the radius of the semicircular rod. The functions $f(I)$ and $f(H)$ are given by

$$\begin{aligned} f(I) = {} & \tfrac{2}{3}\varepsilon_r^2[S(2S+3)/p^2][I_1 + J_0(p)J_1(p)] \\ & + I_1[V^2(\varepsilon_r - 1)/p(p^2+q^2)] \\ & - \tfrac{1}{3}I_0\varepsilon_r^2(S^2-3)/p^3 - \tfrac{1}{3}\varepsilon_r^2 S^2 J_1^2(p)/p^4 \end{aligned} \tag{7.5b}$$

$$\begin{aligned} f(H) = {} & -\tfrac{2}{3}[T(2T+3)/q^3][H_1 + K_0(q)K_1(q)] \\ & + H_1[V^2(\varepsilon_r - 1)/q(p^2+q^2)] \\ & - \tfrac{1}{3}H_0(T^2-3)/q^3 - \tfrac{1}{3}T^2K_1^2(q)/q^4 \end{aligned} \tag{7.5c}$$

$$I_0 = \int_0^p J_0^2(z)\,dz \tag{7.5d}$$

$$I_1 = \int_0^p J_1^2(z)\,dz \tag{7.5e}$$

$$H_0 = \int_q^\infty K_0^2(z)\,dz \tag{7.5f}$$

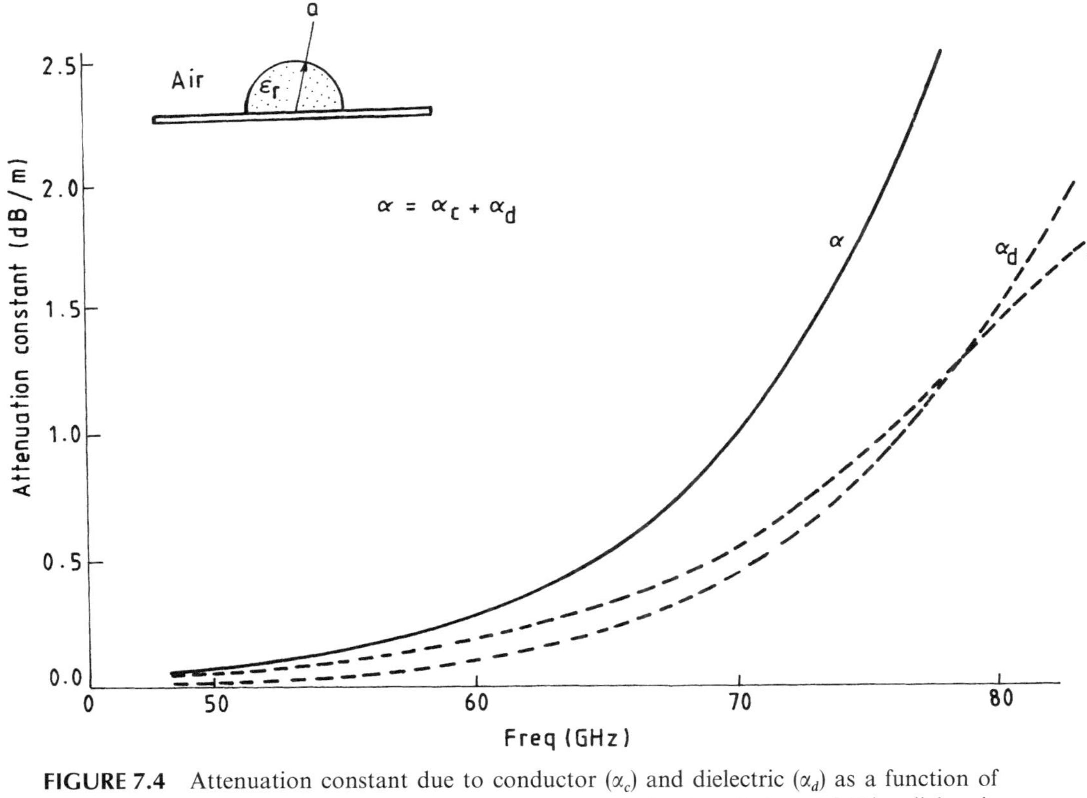

FIGURE 7.4 Attenuation constant due to conductor (α_c) and dielectric (α_d) as a function of frequency for a semicircular dielectric image guide (structure in Fig. 7.3): $2a = 0.047$ in., dielectric (polystyrene); $\varepsilon_r = 2.56$, $\tan\delta = 10^{-3}$; conducting plane (copper): $\sigma = 5.8 \times 10^7$ mhos/m. (From Wiltse [3]. Copyright © 1959 IEEE, reprinted with permission.)

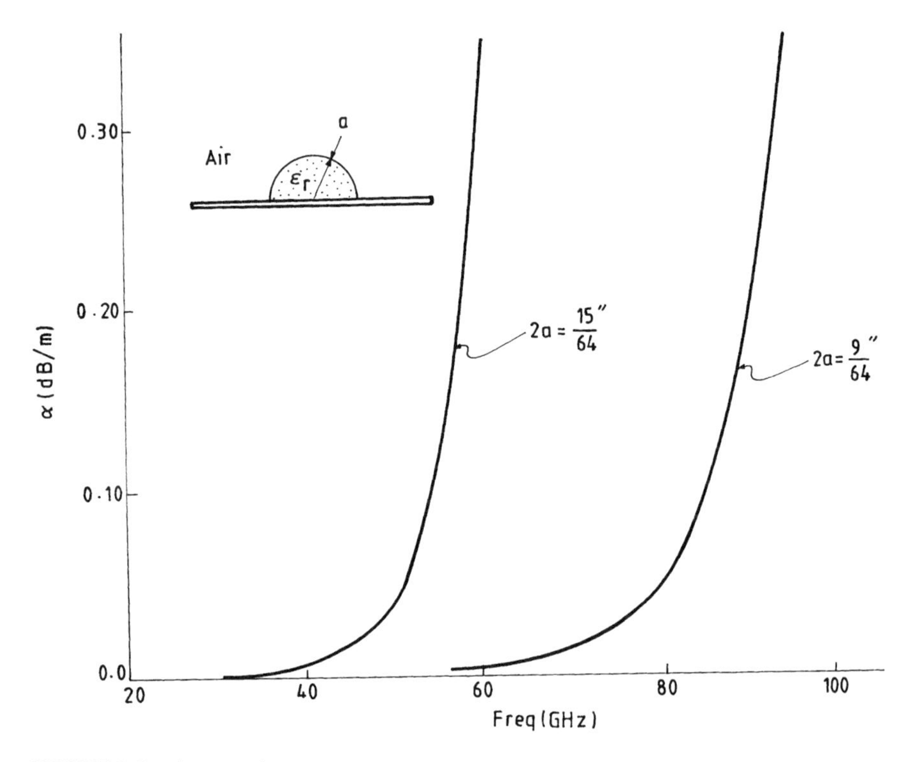

FIGURE 7.5 Attenuation constant α ($=\alpha_c + \alpha_d$) as a function of frequency for a semicircular dielectric image guide of two different diameters. Dielectric (foam polystyrene): $\varepsilon_r = 1.05$, $\tan\delta = 10^{-3}$; conducting plane (copper): $\sigma = 5.8 \times 10^7$ mhos/m. (From Wiltse [3]. Copyright © 1959 IEEE, reprinted with permission.)

$$H_1 = \int_q^{\infty} K_1^2(z)\,dz \tag{7.5g}$$

$$S = (UV/\varepsilon_r) - 1 \tag{7.5h}$$

$$T = UV - 1 \tag{7.5i}$$

For an estimate of the conductor and dielectric losses involved in a semicircular image guide, we reproduce in Figures 7.4 and 7.5 the attenuation characteristics reported by Wiltse [3]. We note from Figure 7.4 that with a dielectric having a diameter $2a = 0.047$ in., $\varepsilon_r = 2.56$, and $\tan\delta = 10^{-3}$, the guide can be operated with reasonably low loss from about 50–65 GHz; beyond this frequency, the attenuation becomes excessive. It has been shown that the attenuation in the guide can be reduced by choosing either a smaller size or smaller ε_r for the rod [3]. The loss tangent must be as small as possible to keep α_d small. Figure 7.5 illustrates the effect of choosing a low dielectric constant in reducing the attenuation in the guide. With a foam polystyrene dielectric rod having a convenient diameter ($2a = \frac{9}{64}$ in.) and low dielectric constant ($\varepsilon_r = 1.05$), the guide attenuation for operation in 50–70 GHz is significantly lower than that for the guide considered in Figure 7.4.

7.3 SEMIELLIPTICAL DIELECTRIC IMAGE GUIDE

An image guide consisting of a dielectric of semielliptical cross-section (Fig. 7.6(a)) has been analyzed by Wakabayashi and Mihara [4, 5]. The analysis is carried out in the elliptical coordinate system by assuming the minor axis length $2b$ to be much smaller than the major axis length $2a$. Figure 7.6(b) shows the elliptical coordinate system, where F_1 and F_2 are the foci of the ellipse and $2h$

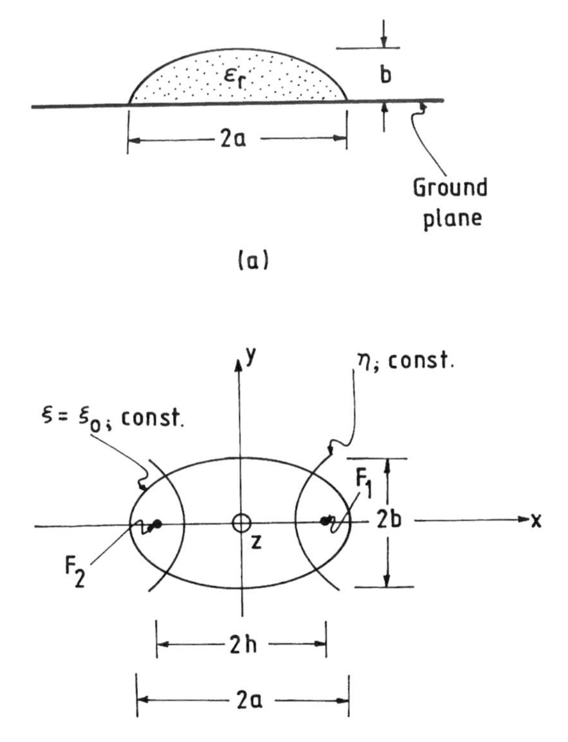

FIGURE 7.6 (a) Cross-sectional view of a semielliptical dielectric image guide. (b) Elliptical coordinate system. (From Wakabayashi and Mihara [5], reprinted with permission of Scripta Publishing Co.)

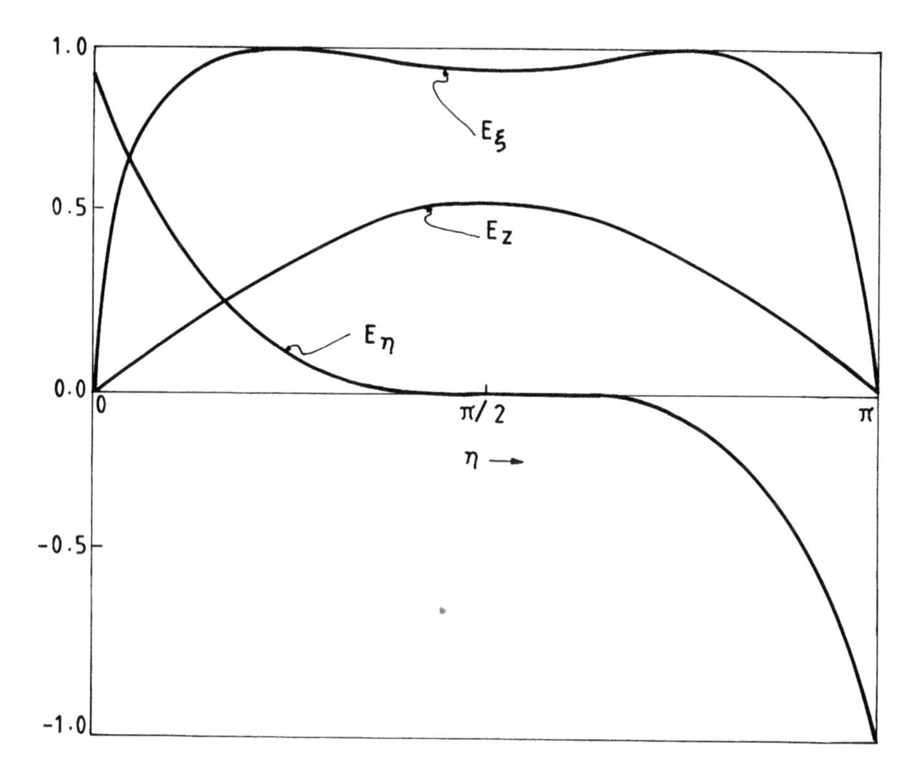

FIGURE 7.7 Distribution of dominant mode electric field components E_ξ, E_η, and E_z in the dielectric region of a semielliptical image guide at $\xi = 0.8\,\xi_0$; $\varepsilon_r = 2.3$, $u_0 = 1.02$, and $2a/\lambda_0 = 0.985$. (From Wakabayashi and Mihara [5], reprinted with permission of Scripta Publishing Co.)

is the distance between the foci. The elliptical (ξ, η, z) and rectangular (x, y, z) coordinates are related by the following:

$$x = h\cosh(\xi)\cos(\eta) \tag{7.6a}$$

$$y = h\sinh(\xi)\sin(\eta) \tag{7.6b}$$

$$z = z; \quad 0 \le \xi \le \infty, \quad 0 \le \eta \le 2\pi \tag{7.6c}$$

The lengths of the major and minor axes are expressed in terms of $2h$ as

$$2a = 2h\cosh(\xi_0) \tag{7.7a}$$

$$2b = 2h\sinh(\xi_0) \tag{7.7b}$$

The eccentricity of the ellipse representing the guide is given by

$$e = h/a = 1/\cosh(\xi_0) \tag{7.8}$$

7.3.1 Fields and Dispersion

Figure 7.7 shows the distribution of the dominant mode electric field components E_ξ, E_η, and E_z inside the semielliptical dielectric region [5]. The field amplitudes, normalized with respect to the maximum value, are plotted for a constant elliptical surface $\xi = 0.8\xi_0$ with η varying from 0 to π. The guide parameters chosen are $\varepsilon_r = 2.3$, $u_0 = \cosh(\xi_0) = 1.02$, and $2a/\lambda_0 = 0.985$, where λ_0 is the free-space wavelength. For the same set of parameters, the computed magnetic field components H_ξ, H_η, and H_z (normalized with respect to the maximum value) are plotted in Figure 7.8 [5]. Because of the elliptical cross-section, the amplitudes of E_ξ and H_η are slightly smaller than the maximum on the y-axis ($\eta = \pi/2$) for the elliptical surface close to the dielectric–air boundary (see Figs 7.7 and 7.8 for $\xi = 0.8\xi_0$), whereas for smaller values of ξ, maximum amplitudes are attained at $\eta = \pi/2$.

Figure 7.9 shows typical dispersion characteristics of a semielliptical dielectric image guide having $\varepsilon_r = 2.5$ [5]. The guide wavelength λ normalized with respect

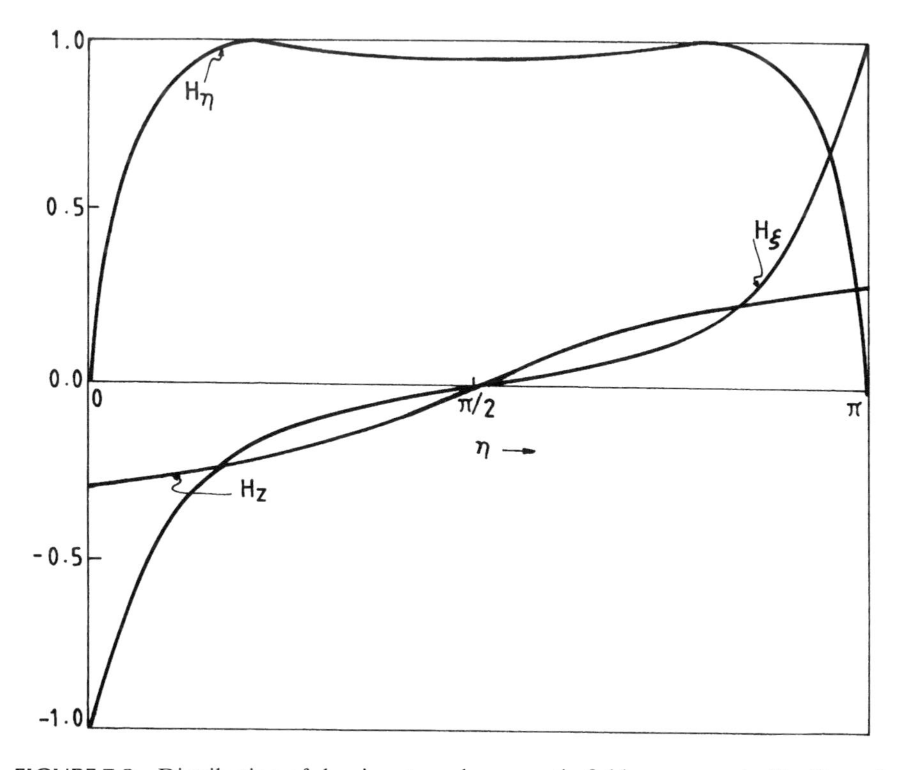

FIGURE 7.8 Distribution of dominant mode magnetic field components H_ξ, H_η, and H_z in the dielectric region of a semielliptical image guide at $\xi = 0.8\xi_0$; $\varepsilon_r = 2.3$, $u_0 = 1.02$, and $2a/\lambda_0 = 0.985$. (From Wakabayashi and Mihara [5], reprinted with permission of Scripta Publication Co.)

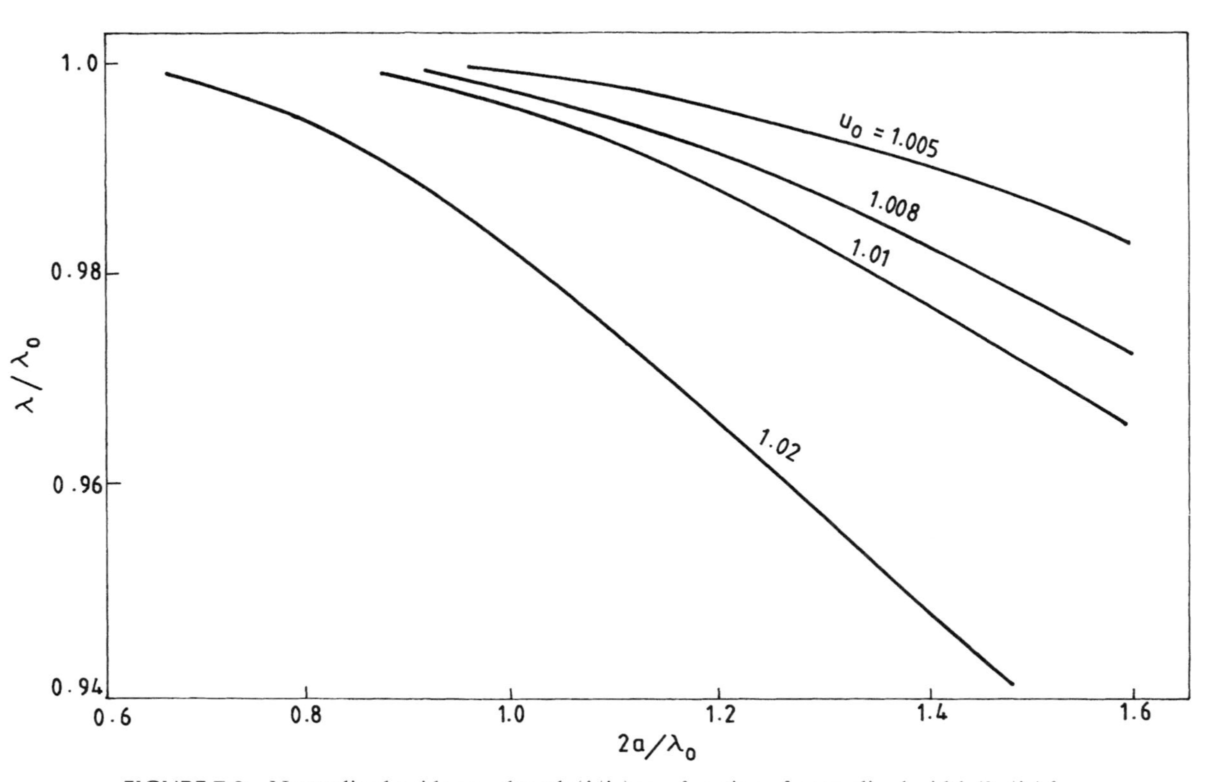

FIGURE 7.9 Normalized guide wavelength (λ/λ_0) as a function of normalized width ($2a/\lambda_0$) for a semielliptical dielectric image guide: $\varepsilon_r = 2.5$. (From Wakabayashi and Mihara [5], reprinted with permission of Scripta Publishing Co.)

to the free-space wavelength λ_0 is plotted as a function of normalized width $2a/\lambda_0$ for different values of u_0, where

$$u_0 = \cosh(\xi_0) = a/h \tag{7.9}$$

The plots show that for a fixed frequency, λ decreases (or equivalently the propagation constant $\beta = 2\pi/\lambda$ increases) either with an increase in the guide

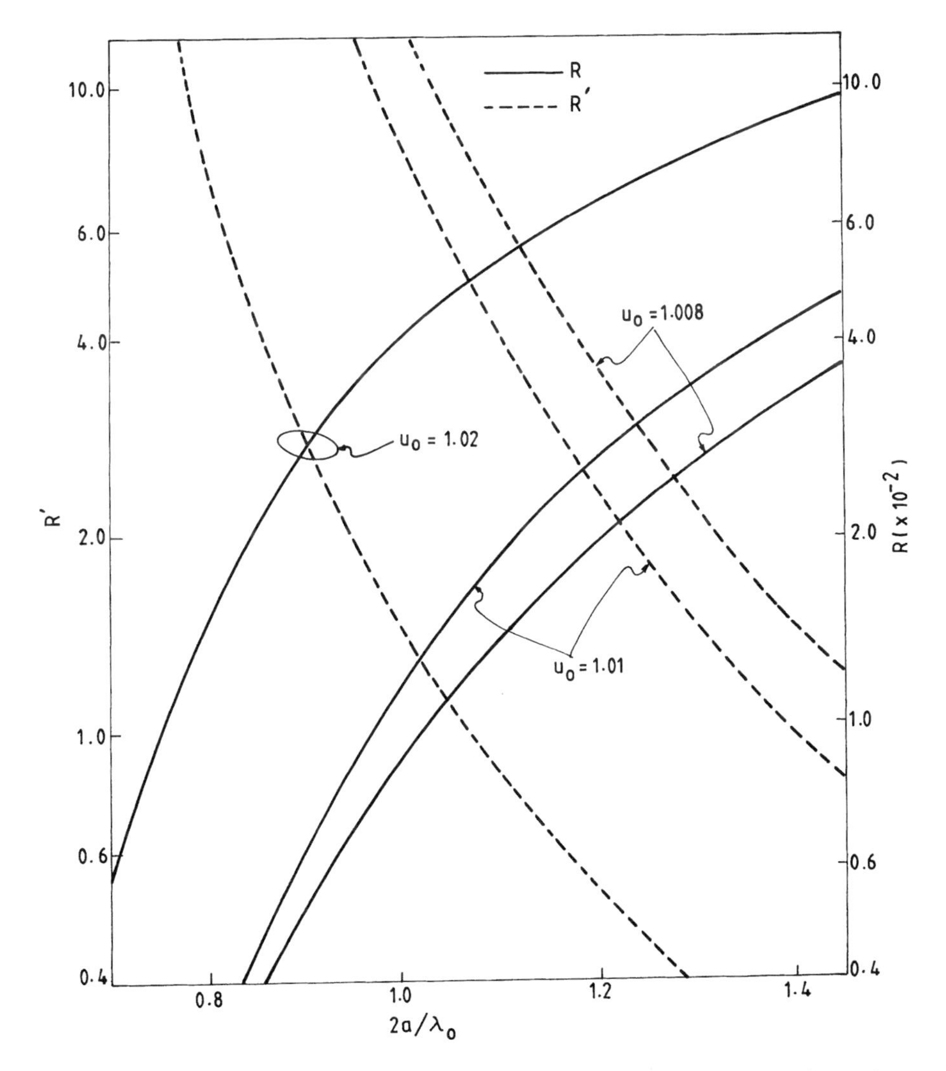

FIGURE 7.10 Attenuation factor R due to dielectric loss and R' due to conductor loss versus normalized width $2a/\lambda_0$ in a semielliptical dielectric image guide: $\varepsilon_r = 2.3$. —— R, ----- R'. (From Wakabayashi and Mihara [5], reprinted with permission of Scripta Publishing Co.)

width $2a$ (h fixed) or with a decrease in h ($2a$ fixed). The increase in β in both cases is because of the increase in the effective dielectric constant with an increase in $2a$ or decrease in h.

7.3.2 Attenuation Characteristics

The attenuation in a straight semielliptical image guide is due to the contribution from dielectric and conductor losses. The expressions for the dielectric attenuation constant α_d and the conductor attenuation constant α_c are given by [5]

$$\alpha_d = 27.3[(\varepsilon_r \tan\delta)/\lambda_0]R \tag{7.10}$$

$$\alpha_c = 34.74(R_s/\eta_0\lambda_0)R' \tag{7.11}$$

where $\tan\delta$ is the loss tangent of the dielectric, R_s is the surface resistivity of the ground plane, and η_0 is the free-space wave impedance. R and R' are the attenuation ratio factors due to the dielectric and conductor losses, respectively. The variation in R and R' as a function of $2a/\lambda_0$ as shown in Figure 7.10 is illustrative of the variation in α_d and α_c [5]. It may be observed that for a fixed u_0, as $2a/\lambda_0$ is increased, the value of R increases and saturates to a certain value. Furthermore, for a fixed $2a/\lambda_0$, R decreases as u_0 approaches unity. Equivalently, the dielectric loss reduces as the thickness b becomes small in comparison with the width $2a$. The conductor loss, however, follows an opposite trend as is evident from the variation in R'. Furthermore, the conductor attenuation is much larger than the dielectric attenuation, which gets accentuated as $2a/\lambda_0$ decreases and as u_0 approaches unity.

7.4 Y-TYPE AND TRIANGULAR DIELECTRIC GUIDES

7.4.1 Y-Dielectric Guide

The Y-dielectric guide (YDG), the cross-section of which is shown in Figure 7.11(a), was proposed by Shinonaga and Kurazono [6] as a low-loss guide for millimeter wave and submillimeter wave applications. The three arms of the YDG are tapered so that the electromagnetic energy is confined to the center part of the guide. The low-loss feature is achieved by allowing a considerable portion of this energy to propagate in the air region near the Y-junction with very little energy existing in the three arms. These arms serve as support to the guide with negligible field perturbation.

The YDG has been analyzed [6] for its characteristics using the generalized telegraphist's method of Schelkunoff [10]. The analysis has been carried out by expanding the transverse electromagnetic fields of the guide in terms of the normal modes (TE_{mn} and TM_{mn} modes) of the metallic rectangular waveguide having the same dimensions as the rectangular hypothetical boundary shown in Figure 7.11(a). Considering the symmetry of the YDG about the vertical plane at

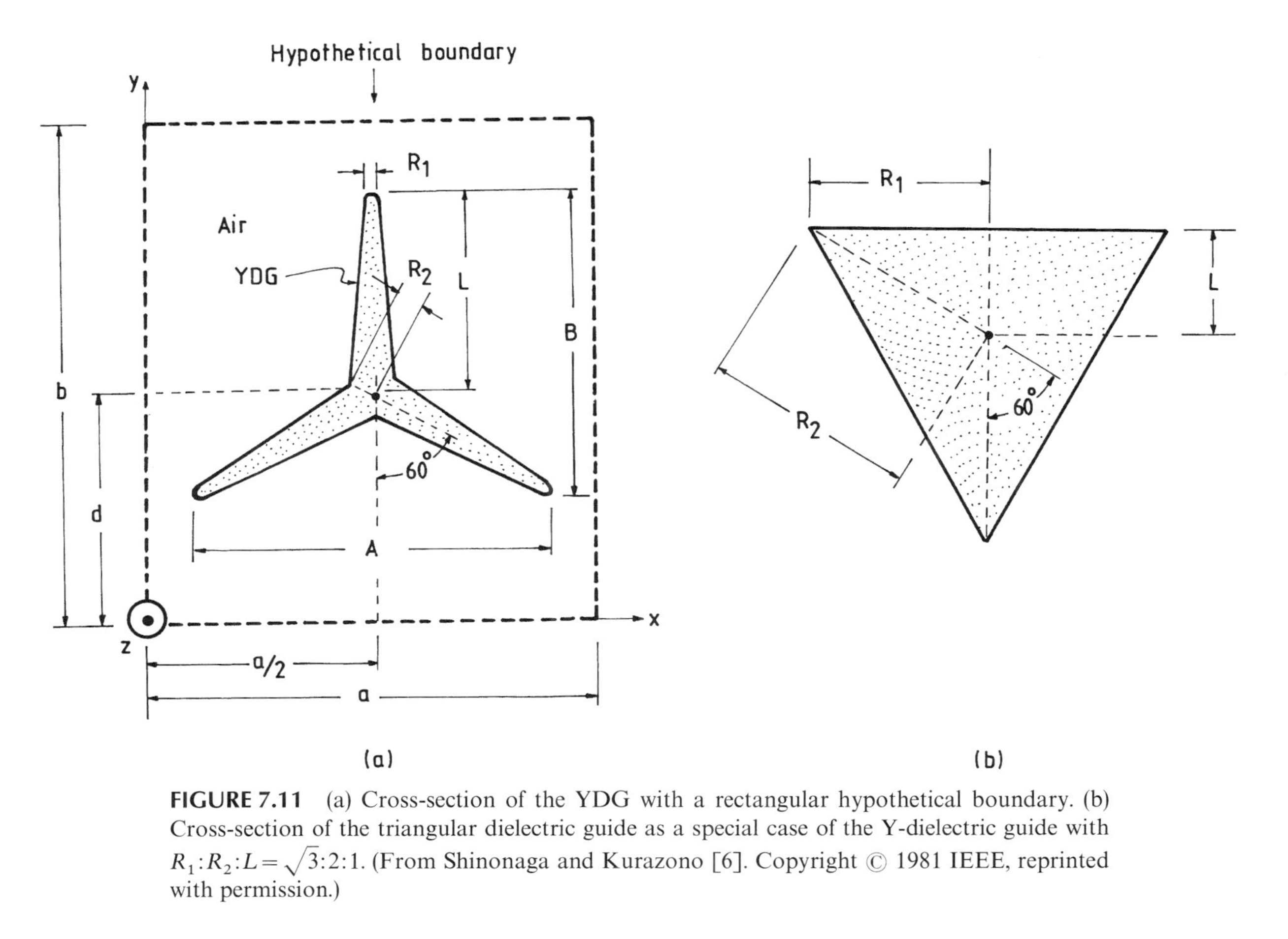

FIGURE 7.11 (a) Cross-section of the YDG with a rectangular hypothetical boundary. (b) Cross-section of the triangular dielectric guide as a special case of the Y-dielectric guide with $R_1:R_2:L=\sqrt{3}:2:1$. (From Shinonaga and Kurazono [6]. Copyright © 1981 IEEE, reprinted with permission.)

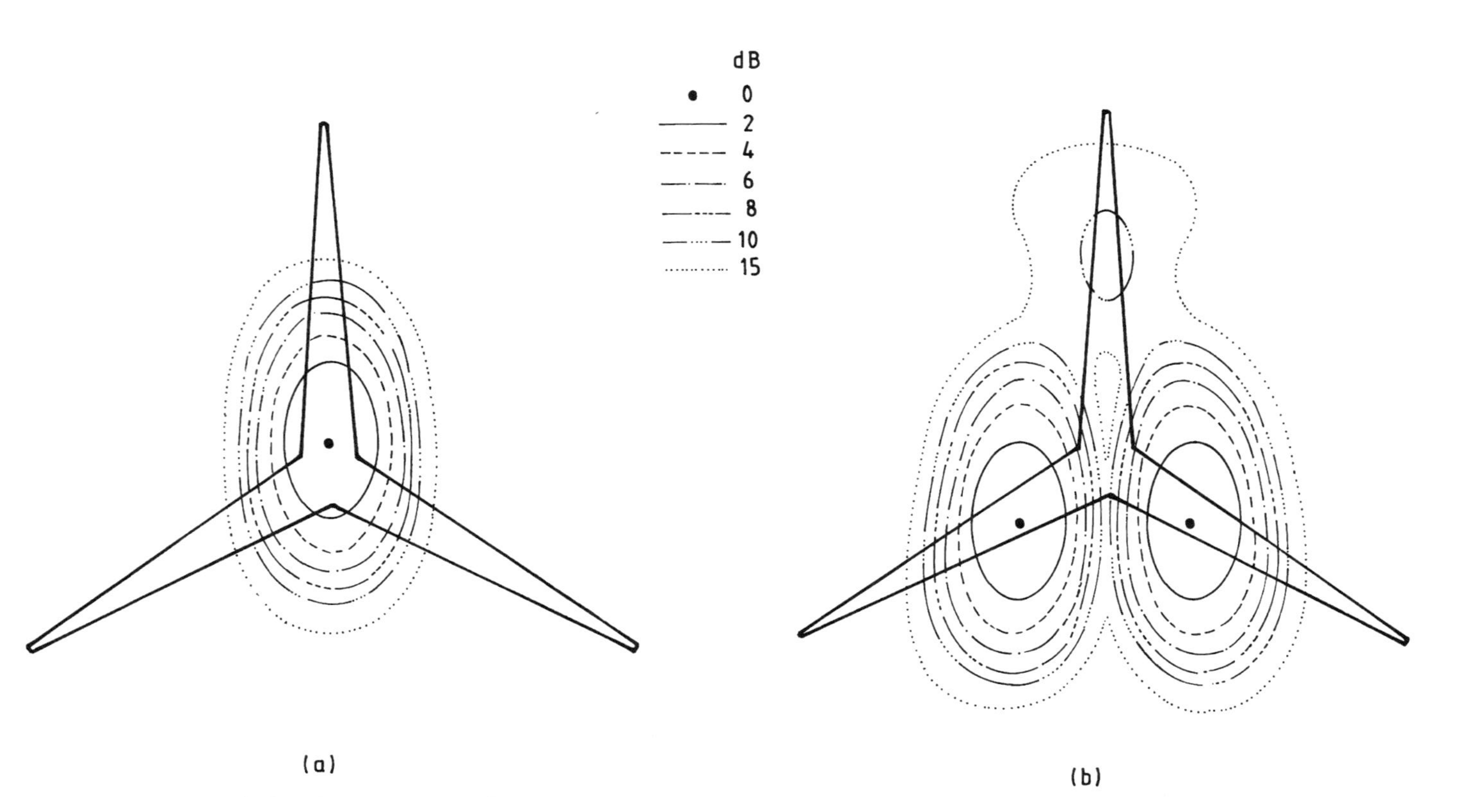

FIGURE 7.12 Power distribution for (a) fundamental mode and (b) second even mode in a YDG (structure in Fig. 7.11(a)). $\varepsilon_r = 2$, $k_0 R_2 = 3.35$, $R_1/R_2 = 0.037$, $L/R_2 = 10.63$, $a/A = 3$, $b/B = 3$, and $d = b/2 - L/4$. (From Shinonaga and Kurazono [6]. Copyright © 1981 IEEE, reprinted with permission.)

$x = a/2$, the propagating modes can be classified into even- and odd-modes depending on whether the plane at $x = a/2$ represents a magnetic wall or an electric wall. For the even mode fields, the expanding modes are TM_{mn} with $m = 1, 3, 5, \ldots$ and $n = 1, 2, 3, \ldots$; and TE_{mn} with $m = 1, 3, 5, \ldots$ and $n = 0, 1, 2, \ldots$; whereas for the odd mode fields, the expanding modes are TM_{mn} with $m = 2, 4, 6, \ldots$ and $n = 1, 2, 3, \ldots$; and TE_{mn} with $m = 0, 2, 4, \ldots$ and $n = 0, 1, 2, \ldots$. Figure 7.12 shows the power distribution for the fundamental mode and the even mode in the guide [6]. As can be seen, energy is strongly confined at and in the vicinity of the Y-junction, particularly for the fundamental mode.

Figure 7.13 shows the typical variation in the normalized propagation constant β/k_0 (where k_0 is the free-space propagation constant) as a function of k_0R_2 for a YDG having parameters $\varepsilon_r = 2$, $R_1/R_2 = 0.037$, $L/R_2 = 10.63$, and

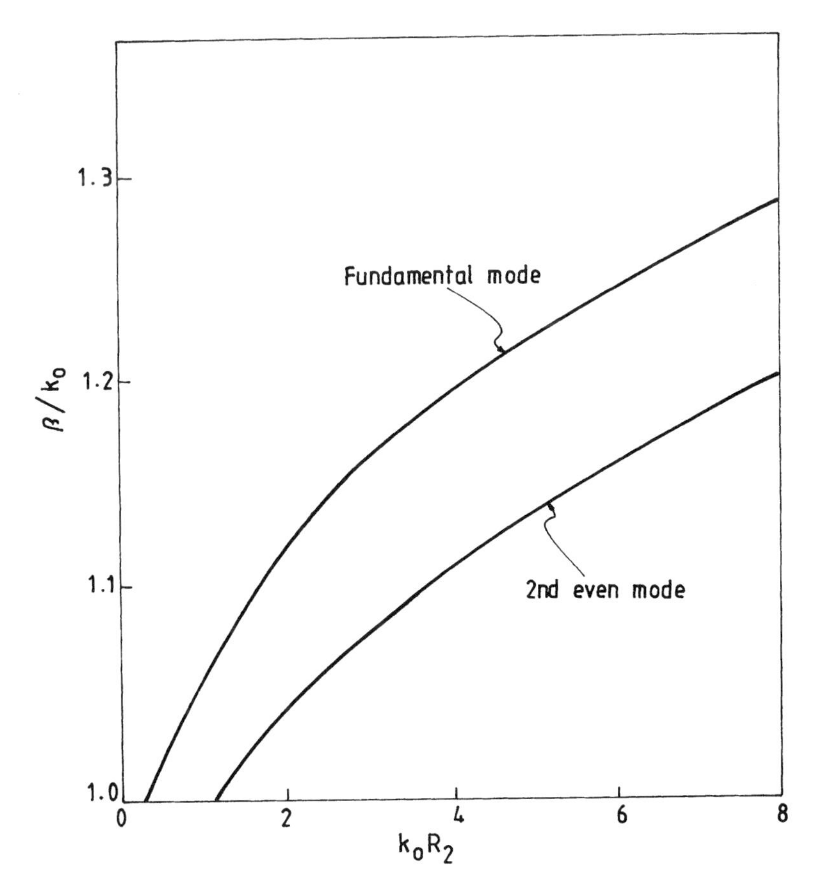

FIGURE 7.13 Dispersion characteristics of a YDG. $\varepsilon_r = 2$, $R_1/R_2 = 0.037$, $L/R_2 = 10.63$, and $d = b/2 - L/4$. (From Shinonaga and Kurazono [6]. Copyright © 1981 IEEE, reprinted with permission.)

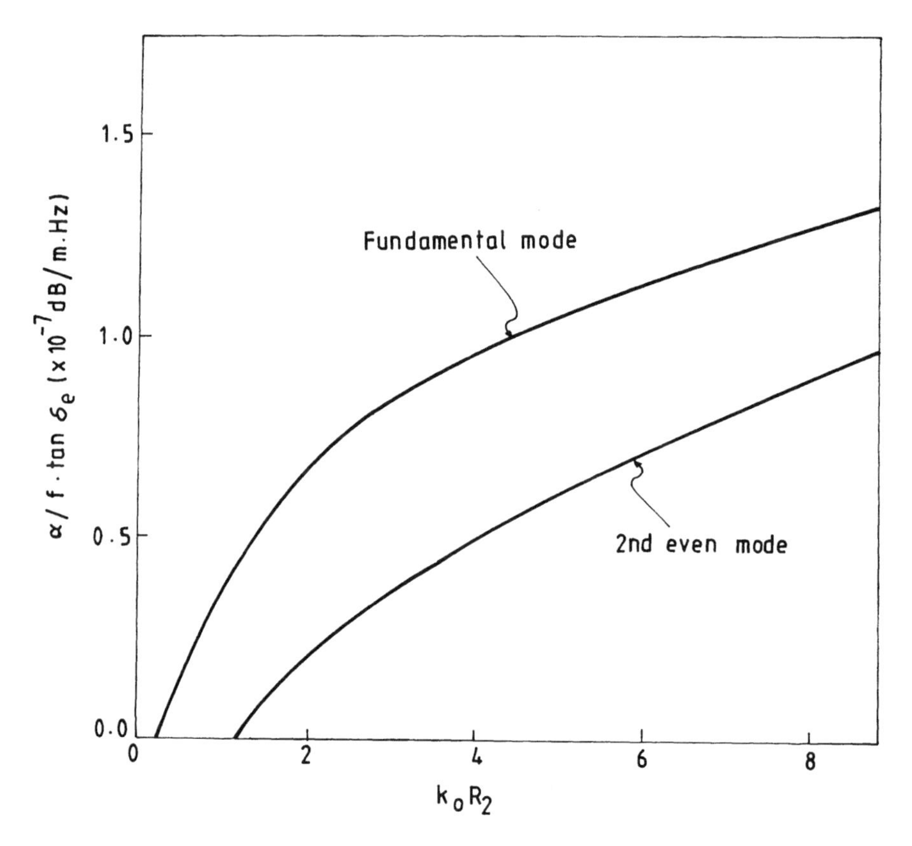

FIGURE 7.14 Transmission losses of a YDG. $\varepsilon_r = 2$, $R_1/R_2 = 0.037$, $L/R_2 = 10.63$, $\tan\delta_e = 10^{-4}$, and $d = b/2 - L/4$. (From Shinonaga and Kurazono [6]. Copyright © 1981 IEEE, reprinted with permission.)

$d = (b/2 - L/4)$ [6]. For the same set of guide parameters, Figure 7.14 shows the variation in the normalized attenuation constant of the YDG as a function of k_0R_2 [6]. The effect of the hypothetical boundary wall on the dispersion and attenuation characteristics is kept small. This is ensured by increasing the values of a/A and b/B suitably as the frequency is decreased.

7.4.2 Triangular Dielectric Guide

The triangular dielectric guide shown in Figure 7.11(b) is a special case of the Y-dielectric guide. The nature of the power distribution in the cross-section of the guide and the dispersion and attenuation characteristics of this guide are similar to those shown in Figures 7.12–7.14 for the YDG. As in the case of the YDG, for the fundamental mode, the E_y component is much greater than the E_x component. As an example of the relative sizes of the two guides, Shinonaga and Kurazono [6] have reported that with polyethylene ($\varepsilon_r = 2$, $\tan\delta = 10^{-4}$) as the dielectric and for equal transmission loss of 0.5 dB/m at 100 GHz, the dimensions of a YDG are $R_1 = 0.023$ mm, $R_2 = 0.621$ mm, and $L = 6.6$ mm, whereas those for a equilateral triangular guide are $2R_1 = 1.9$ mm, $R_2 = 1.1$ mm, and $L = 0.55$ mm (where $2R_1$ represents the length of one side of the triangle). The normalized propagation constant for the YDG is 1.09, whereas that for the triangular guide is 1.05. It is inferred that the YDG offers suitable size, including convenience of support for use at millimeter wave frequencies [6].

7.5 TUBE CONTACTED SLAB GUIDE

The tube contacted slab waveguide proposed by Gomi et al. [7] consists of a ground plane, a substrate of thickness d, a guiding layer of thickness a, and a dielectric tube of thickness t and outer radius r_t, each placed one above the other as shown in Figure 7.15. The relative dielectric constant ε_{r2} of the guiding layer is chosen larger than that of the substrate (ε_{r1}) and the tube (ε_{r3}); so that most of the

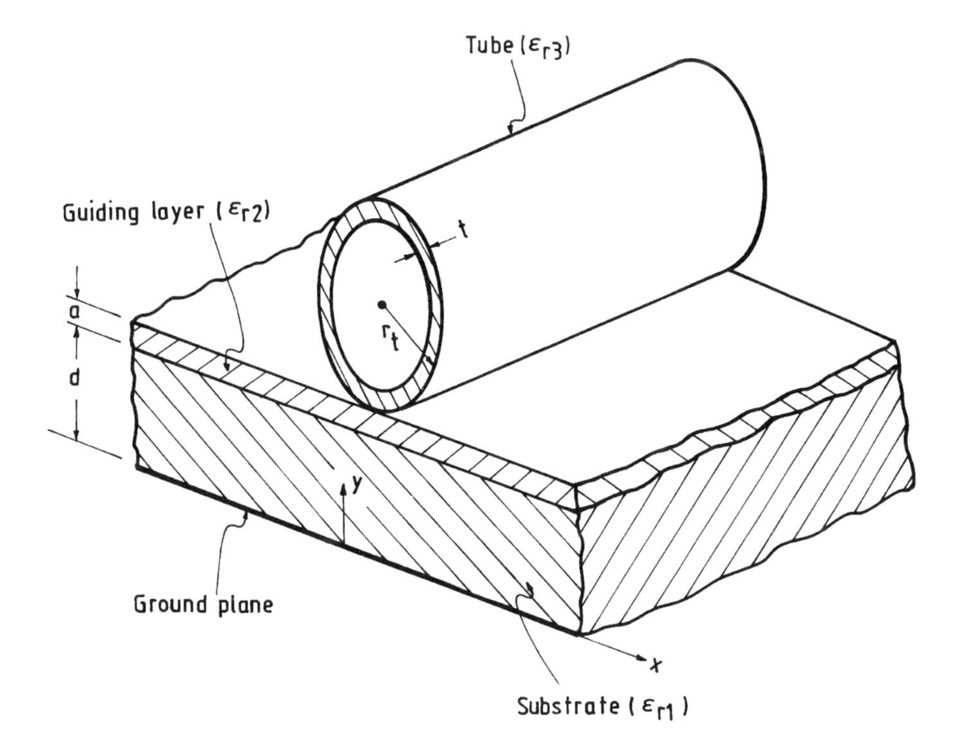

FIGURE 7.15 Geometry of tube contacted slab waveguide. (From Gomi et al. [7], reprinted with permission of IEE.)

propagating energy is confined to the guiding layer. The main advantage of this guide over the planar dielectric strip guide is that its transverse size for monomode operation can be made much larger, thereby making it more suitable for operation at higher millimeter wave and submillimeter wave frequencies. As an illustration of this feature, we reproduce in Figure 7.16 the dispersion characteristic of the tube contacted slab guide reported by Gomi et al. [7]. The dashed lines marked β_L/k_0 and β_H/k_0 in the figure give the square root of the effective

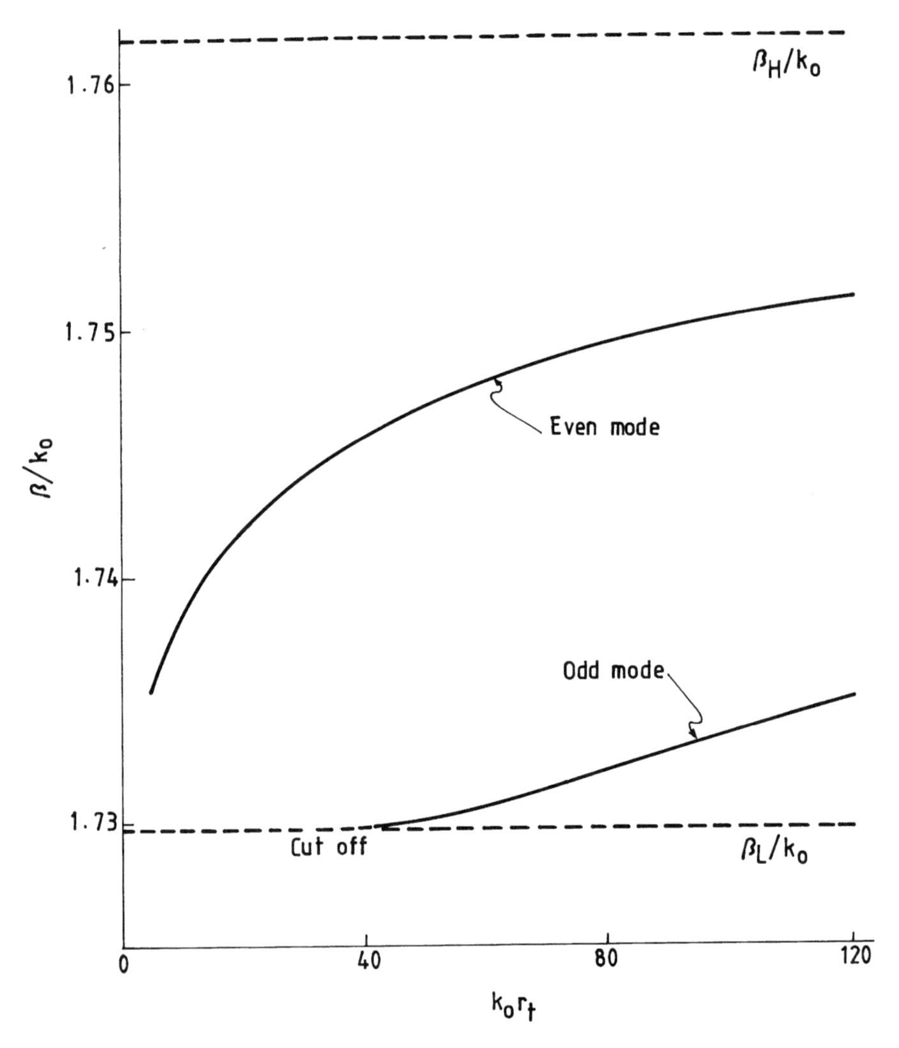

FIGURE 7.16 Normalized propagation constant as a function of normalized tube radius for tube contacted slab waveguide: $\varepsilon_{r1} = 2.54$, $\varepsilon_{r2} = 3.8$, $\varepsilon_{r3} = 2.08$, $k_0 d = 10$, $k_0 a = 2.5$, and $k_0 t = 2$. (From Gomi et al. [7], reprinted with permission of IEE.)

dielectric constants at $x = 0$ and $x = \infty$, respectively. We note from the figure that the cutoff of the first higher order mode occurs at $k_0 r_t = 42$. The radius of the tube at 890 GHz is reported to be about 1 mm, whereas that of the conventional dielectric guide can at most be of the order of a wavelength [7]. The enlarged size of the guide is also favorable from the point of view of fabrication and power handling capability. The tube, if made of flexible dielectric material, can be bent easily, enabling flexibility in component layout.

PROBLEMS

7.1 Draw typical field lines for the first three propagating modes (HE_{11}, TE_{01}, and TM_{01}) in a circular dielectric guide. If this rod is metallized to form a circular waveguide, its first three propagating modes are TE_{11}, TM_{01}, and TE_{21}. Draw typical field lines for these modes and compare.

7.2 The cutoff frequencies of the first three modes in a circular dielectric guide of radius a and relative dielectric constant ε_r are given by

$$f_c|_{HE_{11}} = 0, \quad f_c|_{TE_{01}} = f_c|_{TM_{01}} = \frac{2.405 v_0}{2\pi a \sqrt{\varepsilon_r - 1}}$$

where v_0 is the free-space velocity. The cutoff frequencies of the first three modes in the same guide when fully metallized are given by

$$f_c = \frac{K v_0}{2\pi \sqrt{\varepsilon_r} a}$$

where $K = 1.841$, 2.405, and 3.054 for TE_{11}, TM_{01}, and TE_{21} modes, respectively.

(a) Determine the diameter of a circular dielectric guide having $\varepsilon_r = 1.64$ for operation at 100 GHz. The guide must operate in the dominant mode at 0.9 times the cutoff frequency of the next higher order mode.

(b) If the circular dielectric guide designed above is fully metallized, what would be the recommended frequency band of operation? [Consider the frequency band to be between 1.1 times the cutoff frequency of the dominant mode and 0.9 times the cutoff frequency of the next higher order mode.]

7.3 Draw typical E- and H-field lines in a semicircular dielectric image guide corresponding to the HE_{11}, TE_{01}, and TM_{01} modes of the circular dielectric guide having the same radius and ε_r. Which of these modes will remain unperturbed?

7.4 Explain the principle of wave guidance in a semicircular dielectric image guide. What are its advantages over the circular dielectric guide for millimeter wave circuits?

7.5 **(a)** Figure 7.4 shows variation of conductor and dielectric attenuation constants as a function of frequency for a semicircular dielectric image guide. Draw a graph of total attenuation in dB/m for a corresponding circular dielectric guide having the same radius and ε_r.

(b) What is the effect of lowering the frequency on the radiation loss and overall guidability of the semicircular dielectric image guide?

7.6 In a semicircular dielectric image guide, the total attenuation due to conductor and dielectric losses decreases with a decrease in the radius and/or ε_r of the guide. Discuss the corresponding effect on guidability and radiation loss.

7.7 Draw E- and H-field lines of the dominant mode for the semielliptical dielectric image guide for $e = h/a = 0$, 0.4, and 0.8 with a kept constant.

[*Note*: $e = 0$ corresponds to a semicircular dielectric image guide of radius a.]

7.8 Explain the effect of increase in h/a (with a fixed) on the guide wavelength and the conductor and dielectric losses in a semielliptical dielectric image guide.

7.9 **(a)** Classify the modal fields in a Y-dielectric guide. Distinguish between even- and odd-mode fields. Designate the fundamental modes in the guide.

(b) What are the special features and advantages of a Y-dielectric guide?

7.10 **(a)** Draw power distribution contours for the fundamental mode and the second even mode in a triangular dielectric guide shown in Figure 7.17.

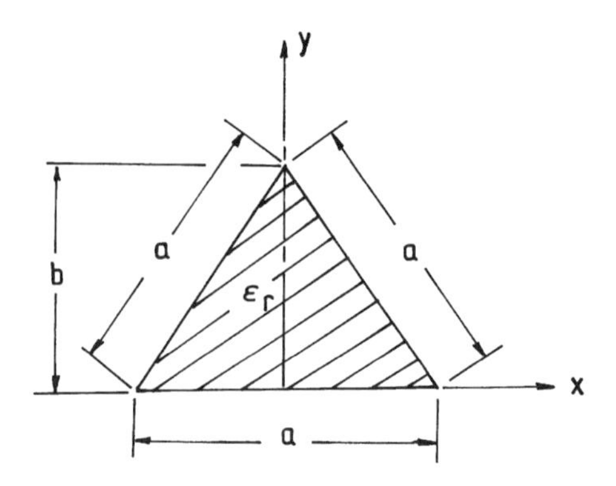

FIGURE 7.17 Cross-section of triangular dielectric guide.

(b) Draw typical field lines for the dominant field component E_y of the fundamental mode in this triangular dielectric guide. Also show the field intensity distribution of E_y as a function of x at $y = b/4$, $b/2$, and $3b/4$.

REFERENCES

1. D. D. King, Properties of dielectric image lines. *IRE Trans. Microwave Theory Tech.*, **MTT-3**, 75–81, Mar. 1955.
2. D. D. King and S. P. Schlesinger, Losses in dielectric image lines. *IRE Trans. Microwave Theory Tech.*, **MTT-5**, 31–35, Jan. 1957.
3. J. C. Wiltse, Some characteristics of dielectric image lines at millimetre wavelengths. *IRE Trans. Microwave Theory Tech.*, **MTT-7**, 65–70, Jan. 1959.
4. T. Wakabayashi and Y. Mihara, Analysis of dielectric tape lines. *Trans. IECE Japan*, **61B(10)**, 880–887, Oct. 1978.
5. T. Wakabayashi and Y. Mihara, Characteristics of the dominant mode in a dielectric tape line. *Electron. Commun. Japan*, **64B(5)**, 70–79, 1981.
6. H. Shinonaga and S. Kurazono, Y dielectric waveguide for millimetre and sub-millimetre waves. *IEEE Trans. Microwave Theory Tech.*, **MTT-29**, 542–546, June 1981.
7. J. Gomi, T. Yoneyama, and S. Nishida, Tube-contacted slab waveguide for millimetre and submillimetre waves. *Electron. Lett.*, **17(3)**, 125–126, Feb. 1981.
8. R. Chatterjee, *Dielectric and Dielectric-Loaded Antennas*, Research Studies Press, London, UK, 1985.
9. W. M. Elsasser, Attenuation in a dielectric circular rod. *J. Appl. Phys.*, **20**, 1192, Dec. 1949.
10. S. A. Schelkunoff, Generalized telegraphist's equations for waveguides. *Bell Syst. Tech. J.*, **31**, 784–801, July 1952.

CHAPTER EIGHT

H-Guides and Groove Guides

8.1 INTRODUCTION

The various dielectric integrated guides discussed in Chapters 2–6 are suitable for millimeter wave applications up to about 100 GHz. Beyond this frequency, these guides pose problems of high attenuation and small physical size. The transmission lines that are better suited for higher millimeter wave frequencies are the H-guide [1, 2] and the groove guide [3–6]. The H-guide, the cross-section of which is shown in Figure 8.1 (a), has the same geometry as the NRD guide except that the plate separation is larger than one-half the free-space wavelength. The emphasis in the case of the H-guide is on achieving low-loss propagation rather than the nonradiative property. The guide relies on surface wave guidance at the dielectric–air interfaces. Electromagnetic energy is confined sideways by parallel metal plates and in the vertical direction by surface wave propagation. The H-guide achieves its low-loss property by virtue of having a large plate separation. However, for achieving loss less than about 0.1 dB/m, the plate separation must be greater than several wavelengths, which is conducive to multimode propagation. The useful frequency range of operation for the H-guide is approximately 100–200 GHz.

The problem of multimode propagation encountered in the H-guide is circumvented in the groove guide. The groove guide, the cross-section of which is shown in Figure 8.1(b), consists of a parallel-plate structure with symmetric rectangular grooves cut uniformly along its length. Because of the wider plate separation in the central groove region, the intrinsic phase velocity in that region is smaller than that in the upper and lower parallel-plate regions. The difference in the two phase velocities creates a surface wave effect identical to that of the dielectric strip region in an H-guide. However, unlike the H-guide, which supports hybrid modes, the propagating modes of the groove guide can be grouped into TE (transverse electric) and TM (transverse magnetic) modes with respect to the direction of propagation (z-direction). The dominant TE mode, designated as TE_{11}, has fields and current distribution similar to those of the dominant

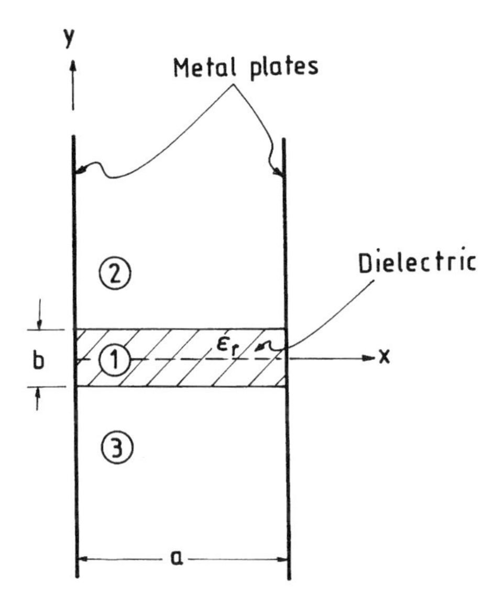

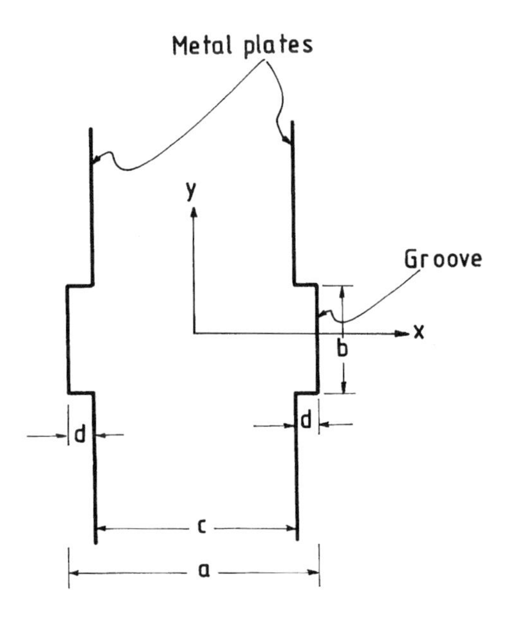

FIGURE 8.1 Cross-sectional views of (a) H-guide and (b) groove guide.

operating mode of the H-guide. The TE_{11} mode therefore forms the low-loss operating mode of the groove guide. Because of the absence of the dielectric strip, the groove guide inherently has much lower loss than the H-guide. Also, since the grooves have the effect of filtering out the higher order modes, the single-mode operating bandwidth is much larger. The groove guide is simple to construct and involves dimensions that are not critical to a fraction of a wavelength. These advantageous features make the groove guide attractive for use at higher millimeter wave frequencies from about 100 GHz up to 300 GHz.

8.2 H-GUIDE

8.2.1 Hybrid Modes and Their Characteristics

The H-guide is basically a parallel-plate guide with a dielectric slab placed perpendicular to the metal plates so as to form the shape of an H (Fig. 8.1 (a)). The hybrid-mode analysis of such a structure has been presented in Chapter 6 (Section 6.2), wherein the propagating modes are classified into TM^y_{mn} and TE^y_{mn} modes. Note that some of the papers on H-guides use the designation PM_{mn} and PE_{mn} for the TM^y_{mn} and TE^y_{mn} modes, respectively [7, 8]. The designation PM refers to the *parallel magnetic* to indicate the fact that the magnetic field lines lie entirely in planes parallel to the air–dielectric interfaces ($H_y = 0$). Similarly, the designation PE refers to *parallel electric* to indicate the fact that the electric field lines lie entirely in planes parallel to the air–dielectric interfaces ($E_y = 0$). The index m gives the number of half-cycles of the sinusoidal variation of the fields in the x-direction and the index n gives the order of the surface wave mode and is indicative of the manner in which the fields vary as a function of y within the strip. The TM^y_{mn} (or PM_{mn}) modes may be considered as resulting from TM surface waves and the TE^y_{mn} (or PE_{mn}) modes may be considered as resulting from TE surface waves, both propagating in an infinite dielectric slab and getting reflected back and forth between the metal walls.

The dominant operating mode of the H-guide is the TM^y_{11} mode, which is the same as the LSM_{11} mode of the NRD guide referred to in Chapter 6. As shown in Figure 8.2, this mode has its electric field primarily parallel to the metal plates and the magnetic field lines parallel to the air–dielectric interfaces. Figure 8.3 shows the nature of variation in the dominant electric field component E_y for the TM^y_{mn} modes for $m = 1, 2, 3$ and $n = 1, 2$. By choosing an appropriate dielectric thickness and a launching device, the H-guide can be excited in only the TM^y_{m1} even modes, where $m = 1, 3$.

Figure 8.4 shows a typical variation in the dielectric attenuation constant α_d as a function of wavelength λ_0 for different values of relative dielectric constant ε_r and thickness b of the dielectric strip [7]. It can be seen that increasing the dielectric constant from a low value can result in lowering the attenuation up to a certain limit, and furthermore, the thinner the dielectric, the more noticeable is the variation in attenuation with ε_r. This is because, in a thin dielectric, a major

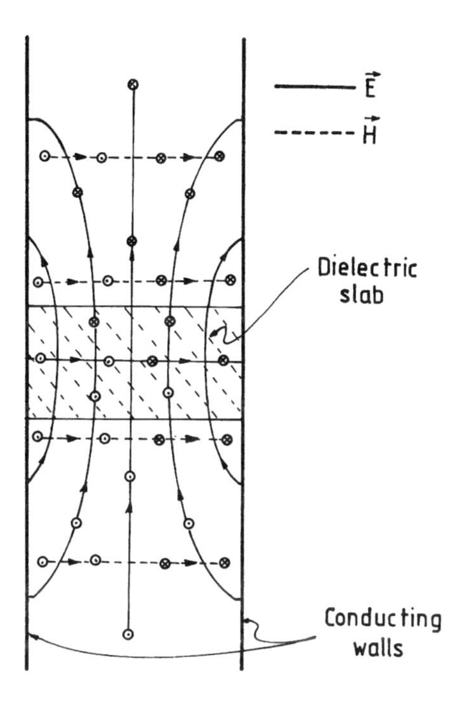

FIGURE 8.2 Field configuration in an H-guide.

portion of the dielectric loss occurs due to the E_y component. Thus, for a given thickness of the dielectric, there is an optimum value of ε_r at which a large part of the electromagnetic energy travels outside the strip, thereby resulting in lowest dielectric loss.

A distinct feature of the H-guide is its low transmission loss, which is achieved by keeping the plate separation more than $\lambda_0/2$. Figure 8.5 illustrates variation in the total attenuation constant α_t, which includes attenuation due to the conductor and dielectric losses ($\alpha_t = \alpha_c + \alpha_d$), as a function of wavelength for three different values of the plate separation distance ($a = 1$ mm, 1.5 mm, 2 mm) [7]. The dielectric loss tangent $\tan\delta$ is 10^{-3} and the effective metal conductivity σ is 10^7 mhos/m. The experimental points of attenuation reported in the literature [9, 10] for rectangular waveguides are superposed for comparison. As can be seen from the graph, the larger the plate separation, the smaller is the attenuation for the TM^y_{11} mode of the H-guide. Since the cutoff of the TM^y_{m1} modes occurs approximately at $a = m\lambda_0/2$, the TM^y_{11} mode becomes radiative for $a > \lambda_0/2$. The level of radiation from the open ends depends mainly on the height of the guide and the exponential decay factor in the air region. For a transmission line, the radiation leakage can be minimized by increasing the height of the metal plates. Another disadvantage of increasing the plate separation is the possibility of multimode propagation. For example, in a guide with a plate separation of

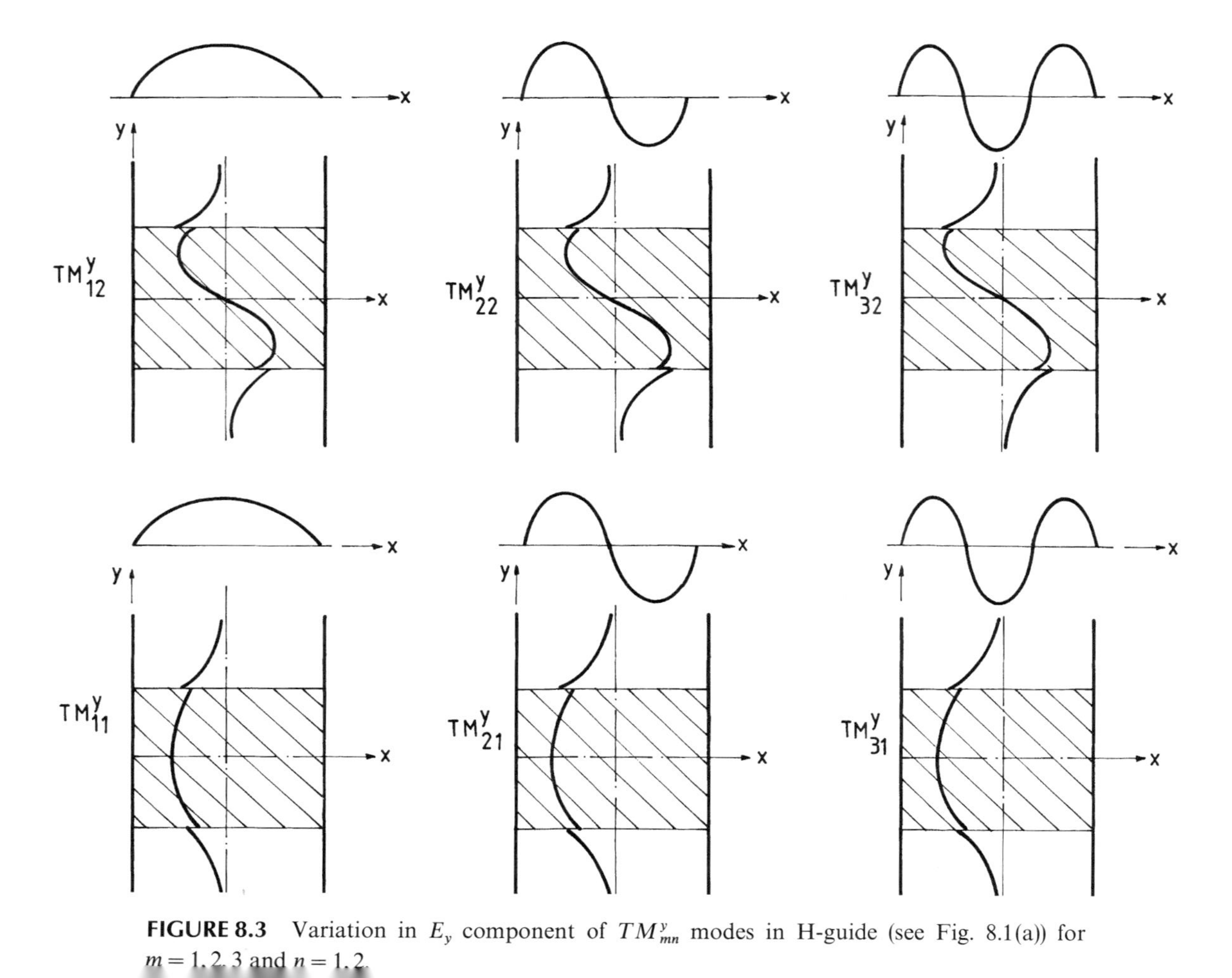

FIGURE 8.3 Variation in E_y component of TM^y_{mn} modes in H-guide (see Fig. 8.1(a)) for $m = 1, 2, 3$ and $n = 1, 2$.

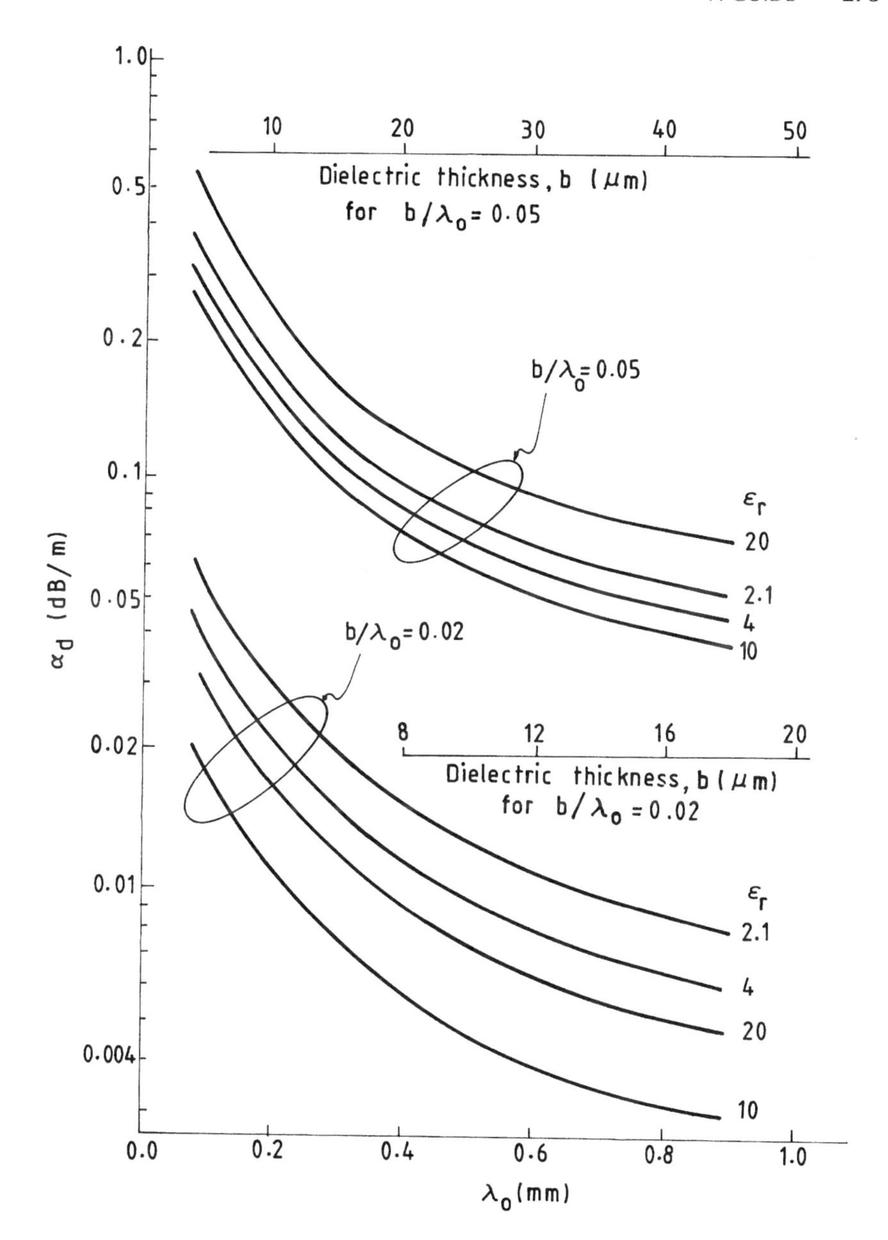

FIGURE 8.4 Variation of dielectric attenuation constant α_d in H-guide (see Fig. 8.1(a)) as a function of wavelength λ_0 with ε_r and b/λ_0 as parameters. (From Doswell and Harris [7]. Copyright © 1973 IEEE, reprinted with permission.)

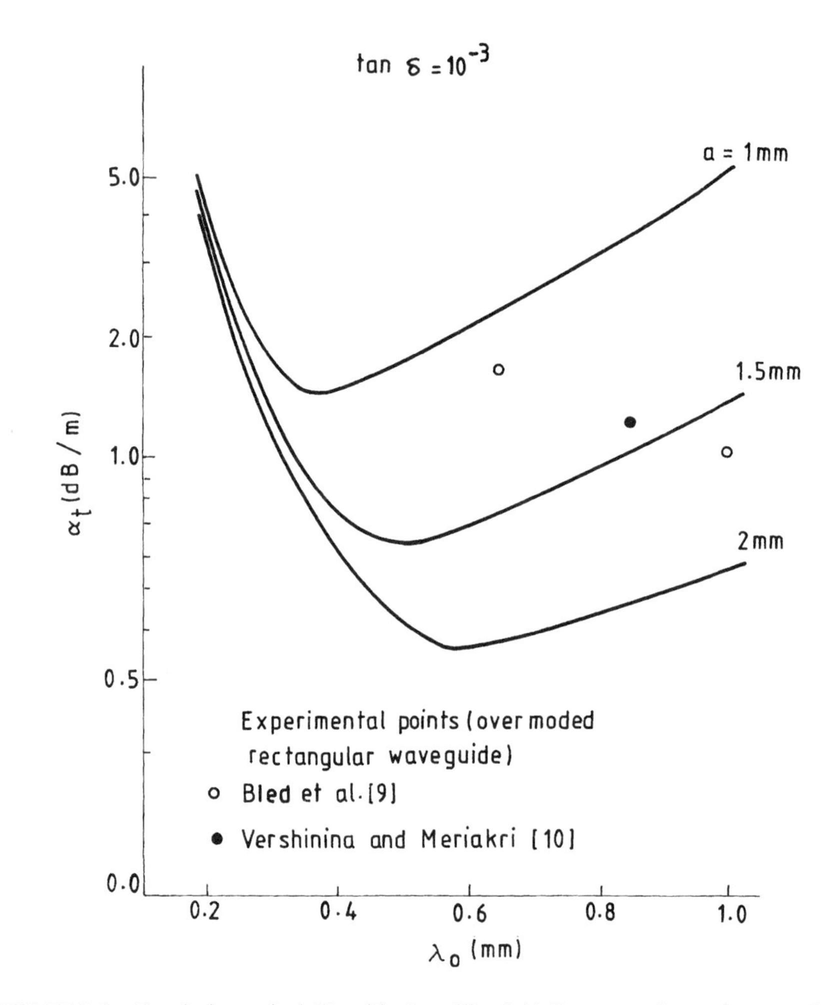

FIGURE 8.5 Total theoretical H-guide (see Fig. 8.1(a)) attenuation $\alpha_t (= \alpha_c + \alpha_d)$ as a function of wavelength λ_0. $\varepsilon_r = 2.1$, $\tan\delta$ (dielectric loss tangent) $= 10^{-3}$, σ (metal conductivity) $= 10^7$ mhos/m, and $b = 15\,\mu$m. (From Doswell and Harris [7]. Copyright © 1973 IEEE, reprinted with permission.)

1.5 mm operating at a wavelength of 0.5 mm, the first higher order even mode—namely, the TM_{31}^y mode—can exist in addition to the TM_{11}^y mode. An effective technique of mode filtering is to introduce grooves in the metal walls and have the dielectric strip penetrate into the walls as shown in Figure 8.6 [7]. The grooves also serve to support the dielectric strip. However, since the groove penetration distance is critical to a fraction of a wavelength, this technique is not quite practical at higher millimeter wave frequencies. A more attractive structure is the groove guide, which permits larger dimensions by eliminating the dielectric strip. This structure is discussed in Section 8.4.

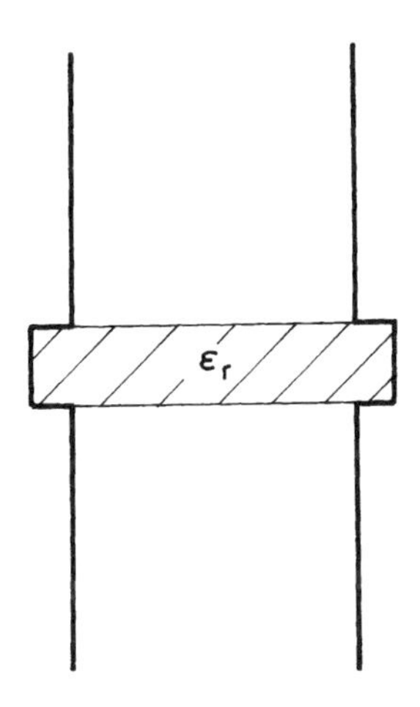

FIGURE 8.6 Cross-section of H-guide with grooves.

8.2.2 Nonhybrid TE_{0n} Modes

The TM^y_{mn} and TE^y_{mn} modes with integers m, $n \geq 1$ are hybrid modes of the H-guide. It is known that this guide cannot support TEM and pure TM modes ($m = 0$). The only nonhybrid modes that it can support are the TE_{0n} modes. The properties of these modes and their utility for device applications have been reported by several investigators [2, 11–13].

The field components of the TE^y_{mn} modes of a dielectric slab loaded parallel-plate guide (H-guide) are derived in Chapter 6. For $m = 0$, the TE^y_{mn} modes reduce to the transverse-electric TE_{0n} modes having E_x, H_y, and H_z as the only nonzero field components. These fields have no variation with respect to the x-direction and can be classified as symmetric or antisymmetric depending on whether E_x is symmetric or antisymmetric about the $y = 0$ plane (see Fig. 8.1(a)). The integer n assumes odd values ($n = 1, 3, 5, \ldots$) for symmetric modes and even values ($n = 2, 4, 6, \ldots$) for antisymmetric modes. The field components for the symmetric modes can be obtained by setting $m = 0$ in Eqs. (6.17) and (6.18) and for antisymmetric modes by setting $m = 0$ in (6.24) and (6.25). Referring to the coordinate system indicated in Figure 8.1(a), the fields may be expressed as follows.

TE_{0n} Symmetric Modes

Region 1 ($|y| < b/2$):

$$E_x = C\omega\mu_0 \cos(\beta_y y) \tag{8.1a}$$

$$H_y = C\beta \cos(\beta_y y) \tag{8.1b}$$

$$H_z = jC\beta_y \sin(\beta_y y) \tag{8.1c}$$

Regions 2 and 3 ($|y| > b/2$):

$$E_x = C\omega\mu_0 \cos(\beta_y b/2)e^{\zeta(b/2 - |y|)} \tag{8.2a}$$

$$H_y = C\beta \cos(\beta_y b/2)e^{\zeta(b/2 - |y|)} \tag{8.2b}$$

$$H_z = \pm jC\zeta \cos(\beta_y b/2)e^{\zeta(b/2 - |y|)} \tag{8.2c}$$

where the upper and lower signs in (8.2c) apply to regions $y > b/2$ and $y < -b/2$, respectively. The common factor $e^{j(\omega t - \beta z)}$ and β are suppressed in (8.1) and (8.2). The propagation constant β, the transverse wavenumber β_y in region 1, and the transverse decay coefficient ζ in regions 2 and 3 are related by

$$\beta^2 = k_0^2 \varepsilon_r - \beta_y^2 = k_0^2 + \zeta^2 \tag{8.3}$$

The characteristic equation for determining β_y (or ζ) is given by (same as (6.20))

$$\beta_y \tan(\beta_y b/2) = \zeta \tag{8.4}$$

TE_{0n} Antisymmetric Modes

Region 1 ($|y| < b/2$):

$$E_x = D\omega\mu_0 \sin(\beta_y y) \tag{8.5a}$$

$$H_y = D\beta \sin(\beta_y y) \tag{8.5b}$$

$$H_z = -jD\beta_y \cos(\beta_y y) \tag{8.5c}$$

Regions 2 and 3 ($|y| > b/2$):

$$E_x = \pm D\omega\mu_0 \sin(\beta_y b/2)e^{\zeta(b/2 - |y|)} \tag{8.6a}$$

$$H_y = \pm D\beta \sin(\beta_y b/2)e^{\zeta(b/2 - |y|)} \tag{8.6b}$$

$$H_z = jD\zeta \sin(\beta_y b/2)e^{\zeta(b/2 - |y|)} \tag{8.6c}$$

where β, β_y, and ζ are related by (8.3). It may be noted that the common factor $e^{j(\omega t - \beta z)}$ and β are suppressed in (8.5) and (8.6). The characteristic equation for determining β_y (or ζ) is given by (same as (6.26))

$$\beta_y \cot(\beta_y b/2) = -\zeta \tag{8.7}$$

Cutoff and Dispersion Characteristics Using the relation (8.3), the characteristic equations (8.4) and (8.7) can be recast in the following forms:

$$\left(\frac{\beta_y b/2}{\cos(\beta_y b/2)}\right)^2 = \pi^2 \left(\frac{b}{\lambda_0}\right)^2 (\varepsilon_r - 1), \quad \text{symmetric modes } (n = 1, 3, \ldots) \tag{8.8}$$

$$\left(\frac{\beta_y b/2}{\sin(\beta_y b/2)}\right)^2 = \pi^2 \left(\frac{b}{\lambda_0}\right)^2 (\varepsilon_r - 1), \quad \text{antisymmetric modes } (n = 2, 4, \ldots) \tag{8.9}$$

The lowest order symmetric mode ($n = 1$) is the dominant TE_{01} mode, the field configuration for which is shown in Figure 8.7. The value of $\beta_y b/2$ for the TE_{01} mode lies in the range $0 \le \beta_y b/2 \le \pi/2$. The lowest order antisymmetric mode ($n = 2$) is the first higher order TE mode (TE_{02}) for which $\beta_y b/2$ lies in the range $\pi/2 \le \beta_y b/2 \le \pi$. The cutoff condition for the TE_{0n} modes can be written as [11]

$$\frac{b}{\lambda_0} = \frac{(n-1)}{2\sqrt{\varepsilon_r - 1}} \tag{8.10}$$

The cutoff condition for the hybrid modes is given by [14]

$$\frac{a_c}{\lambda_0} = 0.5 \left(\frac{1 + \tan^2(\beta_y b/2)}{1 + \varepsilon_r \tan^2(\beta_y b/2)}\right)^{1/2} \tag{8.11}$$

It may be noted that the fields, cutoff wavelength, and guide wavelength of TE modes are independent of the distance between the conducting plates. In particular, the dominant TE_{01} mode has no cutoff and hence can exist down to zero frequency. On the other hand, the hybrid modes depend on the plate separation. Figure 8.8 shows the normalized plate separation a_c/λ_0 corresponding to cutoff of hybrid modes as a function of the normalized slab thickness b/λ_0 for different values of ε_r [11]. Since TEM and pure TM modes cannot exist in the H-guide, only TE modes can exist if we choose the plate separation less than the critical value a_c. It can be seen from Figure 8.8 that the smaller the value of ε_r, the larger is the allowable plate separation. For single-mode operation in only the TE_{01} mode, the higher order TE_{02} mode must be cutoff. This additional condition is given by

$$\frac{b}{\lambda_0} < \frac{1}{2\sqrt{\varepsilon_r - 1}} \tag{8.12}$$

Figure 8.9 illustrates the variation in the normalized guide wavelength of the TE_{01} mode as a function of the normalized slab thickness b/λ_0 with ε_r as the parameter [11]. It can be seen that for small values of b/λ_0, λ/λ_0 approaches unity; and for large values of b/λ_0, its value approaches $1/\sqrt{\varepsilon_r}$. This is an expected result

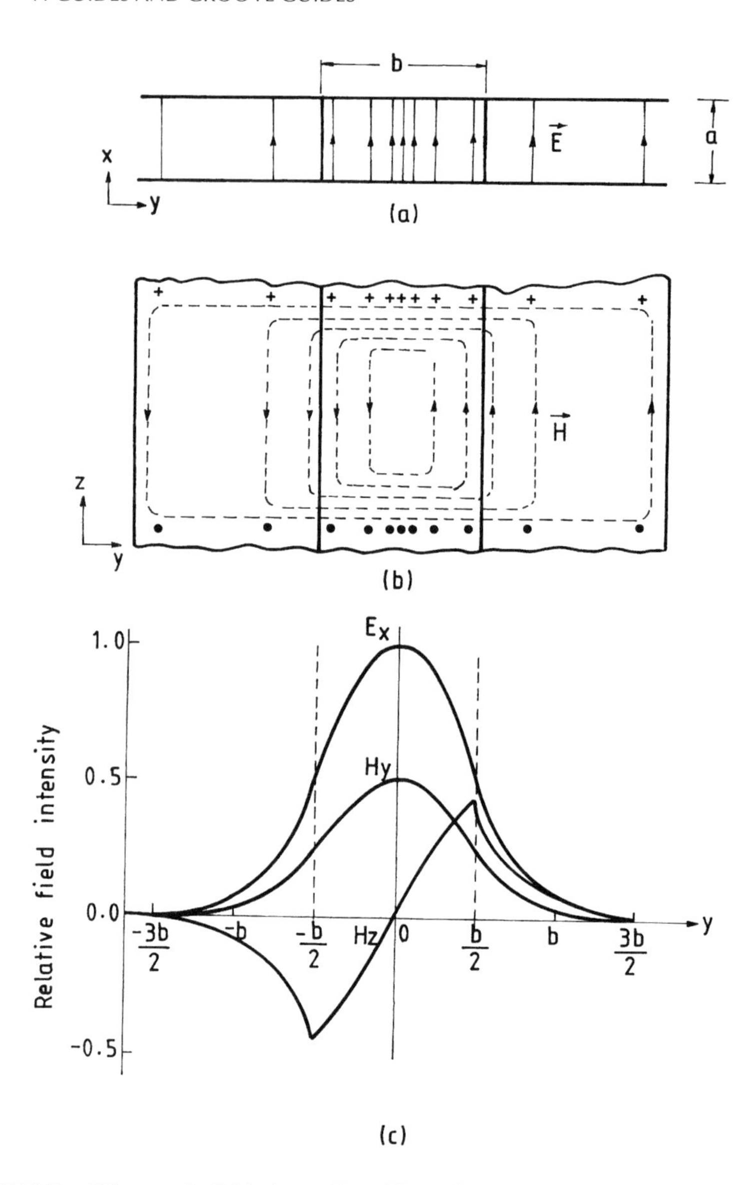

FIGURE 8.7 TE_{01} mode fields in an H-guide: E-lines perpendicular to the plates, (b) H-lines parallel to the plates, and (c) magnitude of the field components. (From Cohn [11]. Copyright © 1959 IEEE, reprinted with permission.)

since for small b/λ_0, most of the energy exists in the air region and with increasing b/λ_0 energy gets increasingly concentrated in the dielectric strip.

When the guide is symmetrically excited (E_x symmetric with respect to the $y = 0$ plane), the lowest order antisymmetric mode may not exist. The practical bandwidth can be taken as the frequency range covering up to the cutoff

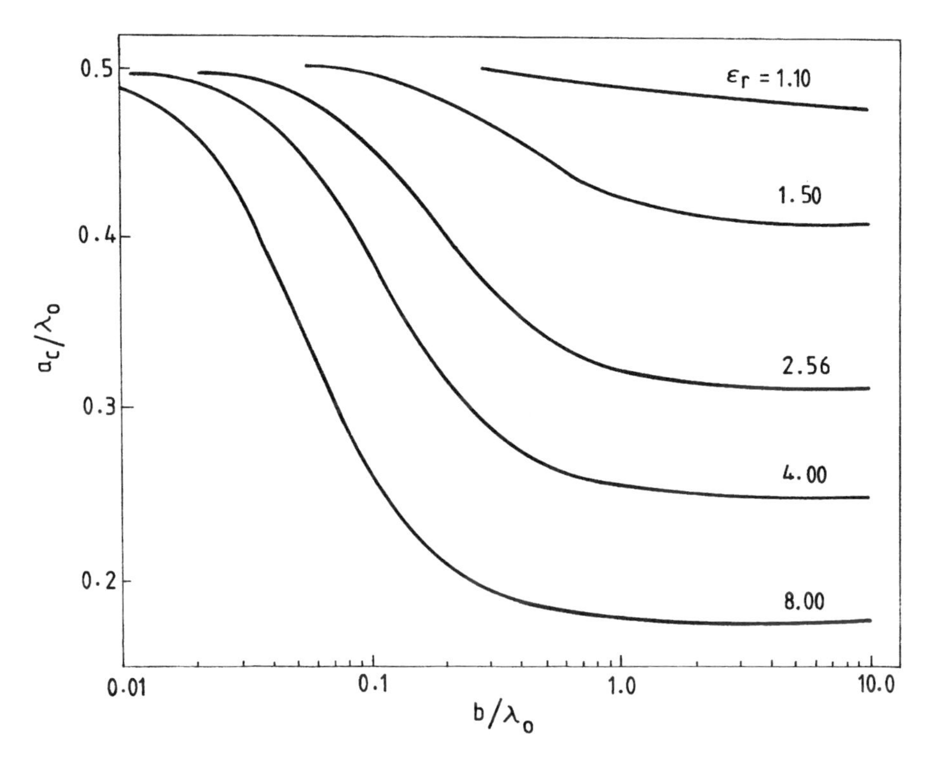

FIGURE 8.8 Cutoff condition for the hybrid modes as a function of the normalized slab thickness b/λ_0 and relative dielectric constant ε_r. (From Cohn [11]. Copyright © 1959 IEEE, reprinted with permission.)

frequency of the TE_{03} mode. The condition that prevents the TE_{03} mode from propagating is given by

$$\frac{b}{\lambda_0} < \frac{1}{\sqrt{\varepsilon_r - 1}} \tag{8.13}$$

The guide can also be excited in only the antisymmetric mode by placing a conducting wall at $y = 0$. With this, the top or bottom half of the H-guide is essentially a trough guide. The dominant mode of this guide is the TE_{02} mode. The ratio of the cutoff wavelengths of the TE_{02} mode and the next higher order antisymmetric mode (TE_{04}) is 3:1. The bandwidth of this trough guide is larger than that of the conventional rectangular waveguide but is smaller than that of the TE_{01} mode of the H-guide [11].

Phase-Shifting Properties A special application of the TE_{01} mode of the H-guide is in phase shifting [12, 13]. The guiding strip is made of ferroelectric

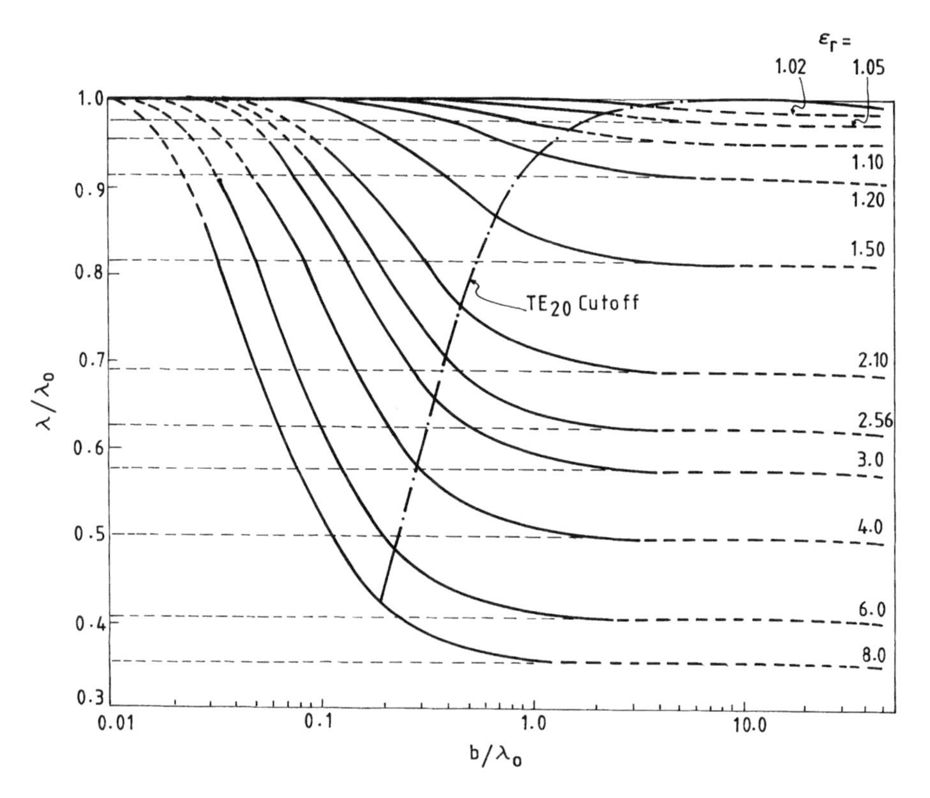

FIGURE 8.9 Normalized guide wavelength λ/λ_0 of the TE_{01} mode as a function of the normalized slab thickness b/λ_0 and relative dielectric constant ε_r. (From Cohn [11]. Copyright © 1959 IEEE, reprinted with permission.)

material such as $LiNbO_3$ (lithium niobate) or $LiTaO_3$ (lithium tantalate). These materials have a property that an applied voltage produces a direct change in their reflection index, thereby causing a change in the phase of the RF signal passing through it. The H-guide enables application of variable dc voltage across the plates. The electric field so produced is parallel to the E-field lines of the TE_{01} mode. The ferroelectric crystal may be oriented with its principal crystalline axis parallel to the E-field lines so that a single tensor component of the electro-optic coefficient contributes to phase change. The phase shift in such a device can be defined as

$$\Delta\phi = l\Delta\beta = l\left(\frac{d\beta}{dn}\right)\Delta n \tag{8.14}$$

where $n = \sqrt{\varepsilon_r}$ is the refractive index of the material, Δn is the induced change in the refractive index, l is the length of the device, and β is the propagation constant given by (8.3). Using (8.3), (8.8), and (8.14), we can obtain the following expression for $\Delta\phi$ [13]:

$$\Delta\phi = \left(\frac{2\pi l\,\Delta n}{\lambda_0}\right)\left(\frac{\lambda}{\lambda_0}\right)\sqrt{\varepsilon_r}\left[\frac{(\beta_y b/2) + \sin(\beta_y b/2)\cos(\beta_y b/2)}{(\beta_y b/2) + \cot(\beta_y b/2)}\right] \tag{8.15}$$

where the tensor notation for n and $\sqrt{\varepsilon_r}$ is suppressed. Depending on the orientation of the crystal axes, $\varepsilon_r = \varepsilon_{\text{or}}$, where $\varepsilon_{\text{or}} = \varepsilon_{11} = \varepsilon_{22} = n_{\text{or}}^2$, or $\varepsilon_r = \varepsilon_{\text{ex}}$, where $\varepsilon_{\text{ex}} = \varepsilon_{33} = n_{\text{ex}}^2$. The subscript "or" and "ex" refer to the "ordinary" and "extraordinary" directions of the crystal. The factor $(2\pi l\,\Delta n/\lambda_0)$ corresponds to the phase shift $(\Delta\phi_0)$ due to a plane wave propagating in the same material of infinite extent. Normalizing $\Delta\phi$ by this factor, we can write

$$\Delta\phi_R = \frac{\Delta\phi}{\Delta\phi_0} = \left(\frac{\lambda}{\lambda_0}\right)\sqrt{\varepsilon_r}\left[\frac{(\beta_y b/2) + \sin(\beta_y b/2)\cos(\beta_y b/2)}{(\beta_y b/2) + \cot(\beta_y b/2)}\right] \tag{8.16}$$

The attenuation constant α_d due to dielectric loss can be derived using (6.40b), (6.41b), and (6.41c) (Chapter 6). The normalized attenuation constant α_R is given by [13]

$$\alpha_R = \frac{\alpha_d}{\alpha_0} = \left(\frac{\lambda}{\lambda_0}\right)\sqrt{\varepsilon_r}\left[\frac{(\beta_y b/2) + \sin(\beta_y b/2)\cos(\beta_y b/2)}{(\beta_y b/2) + \cot(\beta_y b/2)}\right] \tag{8.17}$$

which is identical to the expression for $\Delta\phi_R$. The normalizing factor α_0 is the attenuation constant due to a plane wave in the same material of infinite extent and is given by

$$\alpha_0 = (4\pi/\lambda_0)\sqrt{\varepsilon_r}\tan\delta \tag{8.18}$$

Figure 8.10 illustrates typical phase-shift and attenuation characteristics of the TE_{01} mode of an H-guide using Z-cut $LiNbO_3$ with $\varepsilon_r = \varepsilon_{\text{ex}} = 27$ (crystalline z-axis is normal to the plates) [13]. The plots show the normalized phase shift $\Delta\phi_R$, normalized attenuation coefficient α_R, normalized phase parameter B, and the fractional power confinement factor ρ as a function of normalized frequency. The normalized phase parameter B is defined as

$$B = \frac{(\beta/k_0)^2 - 1}{\varepsilon_r - 1} \tag{8.19}$$

The fractional power confinement factor ρ is the ratio of the power flow in the dielectric (P_d) to the total axial power flow $(P_d + P_a)$.

$$\rho = \frac{P_d}{P_d + P_a} \tag{8.20}$$

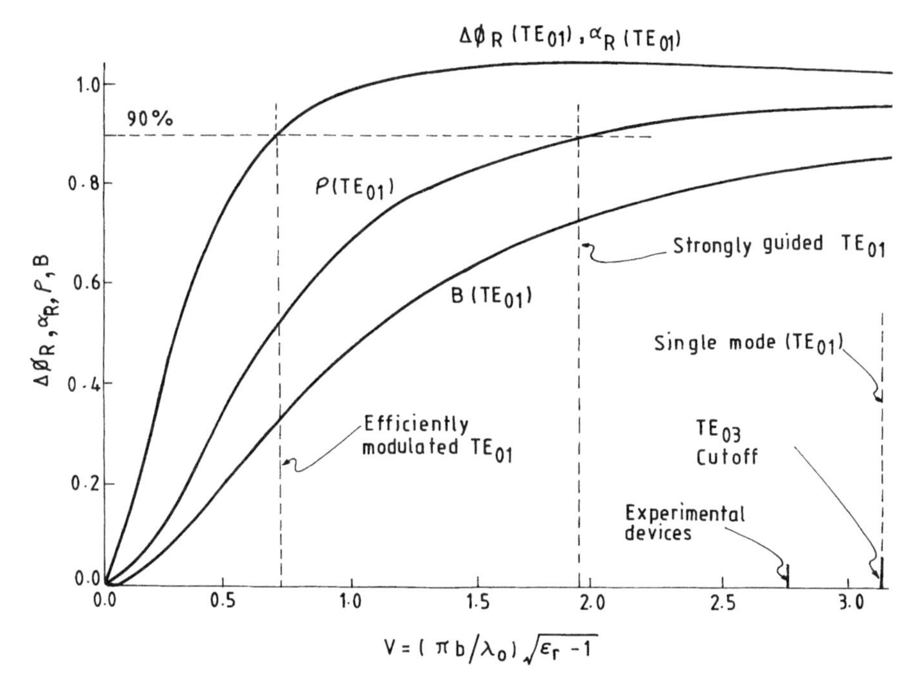

FIGURE 8.10 Characteristics of the TE_{01} mode in the H-guide using Z-cut $LiNbO_3$. $\varepsilon_r = \varepsilon_{ex} = 27$. (From Klein [13], Copyright © 1984 Academic Press Inc., Orlando, FL, reprinted with permission.)

where P_a is the power flow in the air region. The formulas for determining P_d and P_a are

$$P_d = \int_0^a \int_{-b/2}^{b/2} E_{x1} H_{y1}^* \, dx \, dy \tag{8.21}$$

$$P_a = 2 \int_0^a \int_{b/2}^{\infty} E_{x2} H_{y2}^* \, dx \, dy \tag{8.22}$$

where the subscripts 1 and 2 refer to the dielectric and air regions, respectively (see Fig. 8.1(a)). Using the field expressions for the TE_{01} mode, the formula for ρ is obtained as [13]

$$\rho = \frac{(\beta_y b/2) + \sin(\beta_y b/2)\cos(\beta_y b/2)}{(\beta_y b/2) + \cot(\beta_y b/2)} \tag{8.23}$$

The normalized frequency v plotted on the horizontal axis is defined as

$$v = (\pi b/\lambda_0)\sqrt{\varepsilon_r - 1} \tag{8.24}$$

We note from Figure 8.10 that both $\Delta\phi_R$ and α_R increase rapidly as v increases from zero and reaches saturation beyond about $v = 1$. Since the TE_{01} mode has no cutoff and the TE_{02} mode (antisymmetric mode) is not likely to exist, the theoretical frequency range of operation extends from zero up to the TE_{03} mode cutoff at $v = 3.14$. However, for small values of v, most of the energy exists outside the dielectric. If the value of v is chosen corresponding to $\rho = 0.9$ (i.e., 90% of the power being confined within the dielectric), then the range of v for single-mode (TE_{01}) operation is $2 \leq v \leq 3.14$ and the operating bandwidth is then approximately 35%.

8.3 DOUBLE-STRIP H-GUIDE

The double-strip H-guide, the cross-section of which is shown in Figure 8.11, has in general less transmission loss than the H-guide [7, 8, 15, 16]. Further reduction in transmission loss is reported to be feasible by using a laminated dielectric slab [16]. The structure is identical to the coupled NRD guide (see Fig. 6.6) except that the plate separation is larger than half the free-space wavelength. The expressions for the field components and the dispersion relation presented in Chapter 6 (Section 6.4) for the coupled NRD guide are therefore applicable to the double-strip H-guide also. As a single propagating guide operating in the TM^y_{11}

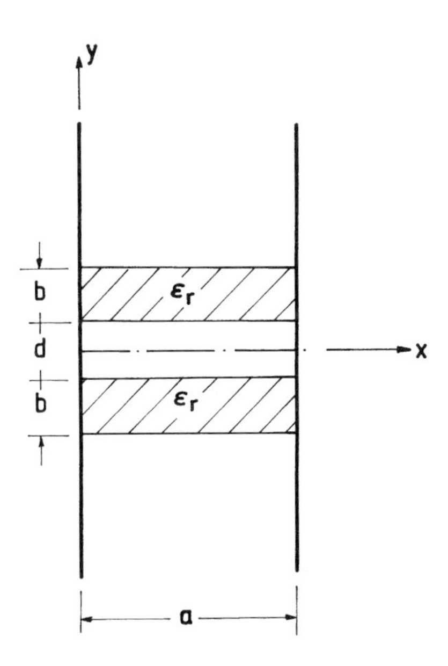

FIGURE 8.11 Cross-section of double-strip H-guide.

mode, the dispersion relation for the double-strip H-guide is given by

$$\frac{\beta_y [\beta_y \tan(\beta_y b) - \varepsilon_r \zeta]}{\varepsilon_r \zeta [\varepsilon_r \zeta \tan(\beta_y b) + \beta_y]} = \tanh\left(\frac{\zeta d}{2}\right) \tag{8.25}$$

where

$$\zeta^2 = k_0^2(\varepsilon_r - 1) - \beta_y^2 \tag{8.26a}$$

$$\beta^2 = k_0^2 \varepsilon_r - \beta_y^2 - (\pi/a)^2 \tag{8.26b}$$

The attenuation coefficients due to the conductor and dielectric losses, denoted as α_c and α_d, respectively, are given by [15]

$$\alpha_c = \left(\frac{k_0\sqrt{2\omega\varepsilon_0}}{a\sqrt{\sigma}}\right)\left(\frac{(\pi/a)^2}{\beta[(\pi/a)^2 + \beta^2]}\right)\left(\frac{A + B_+ + 1/2\zeta}{A + (B_+/\varepsilon_r) + 1/2\zeta}\right) \tag{8.27}$$

$$\alpha_d = \left(\frac{\tan\delta}{2\varepsilon_r\beta}\right)\left(\frac{[(\pi/a)^2 + \beta^2]B_+ + \beta_y^2 B_-}{A + (B_+/\varepsilon_r) + 1/2\zeta}\right) \tag{8.28}$$

where $\tan\delta$ is the dielectric loss tangent and σ is the conductivity of the conductor. The parameters A and $B_\pm$ are given by

$$A = \left(\frac{\beta_y \cos(\beta_y b) + \zeta\varepsilon_r \sin(\beta_y b)}{\beta_y \cosh(\zeta d/2)}\right)^2 \left(\frac{\sinh(\zeta d) + \zeta d}{4\zeta}\right) \tag{8.29a}$$

$$B_\pm = (b/2)[1 + (\zeta\varepsilon_r/\beta_y)^2] \pm (1/4\beta_y)[1 - (\zeta\varepsilon_r/\beta_y)^2] \cdot \sin(2\beta_y b) \pm (\zeta\varepsilon_r/2\beta_y^2)[1 - \cos(2\beta_y b)] \tag{8.29b}$$

Figure 8.12 shows the variation in the total transmission loss in the double-strip H-guide as a function of the normalized separation distance $d/2\lambda_0$ for different values of the normalized strip width a/λ_0 [15]. The graph shows a decrease in the total transmission loss with an increase in the separation distance d, thereby indicating less transmission loss for the double-strip H-guide as compared with the single-strip H-guide having the same parameters ($a \times b$, ε_r, $\tan\delta$). The total transmission loss decreases also with a decrease in the strip thickness b/λ_0. This is because of the decrease in the dielectric loss resulting from the decrease in the volume of the dielectric. As an illustration of the relative contribution to the total transmission loss due to the conductor and the dielectric, Figure 8.13 shows a typical variation in α_c, α_d, and the total attenuation coefficient $\alpha_t = \alpha_c + \alpha_d$ as a function of frequency [8]. At sufficiently low frequencies, where the fields extend far into the air regions, current is induced over a larger surface area of the conductor plates, resulting in large conductor loss. The dielectric loss is also relatively large because the electric field lines are vertically

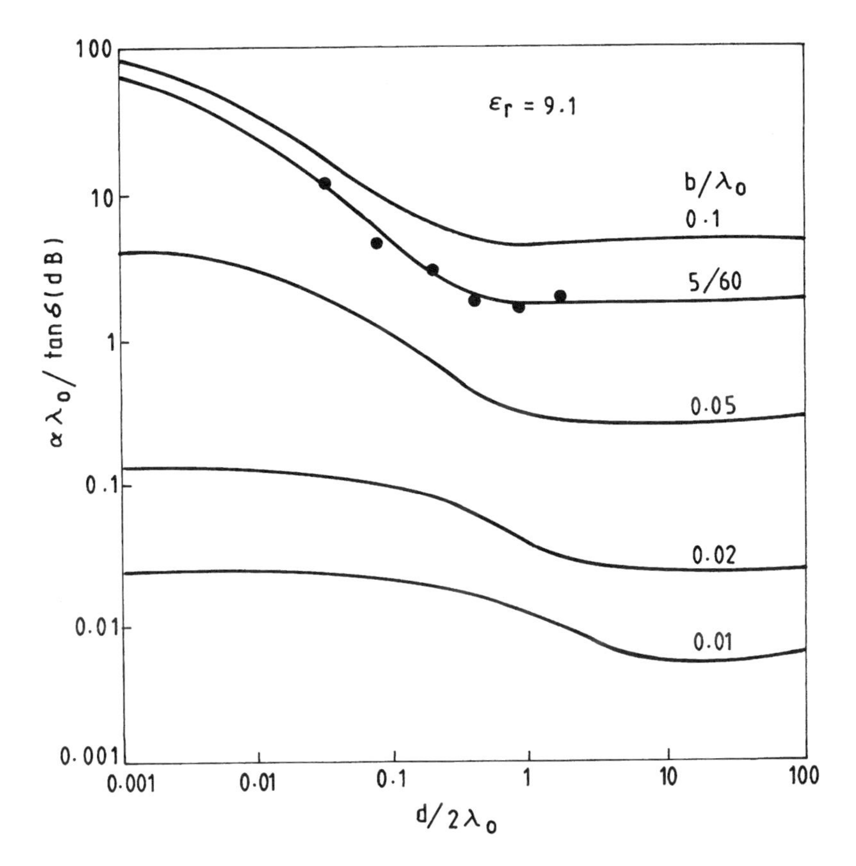

FIGURE 8.12 Calculated transmission loss of double-strip H-guide (see Fig. 8.11) as a function of separation between the two slabs and its comparison with experimental data (••••). (From Kawamura and Kokubo [15]. Copyright © 1980 IEEE, reprinted with permission.)

oriented (parallel to the metal plates) and uniformly distributed in the dielectric. Since the fields decay at a faster rate with frequency in the air region and there is no longitudinal current flow for the TM_{11}^{y} mode, the conductor loss decreases with an increase in frequency. On the other hand, there is first a decrease in the dielectric loss because of the flexing of the E-field lines from the vertical orientation, thereby reducing the effective volume of the dielectric contributing to dielectric loss. A further increase in frequency increases the concentration of electric fields within the dielectric relative to that in the air region, thereby contributing to increased dielectric loss. The magnitude of the dielectric loss is directly proportional to the dielectric loss tangent $\tan\delta$. Thus, unless $\tan\delta$ is very small, the nature of variation in the dielectric loss gets reflected in the total transmission loss. For the same dielectric volume, the total transmission loss in

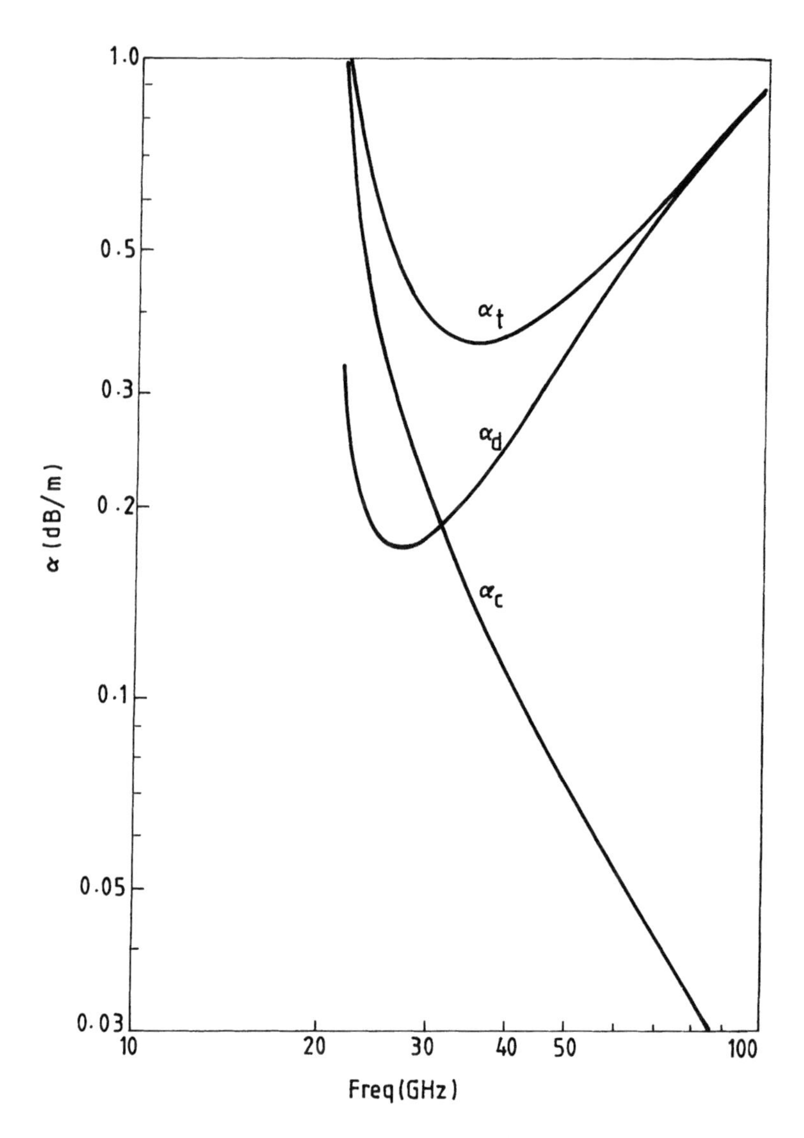

FIGURE 8.13 Attenuation versus frequency for the TM_{11}^y mode of a double-strip H-guide (see Fig. 8.11): $\varepsilon_r = 1.03$, $\tan\delta$ (dielectric loss tangent) $= 10^{-4}$, ρ (metal resistivity) $= 1.73 \times 10^{-8}\,\Omega$-m, $a = 0.712$ cm, $b = 1.27$ cm, and $d/2b = 1$. (From Conlon and Benson [8], reprinted with permission of IEE.)

a double-strip H-guide is less than that in a single-strip H-guide. This feature is illustrated in Figure 8.14 [8]. The figure also shows that the loss in a double-strip H-guide can be much smaller than that in a rectangular waveguide over a certain bandwidth.

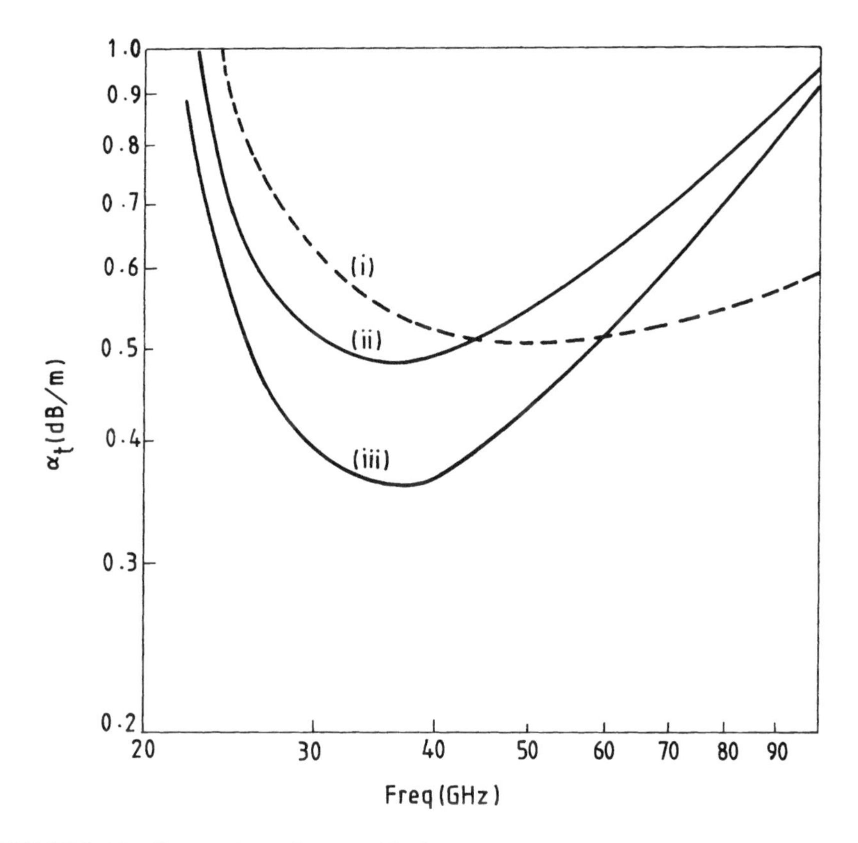

FIGURE 8.14 Comparison of waveguides for 35-GHz operation: (i) conventional rectangular waveguide (WG 22); (ii) single-strip H-guide (Fig. 8.1(a)) with $a = 0.712$ cm and $b = 2.54$ cm; and (iii) double-strip H-guide (Fig. 8.11) with $a = 0.712$ cm, $b = 2.54$ cm, and $d/2b = 1$. (From Conlon and Benson [8], reprinted with permission of IEE.)

8.4 GROOVE GUIDE

8.4.1 Field Analysis

Analysis of the groove guide has been reported by several investigators using the technique of modal analysis [4–6, 17], conformal mapping [18, 19], and transverse equivalent network [20]. In the following we present the modal analysis method reported by Nakahara and Kurauchi [5].

The propagating modes of a groove guide (Fig. 8.15(a)) can be grouped into TE and TM modes with respect to the direction of propagation (z-direction). Using Maxwell's equations, we can conveniently express the transverse E- and H-fields in terms of a single longitudinal component H_z in the case of TE modes ($E_z = 0$) and in terms of E_z in the case of TM modes ($H_z = 0$). Assuming a time and

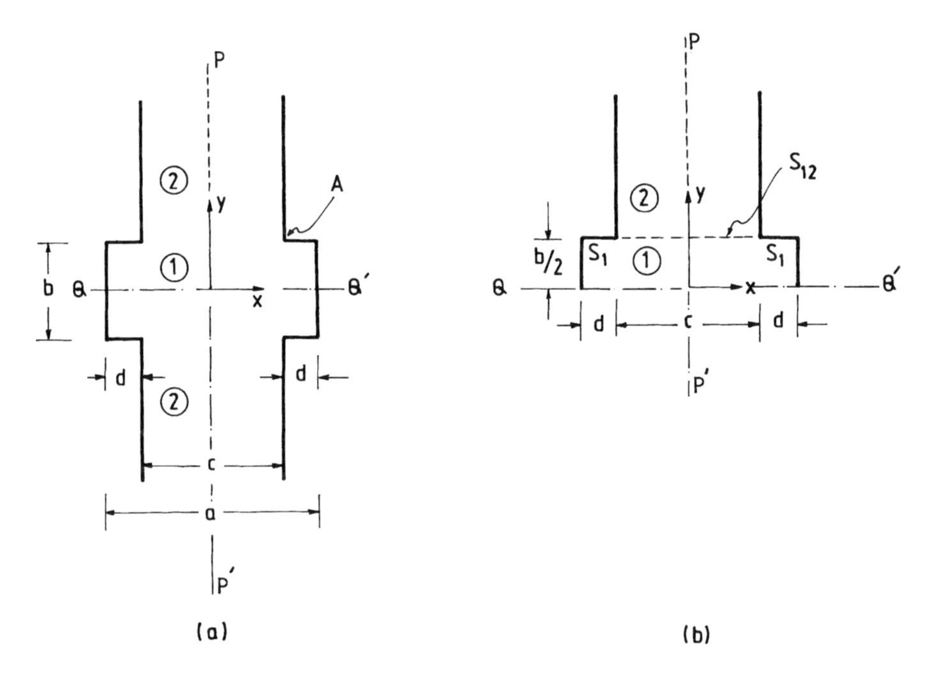

FIGURE 8.15 (a) Cross-sectional view of groove guide. (b) One-half of the structure for the purpose of analysis.

z-dependence of the form $e^{j(\omega t - \beta z)}$, the field relations can be expressed as follows.

***TE Modes* ($E_z = 0$)**

$$E_x = \left(\frac{j\omega\mu_0}{k_0^2 - \beta^2}\right)\frac{\partial H_z}{\partial y} \tag{8.30a}$$

$$E_y = \left(\frac{j\omega\mu_0}{k_0^2 - \beta^2}\right)\frac{\partial H_z}{\partial x} \tag{8.30b}$$

$$H_x = \left(\frac{-j\beta}{k_0^2 - \beta^2}\right)\frac{\partial H_z}{\partial x} \tag{8.30c}$$

$$H_y = \left(\frac{-j\beta}{k_0^2 - \beta^2}\right)\frac{\partial H_z}{\partial y} \tag{8.30d}$$

***TM Modes* ($H_z = 0$)**

$$E_x = \left(\frac{-j\beta}{k_0^2 - \beta^2}\right)\frac{\partial E_z}{\partial x} \tag{8.31a}$$

$$E_y = \left(\frac{-j\beta}{k_0^2 - \beta^2}\right)\frac{\partial E_z}{\partial y} \tag{8.31b}$$

$$H_x = \left(\frac{j\omega\varepsilon_0}{k_0^2 - \beta^2}\right)\frac{\partial E_z}{\partial y} \tag{8.31c}$$

$$H_y = \left(\frac{-j\omega\varepsilon_0}{k_0^2 - \beta^2}\right)\frac{\partial E_z}{\partial x} \tag{8.31d}$$

where

$$k_0 = \omega\sqrt{\mu_0\varepsilon_0} = 2\pi/\lambda_0 \tag{8.32}$$

is the free-space propagation constant. The longitudinal fields E_z and H_z satisfy the wave equation

$$\left(\frac{\partial^2}{\partial x^2} + \frac{\partial^2}{\partial y^2} + (k_0^2 - \beta^2)\right)\begin{Bmatrix} E_z \\ H_z \end{Bmatrix} = 0 \tag{8.33}$$

For ease of field analysis, we divide the groove guide into two regions, marked 1 (groove region $|y| < b/2$) and 2 ($|y| > b/2$). In region 1, the fields are expressed as a summation of normal modes of a rectangular guide, and in region 2, they are expressed in terms of modes of a parallel-plate guide. The modes of the guide structure, however, are represented in terms of the modes of the groove region. That is, they are designated as TE_{mn} and TM_{mn} modes, where the index m refers to the number of half-period sinusoidal variations in the x-direction and n refers to the rank of the mode in the y-direction corresponding to the groove region. These modes can further be classified as symmetric and antisymmetric with respect to each of the two symmetry planes PP' ($x = 0$ plane) and QQ' ($y = 0$ plane)—namely, TE_{mn} (SS), TE_{mn} (SA), TE_{mn} (AS), and TE_{mn} (AA) for TE modes; and TM_{mn} (SS), TM_{mn} (SA), TM_{mn} (AS), and TM_{mn} (AA) for TM modes. The symbols S and A denote *symmetric* and *antisymmetric*, respectively; the first symbol within the parentheses refers to the plane PP' and the second symbol refers to the plane QQ'. The field component that is used for defining the symmetry/asymmetry is E_y for the TE_{mn} mode and E_x for the TM_{mn} mode. For example, for the TE_{mn} (SA) mode, E_y is symmetric about PP' (m odd $= 1, 3, 5, \ldots$) and antisymmetric about QQ' (n even $= 2, 4, 6, \ldots$). We note from (8.30) that $E_y \approx \partial H_z/\partial x$ for TE modes and from (8.31) that $E_x \approx \partial E_z/\partial x$ for TM modes. The expressions for H_z in the case of TE_{mn} modes and those for E_z in the case of TM_{mn} modes can be written as the solutions of the wave equation (8.33). They are

$$\begin{pmatrix} H_z \\ E_z \end{pmatrix}\begin{cases} \approx \sin(m\pi x/a)\cos(\beta_y y), & \text{SS mode; } m \text{ odd, } n \text{ odd} \\ \approx \sin(m\pi x/a)\sin(\beta_y y), & \text{SA mode; } m \text{ odd, } n \text{ even } (2,4,\ldots) \\ \approx \cos(m\pi x/a)\cos(\beta_y y), & \text{AS mode; } m \text{ even, } n \text{ odd} \\ \approx \cos(m\pi x/a)\sin(\beta_y y), & \text{AA mode; } m \text{ even, } n \text{ even } (2,4,\ldots) \end{cases} \tag{8.34}$$

These solutions can be substituted in (8.30) and (8.31) to obtain the transverse fields for the TE_{mn} and TM_{mn} modes, respectively. Referring to the top symmetric half of the guide as shown in Figure 8.15(b), we now provide the complete field expressions in the two regions $0 < y < b/2$ and $y > b/2$ for TE_{mn} modes.

TE_{mn} Modes

TE_{mn} (SS) Modes (m odd, n odd)

Region 1 ($0 < y < b/2$):

$$E_x = jA\omega\mu_0\beta_y \sin(m\pi x/a)\sin(\beta_y y) \tag{8.35a}$$
$$E_y = jA\omega\mu_0(m\pi/a)\cos(m\pi x/a)\cos(\beta_y y) \tag{8.35b}$$
$$E_z = 0 \tag{8.35c}$$
$$H_x = -jA\beta(m\pi/a)\cos(m\pi x/a)\cos(\beta_y y) \tag{8.35d}$$
$$H_y = jA\beta\beta_y \sin(m\pi x/a)\sin(\beta_y y) \tag{8.35e}$$
$$H_z = A(k_0^2 - \beta^2)\sin(m\pi x/a)\cos(\beta_y y) \tag{8.35f}$$

Region 2 ($y > b/2$):

$$E_x = jB\omega\mu_0\gamma_y \sin(p\pi x/c)e^{\gamma_y(b/2-y)} \tag{8.36a}$$
$$E_y = jB\omega\mu_0(p\pi/c)\cos(p\pi x/c)e^{\gamma_y(b/2-y)} \tag{8.36b}$$
$$E_z = 0 \tag{8.36c}$$
$$H_x = -jB\beta(p\pi/c)\cos(p\pi x/c)e^{\gamma_y(b/2-y)} \tag{8.36d}$$
$$H_y = jB\beta\gamma_y \sin(p\pi x/c)e^{\gamma_y(b/2-y)} \tag{8.36e}$$
$$H_z = B(k_0^2 - \beta^2)\sin(p\pi x/c)e^{\gamma_y(b/2-y)} \tag{8.36f}$$

where p is an odd integer and γ_y is the y-directed propagation constant in region 2. $\gamma_y = \zeta_y$ (real) for a decaying type of wave and $\gamma_y = j\xi_y$ with ξ_y real for a propagating type of wave. The parameters β, β_y, and γ_y are related by

$$k_0^2 - \beta^2 = (m\pi/a)^2 + \beta_y^2 = (p\pi/c)^2 - \gamma_y^2 \tag{8.37}$$

TE_{mn} (SA) Modes (m odd, n even, $n \neq 0$) Expressions (8.35)–(8.37) apply except that in (8.35), we replace $\cos(\beta_y y)$ by $\sin(\beta_y y)$ and $\sin(\beta_y y)$ by $-\cos(\beta_y y)$.

TE_{mn} (AS) Modes (m even, n odd, $m \neq 0$)

Region 1 ($0 < y < b/2$):

$$E_x = jA\omega\mu_0\beta_y \cos(m\pi x/a)\sin(\beta_y y) \tag{8.38a}$$

$$E_y = -jA\omega\mu_0(m\pi/a)\sin(m\pi x/a)\cos(\beta_y y) \tag{8.38b}$$

$$E_z = 0 \tag{8.38c}$$

$$H_x = jA\beta(m\pi/a)\sin(m\pi x/a)\cos(\beta_y y) \tag{8.38d}$$

$$H_y = jA\beta\beta_y\cos(m\pi x/a)\sin(\beta_y y) \tag{8.38e}$$

$$H_z = A(k_0^2 - \beta^2)\cos(m\pi x/a)\cos(\beta_y y) \tag{8.38f}$$

Region 2 ($y > b/2$):

$$E_x = jB\omega\mu_0\gamma_y\cos(p\pi x/c)e^{\gamma_y(b/2-y)} \tag{8.39a}$$

$$E_y = -jB\omega\mu_0(p\pi/c)\sin(p\pi x/c)e^{\gamma_y(b/2-y)} \tag{8.39b}$$

$$E_z = 0 \tag{8.39c}$$

$$H_x = jB\beta(p\pi/c)\sin(p\pi x/c)e^{\gamma_y(b/2-y)} \tag{8.39d}$$

$$H_y = jB\beta\gamma_y\cos(p\pi x/c)e^{\gamma_y(b/2-y)} \tag{8.39e}$$

$$H_z = B(k_0^2 - \beta^2)\cos(p\pi x/c)e^{\gamma_y(b/2-y)} \tag{8.39f}$$

where p is an even integer ($p \neq 0$) and $\gamma_y = \zeta_y$ (real) or $j\xi_y$ (imaginary) depending on whether the wave in region 2 is decaying or propagating. The parameters β, β_y, and γ_y satisfy the relation (8.37).

TE_{mn} (AA) Modes (m even, n even, $m, n \neq 0$) Expressions (8.37), (8.38), and (8.39) apply except that in (8.38), we replace $\cos(\beta_y y)$ by $\sin(\beta_y y)$ and $\sin(\beta_y y)$ by $-\cos(\beta_y y)$.

The transverse fields for the TM_{mn} (SS, SA, AS, AA) modes can be written in a similar manner by using the field variation for E_z given by (8.34) in (8.31). The propagation parameters β, β_y, and γ_y for the TM_{mn} modes satisfy the relation (8.37).

Field Configuration The dominant mode of excitation in the groove guide is the TE_{11} mode, which belongs to the TE (SS) group. The E_y and H_x components of this mode are symmetric in both the x- and y-directions. Figure 8.16 shows the cross-sectional view of the TE_{11} mode field pattern. This mode can easily be excited by a TE_{10} mode of a rectangular waveguide through a tapered transition. Figure 8.17 shows the transverse electric field lines of the TE_{mn} and TM_{mn} modes for $m = 1, 3$ and $n = 1, 2$. It may be noted that modes corresponding to m even do not exist in the groove guide.

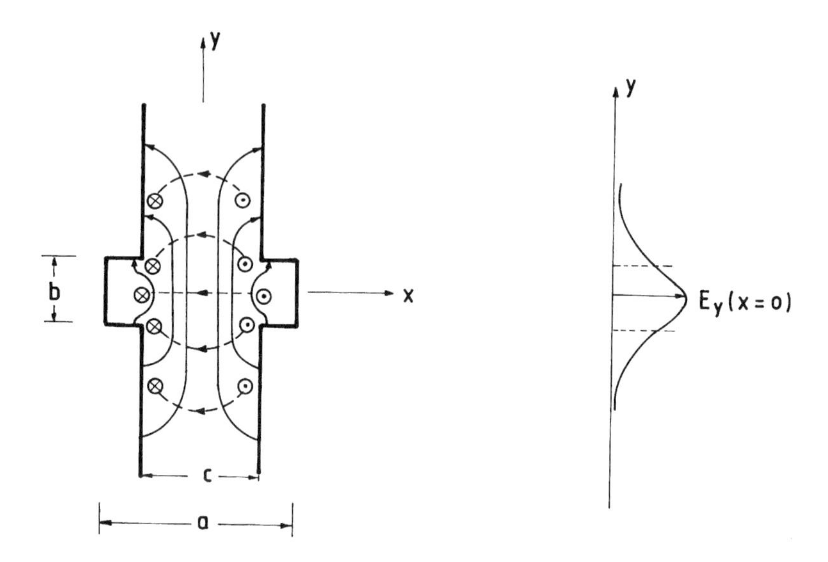

FIGURE 8.16 Field pattern of the dominant TE_{11} mode in groove guide: —— E-lines, ------ H-lines.

Characteristic Equations for TE_{1n} Modes We first derive the characteristic equation for the TE_{1n} (SS) modes (n odd) by applying the boundary conditions at the plane $y = b/2$. The plane $y = b/2$ is divided into two parts—namely, S_{12} on which H_z and E_x must be continuous and the conductor boundary S_1 on which E_x must be zero (refer to Fig. 8.15(b)).

- On S_{12}, H_z must be continuous. Equating (8.35f) and (8.36f) and setting $m = p = 1$ and $y = b/2$, we get

$$A \sin(\pi x/a) \cos(\beta_y b/2) = B \sin(\pi x/c) \tag{8.40}$$

- E_x is continuous on S_{12} and zero on S_1. From (8.35a) and (8.36a), we can write

$$A\beta_y \sin(\pi x/a) \sin(\beta_y b/2) = \begin{cases} B\gamma_y \sin(\pi x/c) & \text{on } S_{12} \\ 0 & \text{on } S_1 \end{cases} \tag{8.41}$$

Multiplying (8.40) by $\sin(\pi x/a)$ and integrating over the boundary S_{12}, we obtain

$$B = A\left(\frac{4\pi \cos(\pi c/2a)}{ac[(\pi/c)^2 - (\pi/a)^2]}\right)\cos\left(\frac{\beta_y b}{2}\right) \tag{8.42}$$

Multiplying (8.41) by $\sin(\pi x/a)$ and integrating over the boundary $S_{12} + S_1$,

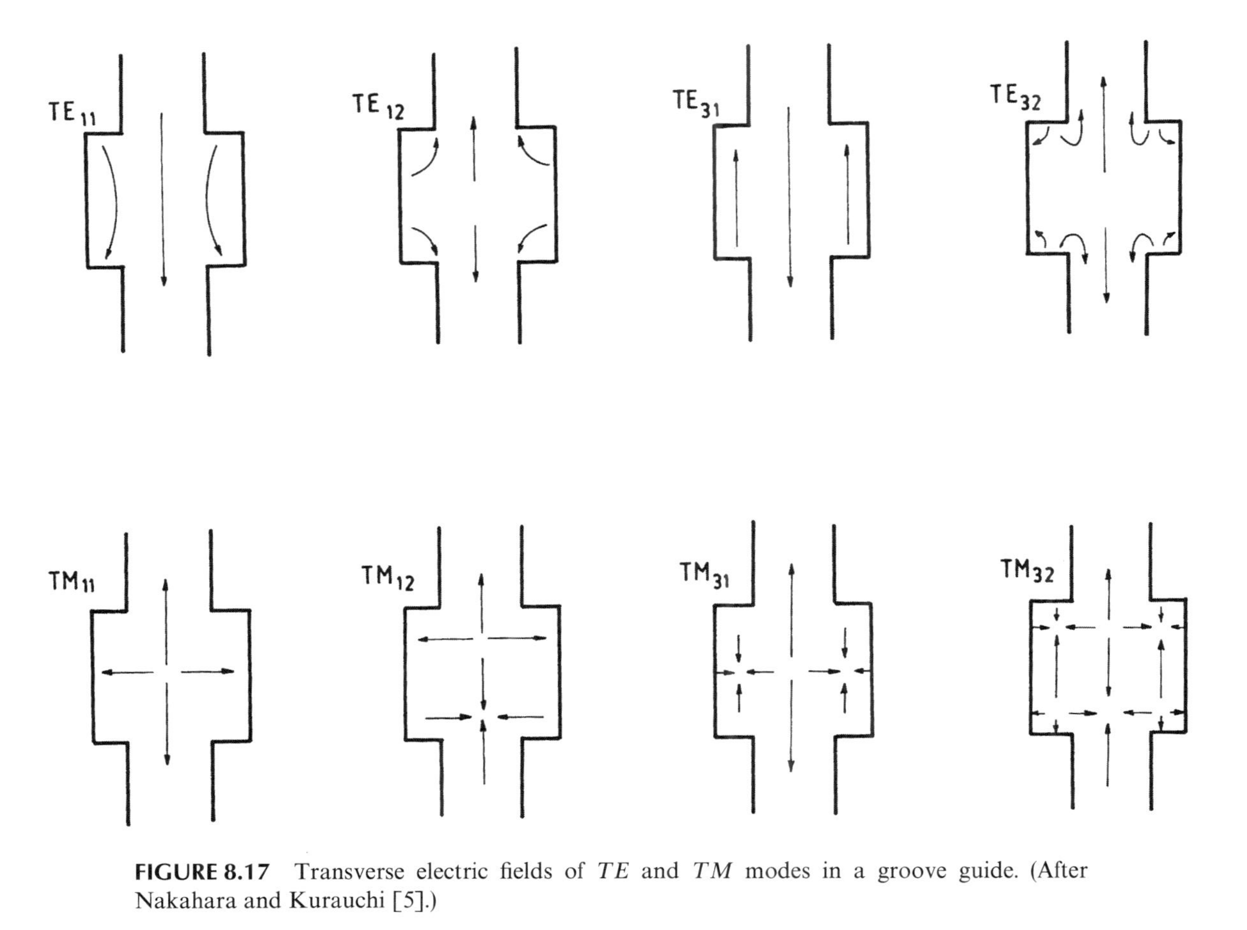

FIGURE 8.17 Transverse electric fields of TE and TM modes in a groove guide. (After Nakahara and Kurauchi [5].)

we get

$$B = A\left(\frac{(a/2)^2\,[(\pi/c)^2 - (\pi/a)^2]}{\pi\cos(\pi c/2a)}\right)\left(\frac{\beta_y}{\gamma_y}\right)\sin\left(\frac{\beta_y b}{2}\right) \tag{8.43}$$

Equating (8.42) and (8.43) yields the following characteristic equation for the TE_{1n} mode with n odd:

$$\gamma_y = \left(\frac{\pi^2 a}{16\,c}\right)\left(\frac{[(a/c) - (c/a)]^2}{\cos^2(\pi c/2a)}\right)\left[\beta_y \tan\left(\frac{\beta_y b}{2}\right)\right] \tag{8.44}$$

The expression for n even is the same as (8.44) with $\tan(\beta_y b/2)$ replaced by $-\cot(\beta_y b/2)$. Defining

$$u = \beta_y b/2 \quad \text{and} \quad v = \gamma_y b/2 \tag{8.45}$$

we can rewrite the characteristic equation for the TE_{1n} mode in the form [5]

$$v = \left(\frac{\pi^2 a}{16\,c}\right)\left(\frac{[(a/c) - (c/a)]^2}{\cos^2(\pi c/2a)}\right)\cdot\begin{cases} u \tan u, & n \text{ odd} \quad (8.46a) \\ -u \cot u, & n \text{ even} \quad (8.46b) \end{cases}$$

Characteristic Equation for TM_{1n} Modes Using the same procedure as outlined above, the characteristic equation for the TM_{1n} modes can be derived. The expression is given by [5]

$$v = \left(\frac{16a}{\pi^2 c}\right)\left(\frac{\cos^2(\pi c/2a)}{[(a/c) - (c/a)]^2}\right)\cdot\begin{cases} u \tan u, & n \text{ odd} \quad (8.47a) \\ -u \cot u, & n \text{ even} \quad (8.47b) \end{cases}$$

where u and v are defined as in (8.45). From (8.37), we note that for both TE_{1n} and TM_{1n} modes, u and v satisfy the relation

$$u^2 + v^2 = (\pi b/2c)^2 - (\pi b/2a)^2 \tag{8.48}$$

Both TE_{1n} and TM_{1n} modes exist in the range $(n-1)\pi/2 < \beta_y b/2 < n\pi/2$. Thus, in the range $\beta_y b/2 < \pi$, only the TE_{11} and TM_{11} modes exist.

8.4.2 Propagation Characteristics

Cutoff Wavelength of TE_{1n} and TM_{1n} Modes At cutoff ($\beta = 0$), the separation equation (8.37) can be rewritten as

$$(2\pi/\lambda_c)^2 = (\pi/a)^2 + (2u'/b)^2 = (\pi/c)^2 - (2v'/b)^2 \tag{8.49}$$

where λ_c is the cutoff wavelength. The values of u' and v' to be substituted in (8.49) are obtained as solutions of (8.46) for TE_{1n} modes and of (8.47) for TM_{1n} modes.

We note that when $c/a = 0$, the TE_{11} mode of the groove guide tends to the TE_{10} mode of the rectangular waveguide of width a, and when $c/a = 1$, it tends to the TE_{11} mode of the parallel-plate guide having plate separation equal to a. Thus, for the groove guide ($0 < c/a < 1$), the cutoff wavelength λ_c has a range $2c < \lambda_c < 2a$ [5].

Condition for Higher Order Mode Cutoff Considering TE-type modes, we note that while TE_{mn} modes are excited in region 1, TE_{pq} modes with $p \leq m$ are induced in region 2. The overall mode category can be considered as TE_{mn}–TE_{pq} mode pair. The y-directed propagation constants β_y for region 1 and γ_y for region 2 are related by

$$(m\pi/a)^2 + \beta_y^2 = (p\pi/c)^2 - \gamma_y^2 \tag{8.50}$$

For low-loss transmission in the groove guide, its dominant mode must be above cutoff in the groove region and the fields must decay in region 2 in the y-direction; that is, $\gamma_y = \zeta_y$ (real). For all higher order modes to diverge out, energy must be lost transversely from the guide. This is characterized by having γ_y imaginary; that is, $\gamma_y = j\xi_y$ with ξ_y real and positive. Equivalently, the condition that must be satisfied is [17]

$$(m\pi/a)^2 - (p\pi/c)^2 + \beta_y^2 \geq 0 \tag{8.51}$$

With β_y real,

$$(m\pi/a)^2 - (p\pi/c)^2 > 0 \tag{8.52a}$$

or

$$c \geq a(p/m) \quad \text{with } m > p \tag{8.52b}$$

will satisfy the condition for transmission of higher order modes transversely out of the guide and the modes will be removed from the system.

From the characteristic equations (8.46) and (8.47), we note that β_y must lie in the range

$$n\pi/b \geq \beta_y \geq (n-1)\pi/b \tag{8.53}$$

Condition (8.51) may be written as

$$\beta_y^2 \geq (p\pi/c)^2 - (m\pi/a)^2 \tag{8.54}$$

Combining (8.53) and (8.54), the condition for filtering out higher order modes can be written in the form

$$(1/b)^2 \geq [1/(n-1)^2][(p/c)^2 - (m/a)^2] \tag{8.55}$$

For the special case $m = p = 1$ and $n \geq 2$, Eq. (8.55) reduces to the following relation [5, 17]:

$$\frac{1}{b^2} = \frac{1}{c^2} - \frac{1}{a^2} \tag{8.56}$$

This is the condition for filtering out the TE_{1n} and TM_{1n} ($n \geq 2$) modes in the transverse direction. For $m > p$, the condition on d can be obtained from (8.52) as

$$d \leq (m/p - 1)\, c/2, \quad m > p \tag{8.57}$$

Figure 8.18 illustrates graphically the condition for filtering out the higher order modes in the transverse direction [5]. The hatched area is bound by the curves corresponding to the condition $c \geq a/3$ (Eq. (8.52) with $m = 3$, $p = 1$) and that given by Eq. (8.56). If the coordinates of the groove edge $A(c/2, d/2)$ (see Fig. 8.15(a)) are in the hatched aréa, the groove propagates only the TE_{11} and TM_{11} modes independent of frequency and the higher order modes are removed transversely out of the guide.

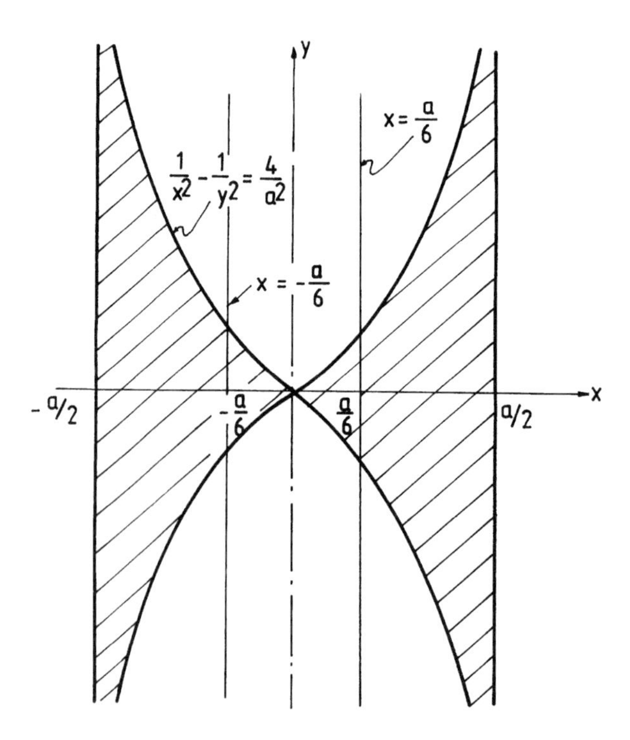

FIGURE 8.18 Conditions of the edge point A for the divergence of higher modes in a groove guide (see Fig. 8.15(a)). (After Nakahara and Kurauchi [5].)

Attenuation of TE_{11} Mode The total power loss in a groove guide is the sum of the wall conductor loss and radiation loss from the open ends. For the TE_{11} mode, since the fields are of decaying type in region 2 toward the openings, the radiation loss can be made negligibly small as compared with the conductor loss by increasing the height of the metal walls on either side of the groove region. For example, the radiation loss coefficient α_r, defined as the ratio of the average power radiated to the incident power, is reported to be less than about 0.002 for $h/c > 5$, where h is the total height of the groove guide. The transmission loss is then dominated by the conductor loss [17].

Figure 8.19 shows the conductor attenuation constant α_c for the TE_{11} mode as

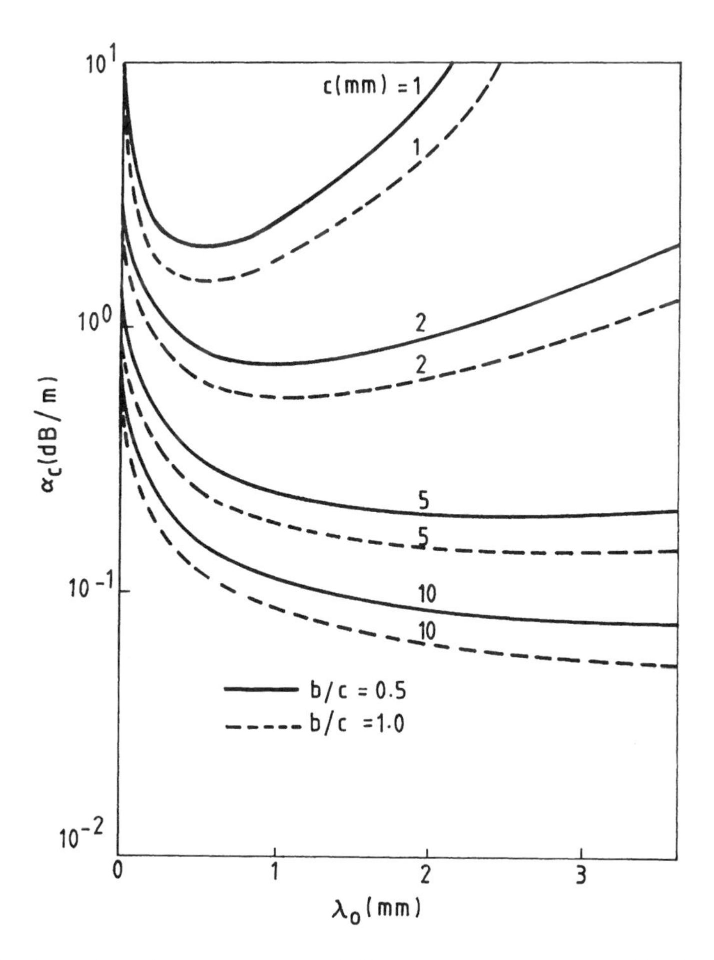

FIGURE 8.19 Theoretical attenuation of TE_{11} mode in a groove guide: $d = c/4$ and σ (metal conductivity) $= 5.8 \times 10^7$ mhos/m. (Reprinted from *J. Infrared Physics*, Vol. 18, Harris et al. [21], Low loss single mode waveguide for submillimeter and millimeter wavelengths, pp. 741–747, Copyright © 1978 with kind permission from Elsevier Science-NL, Sara Burgerhartstraat 25, 1055 KV Amsterdam, The Netherland.)

a function of free-space wavelength λ_0 for different values of the metal plate separation [21]. It can be seen that with $b/c = 1$, $d/c = 0.25$, and $c \approx 3\lambda_0$, the attenuation is less than 0.1 dB/m at 100 GHz, which is an order of magnitude less than in a TE_{10} mode rectangular waveguide. However, such a large plate separation for an H-guide would normally lead to multimode propagation.

Choice of Groove Guide Dimensions The design of a groove guide for low-loss, low-dispersion, single-mode (TE_{11}) propagation requires proper choice of the groove dimensions. The dimensional relations recommended by Choi and Harris [17] are $c = 3$–$3.4\,\lambda_0$, $d = c/3$, $b = 0.4c$, and $h > 5c$. It may be noted that these dimensions are not critical to a fraction of a wavelength, thereby providing relaxed dimensional tolerances for the fabrication of the guide. With a plate separation of about $3\lambda_0$, the guide is reported to have a peak power capability of about 1 MW at 100 GHz and 100 kW at 300 GHz.

8.5 DOUBLE-GROOVE GUIDE

The double-groove guide, the cross-section of which is shown in Figure 8.20, consists of two identical groove sections spaced by a distance $2s$ [22]. The guide supports two modes: the TE_{11} mode, which is symmetrical about the guide centre ($x = 0$, $y = 0$), and the TE_{12} mode, which is antisymmetrical with zero field at the center. Figure 8.21 shows the field configurations of these two modes [22]. The analysis of the guide and the characteristics of these modes [17, 22] are reviewed.

For the purpose of describing the fields, the guide cross-section is divided into three regions: region 1 ($s < |y| < s + b$), region 2 ($|y| > s + b$), and region 3 ($|y| < s$). Since the structure is symmetric about the plane $y = 0$, it suffices to confine the field description to the top symmetric half of the guide. The fields vary sinusoidally in the x-direction, satisfying the boundary condition on the conducting walls. Assuming a time and z-dependence of the form $e^{j(\omega t - \beta z)}$, the expressions for the H_y component of the TE modes in the three regions can be written as follows.

Region $s < y < s + b$:

$$H_z = A(k_0^2 - \beta^2)\sin(m\pi x/a)\cos[\beta_y(y - s - t)] \tag{8.58}$$

Region $y > s + b$:

$$H_z = B(k_0^2 - \beta^2)\sin(p\pi x/c)e^{-\zeta_y(y - s - b)} \tag{8.59}$$

Region $0 < y < s$:

$$H_z = D(k_0^2 - \beta^2)\sin(p\pi x/c)\cdot\begin{cases}\cosh(\zeta_y y), & E_y \text{ symmetric about } y = 0 \quad (8.60a)\\ \sinh(\zeta_y y), & E_y \text{ antisymmetric about } y = 0 \quad (8.60b)\end{cases}$$

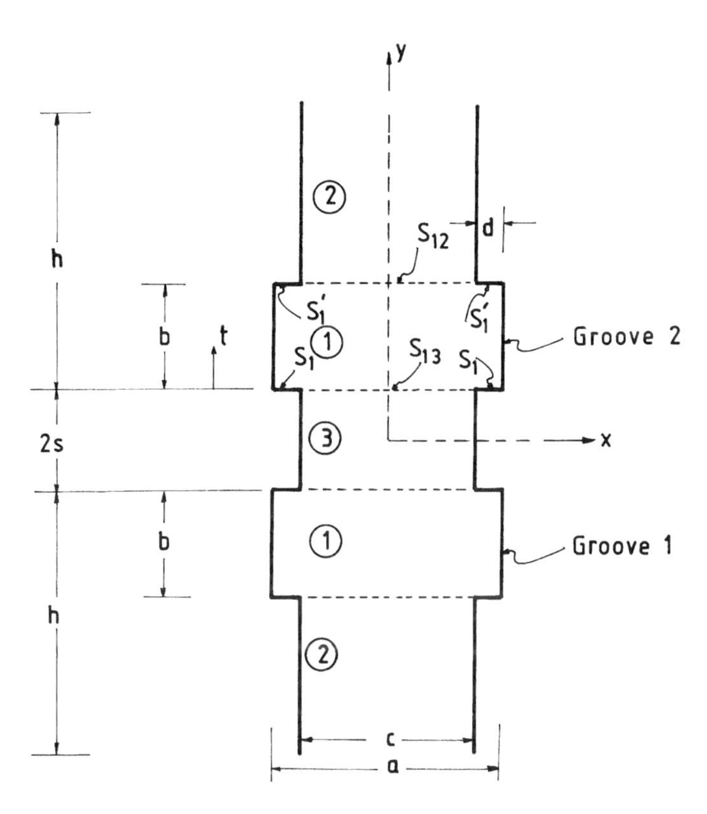

FIGURE 8.20 Cross-section of double-groove guide. (From Harris and Lee [22], reprinted with permission of IEE.)

where A, B, and D are arbitrary constants; t is the value of y for maximum field strength; β_y is the propagation constant in region 1; and ζ_y is the decay constant in regions 2 and 3. The various propagation parameters satisfy the relation

$$k_0^2 - \beta^2 = (m\pi/a)^2 + \beta_y^2 = (p\pi/c)^2 - \zeta_y^2 \tag{8.61}$$

The transverse field components of the TE modes in the three regions can be obtained by substituting (8.58)–(8.60) in (8.30). With $m = p = 1$, Eqs. (8.58), (8.59), (8.60a), and (8.61) are applicable to the TE_{11} mode and Eqs. (8.58), (8.59), (8.60b), and (8.61) are applicable to the TE_{12} mode. The characteristic equations are obtained by satisfying the boundary conditions at the interfaces $y = s$ and $s + b$; namely, E_x (or $\partial H_z/\partial y$) and H_z are continuous at the boundaries S_{12} and S_{13}, and E_x is zero on S_1 and S_1' (refer to Fig. 8.20). The expressions are given by [22]

$$P\left[\frac{P\tan(\beta_y b) - 1}{\tan(\beta_y b) + P}\right] = \begin{cases} \tanh(\zeta_y s) & \text{for } TE_{11} \text{ mode} \qquad (8.62a) \\ \coth(\zeta_y s) & \text{for } TE_{12} \text{ mode} \qquad (8.62b) \end{cases}$$

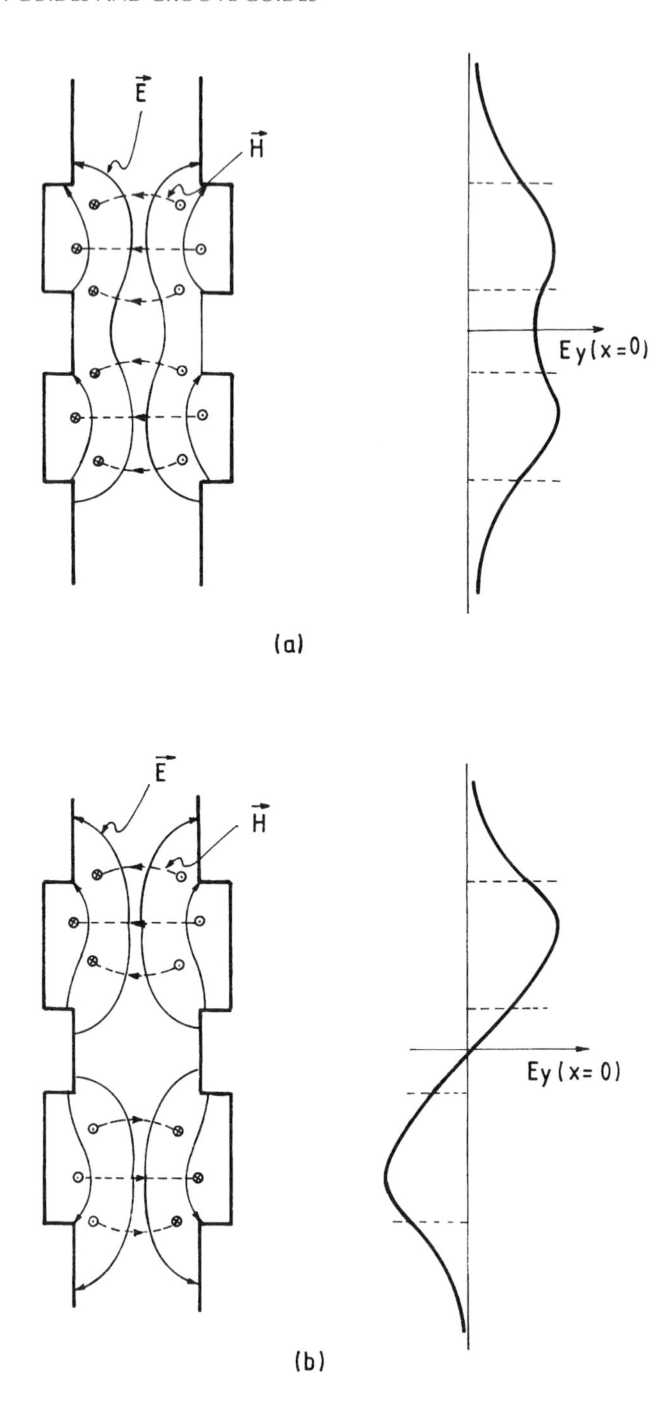

FIGURE 8.21 Field pattern in the double-groove guide and variation of E_y component for (a) TE_{11} mode and (b) TE_{12} mode. (From Harris and Lee [22], reprinted with permission of IEE.)

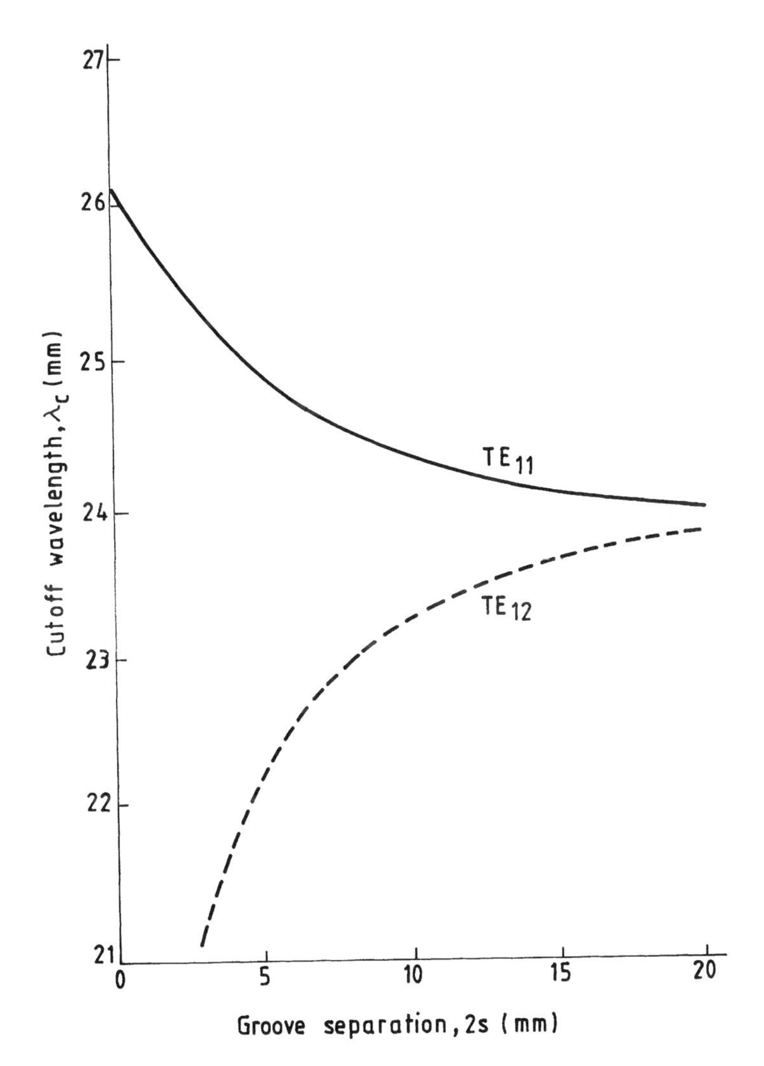

FIGURE 8.22 Variation of cutoff wavelengths of TE_{11} and TE_{12} modes of double-groove guide (see Fig. 8.20) with groove separation: $b = 5\,\text{mm}$, $c = 10\,\text{mm}$, $d = 2\,\text{mm}$. (From Harris and Lee [22], reprinted with permission of IEE.)

where

$$P = \left(\frac{ac}{16}\right)\left(\frac{\beta_y}{\zeta_y}\right)\left(\frac{(\pi/c)^2 - (\pi/a)^2}{(\pi/a)\cos(\pi c/a)}\right)^2 \tag{8.63}$$

Figures 8.22 and 8.23 show variation in the cutoff wavelength λ_c and propagation constant β, respectively, for TE_{11} and TE_{12} modes as a function of groove separation distance $2s$ [22]. It can be seen that with an increase in $2s$, both λ_c and

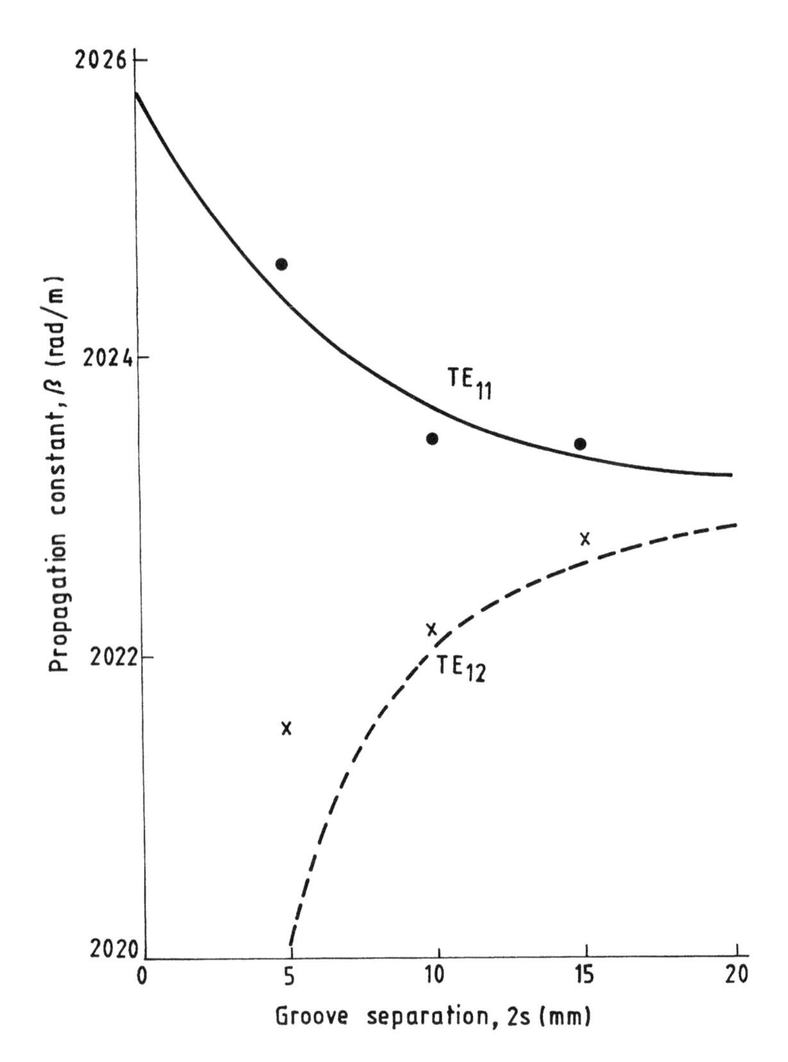

FIGURE 8.23 Variation of propagation constants of TE_{11} and TE_{12} modes with groove separation and experimental results for three separations. Experiments: ● TE_{11}, × TE_{12}, $\lambda_0 = 3.08$ mm, $b = 5$ mm, $c = 10$ mm, $d = 2.5$ mm. (From Harris and Lee [22], reprinted with permission of IEE.)

β decrease for the TE_{11} mode and increase for the TE_{12} mode. Since the two modes have different propagation constants, power transfer takes place from one groove to the other. The difference in the propagation constants decreases with an increase in $2s$. For $2s > 2c$, the two propagation constants are nearly equal and consequently each groove propagates waves rather independent of the other. The attenuation of the TE_{11} mode of the double-groove guide is reported to be less than that of the TE_{11} mode in a single-groove guide by about 20% [22]. But, since the improvement is rather marginal, the single-groove guide, which is simpler to fabricate than the double-groove guide, is preferred as a single-mode propagating structure.

PROBLEMS

8.1 **(a)** What are the essential features of an H-guide that distinguish it from a NRD guide?

(b) Enumerate its advantages and also its limitations.

8.2 **(a)** Explain the properties of PE_{mn} and PM_{mn} modes of an H-guide. Referring to the coordinate system given in Figure 8.24, give the corresponding modal designations in terms of TE and TM modes.

(b) Which is the lowest order hybrid mode that is supported by the H-guide?

8.3 Show that pure TEM and TM waves cannot be propagated along an H-guide.

8.4 **(a)** Derive exact field expressions in the dielectric and air regions of an H-guide for the TM^y_{21} mode. (Refer to the structure shown in Fig. 8.24.)

(b) Obtain the characteristic equation.

(c) Draw typical E- and H-lines for this mode in the cross-sectional plane of the H-guide.

8.5 Using the characteristic equation derived in Problem 8.4, calculate the cutoff frequency of the E^y_{21} mode of the H-guide. Assume $a = 2\,\text{mm}$, $b = 0.5\,\text{mm}$, and $\varepsilon_r = 2.56$ for the H-guide (structure shown in Fig. 8.24).

8.6 The only nonhybrid modes that an H-guide can support are the TE_{0n} modes.

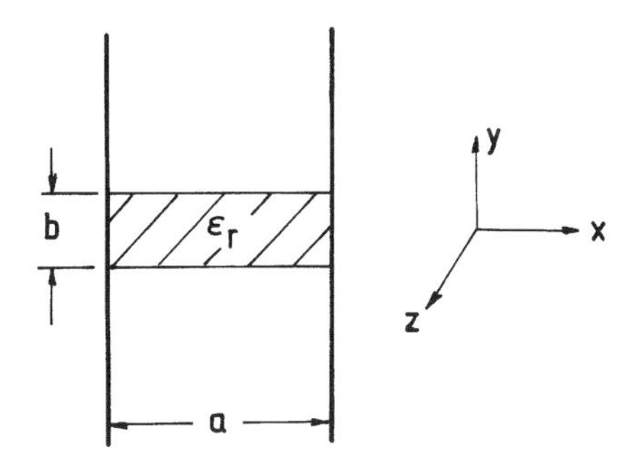

FIGURE 8.24 Cross-section of an H-guide.

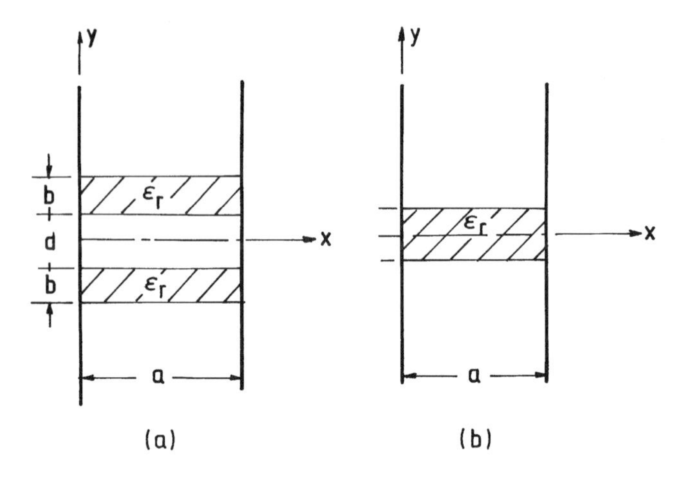

FIGURE 8.25 Cross-section of H-guides: (a) double-strip and (b) single-strip.

(a) What are the special features of these modes?

(b) Draw typical field lines and the field intensity distribution for the TE_{02} mode (similar to those shown in Fig. 8.7 for the TE_{01} mode).

8.7 Consider an H-guide symmetrically excited (E_x symmetric with respect to the $y = 0$ plane; see Fig. 8.24) such that the antisymmetric TE_{0n} modes cannot exist. Determine the plate separation a and the dielectric thickness b such that only the TE_{01} mode can exist up to 140 GHz. Choose $\varepsilon_r = 2.56$ for the dielectric. What would be the dimensions for a lower value of $\varepsilon_r = 1.5$. (Use Fig. 8.8 in addition to the formulas given in Section 8.2.)

8.8 **(a)** Explain the advantages of the double-strip H-guide (Fig. 8.25(a)) over a single-strip H-guide (Fig. 8.25(b)).

(b) Placing an electric wall at the $y = 0$ plane of an H-guide results in a trough waveguide. Specify the modes of the H-guide that can be supported in the trough guide. Identify the dominant hybrid mode and also the dominant nonhybrid mode of the trough guide. Draw field lines for both these modes.

8.9 Explain how the presence of grooves produces a surface wave effect in a groove guide. Compare the properties of modes and nature of wave guidance in a groove guide with those in an H-guide.

8.10 **(a)** Draw typical current flow lines in a groove guide (Fig. 8.26(a)), H-guide (Fig. 8.26(b)), and a hollow rectangular metal waveguide (Fig. 8.26(c)) for the dominant mode in the respective guide.

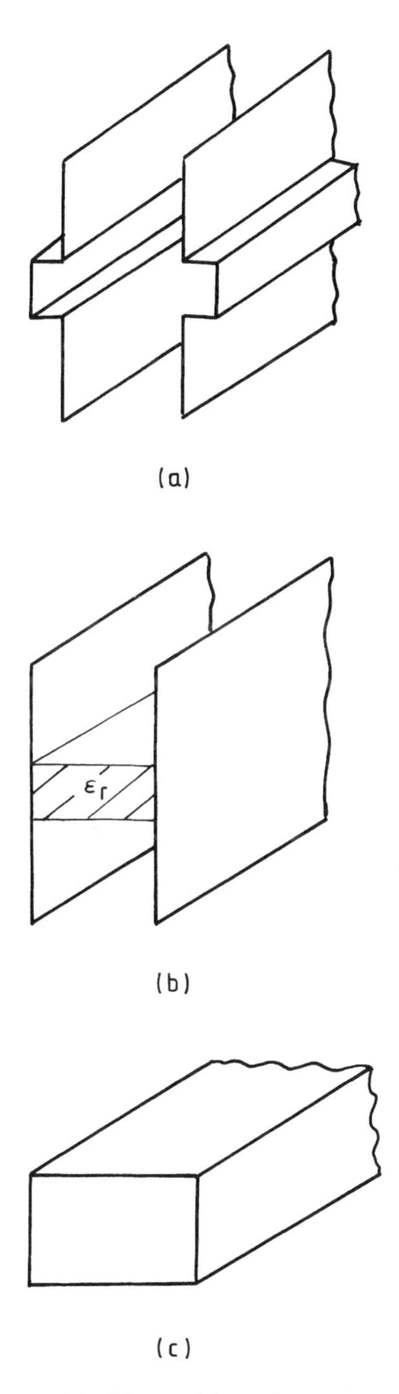

FIGURE 8.26 (a) Groove guide, (b) H-guide, and (c) hollow rectangular metal wave-guide.

(b) Give typical dimensions and approximate attenuation constants for operation (in dominant mode) at 100 GHz, 200 GHz, and 300 GHz. Give reasons for the superiority of the groove guide at these frequencies in offering low loss and largest single-mode bandwidth.

REFERENCES

1. F. J. Tischer, A waveguide structure with low losses. *Arch. Elek. Ubertragung.*, **7**, 592–596, Dec. 1953.
2. F. J. Tischer, Properties of H-guide at microwaves and millimetre waves. *Proc. IEE (Lond.)*, **106B** (Suppl. 13), 47–53, Jan. 1959.
3. F. J. Tischer, The groove guide, a low loss waveguide for millimetre waves. *IEEE Trans. Microwave Theory Tech.*, **MTT-11**, 291–296, 1963.
4. T. Nakahara and N. Kurauchi, Propagation modes in grooved guide. *J. IECE (Japan)*, **47**, 1029–1036, July 1964.
5. T. Nakahara and N. Kurauchi, Transmission modes in the grooved guide. *Sumimoto Electric Tech. Rev.*, **5**, 65–71, Jan. 1965.
6. J. W. E. Griemsmann, Grooved guide. Symposium on Quasi-optics, Polytechnic Institute of Brooklyn, June 1964.
7. A. Doswell and D. J. Harris, Modified H guide for millimeter and submillimeter wavelengths. *IEEE Trans. Microwave Theory Tech.*, **MTT-21**, 587–589, Sept. 1973.
8. R. F. B. Conlon and F. A. Benson, Propagation and attenuation in double-strip H-guide. *IEE Proc.*, **113(8)**, 1311–1320, Aug. 1966.
9. J. Bled et al., Nouvelles techniques d'utilisation des ondes millimetriques et submillimetriques. *Onde Elec.*, **44**, 26–36, 1964.
10. L. N. Vershinina and V. V. Meriakri, A submillimeter waveguide channel. *Radio Eng. Electron. Phys. (USSR)*, **12**, 1698–1700, 1969.
11. M. Cohn, Propagation in a dielectric-loaded parallel plate waveguide. *IRE Trans. Microwave Theory Tech.*, **MTT-7**, 202–208, Apr. 1959.
12. M. Cohn and A. F. Eikenberg, Ferroelectric phase shifters for VHF and UHF. *IRE Trans. Microwave Theory Tech.*, **MTT-10**, 536–548, Nov. 1962.
13. M. B. Klein, Dielectric waveguide electro-optic devices. *Infrared and Millimeter Waves*, Vol. 9, K. J. Button (Ed.), Chap. 3, Academic Press, Orlando, FL, 1984.
14. R. A. Moore and R. E. Beam, A duo-dielectric parallel plane waveguide. *Proc. NEC*, **12**, 689–705, Apr. 1957.
15. M. Kawamura and Y. Kokubo, Transmission loss of the double-strip modified H-guide at 50 GHz. *IEEE Trans. Microwave Theory Tech.*, **MTT-28**, 430–432, Apr. 1980.
16. F. J. Tischer, H-guide with laminated dielectric slab. *IEEE Trans. Microwave Theory Tech.*, **MTT-18**, 9–15, Jan. 1970.
17. Y. M. Choi and D. J. Harris, Groove guide for short millimetric waveguiding systems. *Infrared and Millimeter Waves*, Vol. 11, K. J. Button (Ed.), Chap. 3, Academic Press, Orlando, FL, 1984.
18. G. P. Bava and G. Perona, Conformal mapping analysis of a type of groove guide. *Electron. Lett.*, **2(1)**, 13–15, Jan. 1966.

19. F. J. Tischer, Conformal mapping in waveguide considerations. *Proc. IEEE*, **51**, 1050–1051, 1963.
20. A. A. Oliner and P. Lampariello, Simple and accurate expression for the dominant mode properties of open groove guide. *IEEE MTT-S Int. Microwave Symp. Digest*, 62–64, 1984.
21. D. J. Harris, K. W. Lee, and R. J. Batt, Low-loss single-mode waveguide for submillimetre and millimetre wavelengths. *J. Infrared Phys.* **18**, 741–747, 1978.
22. D. J. Harris and K. W. Lee, Theoretical and experimental characteristics of double-groove guide for 100 GHz operation. *IEE Proc.*, **128(H, 1)**, 6–10, Feb. 1981.

CHAPTER NINE

Dielectric Resonators

9.1 INTRODUCTION

Dielectric resonators in cylindrical and rectangular shapes are commonly employed at microwave and millimeter wave frequencies in the design of filters, oscillators, and as tuning elements. Like the metal cavity resonators, dielectric resonators can be excited in several resonant modes. The basic characteristics defining a dielectric resonator are its resonant frequencies, Q-factors, and field distribution. It is important to know these properties for the lowest order resonant mode and a few higher order modes in order to enable selection of the proper mode for a particular application and incorporation of a suitable coupling mechanism to excite the mode. Knowledge of the field distribution also helps in devising suitable techniques for suppressing or eliminating the undesired modes, which may interfere with the desired mode.

As compared with the rectangular shaped resonator, cylindrical resonators in the form of a pill box (disk) and ring resonators are more often encountered in practice. Cylindrical resonators have received extensive attention in the literature [1–39]. Unlike the metal cavity resonator, the mathematical description of the electromagnetic fields for a dielectric resonator is quite complicated. Therefore approximate methods have traditionally been used for the analysis of cylindrical resonators. These include the *magnetic wall* [1, 2], the *dielectric waveguide-wall* [3, 4], the *mixed model* [5], the *variational* [6], and the *effective dielectric constant* (*EDC*) [7, 8] methods. More sophisticated and rigorous methods have also been reported by several investigators. These include the *mode-matching* [9–14], *finite-element* [15], *finite-difference* [16], and *perturbational-asymptotic series* [17–22] techniques and methods based on *surface or volume integral equations* [23–25]. These rigorous techniques provide accurate description of the electromagnetic fields in and around the resonator and also permit taking into account more accurately the influence of the surrounding structure, such as the dielectric substrate, metallic shield, dielectric post, and metal tuning screws. A fine treatment of the analysis of dielectric resonators using various techniques is available in the book by Kajfez and Guillon [26].

While the rigorous techniques provide for very accurate characterization of resonators, the complexity involved makes their use in practical applications quite difficult. It would be of practical interest to circuit designers to obtain the resonator characteristics through simple formulas, which are still sufficiently accurate. In this chapter, we use the EDC method in conjunction with the *transverse transmission line* technique (described in Chapter 2) to analyze the cylindrical, ring, and rectangular resonators. Resonators in an isolated configuration and in a planar dielectric slab guide environment are analyzed. The latter geometry is compatible with the dielectric integrated guide structures. The EDC technique as applied to cylindrical and ring resonators is essentially an improved version of the dielectric waveguide-wall method (DWM) and is reported to offer quite accurate results for resonant frequencies and Q-factors adequate for most practical purposes [7, 8].

As compared with the cylindrical resonators, the rectangular shaped dielectric resonators are slightly larger in size. The latter type is therefore useful for millimeter wave integrated circuits in view of its advantages in terms of convenience in handling and ease of fabrication. Rectangular resonators have received very little attention in the literature. Approximate methods based on *Marcatili's approximation* for dielectric guides [37] with realistic impedance conditions on the resonator end surfaces [38], *dielectric waveguide-wall* method [39], and a mixture of magnetic walls and dielectric waveguide model [40] have been applied to isolated rectangular resonators. In this chapter, we extend the EDC technique of analysis presented in Chapter 2 (for planar dielectric guides) to study the rectangular resonators.

9.2 ELECTROMAGNETIC FIELDS IN CYLINDRICAL COORDINATES

In a cylindrical coordinate system (r, ϕ, z), the wave equation is given by

$$\frac{1}{r}\frac{\partial}{\partial r}\left(r\frac{\partial\psi}{\partial r}\right)+\frac{1}{r^2}\frac{\partial^2\psi}{\partial\phi^2}+\frac{\partial^2\psi}{\partial z^2}+k^2\psi=0 \tag{9.1}$$

where

$$k=\omega\sqrt{\mu_0\varepsilon} \tag{9.2}$$

is the wavenumber in the medium and ψ stands for either E_z or H_z. We use the method of separation of variables to seek solution in the form

$$\psi(r,\phi,z)=R(r)\,\Phi(\phi)\,Z(z) \tag{9.3}$$

where R, Φ, and Z are functions of r, ϕ, and z, respectively. Using (9.2), the wave equation (9.1) can be divided into three separation equations, each of which determines only one of the functions $R(r)$, $\Phi(\phi)$, and $Z(z)$. They are

$$\left(\frac{1}{Z}\right)\left(\frac{d^2Z}{dz^2}\right) = -\beta_z^2 \tag{9.4}$$

$$\left(\frac{1}{\Phi}\right)\left(\frac{d^2\Phi}{d\phi^2}\right) = -n^2 \tag{9.5}$$

$$\left(\frac{1}{r}\right)\left(\frac{d}{dr}\right)\left(\frac{r\,dR}{dr}\right) + [(k_r r)^2 - n^2]R = 0 \tag{9.6}$$

where

$$k_r^2 = k^2 - \beta_z^2 \tag{9.7}$$

The solution of (9.6) is in the form of Bessel functions, and the solutions of (9.4) and (9.5) are in the form of harmonic functions. The radial wavenumber k_r and the z-directed propagation constant β_z are related through the separation equation (9.7). The choice of β_z and n and solution of (9.4)–(9.6) depend on the structural geometry and the modal pattern to be supported. In cylindrical resonators, k_r, β_z, and n assume discrete values—each combination corresponding to a different mode of resonance.

Once E_z and H_z are obtained as solutions of the wave equation, the other field components can be obtained from Maxwell's equations. The expressions are

$$E_r = \frac{1}{k_r^2}\left(\frac{\partial^2 E_z}{\partial r\,\partial z} - \frac{j\omega\mu_0}{r}\frac{\partial H_z}{\partial \phi}\right) \tag{9.8a}$$

$$E_\phi = \frac{1}{k_r^2}\left(\frac{1}{r}\frac{\partial^2 E_z}{\partial \phi\,\partial z} + j\omega\mu_0\frac{\partial H_z}{\partial r}\right) \tag{9.8b}$$

$$H_r = \frac{1}{k_r^2}\left(\frac{\partial^2 H_z}{\partial r\,\partial z} + \frac{j\omega\varepsilon}{r}\frac{\partial E_z}{\partial \phi}\right) \tag{9.8c}$$

$$H_\phi = \frac{1}{k_r^2}\left(\frac{1}{r}\frac{\partial^2 H_z}{\partial \phi\,\partial z} - j\omega\varepsilon\frac{\partial E_z}{\partial r}\right) \tag{9.8d}$$

9.3 ISOLATED CYLINDRICAL RESONATOR

The resonant fields of a cylindrical resonator (Fig. 9.1(a)) can be divided into constituent modes with azimuthal variation represented by either $\cos n\phi$ or $\sin n\phi$ ($n = 0, 1, 2, \ldots$). For $n=0$ (axisymmetric case), the set of modes can be divided into TE-to-z and TM-to-z modes. These are designated as $TE_{0m\ell}$ and

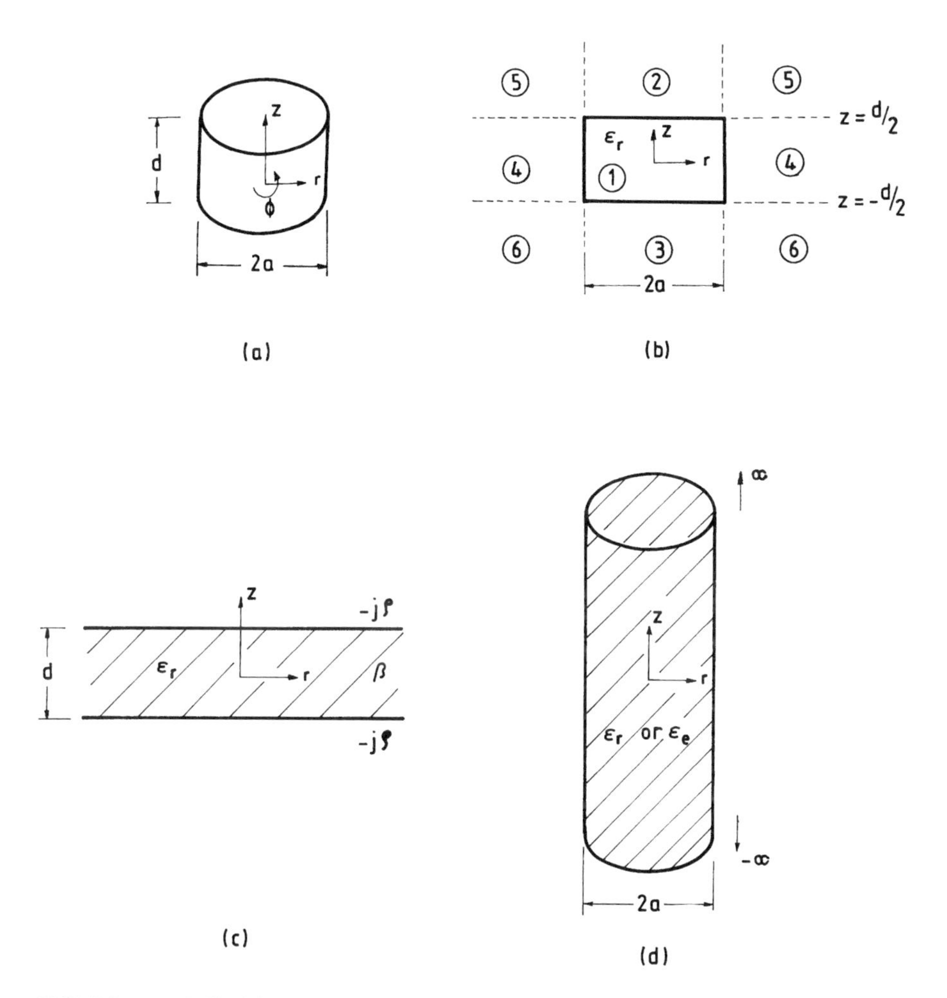

FIGURE 9.1 Cylindrical dielectric resonator and models for EDC analysis: (a) geometry of isolated resonator, (b) cross-sectional view showing various regions, (c) radial slab guide model for determining the z-directed propagation constant, and (d) infinite cylindrical model for determining the radial propagation constant.

$TM_{0m\ell}$ modes. For $m > 0$, the modes are hybrid in nature (having nonzero E_z and H_z) and are designated as $HEM_{nm\ell}$, $HE_{nm\ell}$, and $EH_{nm\ell}$. The designations HE and EH are used when the hydrid modes are predominantly TM-like (E_z dominates over H_z) and TE-like (H_z dominates over E_z), respectively. These mode designations are used because of the strong resemblance of HE_{nm} and EH_{nm} modes with the TM_{nm} and TE_{nm} modes, respectively, in a circular guide with an assumed magnetic wall. The first subscript n refers to the number of circumferential variations (in the ϕ direction). The second subscript m and the third subscript ℓ, refer to the field extrema within the dielectric resonator in the radial (r) and axial (z) directions, respectively. The index ℓ is sometimes replaced by δ with $0<\delta<1$ to indicate fraction of half-cycle variation in the axial direction.

9.3.1 Analysis for $TE_{01\delta}$ Mode

The most commonly used mode of resonance for a cylindrical resonator is the $TE_{01\delta}$ mode. It is a TE-to-z mode ($E_z = 0$) having azimuthal symmetry ($\partial/\partial\phi = 0$) and less than a half-cycle variation in the field in the z-direction. Thus from (9.8), it can be seen that the only nonzero field components are E_ϕ, H_r, and H_z. The wave equation (9.1) in H_z simplifies to

$$\left(\frac{\partial^2}{\partial r^2} + \frac{1}{r}\frac{\partial}{\partial r} + \frac{\partial^2}{\partial z^2} + k^2\right) H_z = 0 \tag{9.9}$$

Once the solution for H_z is obtained, E_ϕ and H_r can be obtained from

$$E_\phi = \frac{j\omega\mu_0}{k_r^2}\frac{\partial H_z}{\partial r} \tag{9.10a}$$

$$H_r = \frac{1}{k_r^2}\frac{\partial^2 H_z}{\partial r\,\partial z} \tag{9.10b}$$

where

$$k_r^2 = k^2 + \left(\frac{\partial^2}{\partial z^2}\right) \tag{9.11}$$

As shown in Figure 9.1(b), we first divide the field region into six subregions. For a high-Q resonator most of the electromagnetic energy is stored in region 1 and the field decays exponentially with distance in the air region outside the resonator. A small amount of energy exists in regions 2, 3, and 4, and even less in regions 5 and 6. We can therefore simplify the analysis by neglecting the fields in regions 5 and 6 and matching the fields only at the four boundary surfaces of region 1.

For the $TE_{01\delta}$ mode, the H_z field in regions 1–4 may be written as

$$H_z = \begin{cases} AJ_0(ur)\cos(\beta z), & \text{region } 1 \\ BJ_0(ur)e^{-\zeta(|z|-d/2)}, & \text{regions } 2,\ 3 \\ CK_0(\xi r)\cos(\beta z), & \text{region } 4 \end{cases} \tag{9.12}$$

where A, B, and C are unknown constants to be determined and J_0 and K_0 are the Bessel and the modified Hankel functions of order zero. The parameters u and β are the radial and z-directed wavenumbers, respectively, in region 1; ζ is the z-directed decay coefficient in regions 2 and 3; and ξ is the radial decay coefficient in region 4. It may be noted that u, β, ζ, and ξ are all real. Using the relation (9.11), we can write

$$u^2 + \beta^2 = k_0^2 \varepsilon_r, \quad \text{region 1} \tag{9.13a}$$

$$u^2 - \zeta^2 = k_0^2, \quad \text{region 2, 3} \tag{9.13b}$$

$$-\xi^2 + \beta^2 = k_0^2, \quad \text{region 4} \tag{9.13c}$$

where

$$k_0 = 2\pi f_0 \sqrt{\mu_0 \varepsilon_0} \tag{9.14}$$

is the free-space resonant wavenumber and f_0 is the resonant frequency. The expression for E_ϕ and H_r can be obtained by substituting for H_z from (9.12) in (9.10). Applying the continuity condition on H_z at the cylindrical surface $r = a, -d/2 \le z \le d/2$, we obtain

$$C = \frac{A J_0(ua)}{K_0(\xi a)} \tag{9.15}$$

Next, applying the continuity condition on E_ϕ at $z = d/2$, $0 \le r \le a$, we obtain

$$B = A \cos(\beta d/2) \tag{9.16}$$

Using the relations (9.15) and (9.16), we can now express all three field components H_z, E_ϕ, and H_r in terms of a single unknown constant A.

$$H_z = \begin{cases} A J_0(ur)\cos(\beta z), & \text{region 1} \\ A\cos(\beta d/2) J_0(ur) e^{-\zeta(|z| - d/2)}, & \text{regions 2, 3} \\ A[J_0(ua)/K_0(\xi a)]\, K_0(\xi r)\cos(\beta z), & \text{region 4} \end{cases} \tag{9.17}$$

$$E_\phi = \begin{cases} -A(j\omega_0\mu_0/u) J_1(ur)\cos(\beta z), & \text{region 1} \\ -A(j\omega_0\mu_0/u)\cos(\beta d/2) J_1(ur) e^{-\zeta(|z| - d/2)}, & \text{regions 2, 3} \\ A(j\omega_0\mu_0/\xi)[J_0(ua)/K_0(\xi a)]\, K_1(\xi r)\cos(\beta z), & \text{region 4} \end{cases} \tag{9.18}$$

$$H_r = \begin{cases} A(\beta/u) J_1(ur)\sin(\beta z), & \text{region 1} \\ A(\zeta/u)\cos(\beta d/2) J_1(ur) e^{-\zeta(z - d/2)}, & \text{regions 2} \\ -A(\zeta/u)\cos(\beta d/2) J_1(ur) e^{\zeta(z + d/2)}, & \text{region 3} \\ -A(\beta/\xi)[J_0(ua)/K_0(\xi a)]\, K_1(\xi r)\sin(\beta z), & \text{region 4} \end{cases} \tag{9.19}$$

Characteristic Equation Using DWM Applying the continuity condition on H_r at $z = d/2$, $0 < r < a$, we obtain the following transcendental equation for determining β:

$$\beta \tan(\beta d/2) = \zeta \tag{9.20a}$$

where

$$\zeta = [k_0^2(\varepsilon_r - 1) - \beta^2]^{1/2} \tag{9.20b}$$

Next, applying the continuity condition on E_ϕ at $r = a$, $-d/2 < z < d/2$, we obtain the following eigenvalue equation for determining the radial wavenumber u:

$$\frac{J_1(ua)}{uJ_0(ua)} + \frac{K_1(\xi a)}{\xi K_0(\xi a)} = 0 \tag{9.21a}$$

where

$$\xi = [k_0^2(\varepsilon_r - 1) - u^2]^{1/2} \tag{9.21b}$$

It may be noted that (9.20a) is nothing but the characteristic equation for the TE_0 mode of a radial slab guide (see Fig. 9.1(c) having the same dielectric constant ε_r and height d as that of the resonator. The characteristic equation (9.21a) is the same as that for the TE_{01} mode of an infinite cylindrical guide (see Fig. 9.1(d)) having the same radius and dielectric constant as that of the resonator. The resonant frequency f_0 is the one at which the values of β and u obtained from (9.20) and (9.21) also satisfy the separation equation (9.13a).

Characteristic Equation Using the EDC Method It may be recognized that the infinitely long homogeneous cylindrical model chosen above for representing regions 1, 2, and 3 having relative permittivities 1, ε_r, and 1, respectively, should have a relative permittivity ε_e lying between 1 and ε_r and not ε_r as assumed in the DWM. A similar argument applies to the radial slab guide chosen for the evaluation of β. In the case of the horizontal radial slab guide, the assumption is justified at least for $d/2a \leq 1$ [6]. It is essentially the radial wavenumber that needs to be evaluated more accurately by assigning a suitable value for the dielectric constant of the infinite dielectric cylinder. The EDC technique [7, 8] that is followed for determining ε_e is as follows.

We first consider an infinite cylindrical guide model of radius a and relative dielectric constant ε_r and solve for its radial wavenumber u', which is now assumed to be different from that of the resonator. Thus u' can be computed from (9.21) with u and ξ replaced by u' and ξ', respectively,

$$\frac{J_1(u'a)}{u'J_0(u'a)} + \frac{K_1(\xi' a)}{\xi' K_0(\xi' a)} = 0 \tag{9.22a}$$

where

$$\xi' = [k_0^2(\varepsilon_r - 1) - (u')^2]^{1/2} \tag{9.22b}$$

is the radial decay coefficient outside the infinite cylinder. The effective dielectric constant ε_e is obtained from

$$\varepsilon_e = (u'/k_0)^2 \tag{9.23}$$

Next, we consider the same infinite cylindrical guide model of radius a but having a relative dielectric constant equal to ε_e. We now assume that the radial wavenumber of this guide is the same as that of the dielectric resonator. The value of u is obtained from

$$\frac{J_1(ua)}{uJ_0(ua)} + \frac{K_1(\xi a)}{\xi K_0(\xi a)} = 0 \tag{9.24a}$$

where

$$\xi = [k_0^2(\varepsilon_e - 1) - u^2]^{1/2} \tag{9.24b}$$

The frequency f_0 at which the equations (9.13a), (9.20), and (9.22)–(9.24) are satisfied simultaneously is the resonant frequency of the $TE_{01\delta}$ mode. The above choice of ε_e is reported [8] to yield resonant frequencies that are in close agreement with the results of rigorous methods [22, 27–29].

9.3.2 Derivation for Radiation Q-Factor

The general definition for the unloaded Q-factor (in the absence of external loading) of a resonator is

$$Q = \omega_0 W/P_L \tag{9.25}$$

where ω_0 is the resonant angular frequency, W is the maximum stored energy, and P_L is the total power loss, which is the sum of conductor loss (P_c), dielectric loss (P_d), and radiation loss (P_r). The total Q-factor can be expressed in terms of the conductor quality factor Q_c, the dielectric quality factor Q_d, and the radiation quality factor Q_r by the relation

$$\frac{1}{Q} = \frac{1}{Q_c} + \frac{1}{Q_d} + \frac{1}{Q_r} \tag{9.26a}$$

where

$$Q_c = \omega_0 W/P_c \tag{9.26b}$$

$$Q_d = \omega_0 W/P_d \tag{9.26c}$$

$$Q_r = \omega_0 W/P_r \tag{9.26d}$$

In the case of an isolated dielectric resonator $Q_c = 0$. The dielectric quality factor Q_d for a homogeneous dielectric is given by

$$Q_d = \omega_0 \varepsilon_0 \varepsilon_r \left[\frac{\displaystyle\int_{\text{vol}} |E|^2 \, dV}{\sigma \displaystyle\int_{\text{vol}} |E|^2 \, dV} \right] = \frac{\omega \varepsilon_0 \varepsilon_r}{\sigma} = \frac{1}{\tan \delta_d} \tag{9.27}$$

where $\tan \delta_d$ is the loss tangent of the dielectric material. Since dielectric materials with $\tan \delta_d$ of the order of 10^{-4} are available, Q_d is much higher than Q_r. Thus the total Q-factor is governed mainly by the value of Q_r. The derivation for Q_r for the $TE_{01\delta}$ mode of a cylindrical resonator is given below.

At resonance, the average electric energy stored is equal to the average magnetic energy. This reactive energy is stored both inside the resonator as well as in the immediate vicinity outside. For a resonator having larger ε_r and operating in the $TE_{01\delta}$ mode, nearly all the electric energy is stored inside the dielectric whereas a substantial portion of the magnetic energy lies outside. Since the fields inside the resonator are known more accurately, we obtain the total stored energy from the stored electric energy rather than from the magnetic energy. The radiation Q-factor can thus be obtained from

$$Q_r = 2\omega_0 W_e / P_r \tag{9.28}$$

The average stored electric energy W_e is given by

$$W_e = \frac{1}{2} \int_{\text{vol}} \varepsilon \, |E|^2 \, dV \tag{9.29}$$

where the integration is performed over the whole volume both inside ($\varepsilon = \varepsilon_0 \varepsilon_r$) and outside ($\varepsilon = \varepsilon_0$) the resonator.

Derivation for Stored Electric Energy W_e For the $TE_{01\delta}$ mode of the resonator, the only nonzero electric field component is E_ϕ. The expressions for E_ϕ in regions 1–4 of the resonator (Fig. 9.1(b)) are given by (9.18a)–(9.18c). For region 5, the approximate expression for E_ϕ can be written as

$$E_\phi = A \left(\frac{j\omega_0 \mu_0}{\xi} \right) \left(\frac{J_0(ua)}{K_0(\xi a)} \right) K_1(\xi a) \cos\left(\frac{\beta d}{2} \right) e^{-\zeta(z - d/2)}, \quad \text{region 5} \tag{9.30}$$

Let W_{ei} denote the average stored electric energy in the ith region. The evaluation of W_{ei} requires the use of certain integrals of Bessel functions. The relevant formulas and Bessel function identities are listed in Appendix 9A [41].

Using the expression for E_ϕ from (9.18a) in (9.29), the average electric energy stored in region 1 is obtained. It is given by

$$W_{e1} = |A|^2 \left(\frac{\pi \varepsilon_0 \varepsilon_r}{2}\right)\left(\frac{\omega_0 \mu_0}{u}\right)^2 \int_0^a r J_1^2(ur)\, dr \int_{-d/2}^{d/2} \cos^2(\beta z)\, dz \tag{9.31a}$$

$$= |A|^2 \left(\frac{\pi \varepsilon_0 \varepsilon_r}{8}\right)\left(\frac{\omega_0 \mu_0}{u}\right)^2 a^2 d\left[1 + \left(\frac{\sin(\beta d)}{\beta d}\right)\right][J_1^2(ua) - J_0(ua) J_2(ua)] \tag{9.31b}$$

The average electric energy stored in region 2 is equal to that in region 3. Using (9.18b) in (9.29), we obtain

$$W_{e2} = W_{e3} = |A|^2 \left(\frac{\pi \varepsilon_0}{2}\right)\left(\frac{\omega_0 \mu_0}{u}\right)^2 \cos^2\left(\frac{\beta d}{2}\right) \int_0^a r J_1^2(ur)\, dr \int_{d/2}^{\infty} e^{-2\zeta(z-d/2)}\, dz \tag{9.32a}$$

$$= |A|^2 \left(\frac{\pi \varepsilon_0}{8\zeta}\right)\left(\frac{\omega_0 \mu_0}{u}\right)^2 a^2 \cos^2\left(\frac{\beta d}{2}\right)[J_1^2(ua) - J_0(ua) J_2(ua)] \tag{9.32b}$$

Substituting for E_ϕ from (9.18c) in (9.29), the average electric energy stored in region 4 is given by

$$W_{e4} = |A|^2 \left(\frac{\pi \varepsilon_0}{2}\right)\left(\frac{\omega_0 \mu_0}{\xi}\right)^2 \left(\frac{J_0^2(ua)}{K_0^2(\xi a)}\right) \int_a^{\infty} r K_1^2(\xi r)\, dr \int_{-d/2}^{d/2} \cos^2(\beta z)\, dz \tag{9.33a}$$

$$= |A|^2 \left(\frac{\pi \varepsilon_0}{8}\right)\left(\frac{\omega_0 \mu_0}{\xi}\right)^2 a^2 d\left(1 + \frac{\sin(\beta d)}{\beta d}\right)\left(\frac{J_0^2(ua)}{K_0^2(\xi a)}\right)[K_0(\xi a) K_2(\xi a) - K_1^2(\xi a)] \tag{9.33b}$$

The average electric energy stored in region 5 is equal to that in region 6. Using (9.30) in (9.29), we obtain

$$W_{e5} = W_{e6} = |A|^2 \left(\frac{\pi \varepsilon_0}{2}\right)\left(\frac{\omega_0 \mu_0}{\xi}\right)^2 \cos^2\left(\frac{\beta d}{2}\right)\left(\frac{J_0^2(ua)}{K_0^2(\xi a)}\right) \int_a^{\infty} r K_1^2(\xi r)\, dr$$

$$\cdot \int_{d/2}^{\infty} e^{-2\zeta(z-d/2)}\, dz \tag{9.34a}$$

$$= |A|^2 \left(\frac{\pi \varepsilon_0}{8\zeta}\right)\left(\frac{\omega_0 \mu_0}{\xi}\right)^2 a^2 \cos^2\left(\frac{\beta d}{2}\right)\left(\frac{J_0^2(\xi a)}{K_0^2(\xi a)}\right)[K_0(\xi a) K_2(\xi a) - K_1^2(\xi a)] \tag{9.34b}$$

The total average stored energy is given by

$$W_e = W_{e1} + 2W_{e2} + W_{e4} + 2W_{e5} \tag{9.35}$$

Derivation for Radiated Power P_r We start with the relation between the displacement density **D** and polarization **P** within a uniform dielectric medium having relative dielectric constant ε_r:

$$\mathbf{D} = \varepsilon_0 \mathbf{E} + \mathbf{P} \tag{9.36a}$$

where

$$\mathbf{P} = \varepsilon_0(\varepsilon_r - 1)\mathbf{E} \tag{9.36b}$$

In the case of the cylindrical dielectric resonator operating in the $TE_{01\delta}$ mode, $\mathbf{E} = \hat{\boldsymbol{\phi}} E_\phi$. Assuming a time dependence of $e^{j\omega t}$, the polarization current density $\mathbf{J}_p$ is given by

$$\mathbf{J}_p = \frac{\partial \mathbf{P}}{\partial t} = \hat{\boldsymbol{\phi}} j\omega\varepsilon_0(\varepsilon_r - 1)E_\phi \tag{9.37}$$

Thus, for the $TE_{01\delta}$ mode, the isolated cylindrical resonator radiates like a magnetic dipole. The magnetic dipole moment $\mathbf{p}_m$ is given by

$$\mathbf{p}_m = \frac{1}{2}\int_{\text{vol}} \mathbf{R} \times \mathbf{J}_p \, dV \tag{9.38}$$

where $\mathbf{R} = \hat{\mathbf{r}} r + \hat{\mathbf{z}} z$ is the radius vector from the origin to the point of observation (within the resonator). Substituting for $\mathbf{J}_p$ from (9.37) in (9.38) and then using the expression for E_ϕ given by (9.18a) with ω set equal to ω_0, we obtain

$$\mathbf{p}_m = Ak_0^2\left(\frac{\varepsilon_r - 1}{2u}\right) 2\pi \int_0^a \int_{-d/2}^{d/2} (r\hat{\mathbf{r}} + z\hat{\mathbf{z}}) \times \hat{\boldsymbol{\phi}}\, J_1(ur)\cos(\beta z) r\, dr\, dz \tag{9.39}$$

The r-component of the integral vanishes to give

$$\mathbf{p}_m = \hat{\mathbf{z}} Ak_0^2(\varepsilon_r - 1)\left(\frac{\pi a^3 d}{u}\right)\left(\frac{J_2(ua)}{ua}\right)\left(\frac{\sin(\beta d/2)}{\beta d/2}\right) \tag{9.40}$$

The power radiated by the magnetic dipole of moment $\mathbf{p}_m$ is given by

$$P_r = 10 k_0^4 |\mathbf{p}_m|^2 \tag{9.41}$$

Substituting for $\mathbf{p}_m$ from (9.40), we obtain

$$P_r = 10|A|^2 (\pi k_0 d/u)^2 (k_0 a)^6 (\varepsilon_r - 1)^2 \left|\left(\frac{J_2(ua)}{ua}\right)\left(\frac{\sin(\beta d/2)}{\beta d/2}\right)\right|^2 \tag{9.42}$$

The final closed form expression for Q_r is obtained by substituting for W_e from (9.35) and P_r from (9.42) in (9.28). If we neglect the small energy stored in regions

4–6, and consider only regions 1–3, the expression for Q_r becomes

$$Q_r = 2\omega_0(W_{e1} + 2W_{e2})/P_r \tag{9.43}$$

Substituting for W_{e1}, W_{e2}, and P_r from (9.31), (9.32), and (9.42), respectively, in (9.43), we obtain

$$Q_r = \frac{[\varepsilon_r d[1 + \sin(\beta d)/\beta d] + (2/\zeta)\cos^2(\beta d/2)][J_1^2(ua) - J_0(ua)J_2(ua)]}{40\pi\omega_0\varepsilon_0(k_0 a)^4(\varepsilon_r - 1)^2 d^2\,[\sin(\beta d/2)/(\beta d/2)]^2[J_2(ua)/ua]^2} \tag{9.44}$$

9.3.3 Closed Form Expressions

The formulas derived above by the application of the EDC technique are far simpler to adopt than those of the rigorous method. However, they are still in the form of transcendental equations, which involve solution by iterative means. In order to enable quick design of cylindrical resonators for the $TE_{01\delta}$ mode of operation, we present below approximate closed form formulas reported by Mongia and Bhat [30]. In these formulas, Bessel functions are replaced by approximate algebraic expressions.

Determination of Resonator Height Over the range $(\varepsilon_r - 1)^{1/2} k_0 a \leq 4$, which normally holds for most of the practical cylindrical resonators, (9.22a) and (9.24a) can be recast to a good approximation into the following simple algebraic equations [30, 31]:

$$u' = c_1 + c_2[(\varepsilon_r - 1)(k_0 a)^2 - c_3]^{1/2} \tag{9.45a}$$

$$ua = c_1 + c_2[(\varepsilon_e - 1)(k_0 a)^2 - c_3]^{1/2} \tag{9.45b}$$

where

$$c_1 = 0.951 p_{01}, \quad c_2 = 0.222, \quad c_3 = 0.951 p_{01}^2 \tag{9.45c}$$

and p_{01} is the first root of the equation $J_0(x) = 0$.

If the resonator parameters ε_r and a and the resonant frequency f_0 are specified, the radial wavenumber u can be computed by using (9.45a), (9.23), and (9.45b). The height d of the resonator can then be obtained from (9.13a) and (9.20).

Determination of Resonant Frequency For a given set of resonator parameters, the following expressions, derived by curve fitting the numerical results of the EDC technique, can be used for obtaining the resonant frequency f_0 [30]. For $0.5 \leq d/2a \leq 1$,

$$k_0 a = [1/(\varepsilon_r + 1)^{1/2}][4.3434 - 2.835458(d/2a) + 1.3014(d/2a)^2] \tag{9.46a}$$

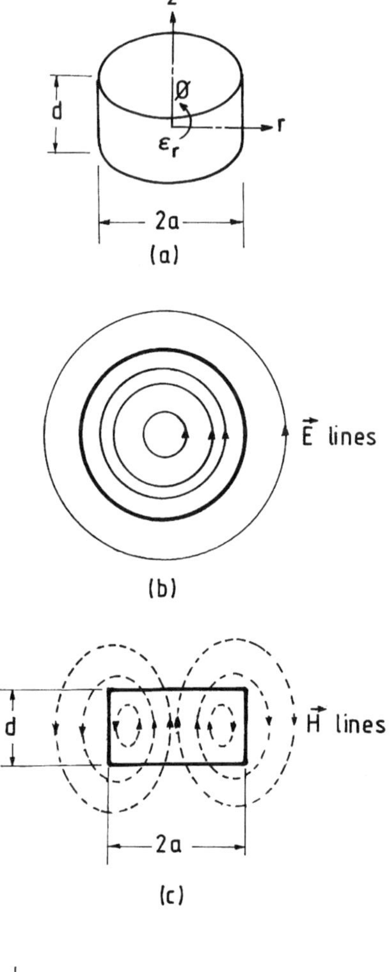

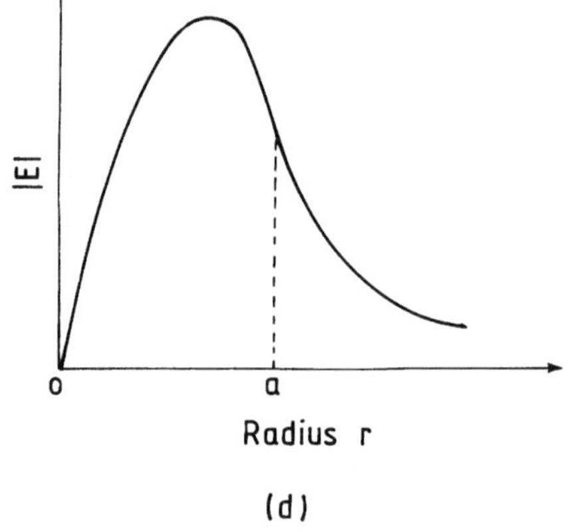

FIGURE 9.2 Typical fields for the $TE_{01\delta}$ mode of a cylindrical resonator: (a) resonator geometry, (b) E-field lines in equatorial plane ($z = 0$), (c) H-field lines in meridian plane, and (d) electric field intensity versus the radial distance.

and for $0.2 \leq d/2a \leq 0.5$

$$k_0 a = [1/(\varepsilon_r + 1)^{1/2}][5.98747 - 10.09767(d/2a) + 9.29892(d/2a)^2] \qquad (9.46b)$$

where

$$k_0 = 2\pi f_0 \sqrt{\mu_0 \varepsilon_0} \qquad (9.46c)$$

The error in using these formulas is reported to be less than 0.5% for values of $\varepsilon_r \geq 25$.

9.3.4 $TE_{01\delta}$ Mode Characteristics

Field Configuration Figure 9.2 shows the typical field distribution for the $TE_{01\delta}$ mode of a cylindrical resonator. Figure 9.2(b) shows the E-field lines in the equatorial plane; that is, the plane passing through the center of the resonator and perpendicular to the z-axis. Figure 9.2(c) shows the H-field lines in the meridian plane; that is, the plane containing the z-axis. The variation in the electric field intensity as a function of radius is shown in Figure 9.2(d). The magnitude of the E_ϕ component is zero at the center of the resonator and has a maximum value around $r = 2a/3$. Outside the resonator, the field decays exponentially. The field variation as a function of radial distance remains nearly the same in different planes parallel to the equatorial plane.

Resonant Frequency and Radiation Q-Factor The accuracy of the EDC technique in determining the resonant frequency of the dominant $TE_{01\delta}$ mode has been well demonstrated in the literature [7, 8]. For a resonator having $\varepsilon_r = 38$ and $a = 1.5$ mm, Table 9.1 shows a comparison of the results of the EDC technique with those of a rigorous method [22] as reported in [8]. It can be seen that the

TABLE 9.1 Resonant Frequency (f_0) and Radiation *Q*-Factor (Q_r) of $TE_{01\delta}$ Mode of Isolated Cylindrical (Pill Box) Resonator: $a = 1.5$ mm, $\varepsilon_r = 38$ (Fig. 9.1(a))

	f_0 (GHz)		Q_r	
$d/2a$	Theory Ref. [22]	EDC Method	Theory Ref. [22]	EDC Method
1/5	21.79	22.13	28.2	22.7
1/3	18.43	18.61	37.4	32.0
1/2	16.48	16.56	42.9	38.2
2/3	15.41	15.44	44.8	40.8
1	14.28	14.24	44.1	41.2

Source: From Mongia and Bhat [8], reprinted from AEU-38, 1987.

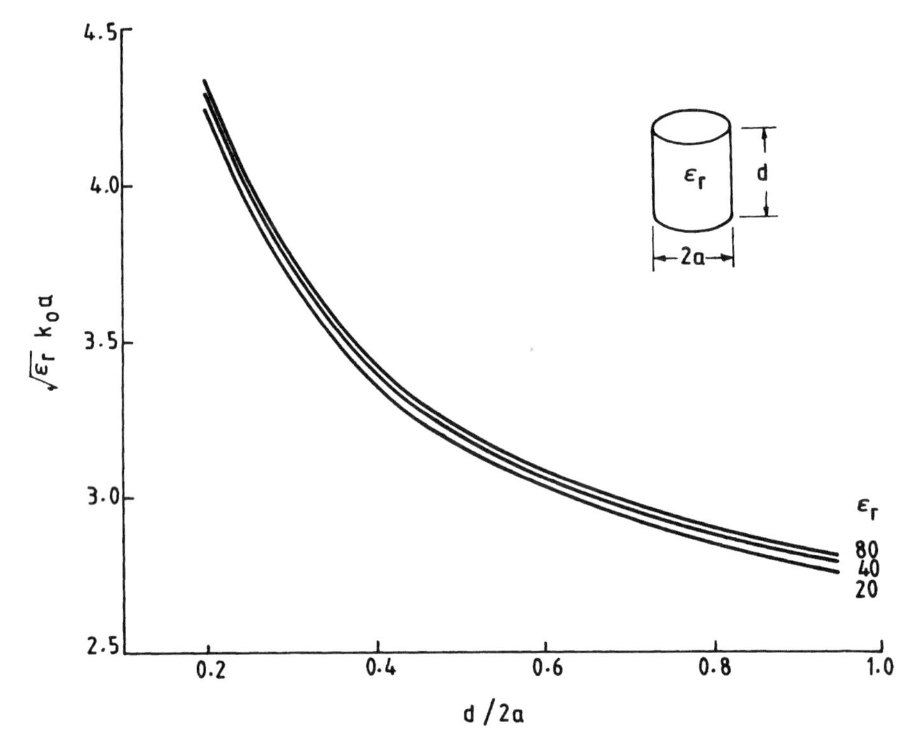

FIGURE 9.3 Normalized resonant wavenumber as a function of aspect ratio $d/2a$ for the $TE_{01\delta}$ mode of a cylindrical resonator (refer to Fig. 9.1(a)).

difference in the resonant frequencies obtained from the two methods is within 1% for aspect ratio in the range $1/3 \leq d/2a \leq 1$. Figure 9.3 shows typical variation of the normalized resonant wavenumber ($k_0\sqrt{\varepsilon_r}a$) as a function of the aspect ratio for three different values of ε_r. It may be noted that for a fixed aspect ratio $d/2a$ and for large ε_r, the value of $k_0\sqrt{\varepsilon_r}a$ increases only marginally with an increase in ε_r. That is, for resonators with large ε_r, both k_0 and ω_0 vary inversely as $\sqrt{\varepsilon_r}$.

The radiation Q-factors calculated from the EDC method are also reported to be in good agreement with the rigorous theories [8]. Table 9.1 shows typical comparison with the results of DeSmedt [22]. Figure 9.4 shows the typical variation of both k_0a and Q_r as a function of ε_r for $d/2a = 0.3, 0.5$, and 1.0. It may be observed that Q_r increases rather rapidly with an increase in ε_r and also with an increase in $d/2a$. Thus for low values of ε_r as well as $d/2a$, the value of Q_r is low, thereby indicating large radiation. Resonators having low values of ε_r and $d/2a$ are therefore useful for antenna applications.

If ε_r is increased beyond a sufficiently large value, both βd and ua approach their asymptotic values. The expression for Q_r given by (9.44) can then be

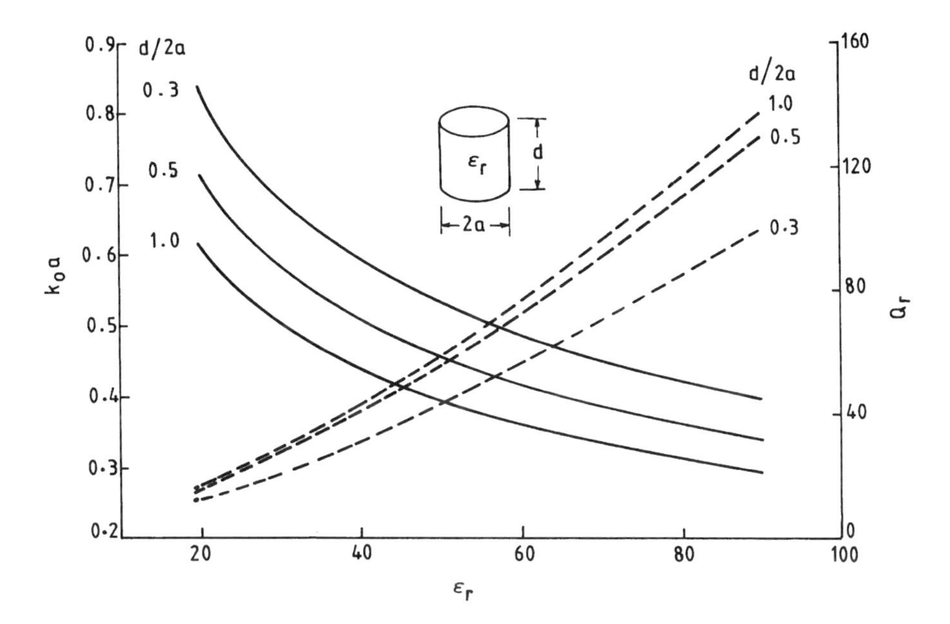

FIGURE 9.4 Normalized resonant wavenumber $k_0 a$ and radiation Q-factor Q_r versus ε_r for the $TE_{01\delta}$ mode of a cylindrical resonator (refer to Fig. 9.1(a)): —— $k_0 a$, ----- Q_r.

simplified to obtain the following proportionality relation:

$$Q_r \propto \frac{1}{\omega_0 \varepsilon_r k_0^4} \tag{9.47}$$

If we use resonators having large ε_r, both ω_0 and k_0 vary as $1/\sqrt{\varepsilon_r}$, and Eq. (9.47) simplifies to

$$Q_r \propto \varepsilon_r^{3/2} \tag{9.48}$$

9.3.5 Higher Order Modes

The higher order modes of an isolated cylindrical resonator that are of interest are $HEM_{11\delta}$ and $TM_{01\delta}$. The $HEM_{11\delta}$ mode is the lowest order hybrid mode. This mode is of special interest because of its application in dual-mode filters [32, 36]. It has also been used in the cylindrical dielectric cavity antenna [42]. When the E_z component dominates over H_z, this hybrid mode is designated as $HE_{11\delta}$ (quasi-TM mode), and when the H_z component dominates over E_z, it is designated as $EH_{11\delta}$ (quasi-TE mode).

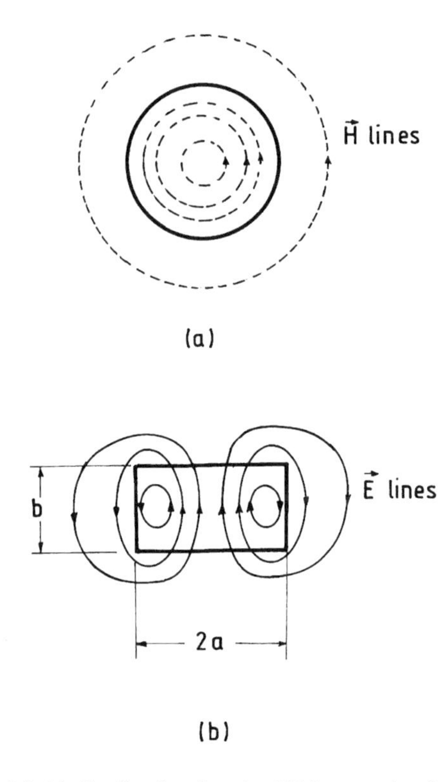

FIGURE 9.5 Typical field distribution for the $TM_{01\delta}$ mode of a cylindrical resonator (refer to Fig. 9.1(a)): (a) H-field lines in equatorial plane ($z = 0$) and (b) E-field lines in meridian plane.

Analysis for $TM_{01\delta}$ Mode The field pattern of the $TM_{01\delta}$ mode of an isolated cylindrical resonator is similar to that of the $TE_{01\delta}$ mode with **E** and **H** lines interchanged. The mode is axisymmetric ($\partial/\partial\phi = 0$). Figure 9.5 shows the typical field pattern. The nonzero field components of this mode are E_z, H_ϕ, and E_r ($H_z = E_\phi = H_r = 0$). E_z satisfies the wave equation given by (9.9). From (9.8), E_r and H_ϕ can be expressed in terms of E_z as

$$E_r = \frac{1}{k_r^2}\frac{\partial^2 E_z}{\partial r\,\partial z} \tag{9.49a}$$

$$H_\phi = -\frac{j\omega\varepsilon}{k_r^2}\frac{\partial E_z}{\partial r} \tag{9.49b}$$

where k_r is the radial wavenumber and ε is the dielectric constant of the medium. Referring to Figure 9.1(b), we can write the following expressions for the fields in regions 1–4:

$$E_z = \begin{cases} AJ_0(ur)\cos(\beta z), & \text{region } 1 \\ A\,\varepsilon_r \cos(\beta d/2) J_0(ur) e^{-\zeta(|z|-d/2)}, & \text{regions } 2,\ 3 \\ A[J_0(ua)/K_0(\xi a)]\,K_0(\xi r)\cos(\beta z), & \text{region } 4 \end{cases} \tag{9.50}$$

$$E_r = \begin{cases} A(\beta/u)J_1(ur)\sin(\beta z), & \text{region } 1 \\ A(\varepsilon_r\zeta/u)\cos(\beta d/2)J_1(ur)e^{-\zeta(z-d/2)}, & \text{regions } 2 \\ -A(\varepsilon_r\zeta/u)\cos(\beta d/2)J_1(ur)e^{\zeta(z+d/2)}, & \text{region } 3 \\ -A(\beta/\xi)[J_0(ua)/K_0(\xi a)]K_1(\xi r)\sin(\beta z), & \text{region } 4 \end{cases} \tag{9.51}$$

$$H_\phi = \begin{cases} A(j\omega_0\varepsilon_0\varepsilon_r/u)J_1(ur)\cos(\beta z), & \text{region } 1 \\ A(j\omega_0\varepsilon_0\varepsilon_r/u)\cos(\beta d/2)J_1(ur)e^{-\zeta(|z|-d/2)}, & \text{regions } 2,\ 3 \\ -A(j\omega_0\varepsilon_0/\xi)[J_0(ua)/K_0(\xi a)]K_1(\xi r)\cos(\beta z), & \text{region } 4 \end{cases} \tag{9.52}$$

where the parameters u, β, ξ, and ζ have the same definition as given in Section 9.3.1 for the $TE_{01\delta}$ mode and they satisfy the relations given by (9.13). The above field expressions are approximate and are obtained by matching the E_z component at $r = a$, $-d/2 < z < d/2$ and the H_ϕ component at $z = d/2$, $0 \le r \le a$. The fields in regions 5 and 6 are neglected.

Applying the continuity conditions on E_r at $z = d/2$, $r < a$, we obtain

$$\beta \tan(\beta d/2) = \zeta\varepsilon_r \tag{9.53}$$

Next, applying the continuity condition on H_ϕ at $r = a$, $-d/2 < z < d/2$, we obtain

$$\frac{\varepsilon_r J_1(ua)}{uJ_0(ua)} + \frac{K_1(\xi a)}{\xi K_0(\xi a)} = 0 \tag{9.54}$$

where

$$\xi = [k_0^2(\varepsilon_r - 1) - u^2]^{1/2} \tag{9.55}$$

We identify (9.53) as the characteristic equation for the TM_0 mode of a radial slab guide having the same height d and relative dielectric constant ε_r as that of the resonator. Similarly, (9.54) is the characteristic equation for the TM_{01} mode of an infinite cylinder of radius a and relative dielectric constant ε_r.

We now consider implementation of the EDC technique for evaluating the resonant frequency. The characteristic equation for evaluating β remains the same as given by (9.53). For evaluation of the radial wavenumber, u is obtained as a solution of (9.54) with ε_r replaced by the effective dielectric constant ε_e. That is,

$$\frac{\varepsilon_e J_1(ua)}{uJ_0(ua)} + \frac{K_1(\xi a)}{\xi K_0(\xi a)} = 0 \tag{9.56a}$$

where

$$\xi = [k_0^2(\varepsilon_e - 1) - u^2]^{1/2} \tag{9.56b}$$

The value of ε_e is obtained from the relation

$$\varepsilon_e = (u'/k_0)^2 \tag{9.57}$$

where u' is the solution of the equation

$$\frac{\varepsilon_r J_1(u'a)}{u' J_0(u'a)} + \frac{K_1(\xi' a)}{\xi' K_0(\xi' a)} = 0 \tag{9.58a}$$

where

$$\xi' = [k_0^2(\varepsilon_r - 1) - (u')^2]^{1/2} \tag{9.58b}$$

The free-space resonant wavenumber $k_0 = \omega_0 \sqrt{\mu_0 \varepsilon_0}$ is the one at which equations (9.53), (9.56)–(9.58), and (9.13a) are satisfied simultaneously.

Closed Form Expressions Van Bladel [17] has shown that, for large ε_r, the surface of the cylindrical resonator acts as a magnetic wall in the case of the $TM_{01\delta}$ mode. Applying the boundary condition for the magnetic wall—namely, $E_z = 0$ at $z = d$, $r \leq a$ and $E_r = 0$ at $r = a$, $-d/2 \leq z \leq d/2$—we get

$$\cos(\beta d/2) = 0 \tag{9.59a}$$

$$J_1(ha) = 0 \tag{9.59b}$$

so that

$$\beta = \pi/d \tag{9.60a}$$

$$h = p_{11}/a \tag{9.60b}$$

where p_{11} is the first root of Eq. (9.59b). Substituting these values in the separation equation (9.13a), we obtain

$$k_0 = (1/\sqrt{\varepsilon_r})[(p_{11}/a)^2 + (\pi/a)^2]^{1/2} \tag{9.61}$$

Characteristic Equations for $HE_{11\delta}$ Mode (Quasi-TM Mode) The hybrid modes of the isolated cylindrical resonator are characterized by the fact that they are axially asymmetric with all six field components present. Referring to Figure 9.1(b), the E_z and H_z components in the four regions at resonance are given by

$$E_z = \begin{cases} AJ_1(ur)\cos(\phi)\cos(\beta z), & \text{region 1} \\ BJ_1(ur)\cos(\phi)e^{-\zeta(|z|-d/2)}, & \text{regions 2, 3} \\ CK_1(\xi r)\cos(\phi)\cos(\beta z), & \text{region 4} \end{cases} \tag{9.62}$$

$$H_z = \begin{cases} DJ_1(ur)\sin(\phi)\sin(\beta z), & \text{region 1} \\ EJ_1(ur)\sin(\phi)e^{-\zeta(|z|-d/2)}, & \text{regions 2, 3} \\ FK_1(\xi r)\sin(\phi)\sin(\beta z), & \text{region 4} \end{cases} \tag{9.63}$$

where u, β, ζ, and ξ satisfy the relations given by (9.13). The expressions for the other field components E_r, E_ϕ, H_r, and H_ϕ are obtained by substituting (9.62) and (9.63) in (9.8). If we apply the continuity conditions on E_z, H_z, E_ϕ, and H_ϕ at the cylindrical surface between regions 1 and 4 ($r = a$, $-d/2 < z < d/2$), we obtain a set of four homogeneous equations in terms of the unknown coefficients A, C, D, and F. They can be written in matrix form as

$$\begin{bmatrix} J_1(ua) & K_1(\xi a) & 0 & 0 \\ 0 & 0 & J_1(ua) & K_1(\xi a) \\ \left(\dfrac{\beta}{u^2 a}\right) J_1(ua) & \left(\dfrac{\beta}{\xi^2 a}\right) K_1(\xi a) & \left(\dfrac{j\omega\mu_0}{u}\right) J_1'(ua) & \left(\dfrac{j\omega\mu_0}{\xi}\right) K_1'(\xi a) \\ \left(\dfrac{-j\omega\varepsilon_0\varepsilon_r}{u}\right) J_1'(ua) & \left(\dfrac{-j\omega\varepsilon_0}{\xi}\right) K_1'(\xi a) & \left(\dfrac{\beta}{u^2 a}\right) J_1(ua) & \left(\dfrac{\beta}{\xi^2 a}\right) K_1(\xi a) \end{bmatrix}$$

$$\cdot \begin{bmatrix} A \\ C \\ D \\ F \end{bmatrix} = 0 \tag{9.64a}$$

where

$$\beta = (k_0^2 \varepsilon_r - u^2)^{1/2} \tag{9.64b}$$

$$\xi = [k_0^2(\varepsilon_r - 1) - u^2]^{1/2} \tag{9.64c}$$

Setting the determinant of the coefficient matrix equal to zero yields the characteristic equation. This characteristic equation is identical to that of the HE_{11} mode of an infinitely long dielectric cylindrical guide [43].

For deriving the characteristic equation in terms of β, we need to match the tangential fields at $z=d/2$. Examination of expressions for E_r, H_r, E_ϕ, and H_ϕ reveals that at $z=\pm d/2$, these tangential fields do not have the same radial dependence. This problem, however, can be circumvented by making use of the feature that for HE modes, the E_z component is much stronger than H_z. If we set $H_z=0$ in regions 1 and 2 ($D=0$, $E=0$ in (9.63)) and apply the continuity conditions on E_ϕ and H_ϕ at $z=d/2$ ($r<a$), we obtain the same characteristic equation as that of a TM_0 mode of radial slab guide of height d and relative dielectric constant ε_r. The expression is given by (9.53).

The procedure for evaluation of the radial wavenumber through the EDC technique follows the same procedure as that outlined above for the $TM_{01\delta}$ mode except that (9.54) gets replaced by the characteristic equation obtained from (9.64a).

Characteristic Equations for $EH_{11\delta}$ Mode (Quasi-TE Mode) The characteristic equation for determining the radial wavenumber u for the $EH_{11\delta}$ mode of the resonator is the same as that of the EH_{11} mode of the corresponding infinitely long dielectric cylinder. The equation can be obtained by replacing β by $-\beta$ in (9.64a) and setting the determinant of the coefficient matrix equal to zero. The EDC technique can be applied to determine first the value of ε_e and then the radial wavenumber u.

The characteristic equation for determining the z-directed wavenumber β is obtained by assuming $E_z=0$ and matching E_ϕ and H_ϕ at $z=d/2$ ($r<a$). The

TABLE 9.2 Resonant Frequencies of Higher Order Modes of Isolated Cylindrical Resonator (Fig. 9.1(a))

Resonator Parameters				Resonant Frequency (GHz)		
					Ref. [27]	
ε_r	a (mm)	d (mm)	Mode	EDC Method	Theory	Experiment
38	5.25	4.60	$HE_{11\delta}$	5.93	6.33	—
38	5.25	4.60	$EH_{11\delta}$	6.79	6.64	6.64
38	5.25	4.60	$TM_{01\delta}$	7.40	7.51	7.60
					Ref. [29]	
					Theory	Experiment
35	10	20	$HE_{11\delta}$	2.11	2.23	—
35	10	20	$TM_{01\delta}$	3.15	3.19	—

Source: From Mongia and Bhat [8], reprinted from AEU-38, 1987.

resulting equation is the same as that for the TE_0 mode of a radial slab guide and is given by (9.20a).

Resonant Frequency Characteristics In order to illustrate the accuracy of the EDC technique in predicting the resonant frequencies of higher order modes, we reproduce in Table 9.2 the results reported in [8]. This table compares the resonant frequencies of $HE_{11\delta}$, $EH_{11\delta}$, and $TM_{01\delta}$ modes due to the EDC technique with those of the rigorous methods [27, 29]. Except for the $HE_{11\delta}$ mode, for which the resonant frequency due to the rigorous method is higher by about 6.5% as compared with the EDC technique, all other results match within 2.4%.

9.4 CYLINDRICAL RESONATOR IN PLANAR DIELECTRIC SLAB GUIDE ENVIRONMENT

9.4.1 EDC Method Combined with Transverse Transmission Line Technique

In this section, we analyze composite resonator structures involving a cylindrical dielectric resonator placed in a planar slab guide environment. Examples of such structures are the insular image cylindrical resonator (Fig. 9.6(a)), cylindrical resonator on a suspended substrate (Fig. 9.6(b)), and cylindrical image resonator with a top dielectric layer (Fig. 9.6(c)). The relative dielectric constant of the substrate is assumed to be very small in comparison with that of the resonator, so that most of the energy is still confined to the dielectric region of the cylindrical resonator.

As in the case of the isolated cylindrical resonator, determination of the resonant frequency of the composite resonator structures shown in Figure 9.6 requires evaluation of radial and axial wavenumbers (u and β) in the dielectric region of the cylindrical resonator. For structures involving layered dielectrics, the characteristic equation for evaluation of the axial wavenumber can be derived easily by applying the *transverse transmission line* technique to the horizontal slab guide model. This horizontal slab guide model is obtained by extending the resonator in the radial direction to infinity. For determining the radial wavenumber u, we adopt the EDC technique of analysis described in the preceding section for the isolated resonator. That is, we ignore the presence of the dielectric substrates and the ground plane and model the resonator as an infinitely long cylindrical dielectric guide. This guide has the same radius a as that of the resonator but an effective (relative) dielectric constant ε_e that is different from the relative dielectric constant ε_r of the resonator. The radial wavenumber of this guide, which is obtained as the solution of its characteristic equation, is taken as the radial wavenumber of the resonator. The value of ε_e required to be substituted is obtained from the relation $\varepsilon_e = (u'/k_0)^2$, were k_0 is the free-space wavenumber and u' is the radial wavenumber of an infinitely long cylindrical guide having the same parameters a and ε_r as those of the resonator.

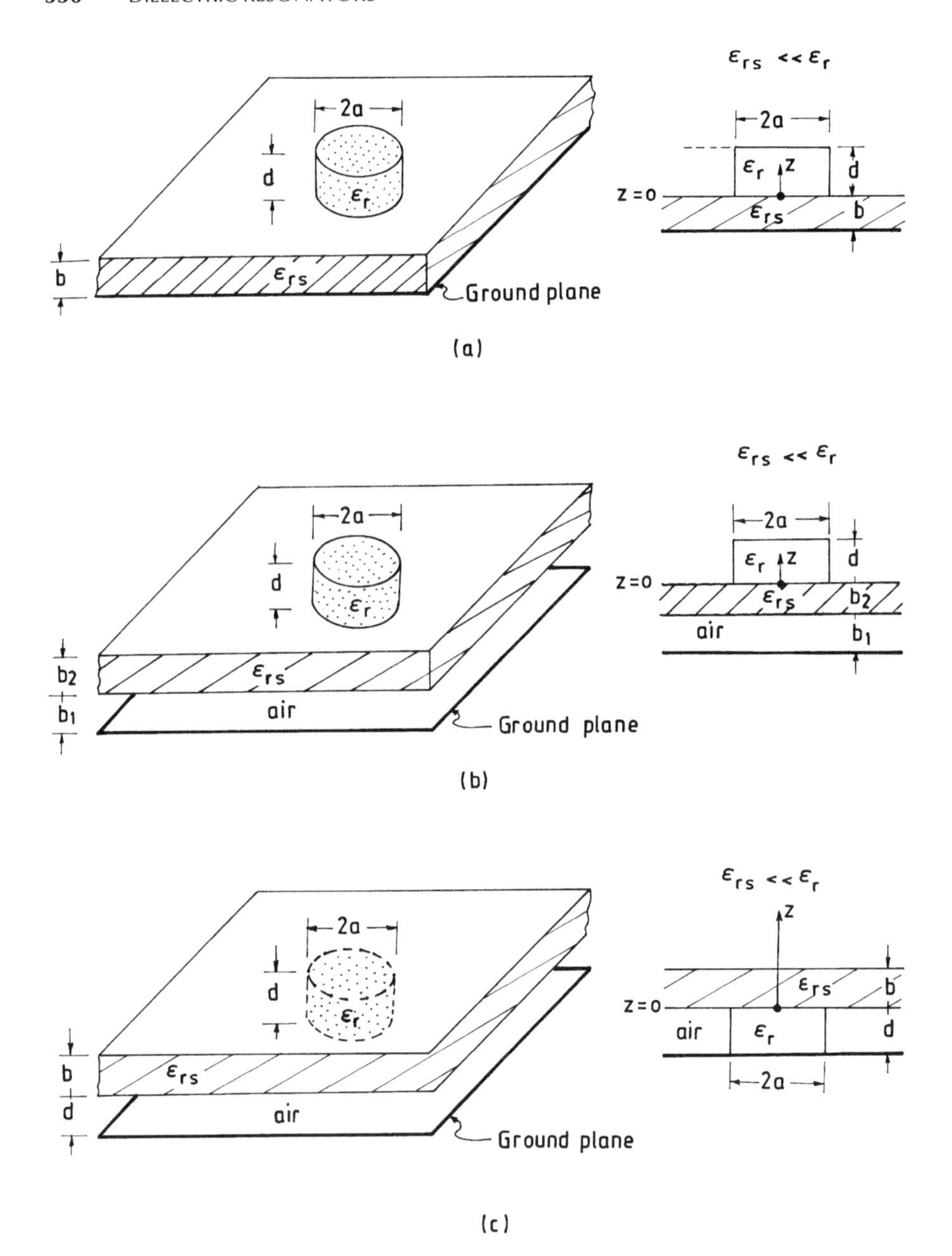

FIGURE 9.6 Cylindrical resonators in a slab guide environment and their sectional views in the meridian plane: (a) insular image cylindrical resonator, (b) cylindrical resonator on a suspended substrate, and (c) cylindrical image resonator with a top dielectric layer.

9.4.2 Insular Image Guide Cylindrical Resonator

Figure 9.6(a) shows the geometry of the insular image cylindrical resonator. The dielectric substrate and the ground plane extend to infinity in the horizontal plane (xy-plane or $r\phi$-plane). It is assumed that the relative dielectric constant ε_{rs} of the substrate is much smaller than that of the resonator (ε_r) and waves are of the decaying type in the substrate as well as in the air region. Let u and β denote the radial and axial wavenumbers, respectively, in the cylindrical resonator (region 1). The separation equation that relates u, β, and the free-space resonant wavenumber k_0 is given by

$$u^2 + \beta^2 = k_0^2\varepsilon_r \tag{9.65}$$

The mode designation for the resonator structure in Figure 9.6(a) corresponds to that defined in Section 9.3 for the isolated resonator. As discussed in Section 9.4.1, for the purpose of determining the radial wavenumber u, we ignore the presence of the dielectric substrate and the ground plane and use the EDC technique as applied to the isolated cylindrical resonator. Thus the characteristic equations derived in Section 9.3 for evaluation of u of the isolated resonator directly apply for the present structure. The presence of the slab guide is taken into account in evaluation of β. In the following, we consider the characteristic equations for determination of the wavenumbers u and β for the $TE_{01\delta}$, $TM_{01\delta}$, $HE_{11\delta}$, and $EH_{11\delta}$ resonant modes.

$TE_{01\delta}$ Mode

Characteristic Equation for Radial Wavenumber u Corresponding to the $TE_{01\delta}$ mode of the resonator, we need to consider the TE_{01} mode of the infinite cylindrical model (Fig. 9.7) for evaluation of the radial wavenumber u. The set of equations are the same as given in (9.22)–(9.24):

$$\frac{J_1(ua)}{uJ_0(ua)} + \frac{K_1(\xi a)}{\xi K_0(\xi a)} = 0 \tag{9.66}$$

where

$$\xi = [k_0^2(\varepsilon_e - 1) - u^2]^{1/2} \tag{9.67a}$$

$$\varepsilon_e = (u'/k_0)^{1/2} \tag{9.67b}$$

and u' is obtained as a solution of (9.66) with u and ξ replaced by u' and ξ', respectively, and ξ' is defined as

$$\xi' = [k_0^2(\varepsilon_r - 1) - (u')^2]^{1/2} \tag{9.67c}$$

Characteristic Equation for Axial Wavenumber β Figure 9.8(a) shows the horizontal slab guide model of Figure 9.6(a). As described in Chapter 2 (Section

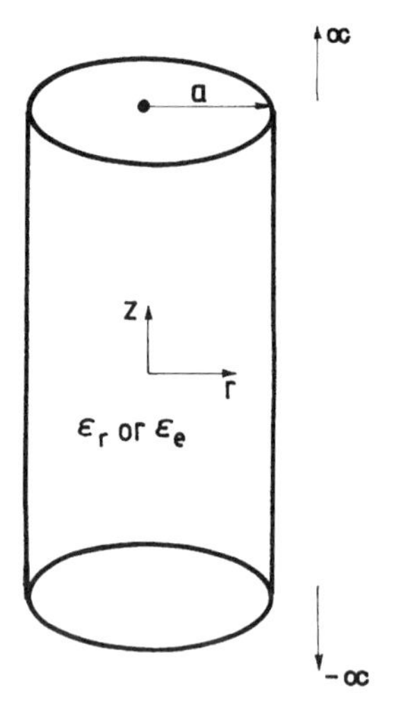

FIGURE 9.7 Infinite cylindrical guide model for determining the radial wavenumber u of structures in Figure 9.6.

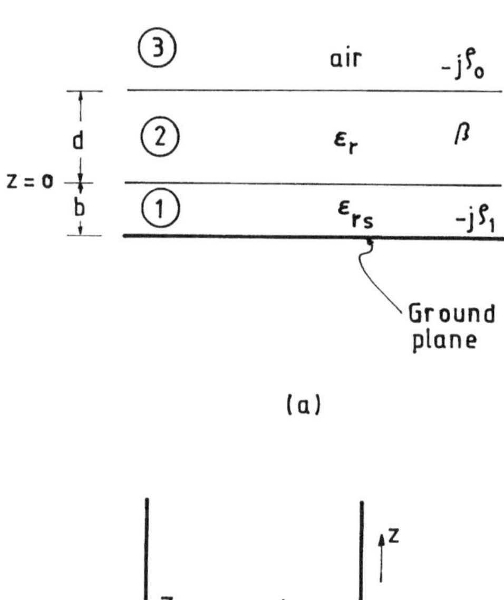

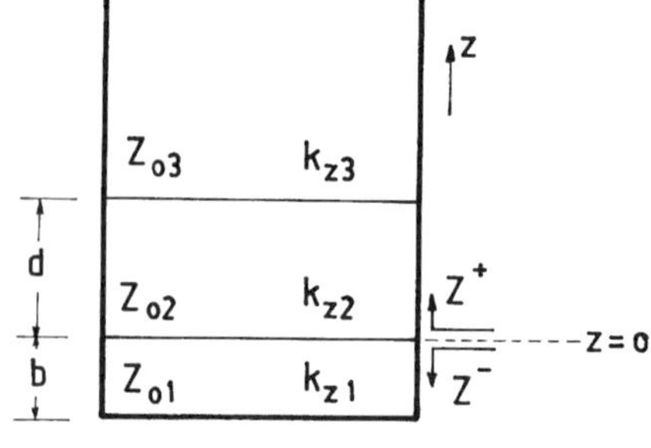

FIGURE 9.8 (a) Horizontal slab guide model of Figure 9.6(a) and (b) its equivalent transmission line network.

2.4.2), the characteristic equation can easily be derived by considering the equivalent transmission line network shown in Figure 9.8(b). For the TE-to-z mode (corresponding to the $TE_{01\delta}$ mode of the resonator) the characteristic impedance Z_{0i} of the ith section is given by (refer to Section 2.4.2)

$$Z_{0i} = \omega\mu_0/k_{zi} \tag{9.68a}$$

where

$$k_{z1} = -j\zeta_1, \quad k_{z2} = \beta, \quad k_{z3} = -j\zeta_0 \tag{9.68b}$$

The parameters ζ_1 and ζ_0 are the decay coefficients in regions 1 and 3, respectively, and β is the propagation constant in region 2, which is assumed to be the same as the axial wavenumber in the cylindrical resonator. Applying the transverse resonance condition—namely, $Z^+ + Z^- = 0$ at $z = 0$ (see Eq. (2.63) of Chapter 2)—we can easily obtain the following characteristic equation:

$$[1 + (\zeta_0/\beta)\tan(\beta d)] + (\beta/\zeta_1)\tanh(\zeta_1 b)[(\zeta_0/\beta) - \tan(\beta d)] = 0 \tag{9.69}$$

where

$$\zeta_1 = [k_0^2(\varepsilon_r - \varepsilon_{rs}) - \beta^2]^{1/2} \tag{9.70a}$$

$$\zeta_0 = [k_0^2(\varepsilon_r - 1) - \beta^2]^{1/2} \tag{9.70b}$$

The resonant frequency for the $TE_{01\delta}$ mode is the one at which (9.65)–(9.67), (9.69), and (9.70) are simultaneously satisfied.

$TM_{01\delta}$ Mode

Characteristic Equation for u In the case of the $TM_{01\delta}$ mode, we consider the infinite cylindrical guide operating in the TM_{01} mode. The set of equations (9.56)–(9.58) derived for the $TM_{01\delta}$ mode of the isolated resonator can therefore be used directly. They are

$$\frac{\varepsilon_e J_1(ua)}{uJ_0(ua)} + \frac{K_1(\xi a)}{\xi K_0(\xi a)} = 0 \tag{9.71}$$

where

$$\xi = [k_0^2(\varepsilon_e - 1) - u^2]^{1/2} \tag{9.72a}$$

$$\varepsilon_e = (u'/k_0)^2 \tag{9.72b}$$

and u' is obtained as a solution of (9.71) with u, ε_e, and ξ replaced by u', ε_r, and ξ', respectively, where

$$\xi' = [k_0^2(\varepsilon_r - 1) - (u')^2]^{1/2} \tag{9.72c}$$

Characteristic Equation for β Referring to the horizontal slab guide model and the transmission line equivalent shown in Figure 9.8, we note that the characteristic impedance Z_{0i} of the ith transmission line section must correspond to that of the TM-to-z mode. It is given by

$$Z_{0i} = k_{zi}/\omega\varepsilon_0\varepsilon_{ri} \tag{9.73}$$

where $\varepsilon_{r1} = \varepsilon_{rs}$, $\varepsilon_{r2} = \varepsilon_r$, $\varepsilon_{r3} = 1$, and the values of k_{zi} for $i = 1$ to 3 are given by (9.68b). Applying the resonance condition to the transmission line equivalent network, we obtain

$$\left[1 + \left(\frac{\zeta_1}{\beta}\right)\left(\frac{\varepsilon_r}{\varepsilon_{rs}}\right)\tanh(\zeta_1 b)\tan(\beta d)\right] + \left(\frac{\beta_y}{\zeta_0\varepsilon_r}\right) \cdot \left[-\tan(\beta d) + \left(\frac{\zeta_1}{\beta}\right)\left(\frac{\varepsilon_r}{\varepsilon_{rs}}\right)\tanh(\zeta_1 b)\right] = 0 \tag{9.74}$$

where ζ_1 and ζ_0 are given by (9.70a) and (9.70b), respectively.

HE$_{11\delta}$ *Mode (Quasi-TM Mode)*

Characteristic Equation for u The characteristic equation for determining the value of u for the resonator is the same as that derived in Section 9.3.5 for the $HE_{11\delta}$ mode of the isolated resonator. The characteristic equation is obtained by setting the determinant of the coefficient matrix in (9.64) equal to zero.

$$\begin{vmatrix} J_1(ua) & K_1(\xi a) & 0 & 0 \\ 0 & 0 & J_1(ua) & K_1(\xi a) \\ \left(\frac{\beta'}{u^2 a}\right) J_1(ua) & \left(\frac{\beta'}{\xi^2 a}\right) K_1(\xi a) & \left(\frac{j\omega\mu_0}{u}\right) J_1'(ua) & \left(\frac{j\omega\mu_0}{\xi}\right) K_1'(\xi a) \\ \left(\frac{-j\omega\varepsilon_0\varepsilon_e}{u}\right) J_1'(ua) & \left(\frac{-j\omega\varepsilon_0}{\xi}\right) K_1'(\xi a) & \left(\frac{\beta'}{u^2 a}\right) J_1(ua) & \left(\frac{\beta'}{\xi^2 a}\right) K_1(\xi a) \end{vmatrix} = 0 \tag{9.75a}$$

where

$$\beta' = (k_0^2\varepsilon_e - u^2)^{1/2} \tag{9.75b}$$

$$\xi = [k_0^2(\varepsilon_e - 1) - u^2]^{1/2} \tag{9.75c}$$

$$\varepsilon_e = (u'/k_0)^2 \tag{9.75d}$$

The value of u' is obtained as a solution of (9.75a) with u, ε_e, β' and ξ replaced by u',

ε_r, β'' and ξ', respectively, where

$$\beta'' = [k_0^2\varepsilon_r - (u')^2]^{1/2} \tag{9.75e}$$

$$\xi' = [k_0^2(\varepsilon_r - 1) - (u')^2]^{1/2} \tag{9.75f}$$

Characteristic Equation for β The characteristic equation for β is the same as that given at (9.74) for the $TM_{01\delta}$ mode.

$EH_{11\delta}$ Mode (Quasi-TE Mode) For the determination of u, the characteristic equation given by (9.75) applies. The characteristic equation for determining β is the same as that given in (9.69) for the $TE_{01\delta}$ mode.

9.4.3 Cylindrical Resonator on a Suspended Substrate

Figure 9.6(b) shows the geometry of a cylindrical resonator on a suspended substrate. The structure incorporates an airgap between the substrate and the ground plane. The relative dielectric constant of the substrate is much less than that of the resonator ($\varepsilon_{rs} \ll \varepsilon_r$) such that the fields decay exponentially inside the substrate as well as in the air regions.

Characteristic Equation for Radial Wave Number u For the purpose of determining the radial wavenumber u inside the cylindrical resonator, we consider the infinite cylindrical guide model as shown in Figure 9.7. Therefore the characteristic equations given in Section 9.4.2 for evaluating u of the insular image cylindrical resonator remain valid; that is, (9.66) and (9.67) for the $TE_{01\delta}$ mode, (9.71) and (9.72) for the $TM_{01\delta}$ mode, and (9.75) for the $HE_{11\delta}$ and $EH_{11\delta}$ mode.

Characteristic Equation for Axial Wavenumber β If the radius of the cylindrical resonator is extended to infinity in the radial direction, a three-layer horizontal slab guide model shown in Figure 9.9(a) results. We note that the dominant TE-to-z mode of the horizontal slab guide corresponds to the $TE_{01\delta}$ and $EH_{11\delta}$ modes of the cylindrical resonator. Similarly, the dominant TM-to-z mode of the horizontal slab guide corresponds to the $TM_{01\delta}$ and $HE_{11\delta}$ modes of the cylindrical resonator.

$TE_{01\delta}$ and $EH_{11\delta}$ Modes We apply the resonance condition for the transmission line equivalent, shown in Figure 9.9(b), and use the formula for Z_{0i} given by (9.68a):

$$Z_{0i} = \omega\mu_0/k_{zi}, \quad i = 1 \text{ to } 4 \tag{9.76a}$$

where

$$k_{z1} = -j\zeta_0, \quad k_{z2} = -j\zeta_2, \quad k_{z3} = \beta, \quad k_{z4} = -j\zeta_0 \tag{9.76b}$$

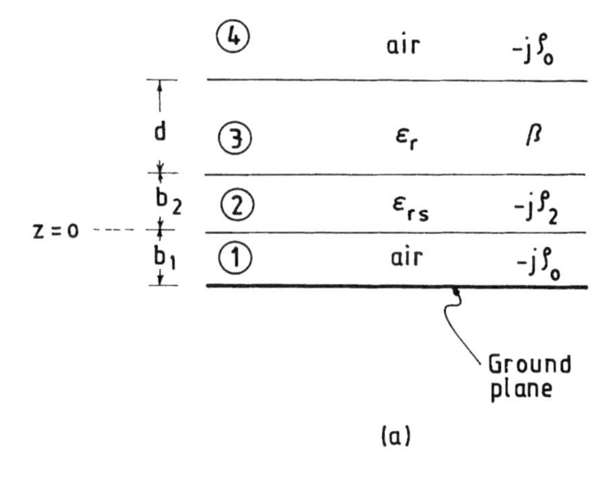

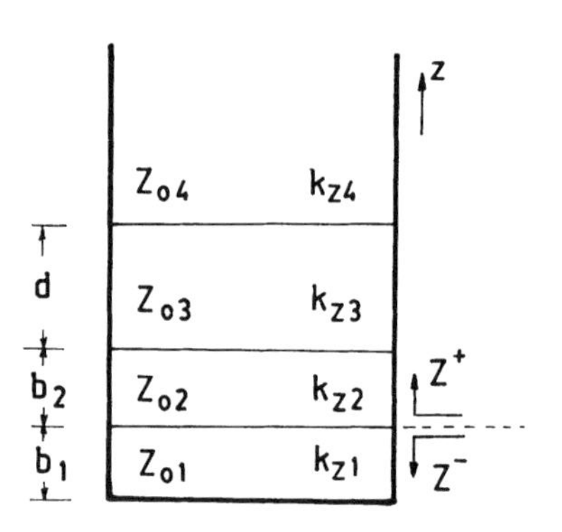

FIGURE 9.9 (a) Horizontal slab guide model of Figure 9.6(b) and (b) its equivalent transmission line network.

The transcendental equation for determination of β of the resonator for the $TE_{01\delta}$ and $EH_{11\delta}$ modes is obtained as

$$[\tanh(\zeta_0 b_1)/\zeta_0]\{[\zeta_0 - \beta\tan(\beta d)] + \zeta_2 \tanh(\zeta_2 b_2)[\zeta_0 + (\tan(\beta d)/\beta)]\}$$
$$+ \{[1 + (\zeta_0/\beta)\tan(\beta d)] + [\tanh(\zeta_2 b_2)/\zeta_2][\zeta_0 - \beta\tan(\beta d)]\} = 0 \qquad (9.77a)$$

where

$$\zeta_0 = [k_0^2(\varepsilon_r - 1) - \beta^2]^{1/2} \qquad (9.77b)$$

$$\zeta_2 = [k_0^2(\varepsilon_r - \varepsilon_{rs}) - \beta^2]^{1/2} \qquad (9.77c)$$

$TM_{01\delta}$ and $HE_{11\delta}$ Modes For these modes, we use the formulas for Z_{0i} given by (9.73) in the transmission line equivalent shown in Figure 9.9(b):

$$Z_{0i} = k_{zi}/\omega\varepsilon_0\varepsilon_{ri}, \quad i = 1 \text{ to } 4 \tag{9.78a}$$

where $\varepsilon_{r1} = \varepsilon_{r4} = 1, \varepsilon_{r2} = \varepsilon_{rs}, \varepsilon_{r3} = \varepsilon_r$, and the parameters k_{zi} for $i = 1$ to 4 are given by (9.76b). Applying the resonance condition, the transcendental equation for the determination of β is obtained as

$$\begin{aligned}&\tanh(\zeta_0 b_1)\{1 + [(\varepsilon_r\zeta_0/\beta)\tan(\beta d)] + [\varepsilon_{rs}\tanh(\zeta_2 b_2)/\varepsilon_r\zeta_2][\beta\tan(\beta d) - \varepsilon_r\zeta_0]\} \\ &\quad + \{1 - [\beta\tan(\beta d)/\varepsilon_r\zeta_0] + [\zeta_2\tanh(\zeta_2 b_2)/\varepsilon_r\zeta_0][1 + (\zeta_0\varepsilon_r/\beta)\tan(\beta d)]\} = 0\end{aligned} \tag{9.79}$$

where ζ_0 and ζ_2 are given by (9.77b) and (9.77c), respectively.

9.4.4 Cylindrical Image Resonator with a Top Dielectric Layer

The structure shown in Figure 9.6(c) makes use of a cylindrical image resonator with a thin dielectric substrate of low dielectric constant. This geometry is compatible for operation with the inverted strip dielectric guide (see Fig. 2.2(d) of Chapter 2).

Characteristic Equation for Radial Wavenumber u Considering the infinite cylindrical guide model (Fig. 9.7), we note that for evaluating the radial wavenumber u inside the resonator, the characteristic equations derived in Section 9.4.2 remain valid; that is, (9.66) and (9.67) for the $TE_{01\delta}$ mode, (9.71) and (9.72) for the $TM_{01\delta}$ mode, and (9.75) for the $HE_{11\delta}$ and $EH_{11\delta}$ mode.

Characteristic Equation for Axial Wavenumber β For determination of β, we use the horizontal slab guide model and its equivalent network shown in Figure 9.10 with $k_{z1} = \beta$, $k_{z2} = -j\zeta_1$, and $k_{z3} = -j\zeta_0$.

$TE_{01\delta}$ and $EH_{11\delta}$ Modes Using the expression for Z_{0i} given by (9.76a) and applying the resonance condition to the network in Figure 9.10(b), we obtain the transcendental equation as

$$[\zeta_0\tan(\beta d)/\beta][1 + (\zeta_1/\zeta_0)\tanh(\zeta_0 b)] + [1 + (\zeta_0/\zeta_1)\tanh(\zeta_1 b)] = 0 \tag{9.80a}$$

where

$$\zeta_0 = [k_0^2(\varepsilon_r - 1) - \beta^2]^{1/2} \tag{9.80b}$$

$$\zeta_1 = [k_0^2(\varepsilon_r - \varepsilon_{rs}) - \beta^2]^{1/2} \tag{9.80c}$$

$TM_{01\delta}$ and $HE_{11\delta}$ Modes Using the expression for Z_{0i} given by (9.78) and

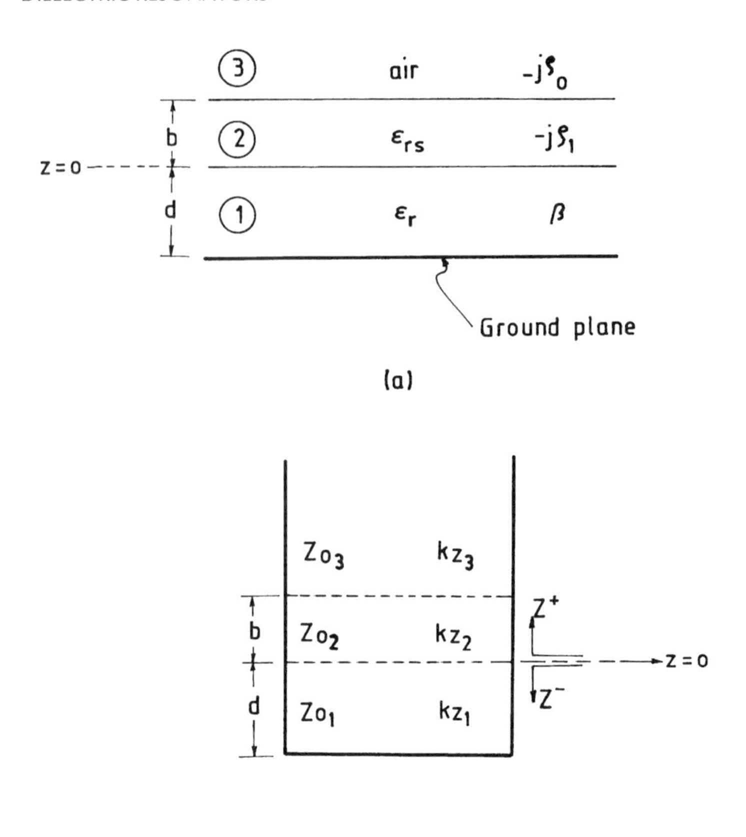

FIGURE 9.10 (a) Horizontal slab guide model of Figure 9.6(c) and (b) its equivalent transmission line network.

applying the resonance condition, we obtain

$$[1-(\varepsilon_{rs}/\varepsilon_r)(\beta/\zeta_1)\tanh(\zeta_1 b)\tan(\beta d)]-(1/\zeta_0\varepsilon_{rs})$$
$$\cdot[-\zeta_1\tanh(\zeta_1 b)+(\varepsilon_{rs}/\varepsilon_r)\beta\tan(\beta d)]=0 \qquad (9.81)$$

where ζ_0 and ζ_1 are given by (9.80b) and (9.80c), respectively.

For each of the three structures considered above and for a given resonant mode, the resonant frequency is the one at which the corresponding characteristic equations for u and β and the separation equation (9.65) are satisfied simultaneously.

9.5 RING RESONATOR STRUCTURES

9.5.1 Isolated Ring Resonator

The operating bandwidth of the $TE_{01\delta}$ mode of a cylindrical resonator depends on its aspect ratio $d/2a$. Maximum modal separation with respect to the next higher order $HE_{11\delta}$ mode occurs when the value of $d/2a$ is approximately 0.4. For this aspect ratio, the resonant frequency of the $HE_{11\delta}$ mode is about 1.35 times the resonant frequency of the $TE_{01\delta}$ mode. For wideband filter applications, the resonator must offer increased modal separation between the $TE_{01\delta}$ and $HE_{11\delta}$ modes [44]. One simple way to achieve this is to remove a cylindrical plug concentric with the axis of the cylindrical resonator, resulting in a coaxial ring resonator [18, 19, 22]. Figure 9.11 shows the geometry of a ring resonator and the typical field lines for the $TE_{01\delta}$ mode. Removing a dielectric plug along the axial length of the cylinder has negligible effect on the E_ϕ component since this field is zero on the axis and is negligible near the axis. For the magnetic field, the presence or absence of the dielectric has practically no effect. Consequently, the resonant frequencies of $HE_{11\delta}$ and $TM_{01\delta}$ modes increase with a removal of dielectric plug from the center of the resonator. Thus, where larger bandwidth of operation is required for the $TE_{01\delta}$ mode, a ring resonator is preferred over a cylindrical resonator. The isolated ring resonator has been analyzed using a rigorous method [22]. This method employs a numerical technique to yield quite accurate results, but at the expense of considerable computational flexibility. In the following, we apply the simple EDC method to analyze the ring resonator.

$TE_{01\delta}$ Mode Let u and β denote the radial and axial wavenumber, respectively, in the dielectric region of the ring resonator (Fig. 9.12(a)). The separation equation relating these parameters is

$$u^2 + \beta^2 = k_0^2 \varepsilon_r \tag{9.82a}$$

where k_0 is the free-space wavenumber.

Determination of Axial Wavenumber β For determination of β, we consider the horizontal slab guide model shown in Figure 9.12(b) for the TE-to-z mode. The transcendental equation is given by (same as (9.20a) of cylindrical resonator)

$$\beta \tan(\beta d/2) = \zeta \tag{9.82b}$$

where

$$\zeta = [k_0^2(\varepsilon_r - 1) - \beta^2]^{1/2} \tag{9.82c}$$

The first root of (9.82b) gives the value of β for the $TE_{01\delta}$ mode of the resonator.

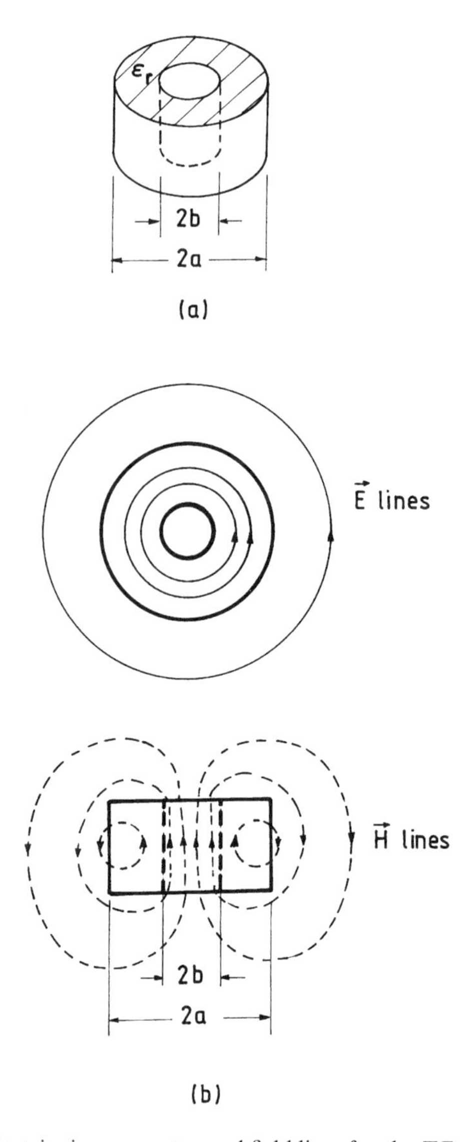

FIGURE 9.11 Dielectric ring resonator and field lines for the $TE_{01\delta}$ mode: (a) geometry of ring resonator and (b) E-field lines in the equatorial plane and H-field lines in the meridian plane.

Determination of Radial Wavenumber u We consider an infinite tubular dielectric guide (Fig. 9.12(c)) having the same cross-section as that of the ring resonator. The relative dielectric constant of the dielectric has an effective value ε_e. Corresponding to the $TE_{01\delta}$ mode of the resonator, the mode of propagation in the tubular guide is TE_{01}. We consider wave propagation in the z-direction

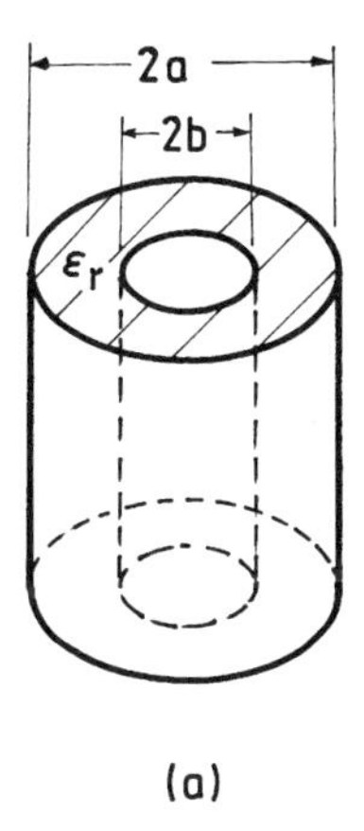

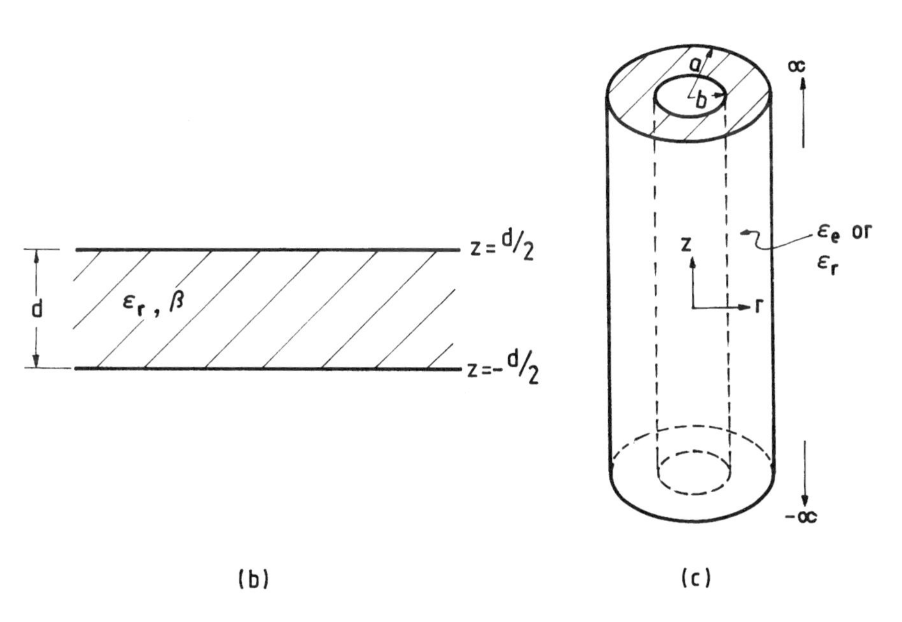

FIGURE 9.12 Ring resonator and models for EDC analysis: (a) ring resonator, (b) horizontal slab guide model for determining β, and (c) vertical tubular guide model for determining the radial wavenumber u.

according to $\exp(-j\beta' z)$. The H_z component in the three regions of the guide can be expressed as

$$H_z = \begin{cases} AJ_0(ur) + BY_0(ur), & \text{region 1} \\ CI_0(\xi r), & \text{region 2} \\ DK_0(\xi r), & \text{region 3} \end{cases} \tag{9.83}$$

where u and ξ satisfy the relations

$$u^2 + (\beta')^2 = k_0^2 \varepsilon_e \tag{9.84a}$$

$$-\xi^2 + (\beta')^2 = k_0^2 \tag{9.84b}$$

The expressions for E_ϕ and H_r in the three regions can be obtained by substituting (9.83) in (9.10). Applying the continuity conditions on H_z and E_ϕ at the cylindrical interfaces $r = b$ and $r = a$, and solving, we obtain the following characteristic equation for the TE_{0n} modes of the tubular guide:

$$\begin{vmatrix} J_0(ub) & Y_0(ub) & -I_0(\xi b) & 0 \\ -J_1(ub)/u & -Y_1(ub)/u & I_1(\xi b)/\xi & 0 \\ J_0(ua) & Y_0(ua) & 0 & -K_0(\xi a) \\ -J_1(ua)/u & -Y_1(ua)/u & 0 & -K_1(\xi a)/\xi \end{vmatrix} = 0 \tag{9.85a}$$

where

$$\xi = [k_0^2(\varepsilon_e - 1) - u^2]^{1/2} \tag{9.85b}$$

The first root of (9.85a) is taken as the value of u for the $TE_{01\delta}$ mode of the ring resonator. The value of ε_e to be substituted in (9.85) is obtained from the relation

$$\varepsilon_e = (u'/k_0)^2 \tag{9.86}$$

where u' is obtained as the first root of (9.85a) with u and ξ replaced by u' and ξ', respectively, and with ξ' given by

$$\xi' = [k_0^2(\varepsilon_r - 1) - (u')^2]^{1/2} \tag{9.87}$$

The free-space resonant wavenumber k_0 is the one at which the values of β and u obtained as above satisfy the separation equation (9.82a).

Radiation Q-Factor The radiation Q-factor (Q_r) for the $TE_{01\delta}$ mode of the ring resonator can be obtained using the same procedure as that described in Section 9.3.2 for the cylindrical resonators. The formula for Q_r is given by (9.28), which is in terms of the stored electric energy (W_e) within and in the vicinity of the resonator and the power radiated (P_r) by the resonator as a magnetic dipole. For evaluation of the power radiated, we need to determine the magnetic dipole moment p_m, which in turn requires the solution for the E_ϕ component in the dielectric region of the resonator. The determination of total stored electric energy requires knowledge of the electric field (E_ϕ) in all regions (see Fig. 9.13). For a high-Q resonator, the bulk of the electric energy gets concentrated in region 1, a small portion in regions 2–4, and an even smaller portion in the remaining regions. We may therefore consider the solution for E_ϕ only in regions 1–4 and

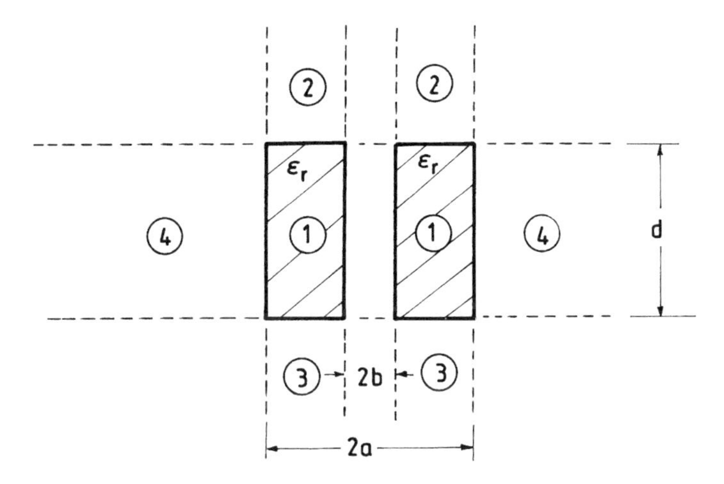

FIGURE 9.13 Cross-section of a cylindrical resonator showing various regions.

neglect the fields in other regions. Referring to Figure 9.13, the solution for the H_z component in regions 1–4 can be expressed as

$$H_z = \begin{cases} A[J_0(ur) + BY_0(ur)]\cos(\beta z), & \text{region } 1 \\ C[J_0(ur) + BY_0(ur)]e^{-\zeta(|z| - d/2)}, & \text{regions } 2,\ 3 \\ DK_0(\xi r)\cos(\beta z), & \text{region } 4 \end{cases} \tag{9.88}$$

Substitution of (9.88) in (9.10) yields the expressions for the E_ϕ component in regions 1–4. By applying the continuity condition on H_z and E_ϕ at the interface $r = a$, $-d/2 < z < d/2$; and on E_ϕ at the interface $z = d/2$, $r < a$; we can eliminate three of the four unknown constants and express the fields in terms of a single unknown. The expressions for E_ϕ are obtained as

$$E_\phi = \begin{cases} -A(j\omega_0\mu_0/u)[J_1(ur) + BY_1(ur)]\cos(\beta z), & \text{region } 1 \\ -A(j\omega_0\mu_0/u)[J_1(ur) + BY_1(ur)]\cos(\beta d/2)e^{-\zeta(|z| - d/2)}, & \text{regions } 2,\ 3 \\ -A(j\omega_0\mu_0/u)[J_1(ua) + BY_1(ua)][K_1(\xi r)/K_1(\xi a)]\cos(\beta z), & \text{region } 4 \end{cases} \tag{9.89}$$

where

$$B = -\left(\frac{uJ_0(ua) + \xi J_1(ua)K_0(\xi a)}{uY_0(ua) + \xi Y_1(ua)K_0(\xi a)}\right) \tag{9.90}$$

The expression for the total energy stored in regions 1–4 can now be derived by substituting for E_ϕ from (9.89) in (9.29) and carrying out the integration over the

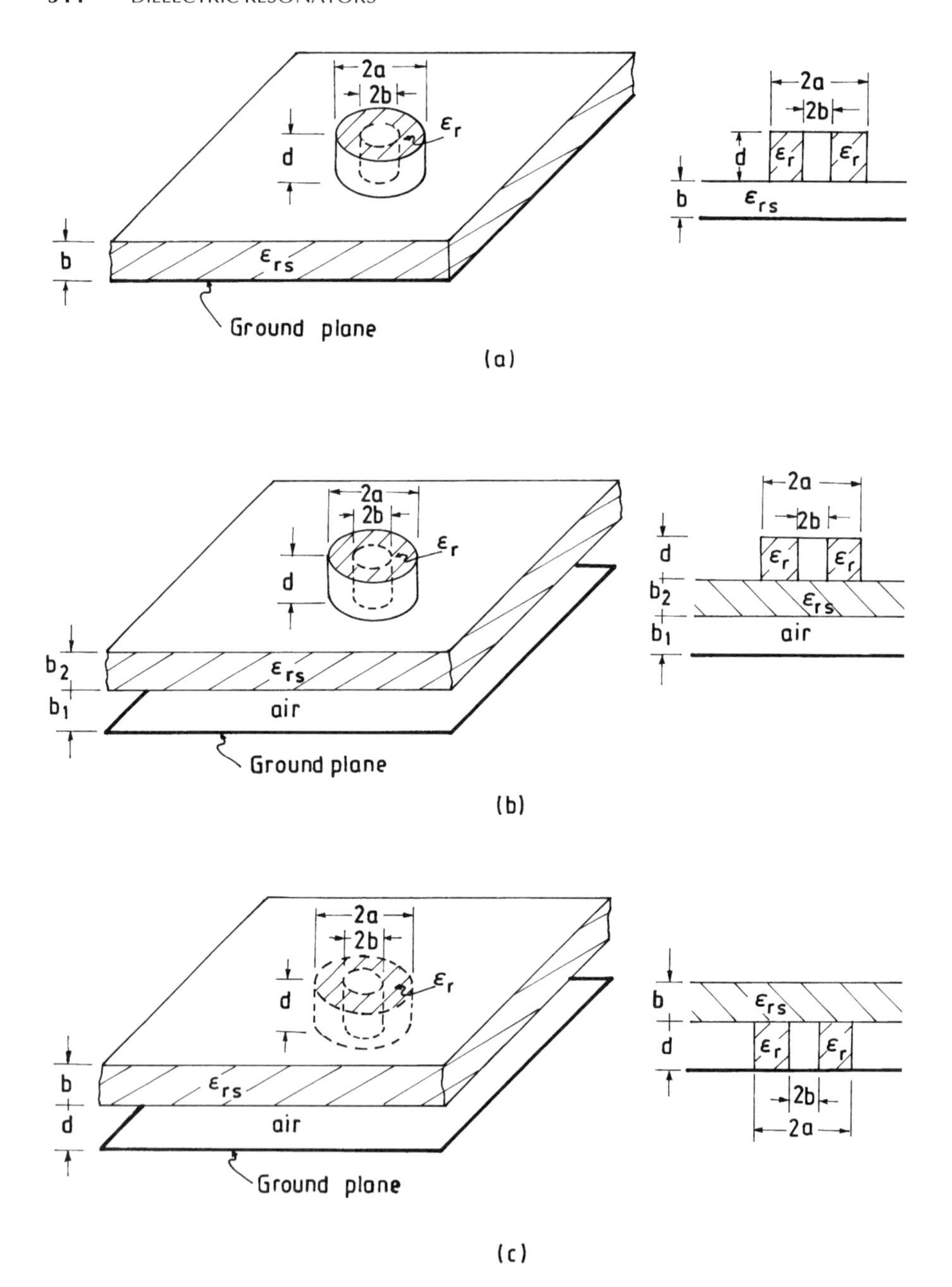

FIGURE 9.14 Ring resonators in a planar slab guide environment: (a) insular image ring resonator, (b) ring resonator on a suspended substrate, and (c) image ring resonator with a top dielectric layer.

volume of the four regions. Using (9.38), the magnetic dipole moment of the ring resonator can be expressed as

$$\mathbf{p}_m = \hat{\mathbf{z}} j\pi\omega\varepsilon_0(\varepsilon_r - 1) \int_b^a \int_{-d/2}^{d/2} r^2 E_{\phi 1}\, dr\, dz \tag{9.91}$$

where $E_{\phi 1}$ refers to the E_ϕ component in region 1 and is given by (9.89a). The expression for the radiated power is obtained by substituting for $|p_m|$ from (9.91) in (9.41).

9.5.2 Ring Resonator in Planar Dielectric Slab Guide Environment

Similar to the composite cylindrical resonator structures shown in Figure 9.11(a)–(c), we consider three ring resonator structures by simply replacing the cylindrical resonator by a ring resonator. The resulting structures are shown in Figure 9.14(a)–(c). The only additional parameter in the present case is the inner radius b of the ring resonator. The analysis of these structures can be carried out using the same technique (EDC method combined with transverse transmission line technique) as followed in Section 9.4.3, for the composite cylindrical resonator structures.

Let u and β denote the radial and axial wavenumbers in the dielectric region of the ring resonator. These parameters then satisfy the separation equation given by (9.82). For high-Q resonator structures, we assume the relative dielectric constant of the slab to be much smaller than that of the resonator ($\varepsilon_{rs} \ll \varepsilon_r$) so that most of the energy is concentrated inside the ring resonator.

For the purpose of determining the radial wavenumber u, we neglect the presence of the planar slab structure and consider an infinite tubular dielectric guide model having the same cross-section as that of the ring resonator. The infinite tubular guide model shown in Figure 9.12(b) applies to all three structures of Figure 9.14. Thus the characteristic equation for determining u is the same for all three ring resonator structures and is identical to that of the isolated ring resonator. For the $TE_{01\delta}$ mode, the characteristic equation is given by (9.85).

For determination of β of three ring resonator structures in Figure 9.14(a)–(c), we consider the horizontal slab guide models obtained by extending the radius of the ring resonator from zero to infinity. Thus the same horizontal slab guide models shown in Figures 9.8–9.10 are applicable to the structures in Figure 9.14(a)–(c), respectively. Considering the $TE_{01\delta}$ mode, the characteristic equations governing β are given by (9.69) for the insular image ring resonator (Fig. 9.14(a)), by (9.77) for the ring resonator on a suspended substrate (Fig. 9.14(b)), and by (9.80) for the image ring resonator with a top dielectric layer (Fig. 9.14(c)).

9.6 RECTANGULAR RESONATORS

9.6.1 EDC Technique of Analysis

The propagating modes in a rectangular dielectric guide extending to infinity in the z-direction can be classified into TE^y_{mn} and TM^y_{mn} modes. A rectangular dielectric resonator may be considered as a section of the dielectric guide truncated at $z = \pm c/2$. Figure 9.15(a) shows the geometry of the resonator. On the basis of the nomenclature used for the dielectric guide, we designate the resonant modes of a rectangular dielectric resonator as $TE^y_{mn\ell}$ and $TM^y_{mn\ell}$ modes. The mode indices m, n, and ℓ refer to the number of extrema of electric and magnetic field components inside the resonator in the x-, y-, and z-directions, respectively. At the resonant frequencies, the fields form standing wave patterns inside the resonator, whereas in the air region they decay exponentially with distance.

In this section, we extend the EDC technique of analysis described in Chapter 2 (for planar dielectric integrated guides) to derive the characteristic equations for rectangular resonators. Let β_x, β_y, and β_z denote the wavenumbers inside the resonator in the x-, y-, and z-directions, respectively. The separation equation that relates these wavenumbers (real) and the free-space resonant wavenumber $k_0 = \omega_0\sqrt{\mu_0\varepsilon_0}$ is given by

$$\beta_x^2 + \beta_y^2 + \beta_z^2 = k_0^2\varepsilon_r \tag{9.92}$$

where ε_r is the relative dielectric constant of the resonator dielectric. In the EDC technique, we derive the characteristic equation in β_y from the horizontal slab guide model obtained by extending the resonator to infinity in the xz-plane (see Fig. 9.15(b)). The relative dielectric constant of the slab guide is assumed to have the same value as that of the resonator. The characteristic equations governing β_x and β_z are derived from the vertical slab guide models, which are obtained by extending the resonator to infinity in the xy-plane and yz-plane, respectively (see Fig. 9.15(c), (d)). The relative dielectric constant of both these vertical slab guides is assumed to be an effective value ε_e, which is obtained from the relation $\varepsilon_e = \varepsilon_r - (\beta_y/k_0)^2$. The resonant frequency is the one at which the characteristic equations derived from the three slab guide models and the separation equation (9.92) are satisfied simultaneously. In the following, we derive the characteristic equations governing β_x, β_y, and β_z for an isolated rectangular dielectric resonator and also for a resonator placed in a planar slab guide environment.

9.6.2 Isolated Rectangular Resonator

Figure 9.15 shows the geometry of a rectangular dielectric resonator along with its three slab guide models and network equivalents. Each of the slab guides can be analyzed in terms of even and odd modes with respect to its plane of symmetry PP'. For the even mode, PP' represents a magnetic wall and for the odd mode it

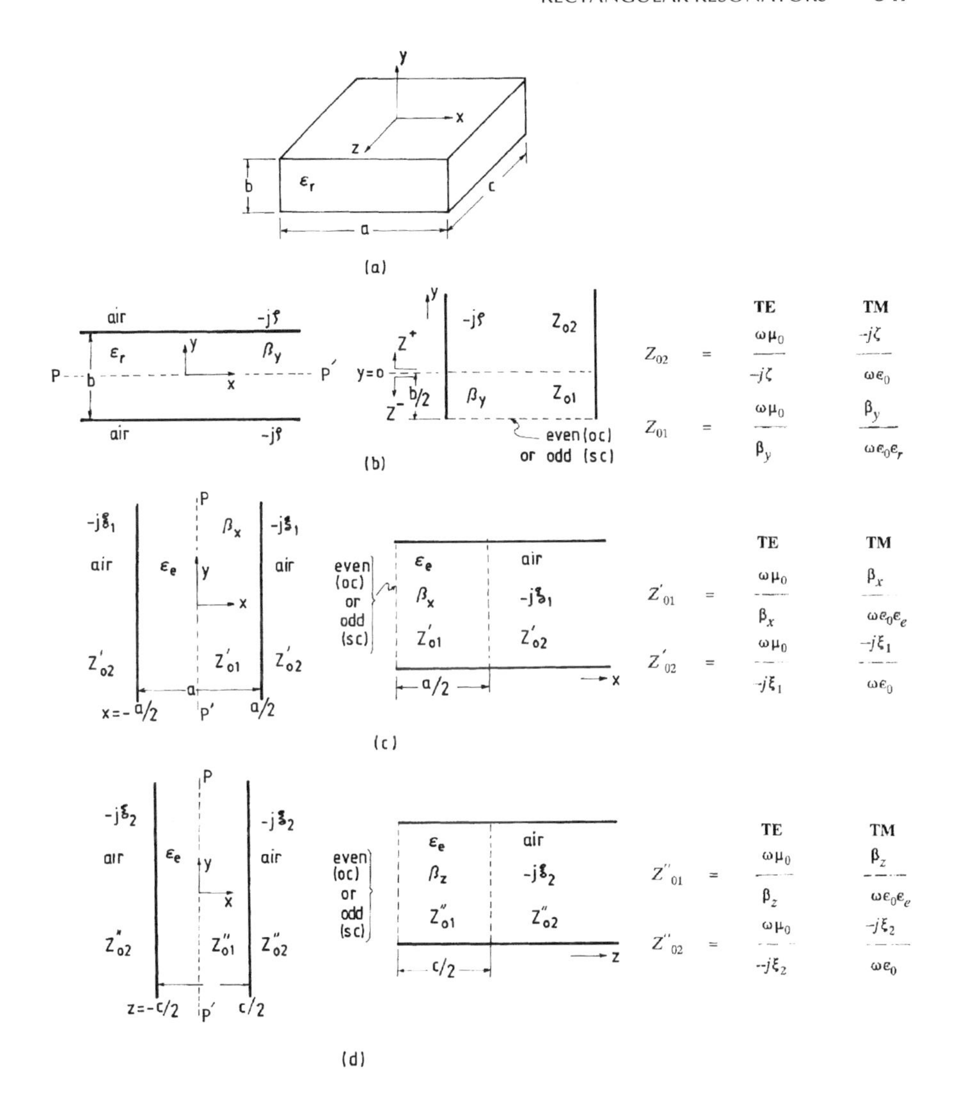

FIGURE 9.15 Rectangular dielectric resonator and models for EDC analysis: (a) geometry of rectangular resonators, (b) horizontal slab guide model and its equivalent network for determining β_y, (c) vertical slab guide model and its equivalent network for determining β_x, and (d) vertical slab guide model and its equivalent network for determining β_z.

represents an electric wall. The equivalent transmission line network for each guide therefore corresponds to one symmetric half of the guide with one end terminated in an open circuit (oc) for the even mode and a short circuit (sc) for odd mode.

***$TM^y_{mn\ell}$* Modes ($H_y = 0$)** Corresponding to the $TM^y_{mn\ell}$ modes of a rectangular resonator, we identify propagation of TM-to-y waves in the horizontal slab guide (Fig. 9.15(b)), TE-to-x waves in the vertical slab guide (Fig. 9.15(c)), and TE-to-z waves in the vertical slab guide (Fig. 9.15(d)).

Characteristic Equation in β_y For deriving the characteristic equation in β_y, we consider the network equivalent of the horizontal slab guide (Fig. 9.15(b)). The wavenumber β_y of the resonator is the y-directed propagation constant of the slab guide and ζ (real) is the decay coefficient outside the slab guide. The characteristic impedances Z_{01} and Z_{02} of the transmission line sections correspond to the TM mode. Applying the transverse resonance condition to the transmission line—namely, $Z^+ + Z^- = 0$ at $y = 0$—the following characteristic equations are easily obtained:

$$(\zeta\varepsilon_r/\beta_y)\tan(\beta_y b/2) + 1 = 0, \quad \text{even mode} \tag{9.93a}$$

$$(\beta_y/\zeta\varepsilon_r)\tan(\beta_y b/2) - 1 = 0, \quad \text{odd mode} \tag{9.93b}$$

where

$$\zeta = [k_0^2(\varepsilon_r - 1) - \beta_y^2]^{1/2} \tag{9.93c}$$

A single characteristic equation that includes both the even- and odd-mode solutions with respect to the xz-plane at $y = 0$ is obtained by multiplying (9.93a) and (9.93b). It is given by

$$2 + (\zeta\varepsilon_r/\beta_y)[1 - (\beta_y/\zeta\varepsilon_r)^2]\tan(\beta_y b) = 0 \tag{9.94}$$

Determination of Effective Dielectric Constant ε_e The effective (relative) dielectric constant of the vertical slab models (Fig. 9.15(c), (d)) is obtained from the relation

$$\varepsilon_e = \varepsilon_r - (\beta_y/k_0)^2 \tag{9.95}$$

Characteristic Equation for β_x Referring to the transmission line network shown in Figure 9.15(c) and using the TE-mode waveguide impedance relation for Z'_{01} and Z'_{02}, we obtain the following characteristic equations:

$$(\beta_x/\xi_1)\tan(\beta_x a/2) - 1 = 0, \quad \text{even mode} \tag{9.96a}$$

$$(\xi_1/\beta_x)\tan(\beta_x a/2) + 1 = 0, \quad \text{odd mode} \tag{9.96b}$$

where

$$\xi_1 = [k_0^2(\varepsilon_e - 1) - \beta_x^2]^{1/2} \tag{9.96c}$$

It may be noted that these even- and odd-mode solutions are with reference to the yz-plane passing through $x = 0$. The combined solution can be given by

$$2 + (\xi_1/\beta_x)[1 - (\beta_x/\xi_1)^2] \tan(\beta_x a) = 0 \tag{9.97}$$

Characteristic Equation in β_z Referring to the transmission line network shown in Figure 9.15(d) and using the TE-mode waveguide impedance relation for Z''_{01} and Z''_{02}, we obtain the characteristic equations for the even and odd modes (with reference to the xy-plane at $z = 0$) as

$$(\beta_z/\xi_2) \tan(\beta_z c/2) - 1 = 0, \quad \text{even mode} \tag{9.98a}$$

$$(\xi_2/\beta_z) \tan(\beta_z c/2) + 1 = 0, \quad \text{odd mode} \tag{9.98b}$$

where

$$\xi_2 = [k_0^2(\varepsilon_e - 1) - \beta_z^2]^{1/2} \tag{9.98c}$$

Multiplying (9.98a) and (9.98b), we obtain a single characteristic equation as

$$2 + (\xi_2/\beta_z)[1 - (\beta_z/\xi_2)^2] \tan(\beta_z c) = 0 \tag{9.99}$$

$TE^y_{mn\ell}$ Modes ($E_y = 0$) Corresponding to the $TE^y_{mn\ell}$ modes of a rectangular resonator, we identify propagation of the TE-to-y waves in the horizontal slab guide (Fig. 9.15(b)), TM-to-x waves in the vertical slab guide (Fig. 9.15(c)), and TM-to-z waves in the vertical slab guide (Fig. 9.15(d)). Accordingly, in the network equivalent of the horizontal slab guide, we use the TE-mode waveguide impedance relation for the characteristic impedances of transmission line sections. Similarly, in the network equivalent of vertical slab guides, we use the TM-mode waveguide impedance relation for the characteristic impedance of transmission line sections. The characteristic equations governing β_x, β_y, and β_z are obtained by using the same procedure as outlined for the $TM^y_{mn\ell}$ modes.

Characteristic Equation in β_y Referring to the network equivalent shown in Figure 9.15(b) and using the TE-mode impedance relations for Z_{01} and Z_{02}, the characteristic equations are obtained as

$$(\beta_y/\zeta) \tan(\beta_y b/2) - 1 = 0, \quad \text{even mode} \tag{9.100a}$$

$$(\zeta/\beta_y) \tan(\beta_y b/2) + 1 = 0, \quad \text{odd mode} \tag{9.100b}$$

where ζ is given by the relation (9.93c).

Combining (9.100a) and (9.100b) gives

$$2 + (\zeta/\beta_y)[1 - (\beta_y/\zeta)^2] \tan(\beta_y b) = 0 \tag{9.101}$$

Determination of β_x Referring to the network equivalent shown in Figure 9.15(c) and using the TM-mode impedance relations for Z'_{01} and Z'_{02}, we obtain the characteristic equations for the even and odd modes as

$$(\beta_x/\xi_1\varepsilon_e)\tan(\beta_x a/2) - 1 = 0, \quad \text{even mode} \tag{9.102a}$$

$$(\xi_1\varepsilon_e/\beta_x)\tan(\beta_x a/2) + 1 = 0, \quad \text{odd mode} \tag{9.102b}$$

where ξ_1 and ε_e are given by (9.96c) and (9.95), respectively. Combining the even- and odd-mode equations, we can write

$$2 + (\xi_1\varepsilon_e/\beta_x)[1 - (\beta_x/\xi_1\varepsilon_e)^2]\tan(\beta_x a) = 0 \tag{9.103}$$

Characteristic Equation in β_z Referring to the network equivalent shown in Figure 9.15(d) and using the TM-mode impedance relation for Z''_{01} and Z''_{02}, we obtain the following characteristic equations for even and odd modes:

$$(\xi_2\varepsilon_e/\beta_z)\tan(\beta_z c/2) + 1 = 0, \quad \text{even mode} \tag{9.104a}$$

$$(\beta_z/\xi_2\varepsilon_e)\tan(\beta_z c/2) - 1 = 0, \quad \text{odd mode} \tag{9.104b}$$

where ξ_2 and ε_e are given by (9.98c) and (9.95), respectively. Combining the even- and odd-mode relations (9.104a) and (9.104b), we can write

$$2 + (\xi_2\varepsilon_e/\beta_z)[1 - (\beta_z/\xi_2\varepsilon_e)^2]\tan(\beta_z c) = 0 \tag{9.105}$$

9.6.3 Insular Image Rectangular Resonator

The EDC technique of analysis described in Section 9.6.2 can easily be applied to composite structures involving rectangular resonators in a planar slab guide environment. As an example, we consider below the analysis of an insular image rectangular resonator. The structure is shown in Figure 9.16(a). The relative dielectric constant of the slab is assumed to be much smaller than that of the resonator ($\varepsilon_{rs} \ll \varepsilon_r$) so that the bulk of the energy is confined to the dielectric region of the resonator. As in the case of the isolated resonator, we classify the resonant modes of this structure into $TM^y_{mn\ell}$ and $TE^y_{mn\ell}$ modes. If β_x, β_y, and β_z denote the wavenumbers inside the resonator with respect to the x, y, and z directions, respectively, then the separation equation (9.92) applies to both sets of resonant modes. The expression is

$$\beta_x^2 + \beta_y^2 + \beta_z^2 = k_0^2\varepsilon_r \tag{9.106}$$

where k_0 is the free-space resonant wavenumber.

Figure 9.16(b) shows the cross-sectional view of the insular image rectangular resonator in the xy-plane at $z = 0$. The structure is divided into two regions: the central vertical region, which includes the rectangular resonator, is marked

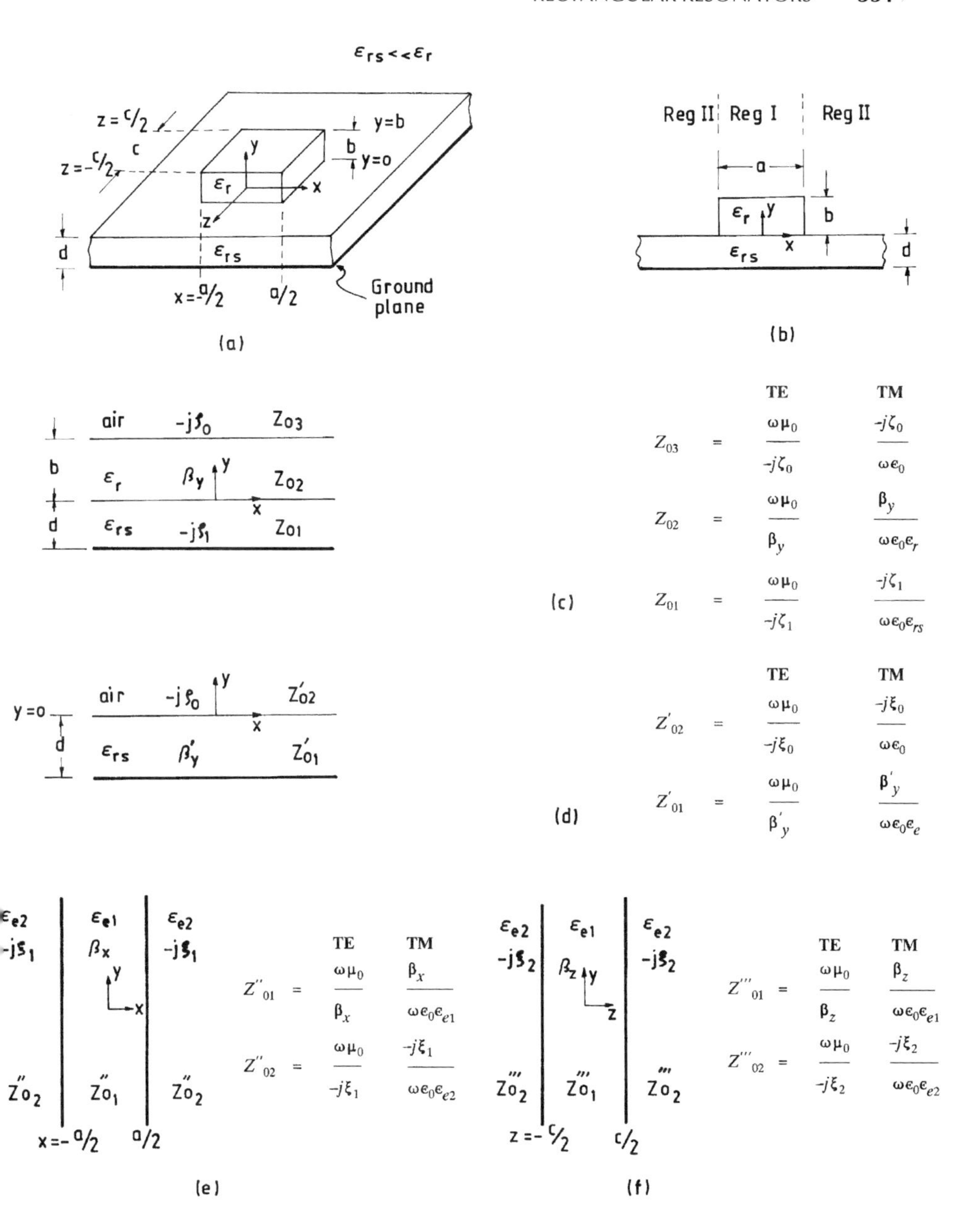

FIGURE 9.16 Insular image rectangular resonators and models for EDC analysis: (a) Cross-sectional view, (b) geometry of insular image resonator, (c) horizontal slab guide model for region I, (d) horizontal slab guide model for region II, (e) vertical slab guide model for determining β_x, and (f) vertical slab guide model for determining β_z.

region I and the remaining portion as region II. The horizontal slab guide model representing region I is shown in Figure 9.16(c). This model is used for deriving the characteristic equation in β_y as well as the effective dielectric constant ε_{e1} corresponding to region I. The horizontal slab guide model representing region II is shown in Figure 9.16(d). This is used for evaluating the effective dielectric constant ε_{e2} of region II. With the effective dielectric constants ε_{e1} and ε_{e2} defined as above, we then use the vertical slab guide as shown in Figure 9.16(e) to derive the characteristic equation in β_x. It is worthwhile to note that the characteristic equations in β_y and β_x would be identical to those of an insular image guide having the same cross-sectional parameters. The only additional analysis involved in the present structure is that of deriving the characteristic equation in β_z by considering another vertical slab guide of infinite extent in the xy-plane and having a thickness equal to c. The characteristic equation in β_z is exactly the same as that for β_x except for a change of symbols.

The above procedure is applied to both $TM^y_{mn\ell}$ and $TE^y_{mn\ell}$ modes. The characteristic equations can be derived by using a transmission line network equivalent for each slab guide and applying the resonance condition. The propagation constants and the characteristic impedances to be used for each model are given in Figure 9.16. While considering the $TM^y_{mn\ell}$ modes of the resonator, we use TM-mode impedances for the horizontal slab guides (Fig. 9.16(c), (d)) and TE-mode impedances for the vertical slab guides (Fig. 9.16(e), (f)). Similarly, while considering the $TE^y_{mn\ell}$ modes of the resonator, we use TE-mode impedances for the horizontal slab guides and TM-mode impedances for the vertical slab guides. The characteristic equations derived using the above procedure are listed below.

$TM^y_{mn\ell}$ Modes ($H_y = 0$)

Characteristic Equation in β_y

$$[1 + (\varepsilon_r/\varepsilon_{rs})(\zeta_1/\beta_y)\tan(\beta_y b)\tanh(\zeta_1 d)]$$

$$- [1/(\zeta_0\varepsilon_r)][\beta_y \tan(\beta_y b) - (\varepsilon_r\zeta_1/\varepsilon_{rs})\tanh(\zeta_1 d)] = 0 \qquad (9.107a)$$

where

$$\zeta_1 = [k_0^2(\varepsilon_r - \varepsilon_{rs}) - \beta_y^2]^{1/2} \qquad (9.107b)$$

$$\zeta_0 = [k_0^2(\varepsilon_r - 1) - \beta_y^2]^{1/2} \qquad (9.107c)$$

Effective Dielectric Constant ε_{e1}

$$\varepsilon_{e1} = \varepsilon_r - (\beta_y/k_0)^2 \qquad (9.108)$$

Effective Dielectric Constant ε_{e2} The effective dielectric constant ε_{e2} is obtained from the propagation constant β'_y of the horizontal slab guide shown in

Figure 9.16(c). The characteristic equation for β'_y is given by

$$(\beta'_y/\zeta_0\varepsilon_{rs})\tan(\beta'_y d) - 1 = 0 \qquad (9.109a)$$

where

$$\zeta_0 = [k_0^2(\varepsilon_{rs} - 1) - (\beta'_y)^2]^{1/2} \qquad (9.109b)$$

The expression for ε_{e2} is

$$\varepsilon_{e2} = \varepsilon_{rs} - (\beta'_y/k_0)^2 \qquad (9.110)$$

Characteristic Equation in β_x

$$2 + (\xi_1/\beta_x)[1 - (\beta_x/\xi_1)^2]\tan(\beta_x a) = 0 \qquad (9.111a)$$

where

$$\xi_1 = [k_0^2(\varepsilon_{e1} - \varepsilon_{e2}) - \beta_x^2]^{1/2} \qquad (9.111b)$$

Characteristic Equation in β_z

$$2 + (\xi_2/\beta_z)[1 - (\beta_z/\xi_2)^2]\tan(\beta_z c) = 0 \qquad (9.112a)$$

where

$$\xi_2 = [k_0^2(\varepsilon_{e1} - \varepsilon_{e2}) - \beta_z^2]^{1/2} \qquad (9.112b)$$

It may be noted that the first root of (9.107a) gives β_y for $n = 1$, the first root of (9.111a) gives β_x for $m = 1$, and the first root of (9.112a) gives β_z for $\ell = 1$.

$TM^y_{mn\ell}$ Modes ($E_y = 0$)

Characteristic Equation in β_y

$$[1 - (\beta_y/\zeta_1)\tan(\beta_y b)\tanh(\zeta_1 d)] + \zeta_0\{[\tan(\beta_y b)/\beta_y] + [\tanh(\zeta_1 d)/\zeta_1]\} = 0 \qquad (9.113)$$

where ζ_1 and ζ_0 are given by (9.107b) and (9.107c), respectively.

Effective Dielectric Constants ε_{e1} and ε_{e2} The formula for ε_{e1} is given by (9.108). For determining ε_{e2}, we first solve the following characteristic equation in β'_y:

$$(\zeta_0/\beta'_y)\tan(\beta'_y d) + 1 = 0 \qquad (9.114)$$

where ζ_0 is given by (9.109b). The value of ε_{e2} is then obtained using (9.110).

Characteristic Equation in β_x

$$2 + (\varepsilon_{e1}/\varepsilon_{e2})(\xi_1/\beta_x)\{1 - [(\varepsilon_{e2}/\varepsilon_{e1})(\beta_x/\xi_1)]^2\} \tan(\beta_x a) = 0 \quad (9.115)$$

where ξ_1 is given by (9.111b).

Characteristic Equation in β_z

$$2 + (\varepsilon_{e1}/\varepsilon_{e2})(\xi_2/\beta_z)\{1 - [(\varepsilon_{e2}/\varepsilon_{e1})(\beta_z/\xi_2)]^2\} \tan(\beta_z c) = 0 \quad (9.116)$$

9.7 OTHER RESONATORS

9.7.1 Dielectric Ring-Gap Resonator

A dielectric ring resonator with a narrow vertical slit called the dielectric ring-gap resonator has been investigated by Hui and Wolff [45]. Figure 9.17 shows the resonator and its dominant mode field pattern. The field configuration closely resembles the TE_{011} mode of a ring resonator and hence is called quasi-TE_{011} mode (or quasi-$TE_{01\delta}$ mode when the height of the resonator is very small). The electric field intensity within the gap is much stronger than in the dielectric and the magnetic field is strong around the ring opposite the gap. Furthermore, the resonant frequencies of the quasi-TE_{0mn} modes are higher than those of the TE_{0mn} modes of the corresponding ring resonator.

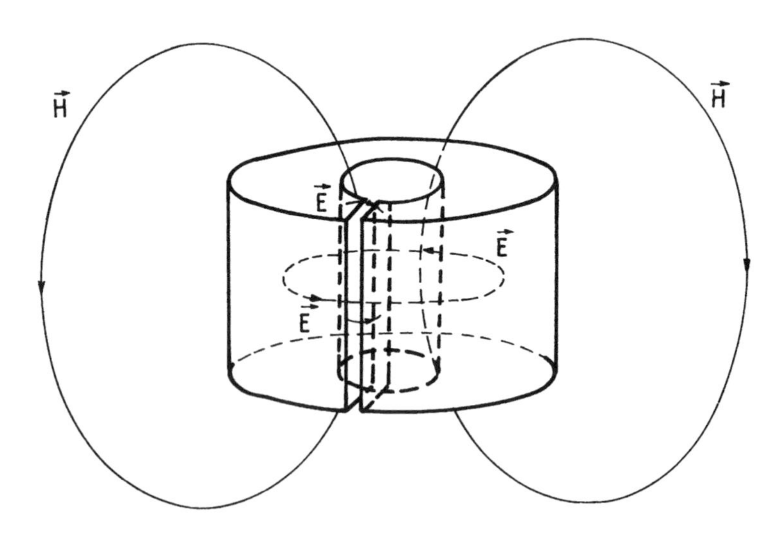

FIGURE 9.17 Dielectric ring-gap resonator and its dominant mode field pattern. (From Hui and Wolff [45]. Copyright © 1991 IEEE, reprinted with permission.)

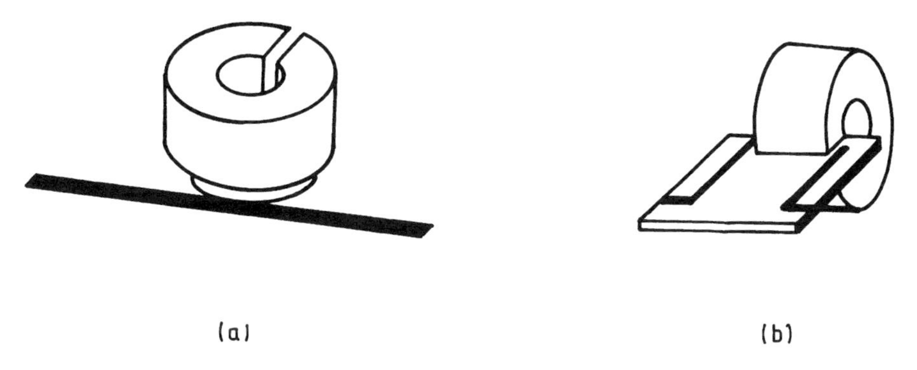

FIGURE 9.18 Excitation of dielectric ring-gap resonator by microstrip through (a) coupling through electric fringe field in the gap and (b) magnetic coupling. (After Hui and Wolff [45].)

The dielectric ring-gap resonator has all the advantages of the dielectric ring resonator. Besides, the ring-gap resonator acquires an important additional feature that it can be excited in the quasi-TE_{011} mode over a wide frequency range with effective suppression of higher order modes. Figure 9.18 shows two techniques of exciting a ring-gap resonator by means of a microstrip. In Figure 9.18(a), the resonator is coupled directly to the microstrip by the electric fringing field and in Figure 9.18(b), the excitation is through magnetic coupling. In the case of magnetic coupling, however, the slit must be suitably oriented with respect to the microstrip for effective suppression of higher order modes. For the orientation considered in Figure 9.18(b), Hui and Wolff [45] have reported that the resonator can be excited in the dominant mode over a frequency range from 45 MHz to 40 GHz without interference from higher order modes.

9.7.2 Optically Controlled Dielectric Resonator

The principle of optical control and the control of propagation characteristics of semiconductor guides with optical illumination are discussed in Chapter 5. The concept of the optically controlled dielectric resonator (OCDR) is illustrated in Figure 9.19 [46]. A thin slab of intrinsic semiconductor is placed on top of a dielectric resonator. When the slab is illuminated from above by light of wavelength shorter than the semiconductor bandgap, a sheet of plasma is formed on its surface. The presence of plasma modifies the dielectric and conductive properties of the slab, which results in a shift of the dielectric resonator center frequency. The depth of the optically induced plasma layer is a function of the semiconductor's optical absorption and also the wavelength of illumination.

Analysis of the OCDR has recently been reported using the modified spectral domain technique by Rong et al. [47], the frequency-dependent finite-difference

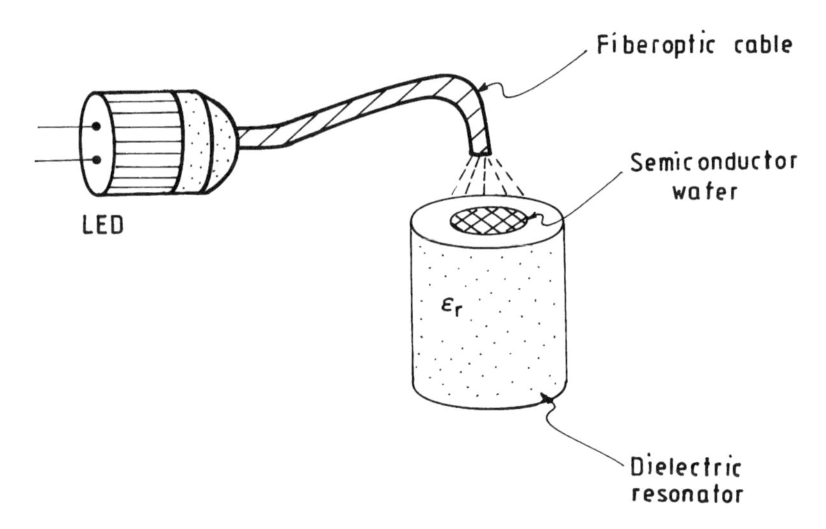

FIGURE 9.19 Concept of optically controlled dielectric resonator. (After Herczfeld et al. [46].)

time domain formulation by Shen et al. [48], and the mode-matching technique by Alphones et al. [49]. Figure 9.20 illustrates typical variations in the $TE_{10\delta}$ resonant frequency of an isolated OCDR as a function of plasma density for various thicknesses of the plasma layer [48]. It is observed that no significant shift in resonance frequency occurs until a critical quantity of photoinduced plasma is created. Beyond this critical plasma density, the resonant frequency shifts rapidly and finally it saturates. Figure 9.21 shows the $TE_{01\delta}$ mode resonance frequencies and intrinsic Q-factors of optically controlled dielectric resonators placed on a dielectric substrate backed by a ground plane [47]. It is observed that for weak and strong light injection, the resonance frequency hardly changes. There is a light-sensitive region, in which the resonance frequencies are strongly affected by the optical illumination. In this region, the intrinsic Q-factor shows a valley.

The optical control technique applies also to resonators made entirely of semiconductors for varying the resonant frequency. Alphones and Tsutsumi [50] have investigated the resonant frequency characteristics of a silicon ridge guide resonator by means of optical control. Figure 9.22 shows the schematic of a resonator illuminated by a xenon arc lamp. The ridge guide resonator consists of a rectangular resonator as a rib of size 2 mm × 0.4 mm × 9 mm on a grounded substrate of size 39 mm × 0.4 mm × 9 mm. As shown in the figure, the optical radiation passing through a slit is focused on to the top surface of the resonator by means of a lens. The illumination intensity on the resonator is varied by adjusting the slit width and the resonator frequency is obtained from the resonance curve.

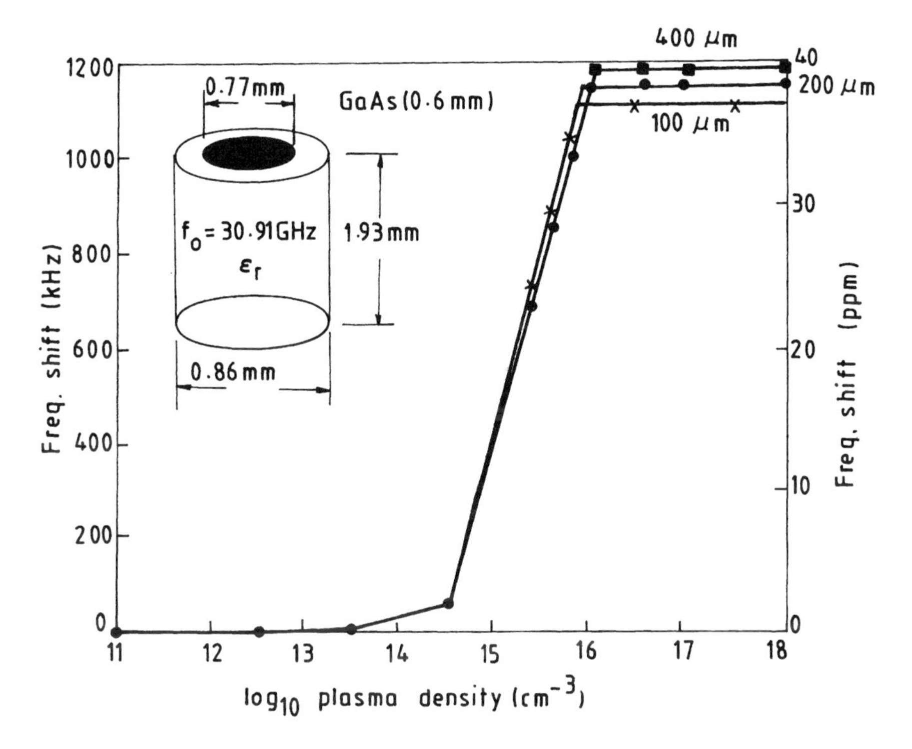

FIGURE 9.20 Variation of the $TE_{01\delta}$ mode resonant frequency shift for a K_a-band resonator ($\varepsilon_r = 29.1$) as a function of plasma density for various plasma depths. (From Shen et al. [48]. Copyright © 1993 IEEE, reprinted with permission.)

Figure 9.23 shows a typical measured response of resonant frequency f_0 and Q-factor as a function of the optical power [50].

9.7.3 Hybrid Dielectric/HTS Resonator

The Q-factor of a dielectric resonator is limited by the losses in the dielectric material of the resonator and the ohmic losses in the metal walls, which form part of the resonator structure. Curtis et al. [51] have proposed the concept of embedding dielectric resonators in a superconducting material for improving the Q-factor. The technique is illustrated in Figure 9.24, which shows a three-pole hybrid dielectric post filter using a high-temperature superconducting (HTS) material. The HTS material is deposited as a thin film on a flat substrate. The three disk-type resonators are thus surrounded by the HTS material. The resonators are optimized so that their TE_{011} mode fields have maximum intensity at the end walls, thereby maximizing the effect of HTS conductivity in lowering the loss. Using ceramics of very low loss tangent and high-quality HTS films, Q-factors of the order of 50,000 at 77 K are predicted [51].

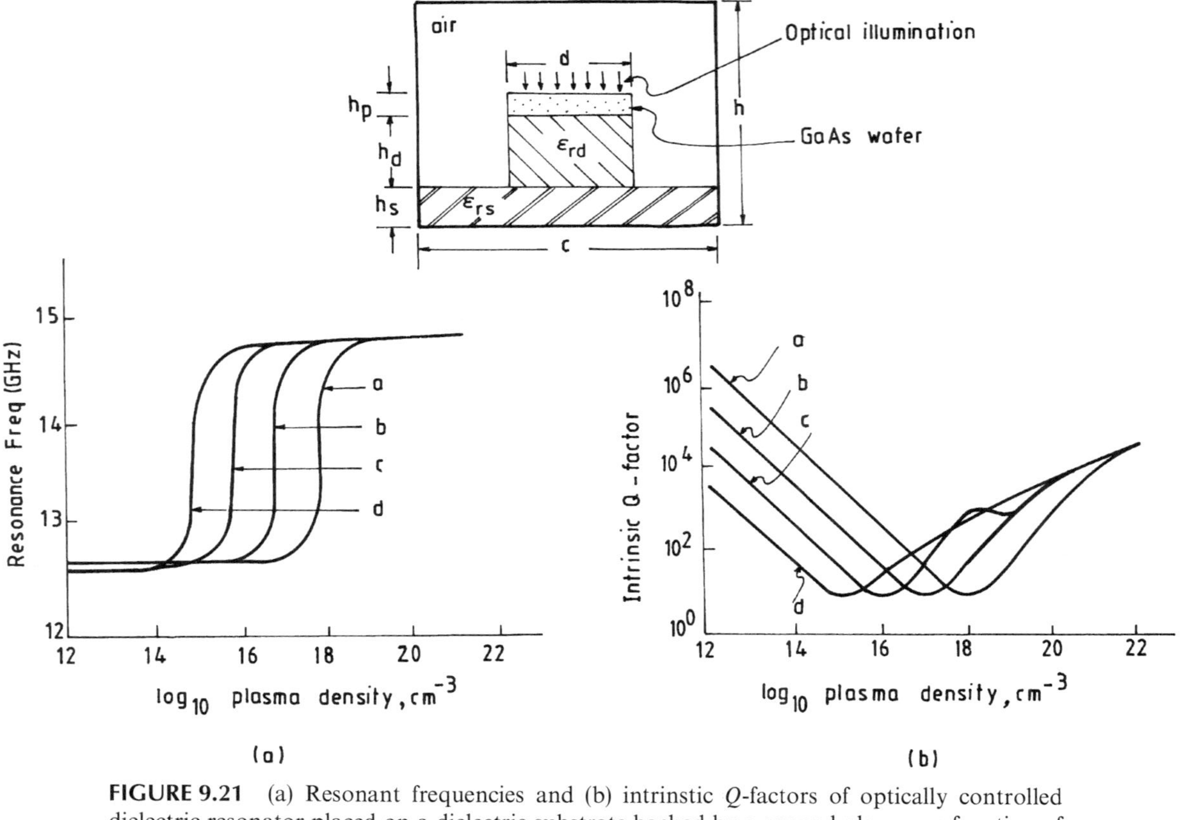

FIGURE 9.21 (a) Resonant frequencies and (b) intrinstic Q-factors of optically controlled dielectric resonator placed on a dielectric substrate backed by a ground plane as a function of optically induced plasma density: $d = 4.86$ mm, $h = 8.0$ mm, $h_d = 1.81$ mm, $h_s = 0.635$ mm, $\varepsilon_{rd} = 35.2$, $\varepsilon_{rs} = 10$; curves a: $h_p = 0.1$ µm, b: $h_p = 1$ µm, c: $h_p = 10$ µm, d: $h_p = 100$ µm. (From Rong et al. [47], reprinted with permission of IEE.)

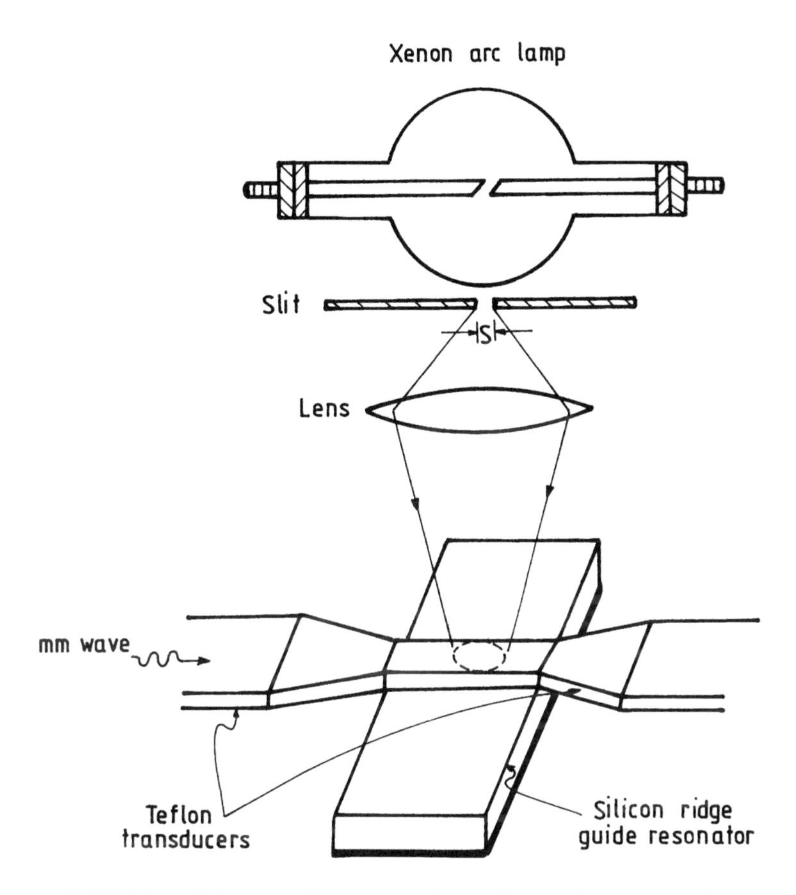

FIGURE 9.22 Schematic diagram of silicon ridge guide resonator illuminated by xenon arc lamp. (From Alphones and Tsutsumi [50], reprinted with permission of Communication Engineers, Denshi Joho Tsushin Gakkai, Tokyo, Japan.)

PROBLEMS

9.1 Compare a cylindrical metal cavity resonator and a cylindrical dielectric resonator for the following:

(a) Basic resonance phenomena.

(b) Classification of modes and the meaning of their modal indices.

(c) Dominant mode designations and their meaning.

(d) Field configurations of the dominant TE and TM modes (draw field lines and discuss).

9.2 Define the Q-factor for a cylindrical metal cavity resonator and also for a cylindrical dielectric resonator. On what factors does the overall Q depend in the two cases? Explain.

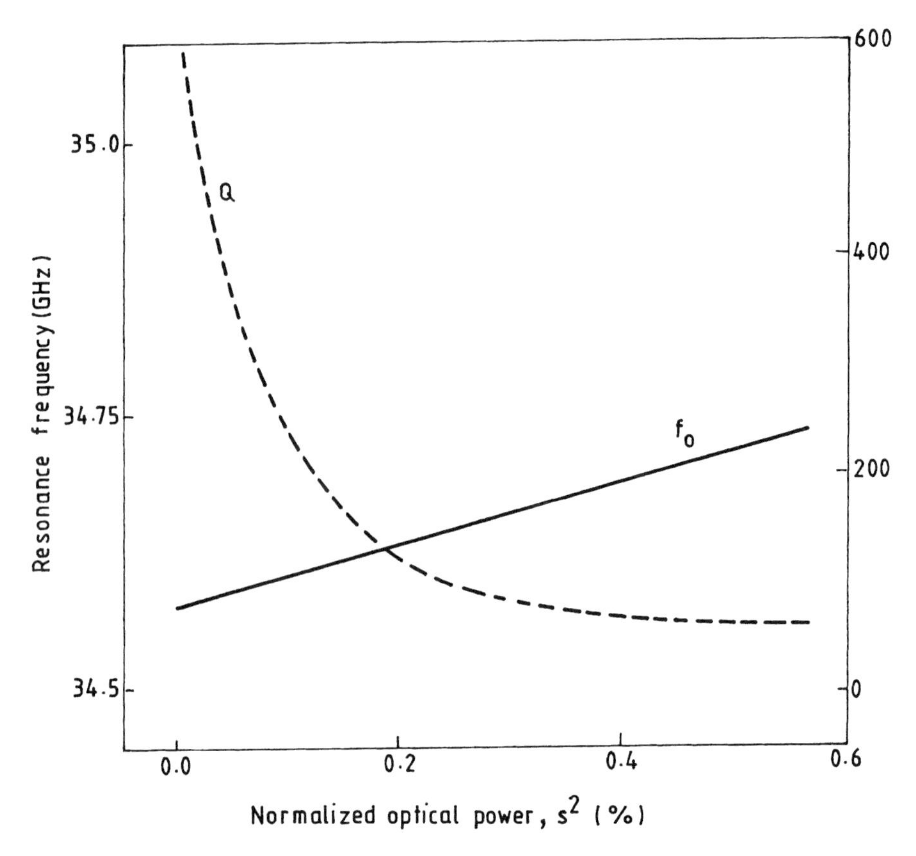

FIGURE 9.23 Variation of resonant frequency and quality factor of ridge guide resonator as a function of optical power. (From Alphones and Tsutsumi [50], reprinted with permission of Communication Engineers, Denshi Joho Tsushin Gakkai, Tokyo, Japan.)

9.3 Calculate the resonance frequency of the $TE_{01\delta}$ mode of a cylindrical dielectric resonator having $\varepsilon_r = 38$, radius $a = 2.5$ mm, and height $d = 3$ mm. (Use approximate closed form expressions.) For the same resonance frequency, determine the size of a cylindrical metal cavity resonator for its dominant mode. (This cavity is to be formed from a circular metal waveguide operating in its dominant mode.)

9.4 The resonant frequency f_0 of the $TE_{01\delta}$ mode of a cylindrical dielectric resonator having $\varepsilon_r = 30$ and radius $a = 0.8$ mm is 30 GHz. Determine the height d of the resonator and its radiation Q-factor Q_r. What would be the value of d and Q_r if only ε_r is increased to 50 (with f_0 and a held fixed)? How do you explain the increase or decrease in Q_r observed?

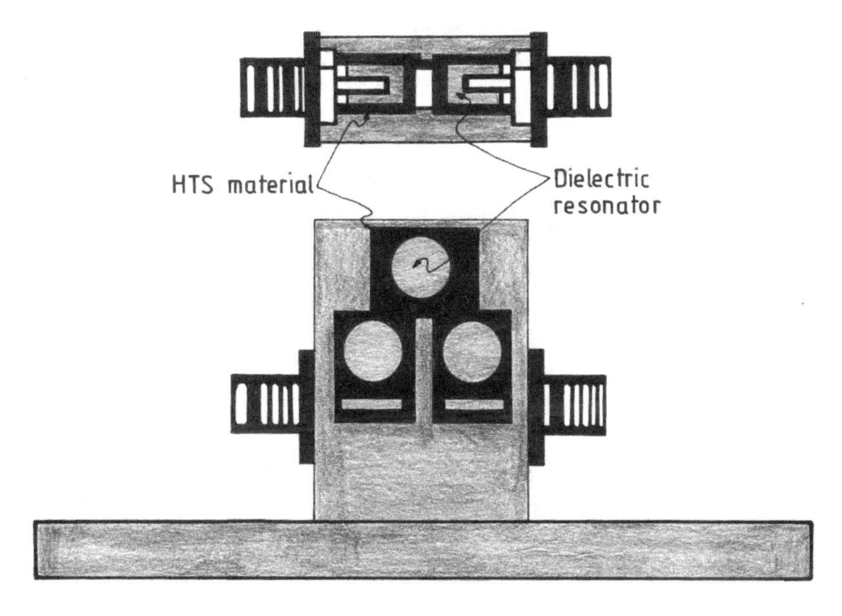

FIGURE 9.24 Single-mode post dielectric resonator HTS filter. (From Curtis et al. [51]. Copyright © 1991 IEEE, reprinted with permission.)

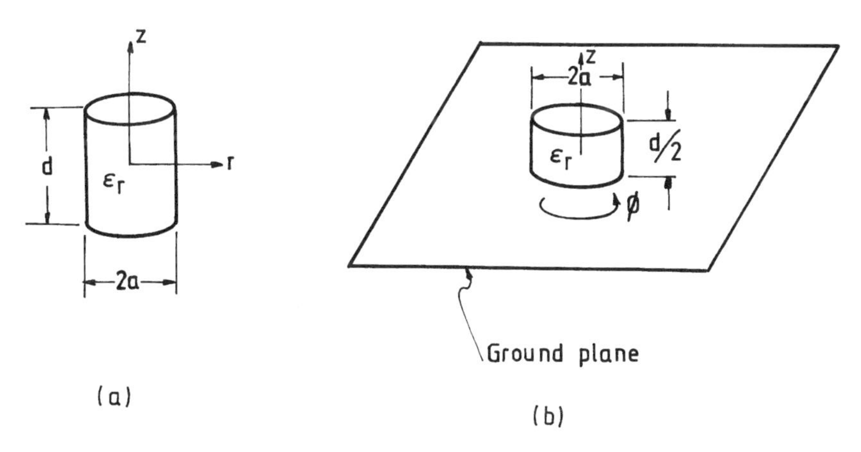

FIGURE 9.25 (a) Cyclindrical dielectric resonator. (b) Cylindrical image resonator formed by placing a ground plane at the $z = 0$ plane of the cylindrical dielectric resonator shown in (a).

9.5 Consider a cylindrical image resonator formed by placing a ground plane at the $z = 0$ plane of a cylindrical resonator (Fig. 9.25(a)) as shown in Figure 9.25(b). For the $TM_{01\delta}$ mode of the cylindrical resonator:

(a) Write expressions for electric and magnetic fields in the dielectric region of the image resonator.

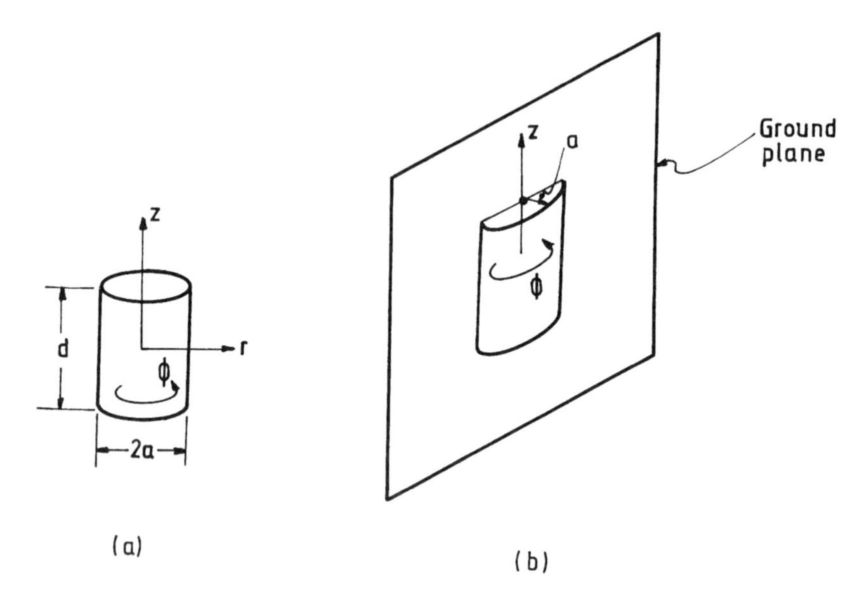

FIGURE 9.26 (a) Cylindrical dielectric resonator. (b) Cylindrical image resonator formed by placing a ground plane passing through the axis of the cylindrical dielectric resonator shown in (a).

(b) Draw current flow lines on the image plane.

(c) Derive a closed form expression for the power dissipated in the ground plane area, which is in contact with the dielectric.

(d) How will the Q-factor get modified with the introduction of the image plane?

9.6 Consider a cylindrical image resonator formed by placing a ground plane passing through the axis of a cylindrical resonator (Fig. 9.26(a)) as shown in Figure 9.26(b). For the $TE_{01\delta}$ mode of the cylindrical resonator:

(a) Write expressions for the electric and magnetic fields in the dielectric region of the image resonator.

(b) Draw current flow lines on the image plane.

(c) Derive a closed form expression for the power dissipated in the ground plane area in contact with the dielectric.

(d) How will the Q-factor get modified with the introduction of the image plane?

9.7 How can $TE_{01\delta}$ and $TM_{01\delta}$ modes be excited in a cylindrical image resonator shown in Figure 9.27. Draw diagrams to show the excitation schemes using coaxial probes.

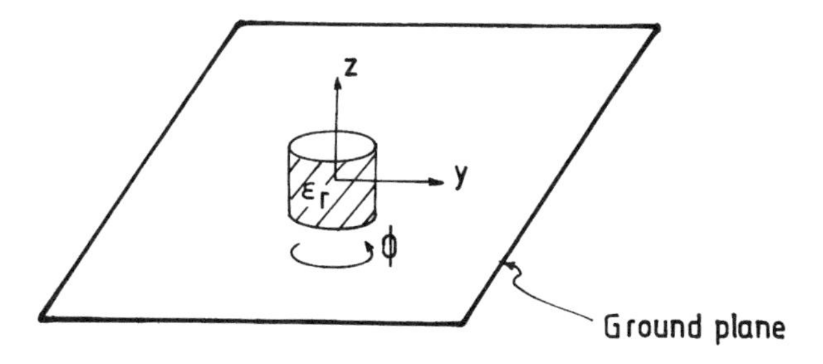

FIGURE 9.27 Cylindrical image resonator.

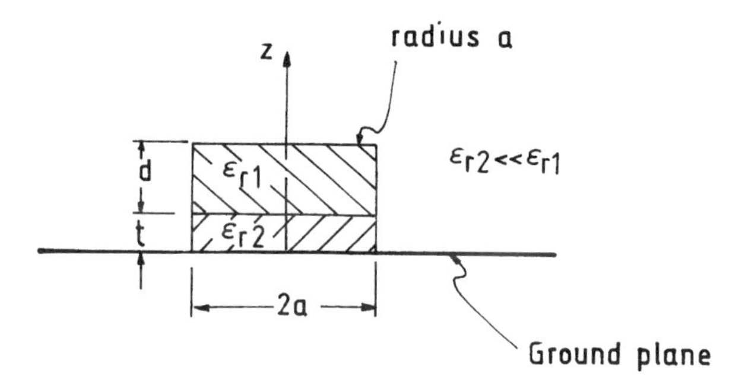

FIGURE 9.28 Cylindrical image resonator with a thin insular disk.

9.8 Consider a cylindrical image resonator with a thin insular disk of low dielectric constant as shown in Figure 9.28.

(a) Derive the characteristic equation for determining the resonant frequency of this resonator.

(b) Explain qualitatively the effect of the thin insular disk on the resonant frequency and Q-factor of the resonator. Assume $t \ll d$ and $\varepsilon_{r2} \ll \varepsilon_{r1}$.

9.9 Draw E- and H-field lines for the $HE_{11\delta}$ mode of a (isolated) cylindrical dielectric resonator.

9.10 **(a)** For a (isolated) cylindrical dielectric resonator having $\varepsilon_r = 38$, radius $a = 5.25$ mm, and height $d = 4.6$ mm, calculate the resonant frequencies for the $TE_{10\delta}$ mode (use closed form expression given in the text).

(b) For the above parameters Table 9.2 gives resonant frequencies of higher order modes. Determine the operating bandwidth of the $TE_{01\delta}$ mode.

9.11 Consider a cylindrical resonator and a ring resonator shown in Figure 9.29 having the same ε_r and aspect ratio $d/2a$ and operating in the $TE_{01\delta}$ mode.

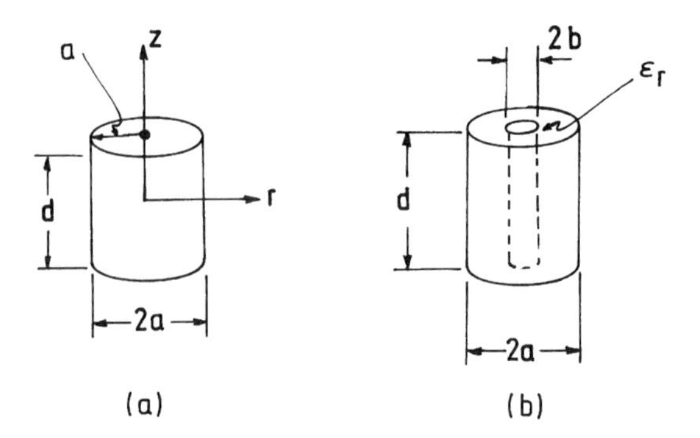

FIGURE 9.29 (a) Cylindrical dielectric resonator and (b) cylindrical ring resonator.

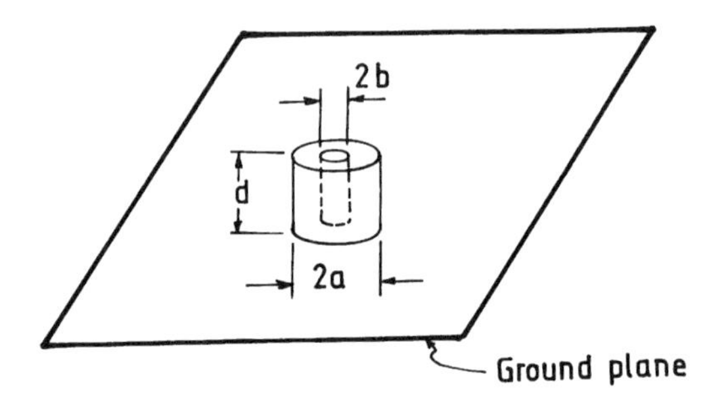

FIGURE 9.30 Cylindrical ring image resonator.

(a) Draw and compare (i) electric field intensity and (ii) magnetic field intensity as a function of radius in the two resonators.

(b) What are the advantages of the ring resonator over the cylindrical resonator.

9.12 For a ring image resonator shown in Figure 9.30, derive characteristic equations for (a) $TE_{01\delta}$, (b) $TM_{01\delta}$, and (c) $HE_{11\delta}$ modes. Use the EDC technique.

9.13 Which is the dominant mode in a rectangular dielectric resonator? Draw its *E*- and *H*-lines. Draw the dominant mode fields for a rectangular metal cavity resonator and compare with those of the dielectric resonator.

9.14 Compare the essential features of the cylindrical ring and rectangular dielectric resonators.

REFERENCES

1. H. Y. Yee, Natural resonant frequencies of microwave dielectric resonators. *IEEE Trans. Microwave Theory Tech.*, **MTT-13**, 256, Mar. 1965.
2. S. B. Cohn, Microwave band pass filters containing high Q dielectric resonators. *IEEE Trans. Microwave Theory Tech.*, **MTT-16**, 218–227, Apr. 1968.
3. K. K. Chow, On the solution and field pattern of cylindrical dielectric resonators. *IEEE Trans. Microwave Theory Tech.*, **MTT-14**, 439, 1966.
4. T. Itoh and R. S. Rudokas, New method for computing the resonant frequencies of dielectric resonators. *IEEE Trans. Microwave Theory Tech.*, **MTT-25**, 52–54, Jan. 1977.
5. P. Guillon and Y. Garault, Accurate resonant frequencies of dielectric resonators. *IEEE Trans. Microwave Theory Tech.*, **MTT-25**, 916–922, Nov. 1977.
6. Y. Konishi, N. Hoshino, and Y. Utsumi, Resonant frequency of a $TE_{01\delta}$ dielectric resonator. *IEEE Trans. Microwave Theory Tech.*, **MTT-24**, 112–114, Feb. 1976.
7. R. K. Mongia and B. Bhat, Accurate resonant frequencies of cylindrical dielectric resonators using a simple analytical technique. *Electron. Lett.*, **21**, 479–480, May 1985.
8. R. K. Mongia and B. Bhat, Effective dielectric constant technique to analyze cylindrical dielectric resonators. *Arch. Elek. Ubertragung.*, **AEU-38**, 161–168, May–June 1987.
9. Y. Kobayashi, N. Fukuoka, and S. Yoshida, Resonant modes for a shielded dielectric rod resonator. *Electron. Commun. Japan*, **64-B(11)**, 46–51, 1981.
10. Y. Komatsu and Y. Murakami, Coupling coefficient between microstrip line and dielectric resonator. *IEEE Trans. Microwave Theory Tech.*, **MTT-31**, 34–40, Jan. 1983.
11. D. Maystre, P. Vincent, and J. C. Mage, Theoretical and experimental study of the resonant frequency of a cylindrical dielectric resonator. *IEEE Trans. Microwave Theory Tech.*, **MTT-31**, 846–848, Oct. 1983.
12. U. S. Hong and R. H. Jansen, Numerical analysis of shielded dielectric resonators including substrate support, support disc and tuning post. *Electron. Lett.* **18**, 1000–1002, Nov. 1982.
13. K. A. Zaki and A. E. Atia, Modes in dielectric-loaded waveguides and resonators. *IEEE Trans. Microwave Theory Tech.*, **MTT-31**, 1039–1045, Dec. 1983.
14. P. Vincent, Computation of the resonant frequency of a dielectric resonator by a differential method. *Appl. Phys.*, **A-31**, 51–54, 1983.
15. P. S. Kooi, M. S. Leong, and A. L. S. Prakash, Finite-element analysis of the shielded cylindrical dielectric resonator. *Proc. IEE Pt. H*, **132**, 7–16, Feb. 1985.
16. J. Delaballe, P. Guillon, and Y. Garault, Local complex permittivity measurement on MIC substrates. *Arch. Elek. Ubertragung.*, **AEU-35(2)**, 80–83, 1981.
17. J. Van Bladel, On the resonances of a dielectric resonator of very high permittivity. *IEEE Trans. Microwave Theory Tech.*, **MTT-23**, 199–208, Feb. 1975.
18. M. Verplanken and J. Van Bladel, The magnetic-dipole resonance of ring resonators of very high permittivity. *IEEE Trans. Microwave Theory Tech.*, **MTT-27**, 328–333, Apr. 1979.

19. M. Verplanken and J. Van Bladel, The electric-dipole resonances of ring resonators of very high permittivity. *IEEE Trans. Microwave Theory Tech.*, **MTT-24**, 108–112, Feb. 1976.

20. R. DeSmedt, Dielectric resonator above an electric or magnetic wall. *Arch. Elek. Ubertragung*, **AEU-37**, 6–14, Jan. 1983.

21. R. DeSmedt, Dielectric resonator inside a circular waveguide. *Arch. Elek. Ubertragung.*, **AEU-38**, 113–120, Mar. 1984.

22. R. DeSmedt, Corrections due to finite permittivity for a ring resonator in free space. *IEEE Trans. Microwave Theory Tech.*, **MTT-32**, 1288–1295, Oct. 1984.

23. M. Jaworski and M. W. Pospieszalski, An accurate solution of the cylindrical dielectric resonator problem. *IEEE Trans. Microwave Theory Tech.*, **MTT-27**, 639–643, July 1979. Also correction in *IEEE Trans. Microwave Theory Tech.*, **MTT-28**, 673, June 1980.

24. A. W. Glisson, D. Kajfez, and J. James, Evaluation of modes in dielectric resonators using a surface integral equation formulation. *IEEE Trans. Microwave Theory Tech.*, **MTT-31**, 1023–1029, Dec. 1983.

25. A. S. Omar and K. Schunemann, Scattering by material and conducting bodies inside waveguides. Part I: Theoretical formulations. *IEEE Trans. Microwave Theory Tech.*, **MTT-34**, 266–272, Feb. 1986.

26. D. Kajfez and P. Guillon (Eds.), *Dielectric Resonators*, Artech House, Norwood, MA, 1986.

27. D. Kajfez, A. W. Glisson, and J. James, Computed modal field distributions for isolated dielectric resonators. *IEEE Trans. Microwave Theory Tech.*, **MTT-32**, 1609–1616, Dec. 1984.

28. M. Tsuji, M. Shigesawa, and K. Takiyama, On the complex resonant frequency of open dielectric resonators. *IEEE Trans. Microwave Theory Tech.*, **MTT-31**, 392–396, May 1983.

29. M. Tsuji, H. Shigesawa, and K. Takiyama, Analytical and experimental investigations on several resonant modes in open dielectric resonators. *IEEE Trans. Microwave Theory Tech.*, **MTT-32**, 628–633, June 1984.

30. R. K. Mongia and B. Bhat, Simple equations quickly design cylindrical DRs. *Microwaves RF*, **26**, 1048–1052, Oct. 1981.

31. T. Hogashi and T. Makino, Resonant frequency stability of the dielectric resonator on a dielectric substrate. *IEEE Trans. Microwave Theory Tech.*, **MTT-29**, 1048–1052, Oct. 1981.

32. S. J. Fiedziuszko, Dual-mode dielectric resonator loaded cavity filters. *IEEE Trans. Microwave Theory Tech.*, **MTT-30**, 1311–1316, Sept. 1982.

33. P. Guillon and Y. Garault, Dielectric resonator dual-mode filters. *Electron. Lett.*, **16**, 646, Aug. 1980.

34. F. Frezza et al., Characterization of the resonant and coupling parameters of dielectric resonators for NRD-guide filtering devices. *IEEE MTT-S Int. Microwave Symp. Digest*, 893–896, June 1993.

35. F. Freeza et al., Theoretical and experimental investigation on rectangular resonators in NRD waveguide. Proceedings of the 1992 Asia–Pacific Microwave Conference (Australia). *Digest*, 833–836, Aug. 1992.

36. F. Freeza et al., NRD waveguide ring resonator. Proceedings of the 1993 Asia–Pacific Microwave Conference (Tokyo). *Digest*, 3–6, Sept. 1993.

37. A. J. Marcatili, Dielectric rectangular waveguide and directional couplers for integrated optics. *Bell. Syst. Tech. J.*, **48**, 2071–2102, Sept. 1969.

38. C. Chang and T. Itoh, Resonant characteristics of dielectric resonators for millimetre-wave integrated circuits. *Arch. Elek. Ubertragung.*, **AEU-33**, 141–144, Apr. 1979.

39. J. F. Legier et al., Resonant frequencies of rectangular dielectric resonators. *IEEE Trans. Microwave Theory Tech.*, **MTT-28**, 1031–1034, Sept. 1980.

40. P. Guillon and Y. Garault, Accurate resonant frequency of dielectric resonators. *IEEE Trans. Microwave Theory Tech.*, **MTT-25**, 916–922, Nov. 1977.

41. M. Abramowitz and I. A. Stegun (Eds.), *Handbook of Mathematical Functions*, Dover Publications, Mineola, NY, 1965.

42. S. A. Long, M. MaAllister, and L. C. Shen, The resonant cylindrical dielectric cavity antenna. *IEEE Trans. Antennas Propag.*, **AP-31**, 406–412, May 1983.

43. E. Snitzer, Cylindrical dielectric waveguide modes. *J. Opt. Soc. Am.*, **51**, 491–498, May 1961.

44. Y. Kobayashi and M. Miura, Optimum design of shielded dielectric rod and ring resonators for obtaining the best mode separation. *IEEE MTT-S Int. Microwave Symp. Digest*, 184–186, June 1984.

45. W. K. Hui and I. Wolff, Dielectric ring-gap resonator for application in MMICs. *IEEE MTT-S Int. Microwave Symp. Digest*, 735–739, 1991.

46. P. R. Herczfeld et al., Optically controlled microwave devices and circuits. *RCA Rev.*, **46**, 528–550, Dec. 1985.

47. A. S. Rong, Y. Cao, and Z. L. Sun, Accurate determination of resonance frequency and intrinsic Q factor for optically-controlled dielectric resonators. *Electron. Lett.*, **27**, 1466–1468, Aug. 1991.

48. Y. Shen, K. Nickerson, and J. Litva, Frequency-dependent FDTD modeling of optically controlled dielectric resonators. *IEEE Trans. Microwave Theory Tech.*, **MTT-41**, 1005–1010, June/July 1993.

49. A. Alphones et al., Determination of resonant frequency of optically controlled dielectric resonator. Private Communication.

50. A. Alphones and M. Tsutsumi, Optical control of millimeter wave in silicon rib guides, Paper of Technical Group on Electronics and Communications, *IEICE Japan*, **MW 90-148**, 19–26, Feb. 1991.

51. J. A. Curtis, S. J. Fiedziuszko, and S. C. Holme, Hybrid dielectric/HTS resonators and their applications. *IEEE MTT-S Int. Microwave Symp. Digest*, 447–450, 1991.

APPENDIX 9A BESSEL FUNCTION FORMULAS USED IN THE ANALYSIS OF CYLINDRICAL RESONATORS

Derivatives and Identities

$$J_0'(x) = -J_1(x)$$

$$J_1'(x) = J_0(x) - J_1(x)/x$$

$$J_2(x) = 2J_1(x)/x - J_0(x)$$

$$K_0'(x) = -K_1(x)$$

$$K_1'(x) = -K_0(x) - K_1(x)/x$$

$$K_2(x) = K_0(x) + 2K_1(x)/x$$

Integrals

$$\int_0^a rJ_0^2(ur)\,dr = (a^2/2)\,[J_1^2(ua) + J_0^2(ua)]$$

$$\int_0^a rJ_1^2(ur)\,dr = (a^2/2)\,[J_1^2(ua) - J_0(ua)J_2(ua)]$$

$$\int_a^\infty rK_0^2(ur)\,dr = (a^2/2)\,[K_1^2(ua) - K_0^2(ua)]$$

$$\int_a^\infty rK_1^2(ur)\,dr = (a^2/2)\,[K_0(ua)K_2(ua) - K_1^2(ua)]$$

$$\int_b^a r^2 J_1(ur)\,dr = (1/u)[a^2 J_2(ua) - b^2 J_2(ub)]$$

CHAPTER TEN

Discontinuities, Transitions, and Measurement Techniques

10.1 INTRODUCTION

The realization of dielectric integrated guide components, both passive and active, invariably involves incorporation of discontinuities in the form of an open end, a gap, an abrupt step, a bend, a curved section, a taper, and so on. Step discontinuities and tapers are used especially in the design of transitions. Corrugated guides involving multiple step discontinuities are used as antenna feeds. Figure 10.1 shows some of the typical transverse discontinuities involving abrupt change in the cross-section of a dielectric guide; (a) open end, (b) gap, (c) abrupt step, (d) notch, (e) ridge, and (f) multiple steps. Such discontinuities may be incorporated in surface wave slab guides, dielectric integrated guides, and also planar optical waveguides. From the point of view of analysis of discontinuities in these guide structures, we may identify two classes of problems, depending on whether the guide structure is shielded or open. Discontinuities in open dielectric guides are far more complex to analyze than those in shielded guide structures. This is because an open dielectric guide supports, besides a discrete set of surface wave modes, a continuum of modes. Specifically, the mode spectra include a discrete spectrum of surface wave modes for $\beta > k_0$, a continuous spectrum of radiated waves for $0 < \beta < k_0$, and a continuous spectrum of attenuated radiation waves for $-j\infty < \beta < j0$, where β is the propagation constant in the direction of propagation. On the other hand, in a shielded guide, the entire mode spectrum is discrete. It involves slow waves for $\beta > k_0$, fast waves for $0 < \beta < k_0$, and evanescent waves for $-j\infty < \beta < j0$.

In view of the analytical complexity involved in solving discontinuity problems in open dielectric guides, most investigations reported in the literature deal with abrupt step discontinuities in surface wave slab dielectric guides [1–14]. Mode-matching methods have been applied in an approximate sense by neglecting the effect of radiation waves on scattering [1–3]. Improved solution

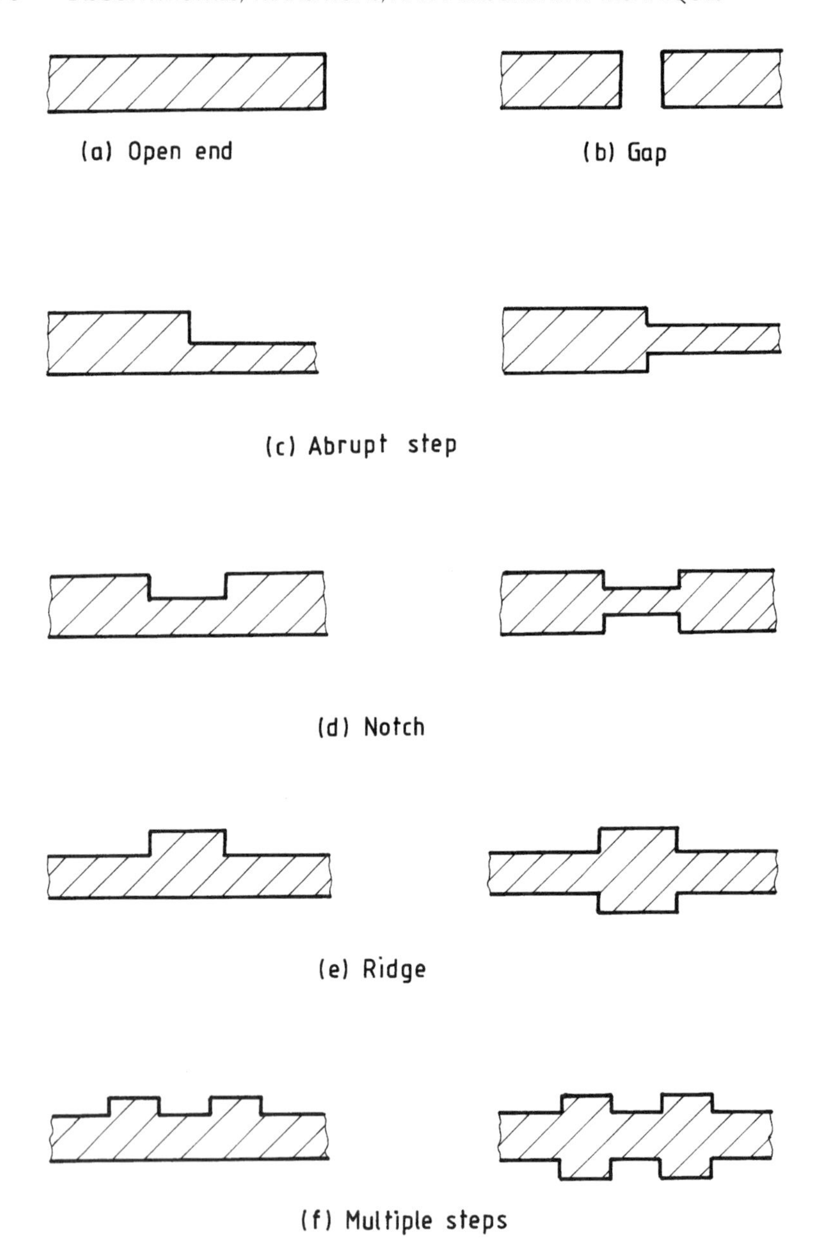

FIGURE 10.1 Typical transverse discontinuities in dielectric guides.

has been reported [4, 5] by transforming the continuous description of radiation modes into a discrete one through expansion of functions in a series of Laguerre polynomials. The mode-matching method has also been applied by introducing a metallic enclosure to the original waveguide [6, 7] and by approximating the guide by a periodic structure [8] in order to discretize the continuous radiation modes. Besides these mode-matching methods, variational approaches have been applied by several investigators. For example, Morishita et al. [9] have made use of the least-squares boundary residual method in conjunction with the normalized Laguerre polynomials. Rozzi [10] and Rozzi and in't Veld [11] have applied the Ritz–Galerkin variational approach to solve the integral equation of transverse field components at the discontinuity interface. Gelin et al. [12, 13] have solved a system of singular coupled integral equations for the discrete and continuous wave amplitudes at the discontinuities by iteration via the Neumann series. Suzuki and Koshiba [14] have developed an approach that uses a combination of the finite-element method for the interior region enclosing the waveguide discontinuities and an analytical method for the exterior region. Chung and Chen [15–17] have applied a partial variational principle for analyzing arbitrary discontinuities in planar dielectric guides. For large multistep discontinuities of arbitrary shape, Liu and Chew [18] have reported a numerically efficient mode-matching method called the eigenmode propagation method.

While the extensive investigations reported on discontinuities in planar dielectric slab guides [1–18] help in the understanding of discontinuities in corresponding dielectric guides having dielectric strips of rectangular cross-section, separate characterization of the latter is important from the point of view of implementation in practical circuits. Discontinuities in open dielectric integrated guides such as the image guide, insular image guide, and π-guide are examples of such structures. The analysis of discontinuities in such guides is considerably simplified if we replace the unbounded configuration by a corresponding bounded one with conducting walls as bounds. Since the entire mode spectrum in such guides is then discrete, the problem can be dealt with using the mode-matching method as applied to shielded guides [19–21].

In this chapter, we first present a general modal analysis technique that can be applied to step discontinuities in shielded dielectric integrated guides. As an example of the application of the mode-matching method, we present a detailed analysis and scattering matrix formulation of a rectangular waveguide containing an E-plane dielectric slab section as a discontinuity. As compared with the literature on step discontinuities, there is little information on curved sections, bends, and junctions in dielectric integrated guides. Since the dielectric image line and the NRD guide are the basic guide structures and are more commonly accepted in practice, studies on such discontinuities and also transitions are mostly confined to these guides. The existing literature on these topics is reviewed in the subsequent sections. The final section presents experimental techniques for the measurement of guide wavelength, fields, and attenuation in dielectric guides.

10.2 STEP DISCONTINUITY AND BENDS

10.2.1 Mode Matching and Generalized Scattering Matrix

The mode-matching method is widely used in the characterization of junctions between dissimilar guides. For a discontinuity in a shielded guide, the method involves expansion of fields in the guide regions located on either side of the discontinuity junction in terms of their respective normal modes. Since the functional form of the normal modes in a shielded guide is known, we apply the continuity conditions on the tangential components of the electric and magnetic fields at the junction in conjunction with the orthogonality property of the normal modes. This procedure leads to an infinite set of linear simultaneous equations for the unknown modal coefficients. Since the exact determination of modal coefficients requires the solution of an infinite set of equations, approximate techniques, such as truncation and iteration, are normally employed in practice. For adequate accuracy, a finite but sufficiently large number of equations is chosen so as to include all the significant higher order modes. With the determination of the modal coefficients, the generalized scattering matrix of the junction is known. This matrix takes into account the scattering phenomena of the dominant as well as the higher-order modes.

Consider a two-step junction as shown in Figure 10.2. When a propagating wave incident from the left encounters the first junction S_1, a part of the energy is reflected back and several higher order modes are generated at the discontinuity. Any of the transmitted propagating or evanescent modes that reach the second junction S_2 will partially reflect and generate a new set of backscattered modes. Some of these will reach the first junction S_1. At junction S_1, the total field consists of the positively directed waves and waves reflected from the second junction S_2. The modal analysis takes into account the interaction effects of the dominant and higher order modes between the two junctions. The amplitudes of the normal modes are chosen such that the boundary conditions are satisfied at the discontinuity. The overall scattering matrix of the two-step junction is obtained by suitably combining the generalized scattering matrices of the two step junctions with the scattering matrix of the guide section connecting the two junctions.

In the case of open-boundary junction problems such as the ones shown in Figure 10.1, the transverse fields on either side of each junction cannot be expanded in terms of the known normal modes. A propagating surface wave incident from the left is scattered by the first junction in all the surface wave modes allowed on either side of the junction as well as in a continuous set of modes. These surface waves and the propagating part of the continuous spectrum, on reaching the next junction, are again scattered. The backscattered waves reach the first junction. The reactive part of the continuous spectrum is nonpropagating and is localized to the vicinity of the junction discontinuity. The interaction between the discontinuity junctions via the propagating continuous modes is an additional feature of the open-boundary problem as compared with a shielded guide problem. The mathematical analysis, however, gets considerably simplified if we provide a shielding boundary to the guide and then compute the results by moving the shielding walls sufficiently away so as to cause negligible effect on the propagation.

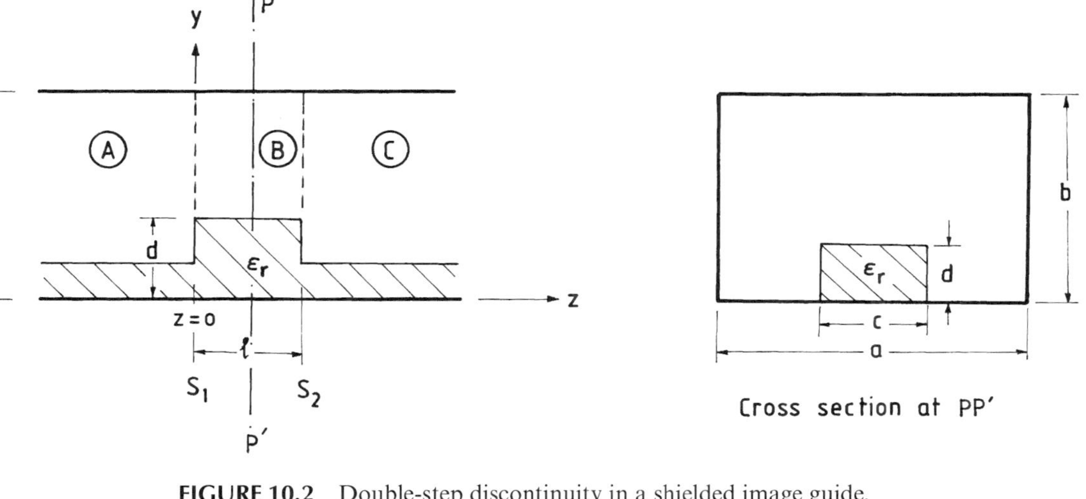

FIGURE 10.2 Double-step discontinuity in a shielded image guide.

10.2.2 General Analysis of Discontinuities in Shielded Dielectric Guides

The general modal analysis technique described by Wexler [19] for solving the waveguide discontinuities can be applied to discontinuities in shielded dielectric guides. In this section, we present the analysis by considering an example of a double-step discontinuity in a shielded image guide (Fig. 10.2).

The electric and magnetic fields of the ith mode in the guiding structure can be expressed in general as

$$\tilde{\mathbf{e}}_i(x, y, z) = a_i \mathbf{e}_i(x, y) e^{\pm \gamma_i z} \tag{10.1a}$$

$$\tilde{\mathbf{h}}_i(x, y, z) = a_i \mathbf{h}_i(x, y) e^{\pm \gamma_i z} \tag{10.1b}$$

where $\mathbf{e}_i$ and $\mathbf{h}_i$ are the transverse vector functions of the electric and magnetic fields, respectively; a_i is the mode coefficient of the ith mode; γ_i is the z-directed propagation constant and the sign in the exponent depends on the direction of propagation. Consider a single propagating mode with mode coefficient $a_1$emanating from a matched source in guide A. At the first junction S_1 located at $z = 0$, let ρ be the reflection coefficient of this mode and let a_i with $i = 2, 3, \ldots$ be the mode coefficients of the scattered modes. The total transverse fields on the left side of the junction S_1 can be expressed in terms of the modes in guide A.

$$\mathbf{E}_A = (1 + \rho) a_1 \mathbf{e}_{A1} + \sum_{i=2}^{\infty} a_i \mathbf{e}_{Ai} \tag{10.2a}$$

$$\mathbf{H}_A = (1 + \rho) a_1 \mathbf{h}_{A1} - \sum_{i=2}^{\infty} a_i \mathbf{h}_{Ai} \tag{10.2b}$$

Just to the right of junction S_1, the fields can be expressed in terms of the propagating modes in guide B and the backscattered modes from the subsequent discontinuity (at junction S_2). Let b_j with $j = 1, 2, \ldots, \infty$ be the mode coefficients of the forward propagating modes in guide B and b_k with $k = 1, 2, \ldots, \infty$ be the mode coefficients of the backscattered modes. Let S_{jk} denote the scattering coefficient of mode k, which is equal to the mode coefficient b_k transformed in amplitude and phase to junction S_1 from junction S_2. By summing up all the forward and backscattered modes, the total transverse electric field and the magnetic field just to the right of junction S_1 can be expressed as

$$\mathbf{E}_B = \sum_{j=1}^{\infty} b_j \left(\mathbf{e}_{Bj} + \sum_{k=1}^{\infty} S_{jk} \mathbf{e}_{Bk} \right) \tag{10.3a}$$

$$\mathbf{H}_B = \sum_{j=1}^{\infty} b_j \left(\mathbf{h}_{Bj} - \sum_{k=1}^{\infty} S_{jk} \mathbf{h}_{Bk} \right) \tag{10.3b}$$

Equations (10.2) and (10.3) are general and can be applied irrespective of the cross-sectional sizes of the guides on either side of the junction. The problem is to evaluate the unknown parameters ρ, a_i, and b_j by matching the boundary conditions at the junction. It is presumed that S_{jk} due to the second junction is known; otherwise it must be evaluated by solving the second discontinuity first. In the case of a single step discontinuity at S_1 and with guide B terminated in a matched load, we have $S_{jk} = 0$. If guide B is terminated in an open or short circuit at a distance ℓ from junction S_1, this termination causes independent reflection of each mode incident upon it irrespective of the amplitude and phase of any other mode. Hence

$$S_{jk} = 0 \quad \text{for } j \neq k \tag{10.4a}$$

and

$$S_{jj} = \pm \exp(-j2\beta_{Bj}\ell) \tag{10.4b}$$

where β_{Bj} is the propagation constant of the jth mode in guide B.

In order to satisfy the boundary conditions at junction S_1, we proceed as follows. We cross-multiply (10.2a) and (10.3a) individually by $\mathbf{h}_{Am}$, integrate over the cross-sectional area of the guide, and equate. The resulting equation is

$$(1+\rho)a_1 \int_A \mathbf{e}_{A1} \times \mathbf{h}_{Am} \cdot \hat{\mathbf{z}}\, dx\, dy + \sum_{i=2}^{\infty} a_i \int_A \mathbf{e}_{Ai} \times \mathbf{h}_{Am} \cdot \hat{\mathbf{z}}\, dx\, dy$$
$$= \sum_{j=1}^{\infty} \left[b_j \int_B \mathbf{e}_{Bj} \times \mathbf{h}_{Am} \cdot \hat{\mathbf{z}}\, dx\, dy + \sum_{k=1}^{\infty} S_{jk} \int_B \mathbf{e}_{Bk} \times \mathbf{h}_{Am} \cdot \hat{\mathbf{z}}\, dx\, dy \right] \tag{10.5a}$$

Next, we take the cross-product of (10.2b) and (10.3b) individually with $\mathbf{e}_{Bn}$, integrate over the cross-section of the guide, and equate to yield

$$(1-\rho)a_1 \int_B \mathbf{e}_{Bn} \times \mathbf{h}_{A1} \cdot \hat{\mathbf{z}}\, dx\, dy - \sum_{i=2}^{\infty} a_i \int_B \mathbf{e}_{Bn} \times \mathbf{h}_{Ai} \cdot \hat{\mathbf{z}}\, dx\, dy$$
$$= \sum_{j=1}^{\infty} \left[b_j \int_B \mathbf{e}_{Bn} \times \mathbf{h}_{Bj} \cdot \hat{\mathbf{z}}\, dx\, dy - \sum_{k=1}^{\infty} S_{jk} \int_B \mathbf{e}_{Bn} \times \mathbf{h}_{Bk} \cdot \hat{\mathbf{z}}\, dx\, dy \right] \tag{10.5b}$$

For a uniform lossless guide, either fully or partially loaded with a dielectric, the following orthogonality relation holds for nondegenerate modes:

$$\int_A (\mathbf{e}_{Ai} \times \mathbf{h}_{Am}) \cdot \hat{\mathbf{z}}\, dx\, dy = 0, \quad \text{if } i \neq m \tag{10.6}$$

This relation holds for both guides A and B. By using the known incident wave amplitudes a_1 as the normalization factor, (10.5) can be written in compact form as

$$\rho I_{A1Am} + \sum_{i=2}^{\infty} \bar{a}_i I_{AiAm} - \sum_{j=1}^{\infty} \bar{b}_j \left(I_{BjAm} + \sum_{k=1}^{\infty} S_{jk} I_{BkAm} \right) = -I_{A1Am} \tag{10.7a}$$

$$\rho I_{BnA1} + \sum_{i=2}^{\infty} \bar{a}_i I_{BnAi} + \sum_{j=1}^{\infty} \bar{b}_j \left(I_{BnBj} - \sum_{k=1}^{\infty} S_{jk} I_{BnBk} \right) = I_{BnA1} \tag{10.7b}$$

where

$$\bar{a}_i = a_i / a_1 \tag{10.8a}$$

$$\bar{b}_j = b_j / a_1 \tag{10.8b}$$

and the notation I_{AiBj} denotes

$$I_{AiBj} = \int \mathbf{e}_{Ai} \times \mathbf{h}_{Bj} \cdot \hat{\mathbf{z}}\, dx\, dy \tag{10.9}$$

with the integration carried over the aperture plane of the discontinuity. If guide B is match terminated, then $S_{jk} = 0$ in (10.7). When the termination is a short or an open, then (10.7) simplifies to

$$\rho I_{A1Am} + \sum_{i=2}^{\infty} \bar{a}_i I_{AiAm} - \sum_{j=1}^{\infty} \bar{b}_j (1 + S_{jj}) I_{BjAm} = -I_{A1Am} \tag{10.10a}$$

$$\rho I_{BnA1} + \sum_{i=2}^{\infty} \bar{a}_i I_{BnAi} + \sum_{j=1}^{\infty} \bar{b}_j (1 - S_{jj}) I_{BnBj} = -I_{BnA1} \tag{10.10b}$$

where S_{jj} is given by (10.4b). Since the higher order modes generated at the junction for large values of i and j are expected to be insignificant, the series may be truncated at $i = M$ and $j = N$. The unknowns ρ, $\bar{a}_i$, and $\bar{b}_j$ then reduce to $(M + N)$ in number. These can be evaluated by solving the $(M + N)$ linear equations generated from (10.7). The transverse field functions $\mathbf{e}_i$ and $\mathbf{h}_i$ appearing in the integrand of the various integrals in (10.7) can be obtained from the modal analysis of dielectric integrated guides covered in Chapter 3.

10.2.3 Example of an *E*-Plane Dielectric Slab Discontinuity in Rectangular Waveguide

As an example, we consider the problem of an E-plane dielectric slab section in a rectangular metal waveguide. The structure is shown in Figure 10.3. Since there is no structural variation in the y-direction, for a TE_{10} mode incident from the rectangular waveguide, the waves excited in the structure can be classified to be of TE_{m0} type. The nonzero field components are E_y^R, H_x^R, and H_z^R, where R denotes any region I to V as marked in Figure 10.3. If we assume a time dependence of the form $e^{j\omega t}$, then from Maxwell's equations, H_x and H_z components can be expressed in terms of E_y as

$$H_x^R = \frac{1}{j\omega\mu_0} \frac{\partial E_y^R}{\partial z} \tag{10.11a}$$

$$H_z^R = -\frac{1}{j\omega\mu_0} \frac{\partial E_y^R}{\partial x} \tag{10.11b}$$

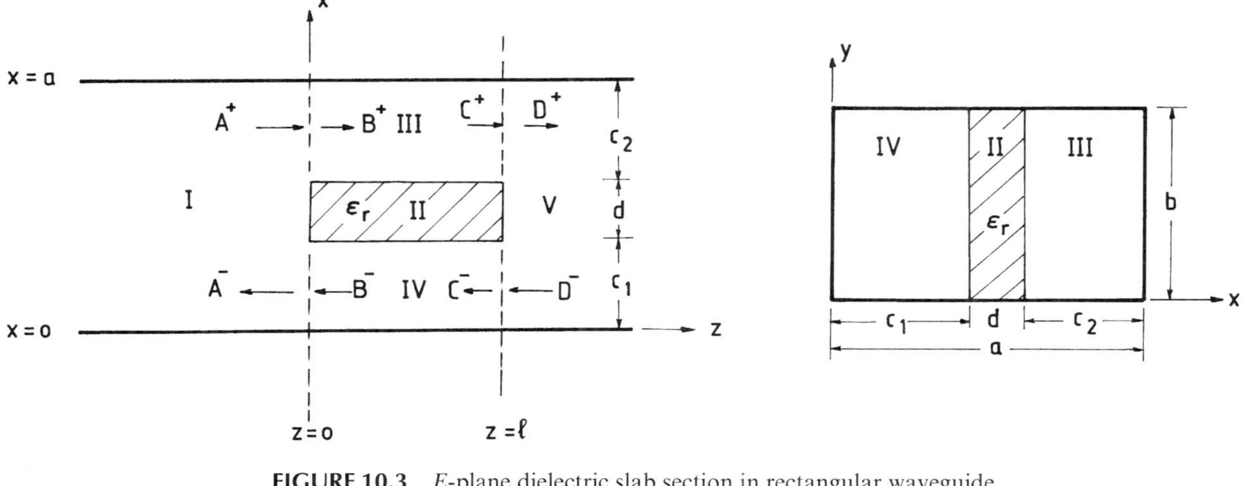

FIGURE 10.3 *E*-plane dielectric slab section in rectangular waveguide.

Let the z-directed propagation constant in the empty guide region I be denoted as β_{zm} and that in regions II, III, and IV of the slab loaded section be denoted as β'_{zm}.

Field Expansion into Eigenmodes Assuming propagation in the z-direction, we first expand the fields in regions I to IV in terms of the eigenmodes.

Region I:

$$E_y^{\mathrm{I}} = -\omega\mu_0 \sum_{m=1}^{M} A_m^{\mathrm{I}} \beta_{zm} \sin(m\pi x/a) e^{-j\beta_{zm}z} \tag{10.12a}$$

$$H_x^{\mathrm{I}} = \sum_{m=1}^{M} A_m^{\mathrm{I}} \beta_{zm}^2 \sin(m\pi x/a) e^{-j\beta_{zm}z} \tag{10.12b}$$

$$H_z^{\mathrm{I}} = -j \sum_{m=1}^{M} A_m^{\mathrm{I}} \beta_{zm}(m\pi/a) \cos(m\pi x/a) e^{-j\beta_{zm}z} \tag{10.12c}$$

where

$$\beta_{zm}^2 = k_0^2 - (m\pi/a)^2 \tag{10.13}$$

Region II:

$$E_y^{\mathrm{II}} = -\omega\mu_0 \sum_{n=1}^{N} \beta'_{zn} [A_n^{\mathrm{II}} \cos(\beta_{xn}^{\mathrm{II}} x) + B_n^{\mathrm{II}} \sin(\beta_{xn}^{\mathrm{II}} x)] e^{-j\beta'_{zn}z} \tag{10.14a}$$

$$H_x^{\mathrm{II}} = \sum_{n=1}^{N} (\beta'_{zn})^2 [A_n^{\mathrm{II}} \cos(\beta_{xn}^{\mathrm{II}} x) + B_n^{\mathrm{II}} \sin(\beta_{xn}^{\mathrm{II}} x)] e^{-j\beta'_{zn}z} \tag{10.14b}$$

$$H_z^{\mathrm{II}} = -j \sum_{n=1}^{N} \beta'_{zn}\beta_{zn}^{\mathrm{II}} [-A_n^{\mathrm{II}} \sin(\beta_{xn}^{\mathrm{II}} x) + B_n^{\mathrm{II}} \cos(\beta_{xn}^{\mathrm{II}} x)] e^{-j\beta'_{zn}z} \tag{10.14c}$$

Region III:

$$E_y^{\mathrm{III}} = -\omega\mu_0 \sum_{n=1}^{N} A_n^{\mathrm{III}} \beta'_{zn} [\sin(\beta_{xn}^{\mathrm{III}} x) - \tan(\beta_{xn}^{\mathrm{III}} a) \cos(\beta_{xn}^{\mathrm{III}} x)] e^{-j\beta'_{zn}z} \tag{10.15a}$$

$$H_x^{\mathrm{III}} = \sum_{n=1}^{N} (\beta'_{zn})^2 A_n^{\mathrm{III}} [\sin(\beta_{xn}^{\mathrm{III}} x) - \tan(\beta_{xn}^{\mathrm{III}} a) \cos(\beta_{xn}^{\mathrm{III}} x)] e^{-j\beta'_{zn}z} \tag{10.15b}$$

$$H_z^{\mathrm{III}} = -j \sum_{n=1}^{N} \beta'_{zn} A_n^{\mathrm{III}} \beta_{xn}^{\mathrm{III}} [\cos(\beta_{xn}^{\mathrm{III}} x) + \tan(\beta_{xn}^{\mathrm{III}} a) \sin(\beta_{xn}^{\mathrm{III}} x)] e^{-j\beta'_{zn}z} \tag{10.15c}$$

Region IV:

$$E_y^{\mathrm{IV}} = -\omega\mu_0 \sum_{n=1}^{N} A_n^{\mathrm{IV}} \beta'_{zn} \sin(\beta_{xn}^{\mathrm{IV}} x) e^{-j\beta'_{zn}z} \tag{10.16a}$$

$$H_x^{\mathrm{IV}} = \sum_{n=1}^{N} A_n^{\mathrm{IV}} (\beta'_{zn})^2 \sin(\beta_{xn}^{\mathrm{IV}} x) e^{-j\beta'_{zn}z} \tag{10.16b}$$

$$H_z^{\mathrm{IV}} = -j \sum_{n=1}^{N} A_n^{\mathrm{IV}} \beta'_{zn}\beta_{xn}^{\mathrm{IV}} \cos(\beta_{xn}^{\mathrm{IV}} x) e^{-j\beta'_{zn}z} \tag{10.16c}$$

where

$$(\beta_{xn}^{\rm II})^2 = k_0^2\varepsilon_r - (\beta'_{zn})^2 \tag{10.17a}$$

$$(\beta_{xn}^{\rm III})^2 = (\beta_{xn}^{\rm IV})^2 = k_0^2 - (\beta'_{zn})^2 \tag{10.17b}$$

Applying the continuity conditions on E_y and H_z at $x = c_1$ between regions II and IV and at $x = (c_1 + d)$ between regions II and III, we can obtain the following set of four algebraic equations in terms of the unknown coefficients $A_n^{\rm II}$, $B_n^{\rm II}$, $A_n^{\rm III}$, and $A_n^{\rm IV}$.

$$A_n^{\rm IV}\sin(\beta_{xn}^{\rm III}c_1) - A_n^{\rm II}\cos(\beta_{xn}^{\rm II}c_1) - B_n^{\rm II}\sin(\beta_{xn}^{\rm II}c_1) = 0 \tag{10.18a}$$

$$A_n^{\rm IV}\beta_{xn}^{\rm III}\cos(\beta_{xn}^{\rm III}c_1) + A_n^{\rm II}\beta_{xn}^{\rm II}\sin(\beta_{xn}^{\rm II}c_1) - B_n^{\rm II}\beta_{xn}^{\rm II}\cos(\beta_{xn}^{\rm II}c_1) = 0 \tag{10.18b}$$

$$A_n^{\rm II}\cos[\beta_{xn}^{\rm II}(c_1+d)] + B_n^{\rm II}\sin[\beta_{xn}^{\rm II}(c_1+d)] + A_n^{\rm III}[\sin(\beta_{xn}^{\rm III}c_2)/\cos(\beta_{xn}^{\rm III}a)] = 0 \tag{10.18c}$$

$$-A_n^{\rm II}\beta_{xn}^{\rm II}\sin[\beta_{xn}^{\rm II}(c_1+d)] + B_n^{\rm II}\beta_{xn}^{\rm II}\cos[\beta_{xn}^{\rm II}(c_1+d)] - A_n^{\rm III}\beta_{xn}^{\rm III}[\cos(\beta_{xn}^{\rm III}c_2)/\cos(\beta_{xn}^{\rm III}a)] = 0 \tag{10.18d}$$

Characteristic Equation for Determining β'_{zm} Eliminating the unknown coefficients from (10.18) yields the following characteristic equation:

$$[\tan(\beta_{xn}^{\rm III}c_1)/\beta_{xn}^{\rm III}] - [\beta_{xn}^{\rm II}/(\beta_{xn}^{\rm III})^2]\tan(\beta_{xn}^{\rm III}c_1)\tan(\beta_{xn}^{\rm III}c_2)\tan(\beta_{xn}^{\rm II}d) \cdot[\tan(\beta_{xn}^{\rm II}d)/\beta_{xn}^{\rm II}] + [\tan(\beta_{xn}^{\rm III}c_2)/\beta_{xn}^{\rm III}] = 0 \tag{10.19}$$

Using (10.17) in (10.19), we can solve for the propagation constant β'_{zn}. Substitution of the value of β'_{zn} in (10.17a) and (10.17b) yields $\beta_{xn}^{\rm II}$ and $\beta_{xn}^{\rm III}$, respectively.

From (10.18a)–(10.18d), we now express $A_n^{\rm II}$, $A_n^{\rm III}$, and $A_n^{\rm IV}$ in terms of the single unknown $B_n^{\rm II}$.

$$A_n^{\rm II} = B_n^{\rm II}p_n \tag{10.20a}$$

$$A_n^{\rm III} = B_n^{\rm II}q_n \tag{10.20b}$$

$$A_n^{\rm IV} = B_n^{\rm II}r_n \tag{10.20c}$$

where

$$p_n = \frac{\beta_{xn}^{\rm II}\tan(\beta_{xn}^{\rm III}c_1) - \beta_{xn}^{\rm III}\tan(\beta_{xn}^{\rm II}c_1)}{\beta_{xn}^{\rm III} + \beta_{xn}^{\rm II}\tan(\beta_{xn}^{\rm II}c_1)\tan(\beta_{xn}^{\rm III}c_1)} \tag{10.21a}$$

$$q_n = -\left[\frac{\cos[\beta_{xn}^{\rm II}(c_1+d)]\cos(\beta_{xn}^{\rm III}a)}{\sin(\beta_{xn}^{\rm III}c_2)}\right]\{p_n + \tan[\beta_{xn}^{\rm II}(c_1+d)]\} \tag{10.21b}$$

$$r_n = [\cos(\beta_{xn}^{\rm II}c_1)/\sin(\beta_{xn}^{\rm III}c_1)[p_n + \tan(\beta_{xn}^{\rm II}c_1)] \tag{10.21c}$$

Scattering Parameters of Junction at z = 0 In order to derive the generalized scattering matrix of the junction discontinuity, assume waves to be incident on the junction from both sides. There will be both forward and reverse traveling waves on either side of the junction. The E- and H-fields in each region can be represented in terms of the wave amplitudes as

$$E_y = \sum_{n=1}^{N} \{V_n^+ e^{-j\beta_{zn}z} + V_n^- e^{j\beta_{zn}z}\} F_{an} \tag{10.22a}$$

$$H_x = \sum_{n=1}^{N} \{V_n^+ e^{-j\beta_{zn}z} - V_n^- e^{j\beta_{zn}z}\} Y_{an} F_{an} \tag{10.22b}$$

where V_n^+ and V_n^- are the forward and reverse wave amplitudes, Y_{an} denotes the wave admittance, and $F_{an}(x)$ represents the nature of the wave in the cross-sectional plane.

Referring to Figure 10.3, we can express the fields at $z = 0$ in terms of the wave amplitudes as follows:

$$E_y^{\mathrm{I}} = -\omega\mu_0 \sum_{m=1}^{M} A_m^{\mathrm{I}} \beta_{zm} (A_m^+ + A_m^-) \sin(m\pi x/a) \tag{10.23a}$$

$$H_x^{\mathrm{I}} = \sum_{m=1}^{M} A_m^{\mathrm{I}} \beta_{zm}^2 (A_m^+ - A_m^-) \sin(m\pi x/a) \tag{10.23b}$$

$$E_y^{\mathrm{II}} = -\omega\mu_0 \sum_{n=1}^{N} \beta'_{zn} (B_n^+ + B_n^-) B_n^{\mathrm{II}} [p_n \cos(\beta_{xn}^{\mathrm{II}} x) + \sin(\beta_{xn}^{\mathrm{II}} x)] \tag{10.24a}$$

$$H_x^{\mathrm{II}} = \sum_{n=1}^{N} (\beta'_{zn})^2 (B_n^+ - B_n^-) B_n^{\mathrm{II}} [p_n \cos(\beta_{xn}^{\mathrm{II}} x) + \sin(\beta_{xn}^{\mathrm{II}} x)] \tag{10.24b}$$

$$E_y^{\mathrm{III}} = -\omega\mu_0 \sum_{n=1}^{N} \beta'_{zn} (B_n^+ + B_n^-) B_n^{\mathrm{II}} q_n [\sin(\beta_{xn}^{\mathrm{III}} x) - \tan(\beta_{xn}^{\mathrm{III}} a) \cos(\beta_{xn}^{\mathrm{III}} x)] \tag{10.25a}$$

$$H_x^{\mathrm{III}} = \sum_{n=1}^{N} (\beta'_{zn})^2 (B_n^+ - B_n^-) B_n^{\mathrm{II}} q_n [\sin(\beta_{xn}^{\mathrm{III}} x) - \tan(\beta_{xn}^{\mathrm{III}} a) \cos(\beta_{xn}^{\mathrm{III}} x)] \tag{10.25b}$$

$$E_y^{\mathrm{IV}} = -\omega\mu_0 \sum_{n=1}^{N} \beta'_{zn} (B_n^+ + B_n^-) B_n^{\mathrm{II}} r_n \sin(\beta_{xn}^{\mathrm{III}} x) \tag{10.26a}$$

$$H_x^{\mathrm{IV}} = \sum_{n=1}^{N} (\beta'_{zn})^2 (B_n^+ - B_n^-) B_n^{\mathrm{II}} r_n \sin(\beta_{xn}^{\mathrm{III}} x) \tag{10.26b}$$

where β_{zm} is given by (10.13), β_{xn}^{II} and $\beta_{xn}^{\mathrm{III}}$ are expressed in terms of β'_{zn} through (10.17), and the constants A_n^{II}, A_n^{III}, and A_n^{IV} are expressed in terms of a single unknown B_n^{II} through (10.20) and (10.21).

In order to determine A_m^{I} and B_n^{II}, we use power normalization by assuming that a power of 1 watt or j watts is incident at the interface $z = 0$ depending on whether the mode is propagating or evanescent, respectively. The expression for the z-directed power in the mth mode in region I is given by

$$W_m^{\mathrm{I}} = -\int_{x=0}^{a}\int_{y=0}^{b} E_{ym}^{\mathrm{I}} H_{xm}^{\mathrm{I}*}\, dx\, dy \tag{10.27}$$

Substituting for E_{ym}^{I} and H_{xm}^{I} from (10.12) and equating $|W_m^{\mathrm{I}}|$ to 1, we obtain

$$A_m^{\mathrm{I}} = \left[\frac{2}{\omega\mu_0 ab|\beta_{zm}|^3}\right]^{1/2} \tag{10.28}$$

For $z > 0$, the total power in the nth mode is obtained by adding the powers in regions II, III, and IV.

$$\begin{aligned} W_n^T &= -\sum_{R=\mathrm{II}}^{\mathrm{IV}} \int_x\int_y E_{yn}^R H_{xn}^{R*}\, dx\, dy \\ &= -\int_{y=0}^{b}\left[\int_0^{c_1} E_{yn}^{\mathrm{IV}} H_{xn}^{\mathrm{IV}*}\, dx + \int_{c_1}^{c_1+d} E_{yn}^{\mathrm{II}} H_{xn}^{\mathrm{II}*}\, dx + \int_{c_1+d}^{a} E_{yn}^{\mathrm{III}} H_{xn}^{\mathrm{III}*}\, dx\right] dy \end{aligned} \tag{10.29}$$

Substituting for E_y^R and H_x^R ($R = \mathrm{II}$ to IV) from (10.14)–(10.16) in conjunction with (10.20) and (10.21) in (10.29), and performing the integration, we obtain W_n^T in terms of B_n^{II}. Setting $|W_n^T| = 1$, we obtain the expression for B_n^{II} as

$$B_n^{\mathrm{II}} = \left(\frac{2}{b\omega\mu_0}\right)^{1/2} \frac{1}{|[(\beta'_{zn})^3(K_n^{\mathrm{II}} + K_n^{\mathrm{III}} + K_n^{\mathrm{IV}})]^{1/2}|} \tag{10.30}$$

where

$$K_n^{\mathrm{II}} = (p_n^2 + 1)d + [(p_n^2 - 1)/2\beta_{xn}^{\mathrm{II}}]\{\sin[2\beta_{xn}^{\mathrm{II}}(c_1 + d)] - \sin(2\beta_{xn}^{\mathrm{II}} c_1)\} \tag{10.31a}$$

$$K_n^{\mathrm{III}} = [q_n^2/\cos^2(\beta_{xn}^{\mathrm{III}} a)]\{c_2 + [\sin(2\beta_{xn}^{\mathrm{III}} c_2)/2\beta_{xn}^{\mathrm{III}}]\} \tag{10.31b}$$

$$K_n^{\mathrm{IV}} = r_n^2\{c_1 - [\sin(2\beta_{xn}^{\mathrm{III}} c_1)/2\beta_{xn}^{\mathrm{III}}]\} \tag{10.31c}$$

where p_n, q_n, and r_n are given by (10.21).

The next step is to apply the continuity conditions on E_y and H_x at $z = 0$. Matching the E_y components at $z = 0$ and integrating with respect to x from 0 to a after multiplying with $\sin(m\pi x/a)$, and noting that the integral on the left hand side vanishes for all terms of the series except for the mth term, we obtain

$$\begin{aligned} 0.5a\beta_{zm}^{\mathrm{I}} A_m^{\mathrm{I}}(A_m^+ + A_m^-) = \sum_{n=1}^{N} \beta'_{zm} B_n^{\mathrm{II}}[(p_n I_{mn}^{\mathrm{II}} + J_{mn}^{\mathrm{II}}) \\ + q_n(-T_n I_{mn}^{\mathrm{III}} + J_{mn}^{\mathrm{III}}) + r_n J_{mn}^{\mathrm{IV}}](B_n^+ + B_n^-) = 0 \end{aligned} \tag{10.32}$$

The coupling integrals I_{mn}, J_{mn} and T_n are given by

$$I^{\mathrm{II}}_{mn} = \int_{c_1}^{c_1+d} \cos(\beta^{\mathrm{II}}_{xn} x) \sin\left(\frac{m\pi x}{a}\right) dx \tag{10.33a}$$

$$J^{\mathrm{II}}_{mn} = \int_{c_1}^{c_1+d} \sin(\beta^{\mathrm{II}}_{xn} x) \sin\left(\frac{m\pi x}{a}\right) dx \tag{10.33b}$$

$$I^{\mathrm{III}}_{mn} = \int_{c_1+d}^{a} \cos(\beta^{\mathrm{III}}_{xn} x) \sin\left(\frac{m\pi x}{a}\right) dx \tag{10.33c}$$

$$J^{\mathrm{III}}_{mn} = \int_{c_1+d}^{a} \sin(\beta^{\mathrm{III}}_{xn} x) \sin\left(\frac{m\pi x}{a}\right) dx \tag{10.33d}$$

$$J^{\mathrm{IV}}_{mn} = \int_{0}^{c_1} \sin(\beta^{\mathrm{III}}_{xn} x) \sin\left(\frac{m\pi x}{a}\right) dx \tag{10.33e}$$

$$T_n = \tan(\beta^{\mathrm{III}}_{xn} a) \tag{10.33f}$$

Denoting

$$D_{Emn} = 0.5\, a\, \beta^{\mathrm{I}}_{zm} A^{\mathrm{I}}_m \tag{10.34a}$$

and

$$L_{Emn} = \beta'_{zn} B^{\mathrm{II}}_n [(p_n I^{\mathrm{II}}_{mn} + J^{\mathrm{II}}_{mn}) + q_n(-T_n I^{\mathrm{III}}_{mn} + J^{\mathrm{III}}_{mn}) + r_n J^{\mathrm{IV}}_{mn}] \tag{10.34b}$$

we can rewrite (10.32) in compact form as

$$D_{Emn}(A^+_m + A^-_m) = \sum_{n=1}^{N} L_{Emn}(B^+_n + B^-_n) \tag{10.35}$$

There will be as many equations as the values of m (m up to M). In matrix form, we can write (10.35) as

$$[D_E][(A^+) + (A^-)] = [L_E][(B^+) + (B^-)] \tag{10.36}$$

Applying the continuity condition on H_x at $z = 0$ and following the same procedure as outlined above for E_y, we obtain

$$D_{Hmn}(A^+_m - A^-_m) = \sum_{n=1}^{N} L_{Hmn}(B^+_n - B^-_n) \tag{10.37}$$

where

$$D_{Hmn} = 0.5\, a\, (\beta^{\mathrm{I}}_{zm})^2 A^{\mathrm{I}}_m \tag{10.38a}$$

$$L_{Hmn} = (\beta'_{zn})^2 B^{\mathrm{II}}_n [(p_n I^{\mathrm{II}}_{mn} + J^{\mathrm{II}}_{mn}) + q_n(-T_n I^{\mathrm{III}}_{mn} + J^{\mathrm{III}}_{mn}) + r_n J^{\mathrm{IV}}_{mn}] \tag{10.38b}$$

In matrix form, (10.37) can be written as

$$[D_H][(A^+)-(A^-)]=[L_H][(B^+)-(B^-)] \tag{10.39}$$

In (10.36) and (10.39), $[D_E]$ and $[D_H]$ are diagonal matrices of order $M \times M$; $[L_E]$ and $[L_H]$ are matrices of order $M \times N$; $[A^+]$ and $[A^-]$ are column matrices of order $M \times 1$; and $[B^+]$ and $[B^-]$ are column matrices of order $N \times 1$.

By matrix manipulation, it is easy to recast (10.36) and (10.39) in the form

$$\begin{bmatrix}(A^-)\\(B^+)\end{bmatrix}=\begin{bmatrix}(S_{11}) & (S_{12})\\(S_{21}) & (S_{22})\end{bmatrix}\begin{bmatrix}(A^+)\\(B^-)\end{bmatrix} \tag{10.40}$$

where

$$(S_{11})=(U)-2(D_H)^{-1}(L_H)(M) \tag{10.41a}$$

$$(S_{12})=2(D_H)^{-1}(L_H)(M)(D_E)^{-1}(L_E) \tag{10.41b}$$

$$(S_{21})=2(M) \tag{10.41c}$$

$$(S_{22})=-2(M)(D_E)^{-1}(L_E)+(U) \tag{10.41d}$$

$$(M)=[(D_E)^{-1}(L_E)+(D_H)^{-1}(L_H)]^{-1} \tag{10.41e}$$

(S_{11}), (S_{12}), (S_{21}), and (S_{22}) are the scattering parameters of the junction at $z=0$.

Overall Scattering Matrix of the Dielectric Slab Section Since the two junction discontinuities considered in Figure 10.3 are identical, the scattering matrix of the junction at $z=\ell$ is the same as that at $z=0$. If C^+ and C^- denote the forward- and backward-wave amplitudes, respectively, on the left side, and D^+ and D^- denote the corresponding quantities on the right side of the junction at $z=\ell$, we can write

$$\begin{bmatrix}(D^+)\\(C^-)\end{bmatrix}=\begin{bmatrix}(S_{11}) & (S_{12})\\(S_{21}) & (S_{22})\end{bmatrix}\begin{bmatrix}(D^-)\\(C^+)\end{bmatrix} \tag{10.42}$$

For the uniform slab loaded section of length ℓ, we can write

$$(C^+)=(R)(B^+) \tag{10.43a}$$

$$(B^-)=(R)(C^-) \tag{10.43b}$$

where R is the diagonal matrix with element

$$R_{nn}=e^{-j\beta'_{zn}\ell} \tag{10.44}$$

The overall scattering matrix $(S)_T$ of the E-plane slab discontinuity in a rectangular waveguide is given by suitably combining (10.40), (10.42), and (10.44):

$$\begin{bmatrix} (A^-) \\ (D^+) \end{bmatrix} = \begin{bmatrix} (S_{11})_T & (S_{12})_T \\ (S_{21})_T & (S_{22})_T \end{bmatrix} \begin{bmatrix} (A^+) \\ (D^-) \end{bmatrix} \tag{10.45}$$

The scattering matrix parameters are given by

$$(S_{11})_T = [(S_{11}) + (S_{12})(R)(W)(S_{22})(R)(S_{21})] \tag{10.46a}$$

$$(S_{12})_T = (S_{12})(R)(W)(S_{21}) \tag{10.46b}$$

$$(S_{21})_T = (S_{12})_T \tag{10.46c}$$

$$(S_{22})_T = (S_{11})_T \tag{10.46d}$$

where

$$(W) = [(U) - (S_{22})(R)(S_{22})(R)]^{-1} \tag{10.46e}$$

10.3 BENDS AND JUNCTIONS

10.3.1 Bends

Bends and curvatures in optical dielectric guides of circular cross-section have been studied by several authors [22–24]. Such open guides are known to radiate tangentially at the bends. An approximate expression for the attenuation due to radiation at a bend is given by [22]

$$\alpha_r \approx \frac{\sqrt{\varepsilon_r}}{4\pi^2} \frac{\lambda_0}{r_0^2} e^{(-1/6\pi^2)(R\lambda_0^2/r_0^3)} \tag{10.47}$$

where R is the radius of curvature, λ_0 is the free-space wavelength, ε_r is the relative dielectric constant of the dielectric, and r_0 is a measure of the extent of transverse field of the dominant HE_{11} mode of the circular dielectric guide. The above formula assumes that the wave is loosely bound to the waveguide $[(r_0/\lambda_0) \gg (1/2\pi)]$ and the rod radius a is small as compared to the quantity r_0 $(a/r_0 \ll 1)$. The reported experimental studies [24] on curved dielectric rods also show that the attenuation of a bend depends mainly on the factor $R\lambda_0^2/r_0^3$. Since bends and curved sections are required in the design of components such as couplers and ring resonators, it is important to implement techniques of minimizing the radiation loss. For bends in open guides, Desai and Mittra [25] have proposed the use of a curved metal shield around the bend (see Fig. 10.4). The shield is about five wavelengths in height and is flared outward at the ends in order to provide a gradual transition from the straight guide to the curved section. The separation distance s between the guide and the shield is optimized

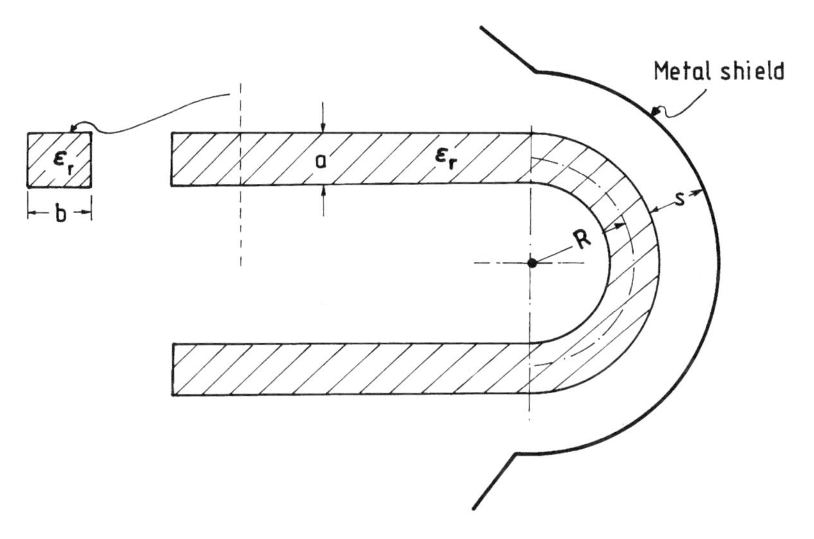

FIGURE 10.4 Dielectric bend with a metallic shield around the curved section. (From Desai and Mittra [25]. Copyright © 1980 IEEE, reprinted with permission.)

to achieve maximum reduction in radiation loss. As an experimental example, for a 180° bend in a Teflon guide of cross-section 2.8 mm × 1.32 mm and operating at 84 GHz, Desai and Mittra [25] have reported an optimum separation of approximately 1 mm between the guide and the shield and a maximum reduction in radiation loss by about 7 dB.

Solbach [26] has reported measured radiation loss for 90° bends in curved dielectric image lines. It is shown that the radiation loss decreases with an increase in frequency as well as the curvature radius R. The radiation loss, however, increases with an increase in frequency if the cross-sectional dimensions of the guide are decreased at higher frequencies so as to allow single-mode operation. It has been pointed out that the minimum curvature radius R to be chosen in practice can be calculated from the approximate formula $R \approx 16\pi^2 r_0^3/\lambda_0^2$, where r_0 is taken as the lateral distance over which the field decays to $1/e$ of the value at the edge of a straight guide. This value of r_0 can be calculated from the field distribution in the image guide.

In the case of a NRD guide, since the structure is basically nonradiative, any loss caused by bends is due to reflection at the transition between the straight guide and the curved guide. Thus the minimum radius of bend permissible in practice is determined by the tolerable reflection. Yoneyama et al. [27] have pointed out that the field maximum shifts outward or inward from the mean path of the curved guide depending on whether the strip is wider or narrower than a certain critical width. This feature of inward shifting of the dominant mode field maximum is specific to the NRD guide. The critical guide width can be chosen for the design of low-loss, very sharp bends. Figure 10.5 shows the configuration of

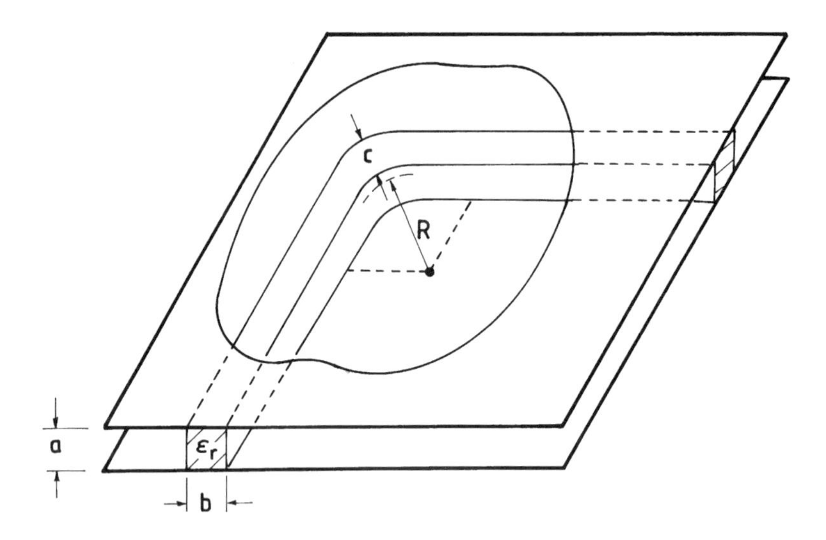

FIGURE 10.5 Bend in NRD guide.

a 90° bend in a NRD guide where R is the bending radius; a and b are the height and width, respectively, of the straight portion of the dielectric strip; and c is the width of the strip at the curved section. Yoneyama et al. [27] have reported measured losses in 90° bends in NRD guides in the frequency range 48–51 GHz. The NRD guide is made of polystyrene ($\varepsilon_r = 2.56$) and has the dimensions $a \times b =$ 2.7 mm × 2.4 mm. For $R = 20$ mm, 16 mm, and 12 mm, the critical width c for minimum loss is reported to be 2.4 mm, 2.2 mm, and 1.9 mm, respectively, showing that for $R \geq 20$ mm, no reduction in the width of the guide at the bend is necessary.

Figure 10.6 shows typical views of right-angle corner bends in an image guide and a trapped image guide. In the image guide bend, a metal reflector is placed in flush with the mitered corner making an angle of 45° with the guide axis (Fig. 10.6(a)). The principle of operation is similar to that of an optical mirror. The dominant E^y_{11} mode incident from the input guide A gets reflected at the metal reflector and is transmitted into the output guide B. Ogusu [28] has shown experimentally that for effective guidance of energy at the bend and hence for minimum bending loss, the reflector must be positioned slightly away from the point of intersection (point P in Fig. 10.6(a)) of the two guide axes. For example, in a dielectric image guide with dimensions $a \times b = 7.2$ mm × 3 mm and operating in the E^y_{11} mode at 22 GHz, the optimum location is reported to be about 1.5 mm away ($s \approx 1.5$ mm) from the intersection of the guide axes. It has also been shown by Ogusu [28] that the bending loss of the corner with reflector is comparable to that of the guide having a large bending radius $R(R > 5a)$. In the case of a trapped image guide, the reflector forms part of the metallic trough as illustrated in Figure 10.6(b).

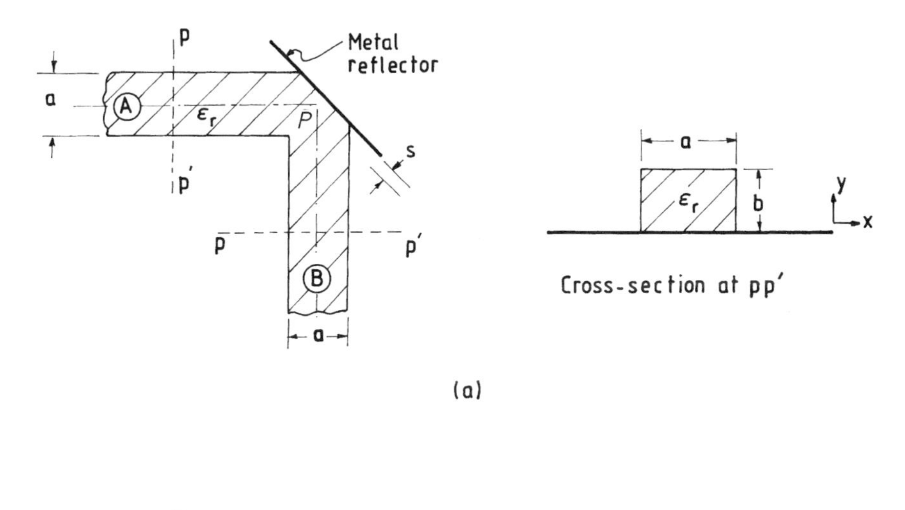

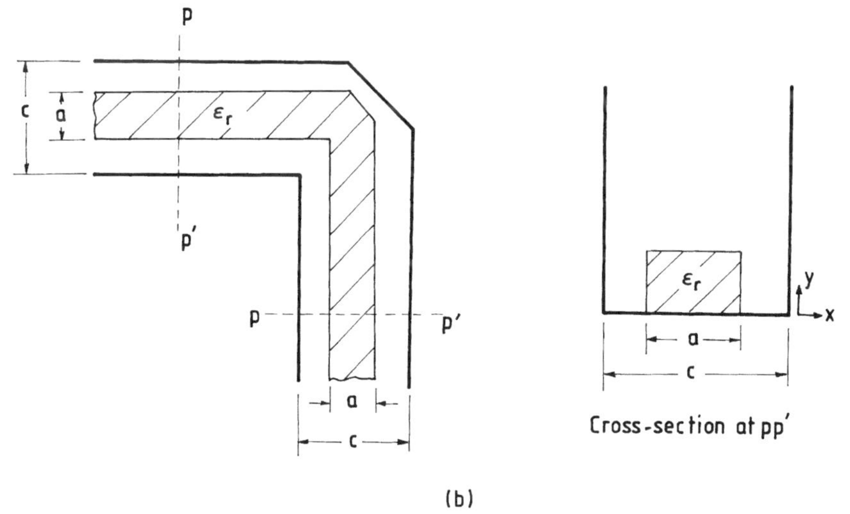

FIGURE 10.6 Right angle bend in (a) image guide and (b) trapped image guide.

10.3.2 Y- and T-Junctions

Dielectric Y- and T-shaped junctions, besides acting as power dividers, find applications in the design of 3-dB couplers, modulators, switches, mixers, and so on. While such junctions have been investigated extensively at optical wavelengths (e.g., see [29–32]), studies reported at millimeter wave frequencies are rather limited [28, 33–35]. Figure 10.7 shows typical configurations of symmetric and asymmetric Y-junctions. At optical wavelengths, the angle θ between the two output guides is usually below 2° so the radiation from the junction can be kept

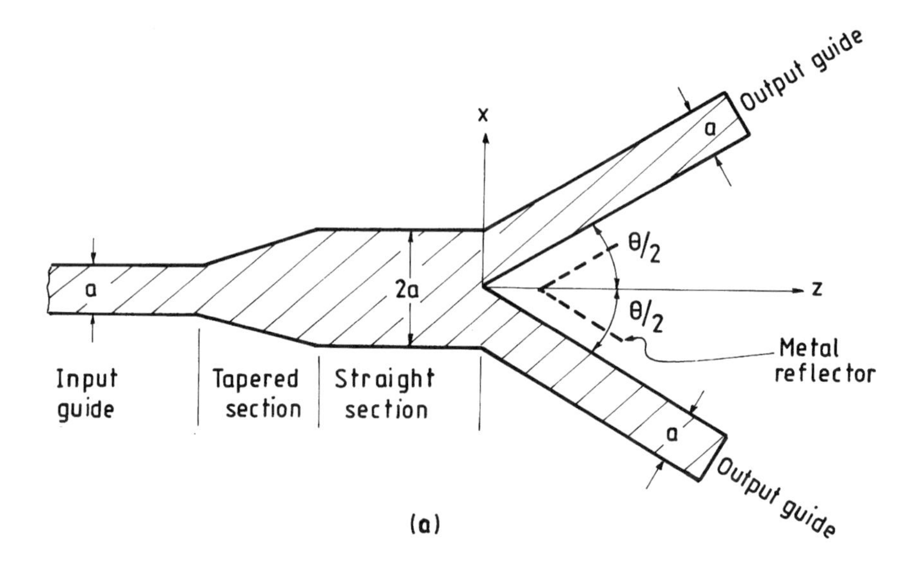

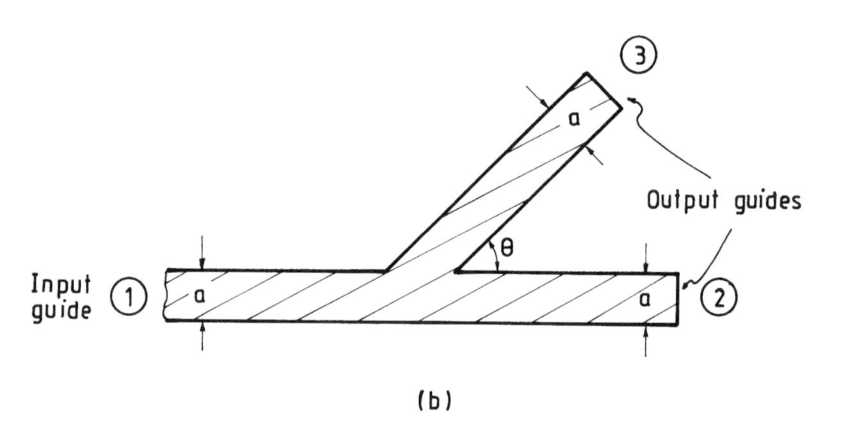

FIGURE 10.7 Typical configurations of (a) symmetric Y-junction and (b) asymmetric Y-junction. ((a) After Prieto et al. [34]; (b) after Ogusu [33].)

negligibly small. However, at millimeter wave frequencies, this angle must be made wider in order to keep the size of the devices incorporating such junctions as small as possible. Experimental results reported by Ogusu [33] on symmetrical and asymmetrical Y-junctions in image lines at K-band have shown high radiation losses for values of θ greater than about 30°.

Prieto et al. [34] have reported theoretical formulas for the determination of transmission coefficient T and reflection coefficient Γ of symmetrical Y-junctions. The formulas are given by

$$T = \left(\frac{\beta_a \beta_b}{(\beta_a + \beta_b)\,\omega \mu_0} \right) \int E_y^i(x,0)\, E_y^{t*}(x,0)\, dx \tag{10.48a}$$

$$\Gamma = \frac{\beta_a - \beta_b}{\beta_a + \beta_b} \tag{10.48b}$$

where β_a and β_b are the propagation constants in the input and output guides, respectively; and E_y^i and E_y^t are the input and output electric fields, respectively, at the junction plane $z = 0$ (refer to Fig. 10.7(a)). These propagation constants and electric fields can be determined by applying either the approximate effective dielectric constant (EDC) method or the rigorous modal analysis method.

The high radiation losses encountered in symmetric Y-junctions in image guides for wide aperture angles are reported to reduce considerably when the image guide on the input side is replaced by a π-guide [34]. This π-guide makes use of a high-permittivity dielectric core in the central region rather than the conventional air window so that there is a larger concentration of the electromagnetic field in the center of the input guide. The improved performance due to the use of the π-guide as reported by Prieto et al. [34] is illustrated in Figure 10.8. The

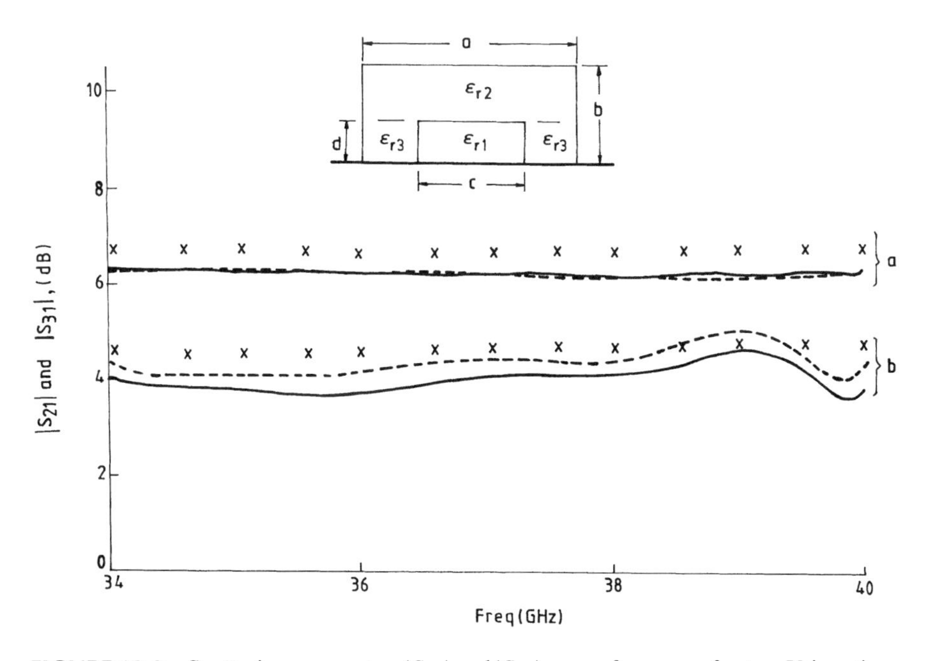

FIGURE 10.8 Scattering parameters $|S_{21}|$ and $|S_{31}|$ versus frequency for two Y-junctions using (a) an image guide ($\varepsilon_{r1} = \varepsilon_{r2} = \varepsilon_{r3} = 2.56$) and (b) a π-guide ($\varepsilon_{r1} = 10$, $\varepsilon_{r2} = \varepsilon_{r3} = 2.56$) in input guide. Dimensions (mm): $a = 3$, $b = 4$, $d = 3$, $c = 1$. —— $|S_{21}|$ Experiment, ----- $|S_{31}|$ Experiment, × × × × Theory. (From Prieto et al. [34], reprinted with permission of IEE.)

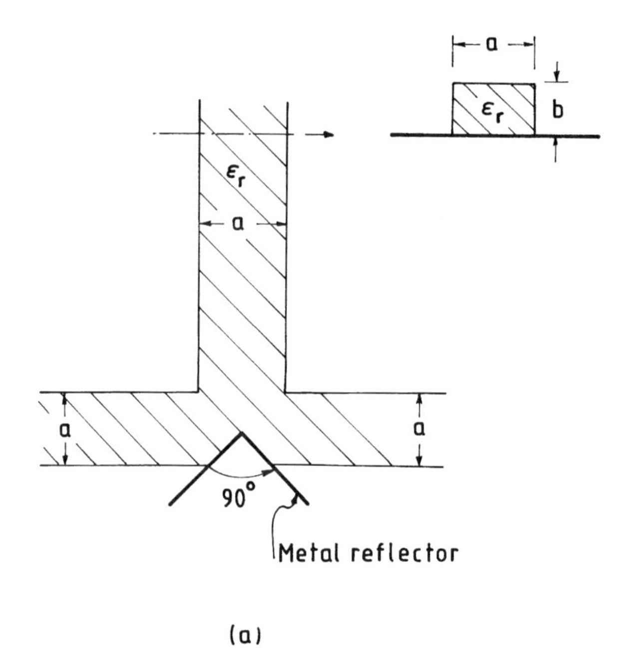

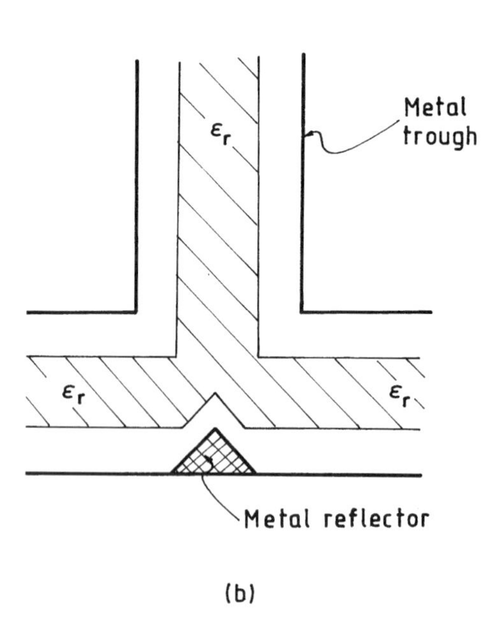

FIGURE 10.9 T-junction configurations in (a) an image guide and (b) a trapped image guide. (After Ogusu [28].)

graph shows theoretical as well as experimental variation in $|S_{12}|$ and $|S_{13}|$ as a function of frequency for two different Y-junctions: one made entirely with an image guide and the other with a π-guide as the input guide. The two Y-junctions have the same outside dimensions and the same output aperture angle $\theta = 120^\circ$. It can be seen that the transmission coefficients $|S_{12}|$ and $|S_{13}|$ are improved by about 2 dB in the Y-junction, which uses a π-guide as the input guide over those obtained in the Y-junction made entirely of an image guide. The use of a wedge-type metal reflector at the output junction as shown in Figure 10.7(a) is also useful in minimizing the radiation.

The asymmetric Y-junction (Fig. 10.7(b)) is useful for unequal power division. The power division ratio can be controlled by changing the aperture angle θ between the output ports. Ogusu [33] has shown that for an asymmetric Y-junction in an image guide having parameters $\varepsilon_r = 2.25$, $a = 8$ mm, and $b = 4$ mm and operating at 21 GHz, equal power division takes place for $\theta \approx 20^\circ$. For $\theta > 20^\circ$ power transmission to port 3 is greater than to port 2 and it increases with an increase in θ. The isolation between the output ports is reported to be better than 38 dB for values of θ in the range 10°–45°.

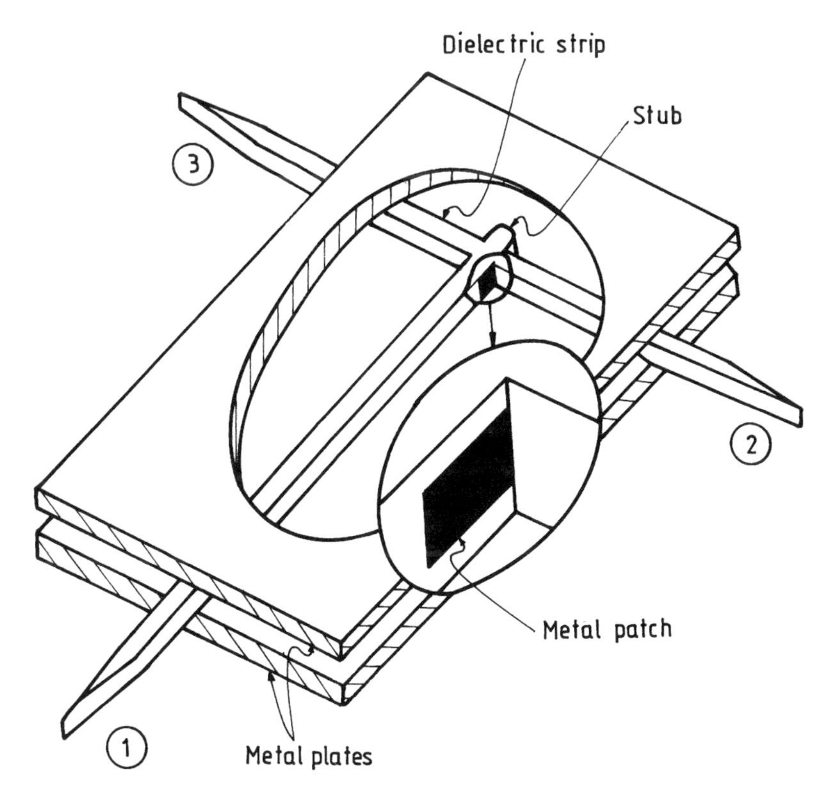

FIGURE 10.10 Sketch of a practical NRD guide T-junction. (From Yoneyama and Nishida [35]. Copyright © 1985 IEEE, reprinted with permission.)

The T-junction is a special case of the Y-junction for $\theta = 180°$. In an image guide T-junction, radiation from the junction can be minimized by placing a wedge-type reflector, as illustrated in Figure 10.9(a)[28]. A wedge-type reflector can also be used in the trapped image guide T-junction (Fig. 10.9(b)) for effective matching. Figure 10.10 shows a NRD guide T-junction reported by Yoneyama and Nishida [35]. The junction incorporates a dielectric stub as an extension of the main arm and the rectangular metal patches on the free surfaces of the main arm as matching elements. A practical T-junction of this type is reported to offer well-balanced power outputs of more than -4 dB over a bandwidth of 2 GHz centered at 35 GHz.

10.4 TRANSITIONS

10.4.1 Transitions to Image Guide

The image guide is commonly excited by a rectangular metal waveguide. The design of the transition must provide for gradual conversion of the TE_{10} mode of the rectangular waveguide to the E^y_{11} mode of the image guide. A simple transition can be formed by inserting a tapered section of the dielectric image guide into the full-height waveguide opening as illustrated in Figure 10.11(a). Both the width and the height of the dielectric strip are tapered over about two wavelengths within the waveguide. Figure 10.11(b) presents a three-section transition similar to the one reported by Dydyk [36]. In the first section of the transition where the dielectric strip enters the launcher, the image guide cross-section does not change but the metal waveguide is tapered down in the E-plane up to about half its height. This half-height waveguide extends over the second section of the transition and in the third section, the waveguide reverts back to its standard height through a linear E-plane taper. In the second and third sections of the transition, the dielectric strip is tapered in the E-plane from the top and in the H-plane from both sides. It may be noted that the wave impedance of the TE_{10} mode is higher than that of the image guide. The wave impedance reduces continuously from the mouth of the horn up to the end of the second section and then increases again to match the standard empty waveguide at the end of the third section. A carefully designed transition of this type with each section having a length equal to one guide wavelength offers a transmission loss of less than 0.2 dB at millimeter wave frequencies.

The image guide can also be excited through a slot cut in its ground plane as illustrated in Figure 10.12 [37]. The slot is fed by a tapered rectangular waveguide so that its feeding end matches the slot aperture. There are two auxiliary slots, one on either side of the driver slot, so that the three slots behave as elements of a Yagi–Uda array and direct energy along the image guide in one direction. The length ℓ_m of the driver slot is at half-wavelength resonance. The reflector slot is spaced by a distance of approximately $\lambda_g/4$ (λ_g is the guide waveguide in the image guide) from the driver slot and its length ℓ_r is approximately 5% longer than ℓ_m.

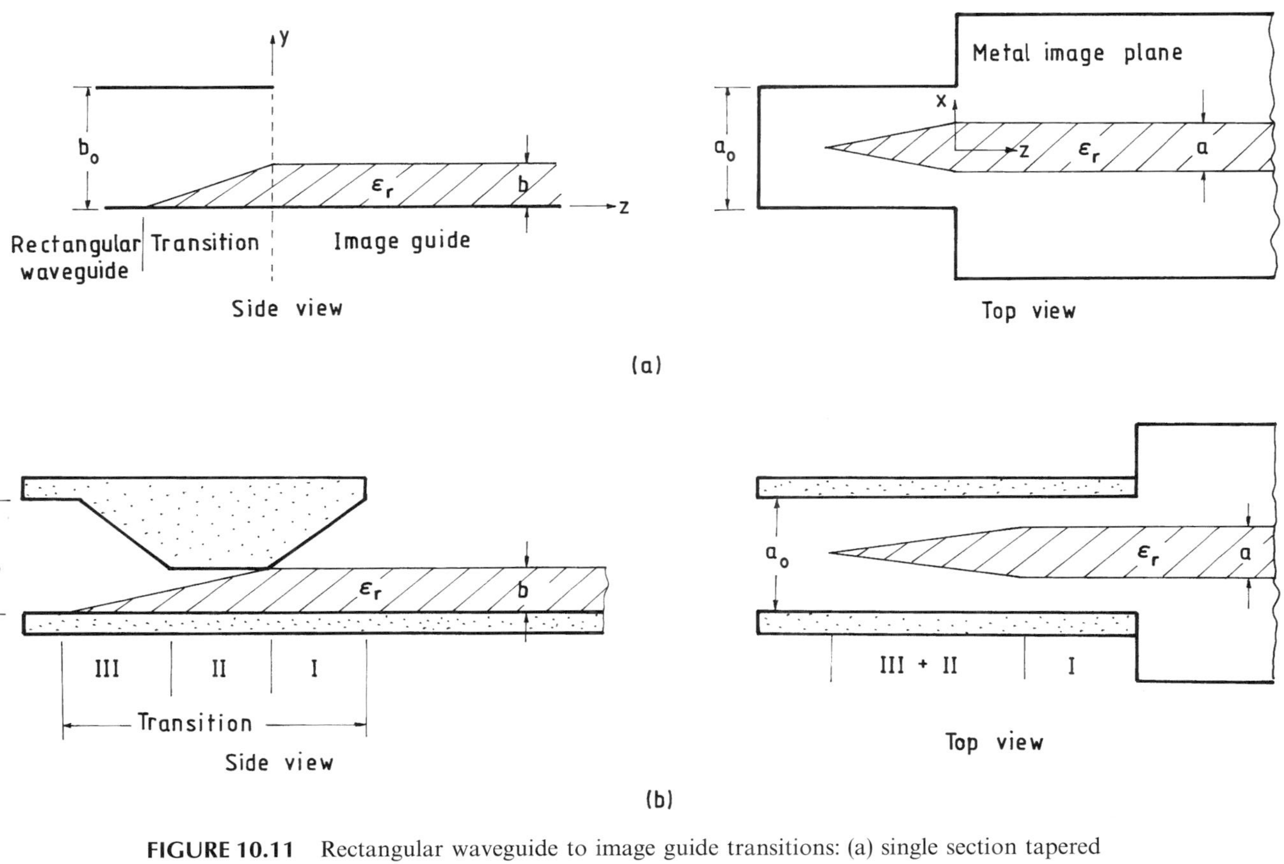

FIGURE 10.11 Rectangular waveguide to image guide transitions: (a) single section tapered transition and (b) composite three-section transition.

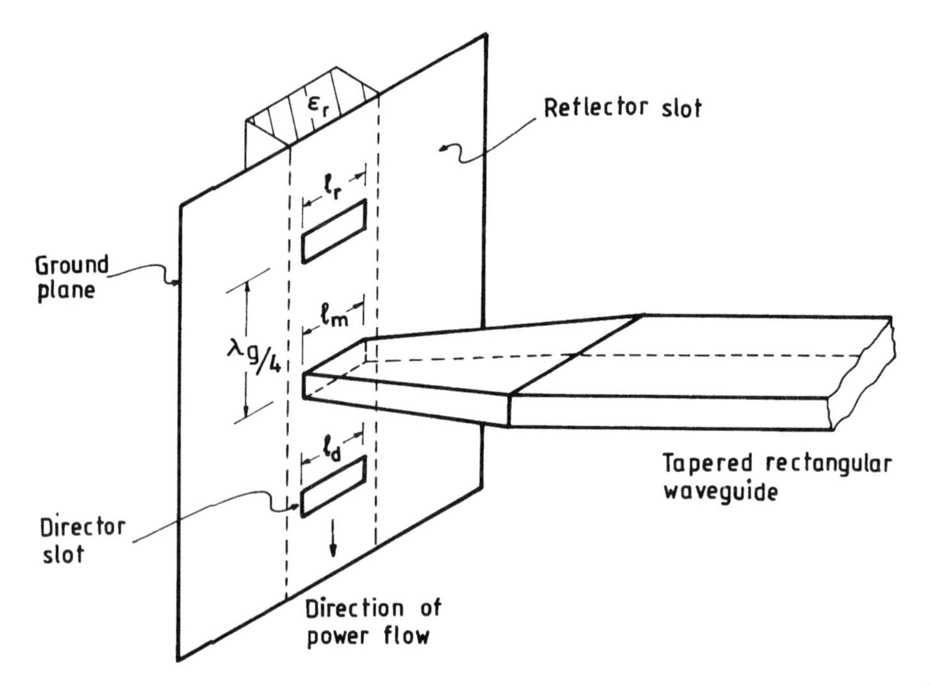

FIGURE 10.12 Image guide fed by rectangular waveguide through a slot in ground plane. (From Shih et al. [37]. Copyright © 1981 IEEE, reprinted with permission.)

The length ℓ_d of the director slot is about 10% shorter than ℓ_m and its distance is adjusted such that minimum energy is directed in one direction.

Instead of using a rectangular waveguide, the slot in the ground plane of the image guide can be fed easily by means of a microstrip line as shown in Figure 10.13. As in any slot resonator excitation, the microstrip is oriented perpendicular to the slot resonator and is terminated in an open circuit at a distance of about one-quarter wavelength ($\lambda_m/4$) in the microstrip line. A single microstrip line can also be used to excite a series of slots cut in the image plane beneath the dielectric strip.

10.4.2 Transitions to NRD Guide

Two different configurations of NRD guide transitions are considered here: one from a standard rectangular waveguide (Fig. 10.14) [38] and the second from a stripline (Fig. 10.15) [39]. The transition shown in Figure 10.14 consists of two sections. The first section is a transition from a standard rectangular waveguide to reduced width dielectric-filled metal waveguide. The metal walls of the waveguide are linearly tapered and the dielectric taper approximates a specified impedance match. The cross-sectional dimensions of the dielectric

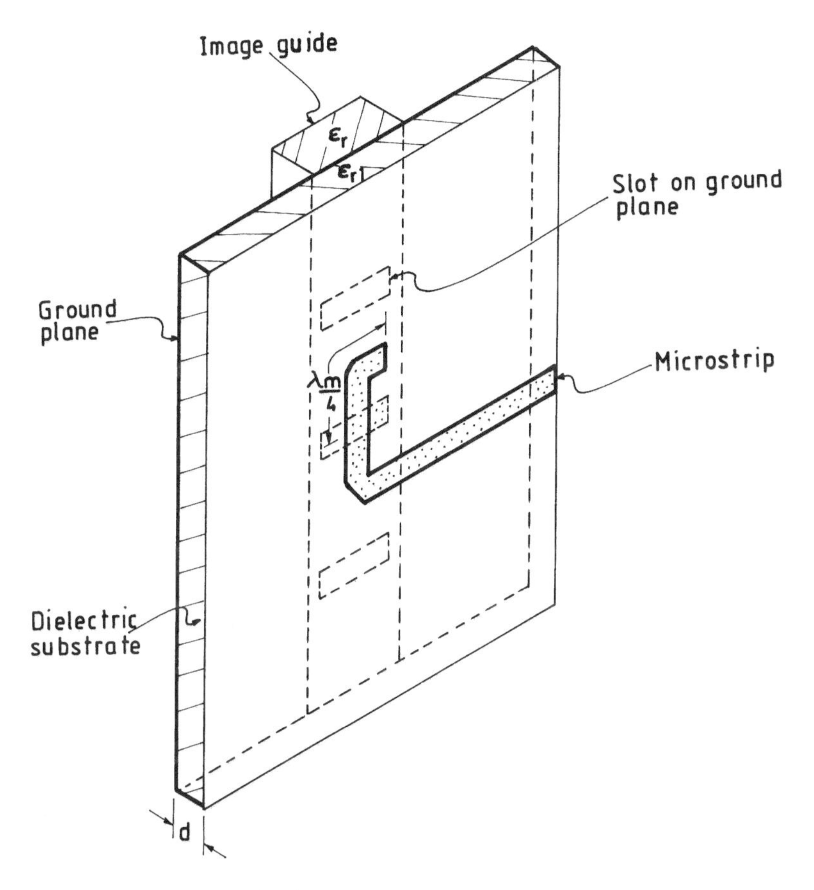

FIGURE 10.13 Image guide fed by a microstrip line through slot in ground plane.

at the end of this first section are the same as those of the dielectric in the NRD guide. The second section of the transition from the dielectric-filled waveguide to NRD guide therefore involves increasing only the metal waveguide height c while keep- ing the dielectric cross-section unaltered. The rate of increase of c is calculated for a specified impedance match while the TE_{10} mode of the dielectric-filled wave-guide is gradually transformed to the dominant mode of the NRD guide.

Stripline to NRD guide transition can find useful applications at lower millimeter wave frequencies. The stripline feed can easily be integrated with the NRD guide by making use of the two ground planes of the NRD guide. The geometry is shown in Figure 10.15 [39]. The NRD guide is shorted at one end in the xy-plane. The strip conductor is oriented parallel to the x-axis at a distance of ℓ from the shorted end and protrudes into the dielectric strip over a short distance

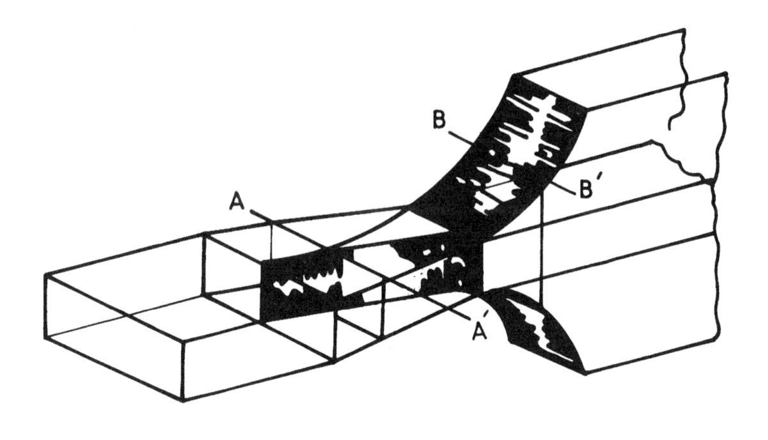

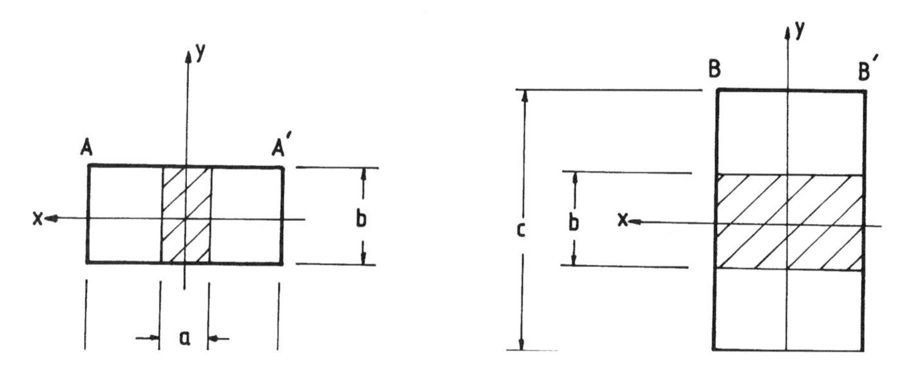

FIGURE 10.14 Composite transition from rectangular metal waveguide to NRD guide. (From Malherbe et al. [38]. Copyright © 1985 IEEE, reprinted with permission.)

d from one of the free sides. The width w of the strip conductor is chosen corresponding to 50 Ω in air stripline and its length L in the air medium is chosen such that the NRD guide fields would have decayed to a sufficiently low value over that length. The spacing ℓ and the penetration depth d into the dielectric are two critical parameters, which are to be adjusted to provide the best match to the 50-Ω stripline impedance. As an example, the dimensions of a theoretically optimized transition reported for a NRD guide having parameters $2a = 2.879$ mm, $b = 2.456$ mm, and $\varepsilon_r = 2.13$ and operating at 54 GHz are $L = 2.492$ mm, $w = 1.8$ mm, $\ell = 1.051$ mm, and $d = 2.51$ mm [39]. The theoretical operating frequency range for $VSWR < 3$ is from 52.24 GHz to 55.01 GHz with $VSWR = 1$ at 54 GHz.

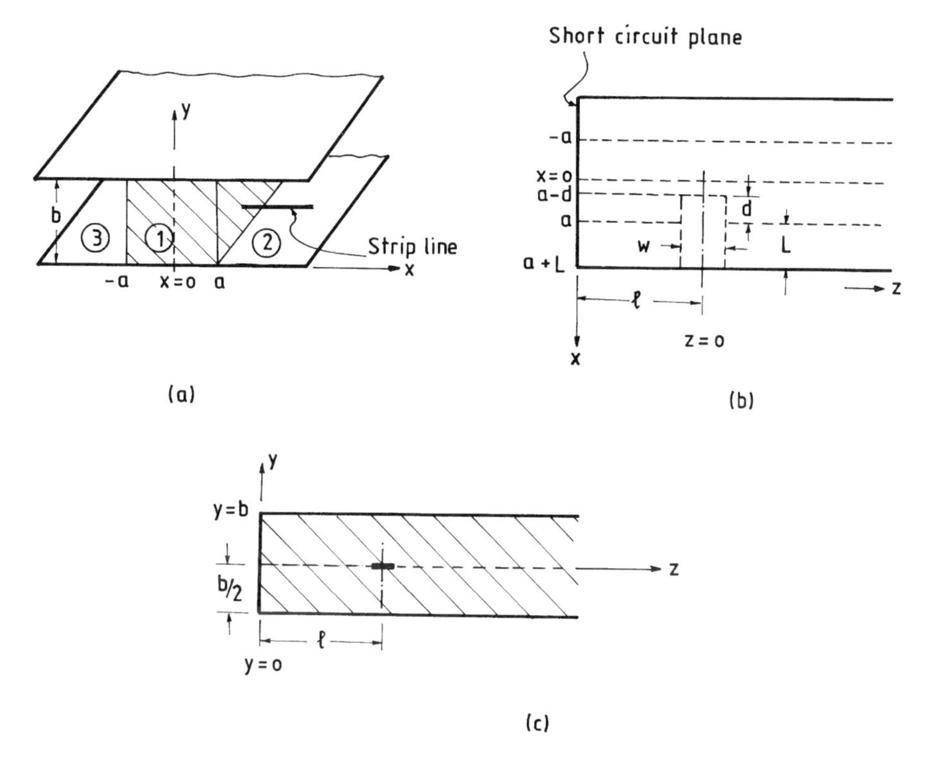

FIGURE 10.15 Stripline to NRD guide transition. (After Dawn and Sachidananda [39].)

10.5 MEASUREMENT TECHNIQUES

10.5.1 Electric Field Probe

The electric field probe is commonly used for measuring the field distribution in the vicinity of a dielectric guide [40]. Figure 10.16 shows the geometry of a typical probe reported for use at K_a-band [41]. The probe is made of a 50-Ω semirigid cable (inner conductor diameter = 0.08 mm, outer conductor diameter = 0.8 mm) of approximate length 80 mm. The outer conductor of the semirigid cable is removed over a length of 1 mm and 2 mm from the two ends. The 1-mm long unipole at one end serves as a field probe to sample the fields of the dielectric guide. The 2-mm long inner conductor at the other end is inserted into a standard rectangular waveguide through a small hole drilled at the center of one of the broad walls of the rectangular waveguide. The outer conductor is connected to the outer position of the waveguide wall. A resistive cotton thread is wound on the semirigid cable so as to avoid any coupling from stray fields. One end of the rectangular waveguide is provided with a movable short while at the other end, a thermistor mount is fitted, which in turn is connected to a power meter. The function of the variable short is to adjust for maximum power output from the thermistor mount.

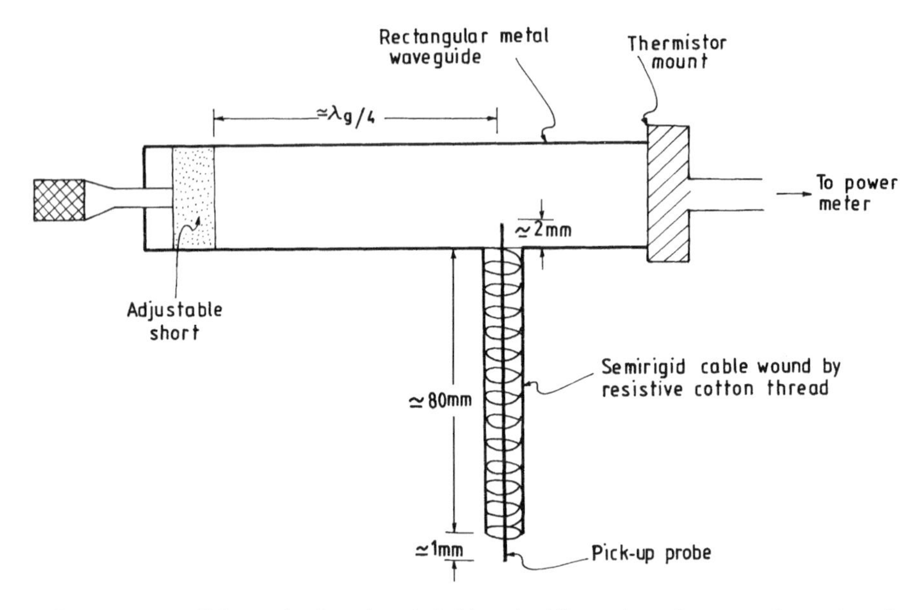

FIGURE 10.16 Schematic of an electric field probe (dimensions shown are for K_a-band).

10.5.2 Measurement of Attenuation Constant and Guide Wavelength

Several techniques for measurement of the attenuation constant of dielectric image guides are reported. These include the insertion loss method [42], the Q-factor method [43], and the $VSWR$ method [40, 44]. The insertion loss method involves measurement of insertion loss of straight image guide sections. This method involves measurement of mode-launcher losses also and hence requires rather long line sections. The Q-factor method requires measurement of loaded Q-factors of short-circuited image guide resonators of variable length. For accurate measurements, this technique requires the fabrication of a separate length of image guide for each frequency. The $VSWR$ technique [40, 44] is the simplest of the three methods and can conveniently be adopted in practice. The setup consists of a section of image guide shorted by a metal reflector at the load end and fed by a rectangular metal waveguide through a mode launcher at the input end (see Fig. 10.17) [40]. A CW signal fed from the matched mode launcher gets reflected from the metal reflector and sets up a standing wave pattern. The resulting field pattern is probed by moving an electric probe along the length of the guide. The $VSWR$ is measured as the ratio of the probe voltages at adjacent minimum and maximum field points just above the image guide. The $VSWR$ on the line can be expressed as [44]

$$VSWR = \frac{1 + e^{-(\tau + 2\alpha z)}}{1 - e^{-(\tau + 2\alpha z)}} \tag{10.49}$$

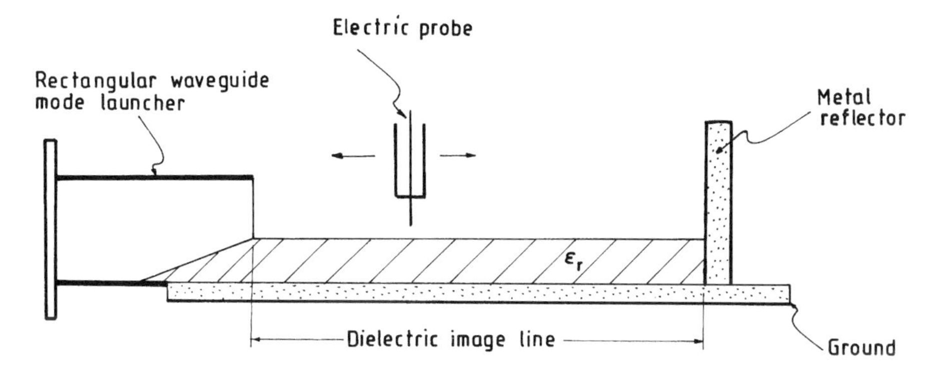

FIGURE 10.17 Arrangement for measuring the attenuation constant of dielectric image guide line using *VSWR* technique. (After Solbach [40].)

where τ is the reflection loss, α is the attenuation constant, and z is the distance between the probe and the reflector plane. It may be noted that the *VSWR* is maximum at the reflector plane and decreases with distance z toward the input. The two unknown quantities τ and α can be determined using the measured values of *VSWR* at different distances.

The wavelength is measured as twice the distance between the successive voltage minima on the line. The electric field intensity variation in the vicinity of the guide can be measured by replacing the reflector by a matched load and moving the electric probe in the lateral direction.

10.5.3 Measurement of Radiation Loss at Bends

The total insertion loss of a curved line section is the sum of the dielectric, conductor, and radiation losses. Solbach [26] has experimentally evaluated the radiation loss in curved dielectric image lines by measuring the total insertion loss of the curved section and subtracting from it the measured conductor and dielectric losses of a straight image guide section of the same length. The curved image guide is excited by a rectangular guide through a mode launcher from one end while the other end is terminated in a matched load. The total insertion loss is determined by measuring the amplitudes of the waves incident on the bend and the waves emerging from the end by means of an electric probe [40]. The dielectric and conductor losses in the curved section are taken to be approximately the same as the losses in the straight section, which are measured separately employing the *VSWR* method [40].

PROBLEMS

10.1 **(a)** Using mode- matching analysis, derive the scattering matrix of an E-plane slab discontinuity in a rectangular waveguide, as shown in

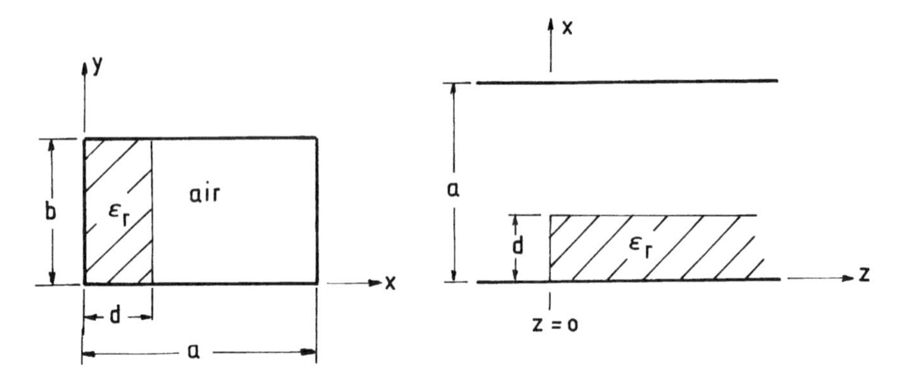

FIGURE 10.18 *E*-plane slab discontinuity in a rectangular waveguide.

Figure 10.18. Assume that only one higher order mode is generated at the discontinuity.

(b) Calculate the reflection and transmission coefficient versus frequency for the fundamental mode (TE_{10}). Assume $a = 22.86$ mm, $b = 10.16$ mm, $d = 4$ mm, and $\varepsilon_r = 2$.

(c) Draw a lumped element equivalent circuit to represent the junction. Determine the element values.

10.2 **(a)** Using the scattering matrix of the single junction derived in Problem 10.1, obtain the scattering matrix of a section of slab of length ℓ in the waveguide (Fig. 10.19).

(b) What should be the length of the dielectric slab to achieve an insertion phase of 180°? Choose the same parameters as in Problem 10.1(b). What should be the length ℓ, if ε_r is increased to 4?

(c) Draw a lumped element equivalent circuit and determine the element values.

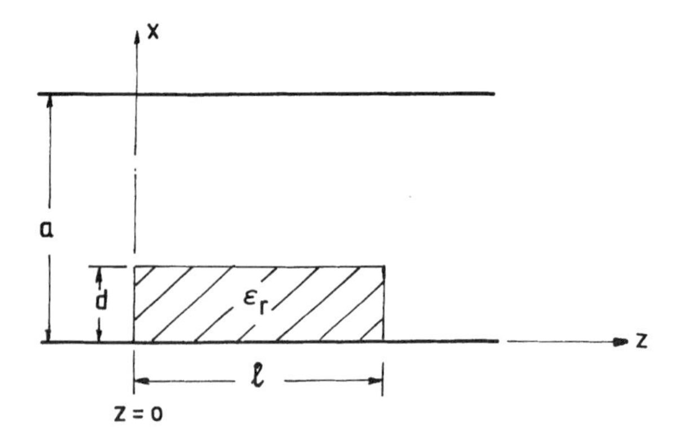

FIGURE 10.19 *E*-plane slab of length ℓ in a rectangular waveguide.

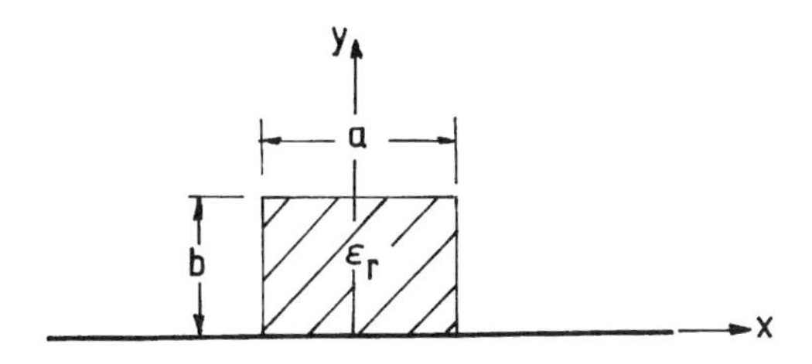

FIGURE 10.20 Cross-section of an image guide.

10.3 Discuss schemes of reducing loss due to bends and curvatures in an image guide, a trapped image guide, and a NRD guide.

10.4 Consider an image guide (Fig. 10.20) operating in the E^y_{11} mode at 20 GHz. The guide parameters are $a \times b = 7.2\,\text{mm} \times 3\,\text{mm}$ and $\varepsilon_r = 2.2$. At a lateral distance of 3.5 mm, the field intensity reduces to $1/e$ of its value at the edge of the guide. If a curved bend is to be designed in this guide, calculated the minimum radius of curvature R of the bend that would give negligible radiation loss.

10.5 What are the critical factors that control radiation loss at a Y-junction discontinuity in an image guide and a trapped image guide? How can this junction loss be reduced in practice in a Y-junction power divider? What must be the impedance level of the output arms with respect to that of the input arm for a Y-junction 3-dB power divider?

10.6 How does a Y-junction 3-dB power divider in a NRD guide differ from that in an image guide? What must be the impedance level of the output arms with respect to that of the input arm? What precautions must be taken to minimize the mismatch loss?

10.7 Draw a schematic of a transition from a microstrip to (a) an image guide and (b) an insular image guide. Show how the quasi-TEM field lines are transformed from microstrip through the transition.

10.8 Draw a schematic of a transition from a rectangular waveguide to **(a)** an H-guide and **(b)** a groove guide. Show how the TE_{10} mode field lines of the rectangular waveguide are transformed through the transition.

10.9 How can a E^x_{11} mode be launched on to an image guide? Assume the ground plane to lie in the xz-plane. What kind of mode suppressor would be required to suppress the E^y_{11} mode?

10.10 Figures 10.12 and 10.13 show schematics of an image guide excited through a slot in its ground plane. Draw a similar schematic for exciting a NRD guide through a slot in the ground plane.

REFERENCES

1. D. Marcuse, Radiation losses of tapered dielectric slab waveguide. *Bell Syst. Tech. J.*, **49**, 273–290, Feb. 1970.
2. P. J. B. Clarricoats and A. B. Sharpe, Modal matching applied to a discontinuity in a planar surface waveguide. *Electron. Lett.*, **8(2)**, 28–29, Jan. 1972.
3. T. Yoneyama and N. Nishida, Approximate solution for step discontinuity in dielectric slab waveguide. *Electron. Lett.*, **17(4)**, 151–153, Feb. 1981.
4. S. F. Mahmood and J. C. Beal, Scattering of surface waves at a dielectric discontinuity on a planar waveguide. *IEEE Trans. Microwave Theory Tech.*, **MTT-23**, 193–198, Feb. 1975.
5. M. Shigesawa and M. Tsuji, Mode propagation through a step discontinuity in dielectric planar waveguide. *IEEE Trans. Microwave Theory Tech.*, **MTT-34**, 205–212, Feb. 1986.
6. G. H. Brooke and M. M. Z. Kharadly, Step discontinuities in dielectric waveguides. *Electron. Lett.*, **12(8)**, 473–475, Sept. 1976.
7. G. H. Brooke and M. M. Z. Kharadly, Scattering by abrupt discontinuities on planar dielectric waveguides. *IEEE Trans. Microwave Theory Tech.*, **MTT-30**, 760–770, May 1982.
8. T. Hosono, T. Hinata, and A. Inoue, Numerical analysis of the discontinuities in slab dielectric waveguides. *Radio Sci.*, **17**, 75–83, Jan.–Feb. 1982.
9. K. Morishita, S. I. Inagaki, and N. Kumagi, Analysis of discontinuities in dielectric waveguides by means of the least squares boundary residual method. *IEEE Trans. Microwave Theory Tech.*, **MTT-27**, 310–315, Apr. 1979.
10. T. E. Rozzi, Rigorous analysis of a step discontinuity in a planar dielectric waveguide. *IEEE Trans. Microwave Theory Tech.*, **MTT-26**, 738–746, Oct. 1978.
11. T. E. Rozzi and G. H. in't Veld, Field and network analysis of interacting step discontinuities in planar dielectric waveguides. *IEEE Trans. Microwave Theory Tech.*, **MTT-27**, 303–309, Apr. 1979.
12. Ph. Gelin, M. Petenzi, and J. Citerne, New rigorous analysis of the step discontinuity in a slab dielectric waveguide. *Electron. Lett.*, **15(12)**, 355–356, June 1979.
13. Ph. Gelin, S. Toutain, and J. Citerne, Scattering of surface waves on transverse discontinuities in planar dielectric waveguide. *Radio Sci.*, **16**, 1161–1165, Nov.–Dec. 1981.
14. M. Suzuki and M. Koshiba, Finite-element analysis of discontinuity problems in a planar dielectric waveguide. *Radio Sci.*, **17**, 85–91, Jan.–Feb. 1982.
15. S. J. Chung and C. H. Chen, Partial variational principle for electromagnetic field problems—theory and applications. *IEEE Trans. Microwave Theory Tech.*, **MTT-36**, 473–479, Mar. 1988.
16. S. J. Chung and C. H. Chen, Analysis of irregularities in planar dielectric waveguide. *IEEE Trans. Microwave Theory Tech.*, **MTT-36**, 1352–1358, Sept. 1988.
17. S. J. Chung and C. H. Chen, A partial variational approach for arbitrary discontinuities in planar dielectric waveguides. *IEEE Trans. Microwave Theory Tech.*, **MTT-37**, 208–214, Jan. 1989.
18. Q. H. Liu and W. C. Chew, Analysis of discontinuities in planar dielectric waveguides: an eigenmode propagation method. *IEEE Trans. Microwave Theory Tech.*, **MTT-39**, 422–430, Mar. 1991.

19. A. Wexler, Solution of waveguide discontinuities by modal analysis. *IEEE Trans. Microwave Theory Tech.*, **MTT-15**, 508–517, Sept. 1967.
20. P. J. B. Clarricoats and K. R. Slinn, Numerical solution of waveguide discontinuity problems. *Proc. IEE*, **114(7)**, 878–886, July 1967.
21. W. K. McRitchie and M. M. Z. Kharadly, Properties of interface between homogeneous and inhomogeneous waveguides. *Proc. IEE*, **112(11)**, 1367–1374, Nov. 1974.
22. E. A. J. Marcatili, Bends in optical dielectric guides. *Bell Syst. Tech. J.*, **48**, 2103–2132, Sept. 1969.
23. L. Lewin, Local form of the radiation condition: application to curved dielectric structures. *Electron Lett.*, **9**, 468–469, Oct. 1973.
24. E. G. Neumann and H. D. Rudolph, Radiation from bends in dielectric and transmission lines. *IEEE Trans. Microwave Theory Tech.*, **MTT-23**, 142–149, Jan. 1975.
25. M. Desai and R. Mittra, A method for reducing radiation losses at bends in open dielectric structures. *IEEE MTT-S Microwave Symp. Digest*, 211–213, 1980.
26. K. Solbach, The measurement of the radiation losses in dielectric image line bends and the calculation of a minimum acceptable curvature radius. *IEEE Trans. Microwave Theory Tech.*, **MTT-27**, 51–53, Jan. 1979.
27. T. Yoneyama, M. Yamaguchi, and S. Nishida, Bends in non-radiative dielectric waveguides. *IEEE Trans. Microwave Theory Tech.*, **MTT-30**, 2146–2150, Dec. 1982.
28. K. Ogusu, Dielectric waveguide corner and power divider with a metallic reflector. *IEEE Trans. Microwave Theory Tech.*, **MTT-32**, 113–116, Jan. 1984.
29. H. Sasaki and I. Anderson, Theoretical and experimental studies on active Y-junctions in optical waveguides. *IEEE J. Quantum Electron.*, **QE-14**, 883–892, Nov. 1978.
30. M. Belanger, G. L. Yip, and M. Haruna, Passive planar multi-branch optical power divider: some design considerations. *Appl. Opt.*, **22**, 2383–2389, Aug. 1983.
31. O. Honaizumi, M. Miyagi, and S. Kawakami, Wide Y-junction with low losses in three-dimensional dielectric optical waveguides. *IEEE Quantum Electron.*, **21(2)**, 168–178, Feb. 1985.
32. H. Sasaki and R. M. De La Rue, Electro-optic Y-junction modulator/switch, *Electron Lett.*, **12**, 459–460, Sept. 1976.
33. K. Ogusu, Experimental study of dielectric waveguide Y-junction for millimetre-wave integrated circuits. *IEEE Trans. Microwave Theory Tech.*, **MTT-33**, 506–509, June 1985.
34. A. Prieto, J. Rodriguez, and M. A. Solano, Y-junctions in dielectric waveguides using very wide-aperture angles. *Electron. Lett.*, **23(4)**, 137–138, Feb. 1987.
35. T. Yoneyama and S. Nishida, Non-radiative dielectric waveguide T-junctions for millimetre-wave applications. *IEEE Trans. Microwave Theory Tech.*, **MTT-33**, 1239–1241, Nov. 1985.
36. M. Dydyk, Image guide: a promising medium for EHF circuits. *Microwaves*, 72–80, Apr. 1980.
37. Y. Shih, J. Rivera, and T. Itoh, Directive planar excitation of an image guide. *IEEE MTT-S Microwave Symp. Digest*, 5–10, 1981.
38. J. A. G. Malherbe, J. H. Cloete, and L. E. Losch, A transition from rectangular to non-rectangular dielectric waveguide. *IEEE Trans. Microwave Theory and Tech.*, **MTT-33**, 539–543, June 1985.

39. D. Dawn and M. Sachidananda, Analysis and design of stripline to NRD guide transition, Proceedings of the 3rd Asia–Pacific Microwave Conference (Tokyo), *Digest*, 15–18, 1990.
40. K. Solbach, Electric probe measurements on dielectric image lines in the frequency range of 26–90 GHz, *IEEE Trans. Microwave Theory Tech.*, **MTT-26**, 755–758, Oct. 1978.
41. A. K. Tiwari, *Investigation on Dielectric Integrated Guides for Millimetre Wave Applications*, Ph.D. Thesis, Indian Institute of Technology, Delhi, 1984.
42. D. D. King, Properties of dielectric image lines. *IRE Trans. Microwave Theory Tech.*, **MTT-5**, 31–35, Mar. 1955.
43. D. D. King and S. P. Schlesinger, Losses in dielectric image lines. *IRE Trans. Microwave Theory Tech.*, **MTT-5**, 31–35, Jan. 1957.
44. K. Solbach, Calculation and measurement of the attenuation constants of dielectric image lines of rectangular cross-section. *Arch. Elek. Ubertragung.*, **32**, 321–328, Aug. 1978.

CHAPTER ELEVEN

Passive Components

11.1 INTRODUCTION

Dielectric integrated guides have been used in the design of a variety of passive and active components, thus facilitating the realization of complete receiver systems. Of the various passive components realized thus far, directional couplers and filters have received maximum attention. The theory and design of these components are discussed in detail in this chapter. Other passive components, namely, power dividers, attenuators, mechanical phase shifters, isolators, and circulators, are briefly reviewed. Components employing semiconductor devices are covered in Chapter 12.

11.2 POWER TRANSFER IN PARALLEL-COUPLED GUIDES

The analysis of parallel-coupled dielectric guides and the formulas for determining the even- and odd-mode propagation parameters are presented in Chapters 2 and 3. In this section, we review the mechanism and formulas governing the power transfer in coupled guides [1–5]. The structures considered are symmetric parallel-coupled guides employing guides of identical cross-sections and asymmetric parallel-coupled guides employing guides of different cross-sections.

11.2.1 Symmetric Parallel-Coupled Guide

Consider two identical dielectric guides placed parallel to each other with one of the guides excited in its dominant mode. Because of the exponentially decaying fields of the feed waveguide, coupling takes place with the second waveguide. The coupled guide structure now supports a new pair of normal modes known as the even (symmetric) and odd (antisymmetric) modes. For the even mode, the transverse electric field components in the two guides point in the same direction, and for the odd mode the fields point in opposite directions. The plane of

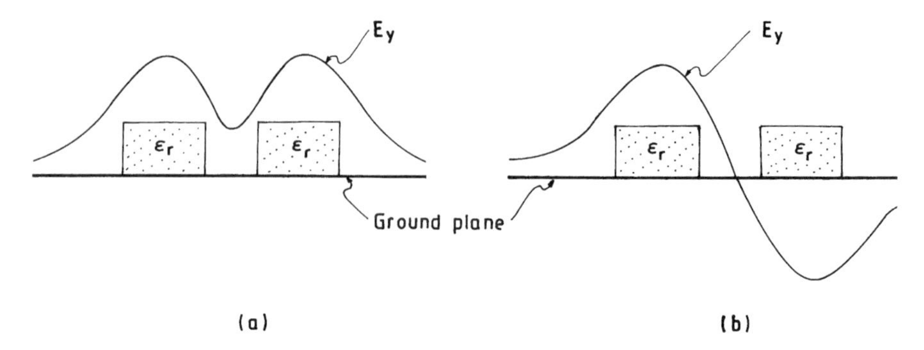

FIGURE 11.1 Modes in symmetric parallel-coupled guide: (a) symmetric or even mode and (b) antisymmetric or odd mode.

symmetry of the coupled structure represents a magnetic wall in the case of the even mode and an electric wall in the case of odd mode. As an illustration, Figure 11.1 shows the typical electric field variation of the dominant even- and odd-modes in a coupled dielectric image guide. Since the two normal modes (even- and odd modes) travel with different phase velocities, regions exist along the length of the structure at which these modes either constructively interfere to produce a power maximum or destructively interfere to produce a null. This aspect is depicted in Figure 11.2.

In order to derive the relation between the length of the coupled guide and coupling, consider a section of coupled guide shown in Figure 11.3. Let port 1 be excited by a generator with voltage V_0. We may consider this excitation as a superposition of even- and odd-mode excitations of the coupled guide with respect to ports 1 and 4. In the case of the even mode, ports 1 and 4 are excited by in-phase voltages each of magnitude $V_0/2$; and in the case of the odd mode, ports 1 and 4 are excited by voltages $V_0/2$ and $-V_0/2$, respectively. If we consider the guides to be lossless and the four ports to be matched, then the voltages $V_A(z)$ and $V_B(z)$ at any distance z along the guides A and B, respectively, can be expressed as

$$V_A(z) = (V_0/2)\,[\exp(-j\beta_e z) + \exp(-j\beta_0 z)] \tag{11.1}$$

$$V_B(z) = (V_0/2)\,[\exp(-j\beta_e z) - \exp(-j\beta_0 z)] \tag{11.2}$$

where β_e and β_o are the even- and odd-mode propagation constants, respectively, of the coupled guide.

The voltage transmission coefficients S_{21} from port 1 to port 2, and S_{31} from port 1 to port 3 are given by

$$S_{21} = V_A(z)/V_0\,|_{z=\ell} \tag{11.3a}$$

$$S_{31} = V_B(z)/V_0\,|_{z=\ell} \tag{11.3b}$$

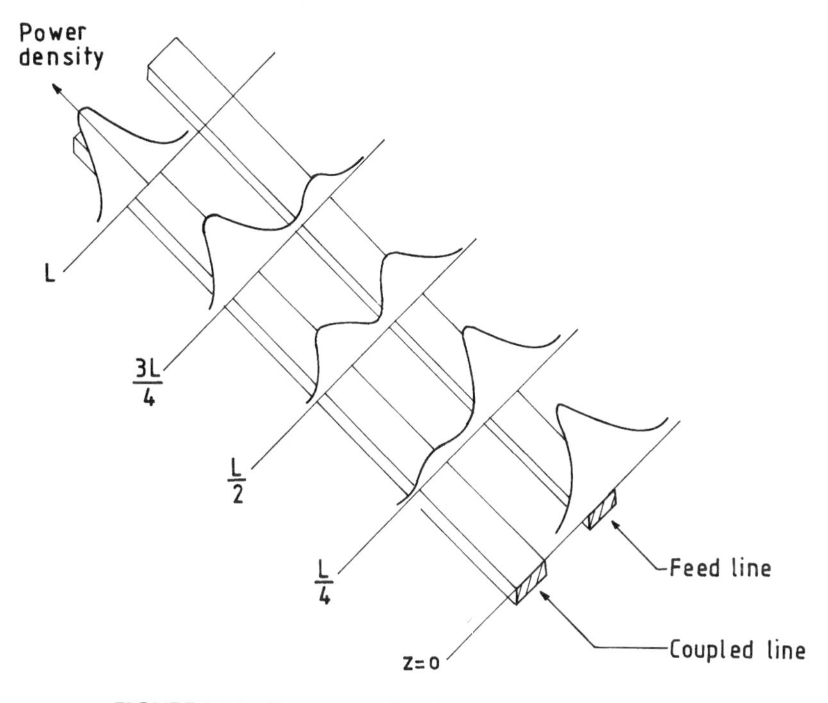

FIGURE 11.2 Power transfer along a pair of coupled lines.

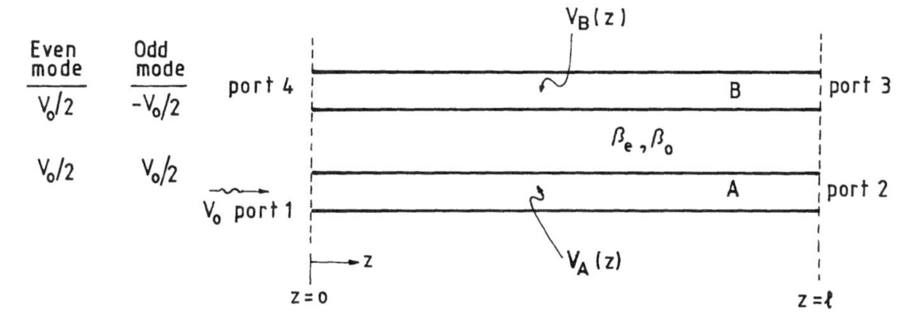

FIGURE 11.3 Parallel symmetrically coupled dielectric guide section.

Because of symmetry, $S_{21} = S_{12}$ and $S_{31} = S_{13}$. Using (11.1) and (11.2) in (11.3), we obtain

$$S_{12} = S_{21} = \exp[-j(\beta_e + \beta_o)\ell/2]\cos[(\beta_e - \beta_o)\ell/2] \tag{11.4a}$$

$$S_{13} = S_{31} = -j\exp[-j(\beta_e + \beta_o)\ell/2]\sin[(\beta_e - \beta_o)\ell/2] \tag{11.4b}$$

Let L be the length at which complete power transfer takes place from guide A to guide B. Setting $S_{12}=0$ at $\ell=L$ gives

$$L=\frac{\pi}{\beta_e-\beta_o} \tag{11.5}$$

Using this result in (11.4), the scattering coefficients can be expressed in terms of the coupling length ℓ as

$$|S_{12}|=|\cos(\pi\ell/2L)|=|\cos[(\beta_e-\beta_o)\ell/2]| \tag{11.6a}$$

$$|S_{13}|=|\sin(\pi\ell/2L)|=|\sin[(\beta_e-\beta_o)\ell/2]| \tag{11.6b}$$

In the unmatched condition, the scattering parameters at each port of the coupled guide section shown in Figure 11.3 are given by [6]

$$S_{11}=\frac{j}{2}\left[\frac{1}{D_e}\left(\frac{\beta}{\beta_e}-\frac{\beta_e}{\beta}\right)\sin(\beta_e\ell)+\frac{1}{D_0}\left(\frac{\beta}{\beta_o}-\frac{\beta_o}{\beta}\right)\sin(\beta_o\ell)\right] \tag{11.7a}$$

$$S_{21}=\frac{j}{2}\left[\frac{1}{D_e}\left(\frac{\beta}{\beta_e}-\frac{\beta_e}{\beta}\right)\sin(\beta_e\ell)-\frac{1}{D_0}\left(\frac{\beta}{\beta_o}-\frac{\beta_o}{\beta}\right)\sin(\beta_o\ell)\right] \tag{11.7b}$$

$$S_{31}=\left(\frac{1}{D_e}-\frac{1}{D_o}\right) \tag{11.7c}$$

$$S_{41}=\left(\frac{1}{D_e}+\frac{1}{D_o}\right) \tag{11.7d}$$

where

$$D_e=2\cos(\beta_e\ell)+j[(\beta/\beta_e)+(\beta_e/\beta)]\sin(\beta_e\ell) \tag{11.8a}$$

$$D_o=2\cos(\beta_o\ell)+j[(\beta/\beta_o)+(\beta_o/\beta)]\sin(\beta_o\ell) \tag{11.8b}$$

and

$$\beta=\frac{\beta_e+\beta_o}{2} \tag{11.9}$$

11.2.2 Nonsymmetric Parallel-Coupled Guide

Dielectric guides of different cross-sections propagate electromagnetic waves with different phase velocities. When two such guides are parallel-coupled (Fig. 11.4(a)), the structure does not possess any clear plane of symmetry. Therefore the even- and odd-mode analysis that is commonly used for symmetric coupled guides is not strictly applicable to the asymmetric coupled guides. We may, however, describe the propagation in the latter in terms of *even-like* and *odd-like*

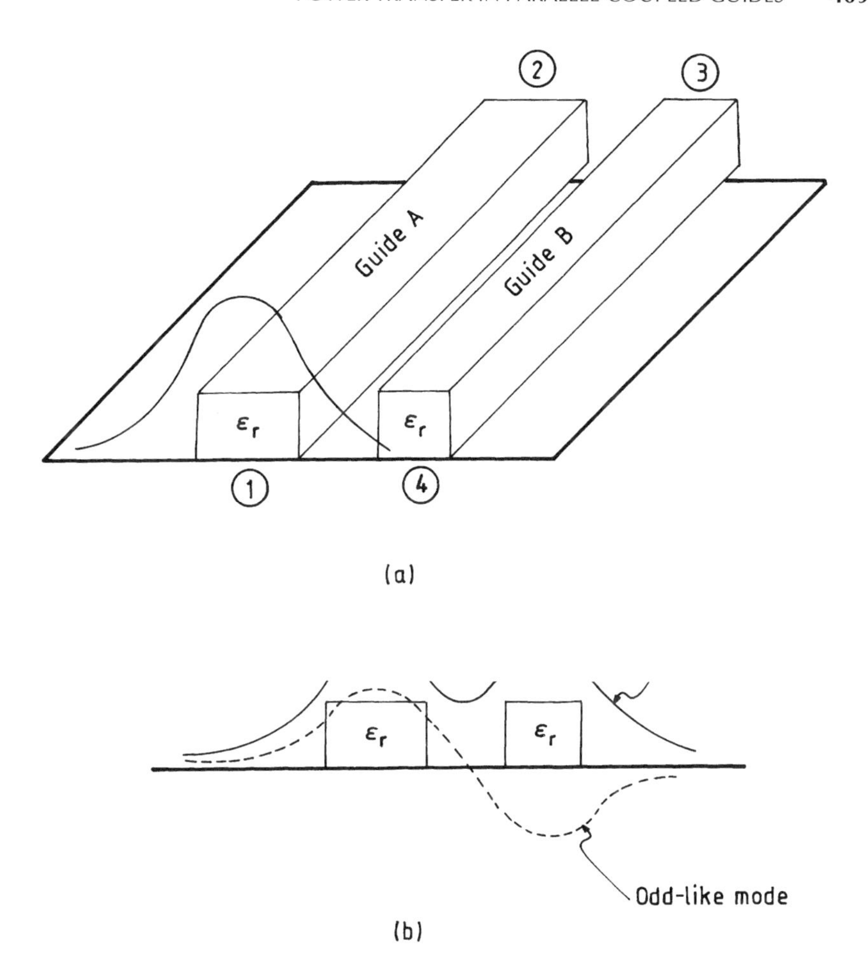

FIGURE 11.4 Nonsymmetrically coupled image guide showing (a) excitation of guide A and (b) its decomposition into even-like and odd-like modes.

modes [7]. Typical transverse mode patterns of an asymmetric coupled image guide corresponding to the E^y_{11} mode excitation of an uncoupled guide are illustrated in Figure 11.4(b). Ikalainen and Matthaei [7] have provided an approximate analysis for the power transfer in such guides, which is summarized below. The power carried by each guide is defined as the square of the magnitude of the total transverse electric field multiplied by a proportionality constant. Referring to Figure 11.4, if we denote E^e_{yi} and E^o_{yi} as the transverse electric fields at the center of guide i (i refers to A or B) for the even-like and odd-like modes, respectively, then the power carried by the guide is given by

$$P_i = G_i |E^e_{yi} + E^o_{yi}|^2 \tag{11.10}$$

where G_i is the proportionality constant for guide i. Since the coupled guide is unsymmetrical, the transverse fields in guide A may be related to those in guide B by another set of proportionality factors, g_e and g_o, which are given by

$$g_{e,o} = E^{e,o}_{yA} / E^{e,o}_{yB} \tag{11.11}$$

where g_o is negative. We now consider that the power fed to guide A at $z = 0$ is unity while that fed to guide B at $z = 0$ is zero. That is, at the center of guide B at $z = 0$, the excitation amplitudes may be taken as $E^e_{yB} = V_0$ and $E^o_{yB} = -V_0$ for the even-like mode and odd-like mode, respectively, so that the total power fed to guide B is zero as per the relation (11.10). Furthermore, we define a normalized wave amplitude $V_i(z)$ for guide i such that the power carried by guide i is given by $|V_i(z)|^2$. The normalized wave amplitude $\bar{V}_A(z)$ and $\bar{V}_B(z)$ can be written as

$$\bar{V}_A(z) = \sqrt{G_A}\,[g_e V_0 \exp(-j\beta_e z) - g_o V_0 \exp(-j\beta_o z)] \tag{11.12a}$$

$$\bar{V}_B(z) = \sqrt{G_B}\,[V_0 \exp(-j\beta_e z) - V_0 \exp(-j\beta_o z)] \tag{11.12b}$$

where

$$V_0 = \frac{1}{\sqrt{G_A[g_e - g_o]}} \tag{11.13}$$

The total power in the coupled guide is given by

$$\begin{aligned} P_T(z) = {} & G_B |V_0 \exp(-j\beta_e z) - V_0 \exp(-j\beta_o z)|^2 \\ & + G_A |g_e V_0 \exp(-j\beta_e z) - g_o V_0 \exp(-j\beta_o z)|^2 \end{aligned} \tag{11.14}$$

For a lossless coupled guide, the total power as a function of z must remain constant. Setting $dP_T/dz = 0$ yields

$$G_B / G_A = -g_e g_o \tag{11.15}$$

Using (11.14) and (11.15), the normalized wave amplitudes $\bar{V}_A(z)$ and $\bar{V}_B(z)$ can be written in the form [7]

$$\bar{V}_A(z) = \{\cos[(\beta_e - \beta_o)z/2] + j[(1 - p_e)/(1 + p_e)] \sin[(\beta_e - \beta_o)z/2]\} \cdot \exp\{-j(\beta_e + \beta_o)z/2\} \quad (11.16a)$$

$$\bar{V}_B(z) = -2j[\sqrt{p_e}/(1 + p_e)] \sin[(\beta_e - \beta_o)z/2] \exp[-j(\beta_e + \beta_o)z/2] \quad (11.16b)$$

where

$$p_e = -g_e/g_o \quad (11.17)$$

For a coupled guide section of length ℓ having all the four ports matched, the scattering parameters can be obtained from (11.16) by setting $z = \ell$. Referring to Figure 11.4(a), we have

$$S_{12} = S_{21} = \{\cos[(\beta_e - \beta_o)\ell/2] + j[(1 - p_e)/(1 + p_e)] \sin[(\beta_e - \beta_o)\ell/2]\} \cdot \exp[-j(\beta_e + \beta_o)\ell/2] \quad (11.18a)$$

$$S_{13} = S_{31} = -2j[\sqrt{p_e}/(1 + p_e)] \sin[(\beta_e - \beta_o)\ell/2] \exp[-j(\beta_e + \beta_o)\ell/2] \quad (11.18b)$$

For a symmetrical coupled guide section employing identical guides, $p_e = 1$. If we set $p_e = 1$ in (11.18), the expressions for S_{12} and S_{13} reduce to those given by (11.4). It is evident from (11.4) that in a coupled line employing identical guides, all the power propagating in one guide can be transferred to the other if the coupling region is long enough. On the other hand, when the two guides have different cross-sections, we note from (11.18) that only partial power transfer can occur from one guide to another. After the maximum possible power transfer has occurred, the power from the second guide starts coupling back to the first guide. As an illustration of the coupling variation, Figure 11.5 shows typical plots of coupled power as a function of distance in an asymmetrical coupled slab guide [7]. It can be seen that the maximum coupling level is higher at lower frequencies, and furthermore, it takes a shorter length for the maximum power transfer to occur. This is because, at lower frequencies, the evanescent fields of the two guides extend laterally over a larger distance so as to cause strong coupling. On the other hand, at higher frequencies, the evanescent fields decay faster with lateral distance, thereby resulting in a weaker coupling level. The above feature can be advantageously utilized to achieve broad bandwidth in couplers. This aspect is illustrated in Figure 11.5 for 3-dB coupling over a band of frequencies. The guide dimensions and spacing are chosen such that, at the lower end of the frequency band, the maximum coupled power is more than half of the input power, whereas the guide length L is chosen longer than that required for maximum power transfer so that the coupled power at the output end reduces to the desired 3-dB value. It can be seen that the curves covering a wide range of frequencies pass close to the 3-dB level for the coupler length L, thereby demonstrating its capability for broadband operation.

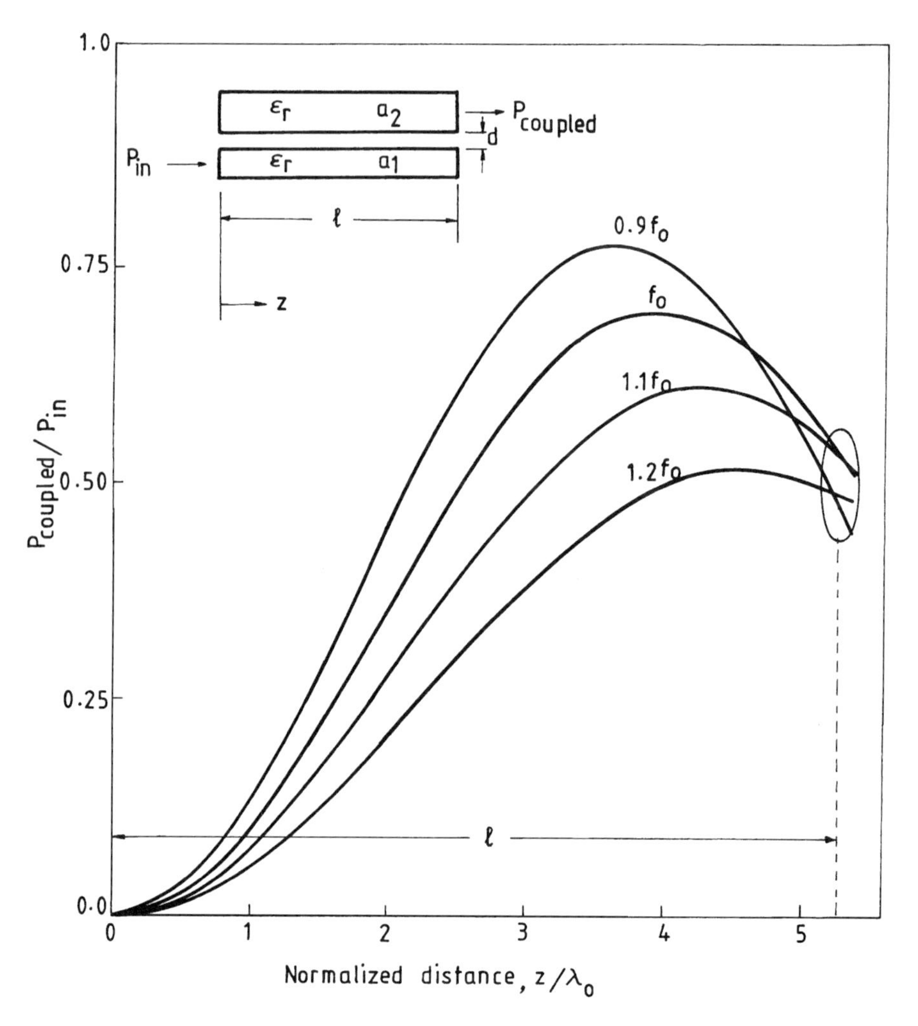

FIGURE 11.5 Coupled power versus normalized distance at various frequencies (for the length ℓ shown, close to 3-dB coupling is achieved over a wide band). $a_1/\lambda_0 = 0.296$, $a_2/\lambda_0 = 0.423$, $d/\lambda_0 = 0.169$, and $\varepsilon_r = 2.25$; λ_0 is the free-space wavelength. (From Ikalainen and Matthaei [7]. Copyright © 1987 IEEE, reprinted with permission.)

11.3 PARALLEL GUIDE DIRECTIONAL COUPLERS

Practical parallel guide couplers incorporate bend sections on either side of the parallel-coupled section in order to achieve decoupling of the four ports. The bend sections form nonuniformly coupled regions with increasing interguide distance. Practical designs must ensure that these bends introduce negligible radiation. Having fixed the bend angle or radius of curvature, the proximity coupling effects must be taken into account in determining the overall coupling.

11.3.1 Symmetric Couplers

The scattering parameters of a uniformly coupled symmetric parallel guide section are given by (11.6). Directional couplers employing such a parallel guide section as the main coupling region and symmetric connecting arms are illustrated in Figure 11.6. The overall scattering parameters of such couplers can be expressed in terms of an effective coupling length ℓ_f. The expressions are given by [8, 9]

$$|S_{12}| = |\cos(\pi \ell_f / 2L)| \tag{11.19a}$$

$$|S_{13}| = |\sin(\pi \ell_f / 2L)| \tag{11.19b}$$

where L is given by (11.5). To a first-order approximation, ℓ_f can be expressed as

$$\ell_f = \ell_0 + [2/(\beta_e - \beta_o)]\Delta\phi = \ell_0 + (2L/\pi)\Delta\phi \tag{11.20a}$$

where

$$\Delta\phi = \int_{z_2}^{z_2} [\beta_e(z) - \beta_o(z)]\, dz \tag{11.20b}$$

where ℓ_0 is the length of the straight coupled section and $(2L/\pi)\Delta\phi$ is the correction factor due to the connecting arms. The integration in (11.20b) is carried out along the axial direction of the coupler from the junction point z_1 (between the uniformly coupled line section and the connecting arms) to the point z_2, where the coupling between the connecting arms reduces to a negligibly small value.

The determination of $\beta_e(z)$ and $\beta_o(z)$ at any distance z in the nonuniformly coupled section requires knowledge of the separation between incremental coupling lengths. For this, we make use of the property that the propagating modes in a dielectric guide are normal to the axial propagation direction, and furthermore, in a coupled guide, the phase fronts can be approximated by cylindrical planes [10]. Referring to Figure 11.6, we may therefore take the separation between the incremental coupling lengths of the two lines in the nonuniformly coupled region as the arc length $2s'$. In the case of a symmetric coupler with straight connecting arms as shown in Figure 11.6(a), the expression for s' is given by

$$s' = \theta\{(s/\sin\theta) + [(z - z_1)/\cos\theta]\} \tag{11.21}$$

where θ is the angle that the connecting arm makes with the axis of symmetry. For a given coupler layout, θ is fixed. Thus the values of $\beta_e(z)$ and $\beta_o(z)$ of the coupled guide at any distance z in the range $z > z_1$ are computed by taking $2s'$ as the spacing in place of $2s$.

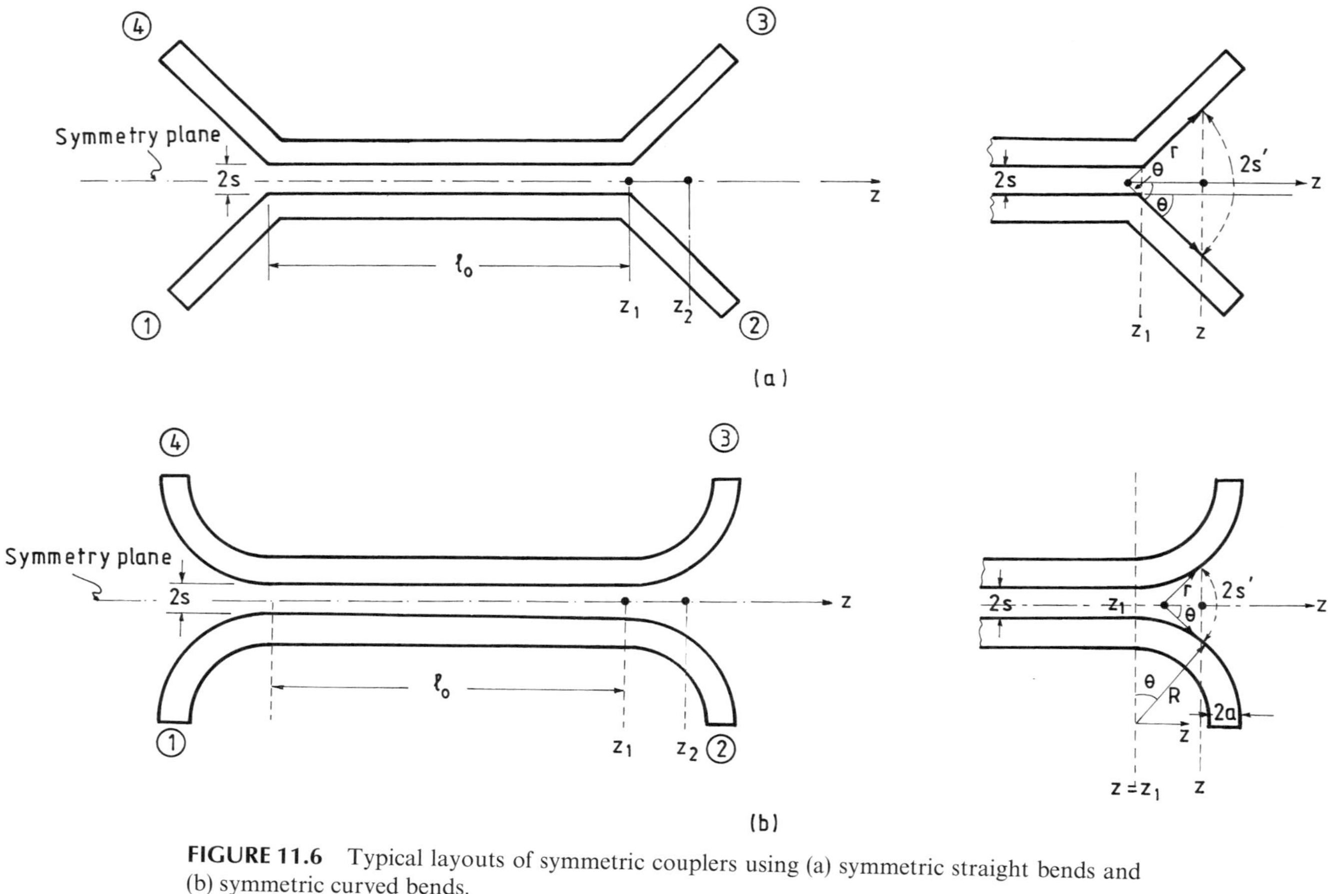

FIGURE 11.6 Typical layouts of symmetric couplers using (a) symmetric straight bends and (b) symmetric curved bends.

In the case of a symmetric coupler using curved connecting arms in the form of circular arcs (Fig. 11.5(b)), the arc length $2s'$ can be obtained from [11]

$$s' = r\theta = \theta[s + (R + a)(1 - \cos\theta)]/\sin\theta \tag{11.22}$$

where r is the radius of the cylindrical phase front, θ is the angle from the incremental coupling length to the symmetry line, and R is the radius of the circular arc. The integral in (11.20b) can be expressed in terms of θ rather than z. Thus

$$\Delta\phi = \int_{\theta=0}^{\pi/2} [\beta_e(\theta) - \beta_o(\theta)]\, d\theta \tag{11.23}$$

The values of $\beta_e(\theta)$ and $\beta_o(\theta)$ for a given angle θ are determined from the same parallel-coupled guide formulas (see Chapter 3) with the spacing between the guide taken as $2s'$.

Figure 11.7 shows typical plots of effective coupling length ℓ_f as a function of frequency for 0-dB and 3-dB couplers in edge-coupled image guide and broadside-coupled image guide configurations [12]. The effective length is computed using (11.20), and the dominant mode propagation constants β_e and β_o of the coupled guides are determined using the mode-matching method. It can be seen from Figure 11.7 that for a fixed set of parameters $b/2a$, ε_r, and s, the coupling length required in the case of a broadside-coupled image guide is considerably less than that required in the case of the edge-coupled image guide. In both cases, a reduction in the gap spacing $2s$ results in a decrease in the variation of ℓ_f with frequency. Thus smaller spacing helps in both reducing the coupler length and widening the bandwidth.

From the scattering coefficient expressions given by (11.19), we note that broadband performance is also possible if the difference in the even- and odd-mode propagation constants (factor $(\beta_e - \beta_o)$) of the parallel-coupled guide is independent of frequency. The even- and odd-mode propagation characteristics of a variety of coupled guide structures are presented in Chapter 5. In the case of the coupled image guide, which is the simplest of the dielectric guide geometries, the value of $(\beta_e - \beta_o)$ for the dominant mode changes sharply with frequency. Consequently, the bandwidth achievable in image guide couplers is quite small (typically 3%). However, couplers using certain modified versions of the coupled image guide are reported to offer broadband performance [13, 14]. One such structure is the directly connected image guide, in which the two guides are tightly coupled by means of an additional dielectric in the coupling region (see inset to Fig. 11.8). With a suitable choice of the guide parameters, the characteristic of $(\beta_e - \beta_o)$ versus frequency for the directly coupled image guide can be made much flatter than that of a coupled image guide. Typical theoretical plots shown in Figure 11.8 illustrate this feature. Using a directly connected image guide, Kim et al. [13] have realized about 28% bandwidth in 3-dB couplers for a coupling tolerance of ± 0.25 dB.

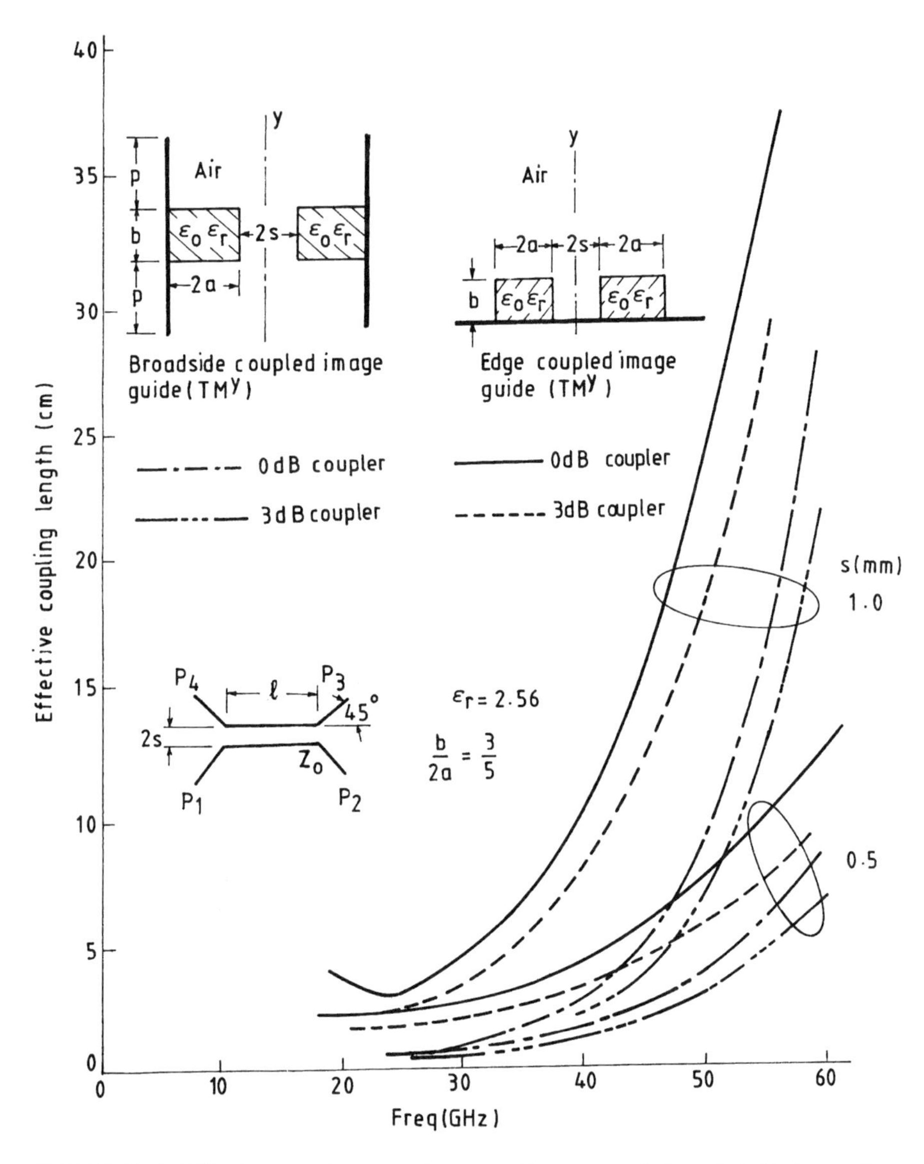

FIGURE 11.7 Effective coupling length versus frequency of 3- and 0-dB broadside-coupled and edge-coupled couplers. (From Bhat and Tiwari [12]. Copyright © 1984 IEEE, reprinted with permission.)

Another structure that is useful for broadband couplers is the coupled π-guide. As an illustration, Figure 11.9(a) shows how the characteristics of $(\beta_e - \beta_o)$ versus frequency of a parallel-coupled π-guide can be flattened by varying the window height. Similar behavior is observed when the width of the window is

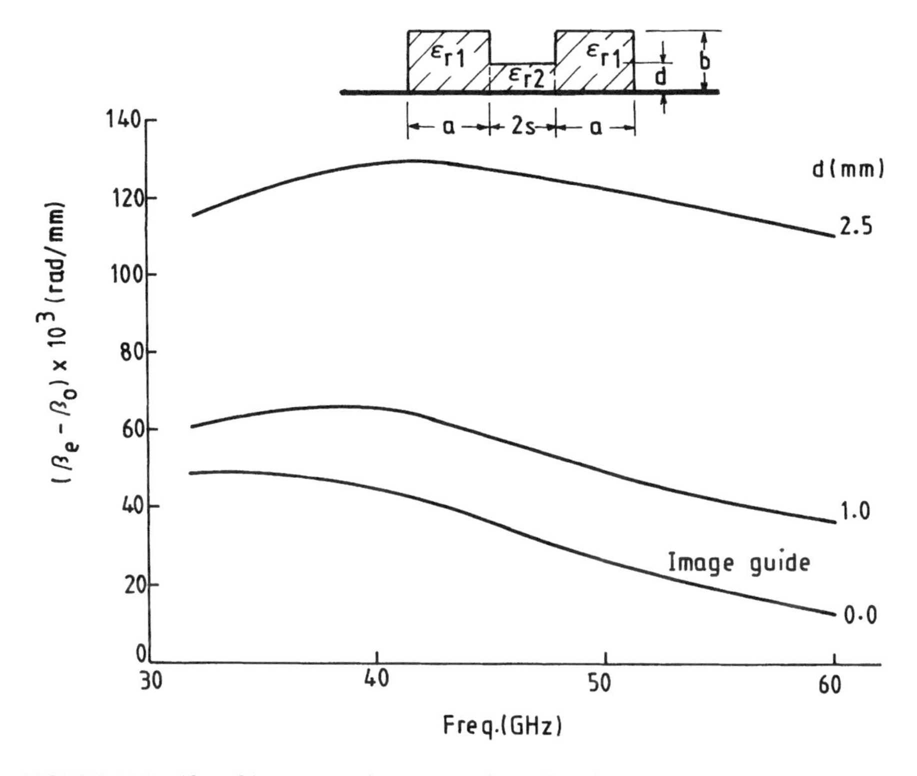

FIGURE 11.8 $(\beta_e - \beta_o)$ versus frequency for directly connected image guide: $\varepsilon_{r1} = \varepsilon_{r2} = 2.1$, $a = 2.5\,\text{mm}$, $b = 2\,\text{mm}$, and $s = 1.5\,\text{mm}$.

varied as shown in Figure 11.9(b). Using coupled π-guides, Rodriguez and Prieto [14] have realized nearly 20% bandwidth in 10-dB couplers for a tolerance of $\pm 0.5\,\text{dB}$.

11.3.2 Nonsymmetric Couplers

Nonsymmetric Couplers Using Identical Guides Figure 11.10(a) shows the typical layout of a parallel guide coupler with nonsymmetric connecting arms. The connecting arms incorporate coupling between a straight guide and a curved guide in the form of a circular arc. As considered in the preceding section, we approximate the phase fronts in the nonuniformly coupled region to cylindrical planes. Referring to Figure 11.10(a), the separation distance s' between incremental coupling lengths in the connecting arm can be written as

$$s' = r\theta = \theta[2s + (R + a)(1 - \cos\theta)]/\sin\theta \tag{11.24}$$

As in the case of the symmetric couplers considered in Section 11.3.1, the effective

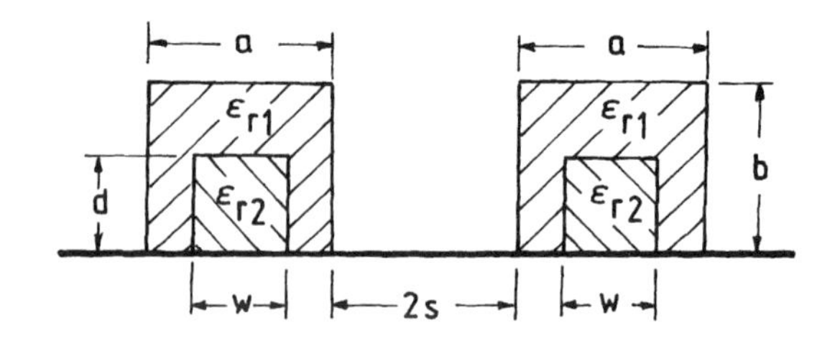

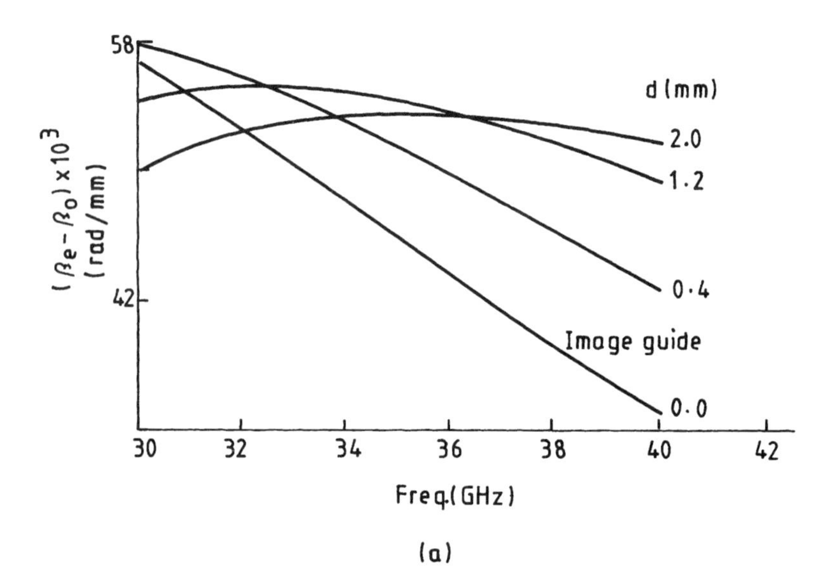

(a)

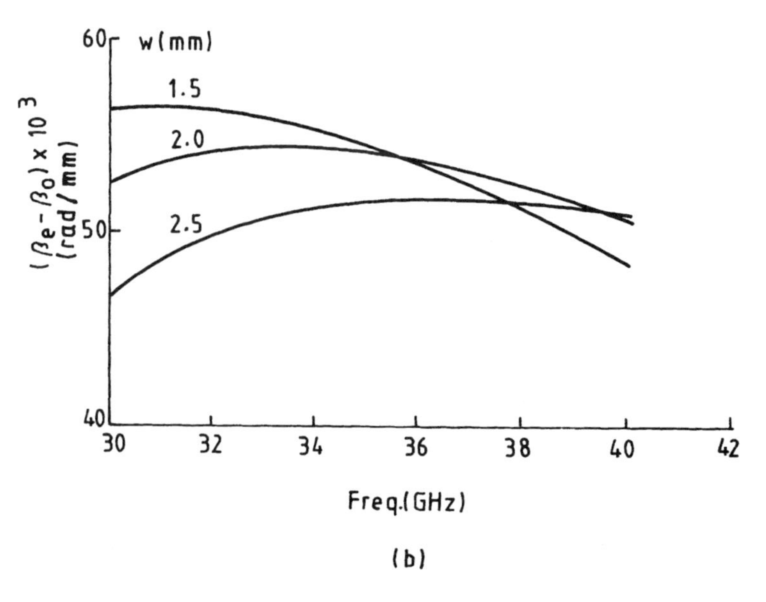

(b)

FIGURE 11.9

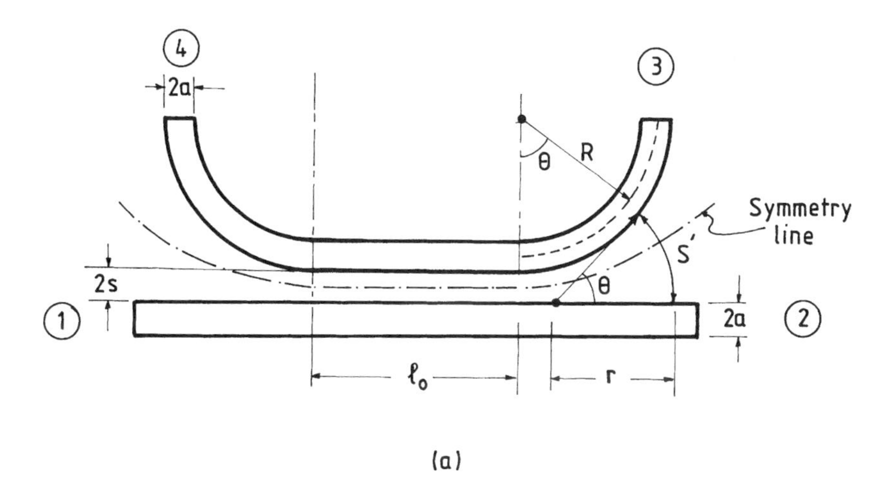

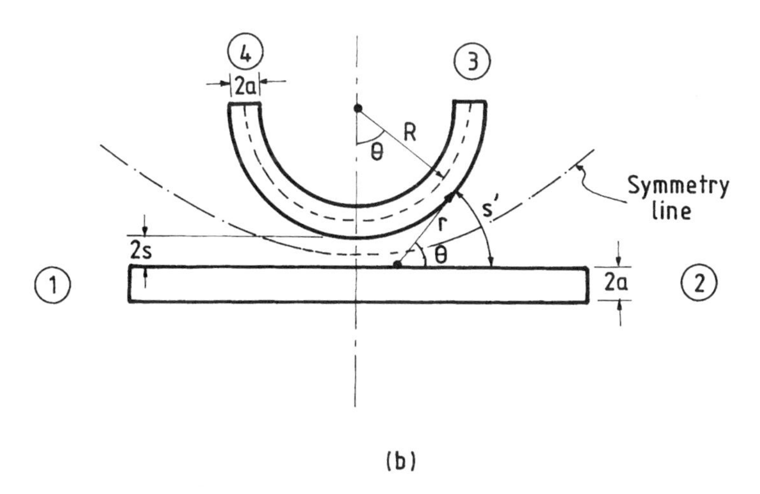

FIGURE 11.10 Nonsymmetric couplers employing (a) parallel-coupled guide section with two nonsymmetric curved arms (b) straight guide coupled to a curved guide.

←

FIGURE 11.9 $(\beta_e - \beta_o)$ versus frequency for coupled π-guide with (a) d as parameter, $w = 2$ mm and (b) w as parameter, $d = 1.5$ mm. $\varepsilon_{r1} = 2.1$, $\varepsilon_{r2} = 1$, $a = 4$ mm, $b = 3$ mm, and $s = 1$ mm. (From Rodriguez and Prieto [14]. Copyright © 1987 IEEE, reprinted with permission.)

coupling length can be obtained from

$$\ell_f = \ell_0 + [2L/\pi]\,\Delta\phi' \tag{11.25a}$$

where

$$\Delta\phi' = \int_{\theta=0}^{\pi/2} [\beta'_e(\theta) - \beta'_o(\theta)]\,d\theta \tag{11.25b}$$

For a given angle θ, the values of $\beta'_e(\theta)$ and $\beta'_o(\theta)$ are obtained from the parallel-coupled guide formulas (see Chapter 3) with s' as the separation distance between the guides. It may be noted that the above formulas assume identical propagation constants for the curved and straight guides. In practical guides, this assumption is not generally valid. In order to account for the asymmetry in coupling, Trinh and Mittra [11] have introduced a correction factor v in the scattering coefficients. The modified formulas are

$$|S_{12}| = |\cos(v\pi\ell_f/2L)| \tag{11.26a}$$

$$|S_{13}| = |\sin(v\pi\ell_f/2L)| \tag{11.26b}$$

For a given frequency, the value of v can be determined experimentally by making a single measurement on $|S_{12}|$ or $|S_{13}|$, and the value of v so obtained remains valid for any arbitrary spacing s'. The same procedure is applicable to the nonsymmetrical coupler shown in Figure 11.10(b) except that $\ell_0 = 0$ in the above set of formulas.

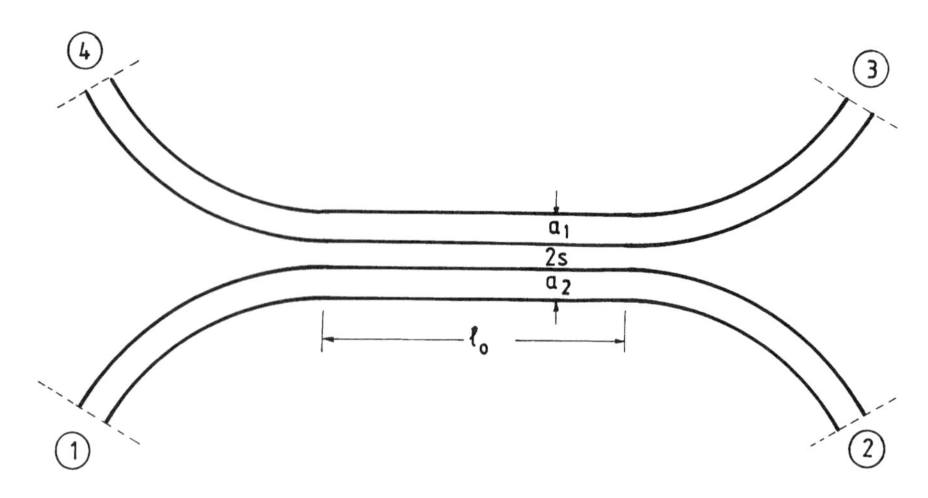

FIGURE 11.11 Nonsymmetric coupler using guides of unequal cross-sections.

Nonsymmetric Coupler Using Guides of Different Cross-Sections The scattering parameters for a section of parallel-coupled guide using guides of nonidentical cross-sections (Fig. 11.11) are given by (11.18). For a practical coupler, which includes the effect of connecting arms, the scattering parameters may be expressed in terms of an effective length ℓ_f as

$$|S_{12}| = [\cos^2(\pi\ell_f/2L) + \{(1 - p_e)/(1 + p_e)\}^2 \sin^2(\pi\ell_f/2L)]^{1/2} \quad (11.27a)$$

$$|S_{13}| = 2[\sqrt{p_e}/(1 + p_e)] \sin(\pi\ell_f/2L) \quad (11.27b)$$

where the expressions for L and ℓ_f are given by (11.15c) and (11.20), respectively. The parameters β_e and β_o refer to the propagation constants of the parallel-coupled guide for the dominant even-like mode and odd-like mode, respectively. The effective length ℓ_f may be determined using the procedure outlined in Section 11.3.1.

11.4 OTHER DIRECTIONAL COUPLERS

11.4.1 Beam-Splitter Type Coupler

The parallel guide couplers considered in the preceding section are of the distributed type. Since these couplers generally employ evanescent field interaction over the coupled length, the bandwidth tends to be narrow. The beam-splitter type directional coupler considered below is based on the principle of the optical beam splitter, where the incident beam splits into a reflected beam and a transmitted beam and is inherently broadband in nature.

Figure 11.12 shows the basic geometry of the beam-splitter coupler. The structure consists of two identical dielectric guides intersecting at right angles with a gap spacing oriented at 45° with respect to each of the four arms of the coupler. An incident wave from port 1 is obliquely incident at 45° at the interface between the dielectric and the air gap. Since the reflected wave is oriented at 90° with respect to the incident wave, the reflected power emerges out of port 2. The transmitted wave, after passing through the gap, is in the same direction as the incident wave and hence emerges out of port 3. No power couples to port 4. Rudokas and Itoh [8] have investigated a beam-splitter coupler in inverted strip dielectric guide where the desired coupling performance is achieved by varying the gap spacing s. Collier and Hjipieris [15] have reported increased bandwidth in such couplers by filling the gap with a dielectric film whose dielectric constant is greater than that of the guide.

Analysis To a first-order approximation, the phase front of the dominant mode in the individual guide may be treated as a plane wave. The coupler is modeled as an infinite dielectric slab of thickness s embedded in another dielectric medium having a different dielectric constant. The incident, reflected, and transmitted rays then correspond to signals at ports 1, 2 and 3, respectively. For a plane wave

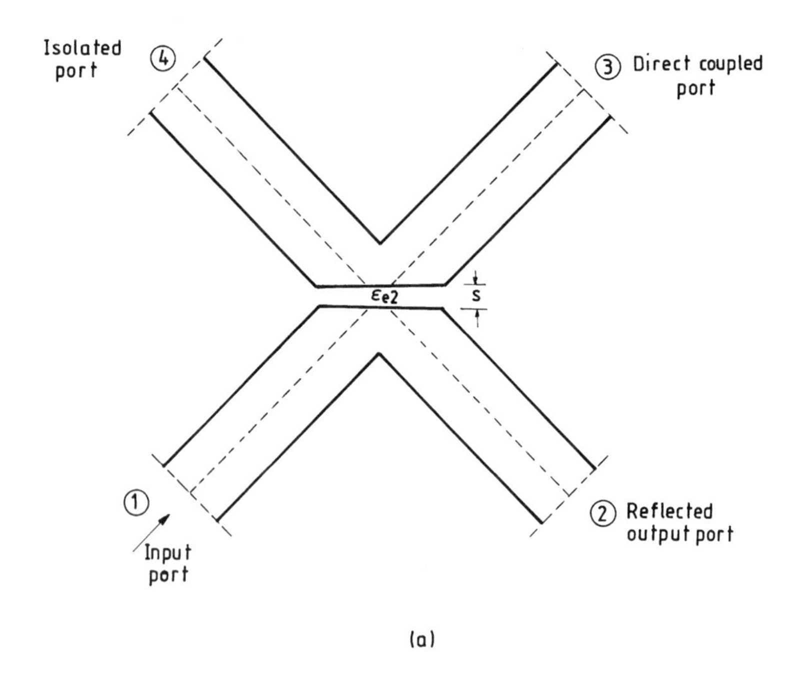

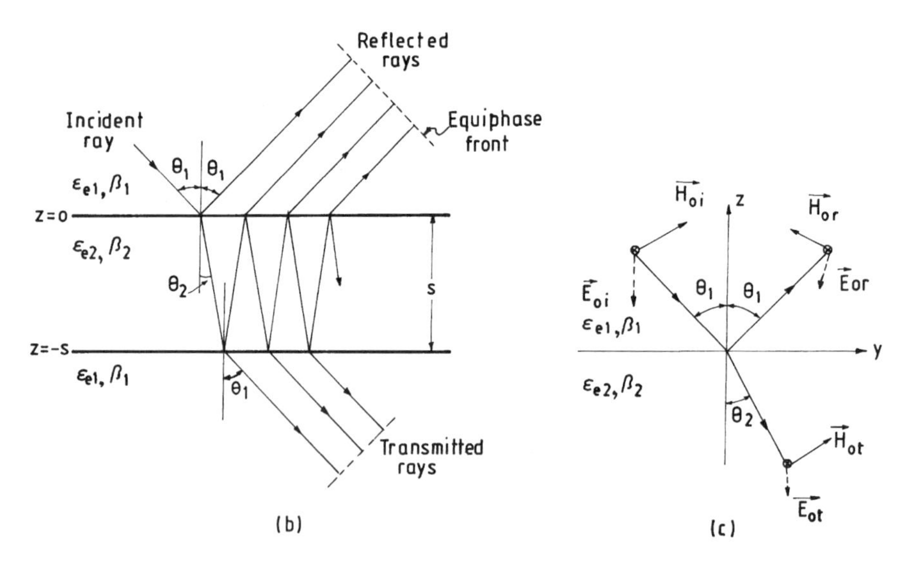

FIGURE 11.12 Beam-splitter coupler: (a) basic geometry, (b) equivalent model showing multiple reflections, and (c) schematic showing **E** and **H** vectors for perpendicular polarization.

incident on the dielectric slab, the overall reflection and transmission coefficients can be derived by considering multiple reflections at the two interfaces (see Fig. 11.12(b)) [15, 16].

Consider a plane wave, represented by a ray, incident on a dielectric interface at an oblique angle θ_1 as shown in Figure 11.12. We assume the wave to be perpendicularly polarized; that is, the **E**-vector is oriented perpendicular to the plane of incidence (yz-plane). This assumption is applicable for an image guide, a π-guide, and other guides, where the **E**-vector for the dominant mode is perpendicular to the ground plane. Referring to Figure 11.12(c), let $\mathbf{E}_i$, $\mathbf{E}_r$, and $\mathbf{E}_t$, respectively, denote the incident, reflected, and transmitted **E**-vectors. These fields can be represented in the form

$$\mathbf{E}_i = \mathbf{E}_{oi} \exp[-j\beta_1 (y \sin\theta_1 - z \cos\theta_1)] \tag{11.28a}$$

$$\mathbf{E}_r = \mathbf{E}_{or} \exp[-j\beta_1 (y \sin\theta_1 + z \cos\theta_1)] \tag{11.28b}$$

$$\mathbf{E}_t = \mathbf{E}_{ot} \exp[-j\beta_2 (y \sin\theta_2 - z \cos\theta_2)] \tag{11.28c}$$

where θ_2 is the angle of refraction and

$$\beta_1 = \omega\sqrt{\mu_0 \varepsilon_1}, \quad \varepsilon_1 = \varepsilon_0\, \varepsilon_{e1} \tag{11.29a}$$

$$\beta_2 = \omega\sqrt{\mu_0 \varepsilon_2}, \quad \varepsilon_2 = \varepsilon_0\, \varepsilon_{e2} \tag{11.29b}$$

The parameters ε_1 and β_1 are the dielectric constant and propagation constant, respectively, of medium 1; and ε_2 and β_2 are the corresponding values for medium 2. The relative dielectric constants of the two dielectric media are denoted as ε_{e1} and ε_{e2} so that they represent the effective relative dielectric constants of the individual guide region and the slab region, respectively, of the coupler made of any dielectric guide geometry.

Applying the continuity conditions on the tangential components of **E** and **H** at $z = 0$, we obtain from (11.28) the following relations:

$$E_{oi} + E_{or} = E_{ot} \tag{11.30a}$$

$$\sqrt{\varepsilon_{r1}}\,(E_{oi} - E_{or}) \cos\theta_1 = \sqrt{\varepsilon_{r2}}\, E_{ot} \cos\theta_2 \tag{11.30b}$$

Solving these equations yields the following expressions for the Fresnel reflection coefficients ρ_0.

$$\rho_0 = \frac{E_{or}}{E_{oi}} = \frac{\sqrt{\varepsilon_{r1}} \cos\theta_1 - \sqrt{\varepsilon_{r2}} \cos\theta_2}{\sqrt{\varepsilon_{r1}} \cos\theta_1 + \sqrt{\varepsilon_{r2}} \cos\theta_2} \tag{11.31}$$

The Fresnel transmission coefficient τ_0 is given by

$$\tau_0 = \frac{E_{ot}}{E_{oi}} = 1 + \rho_0 = \frac{2\sqrt{\varepsilon_{r1}} \cos\theta_1}{\sqrt{\varepsilon_{r1}} \cos\theta_1 + \sqrt{\varepsilon_{r2}} \cos\theta_2} \tag{11.32}$$

We now consider multiple reflections at the two surfaces of the slab as illustrated in Figure 11.12(b). When a ray is incident on the slab surface at $z = 0$ at an angle θ_1, a fraction ρ_0 of the incident amplitude is reflected at angle θ_1 and a fraction $1 + \rho_0$ is refracted into the slab at an angle θ_2. The refracted ray reaches the interface at $z = -s$ with a phase delay of $\beta_2 s \sec\theta_2$. At $z = -s$, a fraction $-\rho_0$ of the refracted ray amplitude is reflected back into the slab toward the first interface and a fraction $1 - \rho_0$ is transmitted out of the slab, and so on. The overall reflection coefficient R is obtained by superposing all the reflected rays at an equiphase front, which is normal to their direction of propagation. Referring to the point of incidence, the expression is given by

$$R = \rho_0 + \sum_{m=1}^{\infty} \rho_m \tag{11.33}$$

where

$$\rho_1 = -\rho_0(1 + \rho_0)(1 - \rho_0)e^{-j2Q} \tag{11.34a}$$

$$\rho_2 = -\rho_0^3(1 + \rho_0)(1 - \rho_0)e^{-j4Q} \tag{11.34b}$$

$$\rho_m = -\rho_0^{2m-1}(1 + \rho_0)(1 - \rho_0)e^{-j2mQ} \tag{11.34c}$$

$$Q = (\beta_2 \sec\theta_2 - \beta_1 \tan\theta_2 \sin\theta_1)s \tag{11.35}$$

According to Snell's law

$$\beta_1 \sin\theta_1 = \beta_2 \sin\theta_2 \tag{11.36}$$

With (11.36), the expression for Q reduces to

$$Q = \beta_2 s \cos\theta_2 \tag{11.37}$$

Substituting (11.34c) in (11.33), the expression for R becomes

$$R = \rho_0 + \frac{(\rho_0^2 - 1)}{\rho_0} \sum_{m=1}^{\infty} \rho_0^{2m} e^{-j2mQ} \tag{11.38}$$

Using the expansion

$$\sum_{m=0}^{\infty} x^m = \frac{1}{1 - x} \tag{11.39}$$

Eq. (11.38) can be recast in the form

$$R = \frac{\rho_0(1 - e^{-j2Q})}{1 - \rho_0^2 e^{-j2Q}} \tag{11.40a}$$

$$= \frac{j(2\rho_0 \tan Q)}{(1 - \rho_0^2) + j(1 + \rho_0^2)\tan Q} \tag{11.40b}$$

The overall transmission coefficient T is similarly obtained by summing up contributions from all the transmitted rays at an equiphase front normal to their direction of propagation. With reference to the point of incidence, the expression for T is obtained as

$$T = \frac{(1-\rho_0^2)\,e^{-jQ}}{1-\rho_0^2\,e^{-j2Q}} \tag{11.41a}$$

$$= \frac{(1-\rho_0^2)\sec Q}{[(1-\rho_0^2)+j(1+\rho_0^2)\tan Q]} \tag{11.41b}$$

Equations (11.40b) and (11.41b) show that the reflection and transmission coefficients differ in phase by $\pi/2$ and this phase difference is independent of frequency and the guide parameters. The power reflection coefficient $|R|^2$ has zeros at

$$Q = n\pi, \quad n = 0, 1, 2, \ldots \tag{11.42a}$$

and maxima at

$$Q = (2n+1)\pi/2, \quad n = 0, 1, 2, \ldots \tag{11.42b}$$

Figure 11.13 shows the variation of $|R|^2$ as a function of Q. It can be seen that around each maximum, the variation of $|R|^2$ as a function of Q (equivalently frequency) is minimum. The maximum value of $|R|^2$ is obtained from (11.40b) as

$$|R|^2 = [2\rho_0/(1+\rho_0^2)]^2 \tag{11.43}$$

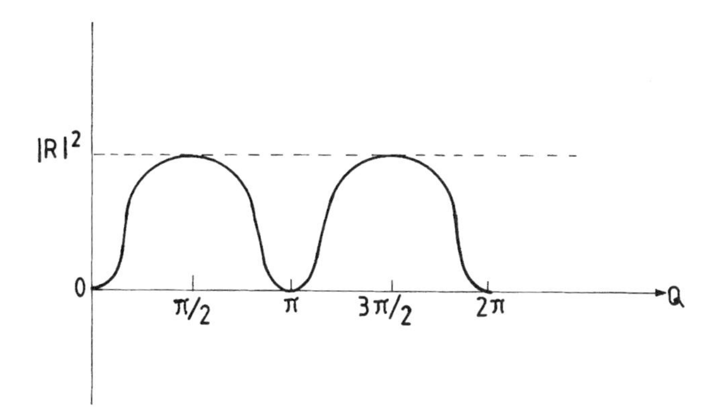

FIGURE 11.13 $|R|^2$ versus $Q = \beta_2 s \cos\theta_2$ for beam-splitter coupler (refer to Fig. 11.12(a) for coupler geometry; R is the reflection coefficient). (From Collier and Hjipieris [15]. Copyright © 1985 IEEE, reprinted with permission.)

Design Formulas [14] Referring to Figure 11.12(a), we define the coupling coefficient C of the coupler as the ratio of the reflected power at port 2 to the incident power at port 1. For broadband performance, the design of the coupler should be centered at $Q = \pi/2$ (or odd multiples of $\pi/2$), where the coupling is

$$C = [2\rho_0/(1 + \rho_0^2)]^2 \tag{11.44}$$

Substituting for ρ_0 from (11.31) for perpendicular polarization and solving yields [15]

$$\frac{\varepsilon_{e2}}{\varepsilon_{e1}} = \frac{(1 + C \pm 2\sqrt{C}) + (1 - C)\tan^2\theta_1}{(1 - C)\sec^2\theta_1} \tag{11.45}$$

For the incidence angle $\theta_1 = \pi/4$, there are two realizable solutions given by

$$\frac{\varepsilon_{e2}}{\varepsilon_{e1}} = \frac{1}{1 + \sqrt{C}} \tag{11.46a}$$

or

$$\frac{\varepsilon_{e2}}{\varepsilon_{e1}} = \frac{1}{1 - \sqrt{C}} \tag{11.46b}$$

Combining (11.46b) with (11.36) and using $Q = \beta_2 s \cos\theta_2 = \pi/2$ and $\theta_1 = \pi/4$, the thickness s of the slab can be expressed in terms of β_2 and C as

$$s = \frac{\pi}{\sqrt{2}\beta_2[(1 + \sqrt{C})]^{1/2}} \tag{11.47}$$

It may be noted that, in practical couplers, the maxima in coupling do not occur exactly at values of Q as defined by (11.42b) because of the frequency dependence of the effective dielectric constant. The frequency shift, however, is quite small to be of any practical significance.

Beam-Splitter Coupler with a Capacitive Diaphragm A modified version of the beam-splitter coupler in an image guide employing a capacitive diaphragm at the airgap has been reported by Birch and Collier [17]. Figure 11.14 shows the geometry of the coupler. The capacitive diaphragm is constructed on a thin film of polythene with an aluminum conductor. The coupling can be varied by changing the height of the capacitive diaphragm (spacing d) above the ground plane.

For a wave incident from port 1, the ratio of the reflected power P_r appearing at port 2 to the incident power P_i at port 1 can be written as [17]

$$\frac{P_r}{P_i} = \frac{A}{1 + A} \tag{11.48}$$

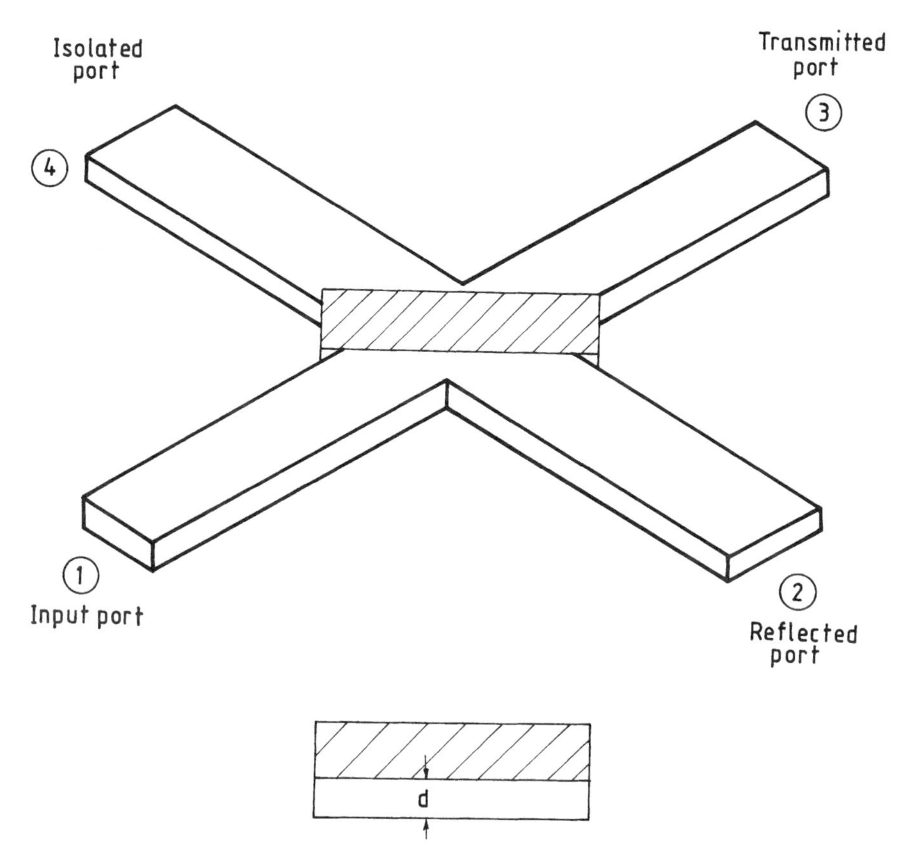

FIGURE 11.14 Image guide coupler with capacitive diaphragm. (From Birch and Collier [17], first presented at the 10th European Microwave Conference, reprinted with permission.)

where

$$A = \left(\frac{\omega C_d}{4Y_0}\right)^2 \tag{11.49}$$

In (11.49), C_d is the capacitance of the diaphragm when viewed obliquely along the image guide and Y_0 is the characteristic admittance of the guide. For a lossless coupler, the ratio of the transmitted power P_t at port 3 to the incident power is

$$\frac{P_t}{P_i} = \frac{1}{1+A} \tag{11.50}$$

An approximate expression for the capacitance of such a diaphragm in waveguide is given by [18]

$$C_d = \left(\frac{8bY_0}{\pi v_p}\right) \ln\left[\operatorname{cosec}\left(\frac{\pi d}{2b}\right)\right] \tag{11.51}$$

where d is the height of the diapragm above the ground plane, b is half the height of the equivalent waveguide, and v_p is the velocity of propagation in the waveguide. Birch and Collier [17] have reported on a 3-dB coupler in an image guide, employing a polystyrene ($\varepsilon_r = 2.5$) guide of cross-section 3.3 mm × 1.6 mm and a polythene film of size 2 cm × 1 cm and thickness 0.1 mm. The coupler is reported to operate over the entire K_a-band with an isolation better than 20 dB and a coupling variation within ± 1dB at the band edges.

11.4.2 Image Guide–Microstrip Coupler

Directional couplers employing coupling between an image guide and a microstrip line through small apertures in the common ground plane have been reported [19, 20]. Figure 11.15(a) shows the schematic of such a coupler and Figure 11.15(b) shows the cross-sectional geometry at the location of an aperture. For a small circular aperture having a diameter much less than the wavelength, coupling from one guide to another can be evaluated from the dipole moments associated with the aperture fields [18, 21]. Let α_e and α_m denote the electric and magnetic polarizabilities, respectively, and E^+, H^+, E^-, and H^- denote the modal fields in the coupled line. The radiated fields in the coupled line due to the electric dipole at the aperture are given by

$$\mathbf{E}_e = A_1 \mathbf{E}^+, \quad \mathbf{H}_e = A_1 \mathbf{H}^+, \quad \text{for } z > 0 \tag{11.52a}$$

$$\mathbf{E}_e = A_2 \mathbf{E}^-, \quad \mathbf{H}_e = A_2 \mathbf{H}^-, \quad \text{for } z < 0 \tag{11.52b}$$

Similarly, the radiated fields in the coupled line due to the magnetic dipole at the aperture are

$$\mathbf{E}_m = B_1 \mathbf{E}^+, \quad \mathbf{H}_m = B_1 \mathbf{H}^+, \quad \text{for } z > 0 \tag{11.53a}$$

$$\mathbf{E}_m = B_2 \mathbf{E}^-, \quad \mathbf{H}_m = B_2 \mathbf{H}^-, \quad \text{for } z < 0 \tag{11.53b}$$

The amplitude constants appearing in (11.52) and (11.53) are given by

$$A_1 = A_2 = -j\omega(\mathbf{E}^- \cdot \mathbf{P})/P_n \tag{11.54a}$$

$$B_1 = B_2 = j\omega\mu_0(\mathbf{H}^- \cdot \mathbf{M})/P_n \tag{11.54b}$$

where

$$P_n = 2\int_S (\mathbf{E}^+ \times \mathbf{H}^+) \cdot \hat{\mathbf{z}}\, ds \tag{11.54c}$$

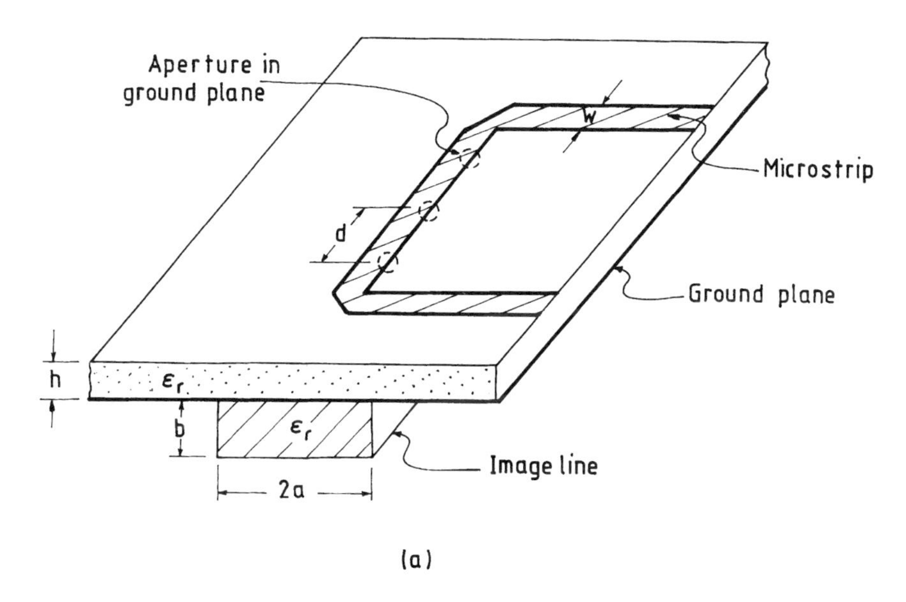

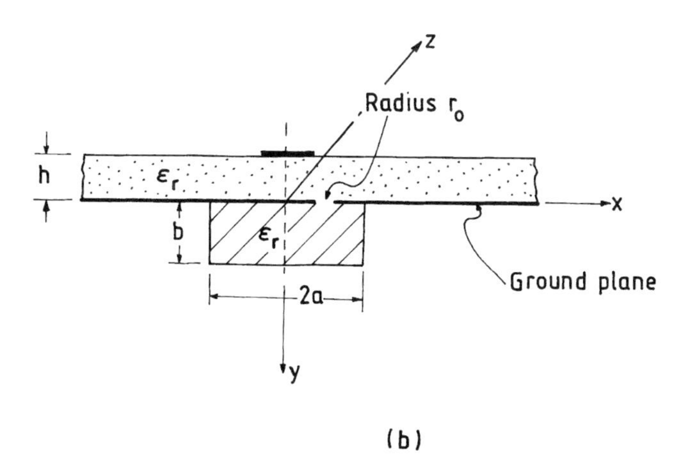

FIGURE 11.15 (a) Schematic of an image guide–microstrip coupler. (b) Cross-section at an aperture. (After Pramanick and Bhartia [19].)

The integration in (11.54c) is carried out over the cross-sectional area a of the primary guide. **P** and **M** are the electric and magnetic polarization vectors, respectively, and are given by

$$\mathbf{P} = -\varepsilon_0 \varepsilon_r \alpha_e \mathbf{E} \tag{11.55a}$$

$$\mathbf{M} = -\alpha_m \mathbf{H} \tag{11.55b}$$

The total radiated fields **E** and **H** in the coupled guide are

$$\begin{aligned} \mathbf{E} &= (A_1 + B_1)\mathbf{E}^+, \quad z > 0 \\ &= (A_2 + B_2)\mathbf{E}^-, \quad z < 0 \end{aligned} \tag{11.56a}$$

$$\begin{aligned} \mathbf{H} &= (A_1 + B_1)\mathbf{H}^+, \quad z > 0 \\ &= (A_2 + B_2)\mathbf{H}^-, \quad z < 0 \end{aligned} \tag{11.56b}$$

Fields in Image Guide Referring to the coordinate system given in Figure 11.15(b), the transverse field components for the E_{11}^y mode of the image guide can be expressed in the form [22]

$$E_y = \begin{cases} E_0 \cos(k_x x)\cos(k_y y)\exp(\pm jk_z z), & |x| \le a,\ y \le b \\ E_0 \cos(k_x a)\cos(k_y y)\exp[-k_{x0}(|x|-a)]\exp(\pm jk_z z), & |x| > a,\ y \le b \\ E_0 \cos(k_x x)\cos(k_y b)\exp[-k_{y0}(y-b)]\exp(\pm jk_z z), & |x| \le a,\ y > b \\ 0, & |x| > a,\ y > b \end{cases} \tag{11.57}$$

$$H_x = -\frac{E_y}{\eta} \tag{11.58}$$

where

$$\eta = \eta_0 \frac{k_0}{k_z} = \frac{\eta_0}{\eta_{\text{eff}}} \tag{11.59a}$$

$$\eta_0 = \sqrt{\frac{\mu_0}{\varepsilon_0}} \tag{11.59b}$$

$$\eta_{\text{eff}} = \frac{k_z}{k_0} \tag{11.59c}$$

$$k_0 = \omega\sqrt{\mu_0 \varepsilon_0} \tag{11.59d}$$

The parameters k_x and k_y are the x- and y-directed propagation constants in the dielectric region, and k_{x0} and k_{y0} are the respective quantities in the air region; k_z is the propagation constant in the direction of propagation and k_0 is the free-space propagation constant. Using the above field expressions in (11.54c), P_n is obtained as

$$P_n = a\frac{b}{\eta}\left(1 + \frac{\sin(2k_x a)}{(2k_x a)}\right)\left(1 + \frac{\sin(2k_y b)}{(2k_y b)}\right) \tag{11.60}$$

Fields in the Microstrip Referring to the microstrip geometry shown in Figure 11.15(b), we note that the main field components contributing to the aperture

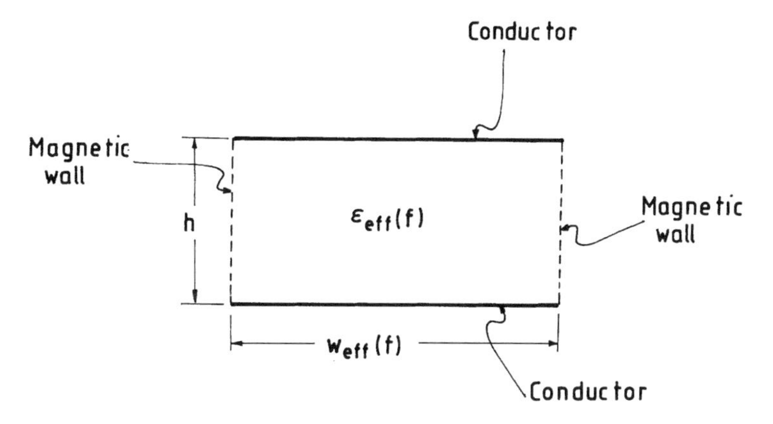

FIGURE 11.16 Parallel-plate waveguide model of microstrip.

coupling are E_y and H_x. Using the parallel-plate waveguide model of the microstrip as shown in Figure 11.16 [23], we can write

$$E_y = E_0 \exp(\pm j\beta_m z) \tag{11.61}$$

$$H_x = \mp \frac{\sqrt{\varepsilon_e(f)}E_y}{\eta_0} \tag{11.62}$$

where

$$\beta_m = k_0\sqrt{\varepsilon_e(f)} \tag{11.63}$$

and $\varepsilon_e(f)$ is the effective dielectric constant of the microstrip. The effective width of the parallel-plate waveguide model is given by [24]

$$w_e(f) = w + \frac{w_{\text{eff}} - w}{1 + (f/f_g)} \tag{11.64}$$

where

$$f_g = \frac{v_0}{2w\sqrt{\varepsilon_r}} \tag{11.65}$$

v_0 is the free-space velocity and w is the width of the microstrip. Using the above expressions for E_y and H_x in (11.54c), the expression for P_n is obtained as [19]

$$P_n = \left(\frac{h}{\eta_0}\right) w_e(f)\sqrt{\varepsilon_e(f)} \tag{11.66}$$

where h is the height of the microstrip substrate. The above expressions are useful for deriving the coupling coefficients of a single aperture.

Coupling Coefficients For a circular aperture of radius r_0, Bhartia and Pramanick [20] have reported the following expressions for the forward and backward coupling coefficients (denoted as C_f and C_b, respectively).

For microstrip to image line:

$$C_f = -20\log|(4r_0^3\omega/3K_1)[\varepsilon_0\varepsilon_r\eta/2 + \mu_0\varepsilon_e(f)/\eta_0]| \tag{11.67}$$

$$C_b = -20\log|(4r_0^3\omega/3K_1)[\varepsilon_0\varepsilon_r\eta/2 - \mu_0\varepsilon_e(f)/\eta_0]| \tag{11.68}$$

where

$$K_1 = \{ab[1 + (\sin(2k_x a)/2k_x a)][1 + (\sin(2k_y b)/2k_y b)]\}^{-1} \tag{11.69}$$

For image guide to microstrip:

$$C_f = -20\log|(4r_0^3\omega/3K_2)[\varepsilon_0\varepsilon_r\eta_0/2 + \mu_0\varepsilon_e(f)/\eta]| \tag{11.70}$$

$$C_b = -20\log|(4r_0^3\omega/3K_2)[\varepsilon_0\varepsilon_r\eta_0/2 - \mu_0\varepsilon_e(f)/\eta]| \tag{11.71}$$

where

$$K_2 = hw_e(f)\sqrt{\varepsilon_e(f)} \tag{11.72}$$

For coupling from either medium to the other, the directivity is given by

$$D = 20\log\left|\frac{\varepsilon_0\varepsilon_r\eta_0\eta + 2\mu_0\varepsilon_e(f)}{\varepsilon_0\varepsilon_r\eta_0\eta - 2\mu_0\varepsilon_e(f)}\right| \tag{11.73}$$

Directional Coupler Design Formulas Practical directional couplers make use of an array of apertures on the common ground to achieve the desired coupling and directivity. For a coupler with symmetrical array of $N+1$ elements, the forward and backward coupled waves are given by

$$C_F = A\exp(-jN\beta_m d)\sum_{n=0}^{N} C_n\exp[-jn(k_z - \beta_m)d] \tag{11.74}$$

$$C_B = A\sum_{n=0}^{N} D_n\exp[-jn(k_z + \beta_m)d] \tag{11.75}$$

For a symmetric array of apertures $r_i = r_{N-i}$. With $C_n = d_n T_f$, $D_n = d_n T_b$, and $\theta = 0.5\,(k_z + \beta_m)d$, the expressions for C_F, C_B, and directivity D are

$$C_F = -20\log|T_f| - 20\log\left|\sum_{n=0}^{M} 2d_n\cos(N-2n)(\theta - \beta_m d)\right| \tag{11.76}$$

$$C_B = -20\log|T_b| - 20\log\left|\sum_{n=0}^{M} 2d_n\cos(N-2n)\theta\right| \tag{11.77}$$

$$D = C_B - C_F = -20\log\left|\frac{T_b}{T_f}\right| - 20\log\left|\frac{\sum_{n=0}^{M} 2d_n\cos(N-2n)\theta}{\sum_{n=0}^{M} 2d_n\cos(N-2n)(\theta - \beta_m d)}\right| \tag{11.78}$$

We note that for $\theta = \pi/2$, the backward coupling $C_B = 0$ and the aperture spacing $d = \pi/(\beta_m + k_z)$. In (11.76) and (11.77) $M = (N-1)/2$ for N odd and $M = N/2$ for N even.

11.4.3 Branch Guide Coupler

The geometry of the branch guide coupler shown in Figure 11.17(a) is similar to the branch line coupler in the microstrip. It consists of a branched coupling region that is connected to uniform input and output guides by curved transition sections. Xu [25] has analyzed this structure from the viewpoint of scattering of incident surface waves by the nonuniform dielectric structure as a whole. Making use of the symmetry, one-quarter of the structure is analyzed by using a staircase approximation to the continuous transition profile. The method combines the building block approach of multimode network theory with a rigorous mode-matching procedure. Figure 11.17(b) shows the variation in the power coupled to the various ports and also the radiated power as a function of transition length [25]. It can be seen that when the transition length L_t is small ($L_t \leq 0.5\lambda$, λ is the wavelength in the guide structure), the power incident from port 1 mostly reaches the direct port 2 and couples lightly to port 4 through the branching coupling region. Furthermore, the radiation loss P_a that is mainly due to the curved transition is high. With an increase in the transition length beyond 0.5λ, the radiation loss decreases sharply and saturates to a low value for w greater than about 1.5λ. The residual part of radiation loss that remains essentially unchanged is contributed by the branched coupling region, and this loss can be minimized by making the separation distance h small. It is interesting to note that in the region $L_t > 1.5\lambda$, where the radiation loss is negligible, the power division between ports 2 and 4 varies continuously and power coupled to port 3 remains negligible. Typical characteristics shown in Figure 11.17(b) indicate that this branching guide configuration can be used for the design of forward wave couplers.

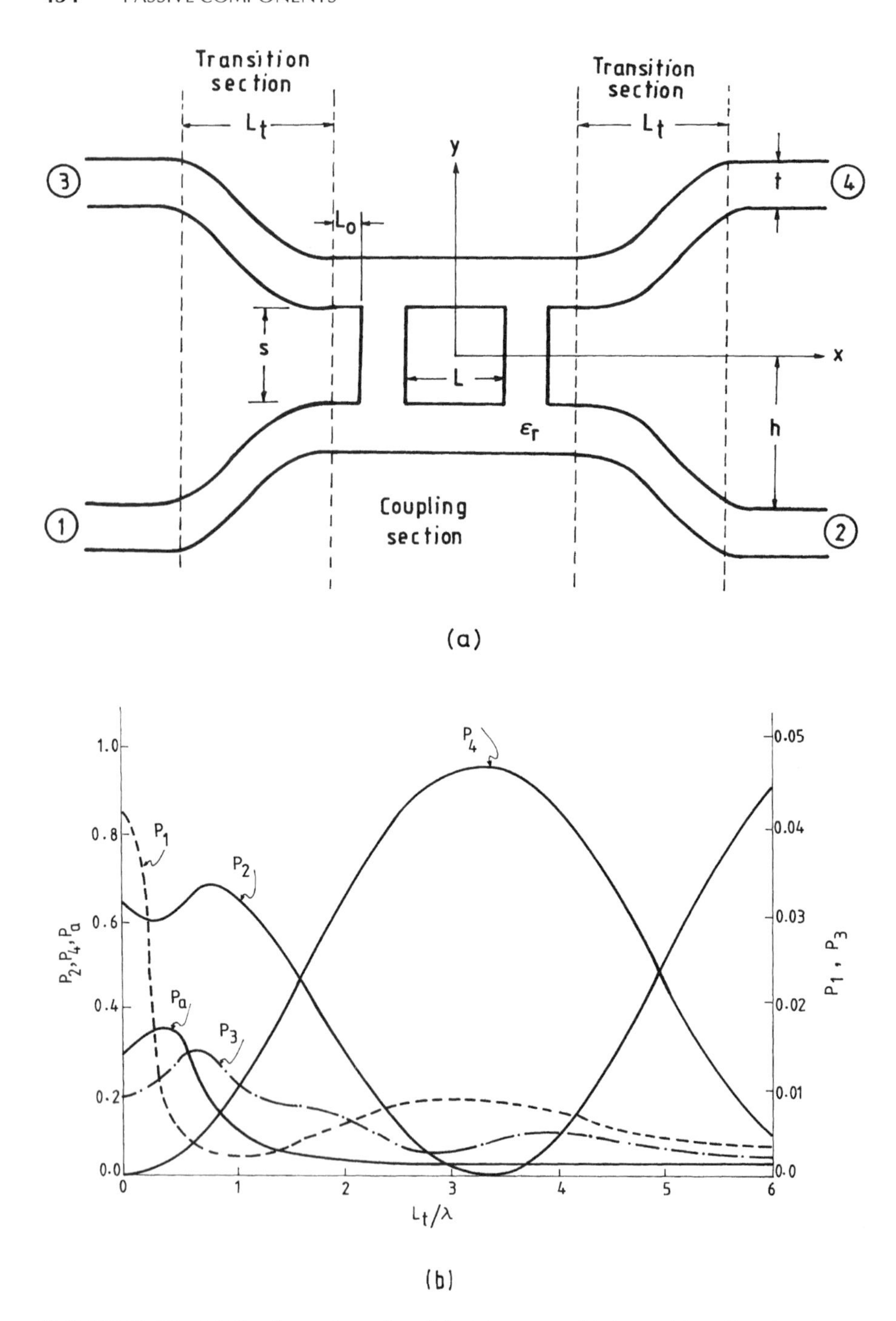

FIGURE 11.17 (a) Configuration of a dielectric waveguide branching directional coupler. (b) Dependence of scattered power on transition length. TE mode: $t = 0.2\lambda$, $s = 0.2\lambda$, $h = 0.3\lambda$, $L = 3\lambda$, $L_0 = \lambda$, and $\varepsilon_r = 2.56$. (From Xu [25], reprinted with permission of IEE.)

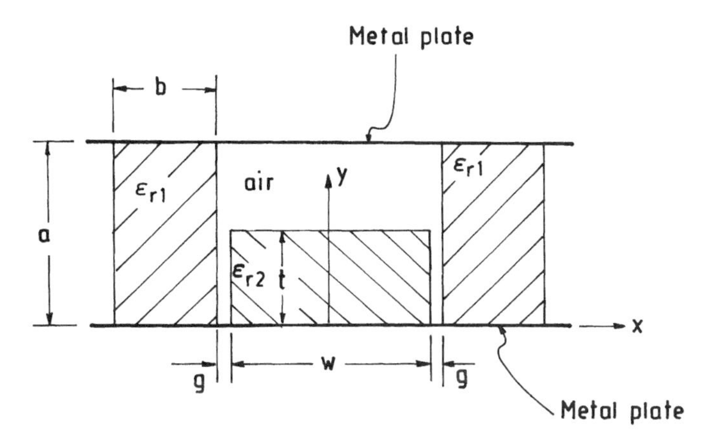

FIGURE 11.18 Cross-sectional view of leaky-wave NRD guide coupler. (After Niu et al. [27].)

11.4.4 Leaky-Wave NRD Guide Coupler

The coupling mechanism in a conventional parallel-coupled guide coupler has been described in Section 11.2. The coupling, which is due to proximity effects, is a function of the spacing between the two guides. At millimeter and optical frequencies, the spacing between the guides becomes prohibitively small especially for tight couplings.

In a leaky-wave coupler, the coupling between the two guides does not primarily depend on the spacing between the guides. The first leaky-wave type coupler using a strip guide structure has been reported by Hue et al. [26] and that using a NRD guide was recently reported by Niu et al. [27]. Figure 11.18 shows a cross-sectional view of NRD guides and a dielectric slab of thickness $t \leqslant a$. The slab guide structure supports several LSE and LSM propagating modes. If the effective refractive index (β_s/k_0) of a particular mode in the slab guide is more than that of the NRD guide (β_n/k_0), leakage takes place. β_s and β_n are the propagation constants of the slab guide and the NRD guide, respectively. The leaky waves are incident on the secondary NRD guide, resulting in field buildup along the guide. Thus power from the primary NRD guide gets coupled to the secondary NRD guide.

11.5 RING RESONATOR FILTERS

The analyses and characteristics of disk, ring, and rectangular shaped resonators are presented in Chapter 9. While all these basic resonator elements have been employed in a variety of millimeter wave devices, the ring resonator is found to be particularly suitable for the realization of bandpass filters. Ring resonators do not have end effects as the rectangular resonators do. Disk resonators need to be necessarily small in order to prevent higher order mode generation, which at millimeter frequencies become inconvenient to handle. Ring resonators do not have this problem since the cross-section of the ring guide can be kept small to permit the presence of only the dominant mode while the ring diameter can be increased to accommodate convenient sizes. The main drawback of ring resonators in image guide and other open dielectric guide configurations is that of radiation. The NRD guide, which is basically nonradiative, is ideal for ring resonator filters. In open guide structures, the ring circumference must be several wavelengths long in order to reduce the radiation to a tolerable limit. This results in numerous closely spaced resonances. The spurious resonances, however, can be suppressed by cascading two or more rings in a filter circuit. The resonance properties of ring circuits have been well studied [28, 29]. Besides being useful in the design of conventional bandpass filters [30–32], the dielectric ring resonator, by virtue of its traveling wave properties, has found applications in directional filters at millimeter wave frequencies [33].

11.5.1 Basic Ring Circuit

As a background to the design of ring resonator filters, we provide below the analysis of a basic ring circuit coupled to a waveguide [29]. Figure 11.19(a) shows the schematic of a single-ring circuit. Directional coupling can be achieved by means of two coupling holes spaced by $\lambda/4$ (quarter wavelength in the guide) apart as indicated in Figure 11.19(b). With this, waves progress only in one direction in the ring. Nondirectional coupling, which provides for waves progressing in both directions, is achieved by a single coupling hole (see Fig. 11.19(c)).

Let V_{ni} and V_{nr} with $n = 1$ to 4 denote the incident and reflected waves, respectively, at each port of a four-port network. Referring to Figure 11.19(b) and considering the symmetry plane AA' as a phase reference, we can write

$$S_{11} = S_{22} = S_{33} = S_{44} \tag{11.79a}$$

$$S_{12} = S_{34}, \quad S_{13} = S_{24}, \quad S_{14} = S_{23} \tag{11.79b}$$

where the scattering parameters S_{13} and S_{14} are the coupling factors. Let port 2 be terminated in a matched load so that $V_{2i} = 0$, and let ports 3 and 4 be connected to form a ring. If ℓ is the mean circumferential length of the ring, we can write

$$V_{3i} = V_{4r} e^{-j\beta\ell} \tag{11.80a}$$

$$V_{4i} = V_{3r} e^{-j\beta\ell} \tag{11.80b}$$

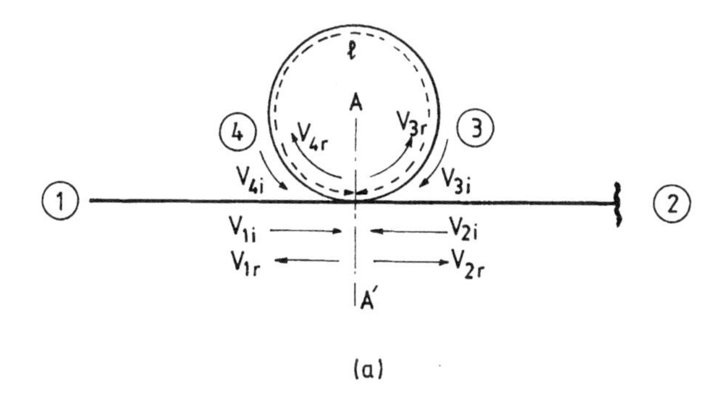

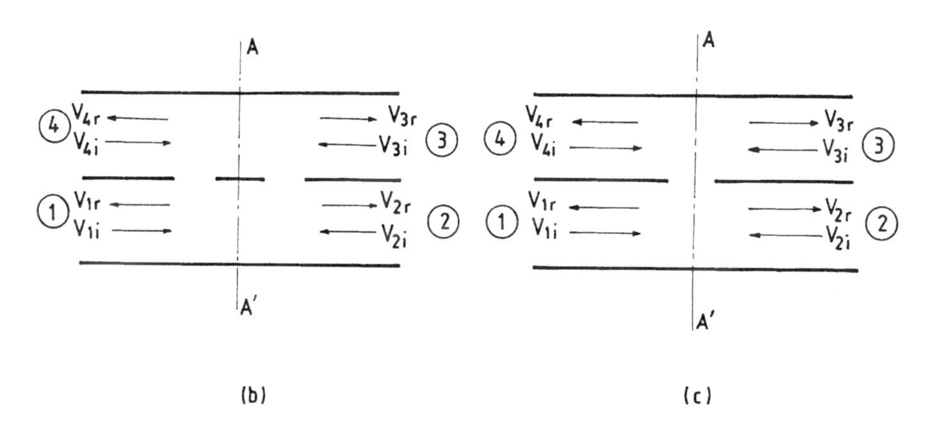

FIGURE 11.19 Schematic of (a) ring circuit, (b) circuit showing directional coupling, and (c) circuit showing nondirectional coupling. (After Tischer [29].)

The reflected voltage waves can be written in terms of the incident waves as

$$V_{1r} = S_{11}V_{1i} + S_{13}V_{3i} + S_{14}V_{4i} \tag{11.81a}$$

$$V_{2r} = S_{21}V_{1i} + S_{23}V_{3i} + S_{24}V_{4i} \tag{11.81b}$$

$$V_{3r} = S_{31}V_{1i} + S_{33}V_{3i} + S_{34}V_{4i} \tag{11.81c}$$

$$V_{4r} = S_{41}V_{1i} + S_{43}V_{3i} + S_{44}V_{4i} \tag{11.81d}$$

Combining (11.80) with (11.81c) and (11.81d), we can express V_{3r} and V_{4r} in terms of V_{1i}.

$$V_{3r} = V_{1i}\left[\frac{S_{13}(1 - S_{12}e^{-j\beta\ell}) + S_{14}S_{11}e^{-j\beta\ell}}{(1 - S_{12}e^{-j\beta\ell})^2 - S_{11}^2e^{-j2\beta\ell}}\right] \tag{11.82a}$$

$$V_{4r} = V_{1i}\left[\frac{S_{14}(1 - S_{12}e^{-j\beta\ell}) + S_{13}S_{11}e^{-j\beta\ell}}{(1 - S_{12}e^{-j\beta\ell})^2 - S_{11}^2e^{-j2\beta\ell}}\right] \tag{11.82b}$$

Using (11.80) and (11.82) in (11.81a) and (11.81b), the expressions for the overall input reflection coefficient ρ_{in} and the transmission coefficient T are obtained as

$$\rho_{in} = \frac{V_{1r}}{V_{1i}} = S_{11} + \left[\frac{2S_{13}S_{14}(1 - S_{12}e^{-j\beta\ell})e^{-j\beta\ell} + (S_{13}^2 + S_{14}^2)S_{11}e^{-j2\beta\ell}}{(1 - S_{12}e^{-j\beta\ell})^2 - S_{11}^2 e^{-j2\beta\ell}} \right] \quad (11.83a)$$

$$T = \frac{V_{2r}}{V_{1i}} = S_{12} + \left[\frac{2S_{13}S_{14}S_{11}e^{-j2\beta\ell} + (S_{13}^2 + S_{14}^2)(1 - S_{12}e^{-j\beta\ell})e^{-j\beta\ell}}{(1 - S_{12}e^{-j\beta\ell})^2 - S_{11}^2 e^{-j2\beta\ell}} \right] \quad (11.83b)$$

These equations describe the property of the ring circuit as a resonant filter. We now consider the cases of directional coupling and nondirectional coupling separately.

Directional Coupling For directional coupling, we consider the waves to progress in one direction only; that is, $V_{4r} = 0$ and hence $S_{14} = 0$. Furthermore, if we assume the input port to be matched ($S_{11} = 0$), the expressions in (11.82) and (11.83) simplify to

$$C = \frac{V_{3r}}{V_{1i}} = \frac{S_{13}}{(1 - S_{12}e^{-j\beta\ell})} \quad (11.84a)$$

$$\rho_{in} = 0 \quad (11.84b)$$

$$T = S_{12} + CS_{13} \quad (11.84c)$$

where C is the voltage coupling coefficient.

Nondirectional Coupling

$$C = \frac{V_{3r}}{V_{1i}} = \frac{V_{4r}}{V_{1i}} = \frac{S_{13}}{1 - (S_{11} + S_{12})e^{-j\beta\ell}} \quad (11.85a)$$

$$\rho_{in} = S_{11} + 2CS_{13}e^{-j\beta\ell} \quad (11.85b)$$

$$T = S_{12} + 2CS_{13}e^{-j\beta\ell} \quad (11.85c)$$

It may be noted that the coupling coefficient C appears in the relation for ρ_{in} and T, indicating that part of the energy in the ring gets coupled back into the main guide. It is the interaction between this coupled wave and the wave in the main guide that determines the properties of the circuit.

11.5.2 Single-Ring Filter

We present below the circuit analysis of a filter consisting of a ring resonator placed between a pair of dielectric guides [28, 32]. Figure 11.20(a) shows the filter configuration. The dielectric ring and the two guides are assumed to be of the

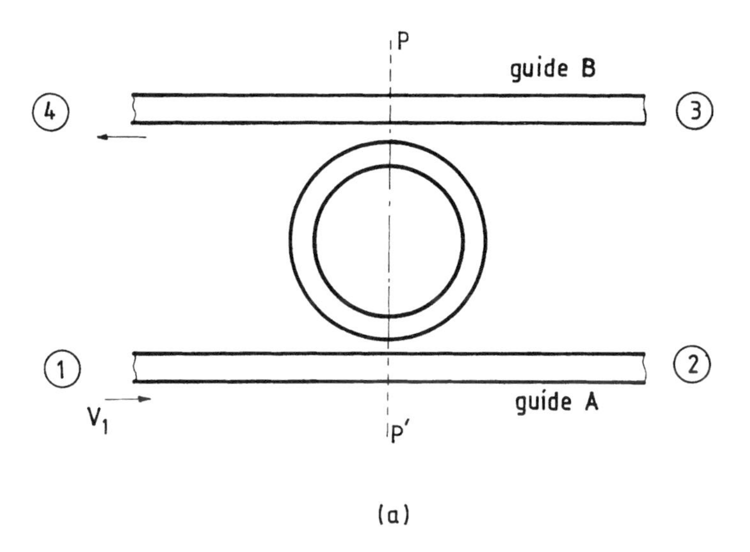

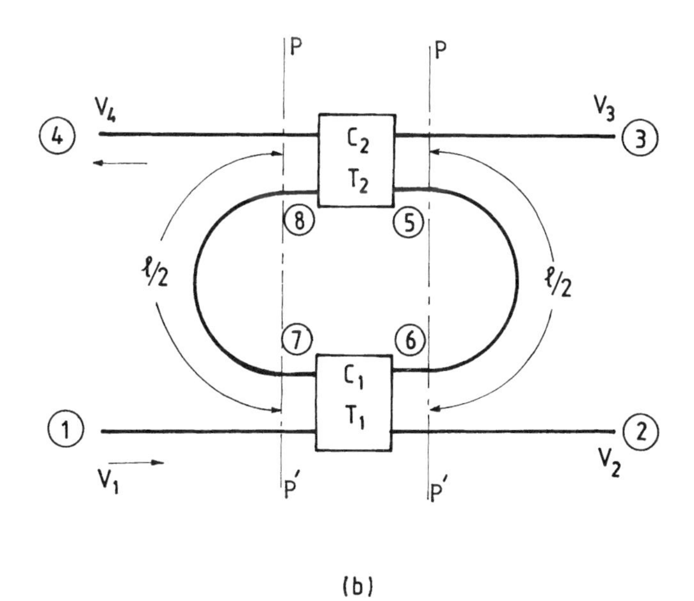

FIGURE 11.20 (a) Single-ring filter and (b) its equivalent circuit representation.

same material and identical cross-sectional dimensions. Let C_1 and T_1 denote the voltage coupling coefficient and transmission coefficient, respectively, of the coupled section between guide A and the ring as illustrated in the equivalent circuit shown in Figure 11.20(b). Similarly, let C_2 and T_2 denote the corresponding quantities for the coupled section between guide B and the ring. If V_1 is the incident voltage wave at port 1, the voltage wave V_5 at port 5 of the filter circuit in Figure 11.20(b) can be expressed as

$$V_5 = V_1 C_1 e^{-\gamma\ell/2}[1 + T_1 T_2 e^{-\gamma\ell} + (T_1 T_2)^2 e^{-2\gamma\ell} + \cdots]$$
$$= V_1 C_1 e^{-\gamma\ell/2}/(1 - T_1 T_2 e^{-\gamma\ell}) \quad (11.86)$$

where $\gamma = \alpha + j\beta$ is the complex propagation constant in the ring and ℓ is the mean circumferential (effective) path around the ring. The voltage wave V_4 at the output port of the second coupled section is given by $C_2 V_5$. Thus

$$V_4 = \frac{C_1 C_2 V_1 e^{-\gamma\ell/2}}{1 - T_1 T_2 e^{-\gamma\ell}} \quad (11.87)$$

If we consider the coupling sections to be identical, we can write

$$C_1 = C_2 = jCe^{-j\phi} \quad (11.88a)$$
$$T_1 = T_2 = Te^{-j\phi} \quad (11.88b)$$

Furthermore, if we consider only light coupling, then $\phi \approx 0$. With these assumptions, we can express $|V_4|$ from (11.87) as

$$|V_4| = \left|\frac{C^2 V_1 e^{-\gamma\ell/2}}{1 - T^2 e^{-\gamma\ell}}\right| \quad (11.89)$$

Thus, for loose coupling, V_4 is maximum when

$$\beta\ell/2 = n\pi, \quad n = 1, 2, 3, \ldots \quad (11.90a)$$

or

$$\ell = n\lambda_g \quad (11.90b)$$

where λ_g is the guide wavelength. The ring gives multiple resonant frequency response corresponding to $\ell = n\lambda_g$ for $n = 1, 2, 3, \ldots$. At resonance

$$|V_4| = \left|\frac{C^2 V_1 e^{-\alpha\ell/2}}{1 - T^2 e^{-\alpha\ell}}\right| \quad (11.91)$$

In order to estimate the loaded Q of the resonator, we consider the half-power point on the resonance curve at which the voltage V_4 falls to $1/\sqrt{2}$ of its value at resonance (center frequency). If $\Delta\beta$ is the change in the phase constant from that at resonance, we can write

$$\frac{1}{\sqrt{2}}\left|\frac{C^2 V_1 e^{-\alpha\ell/2}}{1 - T^2 e^{-\alpha\ell}}\right| = \left|\frac{C^2 V_1 e^{-\alpha\ell/2} e^{-j\beta\ell/2}}{1 - T^2 e^{-\alpha\ell} e^{-j\beta\ell}}\right| \quad (11.92)$$

After some algebraic manipulation, we obtain

$$\cos(\Delta\beta\ell) = 1 - \frac{(1 - T^2 e^{-\alpha\ell})^2}{2T^2 e^{-\alpha\ell}} \tag{11.93}$$

For a high-Q filter, the term $(\Delta\beta\ell)$ is small. Expanding $\cos(\Delta\beta\ell)$ in terms of the power series and retaining only the first two terms, we obtain

$$\Delta\beta\ell = \frac{1 - T^2 e^{-\alpha\ell}}{T e^{-\alpha\ell/2}} \tag{11.94}$$

The loaded Q-factor of the resonator is given by the expression

$$Q_L = \frac{k_0}{2\Delta\beta} \frac{d\beta}{dk_0} \tag{11.95}$$

Substituting for $\Delta\beta$ from (11.94), we obtain

$$Q_L = \frac{k_0\ell}{2} \frac{T e^{-\alpha\ell/2}}{1 - T^2 e^{-\alpha\ell}} \frac{d\beta}{dk_0} \tag{11.96}$$

The unloaded Q-factor of the resonator, denoted as Q_u, may be obtained by setting $C = 0$ or $T = 1$ (since $T^2 + C^2 = 1$) in (11.96)

$$Q_U = \frac{k_0\ell}{2} \frac{e^{-\alpha\ell/2}}{1 - e^{-\alpha\ell}} \frac{d\beta}{dk_0} \tag{11.97}$$

Thus for tight coupling, the ratio Q_U/Q_L is obtained from (11.96) and (11.97) as

$$\frac{Q_U}{Q_L} = \frac{1 - T^2 e^{-\alpha\ell}}{T(1 - e^{-\alpha\ell})} \tag{11.98}$$

The insertion loss of the ring filter at resonance is obtained from

$$IL\,(\text{dB}) = 20\log_{10}\left(\frac{V_4}{V_1}\right) = 20\log_{10}\left(\frac{Q_U}{Q_U - Q_L}\right) \tag{11.99}$$

It may be noted that the above analysis pertains to light coupling. As the coupling level is increased, the phase term ϕ appearing in (11.88) becomes significant. The effective path length in the ring resonator reduces and the resonant frequency increases [32]. Also, the Q-factor decreases with an increase in coupling.

11.5.3 Two-Ring Filter

In order to keep the radiation loss below a tolerable level, ring resonators used in filters must be several wavelengths long in circumference ($\gtrsim 8\lambda_g$). The single-ring filter discussed above produces multiple resonances, which limits its use in many bandpass filter applications. By cascading rings of different diameters, it is possible to realize filters in which the undesired resonances are suppressed.

Consider a two-pole filter employing a cascade of two dielectric rings as illustrated in Figure 11.21. The mean circumferential length of ring 1 is ℓ_1 and that of ring 2 is ℓ_2. The two rings and the dielectric guides are assumed to have identical cross-sections and the same dielectric constant. Referring to Figure 11.21, we can express the coupled output V_3 at port 3 as

$$V_3 = C_3 V_7 \tag{11.100a}$$

where

$$V_7 = \frac{V_5 C_2 e^{-\gamma \ell_2 / 2}}{1 - T_2 T_3 e^{-\gamma \ell_2}} \tag{11.100b}$$

$$V_5 = \frac{V_1 C_1 e^{-\gamma \ell_1 / 2}}{1 - T_1 T_2 e^{-\gamma \ell_1}} \tag{11.100c}$$

The parameters C_1, C_2, and C_3 are the coupling coefficients and T_1, T_2, and T_3 are the transmission coefficients at the three coupling regions as shown in Figure 11.21. V_1 is the incident voltage signal at port 1. Using (11.100b) and (11.100c) in (11.100a), we obtain the overall coupling coefficient between port

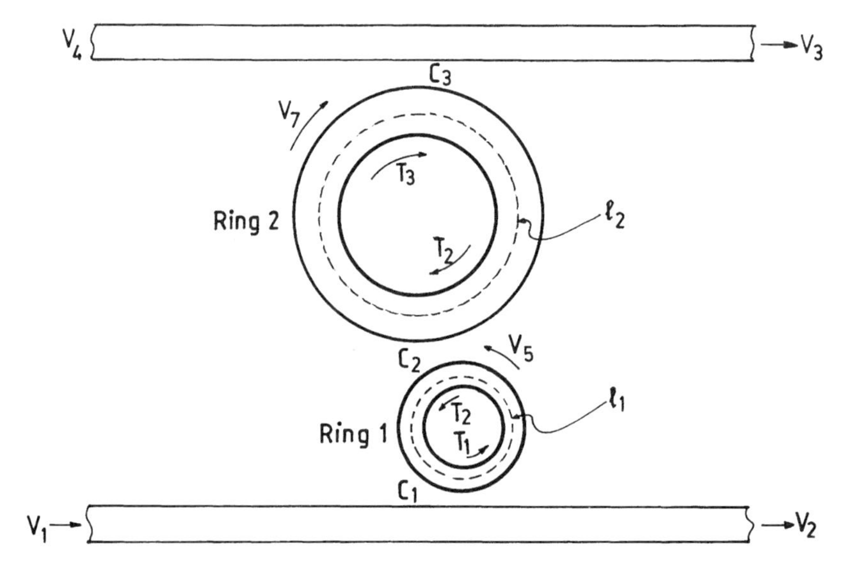

FIGURE 11.21 Geometry of a two-pole ring filter.

1 and 3 as

$$\frac{V_3}{V_1} = \frac{C_1 C_2 C_3 e^{-\gamma(\ell_1 + \ell_2)/2}}{(1 - T_1 T_2 e^{-\gamma \ell_1})(1 - T_2 T_3 e^{-\gamma \ell_2})} \tag{11.101}$$

If we assume the rings to be lossless and set $\alpha = 0$, then for resonance,

$$\beta \ell_1 = 2N_1 \pi, \quad \beta \ell_2 = 2N_2 \pi \tag{11.102a}$$

or

$$\ell_1 = N_1 \lambda_g, \quad \ell_2 = N_2 \lambda_g \tag{11.102b}$$

where N_1 and N_2 are integers and λ_g is the guide wavelength.

The procedure for designing bandpass filters from the lowpass prototype is available in the literature [28]. Using the conductance data on lowpass prototype filters, the coupling coefficients necessary for the realization of maximally flat or Chebyshev responses can be calculated from

$$C_{k,k+1} = \frac{2\omega}{(1 + \omega^2 / g_k g_{k+1}) \sqrt{g_k g_{k+1}}} \tag{11.103}$$

where g_k and g_{k+1} are the lowpass filter element values. The value of ω can be obtained from

$$\omega = \frac{\pi N (\lambda_{g1} - \lambda_{g2})}{\lambda_{g1} + \lambda_{g2}} \tag{11.104}$$

where λ_{g1} and λ_{g2} are the guide wavelengths at the upper and lower ends of the frequency band of interest and N is the number of wavelengths in the ring circumference. For a two-pole filter,

$$N = \sqrt{N_1 N_2} \tag{11.105}$$

It may be noted that the above relations apply only to filters of relatively low bandwidth ($< 10/N\%$). The ring sizes in terms of the number of wavelengths (N_1 and N_2) are chosen on the basis of achieving maximum rejection level at the spurious resonances.

11.6 GRATING AND OTHER TYPES OF FILTERS

11.6.1 Single Guide Grating Filter

Open dielectric guides with grating notches are known to offer very good stopband attenuation [33–35]. Grating type bandstop filters have been realized

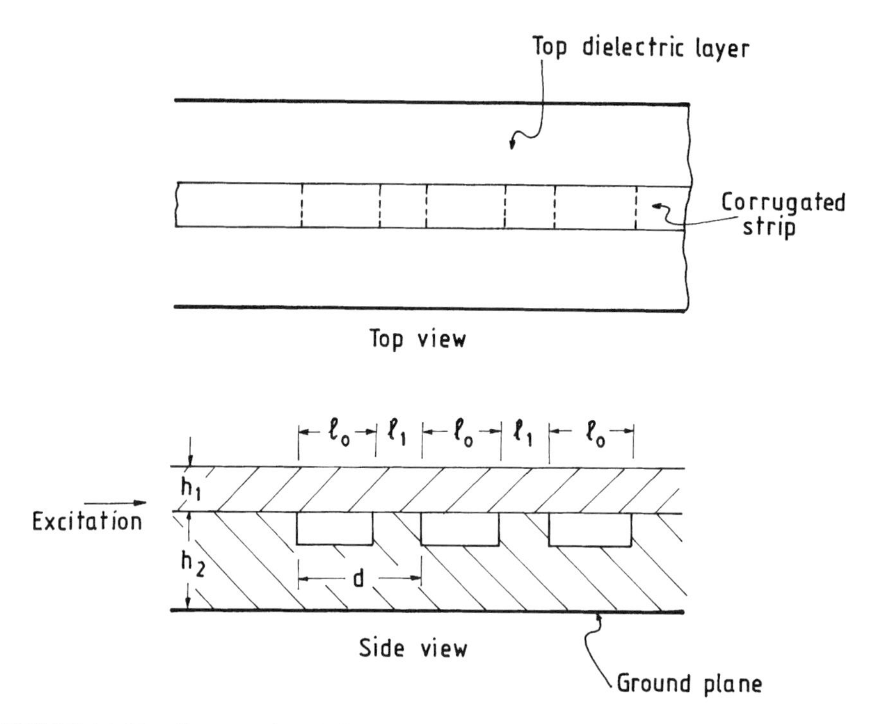

FIGURE 11.22 Cross-sectional view of grating structure in inverted strip guide. (After Itoh [35].)

in inverted strip guide [35] and image guide [35–37]. The application of dielectric guide gratings with construction of bandpass filters has also been reported by several investigators [37–42].

Grating Structure in Inverted Strip Guide Figure 11.22 shows a grating structure in an inverted strip guide wherein the grating is created by placing grooves periodically in the strip while leaving the guide layer intact [35]. This structure can be viewed as conisting of dielectric step discontinuities connected by uniform guide sections. The grating perturbs the transmission characteristics of the guide periodically. The propagating wave in the grating region can be represented in terms of space harmonics whose constants are given by

$$\beta_m = \beta_0 + \left(\frac{2m\pi}{d}\right), \quad m = 0, \pm 1, \pm 2, \ldots \tag{11.106}$$

where $d = \ell_0 + \ell_1$ is the grating period (width of each unit cell) and β_0 is the propagation constant of the dominant ($m = 0$) space harmonic determined by the excitation of the grating. If the perturbation in each unit cell is small, then β_0 is close to the propagation constant β_g of the dominant mode of excitation (E^y_{11}

mode) in the inverted strip guide. For $\beta_0 \approx \beta_g$, if the spacing d is chosen such that

$$\beta_0 d = \pi \tag{11.107}$$

the grating exhibits stopband characteristics resulting from coupling between the fundamental ($m = 0$) forward-traveling and the $m = -1$ harmonic of the fundamental backward-traveling modes. At the frequency at which condition (11.107) is satisfied, an incident wave gets reflected from the grating. The dispersion relation for the grating is given by [35]

$$\cosh(\gamma d) = \cos(\beta_{g0}\ell_0)\cos(\beta_{g1}\ell_1) + 0.5[(\beta_{g1}/\beta_{g0}) + (\beta_{g0}/\beta_{g1})]\sin(\beta_{g0}\ell_0)\sin(\beta_{g1}\ell_1) \tag{11.108}$$

where β_{g0} and β_{g1} are the propagation constants of the grooved and nongrooved sections of the inverted strip guide. It is assumed that the junction effects are negligible and the input and output ends of each unit cell are identical except for a factor that accounts for a complex phase delay. In the stopband, $\gamma = \alpha + j\pi$, where α is real. The dispersion relations for evaluating β_{g0} and β_{g1} as a function of frequency are given in Chapters 2 and 3. Using the values of β_{g0} and β_{g1}, the bandstop characteristic of the grating can be determined from (11.108).

Grating Structure in Image Guide Figure 11.23 shows two types of grating structures in an image guide. For the commonly used E^y_{11} mode of excitation (dominant TM mode), which has its electric field vertically polarized, the grating notches are cut on the two sides of the guide (Fig. 11.23(a)), whereas for the E^x_{11} mode (dominant TE mode), which has its electric field horizontally polarized, notches on the top (Fig. 11.23(b)) are found to be most effective. For excitation in the E^y_{11} mode, cutting notches on the sides is known to yield a strong stopband with no nearby spurious responses [33]. Since the E^y_{11} mode and the grating structure are symmetrical with respect to the x-direction, any TM to TE mode conversion and higher order TM mode generation due to the discontinuities must also have even symmetry with respect to the x-direction. Thus the immediate higher order E^y_{21} and E^x_{11} modes, which have odd symmetry, are not generated. The nearest spurious TM mode that can be generated is the E^y_{31} or E^y_{12} mode and the nearest spurious TE mode is the E^x_{21} mode. These modes occur far away from the desired stopband for the E^y_{11} mode.

For the purpose of filter design, the dielectric grating can be modeled as a transmission line having line sections of equal length (which assumes the same phase velocity throughout). The equivalent circuit is shown in Figure 11.23(c) where Z_0 and Z_1 are the wave impedances of the sections of width w_0 and w_1, respectively. It can be seen that at a frequency f_0, where all the sections are one-quarter wavelength long, the grating structure offers maximum attenuation. If the grating is terminated in Z_0 at both ends, the mid-stopband attenuation L_{max} is given by [37]

$$L_{max} = -10\log\left[\frac{4\gamma^{2n}}{(\gamma^{2n}+1)^2}\right]\text{dB} \tag{11.109}$$

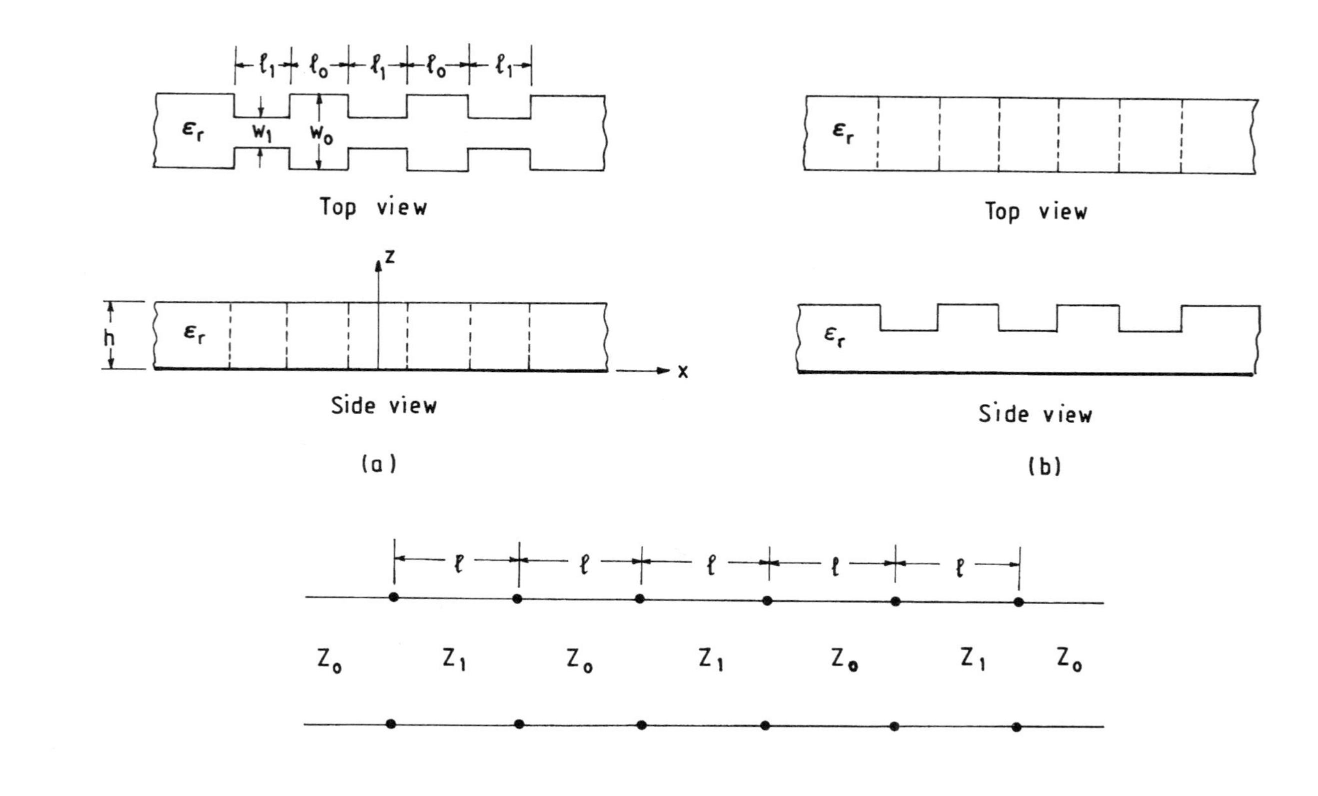

FIGURE 11.23 Image guide grating structure with notches: (a) at the sides of guide, (b) on the top of the guide, and (c) an equal-length equivalent circuit. (After Park et al. [41].)

where

$$\gamma = Z_1/Z_0 > 1 \tag{11.110}$$

and n is the number of sections having the impedance Z_1. The ratio γ is given by the approximate expression

$$\gamma \approx \frac{v_{p1}}{v_{p0}} \tag{11.111}$$

where v_{p1} is the phase velocity in the region having impedance Z_1 and v_{p0} is the phase velocity in the region having impedance Z_0.

Figure 11.24 shows a typical computed bandstop response of an image guide grating consisting of 69 sections with $Z_1/Z_0 = 1.069$ at $f_0 = 34$ GHz, $\varepsilon_r = 2.22$, $\ell_1 = 2.2$ mm, $\ell_0 = 1.82$ mm, $w_1 = 1.5$ mm, and $w_0 = 4$ mm. The grating has 35 identical sections of length $\ell_0 = 1.82$ mm, and 34 identical sections of length $\ell_1 = 2.2$ mm. It can be seen that for 69 sections, the grating offers a stopband attenuation of about 15 dB at the center of the band. For larger attenuation in the stopband, a longer grating will have to be used.

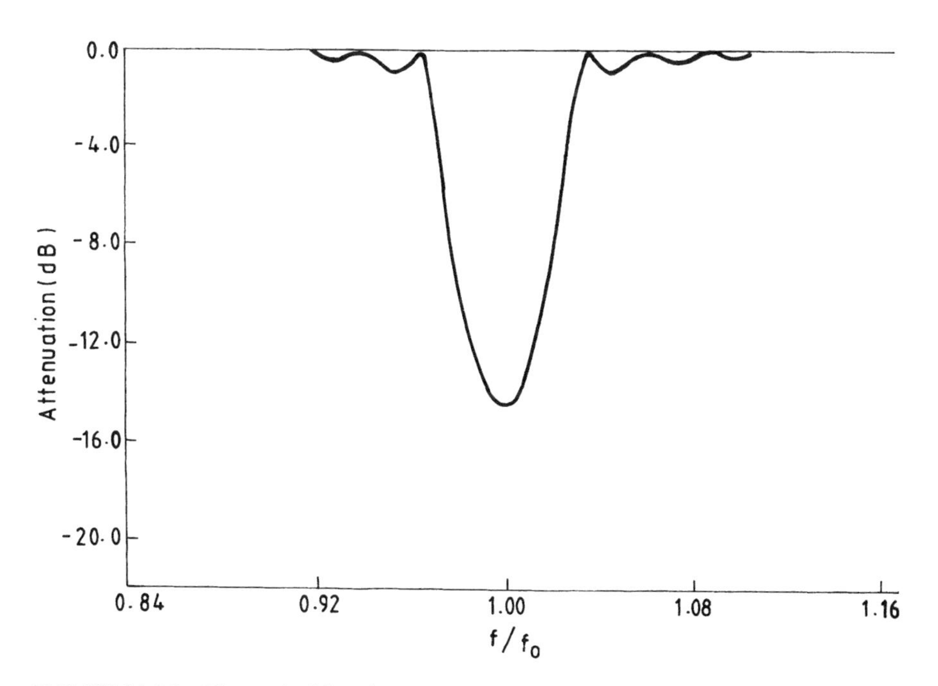

FIGURE 11.24 Theoretical bandstop response of a image guide grating filter (refer to Fig. 11.23(a)). $N = 69$, $f_0 = 34$ GHz, $Z_1/Z_0 = 1.069$, $w_0 = 4$ mm, $w_1 = 1.5$ mm, $\ell_1 = 2.2$ mm, $\ell_0 = 1.82$ mm, and $\varepsilon_r = 2.22$.

Design Formulas for Bandstop Filters Using Long Grating Lines Bandstop filters requiring strong stopband attenuation, of the order of 15 dB or more, employ long gratings. Matthaei et al. [37] have derived design formulas for such filters by modeling an infinite line with uniform gratings. Figure 11.25(a) shows the equivalent circuit of an infinitely long grating and Figure 11.25(b) shows a basic section of the grating. The infinite grating may be considered to be a cascade of basic sections connected end-to-end on an image-matching basis. Referring to Figure 11.25, the image impedances Z_{I0} and Z_{I1} looking from the left and right sides, respectively, of a typical basic section in the grating are given by [28, 37].

$$Z_{I0} = Z_0 \left[\frac{(1+\gamma)\cos\theta - (1-\gamma)}{(1+\gamma)\cos\theta + (1-\gamma)} \right]^{1/2} \tag{11.112}$$

$$Z_{I1} = Z_1 \left[\frac{(1+\gamma)\cos\theta + (1-\gamma)}{(1+\gamma)\cos\theta - (1-\gamma)} \right]^{1/2} \tag{11.113}$$

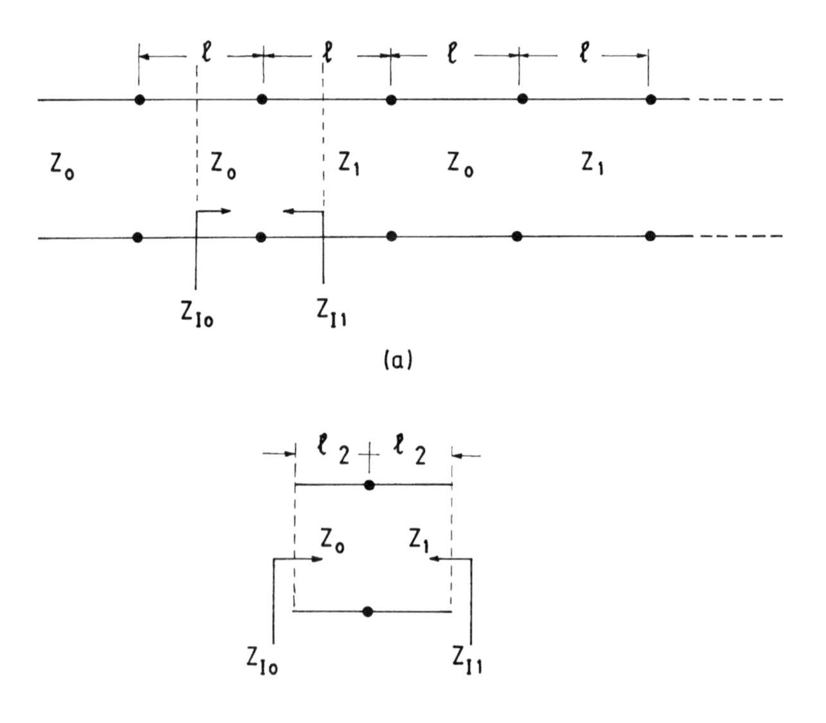

FIGURE 11.25 (a) Equivalent circuit for a grating extending to infinity to the right. (b) A basic section of the infinite grating. (After Matthaei et al. [37].)

where

$$\theta = \pi f / 2 f_0 \tag{11.114}$$

and f_0 is the frequency at which the line length ℓ is a quarter wavelength long. When Z_{I0} and Z_{I1} are imaginary, the sign of the square root is chosen so that the reactance versus frequency curve has a positive slope. From (11.112)–(11.114), the fractional bandwidth of the stopband of the grating can be obtained as

$$\frac{\Delta f}{f_0} = \frac{4}{\pi} \sin^{-1}\left(\frac{\gamma - 1}{\gamma + 1}\right) \tag{11.115}$$

Within the stopband, the image impedance Z_{I0} is purely imaginary. Referring to Figure 11.25(a), we note that the input impedance Z_{in} is obtained by transforming Z_{I0} over a line length $\ell/2$. The reactance slope parameter for the resonance of Z_{in} in Figure 11.25(a) is given by

$$\frac{X_{\text{in}}}{Z_0} = \frac{\pi \gamma}{\gamma - 1} \tag{11.116}$$

This formula assumes dispersion to be negligible. Dispersion effects may be accounted for by multiplying (11.116) by a factor D given by

$$D = -\left[\frac{\ell_0 (d\lambda_{g0}/df) + \ell_1 (d\lambda_{g1}/df)}{\ell_0 (\lambda_{g0}/f_0) + \ell_1 (\lambda_{g1}/f)}\right]\Bigg|_{f = f_0} \tag{11.117}$$

where ℓ_0 and λ_{g0} are the length and guide wavelength of the line section having impedance Z_0; and ℓ_0 and λ_{g1} are the corresponding parameters for the line section of impedance Z_1. The parameter D is greater than 1 for dispersive gratings and is equal to 1 for nondispersive gratings [37].

11.6.2 Coupled Grating Line Filter

As discussed in the preceding section, a single long dielectric image guide grating with periodic notches behaves like a bandstop filter. Two such band-reject gratings can be combined to yield good bandpass filters with broad stopbands [37, 38, 40–42].

Parallel-Coupled Grating Filter [37] Figure 11.26(a) shows the schematic of a simple bandpass filter made of two identical parallel-coupled gratings terminated in distributed absorbing loads [37]. When the gratings operate in their stopbands, power entering at port 1 will emerge out of port 2 yielding a passband. On the other hand, when the gratings are not reflecting, the input power reaches the loads where it will be absorbed and a stopband will result for the filter. In the passband region of the filter, it is important to ensure that coupling exists between the gratings and not directly between the input and output guides.

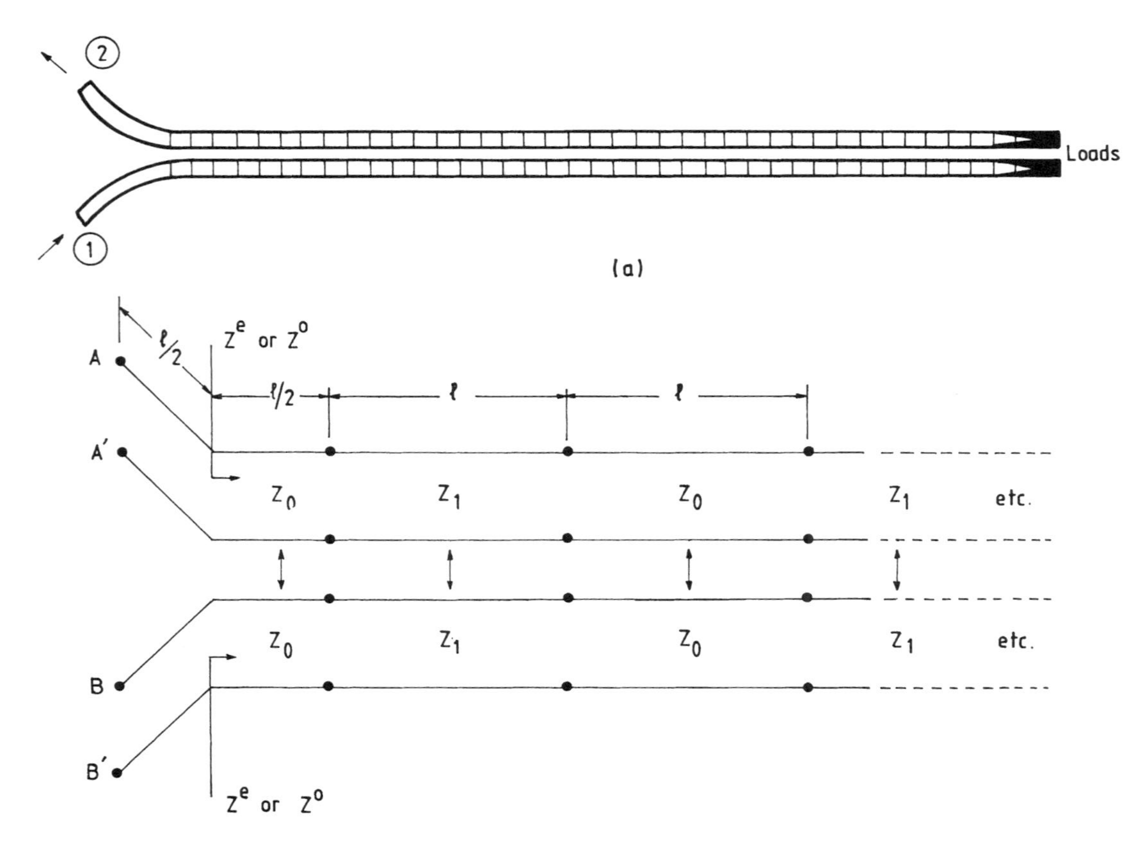

FIGURE 11.26 (a) Schematic of a dielectric waveguide bandpass filter. (b) Equivalent circuit for infinite coupled grating with the first $\ell/2$ portion decoupled. (From Matthaei et al. [37]. Copyright © 1983 IEEE, reprinted with permission.)

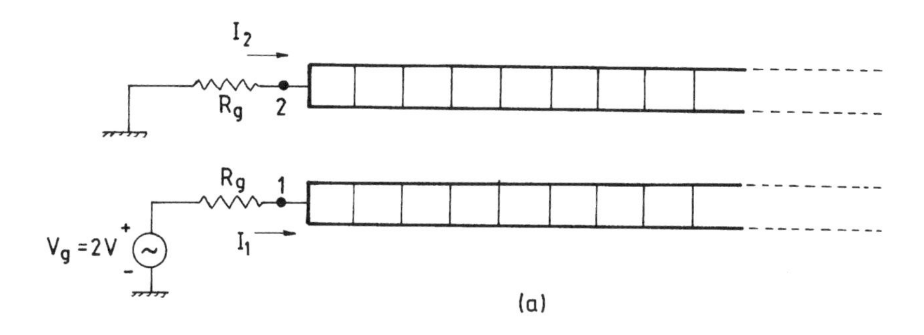

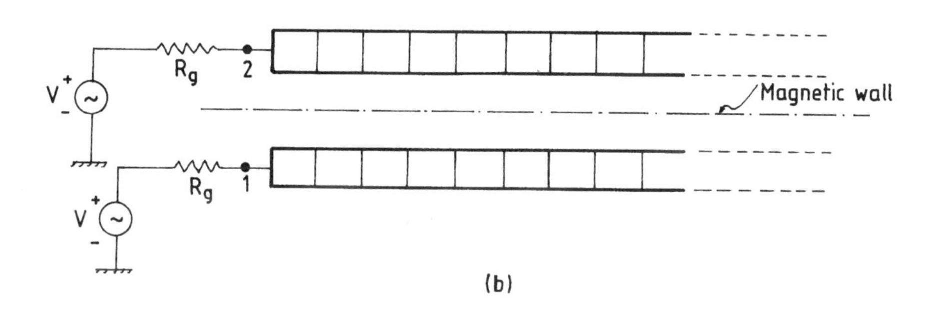

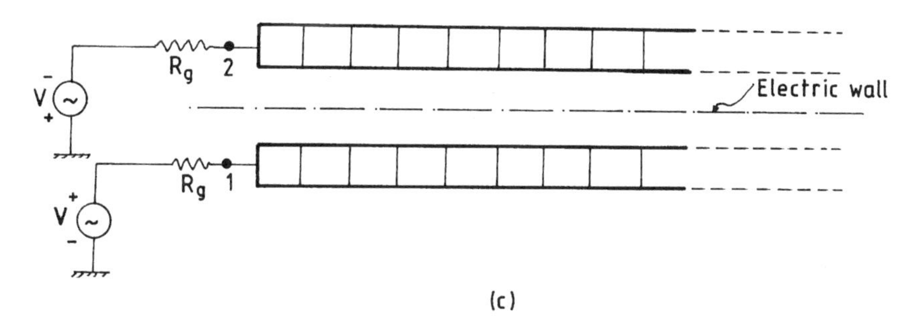

FIGURE 11.27 (a) Parallel-coupled grating excited by a single generator at port 1. (b) Scheme showing even-mode excitation. (c) Scheme showing odd-mode excitation. (After Matthaei et al. [37].)

The analysis of the symmetrical parallel-coupled grating structure can be analyzed in terms of the even and odd modes. Figure 11.27 shows the transmission line representation for such an analysis. The gratings are assumed to be infinitely long (see Fig. 11.26(b)), which is equivalent to placing absorbing terminations at the ends. The original excitation (Fig. 11.27(a)) with port 1 driven by a voltage generator of internal impedance R_g and voltage $V_g = 2\,\text{V}$ can be represented as a superposition of even mode and odd mode excitations as illustrated in Figure 11.27(b) and (c), respectively. Referring to Figure 11.27(b), we note that, for the even mode excitation (magnetic wall at the plane of symmetry), the incident voltages appearing at nodes 1 and 2 are $V/2$ and $V/2$, respectively, and the corresponding reflected voltages are $\Gamma_e V/2$ and $\Gamma_e V/2$, where Γ_e is the even mode reflection coefficient at these ports. Similarly, referring to Figure 11.27(c), for odd mode excitation (electric wall at the plane of symmetry), the incident voltages appearing at nodes 1 and 2 are $V/2$ and $-V/2$, respectively, and the corresponding reflected voltages are $\Gamma_o V/2$ and $-\Gamma_o V/2$, where Γ_o is the odd mode reflection coefficient at the ports. The expressions for Γ_e and Γ_o are

$$\Gamma_e = \frac{Z_{1e} - R_g}{Z_{1e} + R_g} \tag{11.118a}$$

$$\Gamma_o = \frac{Z_{1o} - R_g}{Z_{1o} + R_g} \tag{11.118b}$$

where Z_{1e} and Z_{1o} are the input impedances looking from the left into node 1, for the even- and odd-mode excitations, respectively. For the even mode excitation, the voltages V_1^e and V_2^e appearing at nodes 1 and 2 are

$$V_1^e = V_2^e = V(1 + \Gamma_e)/2 \tag{11.119}$$

Similarly, for the odd mode excitation, the voltages V_1^o and V_2^o appearing at nodes 1 and 2 are

$$-V_1^o = V_2^o = -V(1 + \Gamma_o)/2 \tag{11.120}$$

The total voltage V_2 at node 2 is given by

$$V_2 = V_2^e + V_2^o = V(\Gamma_e - \Gamma_o)/2 \tag{11.121}$$

The desired voltage coupling coefficient between ports 1 and 2 is obtained as

$$V_2/V_g = (\Gamma_e - \Gamma_o)/4 \tag{11.122}$$

We note that for maximum power transfer $|V_2/V_g| = \frac{1}{2}$, which results in the condition

$$\Gamma_e = -\Gamma_o, \quad |\Gamma_e| = |\Gamma_o| = 1 \tag{11.123}$$

Substituting for Γ_e and Γ_o from (11.118) in (11.122), we obtain

$$\frac{V_2}{V_g} = \frac{R_g(Z_{1e} - Z_{1o})}{2(Z_{1e} + R_g)(Z_{1o} + R_g)} \tag{11.124}$$

The impedance Z_{1e} can be computed from the transmission line equivalent circuits using the average even mode velocity v_{pe}. The value of v_{pe} for the grooved and ungrooved sections of the grating will be different and hence an average of the two velocities weighted in proportion to the respective lengths of sections may be used in the equal-line-length equivalent circuit. Similarly, the impedance Z_{1o} can be determined from the average odd mode velocity v_{po}. The effective velocity v_{eff}, which can be used for calculating the center frequency f_0 of the passband of the coupled grating, is

$$v_{\text{eff}} = \frac{2v_{pe}v_{po}}{v_{pe} + v_{po}} \tag{11.125}$$

and the effective propagation constant is given by [37]

$$\beta_{\text{eff}} = \frac{\omega}{v_{\text{eff}}} = \frac{\beta_e + \beta_o}{2} \tag{11.126}$$

With (11.125) and (11.126), the even- and odd-mode electric lengths for each section in the equal-line-length model (see Fig. 11.26(b)) can be expressed as

$$\theta_e = \frac{\omega \ell}{v_{pe}} = \frac{f}{f_0}\left(\frac{\pi\beta_e}{\beta_e + \beta_o}\right) \tag{11.127a}$$

$$\theta_o = \frac{\omega \ell}{v_{po}} = \frac{f}{f_0}\left(\frac{\pi\beta_o}{\beta_e + \beta_o}\right) \tag{11.127b}$$

It may be noted that at $f = f_0$, which is the center of the passband, all the line sections are a quarter wavelength long in terms of the effective wave velocity v_{eff}. Referring to Figure 11.26(a), which uses an infinitely long transmission line model with the leading edge of the coupled region at the center of a Z_0 section, the expression for Z_{1e} and Z_{1o} can be written in the form given by (11.112):

$$Z_{1e} = Z_0 \left[\frac{(1+\gamma)\cos\theta_e - (1-\gamma)}{(1+\gamma)\cos\theta_e + (1-\gamma)}\right]^{1/2} \tag{11.128a}$$

$$Z_{1o} = Z_0 \left[\frac{(1+\gamma)\cos\theta_o - (1-\gamma)}{(1+\gamma)\cos\theta_o + (1-\gamma)}\right]^{1/2} \tag{11.128b}$$

where

$$\gamma = Z_1/Z_0 \tag{11.129}$$

Matthaei et al. [37] have shown that, for complete power transfer at f_0, the gratings must be spaced apart a distance so as to achieve the following velocity ratio:

$$\frac{v_{po}}{v_{pe}} = \frac{\cos^{-1}\,[(1-\gamma)/(1+\gamma)]}{\cos^{-1}\,[(\gamma-1)/(\gamma+1)]} \tag{11.130}$$

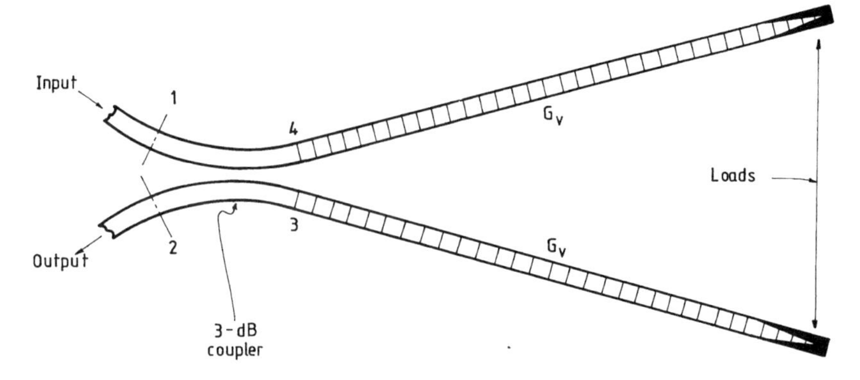

FIGURE 11.28 A bandpass filter formed from two dielectric waveguide bandstrop grating and a 3-dB coupler. (From Park et al. [41]. Copyright © 1984 IEEE, reprinted with permission.)

The parallel-coupled grating bandpass filters are useful for very narrow band applications. The bandwidth is generally of the order of 1% for the reason that the passband width of the filter must be considerably less than the stopband width of the gratings. For a larger bandwidth in the range 5–10%, Part et al. [41] have reported a bandpass filter consisting of a 3-dB coupler with its two output ports connected to two gratings terminated in matched loads. Figure 11.28 shows the configuration. When the gratings are in the stopband, power fed to port 1 of the coupler will emerge at port 2, yielding a passband, and when the gratings are not reflecting, power passes through the gratings and is absorbed in the loads, thus creating a stopband. The passband in this filter is wider because transmission from port 1 to 2 corresponds to the full width of the grating stopband [41].

11.6.3 Gap-Coupled Dielectric Guide Filter

A gap-coupled nonradiative dielectric guide filter reported by Yoneyama et al. [39] is shown in Figure 11.29. It makes use of a cascade of rectangular dielectric strips with alternating airgaps. The operation is based on the property that, except for the dominant mode, all higher order modes decay away from the gap region and can hardly reach the far end of the dielectric strip. This assumption is justified in view of the inherent cutoff nature of the parallel plates spaced by less than a half wavelength. Each gap region can thus be represented by an impedance inverter network with sections of transmission lines connected on both sides as shown in Figure 11.30 [43]. The inverter parameter K and the electrical line length ϕ are given by

$$K = |\tan 0.5(\tan^{-1}\bar{B}_s - \tan^{-1}\bar{B}_o)| \tag{11.131a}$$

$$\phi = -\pi - \tan^{-1}\bar{B}_s - \tan^{-1}\bar{B}_o \tag{11.131b}$$

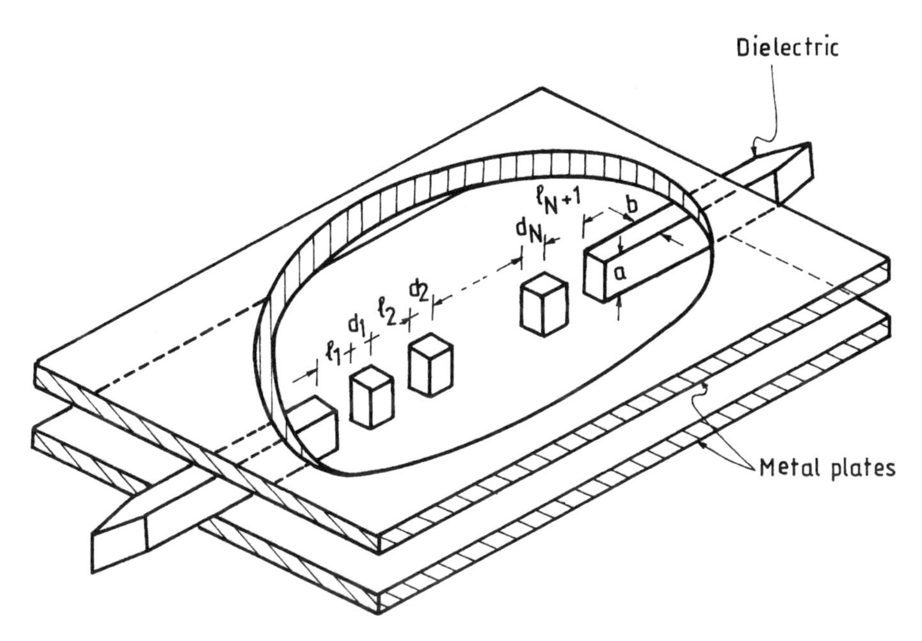

FIGURE 11.29 Gap-coupled NRD guide filter. (From Yoneyama et al. [39]. Copyright © 1984 IEEE, reprinted with permission.)

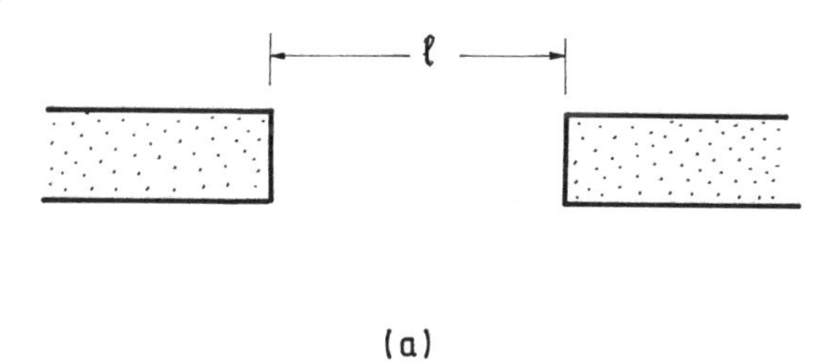

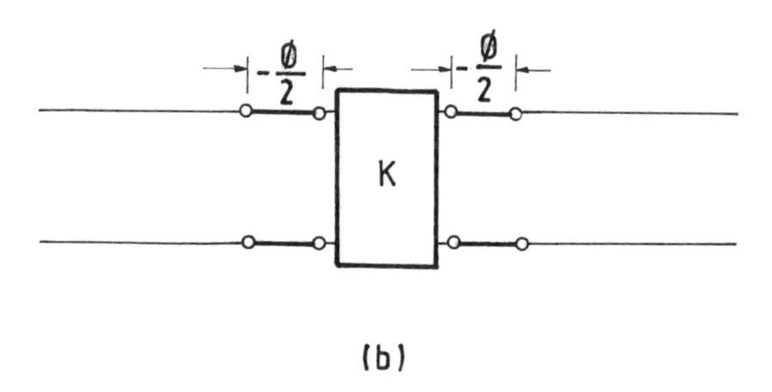

FIGURE 11.30 (a) A pair of gap-coupled semi-infinite dielectric strips and (b) its equivalent circuit with additional transmission line sections on both sides. (From Yoneyama et al. [39]. Copyright © 1984 IEEE, reprinted with permission.)

where $\bar{B}_s$ and $\bar{B}_o$ are the normalized susceptances of the truncated end of the dielectric strip with respect to the transverse symmetry plane in the gap when the plane represents an electric wall and a magnetic wall, respectively. These gap susceptances should be evaluated first by applying, for example, the variational technique to the gap discontinuity problem [39]. By substituting these susceptances in (11.131), the values for K and ϕ as a function of gap length are obtained. The original gap-coupled filter is then replaced by an equivalent bandpass circuit involving impedance inverters connected by transmission line sections (Figure 11.31). The design of the filter is then carried out as follows [39].

First, the inverter parameters are calculated using the available lowpass prototype filter data, the specified fractional bandwidth, and the reactance slope parameter [28]. The reactance slope parameter is given by [39]

$$x = \frac{\pi}{2}\left(\frac{\sqrt{\varepsilon_r}k_0}{\beta_0}\right)^2 \left[\frac{2(q_0^2 + \varepsilon_r p_0^2) + p_0 b(q_0^2 + \varepsilon_r^2 p_0^2)}{2\varepsilon_r(q_0^2 + p_0^2) + p_0 b(q_0^2 + \varepsilon_r^2 p_0^2)}\right] \tag{11.132}$$

In (11.132), b is the width of the dielectric strip, k_0 is the free-space propagation constant, β_0 is the propagation constant of the dominant mode in the NRD guide, and p_0 and q_0 are the lowest eigenvalues of the following characteristic equations:

$$p_0 = (q_0/\varepsilon_r)\tan(q_0 b/2) \tag{11.133a}$$

$$p_0^2 + q_0^2 = (\varepsilon_r - 1)k_0^2 \tag{11.133b}$$

The gap lengths are determined by matching the inverter values computed above with those obtained from (11.131a). The dielectric chip lengths d_i are determined from the relation

$$d_i = [\pi + 0.5(\phi_i + \phi_{i+1})]/\beta_0 \tag{11.134}$$

Yoneyama et al. [39] have designed and tested a three-pole 0.1-dB ripple Chebyshev bandpass filter using Teflon strips ($\varepsilon_r = 2.04$) for 2% bandwidth at a center frequency of 49.5 GHz. The dimensions of the filter are $a = 2.7$ mm, $b = 3.5$ mm, $d_1 = d_3 = 2.04$ mm, $d_2 = 2$ mm, $\ell_1 = \ell_4 = 2.45$ mm, and $\ell_2 = \ell_3 = 4.95$ mm. The filter is reported to have an excess insertion loss of 0.3 dB.

11.6.4 Rectangular Resonator Coupled Filter

Bandstop filters consisting of short-circuited rectangular resonators edge-coupled to a main dielectric guide have been investigated in the image guide [44] and in the nonradiative dielectric guide [45]. Figure 11.32 shows the basic geometry of the filter.

The analysis of a parallel-coupled infinitely long dielectric guide is presented in Section 11.2.1. For a unit incident wave amplitude, the wave amplitudes at a distance z along the two guides A and B shown in Figure 11.3 can be obtained

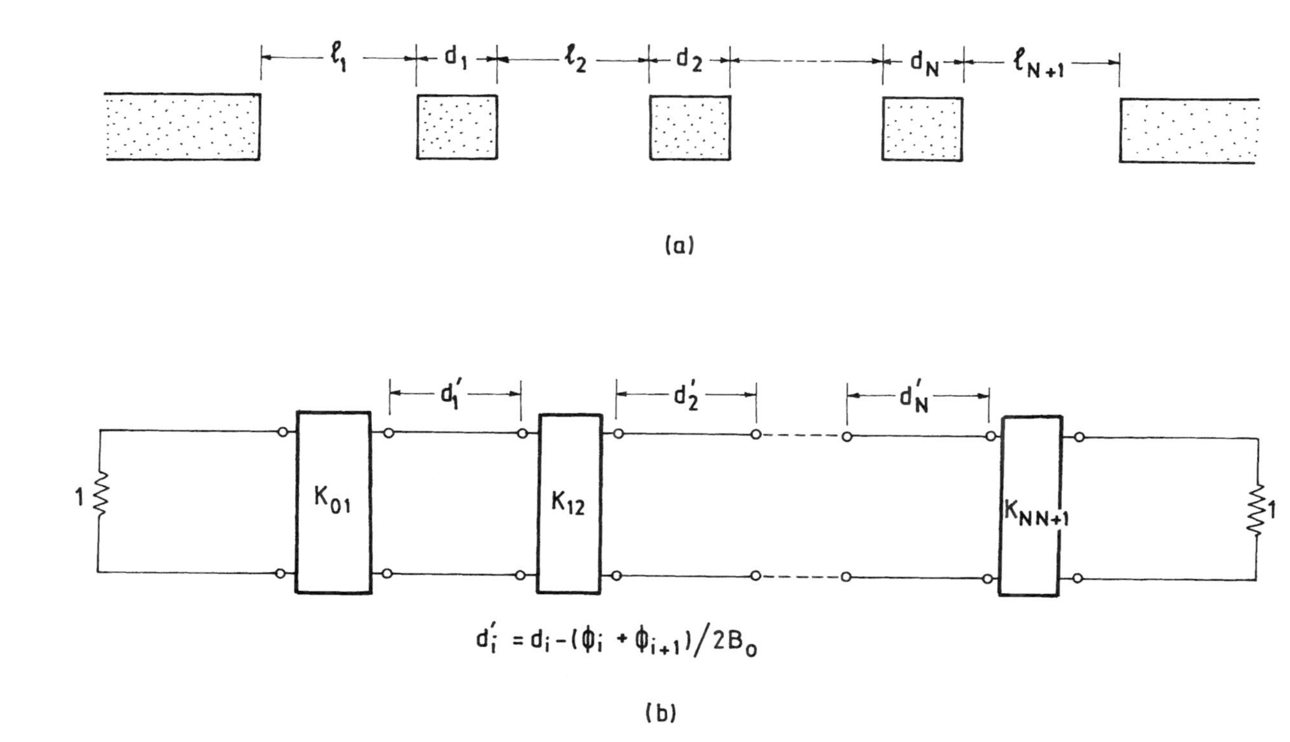

FIGURE 11.31 (a) Gap-coupled filter and (b) its equivalent circuit representation. (From Yoneyama et al. [39]. Copyright © 1984 IEEE, reprinted with permission.)

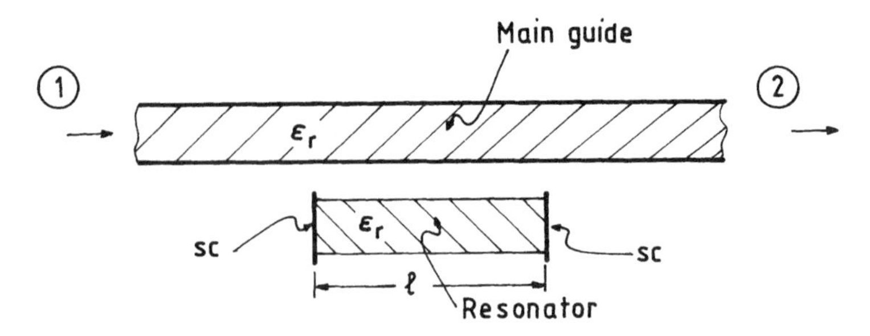

FIGURE 11.32 Short-circuited dielectric resonator coupled to a dielectric guide. (After Malherbe and Olivier [45].)

from (11.1) and (11.2) as

$$\bar{V}_A(z) = e^{-j\beta z}\cos(cz) \tag{11.135a}$$

$$\bar{V}_B(z) = je^{-j\beta z}\sin(cz) \tag{11.135b}$$

where

$$\beta = (\beta_e + \beta_o)/2 \tag{11.136a}$$

$$c = (\beta_e - \beta_o)/2 \tag{11.136b}$$

β_e and β_o are the even- and odd-mode propagation constants of the coupled guide. If one of the guides is replaced by a short-circuited (sc) resonator, the configuration shown in Figure 11.32 results. Within the resonator, the waves undergo multiple reflections due to the shorted ends. The reflection and transmission coefficients for the structure shown in Figure 11.32 are [44]

$$S_{11} = -\left(\frac{e^{-j2\beta\ell}\ \sin^2(c\ell)\ \cos(c\ell)}{1 - e^{-j2\beta\ell}\ \cos^2(c\ell)}\right) \tag{11.137}$$

$$S_{21} = e^{-j\beta\ell}\cos(c\ell)\left(1 - \frac{e^{-j2\beta\ell}\ \sin^2(c\ell)}{1 - e^{-j2\beta\ell}\ \cos^2(c\ell)}\right) \tag{11.138}$$

where ℓ is the length of the resonator section.

For the purpose of filter design incorporating multiple resonators, it is convenient to represent each coupled section by an equivalent circuit. Three different equivalent circuits reported by Malherbe and Olivier [45] are shown in Figure 11.33. Equating the transmission phase of the dielectric coupled guide section to that of the equivalent circuits yields the following relation between the coupling parameter c and admittance Y_s appearing in the equivalent circuits [45]:

$$Y_s = 4\left(\frac{1 - \cos^2(c\ell)}{1 + \cos^2(c\ell)}\right) \tag{11.139}$$

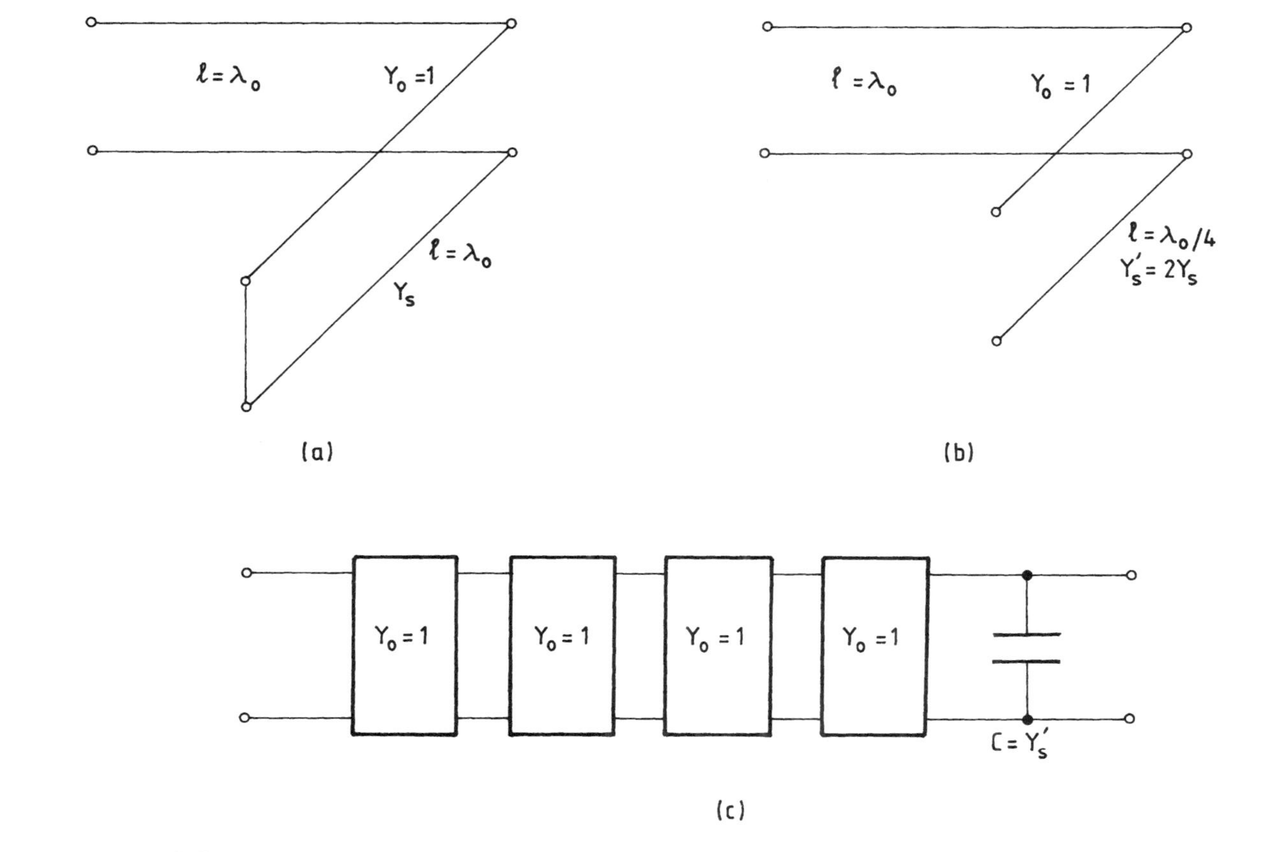

FIGURE 11.33 Equivalent circuits of the structure in Figure 11.31: (a) full-wave shunt stub, (b) quarter-wave shunt stub for narrow band equivalence to circuit in (a), and (c) Richard's equivalent circuit. (From Malherbe and Olivier [45]. Copyright © 1986 IEEE, reprinted with permission.)

In deriving (11.139), it is assumed that the propagation constant of the uncoupled transmission line in Figure 11.32 is equal to β given by (11.136a).

11.6.5 Periodic Branching Filter

Periodic branching filters whose amplitude transmission characteristics vary periodically as a function of frequency find application in satellite communication systems. Figure 11.34 shows the schematic of a filter composed of three directional couplers, a traveling wave resonator, and connecting guide sections of length ℓ_1 and ℓ_2 for phase adjustment [46,47]. Out of the three directional couplers, two are 3-dB hybrids and the third one couples the resonator to one of the connecting guides. Referring to Figure 11.34, the transmission coefficient S_{12} between ports 1 and 2 is given by [46]

$$S_{12} = -A_{12} e^{-j\phi_{12}} \tag{11.140}$$

where

$$A_{12} = (1\sqrt{2})[1 - \cos(\phi - \beta\Delta\ell)]^{1/2} \tag{11.141a}$$

$$\phi_{12} = \tan^{-1}(\cot\{0.5[\phi + \beta(\ell_1 + \ell_2)]\}) \tag{11.141b}$$

Similarly, the transmission coefficient S_{13} between ports 1 and 3 is given by [46]

$$S_{13} = -A_{13} e^{-j\phi_{13}} \tag{11.142}$$

$$A_{13} = (1/\sqrt{2})[1 + \cos(\phi - \beta\,\Delta\ell)]^{1/2} \tag{11.143a}$$

$$\phi_{13} = \tan^{-1}(\cot\{0.5[\phi + \beta(\ell_1 + \ell_2)]\}) \tag{11.143b}$$

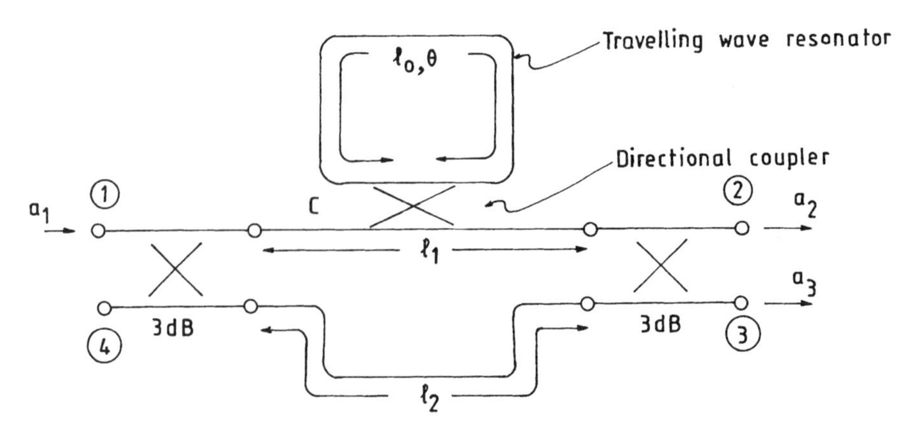

FIGURE 11.34 Schematic of a periodic branching filter with a traveling wave resonator. (From Kumazawa and Ohtomo [46]. Copyright © 1977 IEEE, reprinted with permission.)

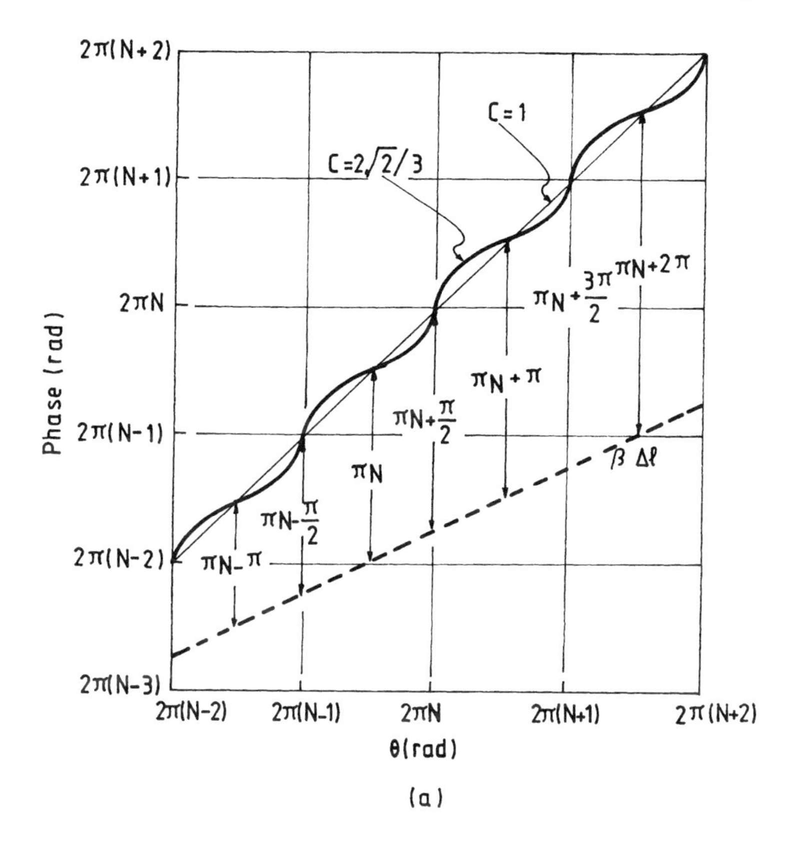

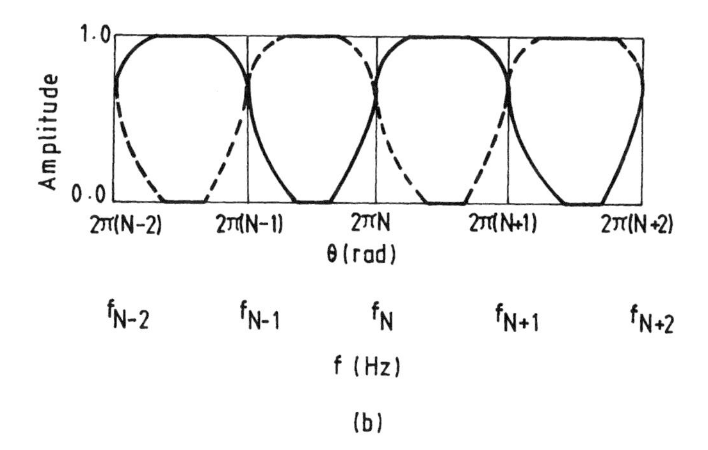

FIGURE 11.35 Phase and amplitude characteristics of a periodic branching filter: (a) phase and (b) amplitude. (From Kumazawa and Ohtomo [46]. Copyright © 1977 IEEE, reprinted with permission.)

The above relations assume that the couplers are ideal and the guide is lossless. β is the propagation constant of the guide, $\Delta\ell = \ell_2 - \ell_1$, and ϕ is the phase lag of the traveling wave resonator, which is given by

$$\phi = \tan^{-1}\left(\frac{(p^2-1)\ \sin\theta}{2p-(p^2+1)\ \cos\theta}\right) \tag{11.144}$$

where

$$p = \sqrt{1-C^2} \tag{11.145}$$

The parameters θ and C denote the electrical length of the resonator and

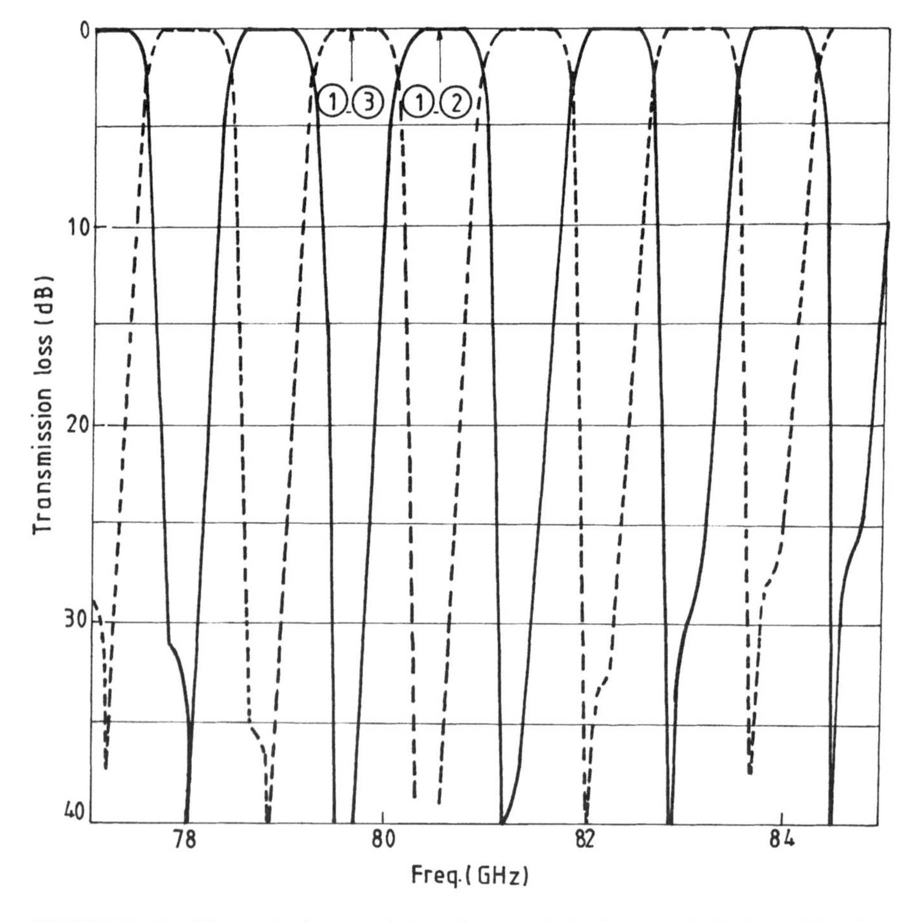

FIGURE 11.36 Theoretical transmission characteristics for a periodic branching filter (see Fig. 11.34); $c = 2\sqrt{2}/3$. (From Itanami [47]. Copyright © 1981 IEEE, reprinted with permission.)

amplitude (voltage) coupling coefficient of the directional coupler, respectively. Figure 11.35 shows the amplitude and phase characteristics of the filter as a function of the electrical length θ of the resonator for two different coupling factors: $C = 1$ and $2\sqrt{2}/3$ [46]. It is shown that $C = 2\sqrt{2}/3$ corresponds to critical coupling for which the filter gives a flat amplitude characteristic at the center frequency. For this coupling, the slopes of resonator phase ϕ and $\beta\Delta\ell$ coincide ($d\phi/d\omega = \Delta\ell\, d\beta/d\omega$) and the value of $\phi - \beta\Delta\ell$ increases periodically by π at successive antiresonance points as θ is increased. The resonance point of the resonator is placed at the edge of each channel frequency band, thereby yielding low-loss characteristics for the filter. If the center frequency f_0 and the 3-dB bandwidth Δf are given, the number of wavelengths (N) along the resonator and the resonator length ℓ_0 can be obtained from [47]

$$N = \beta_\ell/(\beta_u - \beta_\ell) \tag{11.146}$$

$$\ell_0 = N(2\pi/\beta_\ell) \tag{11.147}$$

where β_ℓ and β_u are the propagation constants of the resonator guide at frequencies $f_0 - \Delta f/2$ and $f_0 + \Delta f/2$, respectively. The differential length $\Delta\ell$

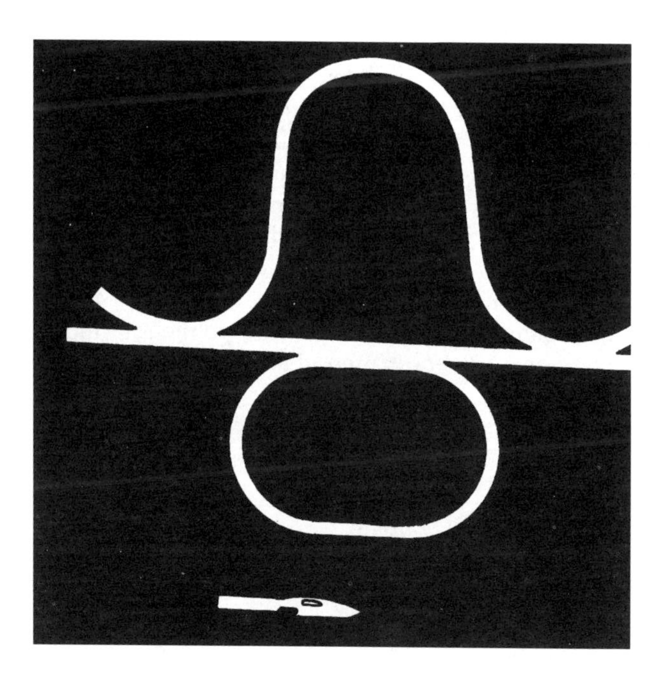

FIGURE 11.37 Practical periodic branching filter in rectangular dielectric guide. (From Itanami [47]. Copyright © 1981 IEEE, reprinted with permission.)

between the connecting guides is given by

$$\Delta\ell = \ell_0/2 - (\pi/2\beta_\ell) \tag{11.148}$$

Using these design equations, Itanami [47] has reported a periodic branching filter designed at $f_0 = 80.425$ GHz with a 3-dB bandwidth of 850 MHz. The dimensions of the filter are reported to be $\ell_0 = 234.22$ mm, $\ell_1 = 123.62$ mm, $\ell_2 = 240$ mm, and $N = 80$. The theoretical amplitude transmission characteristics of the designed filter by considering the couplers to be ideal and $C = 2\sqrt{2}/3$ are shown in Figure 11.36. Figure 11.37 shows a practical filter fabricated in rectangular dielectric ($\varepsilon_r = 2$, $\tan\delta = 2 \times 10^{-4}$) guide having a cross-section 3.5 mm × 3.5 mm. The measured characteristics are reported to match the theoretical response but with an insertion loss of less than 1 dB at each center frequency [47].

11.7 OTHER COMPONENTS

Besides directional couplers and filters, the realization of several other dielectric guide components has been reported in the literature. This section reviews some of these components.

11.7.1 Power Divider

A directional coupler with its isolated port terminated in a matched load functions as a power dividing three-port device. The two output signals, however, have a phase difference between them. For in-phase power division, the T- and Y-type junctions discussed in Chapter 10 may be used.

A compact broadband Y-type power divider composed of three identical dielectric guides has been reported by Zheng-he [48]. The configuration is shown in Figure 11.38. The three guides are parallel-coupled over a certain length with no gap between them. In the coupling region, the transverse field distribution may be decomposed into a fundamental and a third mode as indicated in Figure 11.39. The E_y component can be expressed as [48]

$$E_y(x,z) = AE_{y1}(x)e^{-j\beta_1 z} + BE_{y3}(x)e^{-j\beta_3 z} \tag{11.149}$$

where β_1 and β_3 are the propagation constants of the fundamental mode and the third mode, respectively. The coupling factor is given by

$$C = (\beta_1 - \beta_3)/2 \tag{11.150}$$

and the coupling length L required for 3-dB power division is

$$L = 2\pi/C = 4\pi/(\beta_1 - \beta_3) \tag{11.151}$$

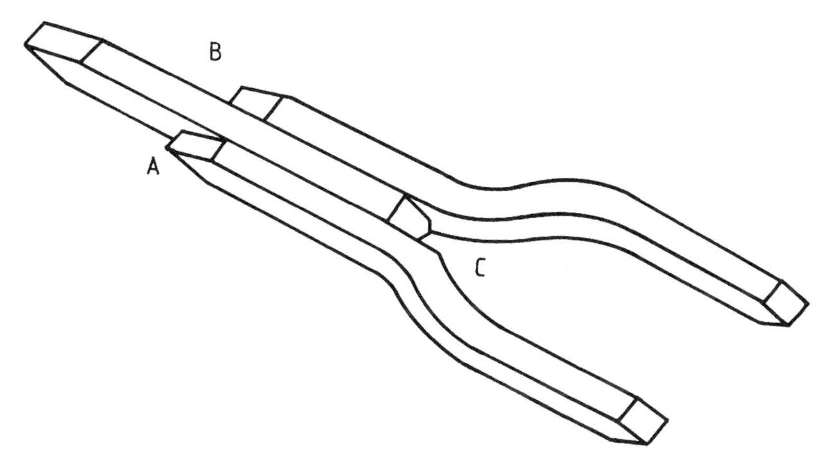

FIGURE 11.38 Dielectric waveguide power divider. (From Zheng-he [48]. Copyright © 1986 IEEE, reprinted with permission.)

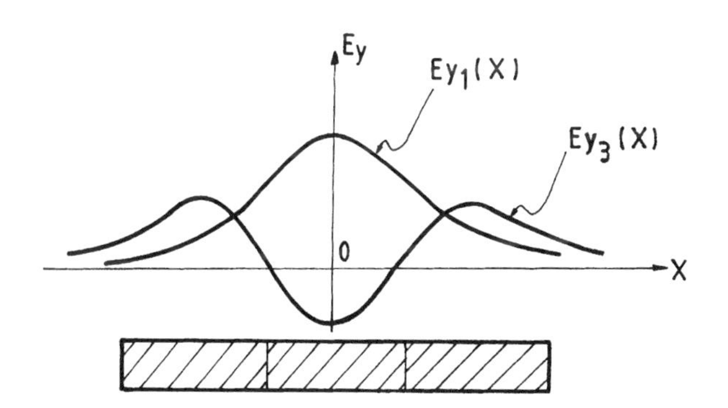

FIGURE 11.39 Fundamental and third mode of coupled dielectric guide in Figure 11.38. (From Zheng-he [48]. Copyright © 1986 IEEE, reprinted with permission.)

This device offers broadband performance because strong coupling takes place in the directly connected guide region and not through evanescent fields.

The configuration of a four-way power divider in the NRD guide is shown in Figure 11.40 [49, 50]. The dielectric strips in this structure are not in touch with each other at the junctions but are coupled electromagnetically via an airgap. As discussed in Chapter 6, the dominant mode in the NRD guide is LSM_{11} (referred to as LSM_{01} in [49, 50]). By suitably choosing the width of the dielectric strip, the guide can support parasitic LSE_{11} (referred to as LSE_{01} in [49, 50]) mode. In the configuration shown in Figure 11.40, the incident LSM_{11} mode is divided into two LSE_{11} modes, each of which is further divided into the LSM_{11} modes, resulting in a four-way power divider network. In order to avoid mixing of LSE_{11}

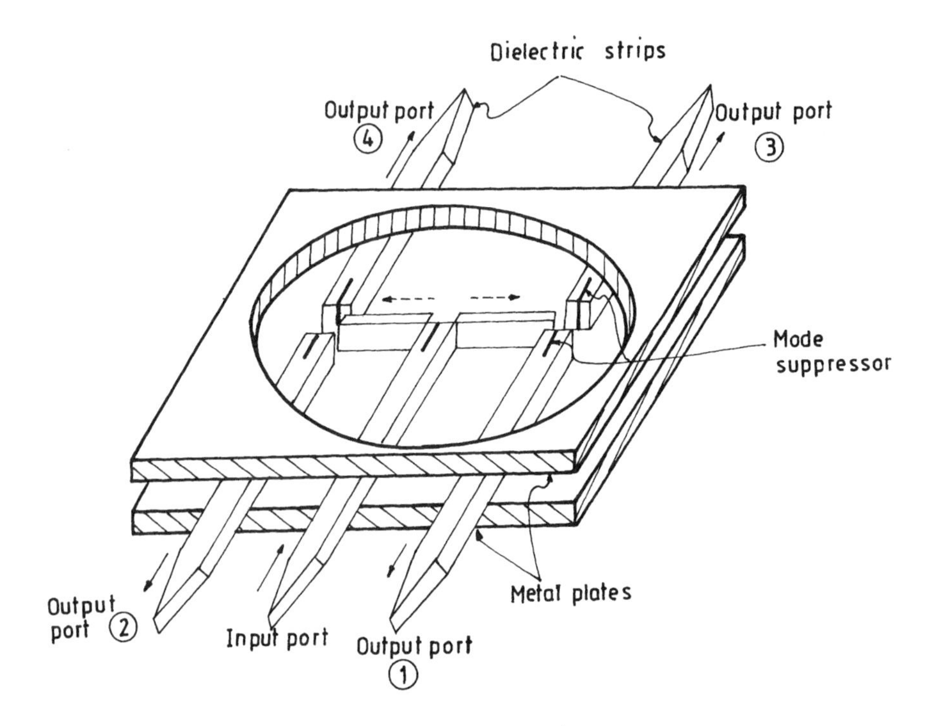

FIGURE 11.40 Configuration of a four-way power divider in the NRD guide: ⟶ LSM_{11} mode, ----→ LSE_{11} mode. (After Yoneyama [49].)

and LSM_{11} modes, mode suppressors are used on the guides supporting the LSM_{11} mode as shown in Figure 11.40. Yoneyama et al. [49, 50] have reported a four-way power divider at 50 GHz with almost equal power division over a wide frequency range.

11.7.2 Variable Attenuator and Phase Shifter

A dielectric sheet coated with a resistive material forms a dissipative element. One simple technique of realizing a variable attenuator in a dielectric guide is to place a resistive vane parallel to the dielectric guide and to use a support mechanism to vary the spacing between them [51]. Figure 11.41 shows a simple schematic of a variable attenuator in an image guide. The resistive vane can be formed, for example, from fiberglass with its ends tapered to provide smooth variation in impedance along the length of the component. The length of the resistive vane is chosen depending on the maximum attenuation required. The attenuation of the device decreases monotonically as the spacing between the dielectric guide and the vane is increased. For a fixed spacing, the attenuation decreases with an increase in frequency as a result of higher concentration of energy within the dielectric.

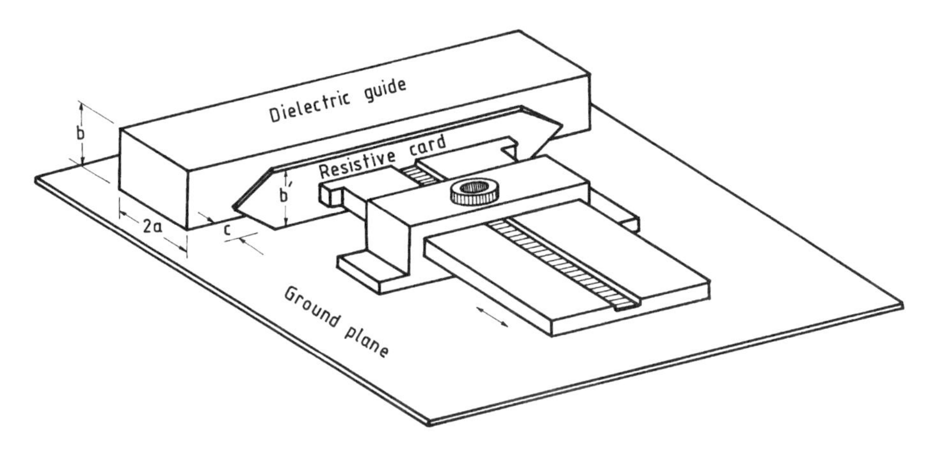

FIGURE 11.41 Variable attenuator in image guide.

The same arrangement as shown in Figure 11.41 can be used as a mechanical phase shifter by replacing the resistive card by a metal sheet with tapered ends. Since the tangential component of the electric field must be zero at the conductor surface, the image guide field changes to that of the odd mode in a corresponding coupled guide with a spacing equal to $2c$. The propagation in the guide is governed by the odd mode propagation constant β_o. As the position of the metallic vane is varied, the value of β_o changes, thereby causing a change in the insertion phase of the device. If L is the length of the metallic vane, then the phase shift $\Delta\phi$ introduced by moving the metal wall from a large c where the image field is negligible to a point where it perturbs the field is given by

$$\Delta\phi = (\beta_e - \beta_u)L \tag{11.152}$$

where β_u is the propagation constant of the unperturbed line. It may be noted that the variation in $\Delta\phi$ as a function of spacing is rather nonlinear. For a fixed spacing, the phase shift decreases with an increase in frequency.

Figure 11.42 shows an electrically controlled dielectric guide millimeter wave phase shifter reported by Tao et al. [52]. The phase-shifting principle is the same as that of the mechanically variable dielectric guide phase shifter, but the mechanically movable metallic sheet is replaced by an electrically controlled metallized piezoelectric bimorph actuator. In Figure 11.42, two bimorph actuators, one on each side of the dielectric guide, are shown. Under an applied bias, the piezoelectric material undergoes mechanical deformation, thereby changing the distance between the guide and the conductor plane. Thus, by appropriate biasing of the actuator, the propagation constant of the guide and hence the phase shift can be controlled.

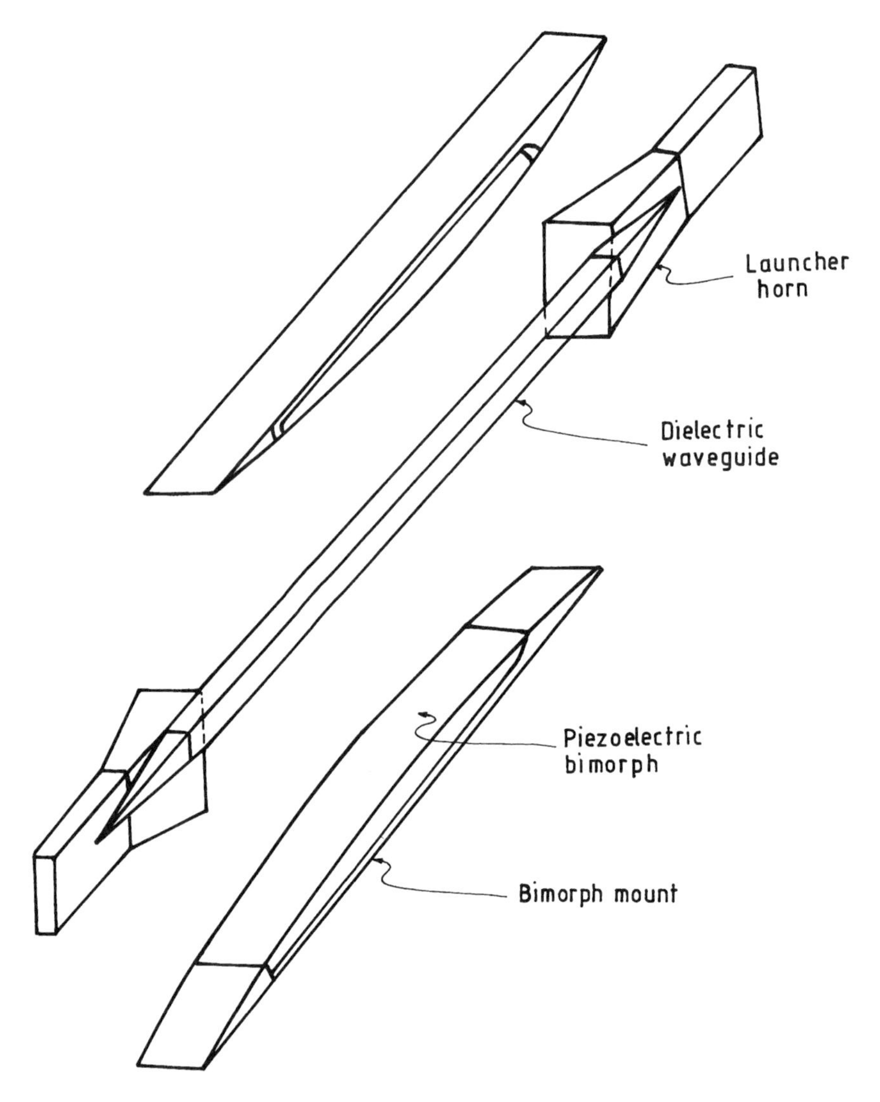

FIGURE 11.42 Montage of electrically controlled phase shifter. (From Tao et al. [52]. Copyright © 1981 IEEE, reprinted with permission.)

11.7.3 Reflectometers

Dielectric guide reflectometers [53–56] employing a six-port technique [57–59] and four-port multistate technique [59] have been reported for operation at millimeter wave frequencies. These reflectometers offer an attractive approach for the measurement of complex scattering parameters of a device without the need

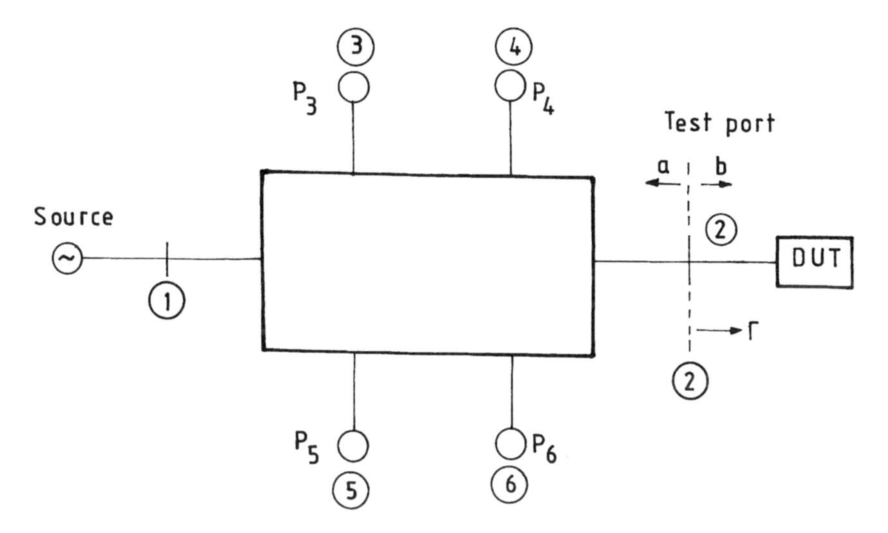

FIGURE 11.43 Block diagram of a six-port network for measuring complex reflection coefficient. (After Engen and Hoer [58].)

for frequency conversion, as is required in many of the automated measurement schemes.

Six-Port Reflectometers The six-port reflectometer technique developed by Engen and Hoer [57–59] is basically a six-port junction that enables determination of the complex reflection coefficient of an unknown load. Figure 11.43 shows the block diagram of a six-port junction, where port 1 is connected to a source, port 2 to the device under test (DUT), and ports 3 to 6 to four different power detectors [58]. The six-port technique permits determination of the complex reflection coefficient Γ at the test port in terms of the measured power readings P_3 to P_6. If a and b are the reflected and incident waves at test port 2, then the powers P_3 to P_6 can be expressed as [59]

$$P_3 = |aA + bB|^2 \tag{11.153a}$$

$$P_4 = |aC + bD|^2 \tag{11.153b}$$

$$P_5 = |aE + bF|^2 \tag{11.153c}$$

$$P_6 = |aG + bH|^2 \tag{11.153d}$$

where $A, \ldots, H$ are complex constants that are known for a given six-port circuit configuration. In a design reported by Engen [59], C is chosen to be zero so that the power reading at port 4 is proportional to $|b|^2$. We set $C = 0$ in (11.153b), and rewrite (11.153) in the forms

$$P_3 = |A|^2\,|b|^2\,|\Gamma - q_3|^2 \quad (11.154a)$$

$$P_4 = |D|^2\,|b|^2 \quad (11.154b)$$

$$P_5 = |E|^2\,|b|^2\,|\Gamma - q_5|^2 \quad (11.154c)$$

$$P_6 = |G|^2\,|b|^2\,|\Gamma - q_6|^2 \quad (11.154d)$$

where $\Gamma = a/b$ is the complex reflection coefficient, $q_3 = -B/A$, $q_5 = -F/E$, and $q_6 = -H/G$. Eliminating $|b|^2$ from (11.154a)–(11.154d) yields

$$|\Gamma - q_3|^2 = |D/A|^2(P_3/P_4) \quad (11.155a)$$

$$|\Gamma - q_5|^2 = |D/E|^2(P_5/P_4) \quad (11.155b)$$

$$|\Gamma - q_6|^2 = |D/G|^2(P_6/P_4) \quad (11.155c)$$

For a given six-port network, the parameters q_3, q_5, and q_6 are known and with the measured powers P_3 to P_6, the right-hand sides of (11.155) are known. The complex reflection coefficient Γ is then calculated as the intersection of three circles in the Γ-plane with centers located at q_3, q_5, and q_6 and radius given by $|\Gamma - q_3|$, $|\Gamma - q_5|$, and $|\Gamma - q_6|$, respectively. As an example, consider the schematic of a six-port circuit shown in Figure 11.44, which employs one 180° hybrid and three 90° hybrids [59]. From the outputs at ports 3 to 6, we note that $|q_3| = \sqrt{2}$, $|q_5| = |q_6| = 2$, and the phase angles between q_3, q_5, and q_6 are 135°, 90°, and 135°, respectively. With these parameters and the power meter readings P_3 to P_6, the value of Γ for a device connected at that port can be determined. It may be noted that, in practice, the circuit needs to be first calibrated with certain known standards.

Using rectangular dielectric guides, six-port reflectometers have been realized for operation at K_a- and W-bands [53–55]. As an illustration, Figure 11.45 shows a layout of a practical W-band dielectric guide six-port reflectometer reported by Radovich and Paul [54]. The circuit consists of a Y-junction in-phase power divider and three parallel-coupled directional couplers terminating in output ports plus two external couplers of the same type for interfacing with the input ports of the network.

Multistate Reflectometer Figure 11.46 shows a four-port multistate reflectometer implemented by Collier and D'Souza [56] for operation in the frequency range 75–140 GHz. The main difference with respect to the six-port reflectometer is that it requires only two power detectors instead of four. The circuit employs two broadband directional couplers (marked C_1 and C_2) and a phase shifter. The couplers are realized by positioning a dielectric film at 45° to the direction of propagation of the wave. The phase shifter consists of a section of dielectric guide with a movable conducting plate parallel to the guide. The circuit has four

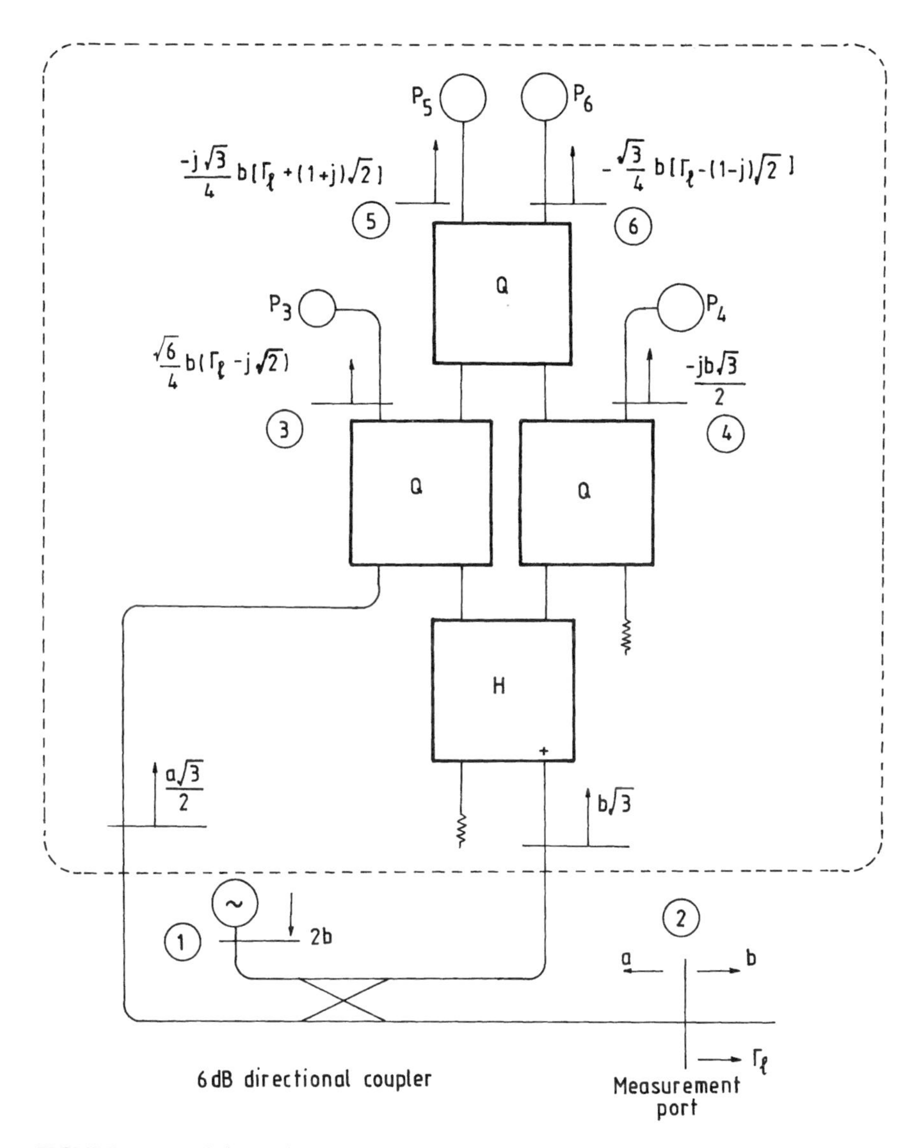

FIGURE 11.44 Schematic of a typical six-port reflectometer: $H = 180^\circ$ hybrid, $Q = 90^\circ$ hybrid. (From Engen [59]. Copyright © 1977 IEEE, reprinted with permission.)

accessible ports: port 1 is the input port, port 2 is the test port, and ports 3 and 4 are connected to power detectors. At each of these ports, the E^y_{11} mode of the rectangular dielectric guide is converted to the TE_{10} mode of the standard rectangular metal waveguide by means of a horn transition. Of the remaining two ports, one is terminated in a matched load and the other is connected to a variable phase shifter. It has been shown [60] that the ratio of the powers at ports 3 and

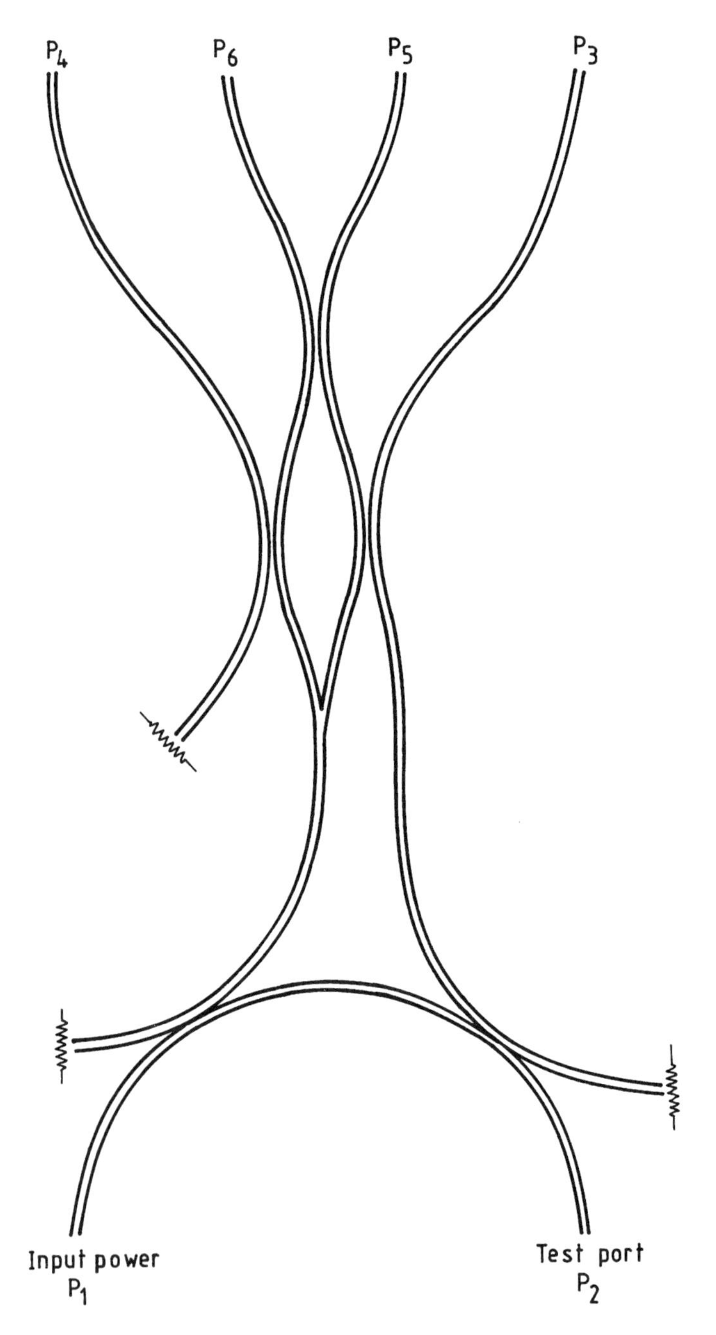

FIGURE 11.45 Layout of a six-port reflectometer in rectangular dielectric guide. (From Radovich and Paul [54]. Copyright © 1982 IEEE, reprinted with permission.)

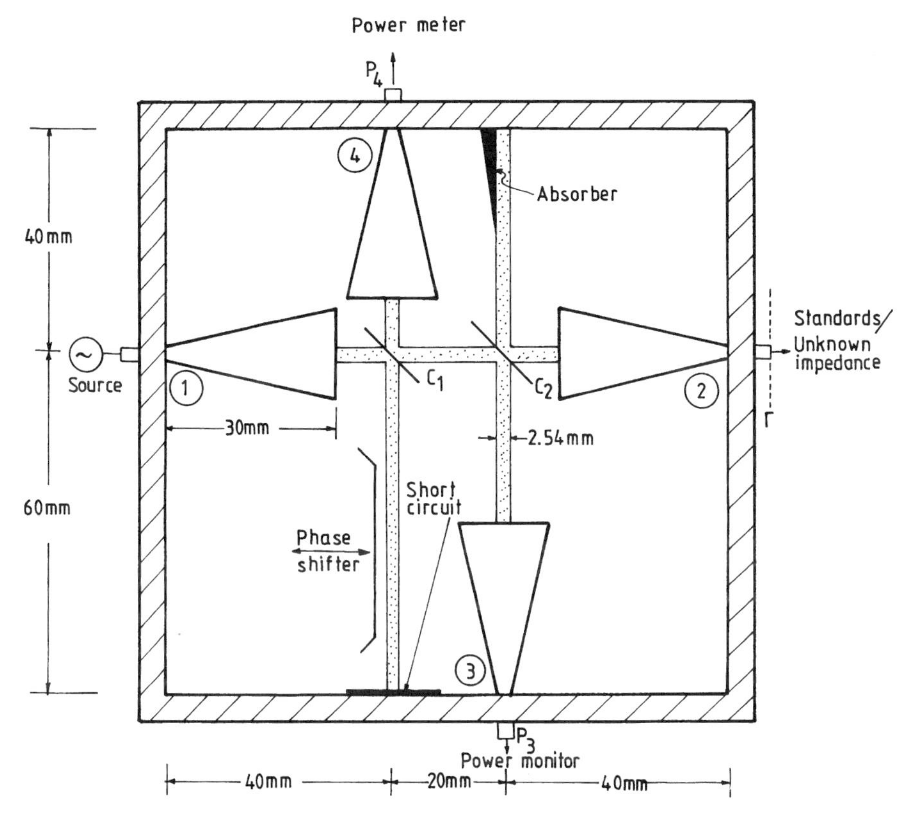

FIGURE 11.46 Dielectric guide multistate reflectometer. (From Collier and D'Souza [56]. Copyright © 1991 IEEE, reprinted with permission.)

4 for each phase shift state k is given by the following bilinear relation:

$$\left[\frac{P_4}{P_3}\right]_k = \left|\frac{d_k\Gamma + e_k}{c\Gamma + 1}\right|^2 \tag{11.156}$$

where c, d_k, and e_k are the calibration constants to be determined and Γ is the reflection coefficient of the device. It may be noted that the calibration constant c is independent of the phase shift state k. The above bilinear relation describes three circles in the Γ-plane. For given frequency, the centers of the circles are given by the scattering parameters of the six-port circuit and their radii are proportional to the square root of the power ratios. The phase shifter is adjusted so that the centers of the three circles are distributed rather evenly around the origin so as to remove any ambiguity. The complex reflection coefficient Γ of the device under test is given by the intersection of the three circles [56].

11.7.4 Isolator

At millimeter wave frequencies, the commonly employed resonant type ferrite isolators are not quite practical because of the extremely high dc magnetic field required for saturating the ferrite material. The *edge-loaded isolator* in an image guide [61] offers an attractive alternative. The device consists of an element of transversely magnetized ferrite introduced into the dielectric image guide from one side as shown in the cross-sectional views in Figure 11.47. In Figure 11.47(a) is shown the dominant mode (E^y_{11}) field distribution in the image guide away from the ferrite loaded region. In the ferrite loaded region, the magnetically biased ferrite modifies the electric field distribution in the guide system, leading to a pronounced nonreciprocal effect. The modified field distributions for the forward and reverse transmission are different. This is illustrated in Figure 11.47(b) and (c). By an appropriate selection of the parameters, the location of small field amplitudes for forward transmission may be made to coincide with that of larger field amplitudes for the reverse transmission case. By introducing a resistive vane into the structure at that location (plane at x_1 in Fig. 11.47(b) and (c)), energy may be transmitted with negligible loss in the forward direction, whereas it would be heavily attenuated for transmission in the reverse direction.

11.7.5 Circulator

Millimeter wave circulators incorporated in an image guide and NRD guide have been reproted [62, 63]. The NRD guide with its E-plane parallel to the top and bottom plates has the advantage of an E-plane circuit. The NRD guide circulator reported by Yoshinaga and Yoneyama [63] has a form similar to the waveguide E-plane counterpart. The structure is shown in Figure 11.48. The operating mode of the NRD guide may be designated as the LSM_{11} (referred to as LSM_{01} in [63]) mode. It resembles the TE_{10} mode of the rectangular waveguide. Another non-radiative mode that can exist in the NRD guide is the LSE_{11} (referred to as LSE_{01} in [63]) mode, which has its field configuration resembling the TM_{11} mode of the waveguide. The circulator shown in Figure 11.48 incorporates mode suppressors for eliminating this parasitic mode. This mode suppressor is in the form of a shaped metal strip inserted at the midplane of the dielectric strip such that it is perpendicular to the metal plates of the NRD guide. The LSM_{11} mode is not affected by this metal strip since its E-field is perpendicular to the strip surface whereas the LSE_{11} mode, which has a component of its E-field parallel to the strip, is suppressed. In order to ensure that the metal strip does not induce any TEM wave, it is shaped into a $\lambda/4$ choke structure for the TEM wave. Figure 11.49 shows the measured performance of a practical NRD guide circulator as reported by Yoshinaga and Yoneyama [63]. The circulator designed at 50 GHz uses a pair of YIG ferrite disks ($4\pi M_s = 1800$ G, $\varepsilon_r = 15$) having diameter 3.37 mm and thickness 0.342 mm. The ferrite disks are supported by a foamed polystyrene block ($\varepsilon_r = 1.03$). The mode suppressor is a four-step choke (4.8 mm in length)

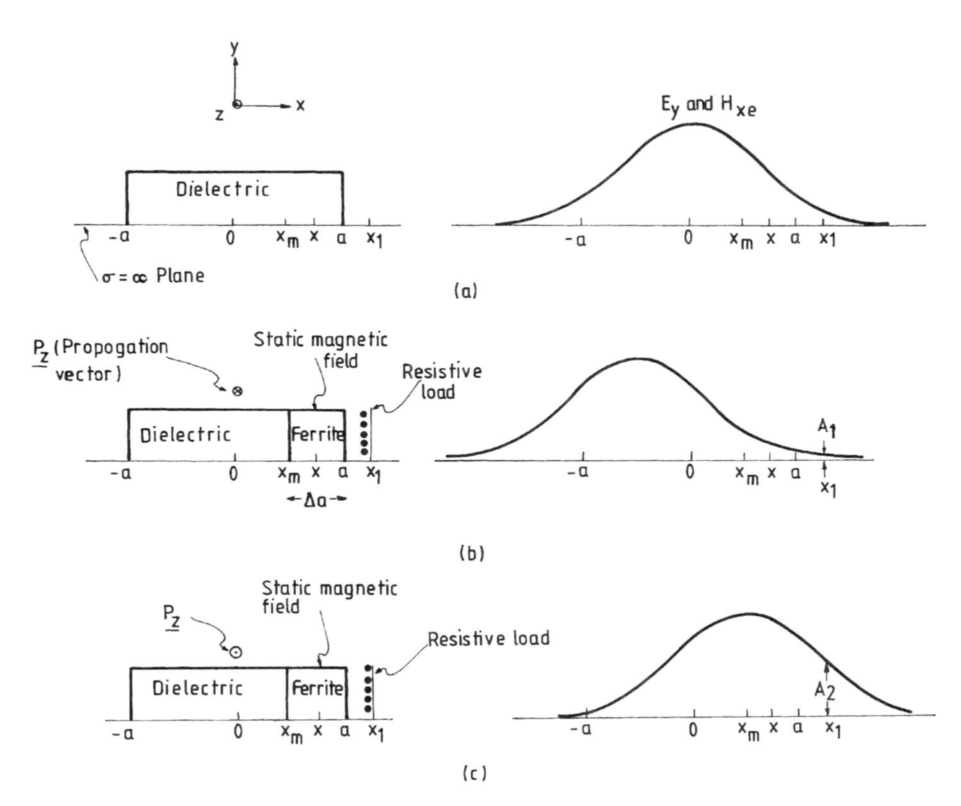

FIGURE 11.47 Transverse field representation for various cases of edge-loaded dielectric guide: (a) typical dielectric guide, (b) forward transmission case of ferrite loaded dielectric guide, and (c) reverse transmission case of ferrite loaded dielectric guide. (From Nanda [61]. Copyright © 1976 IEEE, reprinted with permission.)

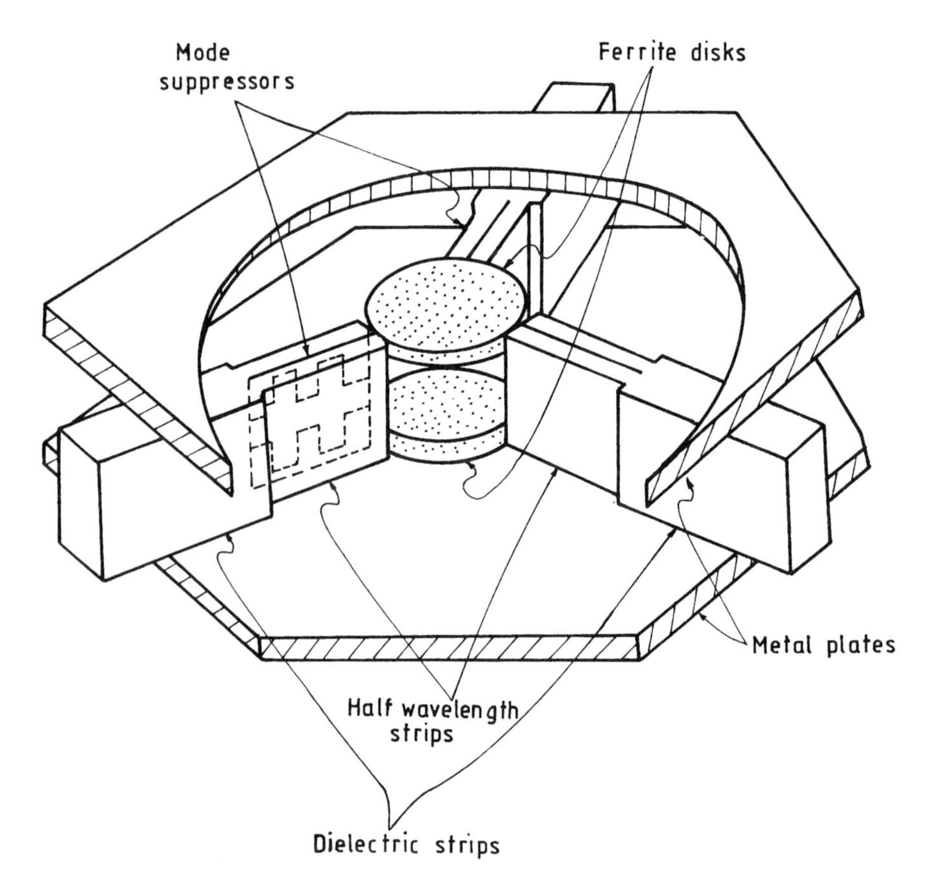

FIGURE 11.48 NRD guide circulator with half-wavelength step transformer and mode suppressor. (From Yoshinaga and Yoneyama [63]. Copyright © 1988 IEEE, reprinted with permission.)

designed for use at 50 GHz and is realized on a 0.13-mm thick dielectric substrate ($\varepsilon_r = 2.6$). A magnetic bias field of 1320 Oe applied perpendicular to the ferrite resonator is reported to give optimum circulation. The circulator is reported to offer a low insertion loss of about 0.3 dB over a 20-dB isolation bandwidth of 1.4 GHz [63].

The Y-juction circulator can also be realized using a groove NRD (GNRD guide [64, 65]. Since, in this configuration, dielectric strips are precisely assembled in the grooves, the GNRD guide circulator offers good stability and symmetry. Lanfen and Yong Xiang [64] have reported experimental results on a GNRD circulator at K_a-band. The insertion loss reported including two transitions is less than 1.5 dB, the isolation is better than 20 dB, and the bandwidth is 1% at the center frequency of 34.5 GHz.

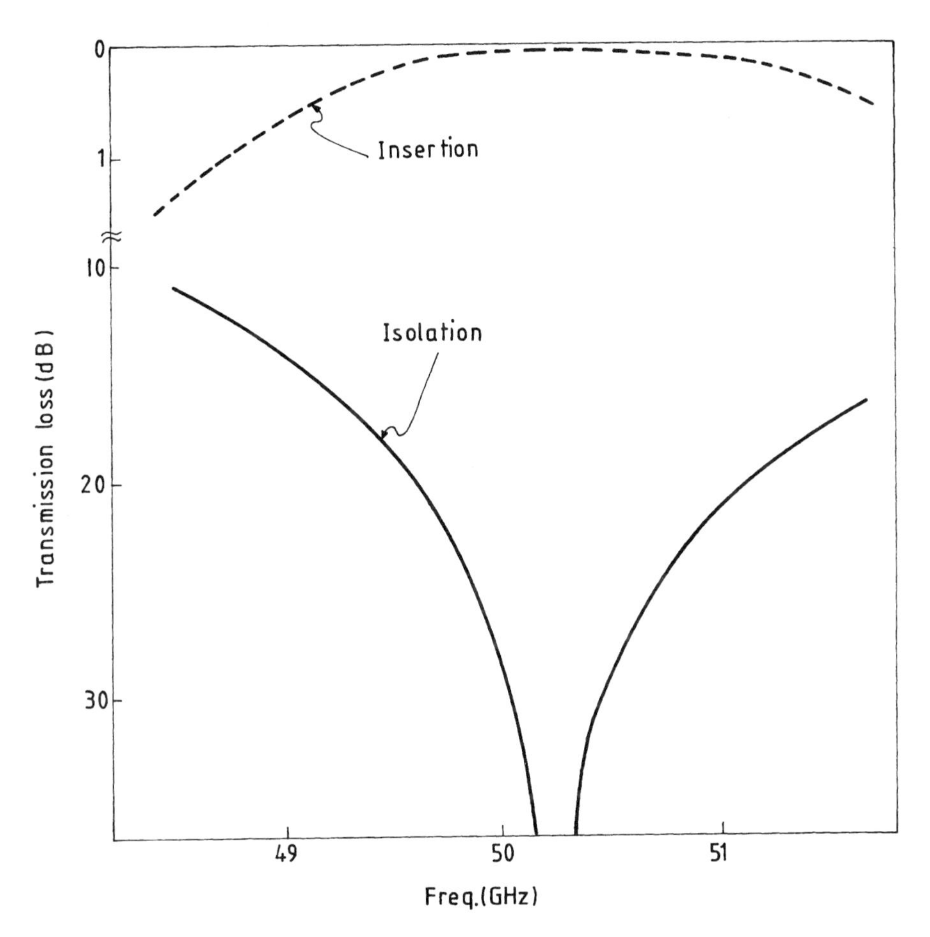

FIGURE 11.49 Measured insertion loss and isolation of NRD guide circulator with mode suppressor (refer to Fig. 11.48). (From Yoshinaga and Yoneyama [63]. Copyright © 1988 IEEE, reprinted with permission.)

PROBLEMS

11.1 Consider a coupled image guide shown in Figure 11.50. The ratio of the voltage wave amplitude in guide B with respect to that in guide A at a distance $z = \ell$ is given by

$$\frac{V_B(z=\ell)}{V_A(z=\ell)} = \tan[(\beta_e - \beta_o)\ell]$$

where β_e and β_o are the even- and odd-mode propagation constants in the coupled guide.

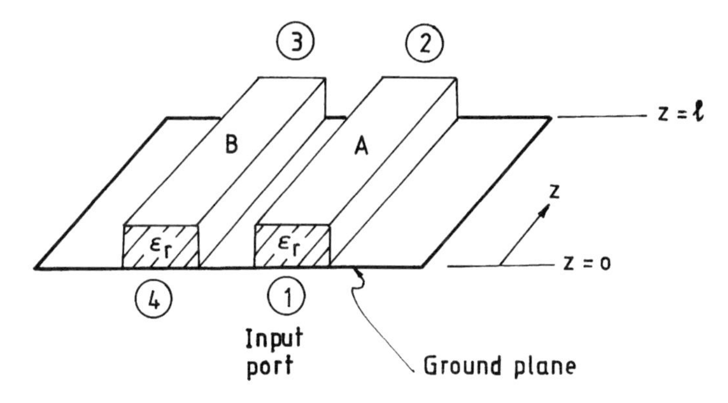

FIGURE 11.50 Coupled image guide.

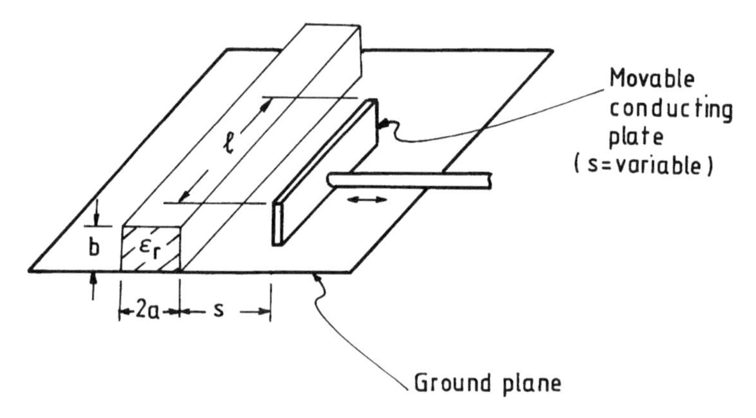

FIGURE 11.51 Schematic of a mechanically variable phase shifter in image guide.

(a) Using the above equations, derive expressions for the scattering parameters $|S_{21}|$ and $|S_{31}|$.

(b) If for a particular coupled guide $\beta_e/k_0 = 1.4$ and $\beta_o/k_0 = 1.1$ at $f = 30\,\text{GHz}$, calculate the length of the uniformly coupled section for 10-dB coupling between ports 1 and 3. What is the length for 0-dB coupling? [*Note*: $k_0 = 2\pi/\lambda_0$ and λ_0 is the free-space wavelength.]

11.2 Figure 11.51 shows the schematic of a mechanically variable phase shifter in an image guide. The image guide is fed by a TE_{10} mode rectangular waveguide so as to excite the E^y_{11} mode. The parameters of the image guide are $\varepsilon_r = 2.56$, $b = 2\,\text{mm}$, $a/b = 0.5$, $s/a \geqslant 0.5$. Determine the length ℓ of moving conducting plate for achieving a continuously variable phase shift (w.r.t $s/a = \infty$) of $180°$ at $f = 40\,\text{GHz}$. Neglect end/edge effects of the conducting plate. [*Note*: Use EDC formulas to calculate the propagation constants of the isolated image guide and also of the coupled image guide (odd mode) with s/a varying from 0.5 to ∞.]

11.3 Figure 11.9(a) shows variation of $\beta_e - \beta_o$ versus frequency for a π-guide. Using this data, calculate and draw corresponding graphs of coupling length versus frequency for 10-dB coupling. Consider a uniformly coupled guide.

11.4 From the graphical plots of coupling length versus frequency of Problem 11.3, design a 10-dB coupler in a π-guide at a center frequency of 35 GHz and plot the coupling versus frequency response for each of the four values of d. Consider only a uniformly coupled guide section and neglect end effects. Determine the bandwidth of the 10-dB coupler for a coupling variation within 0.5 dB. Compare the bandwidth of the image guide coupler ($d = 0$) with that of the π-guide and explain the variation.

11.5 Design a beam-splitter type of coupler (Fig. 11.12(a)) for 10-dB coupling. Assume an airgap at the coupling region ($\varepsilon_{e2} = 1$). Determine the relative dielectric constant ε_{e1} of the guide and the gap dimension s for an incidence angle $\theta_1 = \pi/4$. What would be the values of ε_{e1} and s if $\theta_1 = \pi/5$?

11.6 Design a 3-dB branched NRD guide coupler at 30 GHz. Assume $\varepsilon_r = 2$ for the dielectric and ground plane spacing $a = 3$ mm. [*Note*: Use the formula for the characteristic impedance derived in Problem 6.4.]

11.7 What are the advantages of a two-ring filter over a single-ring filter? What are the criteria for choosing ring sizes?

11.8 Consider an image guide grating of the type shown in Figure 11.23(a). The grating structure is made of a dielectric having $\varepsilon_r = 2.22$. It has n identical sections of characteristic impedance Z_0, length $\ell_0 = 1.82$ mm, and width $w_0 = 4$ mm; and $n - 1$ identical sections of characteristic impedance Z_1, length $\ell_1 = 2.2$ mm, and width $w_1 = 1.5$ mm. The stopband center frequency f_0 is 34 GHz at which $Z_1/Z_0 = 1.069$.

(a) Calculate and plot the attenuation at the center of the stopband as a function of the total number of sections $N = 2n - 1$ for N in the range 20–100.

(b) Determine the fractional bandwidth of this filter for $n \to \infty$.

(c) How can such gratings be used to construct bandpass filters?

11.9 Design a three-pole 0.15-dB ripple Chebyshev bandpass filter with 5% bandwidth at a center frequency of 35 GHz. Use a gap-coupled NRD guide configuration with Teflon ($\varepsilon_r = 2.1$) as the dielectric.

11.10 Compare the relative advantages of a six-port reflectometer and a multi-state reflectometer.

11.11 Draw the layout of a six-port reflectometer using beam-splitter type couplers.

11.12 Draw the layout of a multistate reflectometer using parallel-coupled dielectric couplers.

REFERENCES

1. S. E. Miller, Coupled wave theory and waveguide applications. *Bell Syst. Tech. J.*, **33(3)**, 661–719, May 1954.
2. W. H. Louisell, Analysis of the single tapered mode coupler. *Bell Syst. Tech. J.*, **34(4)**, 853–870, July 1955.
3. D. Marcuse, The coupling of degenerate modes in two parallel dielectric waveguides. *Bell Syst. Tech. J.*, **50(6)**, 1791–1816, 1971.
4. E. A. J. Marcatili, Dielectric rectangular waveguide and directional coupler for integrated optics. *Bell Syst. Tech. J.*, **48(7)**, 2071–2102, 1969.
5. I. Anderson, On the coupling of degenerate modes on non-parallel dielectric waveguides. *IEE J. Microwaves Opt. Acoust.*, **3(2)**, 56–58, Mar. 1979.
6. C. Kaizhou et al., An optimal design of directional coupler of millimetre wave integrated dielectric image lines. *Int. Symp. Electromagnetic Compatibility*, **2**, 642–645, Oct. 1984.
7. P. K. Ikalainen and G. L. Matthaei, Design of broadband dielectric waveguide 3 dB-couplers. *IEEE Trans. Microwave Theory Tech.*, **MTT-35**, 621–628, July 1987.
8. R. Rudokas and T. Itoh, Passive millimetre-wave IC components made of inverted strip dielectric waveguides. *IEEE Trans. Microwave Theory Tech.*, **MTT-24**, 978–981, Dec. 1976.
9. K. Solbach, The calculation and measurement of the coupling properties of dielectric image lines of rectangular cross-sections. *IEEE Trans. Microwave Theory Tech.*, **MTT-27**, 54–58, Jan. 1979.
10. E. G. Neumann and H. D. Rudolph, Radiation from bends in dielectric rod transmission lines. *IEEE Trans. Microwave Theory Tech.*, **MTT-23**, 142–149, Jan. 1975.
11. T. Trinh and R. Mittra, Coupling characteristics of planar dielectric waveguides of rectangular cross-section. *IEEE Trans. Microwave Theory Tech.*, **MTT-29**, 875–880, Sept. 1981.
12. B. Bhat and A. K. Tiwari, Analysis of low-loss broadside-coupled dielectric image guide using the mode-matching technique. *IEEE Trans. Microwave Theory Tech.*, **MTT-32**, 711–717, July 1984.
13. D. I. Kim et al., Directly connected image guide 3-dB couplers with very flat couplings. *IEEE Trans. Microwave Theory Tech.*, **MTT-32**, 621–627, June 1984.
14. J. Rodriguez and A. Prieto, Design and performance of π-guide directional couplers with very flat couplings. *Electron. Lett.*, **21(7)**, 292–293, Mar. 1985.
15. R. J. Collier and G. Hjipieris, A broadband directional coupler for both dielectric and image guides. *IEEE Trans. Microwave Theory Tech.*, **MTT-33**, 161–163, Feb. 1985.
16. W. D. Burnside and K. W. Burgener, High frequency scattering by a thin lossless dielectric slab. *IEEE Trans. Antennas Propag.*, **AP-31**, 104–110, Jan. 1983.
17. R. D. Birch and R. J. Collier, A broadband image guide directional coupler. *10th Eur. Microwave Conf. Digest*, 295–297, 1980.

18. R. E. Collin, *Field Theory of Guided Waves*, McGraw-Hill, New York, 1960.
19. P. Pramanick and P. Bhartia, Design of image guide–microstrip line couplers. *Microwave J.*, 247–255, May 1984.
20. P. Bhartia and P. Pramanick, Image guide–microstrip line directional couplers. *IEEE Int. Microwave Symp. Digest*, 265–269, 1983.
21. J. F. Miao and T. Itoh, Coupling between microstrip line and image guide through small apertures in common ground plane. *IEEE Trans. Microwave Theory Tech.*, **MTT-31**, 361–363, 1983.
22. R. M. Knox and P. P. Toulios, Integrated circuits for millimetre through optical frequency range, *Proceedings of the Symposium on Sub-millimetre Waves*, Mar. 1970.
23. G. Kompa and R. Mehran, Planar waveguide model for calculating microstrip components. *Electron. Lett.*, **11**, 459–460, 1975.
24. W. J. Getsinger, Microstrip dispersion model. *IEEE Trans. Microwave Theory Tech.*, **MTT-21**, 34–39, Jan. 1973.
25. S. J. Xu, Dielectric waveguide branching directional coupler. *IEE Proc.*, **135(H-4)**, 282–284, Aug. 1988.
26. E. W. Hue, S. T. Peng, and A. A. Oliner, A novel leaky-wave strip waveguide directional coupler, *Topical Meet Integrated and Guided Wave Optics*, Paper No. WD2 (Salt Lake City, UT), Jan. 1978.
27. Dow-Chih Niu, T. Yoneyama, and T. Itoh, Analysis and measurement of NRD-guide leaky wave coupler in Ka-band. *IEEE Trans. Microwave Theory Tech.*, **MTT-41**, 2126–2132, Dec. 1993.
28. G. L. Matthaei, L. Young, and E. M. T. Jones, *Microwave Filters, Impedance Matching Network and Coupling Structures*, McGraw Hill, New York, 1964.
29. F. J. Tischer, Resonance properties of ring circuits. *IRE Trans. Microwave Theory Tech.*, **MTT-5**, 51–56, Jan. 1957.
30. A. R. Kaurs, A tunable bandpass ring filter for rectangular dielectric waveguide integrated circuits. *IEEE Trans. Microwave Theory Tech.*, **MTT-24**, 875–876, Nov. 1976.
31. T. Itanami and S. Shindo, Channel dropping filter for millimetre-wave integrated circuits. *IEEE Trans. Microwave Theory Tech.*, **MTT-26**, 759–765, Oct. 1978.
32. R. V. Gelsthorpe and N. Williams, *The Dielectric Waveguide Handbook*, ERA Report No. 81–85, ERA Technology Ltd., UK, 1981.
33. G. L. Matthaei, A note concerning modes in dielectric waveguide gratings for filter applications. *IEEE Trans. Microwave Theory Tech.*, **MTT-31**, 309–312, Mar. 1983.
34. H. Shigesawa and M. Tsuji, A completely theoretical design method of dielectric image guide gratings in the Bragg reflection region. *IEEE Trans. Microwave Theory Tech.*, **MTT-34**, 420–426, Apr. 1986.
35. T. Itoh, Application of gratings in a dielectric waveguide for leaky wave antennas and band-reject filters. *IEEE Trans. Microwave Theory Tech.*, **MTT-25**, 1134–1138, Dec. 1977.
36. D. C. Park et al., Bandstop filter design using a dielectric waveguide grating. *IEEE Trans. Microwave Theory Tech.*, **MTT-33**, 693–702, Aug. 1985.
37. G. L. Matthaei et al., A study of the filter properties of single and parallel coupled dielectric waveguide gratings. *IEEE Trans. Microwave Theory Tech.*, **MTT-31**, 825–835, Oct. 1983.

38. G. L. Matthaei et al., Dielectric waveguide filters using parallel-coupled grating resonators. *Electron. Lett.*, **18**, 509–510, June 1982.
39. T. Yoneyama, F. Kuroki, and S. Nishida, Design of non-radiative dielectric waveguide filters. *IEEE Trans. Microwave Theory Tech.*, **MTT-32**, 1659–1662, Dec. 1984. [Also correction, *IEEE Trans. Microwave Theory Tech.*, **MTT-33**, 741, Aug. 1985.]
40. P. K. Ikalainen and G. L. Matthaei, Design of dielectric waveguide bandpass filters using parallel-coupled gratings. *IEEE Trans. Microwave Theory Tech.*, **MTT-34**, 681–686, June 1986.
41. D. C. Park et al., Dielectric waveguide grating design for bandstop and bandpass filter applications. *IEEE MTT-S Int. Microwave Symp. Digest*, 202–204, 1984.
42. P. K. Ikalainen et al., Dielectric waveguide band-pass filters with broad stop bands. *IEEE MTT-S Int. Microwave Symp. Digest*, 277–279, 1985.
43. D. F. Williams and S. E. Schwarz, Design and performance of coplanar waveguide bandpass filters. *IEEE Trans. Microwave Theory Tech.*, **MTT-31**, 558–566, July 1983.
44. J. S. Kot, S. R. Pennock, and T. Rozzi, Millimetre wave filter synthesis using image guide short circuit resonators, *18th Eur. Microwave Conf. Digest*, 675–679, Sept. 1988.
45. J. A. G. Malherbe and J. C. Olivier, A bandstop filter constructed in coupled non-radiative dielectric waveguide. *IEEE Trans. Microwave Theory Tech.*, **MTT-34**, 1408–1412, Dec. 1986.
46. H. Kumazawa and I. Ohtomo, 30 GHz-band periodic branching filter using a travelling-wave resonator for satellite applications. *IEEE Trans. Microwave Theory Tech.*, **MTT-25**, 683–687, Aug. 1977.
47. T. Itanami, A periodic branching filter for millimeter-wave integrated circuits. *IEEE Trans. Microwave Theory Tech.*, **MTT-29**, 971–978, Sept. 1981.
48. F. Zheng-he, Broadband dielectric waveguide coupler and six-port network. *IEEE MTT-S Int. Microwave Symp. Digest*, 237–240, 1986.
49. T. Yoneyama, Millimetre wave integrated circuits using non-radiative dielectric waveguide, Yagi Symposium on Advanced Technology Bridging the Gap Between Light and Microwaves, *Proceedings of the Second Sendai International Conference*, Japan, pp. 57–66, Sept. 1990.
50. F. Kuroki and T. Yoneyama, NRD-guide center-fed H-shaped five ports at millimetre-wave lengths, *Proceedings of the 1990 Asia–Pacific Microwave Conference* (Tokyo), pp. 7–14, 1990.
51. M. J. Aylward and N. Williams, Feasibility studies of insular guide millimetre wave integrated circuits, AGARD Conference on Millimetre and Sub-millimetre Wave Propagation and Circuits, *Conf. Proc.*, **245**, 30.1–30.11, Sept. 1978.
52. J. W. Tao, B. Chau, H. Baudraud, and J. Atechian, Novel type of electrically-controlled phase shifter for millimeter-wave use: theory and experiment. *IEEE MTT-S Int. Microwave Symp Digest*, 671–674, 1991.
53. J. A. Paul and P. C. H. Yen, Millimeter wave passive components and six-port network analyser in dielectric waveguide. *IEEE Trans. Microwave Theory Tech.*, **MTT-29**, 948–953, Sept. 1981.
54. D. Radovich and J. Paul, Phase and amplitude characteristics of dielectric waveguide couplers and six-port network. *IEEE MTT-S Int. Microwave Symp. Digest*, 322–324, 1982.

55. G. Hjipieris, R. J. Collier, and E. J. Griffin, A millimeter-wave six-port reflectometer using dielectric waveguide. *IEEE Trans. Microwave Theory Tech.*, **MTT-38**, 54–61, Jan. 1990.
56. R. J. Collier and M. F. D'Souza, A multistate reflectometer in dielectric waveguide for the frequency range 75–140 GHz. *IEEE MTT-S Int. Microwave Symp. Digest*, 1027–1030, 1991.
57. C. A. Hoer, A network analyzer incorporating two six-port reflectometers. *IEEE Trans. Microwave Theory Tech.*, **MTT-25**, 1070–1074, Dec. 1977.
58. G. F. Engen and C. A. Hoer, Application of an arbitrary 6-port junction to power measurement problems. *IEEE Trans. Instrum. Meas.*, **IM-21**, 470–474, Nov. 1972.
59. G. F. Engen, The six-port reflectometer: an alternative network analyser. *IEEE Trans. Microwave Theory Tech.*, **MTT-25**, 1075–1080, Dec. 1977.
60. L. C. Oldfield, J. P. Ide, and E. J. Griffin, A multistate reflectometer. *IEEE Trans. Instrum. Meas.*, **IM-34(2)**, 198–201, June 1985.
61. V. P. Nanda, A new form of ferrite device for millimetre wave integrated circuits. *IEEE Trans. Microwave Theory Tech.*, **MTT-24**, 876–879, Nov. 1976.
62. R. A. Stern and R. W. Babbitt, Dielectric waveguide circulator. *Int. J. Infrared Millimetre Waves*, **3(1)**, 11–18, Jan. 1982.
63. H. Yoshinaga and T. Yoneyama, Design and fabrication of a non-radiative dielectric waveguide circulator. *IEEE Trans. Microwave Theory Tech.*, **MTT-36**, 1526–1529, Nov. 1988.
64. Q. Lanfen and L. Yong Xiang, Development of groove NRD waveguide circulator, *Proceedings of the 1990 Asia–Pacific Microwave Conference* (Tokyo), pp. 19–22, 1990.
65. S. Zhong Xiang, L. Xiaoming, and Li. Sifan, Dispersion characteristics of groove nonradiative dielectric waveguide with anisotropic substrate, *Proceedings of the 1990 Asia–Pacific Microwave Conference* (Tokyo), pp. 19–22, 1990.

CHAPTER TWELVE

Components Using Semiconductor Devices and Optical Control

12.1 INTRODUCTION

Semiconductor device-based circuit components—namely, detectors, oscillators, mixers, and phase shifters—have been realized using image guide technology. Except in the case of phase shifters, which mostly use bulk semiconductor guides for their operation, other components use concentrated, lumped semiconductor elements, commonly in package form. The diode is normally inserted by drilling a hole through the guide. One terminal of the diode is screwed on to the ground plane, whereas the other is accessible from the top for biasing. Some sort of metal shielding is normally incorporated in the vicinity of the diode mount for suppressing any radiation.

The literature on semiconductor device-based circuit components, particularly from the point of view of analysis and design, is very limited. Several circuit configurations including diode-mounting techniques, however, are reported. These are reviewed in this chapter.

12.2 IMAGE GUIDE DETECTOR CIRCUITS

Detector circuits have been realized mostly in image guide configuration [1–4]. The most important part of a detector circuit is the diode mount. Schematics of typical image guide circuits employing coaxial detector diodes and beam lead diodes are shown in Figures 12.1–12.3. In circuits employing coaxial diodes, the diode is mounted in the ground plane with the center conductor of the diode protruding into the dielectric as shown in Figure 12.1(a). The conductor, which is approximately $\lambda_g/4$ in length, serves as a probe coupler to the guide. The top and

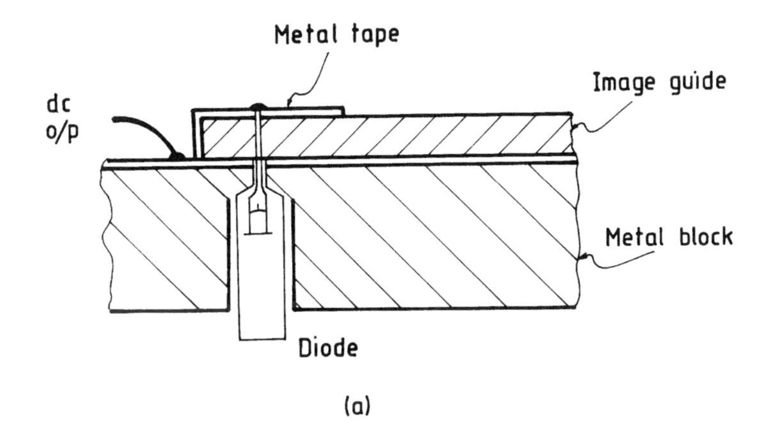

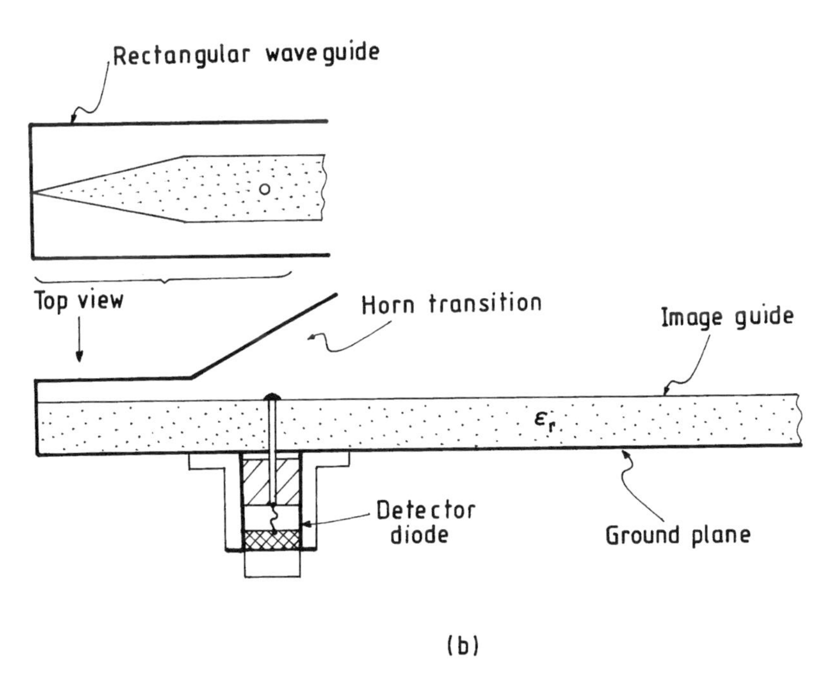

FIGURE 12.1 Coaxial detector diode mounted in image guide with (a) dielectric guide metallized on all sides in the vicinity of the diode and (b) metal horn transition. ((a) After Aylward and Williams [2].)

end faces of the guide around the diode mount are metallized to reduce radiation from the diode [2]. Heavy radiation loss can still occur due to the abrupt transition from the metallized guide to the open guide. A careful choice of the guide dimensions and dielectric constant is necessary to reduce this loss. An alternative technique of reducing the transition loss is to use a metal horn as

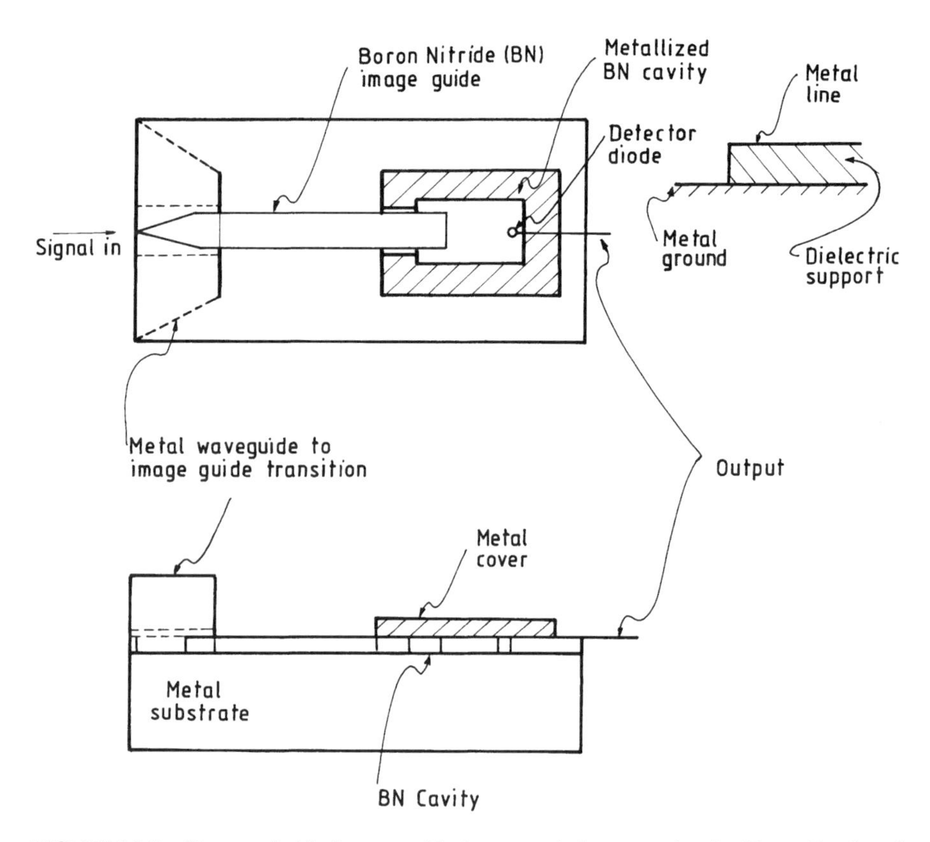

FIGURE 12.2 Boron nitride image guide integrated detector circuit. (From Paul and Chang [3]. Copyright © 1978 IEEE, reprinted with permission.)

illustrated in Figure 12.1(b). With a suitable taper in the image guide dielectric, this transition can be optimized to yield excellent impedance match with improved circuit efficiency.

Figure 12.2 shows the configuration of a boron nitride image guide integrated detector circuit employing a beam lead Schottky diode as reported by Paul and Chang [3]. The beam lead diode is placed in a metallized boron nitride cavity with a metal cover and the detected output is taken out at the rear of the cavity. Millimeter wave detectors in this configuration are reported to operate over nearly the full waveguide band with a tangential sensitivity on the order of − 40 dBm.

Figure 12.3 shows a planar form of an image guide integrated detector circuit reported by Solbach [4]. The basic geometry involves an image guide with a slot resonator pattern etched on the ground plane, which in turn is backed by a metallized dielectric substrate (Fig. 12.3(a)). The slot serves as a two-terminal source for mounting a beam lead diode and the dielectric guide of high dielectric

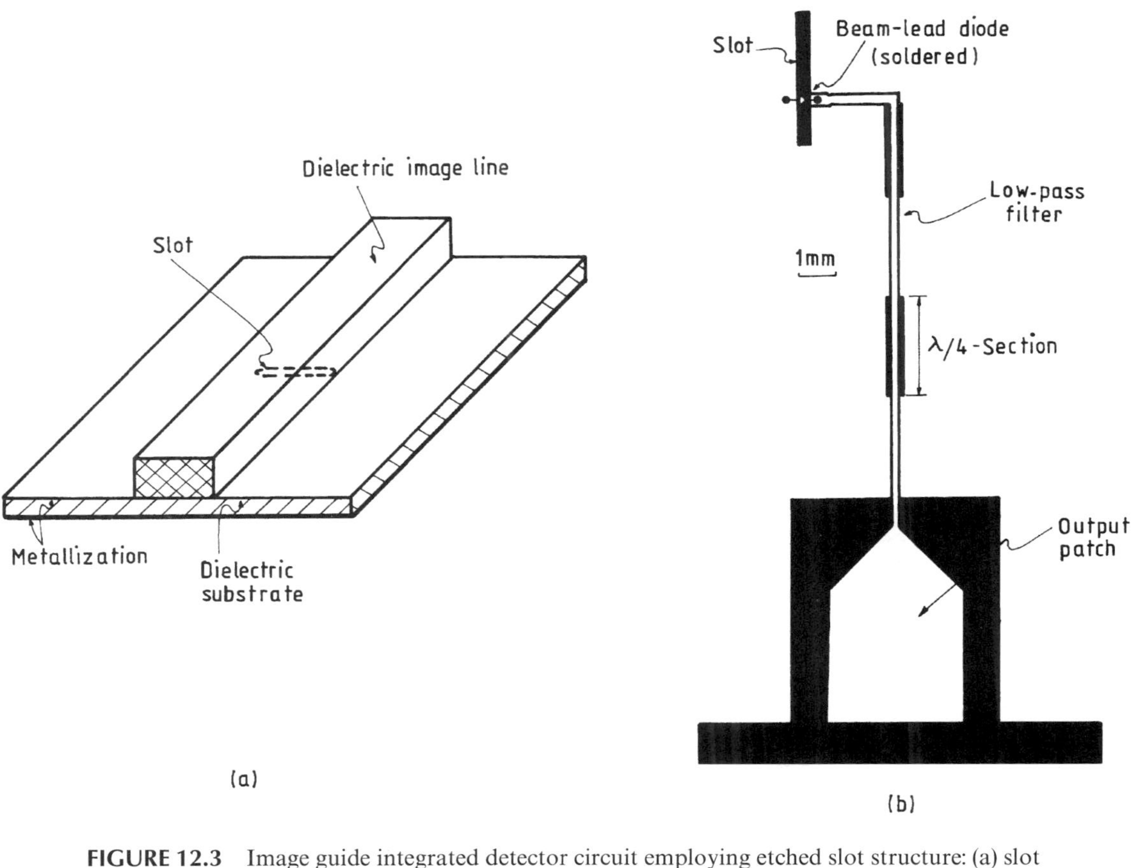

FIGURE 12.3 Image guide integrated detector circuit employing etched slot structure: (a) slot in the ground plane of image guide backed by a metallized substrate and (b) layout of metallization pattern of the detector circuit. (From Solbach [4]. Copyright © 1981 IEEE, reprinted with permission.)

constant acts as a shield to reduce radiation losses in the circuit. The slot resonator is chosen to be a half-wavelength long at the center frequency of operation so that a low-impedance beam lead diode can be tap-matched to a high source impedance across the slot. Where the diode is to be soldered, a narrow groove may be made in the guide so as to conveniently accomodate it without breakage. The guide is terminated in an open circuit at about a quarter wavelength distance from the slot so as to introduce a current maximum at the location of the slot. The dc current path for the diode is provided through a coplanar lowpass filter etched on the bottom ground plane of the substrate (Fig. 12.3(b)).

With the diode perfectly matched, the coupling efficiency of the circuit is determined by the slot radiation resistance, the substrate wave impedance, and the reflection loss at the dielectric guide termination [5]. For achieving high circuit efficiency, a combination of high-permittivity dielectric image line with a low-permittivity substrate is to be used. As an example, for a K_a-band detector employing an image guide made of Epsilam-10 ($\varepsilon_r = 10$) over a RT-duroid substrate ($\varepsilon_r = 2.2$), an efficiency on the order of -1 dB has been reported [5].

12.3 OSCILLATORS

Dielectric integrated guide oscillators have been reported in image guide configuration using Gunn and IMPATT diodes [2, 3, 6–9] and also in NRD guide using a Gunn diode [10]. The most important part of the oscillator is the cavity, which must be designed for high Q-factor. Oscillators employing high-Q metal cavities—namely, rectangular waveguide cavity and coaxial line cavity with transition to image guide-have been reported [2, 6, 7]. More compact versions incorporate dielectric guide cavities with or without surface metallization [3, 8]. The use of a semiconductor image guide employing Si or GaAs permits monolithically integrated Gunn oscillators to be realized [8]. Several guide materials have been reported for use in oscillators at millimeter wave frequencies. These include fused quartz, silica, semi-insulating Si, alumina, and hot-pressed boron nitride. When dielectric guide cavities are used for housing the diode, the guide material, besides having a low-loss tangent, must possess low thermal expansion coefficient and high-temperature characteristics so as to be able to withstand heat and thermal shock. Materials such as fused quartz and silica, which have these properties, have been used for realizing a W-band Gunn oscillator.

12.3.1 Image Guide Oscillators

Coaxial Cavity-Coupled Image Guide Oscillator Figure 12.4 shows the configuration of a coaxial cavity-coupled image guide oscillator reported by Horn et al. [6, 7]. The cross-sectional view of the cavity showing the arrangement of a top disk tuner and transition to alumina image guide is shown in Figure 12.5. As

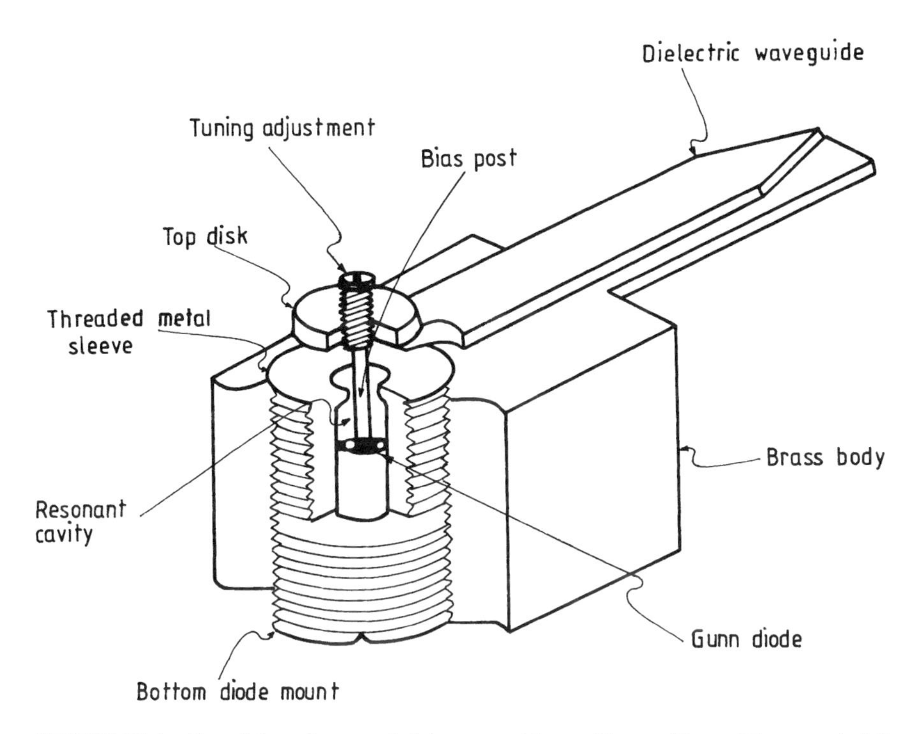

FIGURE 12.4 Coaxial cavity-coupled image guide oscillator. (From Horn et al. [6]. Copyright © 1984 IEEE, reprinted with permission.)

can be seen from the illustration, the Gunn diode is mounted at the bottom of the cavity and the output to the image guide is taken from the top through an iris. The dc bias to the diode is provided through a brass rod, which also forms the center conductor of the cavity. The top metal disk acts as a transformer for coupling the metal cavity to the image guide. The metal disk enables the horizontally oriented electric field lines of the TEM mode of the coaxial cavity to gradually bend and orient in the vertical direction. As a consequence, at the output iris, energy enters the dielectric with vertically polarized electric field, thereby causing the E^y_{11} mode to propagate in the image guide.

The equivalent circuit for the oscillator can be represented as shown in Figure 12.6, where R, C_j, L_p, and C_p are the negative resistance, junction capacitance, package inductance, and package capacitance, respectively, of the diode. $R_{\ell t}$ is the load resistance appearing at the diode package as transformed by the image guide and the cavity. Referring to Figure 12.6, the condition for oscillation can be written as [6]

$$|\mathrm{Re}(Z_g)| \geq |\mathrm{Re}(Z_t)| \tag{12.1}$$

$$\mathrm{Im}(Z_g) = -\mathrm{Im}(Z_t) \tag{12.2}$$

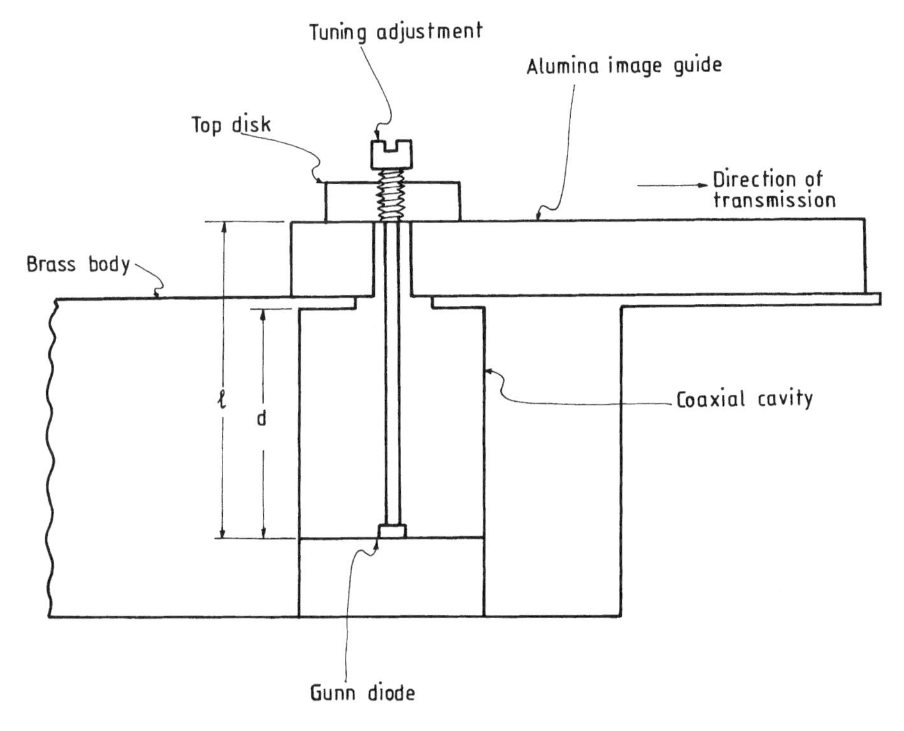

FIGURE 12.5 Coaxial cavity to image guide launch (TEM mode to E_{11}^y mode). (From Horn et al. [7]. Copyright © 1986 IEEE, reprinted with permission.)

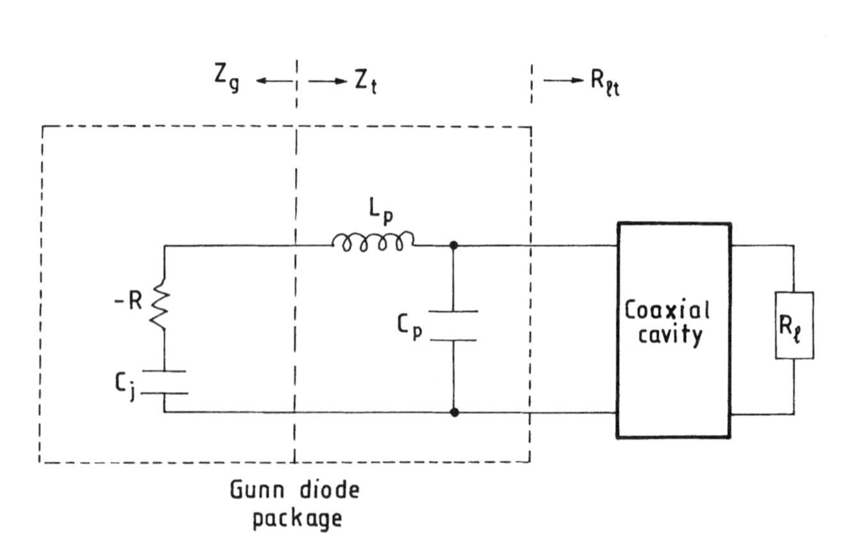

FIGURE 12.6 Equivalent circuit for cavity-coupled Gunn oscillator. (After Horn et al. [6].)

where

$$Z_g = -R - \frac{1}{\omega C_j} \tag{12.3a}$$

$$Z_t = j\omega L_p + \frac{1}{\left(\frac{1}{R_{\ell t}} + j\omega C_p\right)} \tag{12.3b}$$

It is assumed that for optimum power transfer, the load $R_{\ell t}$ is real. Substitution of (12.3) in (12.1) and (12.2) yields the following relations [6]:

$$R \geq \left| \frac{R_{\ell t}}{1 + \omega^2 C_p^2 R_{\ell t}^2} \right| \tag{12.4a}$$

$$C_j = \frac{1 + \omega^2 C_p^2 R_{\ell t}^2}{\omega^2(-C_p R_{\ell t}^2 + L_p + \omega^2 C_p^2 R_{\ell t}^2 L_p)} \tag{12.4b}$$

It can be seen from (12.4a) that for the onset of oscillations $R_{\ell t}$ must be large so as to make $\text{Re}(Z_t)$ less than R. The negative resistance of practical diodes is of the order of $-5\,\Omega$. At resonance, the cavity height corresponds to half wavelength so that $R_{\ell t} = R_\ell$. For a matched image guide, R_ℓ corresponds to the wave impedance of the guide, which is on the order of $200\,\Omega$. As an example, if we consider $f =$ 60 GHz, $C_p = 0.1$ pF, and $R_{\ell t} = 200\,\Omega$, we obtain $\text{Re}(Z_t) = 3.5\,\Omega$, which satisfies the condition for oscillation.

Integrated Image Guide Gunn Oscillator An image guide oscillator that incorporates a dielectric cavity as an integral part of the guide structure offers considerable compactness. The dielectric cavity can be modeled as a section of transmission line with suitable terminal conditions. The equivalent circuit of a dielectric cavity with a Gunn diode mounted in it is shown in Figure 12.7, where Y_0 and β are the characteristic admittance and propagation constant, respectively, of the dielectric guide; $-G$ and B are the negative conductance and susceptance, respectively, of the Gunn diode, and $\ell_1 + \ell_2$ is the length of the cavity. One end of the cavity is assumed to be terminated in a load admittance Y_ℓ and the other end in a short circuit ($Y_s = \infty$). Furthermore, assuming the transmission line sections to be lossless, the expressions for the input admittances $Y_{\text{in}1}$ and $Y_{\text{in}2}$ as referred in Figure 12.7 can be written as

$$Y_{\text{in}1} = -jY_0 \cot \beta\ell_1 \tag{12.5}$$

$$Y_{\text{in}2} = Y_0 \left[\frac{Y_\ell + jY_0 \tan \beta\ell_2}{Y_0 + jY_\ell \tan \beta\ell_2} \right] \tag{12.6}$$

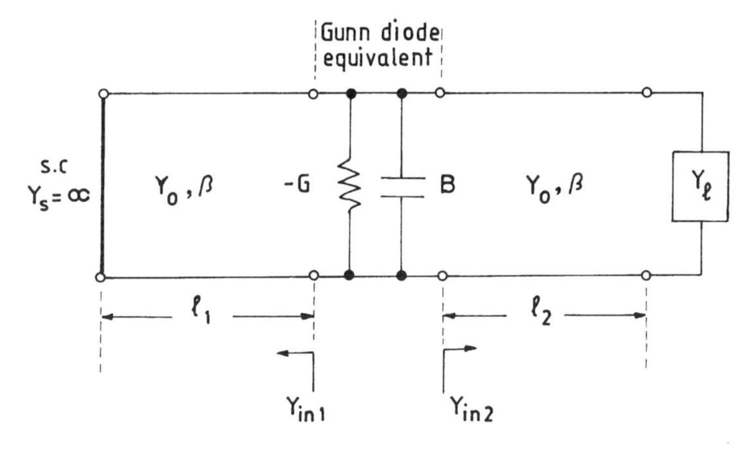

FIGURE 12.7 Equivalent circuit of dielectric cavity with Gunn diode.

The condition for resonance is given by

$$Y_{\text{in}1} + Y_{\text{in}2} - G + jB = 0 \tag{12.7}$$

Separating this equation into real and imaginary parts gives

$$\text{Re}(Y_{\text{in}2}) - G = 0 \tag{12.8}$$

$$\text{Im}(Y_{\text{in}2}) - Y_0 \cot \beta\ell_1 + B = 0 \tag{12.9}$$

The frequency of oscillation is governed by (12.9). Equation (12.8) determines the power output and whether the oscillation can be sustained or not. For an oscillator the load admittance Y_ℓ (real) is chosen to be small in comparison with Y_0. With $Y_\ell/Y_0 \ll 1$ and if $\beta\ell$ is not close to $\pi/2$, Eq. (12.6) can be approximated to

$$Y_{\text{in}2} \simeq Y_\ell + jY_0 \tan\beta\ell_2 \tag{12.10}$$

and (12.9) can be written as

$$\tan\beta\ell_2 - \cot\beta\ell_1 + B/Y_0 = 0 \tag{12.11}$$

It may be noted that if the dielectric cavity is terminated at one end in an open circuit ($Y_s = 0$) instead of the short circuit ($Y_s = \infty$) as considered in Figure 12.7, then the term $\cot\beta\ell_1$ in (12.11) gets replaced by $-\tan\beta\ell_1$. The condition for oscillation then becomes

$$\tan\beta\ell_2 + \tan\beta\ell_1 + B/Y_0 = 0 \tag{12.12}$$

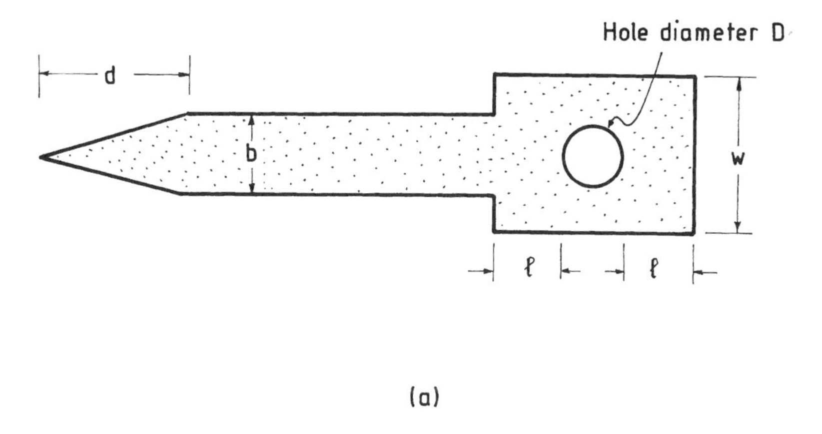

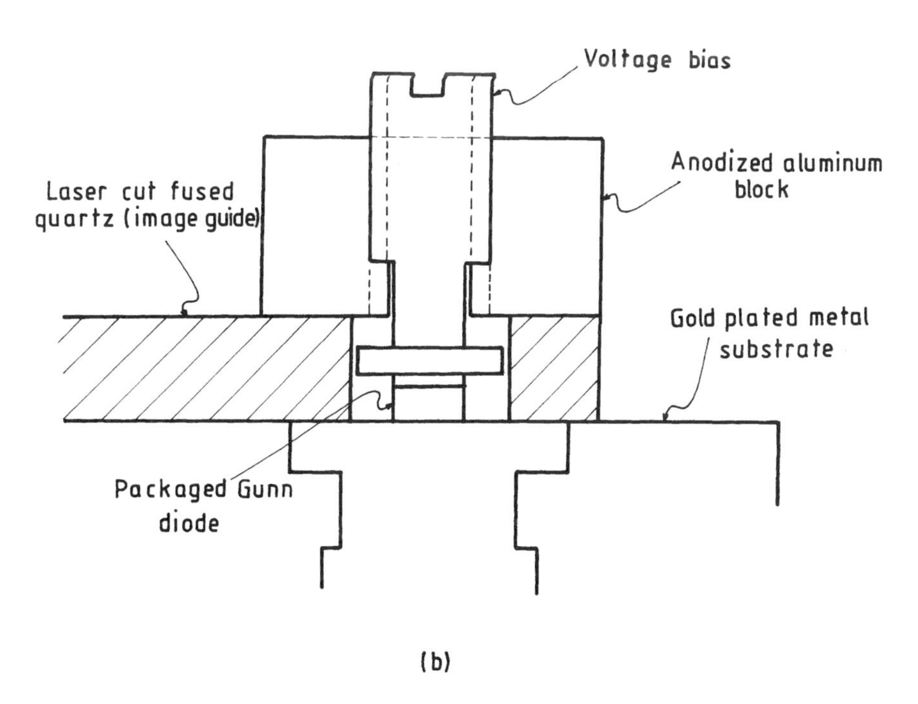

FIGURE 12.8 Image guide integrated Gunn oscillator: (a) quartz image guide dimensions for W-band Gunn oscillator and (b) cross-sectional views of diode mount area. (From Chang [8]. Copyright © 1983 IEEE, reprinted with permission.)

As an example of a dielectric cavity integrated image guide oscillator, we show in Figure 12.8 the schematic of an oscillator reported by Chang [8]. The image guide is tapered on one side to form a suitable transition to rectangular metal waveguide and is increased in width on the other to form a cavity. The cavity is provided with a circular hole for mounting the diode and is metallized on all sides to prevent radiation. Figure 12.8(a) shows the geometry of the guide and its dimensions. The cross-sectional view in Figure 12.8(b) shows the details of the diode mount. The Gunn diode is inserted into the hole and is threaded to a metal base substrate to achieve good thermal contact. A small radial metal disk on the top of the diode and a metal bias pin threaded through a metal block as depicted in Figure 12.8(b) form a series *LC* resonance circuit. In this arrangement, the oscillator frequency can be adjusted by properly choosing the dimensions of the disk and the pin, and the power output of the oscillator can be optimized by varying the cavity length. With metallization on the top and side surfaces of the cavity region, the structure resembles a conventional metal waveguide cavity with a dielectric image guide inserted into the cavity from one side. An experimental oscillator of this type is reported to offer a power output of about 5 mW at 94 GHz with a bias tuning range of 350 MHz [8].

Distributed Bragg Reflector Gunn Oscillator Another example of a dielectric cavity integrated oscillator is the distributed Bragg reflector (DBR) Gunn oscillator reported by Itoh and Hsu [9]. The device consists of a Gunn diode mounted in an image guide grating structure, as illustrated in Figure 12.9.

As discussed in Chapter 11, a dielectric guide with periodic gratings exhibits good stopband characteristics. Stopband occurs at a frequency at which the condition $\beta d = \pi$ is satisfied, where β is the propagation constant of the unperturbed grating and d is the grating period. Thus in the DBR oscillator (Fig. 12.9), the two grating guide sections behave like frequency-selective reflectors for the diode. The structure thus behaves like a diode-mounted cavity, which has a high Q only in the stopband of the grating. By appropriately choosing the impedance of the diode, the frequency of oscillation can be chosen to fall in the stopband of the grating.

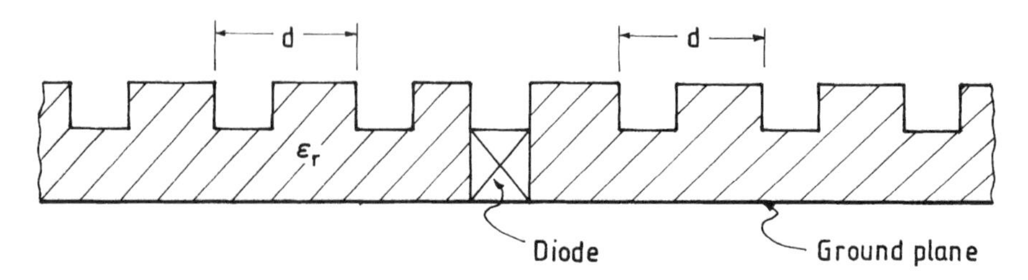

FIGURE 12.9 Schematic of distributed Bragg reflector oscillator. (After Itoh and Hsu [9].)

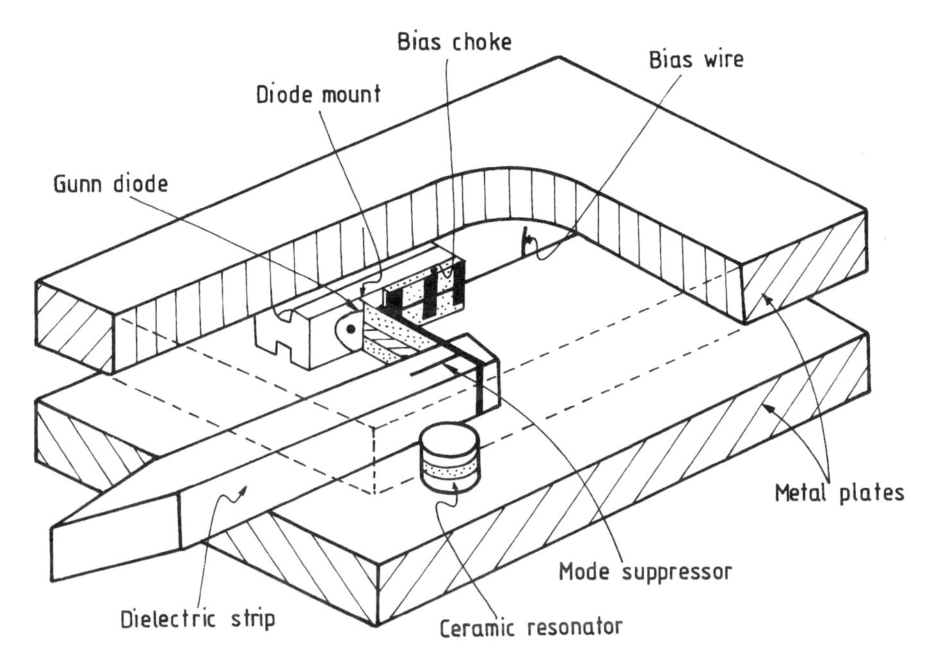

FIGURE 12.10 Structure of NRD guide oscillator. (After Yoneyama [11].)

12.3.2 NRD Guide Oscillator

Figure 12.10 illustrates implementation of a Gunn diode oscillator in NRD guide as reported by Takadaand and Yoneyama [10, 11]. The Gunn diode is fixed on a metal block with a bias choke on it and the output is fed to the dielectric guide via a copper strip etched on a thin Teflon substrate. The substrate is oriented such that the electric field lines are aligned parallel to the metal plates of the NRD guide. The copper strip acts as a resonator and its length determines the oscillator frequency. A temperature-stabilized ceramic resonator is used for phase locking the oscillator.

12.4 ELECTRONIC PHASE SHIFTERS

Integrated dielectric guide electronic phase shifters employing high-purity semiconductors as the guide material have been reported by several investigators [12–17]. The basic mechanism involves changing the phase of the RF propagating wave by creating a plasma into the guide and controlling its dielectric constant. Two techniques have been reported for the generation of plasma: one requiring the injection of electrons into the semiconducting medium via contacts [12] and the other through optical illumination [13–17].

12.4.1 Phase Shifter Using Contact Injection Technique

Figure 12.11 shows a millimeter wave phase shifter due to Jacobs and Chrepta [12], which employs the contact injection technique. The device makes use of plasma injection into a silicon dielectric guide by means of a triangular shaped p-i-n diode placed on its broad wall. When a forward bias is applied to the p-i-n diode, mobile charge carriers are injected into the I-region. Conduction starts at the upper edge of the triangular piece and, as the current is increased, an increasing portion of the I-region sets into conduction in a downward motion. This results in a change in the effective height of the dielectric guide with a consequent increase in the propagation constant of the guide in the direction of propagation (z-direction). If β_z and β'_z, respectively, denote the propagation constants corresponding to zero bias and forward bias, and ℓ is the length of the p-i-n diode structure in the z-direction, then the differential phase shift $\Delta\phi$ is given by

$$\Delta\phi = (\beta_z - \beta'_z)\ell \tag{12.13}$$

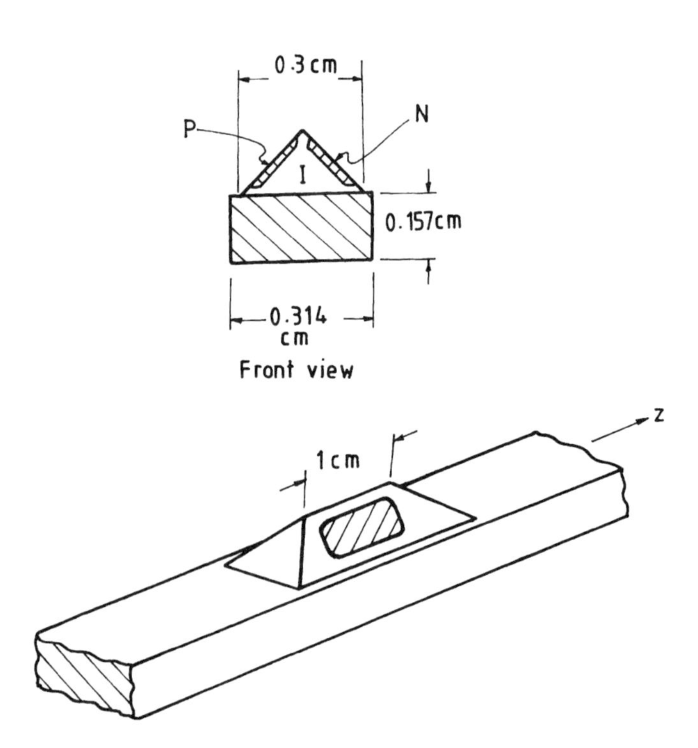

FIGURE 12.11 Dielectric image line phase shifter using distributed p-i-n diode. (From Jacobs and Chrepta [12]. Copyright (©) 1974 IEEE, reprinted with permission.)

12.4.2 Phase Shifter Using Optical Injection Technique

The contact injection technique requires metallization for contact to the junction area, owing to which the device incurs considerable loss. This problem is circumvented in the optical injection technique by creating a surface layer plasma rather than a bulk plasma. Optical control also offers other advantages, which include excellent isolation between the source and the guide, ultrafast response for switching of phase, and high-power handling capability. In this technique, the semiconductor guide is illuminated on its broad surface with above-bandgap radiation from a laser beam. The semiconductor material for the guide can be silicon, GaAs, silicon-on-sapphire, or silicon-on-alumina. The guides that have been experimentally studied for optically controlled phase shifter characteristics are rectangular silicon/GaAs guides [13–15], the silicon image guide [16], and the silicon rib guide [17]. The theory of dielectric guides with optically induced plasma layer and the phase shift characteristics of the above guides are presented in Chapter 5 (sections 5.3 and 5.4) and hence are not discussed here.

12.5 BALANCED MIXER

An integrated balanced mixer incorporating identical mixer diodes in the two output arms of a boron nitride image guide 3-dB coupler has been reported by Paul and Chang [3]. The circuit details are illustrated in Figure 12.12. Each mixer diode is housed in an air-filled cavity so as to provide approximately a quarter-wavelength termination behind the diode. The RF and LO signals are fed to the two input arms of the coupler through waveguide-to-image guide transitions. The IF output is fed to an amplifier through printed lines on a boron nitride substrate. A 60-GHz balanced mixer utilizing mixer diodes with a cutoff frequency of 750 GHz is reported to offer a DSB noise figure of about 12 dB [3].

Figure 12.13 shows the cross-sectional view of an image guide self-oscillating mixer circuit reported by Dixon and Jacobs [18]. Unlike the conventional mixers, the circuit eliminates the need for mixer diodes. The indium phosphide (InP) Gunn diode used in the circuit serves both as a local oscillator and a mixing element. The Gunn diode is mounted in a Al_2O_3 image guide such that the resonant section at the back of the diode is approximately $(2n+1)\lambda_g/2$ in length, where λ_g is the guide wavelength and n is an integer. The other end of the guide is tapered to couple to a metal waveguide. As illustrated in the figure, the Gunn diode protrudes into a circular hole made in the image guide, with a metal disk on the top serving as a tuning element. The signal is introduced from the left directly into the oscillator and the IF power is extracted through an IF probe from the top of the image guide. A practical self-oscillating mixer designed at 60 GHz is reported to have a peak sensitivity on the order of -79 dBm in a 100-MHz bandpass window.

The structure of a beam lead diode mount in a NRD guide for application in a mixer circuit is shown in Figure 12.14 [11, 19]. The circuit consisting of

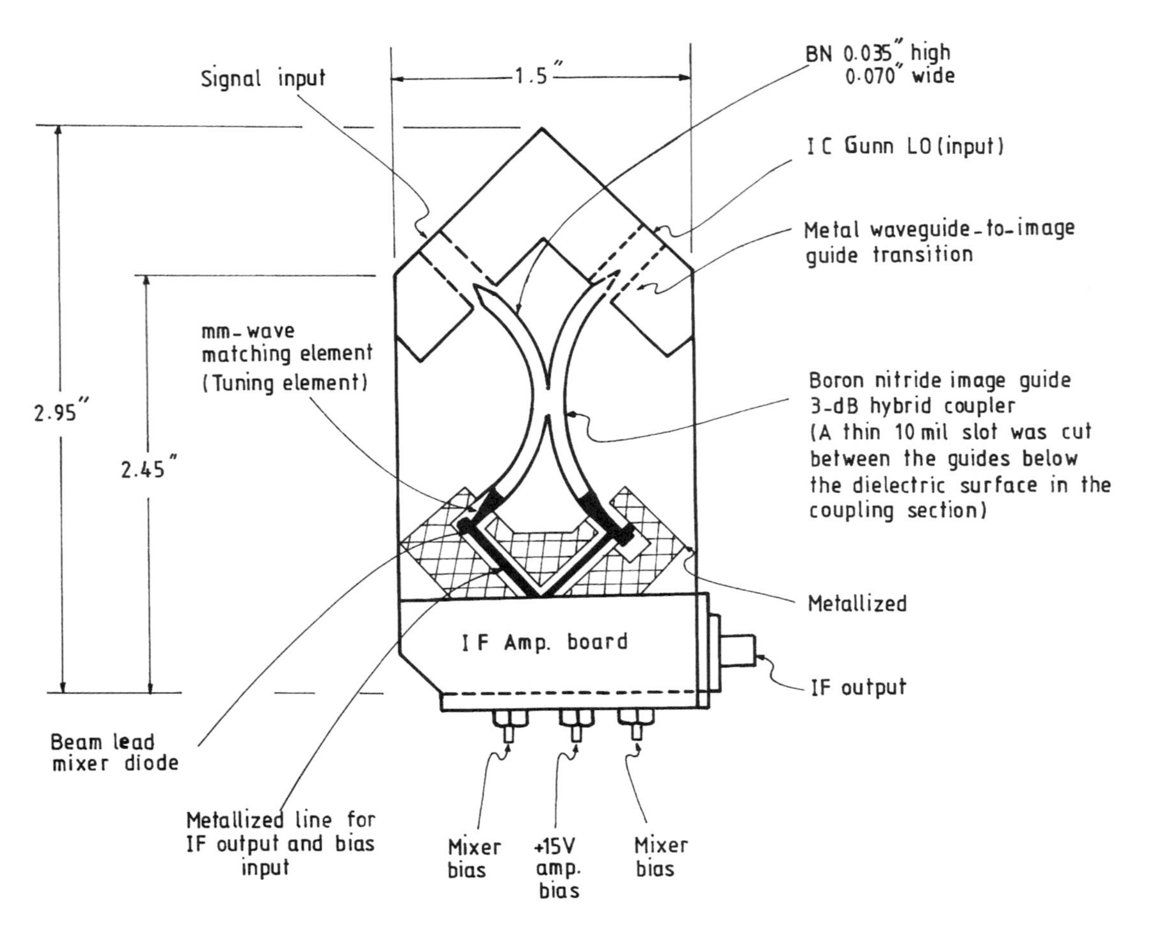

FIGURE 12.12 Integrated image guide balanced mixer. (From Paul and Chang [3]. Copyright © 1978 IEEE, reprinted with permission.)

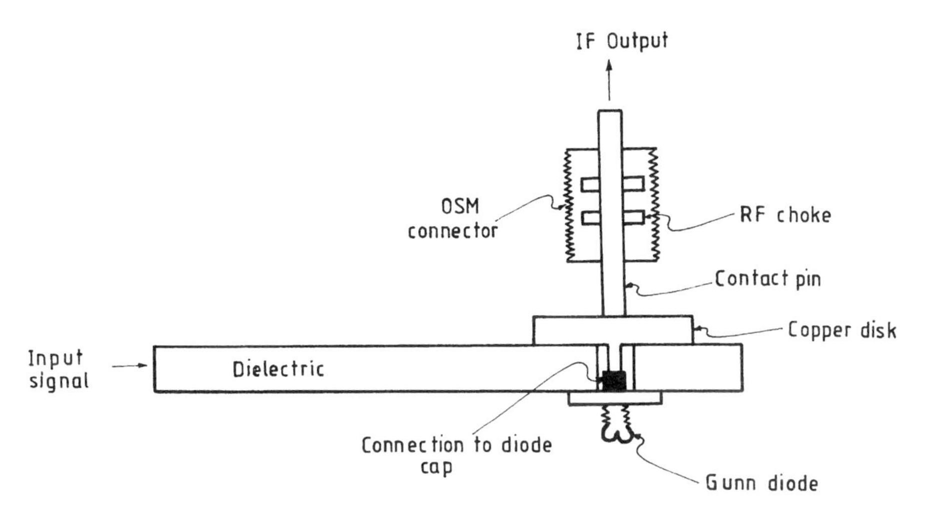

FIGURE 12.13 Cutaway view of image guide self-oscillating mixer. (From Dixon and Jacobs [18]. Copyright © 1981 IEEE, reprinted with permission.)

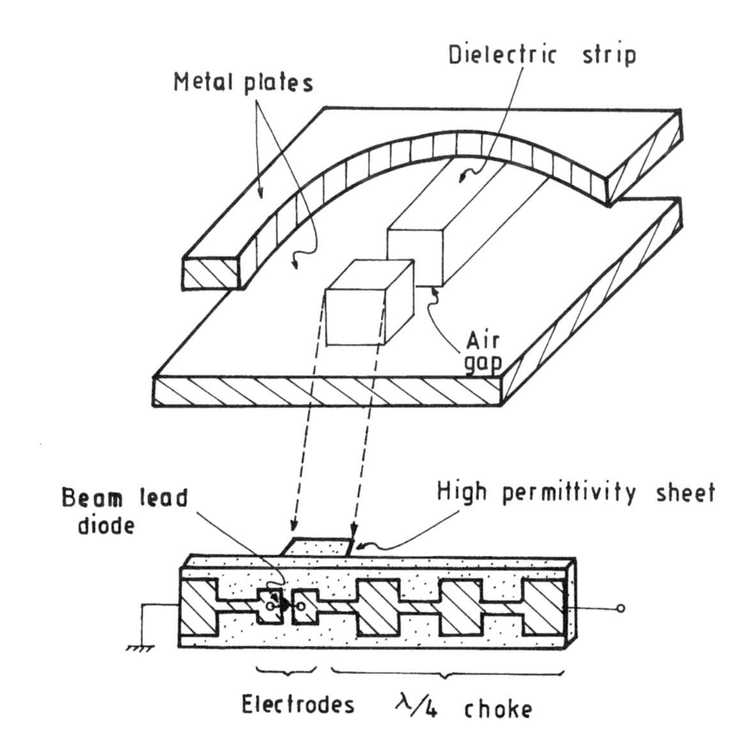

FIGURE 12.14 Structure of a beam lead diode mount. (After Yoneyama [11].)

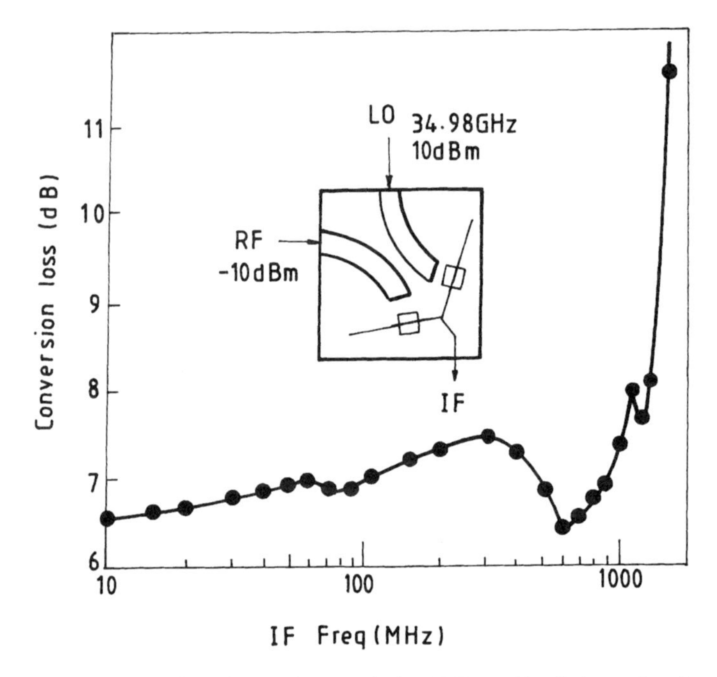

FIGURE 12.15 Conversion loss of a typical NRD guide balanced mixer. (From Yoneyama [11], reprinted with permission.)

rectangular electrodes and a quarter-wave choke (lowpass filter) printed on a single Teflon substrate. This mount is attached to the end of the NRD dielectric strip. The beam lead diode is bonded across the gap between the electrodes. The matching of the diode is achieved by placing a high dielectric thin plate in front of the electrodes and creating an airgap in the dielectric strip as shown in Figure 12.14. Figure 12.15 shows the conversion loss of a NRD guide balanced mixer using the beam lead diode mount as reported by Yoneyama [11]. For an LO power of 10 mW, a conversion loss of 7 dB on the average has been reported over the IF bandwidth of about 800 MHz [11].

12.6 AMPLIFIER

The layout of a HEMT amplifier as reported by Artuzi and Yoneyama [20] is shown in Figure 12.16. It consists of input and output NRD guides and the amplifier circuit realized in the laterally shielded coplanar waveguide. Figure 12.17 shows circuit details of the HEMT chip carrier. Besides supporting the HEMT chip, the circuit consists of matching stubs, isolation choke, bias choke, stabilization resistor, bias pad, chip condensor, and the earth pad. The chokes and the matching elements are implemented with short-circuited slotline stubs placed in the coplanar waveguide ground planes. Two metal strip probes are

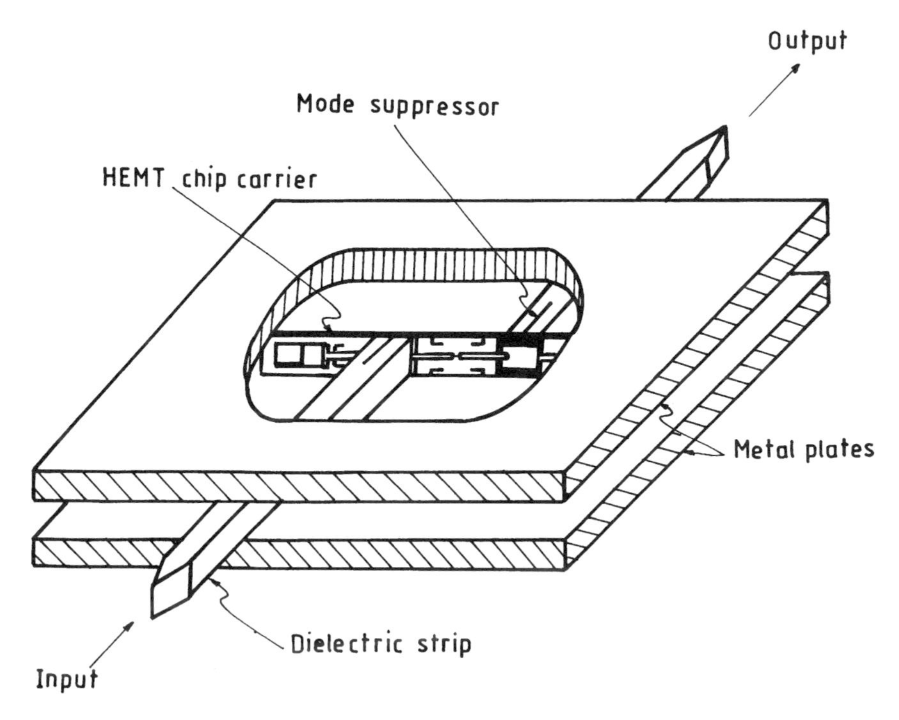

FIGURE 12.16 Structure of a HEMT amplifier for NRD guide. (From Artuzi and Yoneyama [20], reprinted with permission of Communication Engineers, Denshi Joho Tsushin Gakkai, Tokyo, Japan.)

placed at the location of maximum electric field strength so as to provide efficient transitions between the coplanar waveguide and the NRD input and output guides. Using the Toshiba JS-8900-AS low-noise HEMT, Artuzi and Yoneyama [20] have reported measured maximum gain of about 5 dB at 34.3 GHz as compared with the theoretically calculated maximum available gain of about 7 dB.

12.7 CIRCUIT MODULES

12.7.1 Oscillator-cum-Power Combiner

Figure 12.18 illustrates an arrangement for combining power outputs from two Gunn oscillators in an image guide configuration [21]. The device makes use of two rectangular shaped high-resistivity silicon resonators in each of which is mounted a Gunn diode with its lower end screwed to the ground plane. The bias pins located on the top metal plate make pressure contact with the top terminals of the diodes. The power outputs from these two diode-mounted resonators are coupled into another rectangular dielectric guide section (marked G), which

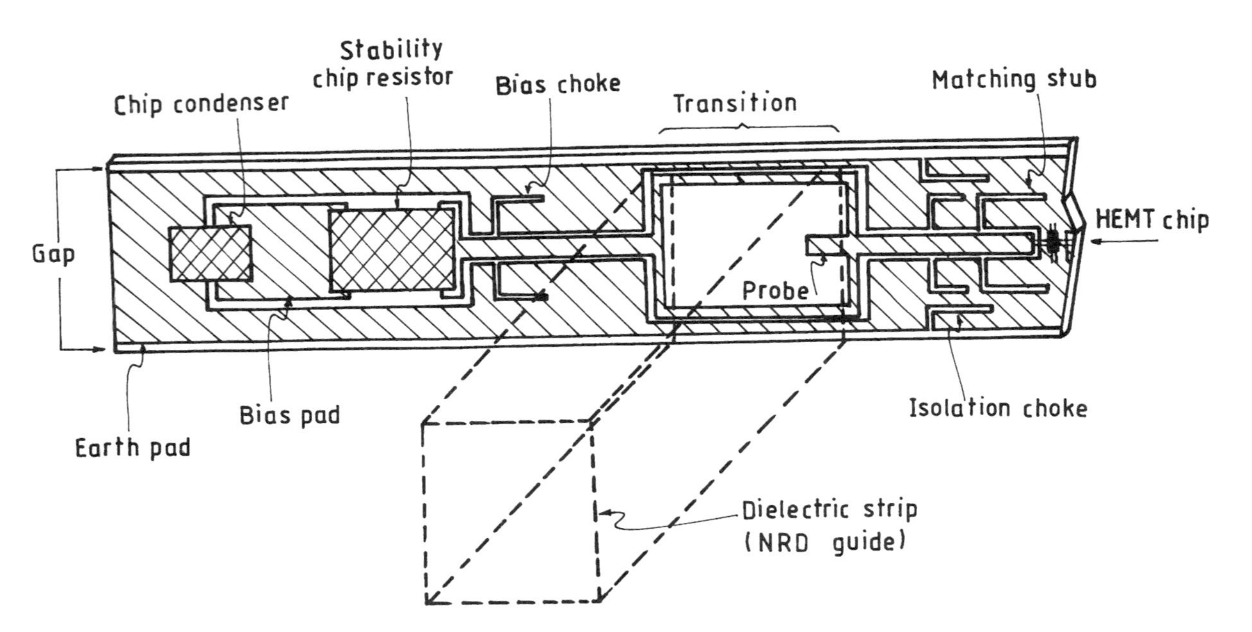

FIGURE 12.17 Layout of the HEMT chip carrier. (From Artuzi and Yoneyama [20], reprinted with permission of Communication Engineers, Denshi Joho Tsushin Gakkai, Tokyo, Japan.)

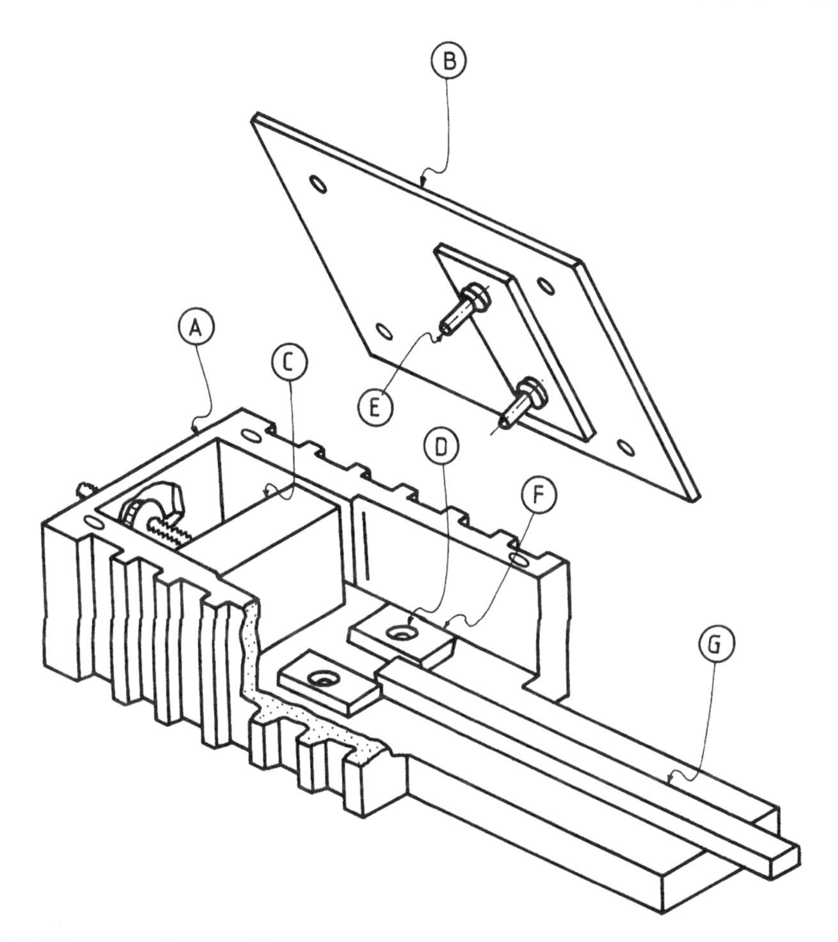

FIGURE 12.18 Gunn oscillator-cum-power combiner: A, metal enclosure; B, top metal plate; C, sliding short; D, Gunn diode; E, bias pin; F, silicon resonator; and G, central resonator (Al_2O_3). (From Potoczniak et al. [21]. Copyright © 1982 IEEE, reprinted with permission.)

serves as a central resonator. The position of this central cavity is adjusted to give optimum match for maximum output coupling. The sliding short behind the diode-mounted resonators serves as a fine tuner and the metal enclosure suppresses any radiation from the mounts. With this two-diode configuration, Potoczniak et al. [21] have reported power output that is nearly four times that obtainable from a single diode oscillator.

12.7.2 Transceiver Circuit

Figure 12.19 shows photographs of a 35-GHz practical NRD guide transmitter and receiver reported by Yoneyama [11]. The transmitter consists of a Gunn

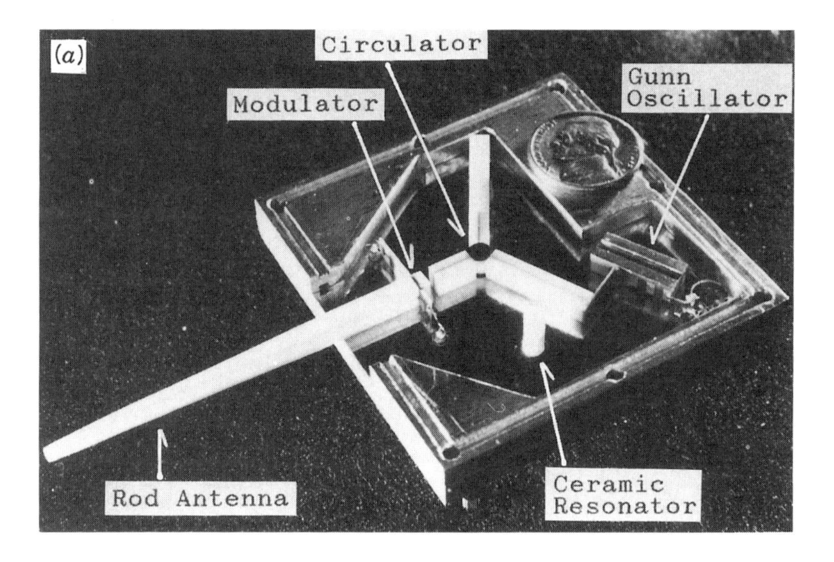

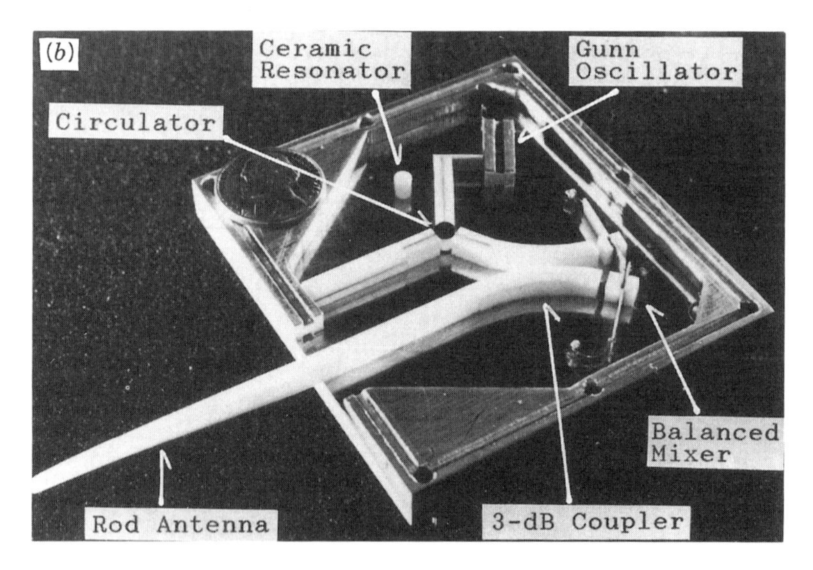

FIGURE 12.19 Photographs of (a) transmitter and (b) receiver in NRD guide (top cover removed). (After T. Yoneyama, private communication.)

diode oscillator, a circulator, a p-i-n diode pulse modulator, and a dielectric rod antenna all contained in a housing of size 6.8 cm × 6.6 cm. The Gunn diode oscillator circuit is the same as that shown in Figure 12.10 [10, 11]. The oscillator is locked in phase by means of a temperature-stabilized ceramic resonator. The output of the Gunn diode oscillator is fed to one of the three ports of the

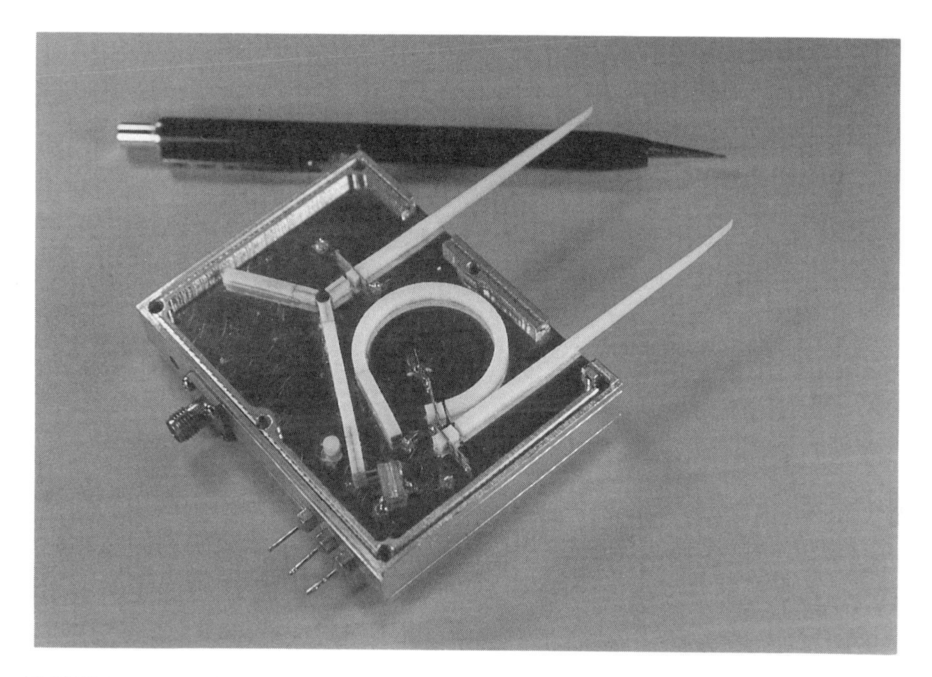

FIGURE 12.20 Photograph of a 60-GHz transreceiver in NRD guide for digital wireless LAN application. (After F. Kuroki and T. Yoneyama, private communication.)

circulator. The p-i-n diode modulator is located in another port of the circulator leading to the dielectric antenna.

The receiver consists of an assembly of Gunn diode oscillator, a circulator, a balanced mixer, and a receiving dielectric rod antenna, all contained in a housing of size 8.5 cm × 7.5 cm. A ceramic resonator is used for phase locking the receiver to yield a frequency stability of -0.76 ppm/°C. The measured noise figure of this receiver is reported to be about 10.6 dB [11].

For wireless digital communication between computers, Kuroki and Yoneyama [22] and Yoneyama [23] have reported a NRD guide digital transreceiver at 60 GHz. The transreceiver consists of a circulator, a self-injection locked Gunn oscillator, a pulse modulator, two directional couplers, a balanced mixer, and a horn antenna. Figure 12.20 shows a photograph of a NRD guide digital transreceiver at 60 GHz for wireless digital communication between computers [F. Kuroki and T. Yoneyama, private communication]. This transreceiver uses separate antennas for transmission and reception of the digital signals.

PROBLEMS

12.1 Discuss techniques of mounting packaged semiconductor devices in an image guide. What precautions are to be taken to minimize the radiation loss?

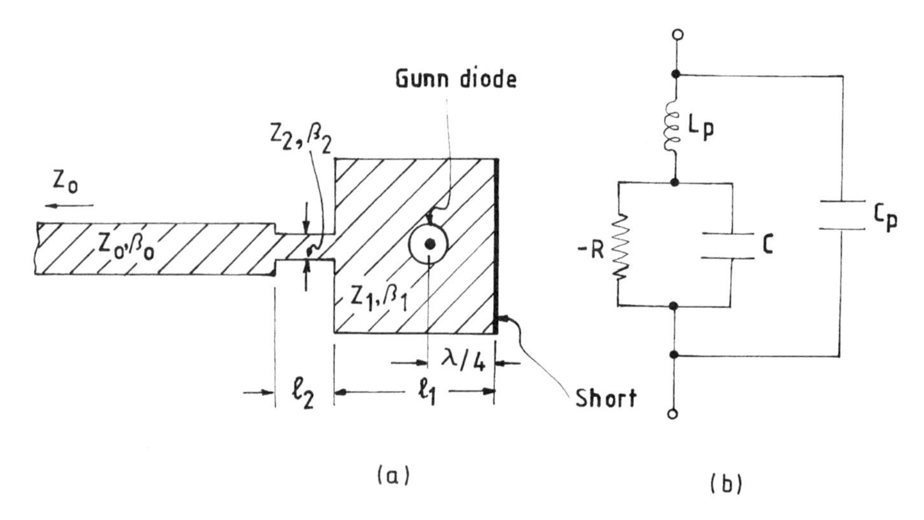

FIGURE 12.21 (a) Top view of an integrated image guide Gunn oscillator. (b) Gunn diode equivalent circuit.

12.2 Figure 12.21 shows the top view of an integrated guide Gunn oscillator along with the Gunn diode equivalent circuit.

(**a**) Draw the equivalent circuit of the Gunn oscillator.

(**b**) Derive conditions for oscillation and also for optimum power transfer.

12.3 Figure 12.10 shows the schematic of a Gunn diode oscillator in a NRD guide.

(**a**) Draw the equivalent circuit of the Gunn oscillator. For the Gunn diode, assume the equivalent circuit given in Problem 12.2.

(**b**) Explain the necessity of using the mode suppressor (shown in Fig. 12.10). Which modes does it suppress?

12.4 Compare the relative advantages of the contact injection and optical injection technique in achieving phase change in a semiconductor guide.

12.5 Draw a schematic of an electronic phase shifter for digital phase shifting.

12.6 Describe a scheme of realizing an optically controlled SPST switch. Draw a schematic diagram.

12.7 Compare the principles of operation and relative advantages of a balanced mixer of the type shown in Figure 12.12 and a self-oscillating mixer shown in Figure 12.13.

12.8 Draw a schematic of a balanced mixer in a NRD guide using beam lead diodes. Discuss design guidelines.

12.9 Draw a schematic diagram of a SPST switch in a NRD guide using beam lead p-i-n diodes. Show the biasing arrangement.

12.10 Draw a schematic diagram of a SPDT switch in a NRD guide using beam lead p-i-n diodes. Draw the equivalent circuit of the switch and explain its operation.

REFERENCES

1. J. C. Wiltse, Some characteristics of dielectric image lines at millimetre wavelengths. *IRE Trans. Microwave Theory Tech.*, **MTT-7**, 60–65, Jan. 1959.
2. M. J. Aylward and N. Williams, Feasibility studies of insular guide millimetre wave integrated circuits, Proceedings of the AGARD Conference of Millimetre and Sub-millimetre Wave Propagation and Circuits, *Conf. Reprint*, **245**, 30.1–30.11, Sept. 1978.
3. J. A. Paul and Y. W. Chang, Millimetre-wave image-guide integrated passive devices, *IEEE Trans. Microwave Theory Tech.*, **MTT-26**, 751–754, Oct. 1978.
4. K. Solbach, Millimetre-wave dielectric image line detector circuit employing etched slot structure. *IEEE Trans. Microwave Theory Tech.*, **MTT-29**, 953–957, Sept. 1981
5. K. Solbach and I. Wolff, Slot as new circuit-elements in dielectric image line. *IEEE MTT-S Int. Microwave Symp. Digest*, 8–10, 1981.
6. R. E. Horn et al., Integrated tunable cavity Gunn oscillator for 60 GHz operation in image line waveguide. *IEEE Trans. Microwave Theory Tech.*, **MTT-32**, 171–176, Feb. 1984.
7. R. E. Horn et al., Millimetre-wave oscillator using image-line or microstrip waveguides. *IEEE Trans. Microwave Theory Tech.*, **MTT-34**, 285–288, Feb. 1986.
8. Y. W. Chang, Millimetre-wave (W-band) quartz image guide Gunn oscillator. *IEEE Trans. Microwave Theory Tech.*, **MTT-31**, 194–199, Feb. 1983.
9. T. Itoh and F. J. Hsu, Distributed Bragg reflection Gunn oscillator for dielectric millimetre-wave integrated circuits. *IEEE Trans. Microwave Theory Tech.*, **MTT-27**, 514–518, May 1979.
10. T. Takadaand and T. Yoneyama, Self-injection locked Gunn-oscillator using non-radiative dielectric waveguide. *IEIEC Trans. (CI)*, **J72-CI (1)**, 53–58, Jan. 1989.
11. T. Yoneyama, Millimetre wave integrated circuits using non-radiative dielectric waveguides, Yagi Symposium on Advanced Technology, Bridging the Gap Between Light and Microwave, *Proceedings of the Second Sendai International Conference*, (Japan), pp. 57–66, Sept. 1990.
12. H. Jacobs and M. M. Chrepta, Electronic phase shifter for millimetre wave semiconductor dielectric integrated circuits. *IEEE Trans. Microwave Theory Tech.*, **MTT-22**, 411–417, Apr. 1974.
13. C. H. Lee et al., Optical control of millimetre wave propagation in dielectric waveguide. *IEEE J. Quantum Electron.*, **QE-16**, 277–288, Mar. 1980.
14. A. M. Vaucher, C. D. Striffler, and C. H. Lee, Theory of optically controlled millimetre-wave phase shifters. *IEEE Trans. Microwave Theory Tech.*, **MTT-31**, 209–216, Feb. 1983.

15. C. H. Lee, Optical generation and control of microwaves and millimetre waves. *IEEE MTT-S Int. Microwave Symp. Digest*, 811–814, 1987.

16. A. Alphones and M. Tsutsumi, Optical control of image lines at 140 GHz. *Symp. Radio Sci.*, **RS 90**, 13, Dec. 1990.

17. A. Alphones and M. Tsutsumi, Optical control of millimetre waves in silicon rib guides, paper of Technical Group on Electronics and Communications, *IEIEC (Japan)*, **MW 90–148**, 19–26, Feb. 1991.

18. S. Dixon and H. Jacobs, Millimetre-wave InP image guide line self-mixing Gunn oscillator. *IEEE Trans. Microwave Theory Tech.*, **MTT-29**, 958–961, Sept. 1981.

19. F. Kuroki and T. Yoneyama, Non-radiative dielectric waveguide circuit components using beam-lead diodes. *IEICE Trans. (CI)*, **J 73-C-I (2)**, 71–76, Feb. 1990.

20. W. A. Artuzi and T. Yoneyama, A HEMT amplifier for non-radiative dielectric waveguide integrated circuits. *IEICE Trans.*, **E-74 (5)**, 1185–1190, May 1991.

21. J. J. Potoczniak et al., Power combiner with Gunn-diode oscillators. *IEEE Trans. Microwave Theory Tech.*, **MTT-30**, 724–728, May 1982.

22. F. Kuroki and T. Yoneyama, Non-radiative waveguide digital transreceiver at 60 GHz band, *19th IR and MM-Wave Conference Digest*, pp. 437–438, 1994.

23. T. Yoneyama, Principles and applications of non-radiative dielectric waveguide, *19th IR and MM-Wave Conference Digest*, pp. 222–223, 1994.

CHAPTER THIRTEEN

Antennas

13.1 INTRODUCTION

It is well known that any discontinuity placed in a dominant mode guiding structure not only reflects the incident dominant mode partially but also excites higher order modes in its vicinity. In a shielded guide such as a rectangular metal waveguide, these higher order modes represent stored energy, whereas in an open dielectric guide, they cause radiation loss. Dielectric guides with discontinuities such as tapers, notches, and conducting strips introduced along the propagation path find extensive applications in antennas.

A major advantage of dielectic antennas is that they can easily be integrated with dielectric integrated circuits to form transmitter/receiver front ends. Dielectric integrated guide antennas are particularly attractive at millimeter wave frequencies because of their compact size, high reliability, and ease of manufacture. Furthermore, at these frequencies, the conventional microwave techniques can be combined with optical techniques to yield a variety of antenna configurations.

Most dielectric and dielectric integrated guide antennas belong to a general class of traveling wave antennas. In a traveling wave antenna, excitation is from a source at the feed end or from a feed line so that the wave propagates along the antenna illuminating the entire radiating aperture. Traveling wave antennas can be further classified as surface wave and leaky-wave antennas. Surface waves are slow waves; that is, the wave velocity along the structure is less than the free-space velocity. Dielectric rods of uniform cross-section are examples of surface wave structures. Such antennas radiate in the end-fire or off-broadside direction depending on the mode of excitation. Leaky waves are produced by altering the geometry of a waveguide in such a way that the initially bound mode is converted into a leaky mode. Leaky waves radiate all along the antenna and form a narrow beam that is frequency scannable. Most leaky-wave antennas at microwave frequencies are based on closed waveguides with selected openings to produce the leakage. On the other hand, dielectric guides used in millimeter wave antennas are open structures. Leaky-wave radiation from open guides may be produced in several ways, such as by (i) tapering a dielectric rod, (ii) introducing periodic gratings or certain asymmetry in the longitudinal direction, (iii) foreshortening one side of the open waveguide, and (iv) permitting a leaky higher order mode to propagate. Of these, the grating type antennas are more popular. In this chapter, we present the theory and design aspects of grating antennas in detail. Antennas other than the grating type in various dielectric integrated guide configurations, semiconductor antennas including the optically controlled antennas, and their frequency-scanning capabilities are also covered.

13.2 DIELECTRIC ROD ANTENNAS

Studies on dielectric rods of circular and rectangular cross-sections have been reported extensively in the literature [1–6]. In a dielectric rod of uniform cross-section, radiation takes place primarily from the feed region and the terminated open end. Such a structure is basically a low-gain surface wave antenna radiating a broad beam in the end-fire direction. While the gain of the antenna can be increased by increasing its length, the side lobe also increases simultaneously, often exceeding the tolerable limit.

Longitudinally tapered dielectric rods offer considerable improvement in the radiation characteristics over those of uniform rods [4–6]. Tapering reduces the reflection and radiation from the terminated open end. More importantly, the antenna produces leaky-wave radiation from the entire tapered length. The surface wave radiation from the feed end, which normally gives rise to high side lobes, can be suppressed by employing a matching transition to the feed guide. The design of an integrated antenna therefore involves the design of a tapered section and a matching transition to the feed guide [7].

For a given frequency and choice of dielectric, the design of the tapered section involves the determination of the cross-sectional dimensions at the feed and open ends, and the taper length. For high gain and low side lobes, the antenna structure must support only the dominant mode. It has been pointed out [1, 5] that for optimum radiation characteristics, the ratio of the free-space wavelength λ_0 to the guide wavelength λ should be about 1.1 ($\lambda_0/\lambda = 1.1$). It is assumed that the rod is matched to free space at the terminal end so that no standing waves are formed on the rod. The diameter at the end of the rod is chosen for the condition $\lambda_0/\lambda = 1.0$. Based on the above criteria, approximate formulas have been reported for the design of the tapered rod antennas [5].
They are

$$d_1 \approx \frac{\lambda_0}{[\pi(\varepsilon_r - 1)]^{1/2}} \tag{13.1a}$$

$$d_0 \approx \frac{\lambda_0}{[2.5\pi(\varepsilon_r - 1)]^{1/2}} \tag{13.1b}$$

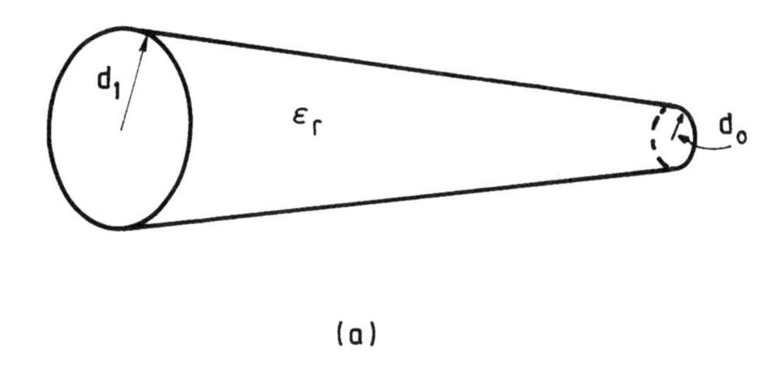

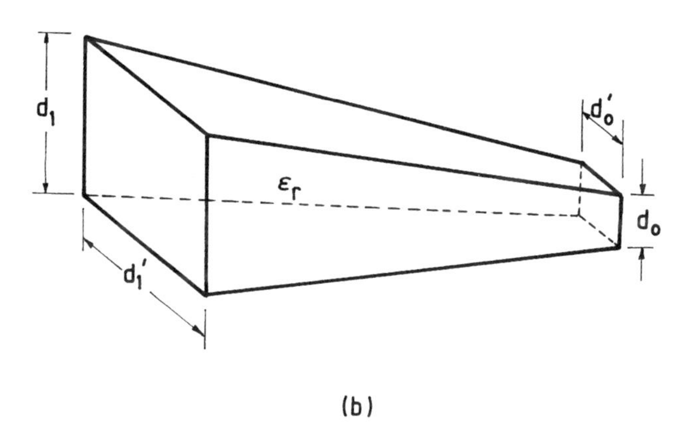

FIGURE 13.1 Tapered dielectric rod antenna of (a) circular cross-section and (b) rectangular cross-section.

where d_1 and d_0 are the diameters of the tapered rod at the feed end and open end, respectively, and ε_r is the relative dielectric constant of the dielectric material (see Fig. 13.1(a)). For a rod of square cross-section, d_1 and d_0 may be taken as the side dimensions at the feed and open ends, respectively (Fig. 13.1(b) with $d'_1 = d_1$ and $d'_0 = d_0$). As an example, for an alumina rod ($\varepsilon_r = 9.8$) of square cross-section, Shiau [5] has reported maximum gain with a dimension of $d_1/\lambda \approx 0.2$ (which is close to the theoretical value of 0.19 as calculated from (13.1a)). The gain increases initially with an increase in the length until it reaches a maximum value and then saturates. With a tapered alumina rod of square cross-section having dimensions $d'_1 = d_1 = 160\,\text{mil}$ and $d'_0 = d_0 = 110\,\text{mil}$, the measured values of gain at the K_u-band are 14.5 dB for a rod length $L = 5\lambda_0$, 16.2 dB for $L = 8\lambda_0$, with no increase in gain for $L > 8\lambda_0$ [5].

As an example of an integrated antenna, Figure 13.2(a) shows the view of a X-band tapered rod of rectangular cross-section with transition to non-

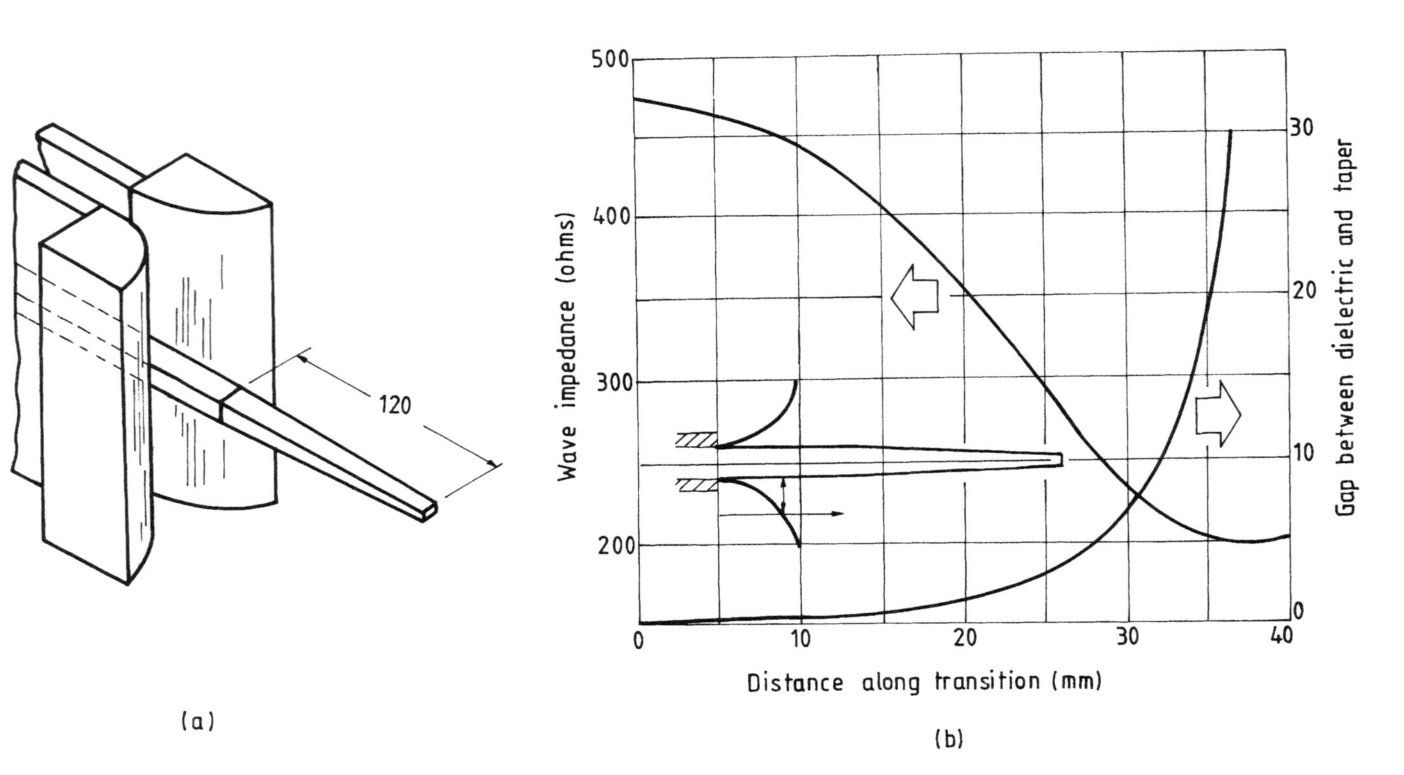

FIGURE 13.2 Integrated tapered dielectric rod antenna fed by NRD guide: (a) antenna construction and (b) wave impedance versus distance along the transition. (From Malherbe [8]. Copyright © 1985 IEEE, reprinted with permission.)

radiative dielectric guide [8]. Throughout the length of the transition, the cross-section of the dielectric is held constant at 15 mm × 10.16 mm, whereas the gap between the dielectric and the metal wall on either side is increased from zero to infinity. The tapered rod connected at the end of the transition has a length of 4.7533 radians at 9 GHz. The transition converts the NRD guide mode to the dominant mode of the open rectangular dielectric guide and also provides the necessary impedance transformation (see Fig. 13.2(b)). This antenna is reported to offer a gain of 15.9 dB (close to the theoretical maximum) and a return loss better than 15 dB over a frequency range of 9–10 GHz [8].

13.3 IMAGE GUIDE ANTENNAS

The image line offers a convenient dielectric guide medium for realizing integrated antennas for millimeter wave applications [9–13]. It can be configured as a grating antenna by introducing periodic notches (Fig. 13.3(a)), conducting patches on the dielectric surface (Fig. 13.3(b), (c)), and grooves in the ground plane (Fig. 13.3(d)). These structures belong to a class of leaky-wave grating antennas. It may be noted that, without the ground plane, the structures shown in Figure 13.3(a)–(c) reduce to dielectric rod grating leaky-wave antennas [14–15]. The advantage of the image guide counterpart is that the ground plane redirects the entire radiated energy, which would otherwise exist on both sides of the dielectric to only one side of the dielectric.

13.3.1 Theory of Leaky-Wave Grating Antenna

In order to understand the theory of grating antennas, we consider the example of an image guide with grooves in the dielectric as shown in Figure 13.3(a). When a wave propagating in the guide encounters a discontinuity, the incident power, besides suffering partial reflection and transmission, is partly radiated into the space above the dielectric. The amplitude of the wave decays exponentially as it propagates along the structure. Radiation takes place in certain preferred directions determined primarily by the phase of the unperturbed structure and the grating period d. The width of the radiated beam is proportional to the decay constant of the guided wave. Alternatively, the attenuation of the wave is a measure of the energy leakage or radiation rate. The design of such antennas therefore requires knowledge of both the attenuation constant α and the phase constant β of the wave propagating along the structure. When the values α and β are known, the field distribution on the antenna aperture is known in principle. The far field of the antenna can then be calculated using the spatial Fourier transform.

From the study of wave propagation in a grating structure of finite width, it has been shown [16, 17] that, for sufficiently large guide widths, the phase constant β of the propagating wave is essentially that of a corresponding unperturbed guide ($\beta \approx \beta_u$) and the attenuation constant α can be approximated

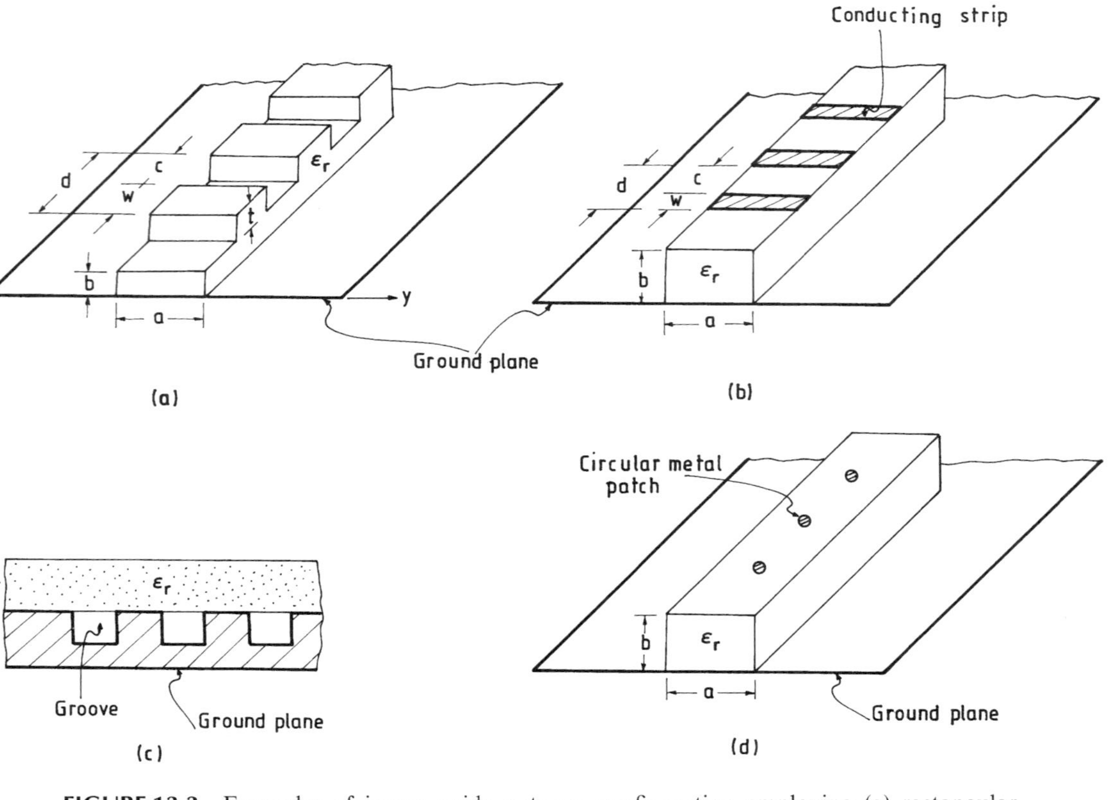

FIGURE 13.3 Examples of image guide antenna configuration employing (a) rectangular grooves in dielectric, (b) conducting strips on dielectric, (c) rectangular grooves in the ground plane, and (d) circular metal patches on dielectric.

by that of a two-dimensional grating antenna at normal incidence. Based on this assumption, Schwering and Peng [10] have reported an approximate procedure for the design of grating antennas.

Phase Constant The phase constant β of the unperturbed guide is determined using the EDC techniques, the procedure for which is described in Chapter 2. In the case of an image guide grating antenna, the effect of the periodic grooves along the guide is taken into account by adding a uniform dielectric layer of thickness t (equal to the groove thickness) and dielectric constant equal to the average dielectric constant ε_{av} on the top of the uniform guide of height b. The modified uniform image guide and its EDC model for determining β are shown in Figure 13.4(a).

Attenuation Constant α For determining the attenuation constant α, the dielectric and grooves are assumed to extend to infinity in the lateral direction. The electromagnetic field in the resulting two-dimensional periodic grating

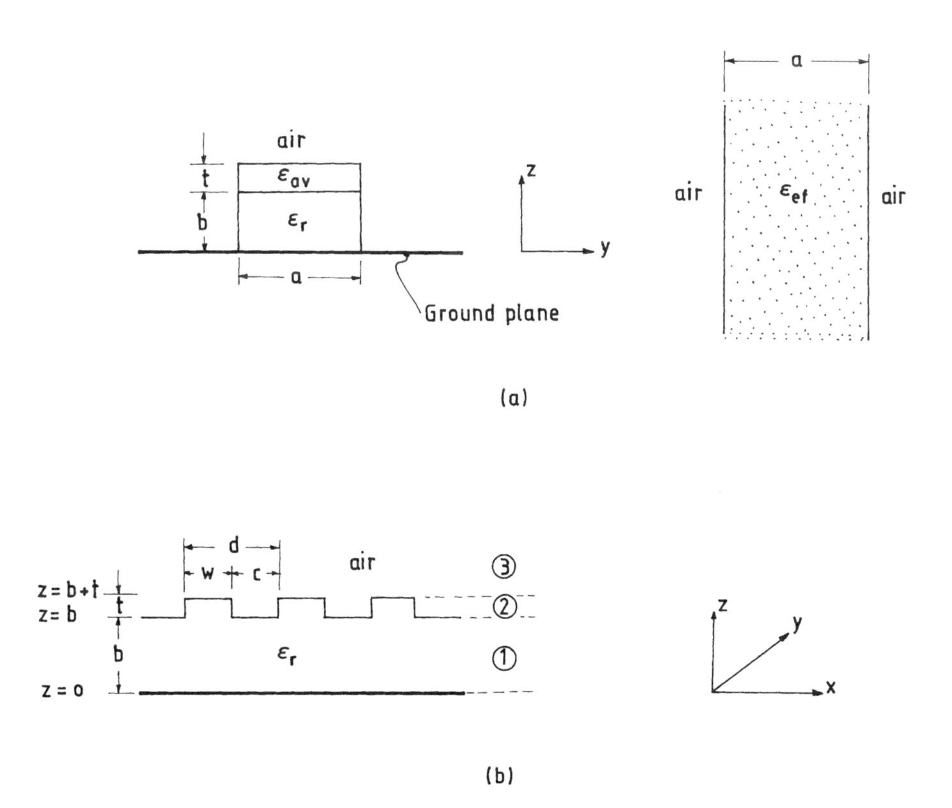

FIGURE 13.4 (a) Unperturbed image guide equivalent of corrugated image guide in Figure 13.3(a) and its EDC model for determining β. (b) Two-dimensional corrugated slab image guide model for determining α. (After Schwering and Peng [10].)

structure (Fig. 13.4(b)) can be expressed in terms of infinitely many harmonics. The magnetic field for the TM modes in each of the three subregions (region 1, $0 \leq z \leq b$; region 2, $b \leq z \leq b+t$; and region 3, $z > b+t$) can be expressed as [10]

$$H_y(x, z) = \sum_{m=-\infty}^{\infty} I_m^j(z) e^{jk_{xm}x} \tag{13.2}$$

where $j = 1, 2, 3$ refers to the three subregions, I_m^j is the magnetic field amplitude of the mth space harmonic, and k_{xm} is the x-directed complex wave number for this harmonic.

$$k_{xm} = k_{x0} + (2\pi m/d) \tag{13.3a}$$

$$= \beta - j\alpha + (2\pi m/d) = \beta_m - j\alpha \tag{13.3b}$$

where β and α are the propagation and attenuation constants of the dominant space harmonic ($m = 0$). It may be noted that the various space harmonics have different phase constants but the same attenuation constant α.

Leaky-Wave Beam When the grating period d is chosen such that the propagation constant k_{xm} satisfies the condition

$$|(\beta_m/k_0)| < 1, \quad k_0 = 2\pi/\lambda_0 \tag{13.4}$$

where λ_0 is the free-space wavelength, the grating region supports a mth leaky wave. If the perturbation by each unit cell of the grating is small, the leakage radiation would be small. Consequently, k_{x0} is almost real and is close to the propagation constant of the unperturbed guide. The direction of radiation of the mth leaky-wave beam with respect to the broadside direction is given by [18]

$$\theta_m = \sin^{-1}[(\beta + 2m\pi/d)/k_0], \quad m = 0, \pm 1, \pm 2, \ldots \tag{13.5}$$

In most practical cases, it is desirable to have single-beam radiation. Single-beam operation implies that only the $m = -1$ harmonic radiates. Setting $m = -1$ and $\beta = 2\pi/\lambda$, where λ is the guide wavelength in the structure, (13.5) can be written as

$$\theta_m|_{m=-1} = \sin^{-1}[(\beta/k_0)(1 - \lambda/d)], \quad |(\lambda_0/\lambda)(1 - \lambda/d)| < 1 \tag{13.6}$$

This expression is the same as that of the conventional linear array of spacing d with a linear phase taper. Correspondingly, for a spacing $d = \lambda$, the beam points in the broadside direction.

Radiation Pattern Schwering and Peng [10] have derived an expression for the radiation pattern of a corrugated image guide antenna from the field distribution

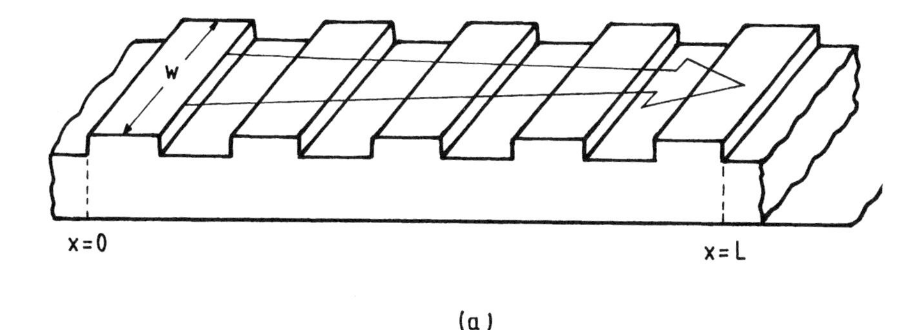

(a)

(b)

FIGURE 13.5 (a) Grating antenna of finite length L and width a. (b) Antenna aperture and coordinate system. (From Schwering and Peng [10]. Copyright © 1983 IEEE, reprinted with permission.)

at the antenna aperture. Referring to Figures 13.4 and 13.5, the magnetic field distribution for the harmonic $m = -1$ at the aperture $z = b + t$ is expressed as

$$H_y(x, y, b + t) = Ie^{-\alpha x}e^{-(jk_0 x \sin\theta_{-1})} \cos(\pi y/a), \quad 0 \leq x \leq L, |y| \leq b/2 \qquad (13.7)$$

where I is a constant and the relation $\beta_- = \beta(1 - \lambda/d) = k_0 \sin\theta_{-1}$ has been made use of from (13.6). The above distribution assumes that the dielectric strip is sufficiently wide ($a > \lambda$) and the length L of the antenna is large compared with λ. The power radiation pattern of the antenna obtained from the formula (13.7) is [10]

$$G(\theta, \phi) = G_d T(\sin\theta \sin\phi)\, S(\cos\phi) \tag{13.8}$$

where G_d is the directivity, $T(\sin\theta\sin\phi)$ is the E-plane pattern, and $S(\cos\phi)$ is the H-plane pattern. With S and T normalized to unity in the main beam direction $\phi = 90°$, $\theta = \theta_{-1}$, we have

$$G_d = (16/\pi^3)(k_0^2 a/\alpha)\tanh(\alpha L/2)\cos\theta_{-1} \tag{13.9a}$$

$$T(\sin\theta\sin\phi) = \left(\frac{\alpha L}{1 - e^{-\alpha L}}\right)^2 \left[\frac{1 - 2e^{-\alpha L}\cos[k_0 L(\sin\theta\ \sin\phi - \sin\theta_{-1})] + e^{-2\alpha L}}{(\alpha L)^2 + (k_0 L)^2\ (\sin\theta\ \sin\phi - \sin\theta_{-1})^2}\right] \tag{13.9b}$$

$$S(\cos\phi) = \left(\frac{\pi}{2}\right)^4 \left[\frac{\cos^2[(k_0 a\ \cos\phi)/2]\sin^2\phi}{\{(\pi/2)^2 - [(k_0 a\ \cos\phi)/2]^2\}^2}\right] \tag{13.9c}$$

The E-plane pattern corresponds to an exponentially tapered source distribution and the H-plane pattern is that of a cosine tapered source distribution. The 3-dB beamwidth in the H-plane is $\Delta\phi = 2\pi/k_0 a$. For $\alpha L \ll 1$, the directivity gain G_d is proportional to the antenna aperture area aL and the antenna efficiency is rather low. For $\alpha L \gg 1$, G_d is proortional to $2a/\alpha$ and hence becomes independent of the length L. For a grating antenna to be efficient, L must be sufficiently large so that most of the guided power is radiated before it reaches the terminating end. The length required to achieve a given efficiency η is given by

$$L = -(1/2\alpha)\ln(1 - \eta) \tag{13.10}$$

If we approximate the grating antenna as a linear array of isotropic elements, the radiation pattern takes the well known form

$$|f(\theta)|^2 = (1/N^2)|\sin(N\psi/2)/\sin(\psi/2)|^2 \tag{13.11}$$

$$\psi = k_0 d\sin\theta - \beta d \tag{13.12}$$

where N is the total number of grating elements and θ is measured from the broadside direction.

13.3.2 Antennas with Grooves in Dielectric

Figure 13.6 illustrates typical radiation characteristics of a broad image line ($a > \lambda$) grating structure [10]. Plots show the effect of normalized guide height b/λ and normalized thickness t/λ on the attenuation constant (radiation rate) and the scan angle θ_{-1} (for the space harmonic $m = -1$). The fixed parameters are the relative dielectric constant ($\varepsilon_r = 12$), grating period ($d = 0.25\lambda$), and groove spacing ($c = 0.5d$). It can be seen from Figure 13.6(a) that for a fixed groove depth ($t = 0.05\lambda$), the radiation rate is a strong function of the dielectric height b exhibiting a peak around $b = 0.21\lambda$. This indicates that the normalized guide height b/λ has a strong influence on the field distribution of the guided wave. At low frequencies (near cutoff), most of the field exists in the air region and at high frequencies, the field tends to concentrate within the dielectric, thereby leaving very weak fields to exist in the groove region for radiation. Thus for a given value of b, the field in the groove region (and hence radiation) peaks at some intermediate frequency [10]. Equivalently, for a given frequency, there is an optimum height b that yields maximum radiation. On the other hand, the beam angle θ_{-1} shows a sharp change with an increase in b only for small values of b. For the near optimum value of $b/\lambda(= 0.2)$, Figure 13.6(b) shows that the radiation rate increases sharply with an increase in t for small values of t and reaches saturation for large t.

13.3.3 Antennas with Periodic Conducting Patches

Image guide leaky-wave antennas with periodic conducting patches [13, 19, 20] are easier to fabricate than groove type structures. For millimeter wave applica-

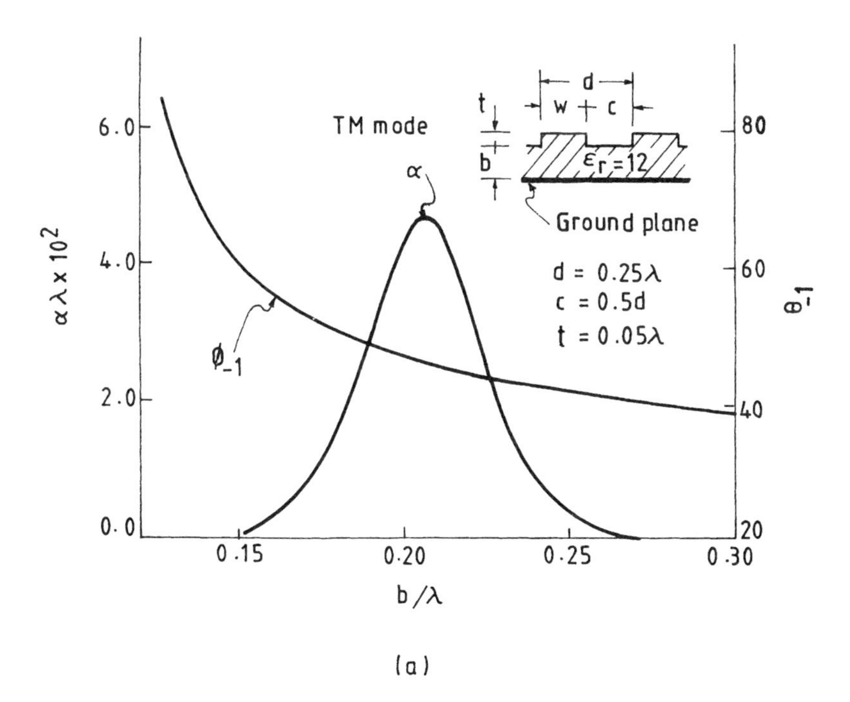

FIGURE 13.6 Theoretical variation in radiation rate and radiation angle θ_{-1} in a broad dielectric image guide grating structure. Effect of (a) guide height and (b) grating thickness. (From Schwering and Peng [10]. Copyright © 1983 IEEE, reprinted with permission.)

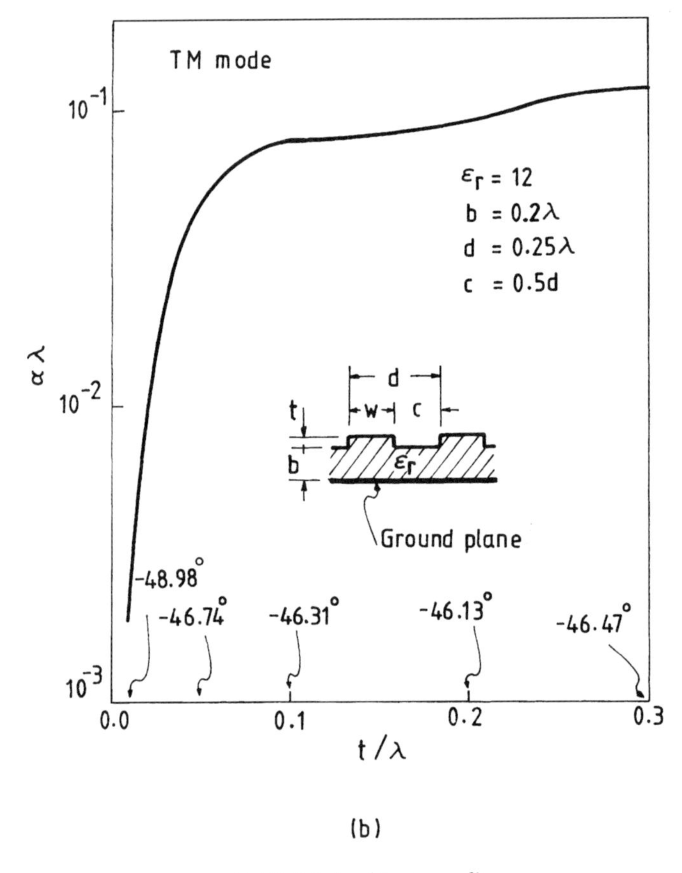

(b)

FIGURE 13.6 *(Continued)*

tions, the guide may be cut from copper plated RT-duroid substrate with strips or disks etched on one side using the photolithographic technique. The dielectric strip is then fixed to a metal plate using film adhesives. A schematic of a grating antenna with periodic conducting strips is shown in Figure 13.3(b) and that with periodic conducting disks is shown in Figure 13.3(c).

It is desirable that the power radiated at each conducting patch is small so that the incident energy is guided over the entire length of the antenna. It is also important that the entire guided energy is nearly totally radiated by the patches so that the residual energy at the end of the antenna is negligible. For keeping the radiation from each element small, the strip width must be small. This in turn would necessitate increasing the length of the antenna so as to effectively radiate the incident energy. Using strips of large width reduces the effective aperture area, increases mismatch at the first strip, and results in high side lobes. Trinh et al. [20] have experimentally inferred that, for antennas employing uniform strip widths, the optimum strip width w for achieving high gain is approximately 0.4λ.

Horn Image Guide An improved design of a image guide leaky-wave antenna that employs strips linearly tapered in width and a flared horn in the H-plane has been reported by Trinh et al. [19, 20]. Figure 13.7(a) shows the antenna geometry. Using strips of smaller width at the input end improves the impedance match in the transition region and lowers the side lobe level of the antenna. Strips of larger width located toward the end of the guide radiate more efficiently and minimize the residual power. The width of the nth strip is given by the empirical relation [20]

$$\begin{aligned} w_n &= 0.15 + 0.015(n-1)\lambda, \quad && n \leq 18 \\ &= 0.4\lambda, && n > 18 \end{aligned} \tag{13.13}$$

where λ is the guide wavelength. The antenna is constructed by embedding the dielectric strip with metal patches in a metallic trough and with metal flares attached to the two sides. For this trough guide fed by a standard rectangular metal waveguide, the dominant mode in the trough region has its major magnetic field component in the x-direction (see Fig. 13.7(a)). The current excited in each strip is in the z-direction and consequently the principal aperture field at the metal flare region is E_z. The metal flares can therefore be modeled as an H-plane sectoral horn. The flare angle and the length of the metal horn are suitably chosen to yield maximum gain in the H-plane. Figure 13.7(b) shows the measured radiation pattern of the horn image guide antenna designed at 81.5 GHz [19]. The various dimensions of the antenna are indicated in Figure 13.7(a). The relative dielectric constant ε_r of the guide material is 2.47. The antenna uses 32 strips with grating period $d = 2.73$ mm, with widths calculated as per formula (13.13). This antenna is reported to offer a half-power beam width of 4° and 13° in the E-plane and H-plane, respectively, and an overall gain of 26 dB with a side lobe level better than 25 dB [20].

Two-Dimensional Array Another technique of narrowing the H-plane beam is to form a two-dimensional array by stacking several single-guide leaky-wave antennas. As an example, a two-dimensional array with etched disk radiators as reported by Solbach [9] is shown in Figure 13.8. The image guides are fed by a rectangular metal waveguide broadwall slot array. A common horn transition is used at the feed end for efficient launching of waves on to the image guides. The slots in the broadwall of the rectangular metal waveguide are offset by a fixed distance from the guide axis so as to produce constant-amplitude constant-phase aperture distribution.

13.3.4 Antennas with Grooves in Ground Plane

Figure 13.3(d) shows the schematic of a leaky-wave image guide with grooves in the ground plane. Two types of grooves have been adopted in practice: (i) narrow rectangular transverse grooves and (ii) circular grooves. A single groove discontinuity in the ground plane of an image guide (Fig. 13.9(a)) can be modeled as

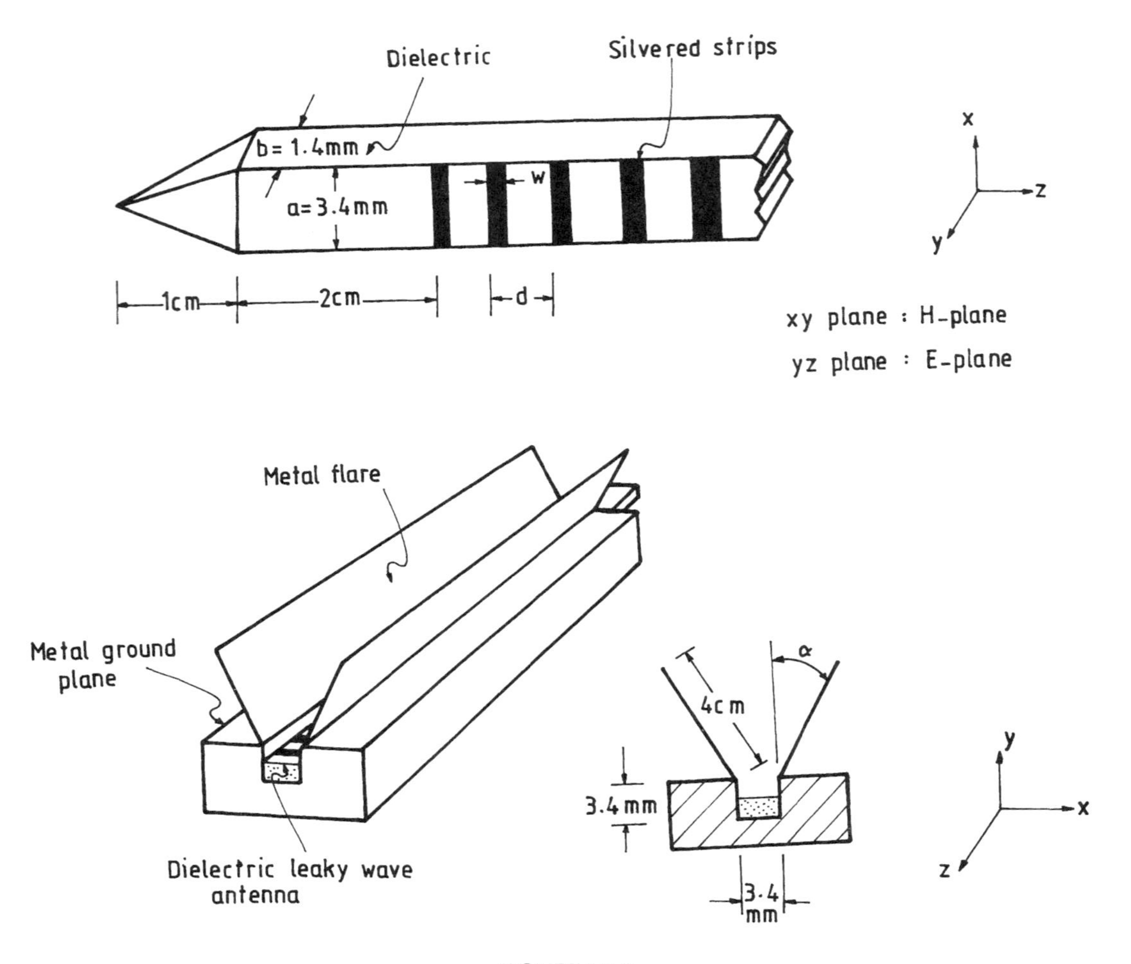
Dielectric
Silvered strips
b= 1.4mm
a=3.4mm
w
1cm
2cm
d
x
z
y
xy plane : H-plane
yz plane : E-plane
Metal flare
Metal ground plane
Dielectric leaky wave antenna
4cm
α
3.4mm
3.4 mm
y
x
z

FIGURE 13.7

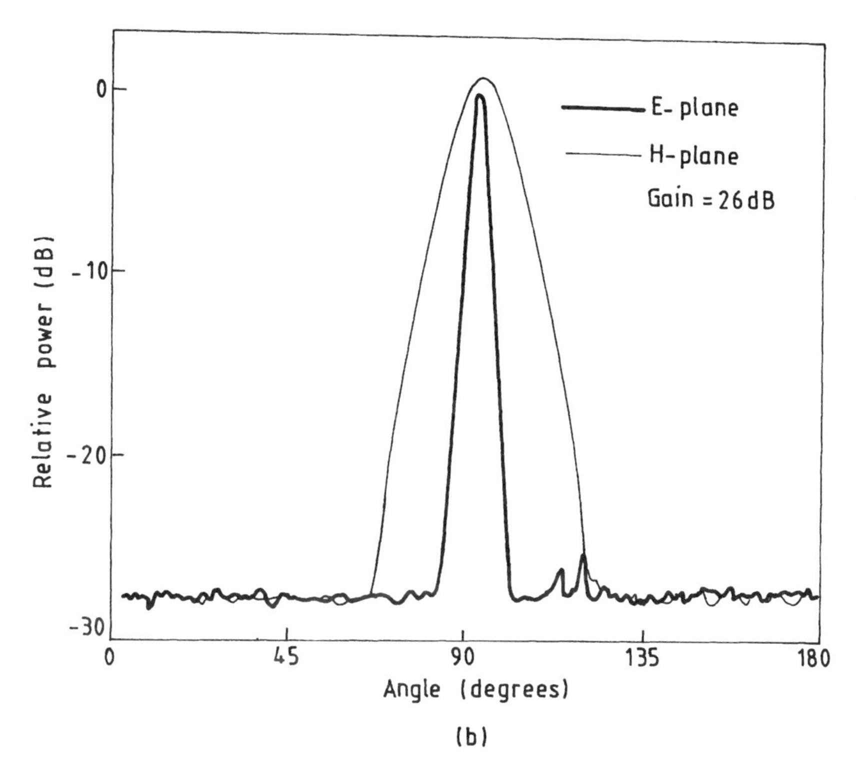

FIGURE 13.7 (a) Horn image guide leaky-wave antenna and (b) its *E*- and *H*-plane radiation patterns. (From Trinh et al. [19]. Copyright © 1981 IEEE, reprinted with permission.)

a slot backed by a short-circuited waveguide of length equal to the groove depth g. Figure 13.9(b) shows the approximate equivalent circuit of the discontinuity, where R_r is the radiation resistance, the capacitance C_d represents the energy stored near the discontinuity, and β_g and Z_g are the propagation constant and wave impedance, respectively, of the waveguide created by the groove [12]. The radiation resistance is primarily a function of the height and the dielectric constant of the image guide. Varying the groove depth changes the reactance shunting the radiation resistance.

As an example, Figure 13.10 shows a grating antenna employing transverse rectangular grooves in the ground plane of an image guide made of RT-duroid [9]. The antenna uses 31 grooves of varying depth and width so as to provide a tapered source distribution along the length of the antenna. Any residual energy reaching the end of the antenna is absorbed in the terminating load. As shown in the figure, the antenna has a narrow beam in the *E*-plane but the *H*-plane beamwidth is large. Increasing the *H*-plane directivity would require the use of longer grooves and a dielectric guide of larger width with the possibility of

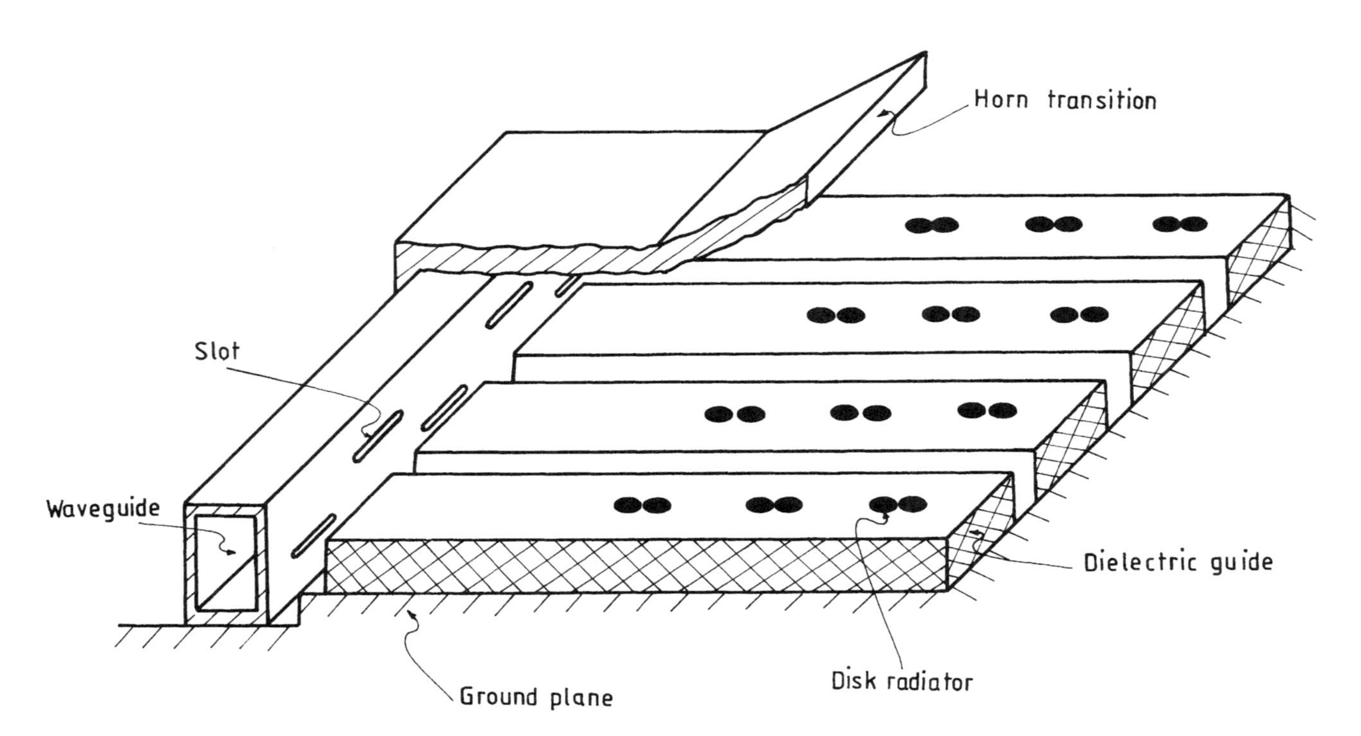

FIGURE 13.8 Two-dimensional dielectric image guide antenna array with etched dielectric radiators. (From Solbach [9]. Copyright © 1986 Academic Press Inc. Orlando, FL, reprinted with permission.)

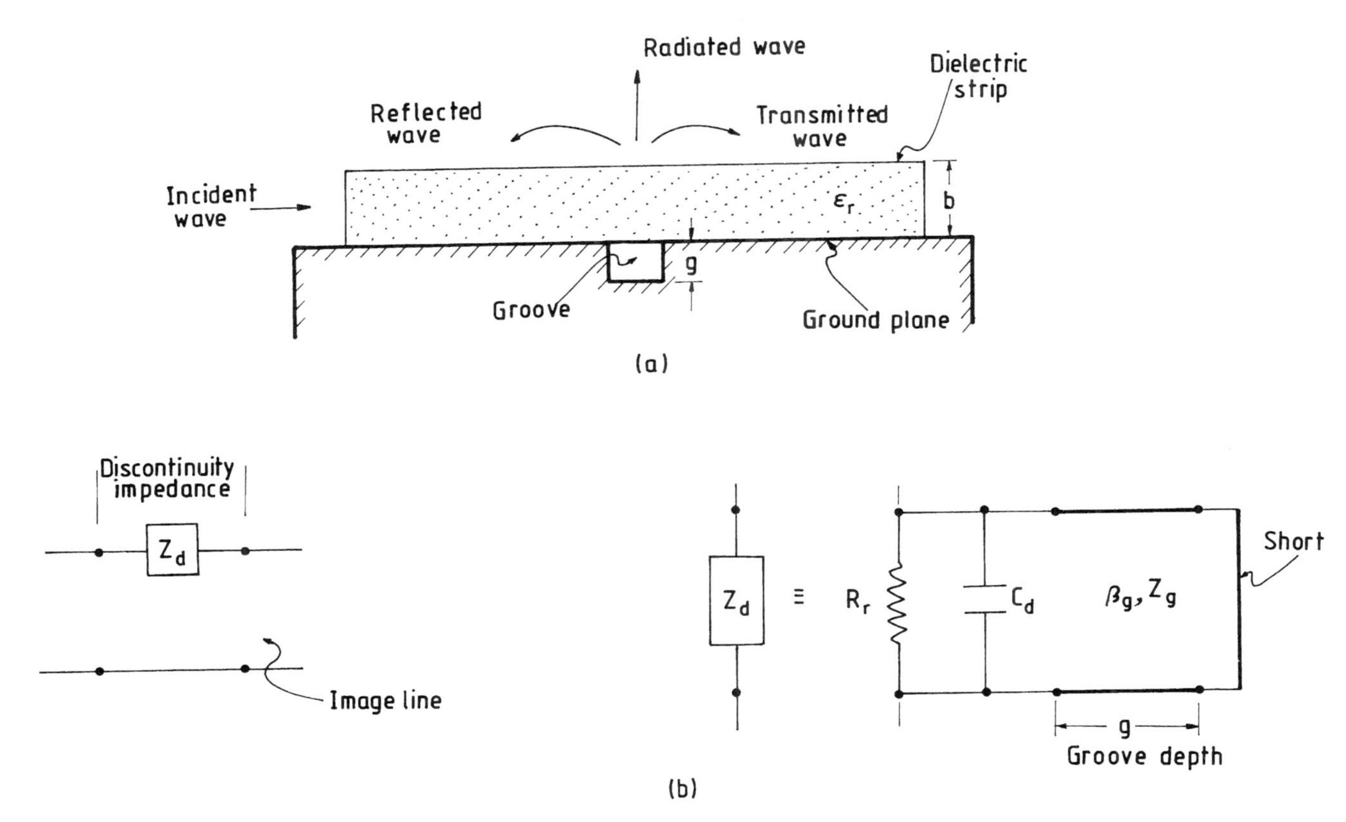

FIGURE 13.9 (a) Groove in the ground plane of a dielectric image guide and (b) its equivalent circuit. (From Wolff and Solbach [12]. Copyright © 1985 IEEE, reprinted with permission.)

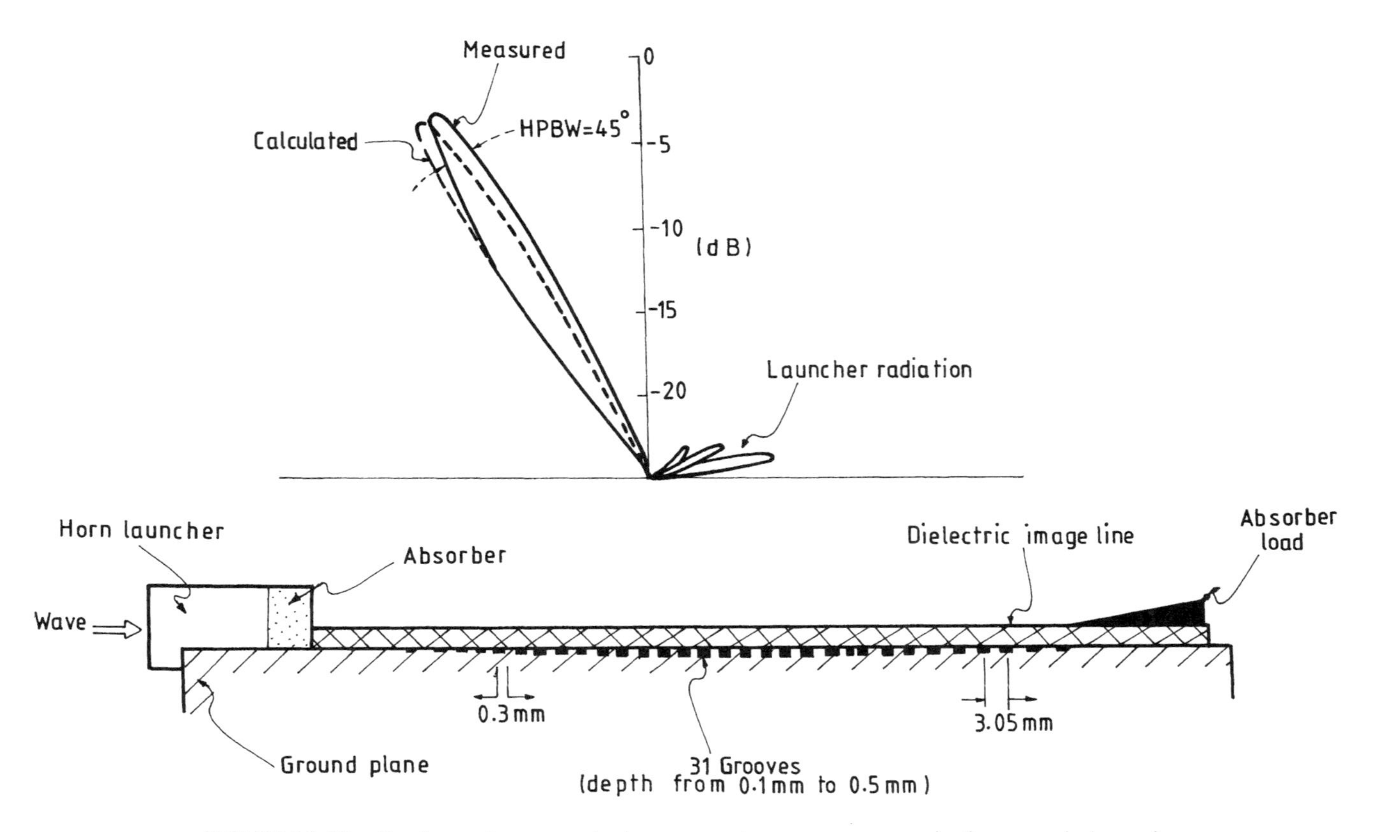

FIGURE 13.10 Grating antenna employing narrow transverse grooves in the ground plane of a dielectric image guide made of RT-duroid. (From Solbach [9]. Copyright © 1986 Academic Press Inc., Orlando, FL, reprinted with permission.)

exciting higher order slot modes. The problem of higher order modes can be circumvented by employing circular grooves instead of rectangular grooves in the ground plane. The diameters of the holes are chosen such that they can be considered as circular waveguides far below cutoff of the first waveguide mode. Solbach [9] has reported on an antenna incorporating a two-dimensional array of 5 × 30 circular grooves in the ground plane of the image guide. An extra broad dielectric guide having width to height ratio equal to 10 is used to cover the entire array of grooves, which are tapered in diameter in the *E*-plane and constant in diameter in the *H*-plane. This antenna is reported to offer a half-power beam width of 36° × 6° with side lobes well below −25 dB and an overall gain of about 20 dBi [9].

13.4 IMAGE AND INSULAR IMAGE GUIDE-FED ANTENNAS

In contrast to periodically modulating the dielectric guide for realizing antennas, the guide may be used as a low-loss feeder to an array of discrete radiating elements [21–24]. Figure 13.11 shows basic schematics of arrays of radiating resonant rectangular slots and circular holes cut on the ground plane of an image guide. The image guide is excited in its dominant (E^y_{11}) mode at one end and as the wave propagates along the guide, the slots/holes are excited, causing radiation. The dimensions of slots/holes are chosen such that no higher order modes are excited. The excitation coefficients of the radiating elements may be chosen by suitably displacing them from the centerline.

Figure 13.12 shows a circularly polarized resonant slot array fed by an image guide [21]. The antenna employs several pairs of inclined slots. The slots in each pair are oriented at angles $\pm\phi°$ with respect to the center line and are displaced from it by distance $\pm\delta$. The antenna is similar to the rectangular metal waveguide broadwall slot array with inclined slots. Referring to the coordinate system shown in Figure 13.12, the main beam is in the *xz*-plane and makes an angle θ_0 with respect to the broadside given by

$$\theta_0 = \sin^{-1}[(\lambda_0/\lambda)(1 - \lambda/d_2)] \tag{13.14}$$

where d_2 is the distance between adjacent pairs of slots and λ is the guide wavelength. The axial ratio, defined as the ratio of the maximum to minimum power, depends on the angle ϕ and the spacing d_1 between the two slots in a pair. The condition for achieving circular polarization in the main beam is given by [21]

$$\tan\phi = -\cos\theta_0 \cot\{(k_0 d_1/2)[\sin\theta_0 - (\lambda_0/\lambda)]\} \tag{13.15}$$

where $k_0 = 2\pi/\lambda_0$ is the free-space propagation constant. If $d_1 = \lambda/4$ and $\phi = \pi/4$, then slots 1 and 2 are fed with a phase difference of $\pi/2$, and a circularly polarized main beam appears at broadside. This type of slot array antenna is reported to

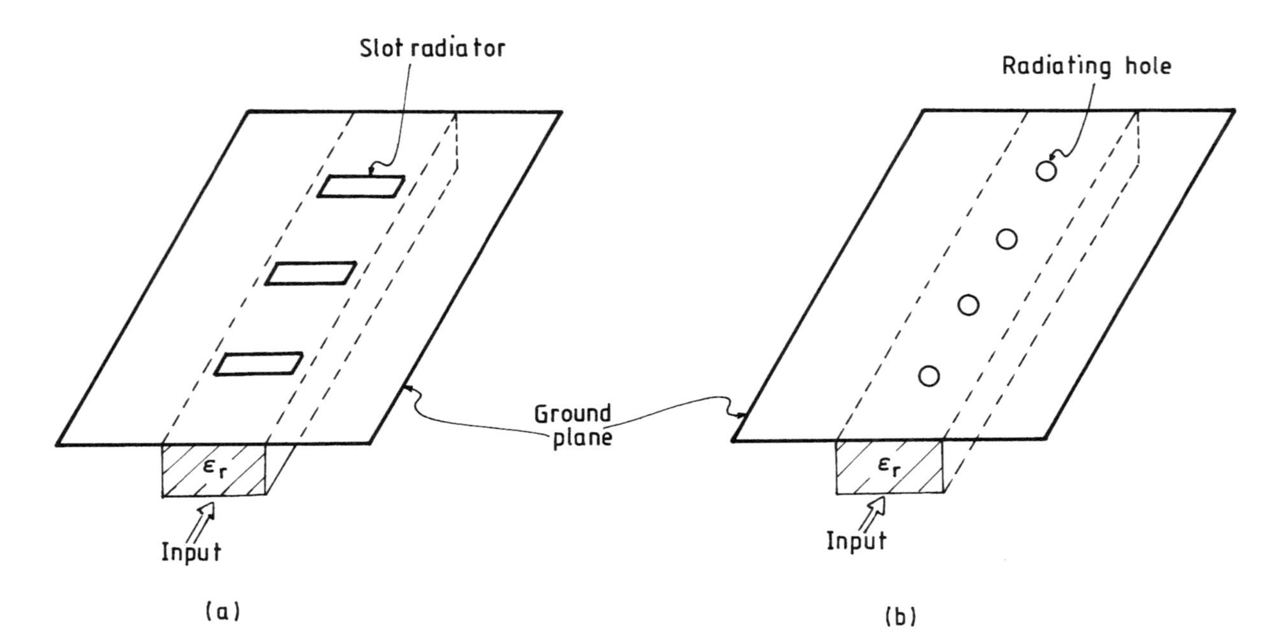

FIGURE 13.11 Schematic of arrays of radiating resonant (a) rectangular slots and (b) circular holes cut on the ground plane of an image guide.

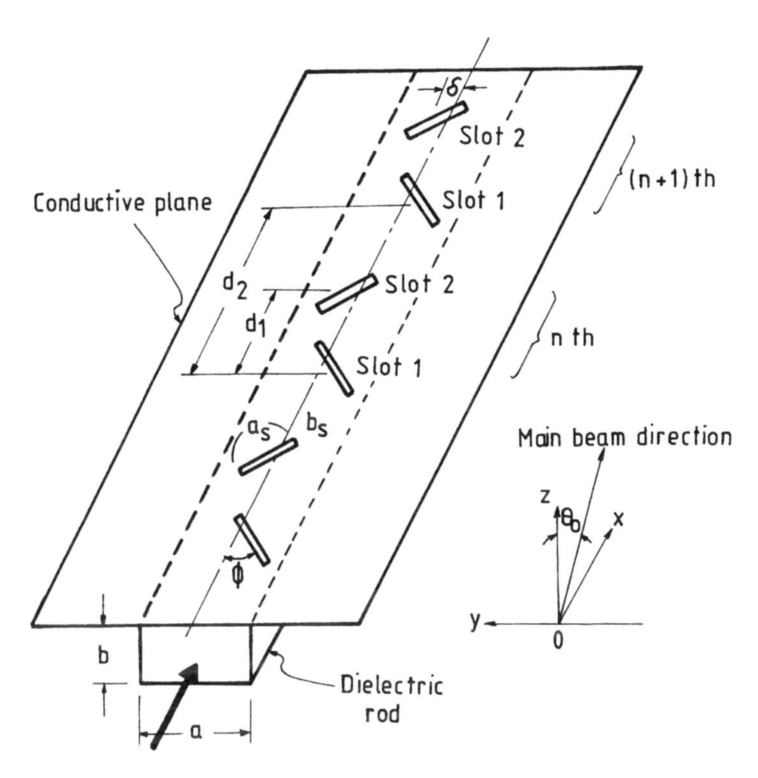

FIGURE 13.12 Circularly polarized array antenna using slots in the ground plane of a dielectric image line. (From Hori and Itanami [21]. Copyright © 1981 IEEE, reprinted with permission.)

offer lower loss, wider bandwidth in the axial ratio, and better circular polarization characteristics over wide angles as compared with microstrip line-fed slot array antennas [21, 25].

Figure 13.13 shows a millimeter wave printed dipole array employing an insular image guide as the feeder [22, 23]. The antenna employs pairs of printed half-wave dipoles on a thin dielectric sheet with each successive pair spaced by one guide wavelength ($d = \lambda$). The dipoles are excited by the H_x component of the dominant mode of the insular image guide. The level of coupling between the insular guide and the dipoles can be varied by either adjusting the dielectric sheet thickness t or by displacing the dipoles in the x-direction. The H_x component at the surface $y = h + b + t$ may be written as [23]

$$H_x = A \cos(k_{x1}x), \qquad -a/2 \leq x \leq a/2 \tag{13.16a}$$

$$= A \cos(k_{x1}a/2)\, e^{-k_{x2}(x-a/2)}, \quad |x| > a/2 \tag{13.16b}$$

where k_{x1} and k_{x2} are the x-directed propagation constants in the regions

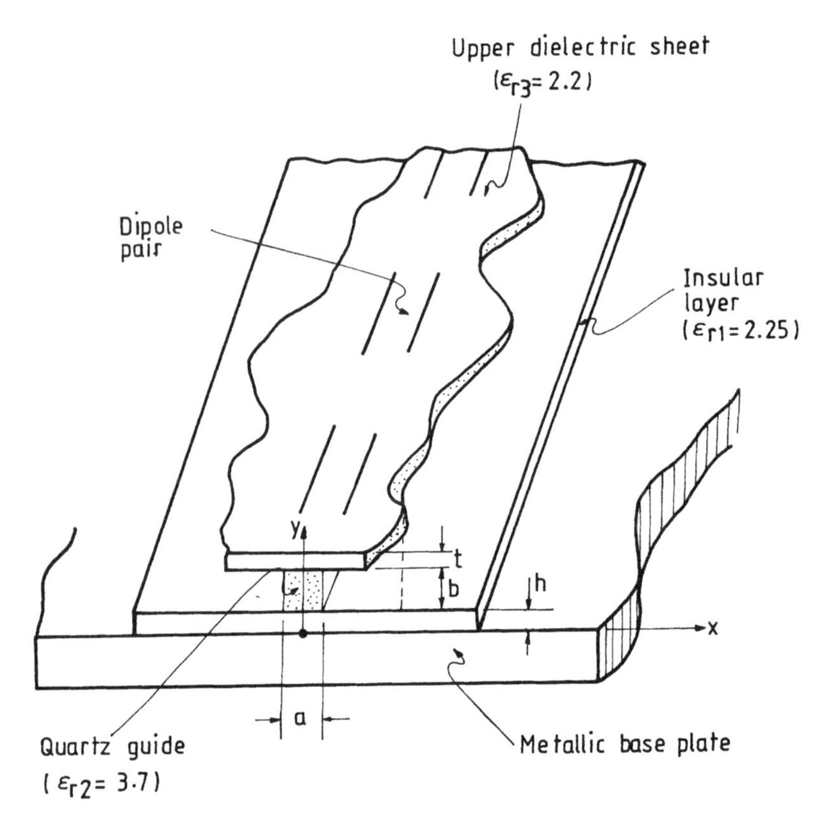

FIGURE 13.13 Linear array of dipole pairs fed by dielectric waveguide. Typical parameters for operation at 30 GHz: $a = b = 2$ mm, $h = 0.25$ mm, $t = 0.51$ mm, $\varepsilon_{r1} = 2.25$, $\varepsilon_{r2} = 3.7$, and $\varepsilon_{r3} = 2.2$. (From Inggs et al. [22], reprinted with permission of IEE.)

$-a/2 \le x \le a/2$ and $|x| > a/2$, respectively, and A is a constant. The propagation parameters k_{x1} and k_{x2} can be calculated by applying the approximate EDC technique to the three-layer dielectric guide model (without dipoles). The relative current excitation coefficients of the dipoles are then obtained as proportional to the calculated H-field values at the element locations. The electric field in the far-field zone can be expressed as

$$E(r, \theta, \phi) \approx \hat{\boldsymbol{\theta}}(I_0 \eta_0 / 2\pi r) f_e(\theta) f(\theta, \phi) e^{-jk_0 r} \tag{13.17}$$

where I_0 is the current excitation coefficient of the reference element, η_0 is the free-space impedance, $f_e(\theta)$ is the element pattern, and $f(\theta, \phi)$ is the array factor. The array factor may be expressed as [23]

$$f(\theta, \phi) = \cos(k_0 a \sin\theta \cos\phi) \sum_{n=1}^{N} a_n \exp[jnk_0 d(\cos\theta - \lambda_0/\lambda)] \tag{13.18}$$

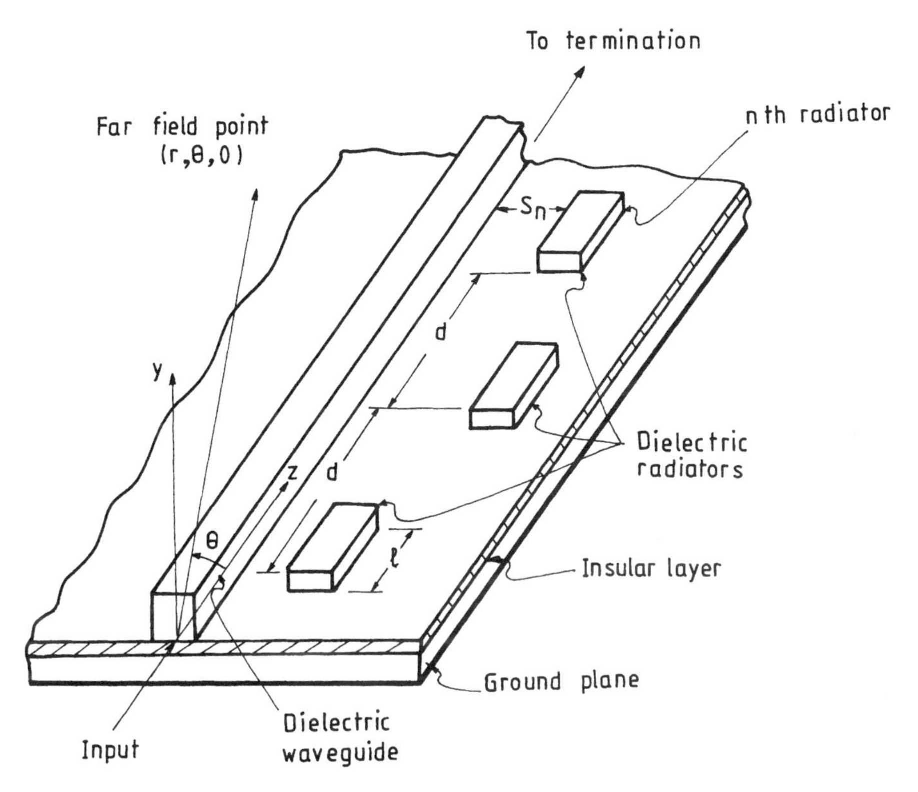

FIGURE 13.14 Array antenna using dielectric resonators as radiating elements. (From Birand and Gelsthorpe [24], reprinted with permission of IEE.)

where a_n is the relative amplitude excitation coefficient of the nth element. The element pattern $f_e(\theta)$ may be approximated by the pattern of a printed dipole over a dielectric slab backed by a ground plane [26]. The direction of the mth grating lobe can be obtained from the relation (13.5). For the guide parameters shown in Figure 13.13 and for a linear array of ten dipole pairs spaced by one guide wavelength apart, Inggs et al. [22] have reported frequency scanning from 28 to 32 GHz with low side lobe levels (better than about 12 dB).

Another example of an insular image guide-fed array is illustrated in Figure 13.14 wherein the rectangular dielectric resonators placed adjacent to the guide act as radiators [24]. The resonators are chosen to be half-wavelength long in insular guide geometry (see Chapter 9). The energy coupled to each resonator is a function of the distance between the resonator and the main feed guide. The element excitation coefficients can therefore be controlled easily by adjusting the positions of the resonators with respect to the feed guide. This type of antenna is reported to offer high gain in the broadside direction [24].

13.5 ANTENNAS BASED ON NRD AND GROOVE GUIDES

Besides the image guide, other dielectric guide structures and the groove guide can be used for realizing millimeter wave antennas. Among the other dielectric guide structures, the inverted strip guide with periodic corrugations on the dielectric strip [27, 28] and the trapped image guide with periodic metal strips on the dielectric guide [29] have been investigated as grating type leaky-wave antennas. The basic schematics of these antennas are shown in Figure 13.15. The theory of operation of these antennas is similar to that of the image guide grating leaky-wave antennas. The direction of the main beam measured from the broadside direction in the E-plane (xz-plane in Fig. 13.15) is given by (13.6) and the radiation pattern in the E-plane can be obtained from (13.11). One specific advantage of these structures over the image guide is that they are amenable for flush mounting on large conducting bodies. The inverted strip guide has the additional advantage that its top dielectric layer can serve as a radome.

13.5.1 NRD Guide-Based Antennas

Among the various dielectric guide structures, the NRD guide has assumed special significance in the realization of millimeter wave integrated circuits because of its nonradiative nature. The NRD guide is an open structure with its magnetic field lines lying in a plane perpendicular to the side metallic walls and its electric field oriented predominantly parallel to the plates (see Fig. 6.2, Chapter 6). The open nature of the guide and its field structure have been exploited in configuring different types of antennas [30–33]. These include antennas employing asymmetric airgap [30], foreshortened side [30, 31], slots on side metallic wall [32–33], and a certain asymmetry along the guide [34].

The leakage mechanism in a NRD guide employing an asymmetric airgap is illustrated in Figure 13.16(a) [30]. The asymmetry produces a net horizontal electric field, which creates a TEM field in the air region on either side of the dielectric. Since the TEM mode has no cutoff, power leaks away from the open end. The structure thus operates as a leaky-wave antenna.

In the NRD guide with a foreshortened side (see Fig. 13.16(b)), the dielectric strip is displaced to one side of the symmetry plane so as to make the distance from one of the open ends small. The fields then would not decay to negligible value at that open end and some power would leak away. This open end from where the leakage power radiates forms the antenna aperture. Sanchez and Oliner [31] have evaluated the complex propagation constant ($\beta - j\alpha$) of this antenna by considering a transverse equivalent network corresponding to LSE and LSM modes (with respect to the xy-plane in Fig. 13.16(b)). Figure 13.17 shows a typical variation of the attenuation (leakage) constant α and phase constant β of a NRD guide antenna as a function of the distance d [31]. Since the field decays exponentially away from the dielectric region, the value of α and hence the leakage radiation increase with a decrease in d. Furthermore, with a decease in d, the value of α varies over a fairly wide range, whereas β remains

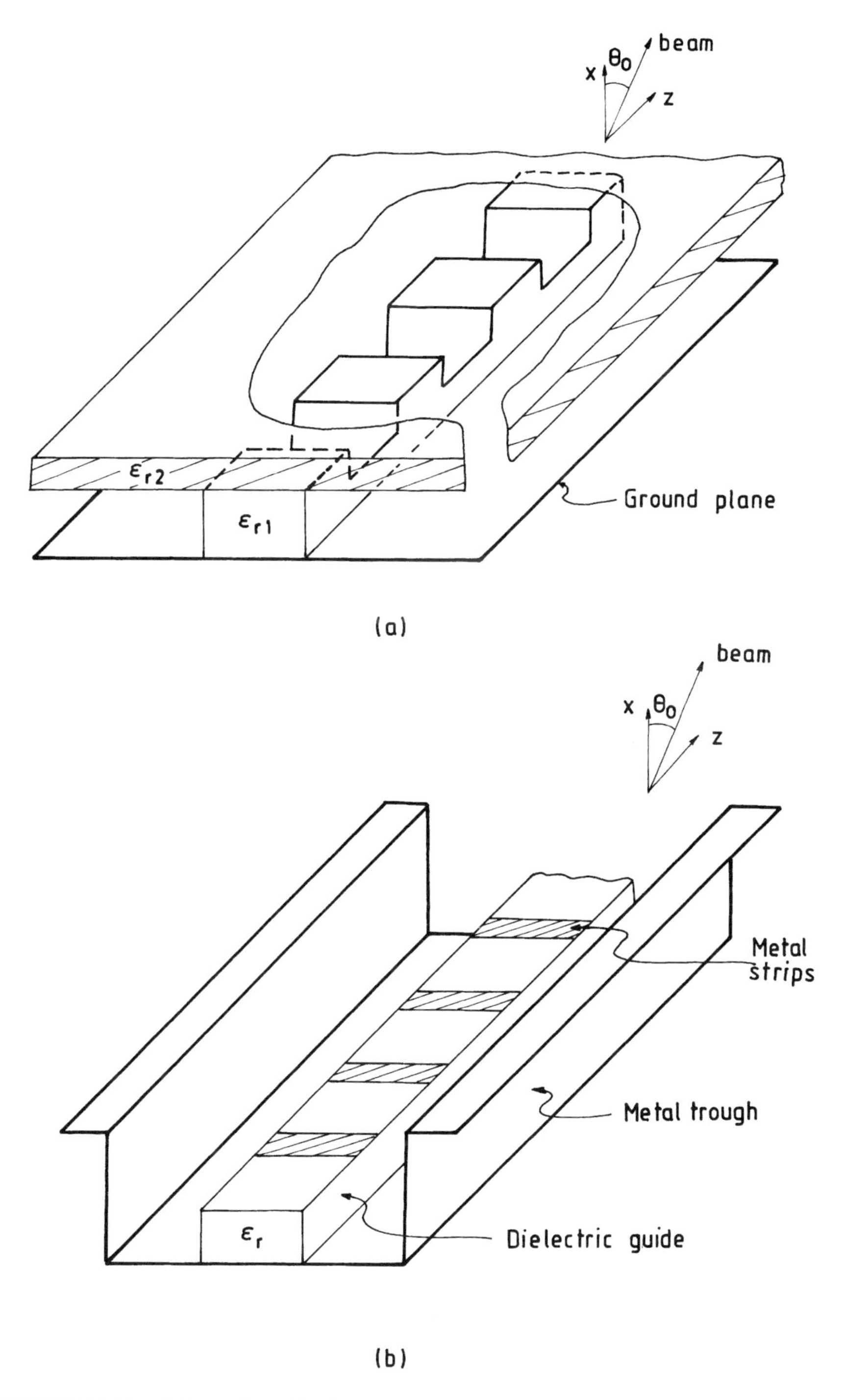

FIGURE 13.15 Schematics of leaky wave antenna in (a) inverted strip guide and (b) trapped image guide.

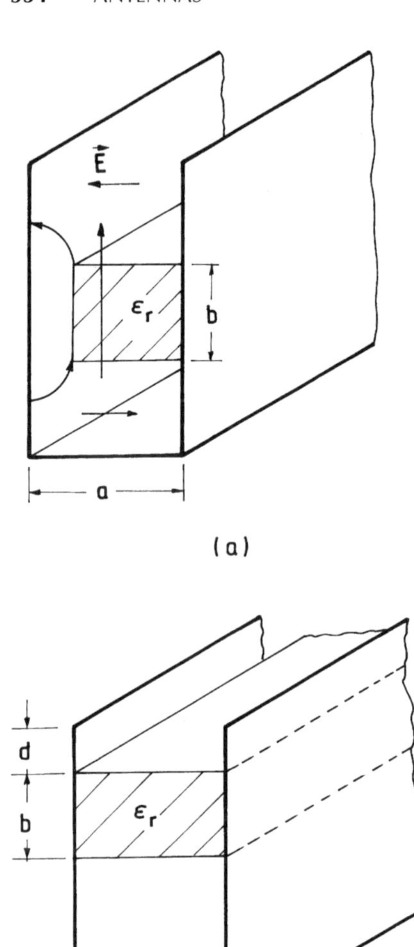

FIGURE 13.16 Leaky-wave antennas based on NRD guide using (a) asymmetric airgap and (b) foreshortened side. (After Oliner [30].)

essentially constant until d becomes very small. The antenna aperture amplitude distribution can therefore be tapered by varying d along the length of the antenna. Figure 13.18 shows the end view of such an antenna with the strip terminated in an absorbing load. The input end of the antenna can easily be connected to the rest of the millimeter wave integrated circuit realized in a NRD guide.

The field configuration of the NRD guide enables a slot array to be realized by cutting rectangular slots on the side metallic wall (Fig. 13.19). Since the tangential magnetic field lines on the inner surfaces of the side walls are z-directed, the

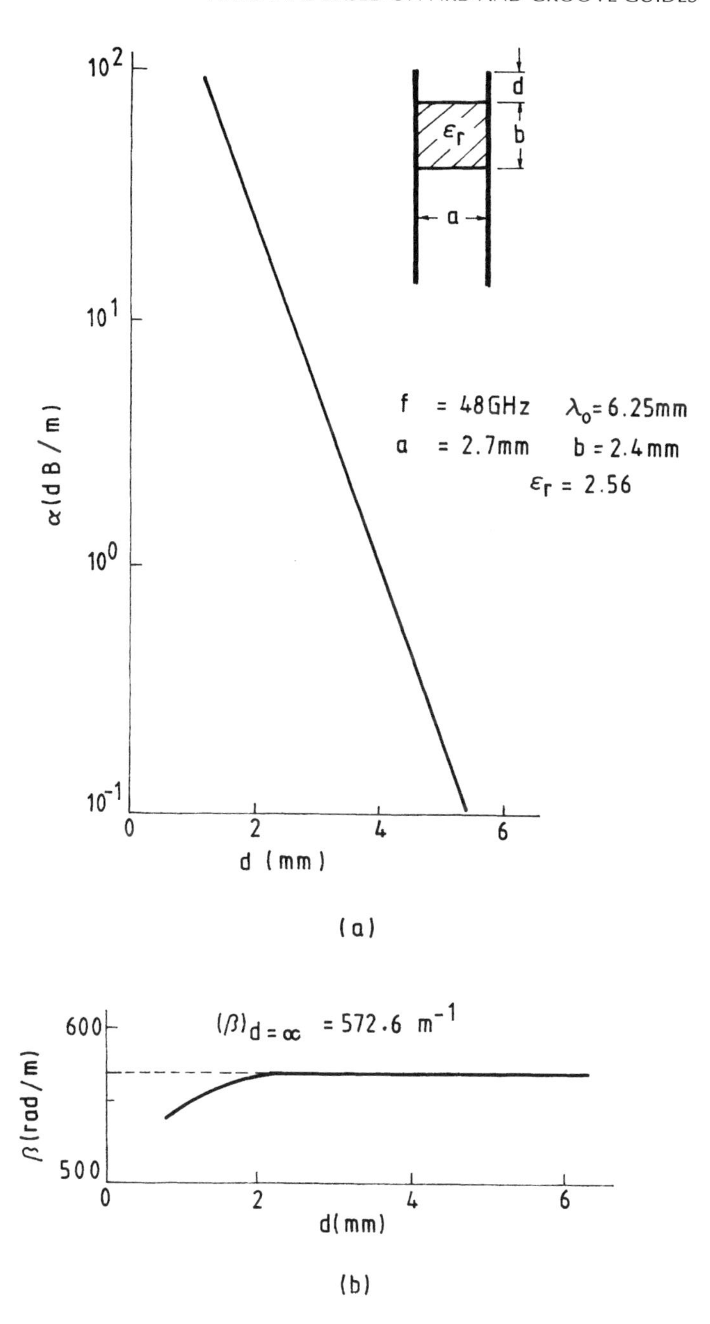

FIGURE 13.17 (a) Leakage constant α and (b) phase constant β of a NRD guide leaky-wave antenna as a function of distance d from the dielectric strip to the open end of the vertical metal well. (From Sanchez and Oliner [31]. Radio Science, Vol. 19, pp. 1225–1228, 1984, Copyright © by the American Geophysical Union.)

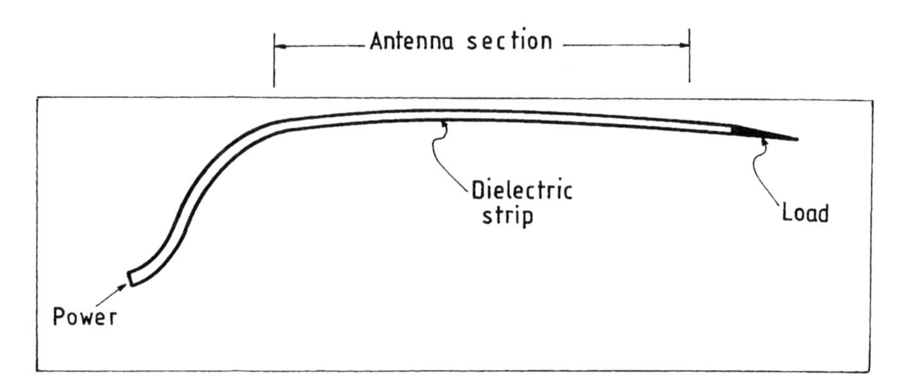

FIGURE 13.18 Side view of NRD guide antenna (refer to inset in Fig. 13.17(a)). (From Sanchez and Oliner [31]. Radio Science, Vol. 19, pp. 1225–1228, 1984, Copyright © by the American Geophysical Union.)

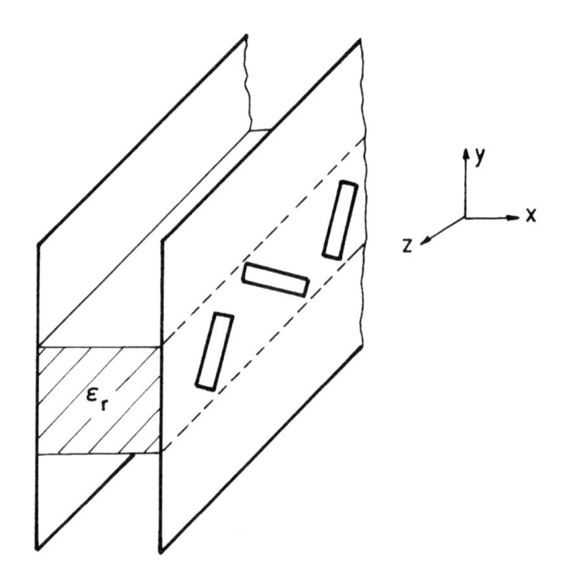

FIGURE 13.19 NRD guide slot array.

surface current flows mostly in the y-direction and is an even function about the centerline. Rectangular slots cut in the axial direction need to be spaced by one guide wavelength for a broadside array. With this spacing, however, unwanted grating lobes appear in the radiation pattern. By inclining the slots, the slot spacing can be reduced to half wavelength and the coupling can be chosen by adjusting the angle of inclination. Malherbe [32] and Malherbe et al. [33] have investigated a NRD guide slot array with inclined slots. The analysis and design procedure are similar to that of the standard rectangular waveguide narrow-wall slot array reported by Elliot [35]. Malherbe [32] has reported the design of a ten-element broadside array with uniform illumination and has shown that the effect of mutual impedance in the NRD guide antenna is far less than that in a rectangular waveguide slot array.

A different type of leaky-wave antenna that achieves radiation by asymmetrically perturbing the NRD guide has been reported by Yoneyama et al. [34]. In this antenna, the dielectric strip of the NRD guide is gradually changed in cross-section from a rectangular shape to a trapezoid shape as illustrated in Figure 13.20. When excited by a rectangular waveguide at the input end, the electric field of the operating NRD guide is predominantly parallel to the metal plates. In the portion of the guide containing the trapezoid dielectric strip, it is required to have a component of electric field prependicular to the metal plates in

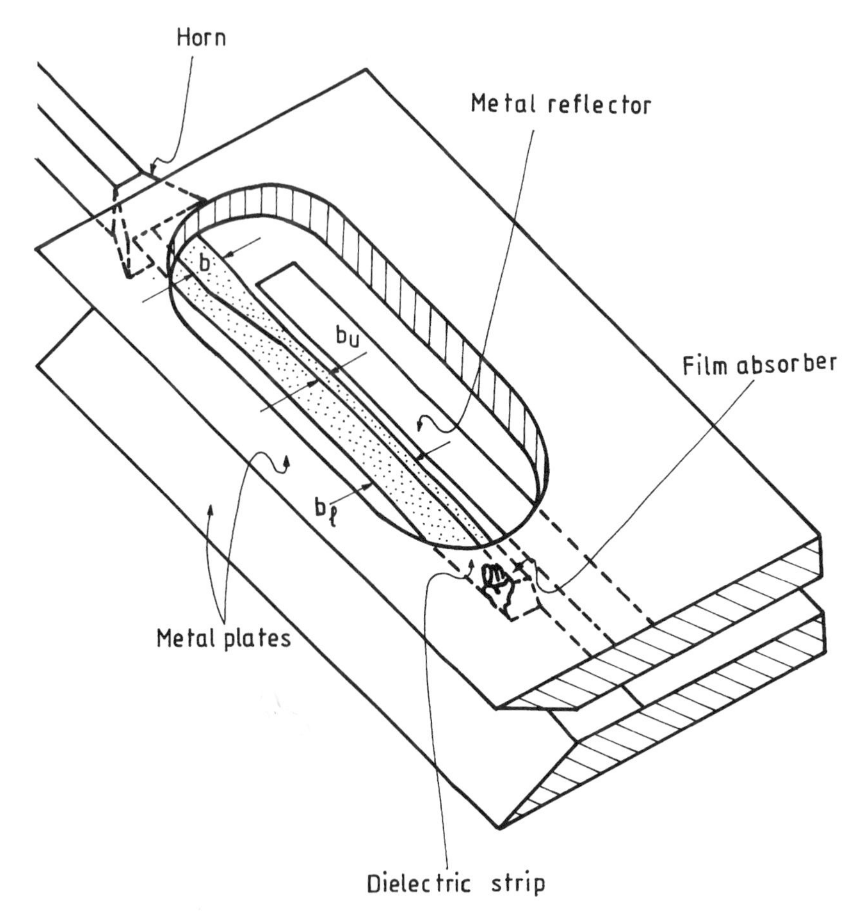

FIGURE 13.20 Structure of a NRD guide leaky-wave antenna (b_u and b_ℓ are the upper and lower widths of trapezoid section). (From Yoneyama et al. [34]. Copyright © 1985 IEEE, reprinted with permission.)

order to satisfy the continuity condition at the air–dielectric interfaces. This perpendicularly oriented field component leaves the waveguide as leaky-wave radiation. The metal bar of rectangular cross-section located on one side of the dielectric strip acts as a reflector to produce unidirectional radiation. On the radiating side, the top and bottom plates are shaped into a two-dimenisional horn to improve the radiation in the vertical plane. The film absorber located at the end of the dielectric strip absorbs the residual power. From the measured amplitude and phase profile along the dielectric strip of a NRD guide antenna designed at 50 GHz, Yoneyama et al. [34] have inferred that the amplitude profile can be approximated by

$$\begin{aligned} f(z) &= \sin(\pi z/2\ell_0), \quad 0 < z < \ell_0 \\ &= e^{-\alpha(z-\ell_0)}, \quad \ell_0 < z < \ell \end{aligned} \tag{13.19}$$

where ℓ_0 is a length along the dielectric strip (including the transition) and α is the attenuation constant. From the measured phase profile, it is shown that if the width b of the uniform rectangular strip is equal to the geometric mean of the widths b_u and b_ℓ corresponding to the upper and lower sides, respectively, of the trapezoidal strip, the phase constant β is nearly uniform along the strip. Thus the condition

$$b = 0.5(b_u + b_\ell) \tag{13.20}$$

forms a basic design formula for obtaining uniform phase distribution, which is essential for achieving a sharp radiation pattern. From the measured near-field distribution, the radiation pattern can be calculated using the formula

$$F(\theta) = \int_0^\ell f(z)\exp[-j(\beta - k_0 \sin\theta)z]dz \tag{13.21}$$

where k_0 is the free-space propagation constant, β is the phase constant of the leaky wave, θ is the azimuthal angle measured from the broadside direction, and ℓ is the total length of the antenna including the transition. This type of leaky-wave antenna is shown to exhibit very low side lobes and low cross-polarization (better than -30 dB) [34].

13.5.2 Groove Guide Antennas

Two types of groove guide antennas that employ asymmetric leakage mechanisms have been reported in the literature [30, 36, 37]. The cross-sectional views of these structures are shown in Figure 13.21 [30]. In the first structure, asymmetry is introduced by placing a continuous metallic strip of small width δ along one side wall. In the absence of the strip, the field of the basic symmetric guide, with top and bottom sides open, decays exponentially away from the groove region in

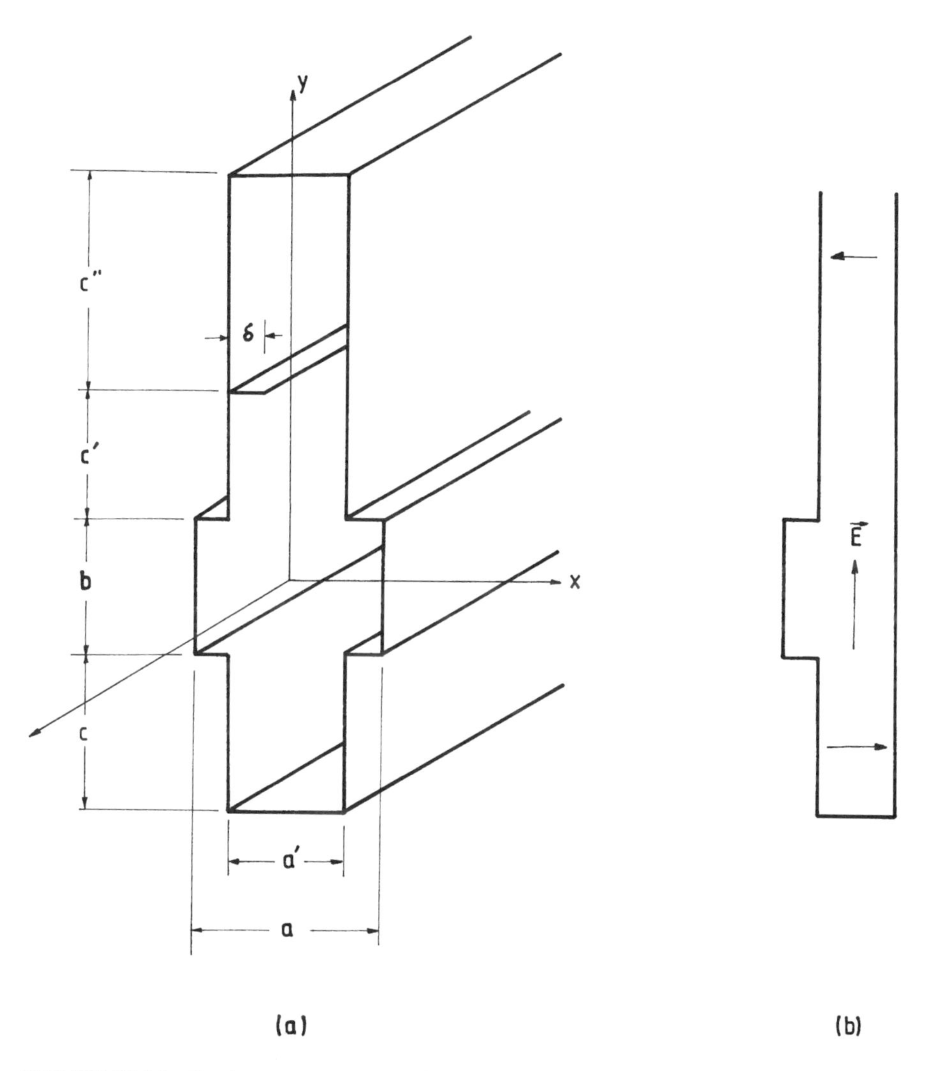

FIGURE 13.21 Leaky-wave groove guide antennas: (a) asymmetric continuous metal strip in the groove guide and (b) bisected groove guide. (From Oliner [30]. Copyright © 1985 IEEE, reprinted with permission.)

the vertical (y)-direction. The asymmetrically placed strip produces a net horizontal electric field, thereby creating a new transverse mode similar to the TEM mode of a parallel-plate guide. By virtue of the strip, this new mode and the original transverse mode get coupled. The combined effect is to produce a TE longitudinal mode (in the x-direction) with a complex propagation constant $(\beta - j\alpha)$. While the original mode of the groove guide has a half sinusoid variation with respect to the x-direction, the new transverse mode has no variation with x.

In the groove region, both these modes are above cutoff and in the outer regions, the original mode is below cutoff, whereas the new transverse mode remains above cutoff, leading to a standing wave effect in those regions and radiation from the open end. Lampariello and Oliner [37] have evaluated the propagation constant of the structure by modeling it as an equivalent network and applying the transverse resonance technique. For designing the structure as an efficient antenna, it is required to taper the aperture amplitude distribution while keeping the phase linear along the aperture length. Equivalently, α must vary along the aperture length while β remains constant. The value of β is determined essentially by the original transverse mode (of the symmetric groove guide) and hence is governed primarily by the dimensions a, a', and b. Thus for a given frequency the design involves choosing first the width a and adjusting a' and b to achieve the desired value of β. It may be noted that the beam pointing direction is fixed mainly by the value of β/k_0. The attenuation constant increases with an increase in δ and also with a decrease in c'. Figure 13.22(a) shows the nature of decrease in α/k_0 with an increase in c'/a, and Figure 13.22(b) illustrates that $(c + c')/a$ is nearly constant for a wide range of values of α/k_0 [37]. Once the dimension c' is determined for an optimized value of α/k_0 from Figure 13.22(a), the dimension c can be obtained from Figure 13.22(b). This type of leaky-wave antenna is suitable for higher millimeter wave frequencies and is capable of offering a wide range of pointing angles and beam widths [37].

Leaky-wave groove guide antennas can be realized by making use of higher order modes that are leaky. For higher order modes to exist, the guide width must be increased in the central region. When the first higher order even mode ($n = 2$) is excited, the leakage is in the form of TEM mode and the polarization of the leaky-wave radiation is horizontal. When the first higher order odd mode ($n = 3$) is incident, the leakage radiation corresponds to vertical polarization. Figure 13.21(b) shows the cross-section of an antenna for yielding horizontally polarized leaky-wave radiation [30]. Only one-half of the groove guide serves as the antenna because, for the even mode ($n = 2$), the vertical symmetry plane corresponds to an electric wall and hence can be replaced by a metal wall. The antenna can thus be viewed as a parallel-plate guide with a stub. This type of antenna is reported to have the capability of offering narrow as well as broad beams.

13.5.3 NRD Guide-Fed Planar Antennas

Yoneyama et al. [38, 39] have proposed a new type of leaky NRD guide that can be used to feed microwave planar antennas. This guide consists of a periodic array of notches on the upper surface of the dielectric strip as shown in Figure 13.23. When the periodicity of the notches is very small compared to the guide wavelength (λ_g), the operation of the periodic leaky NRD guide resembles its uniform counterpart described in Section 13.5.1 [40, 41]. For the case when notches are separated by a guide wavelength as shown in Figure 13.24(a), broadside radiation occurs. This structure, however, suffers from grating lobe radiation, which is undesirable for practical applications. These grating lobes can

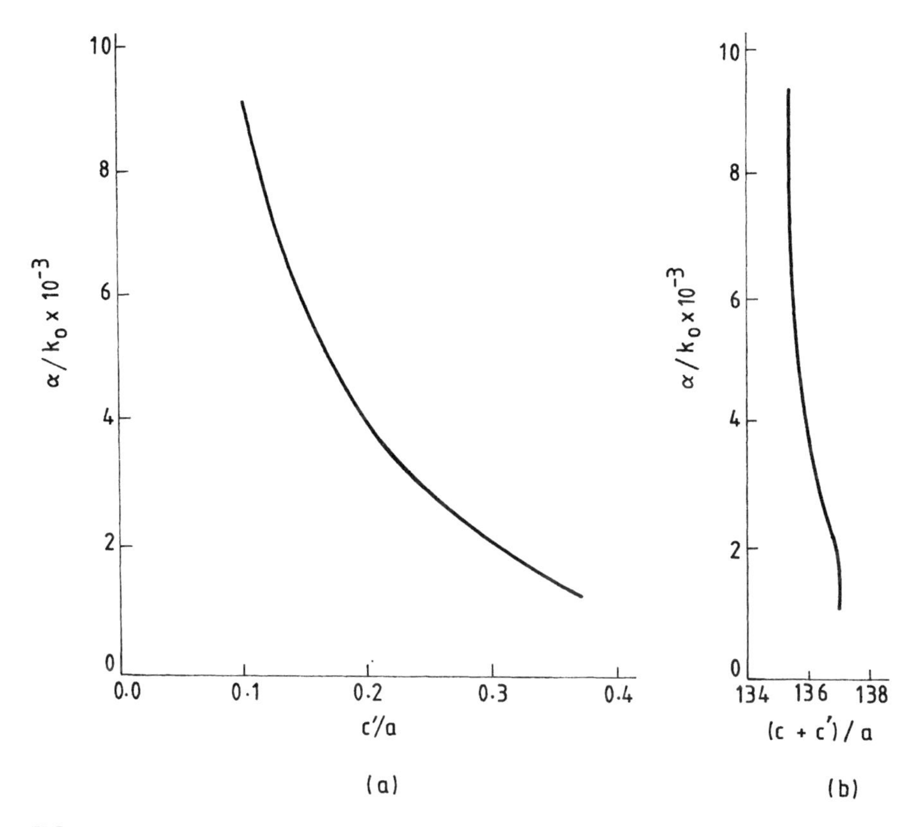

FIGURE 13.22 (a) Leakage constant α/k_0 of leaky-wave antenna shown in Figure 13.21(a) as a function of distance c'/a of perturbing strip from step junction. (b) Optimum value of leakage constant α/k_0 as a function of $(c + c')/a$ required for optimization. (From Lampariello and Oliner [37], reprinted with permission of IEE.)

be eliminated by creating the same notch array shifted in position by half a guide wavelength on the lower surface of the dielectric strip as indicated in Figure 13.24(b). The reflections from each notch in this structure add in phase at the input port. To reduce these reflections, Yoneyama and co-workers [38, 42] devised a matched element that consists of a pair of symmetric notches on the upper and lower sides of the dielectric strip as shown in Figure 13.25(a). The matching element does not radiate at all. Arranging the radiating and matching elements as shown in Figure 13.25 reduces the reflections considerably [42]. The configuration of a broadside leaky NRD guide with matching element is shown in Figure 13.26 [38, 42]. This leaky-wave structure can be used to feed a slot array in a planar antenna as shown in Figure 13.27 [38, 42]. Figure 13.28 shows the radiation pattern of the planar antenna fabricated at 23 GHz as reported by Maamria et al. [42]. The pattern shows narrow beamwidth and low side lobes, which are less than 20 dB from the main lobe level. The photograph of the

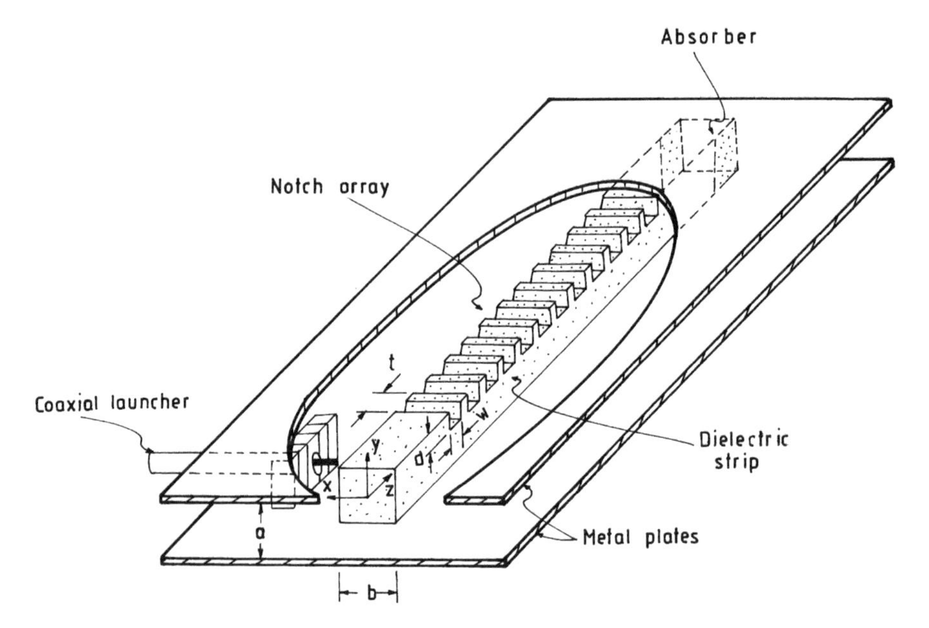

FIGURE 13.23 Periodic leaky NRD guide. (After Yoneyma [39].)

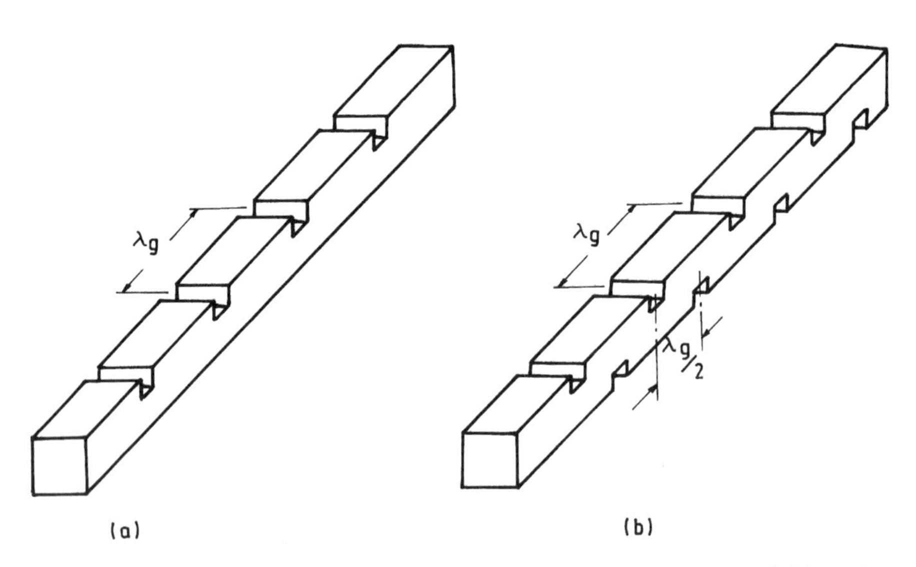

FIGURE 13.24 Broadside leaky NRD guide: (a) grating lobes appear and (b) grating lobes disappear. (From Yoneyama [38], reprinted with permission.)

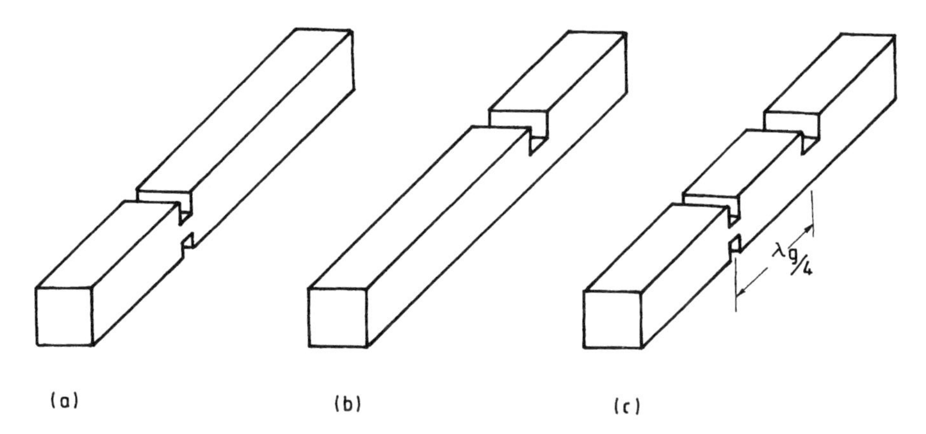

FIGURE 13.25 Matching and radiating elements: (a) matching element, (b) radiating element, and (c) radiating and matching elements. (After Yoneyama [38].)

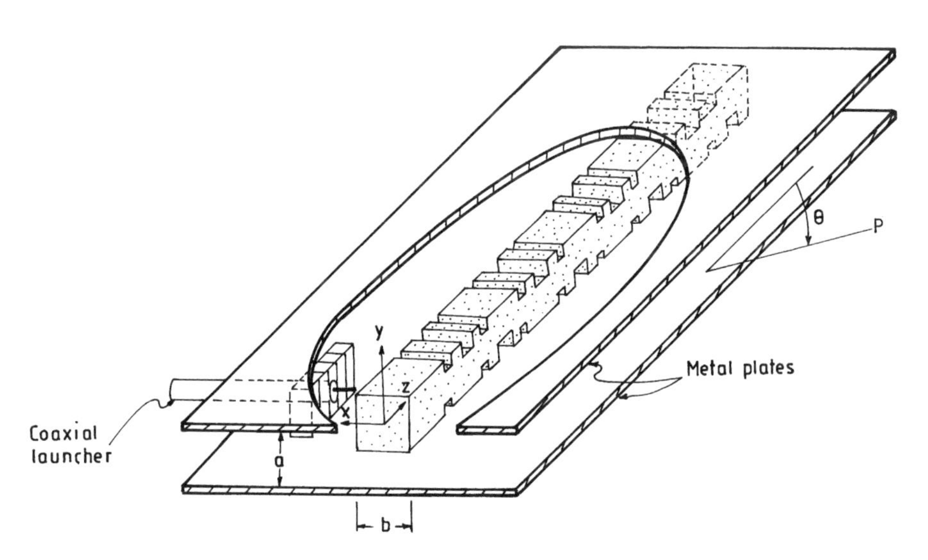

FIGURE 13.26 Broadside leaky NRD guide with matching element. (From Yoneyama [38], reprinted with permission.)

fabricated antenna is shown in Figure 13.29 (T. Yoneyama, private communication.)

A leaky NRD guide-fed slot array antenna that can produce circular polarization has also been reported. The antenna structure reported by Maamria et al. [42] is shown in Figure 13.30. Two leaky waveguides are connected to a NRD guide directional coupler, which divides equally the power fed between the two leaky NRD guides with 90° phase shift. This feeding arrangement ensures that the

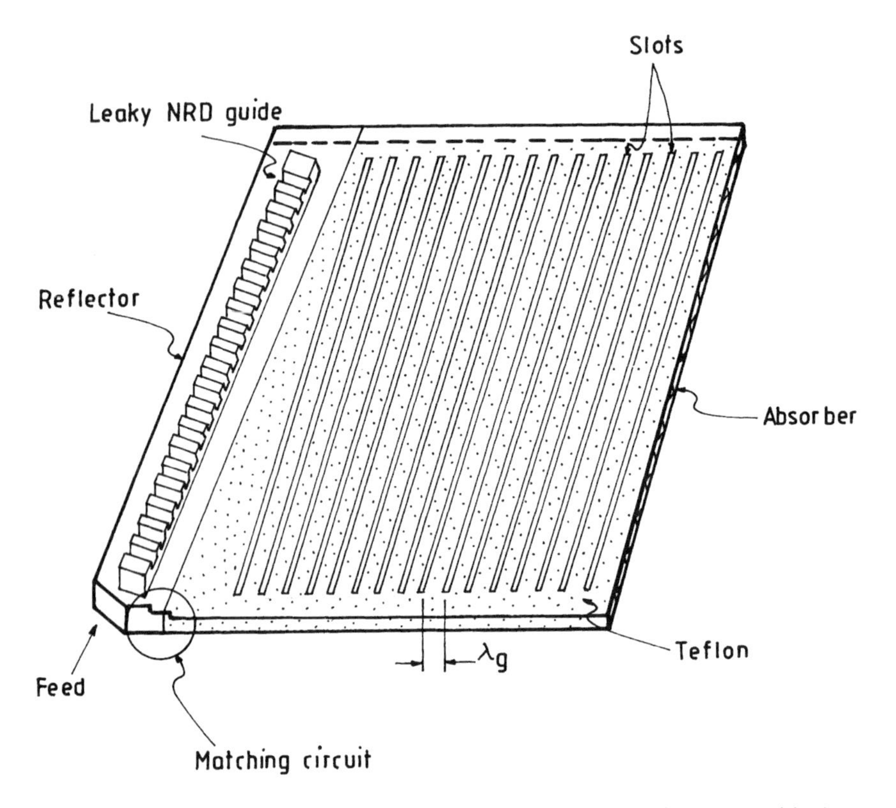

FIGURE 13.27 Structure of planar antenna fed by broadside leaky NRD guide. (From Maamria et al. [42]. Copyright © 1993 IEEE, reprinted with permission.)

radiated fields from the two sets of slots are equal in amplitude but in phase quadrature. In addition, the fields are perpendicular to each other, thereby generating circularly polarized waves.

13.6 SEMICONDUCTOR GUIDE ANTENNAS

Semiconductor guides differ from other dielectric guides in that the conductivity of the guide can be varied over a wide range ($\sigma \approx 10^{-4}$–10^{-6} mhos/m) either by suitable doping or carrier injection. Semiconductor guides also permit permittivity modulation through optical control. Antennas based on semiconductor guides have the advantage that their radiation characteristics can be controlled during fabrication as well as operation. Furthermore, these antennas are compatible for integration with semiconductor devices and other monolithic integrated circuits to yield fully integrated T-R modules for millimeter wave radars [43].

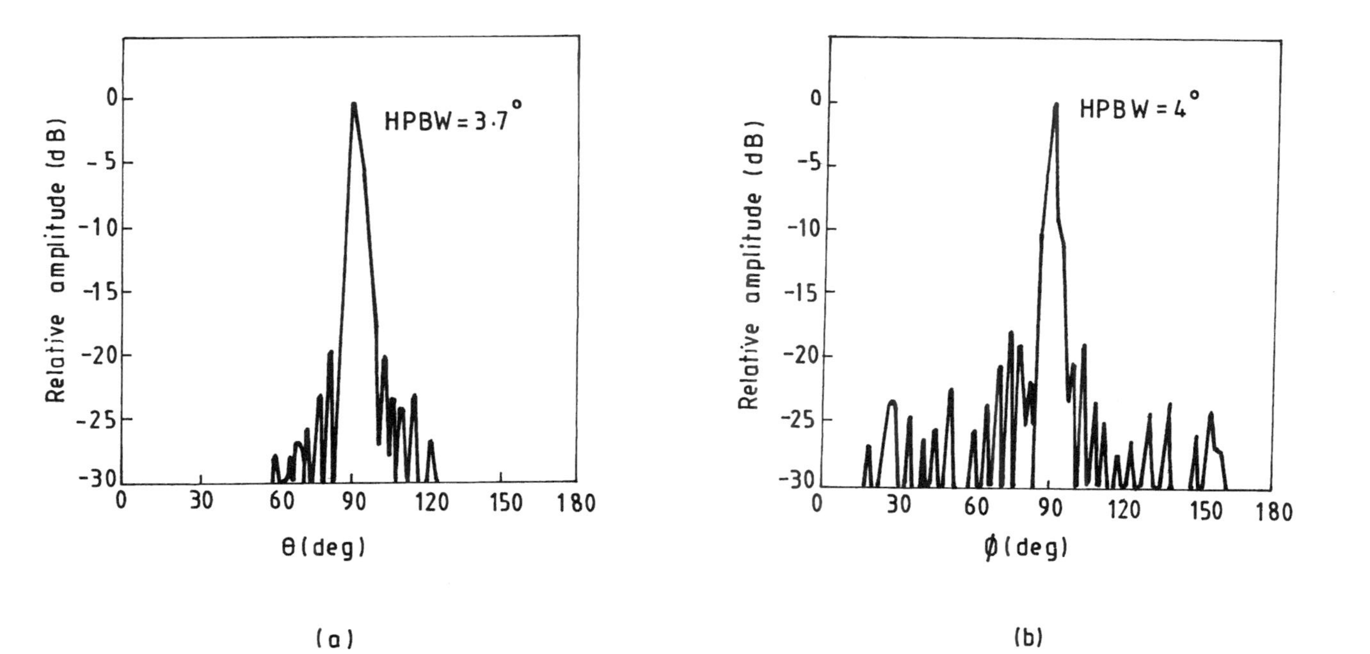

FIGURE 13.28 Radiation pattern of planar leaky NRD guide-fed antenna: (a) in plane parallel to the slots and (b) in plane normal to the slots. (From Maamria et al. [42]. Copyright © 1993 IEEE, reprinted with permission.)

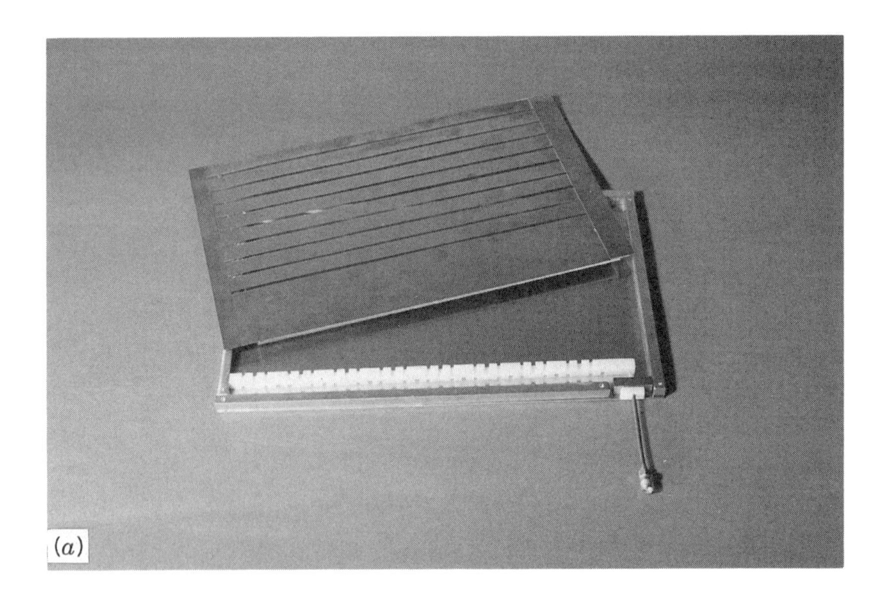

FIGURE 13.29 Photograph of planar antenna fed by broadside leaky NRD guide: (a) top plate removed and (b) completely assembled. (From T. Yoneyama, private communication.)

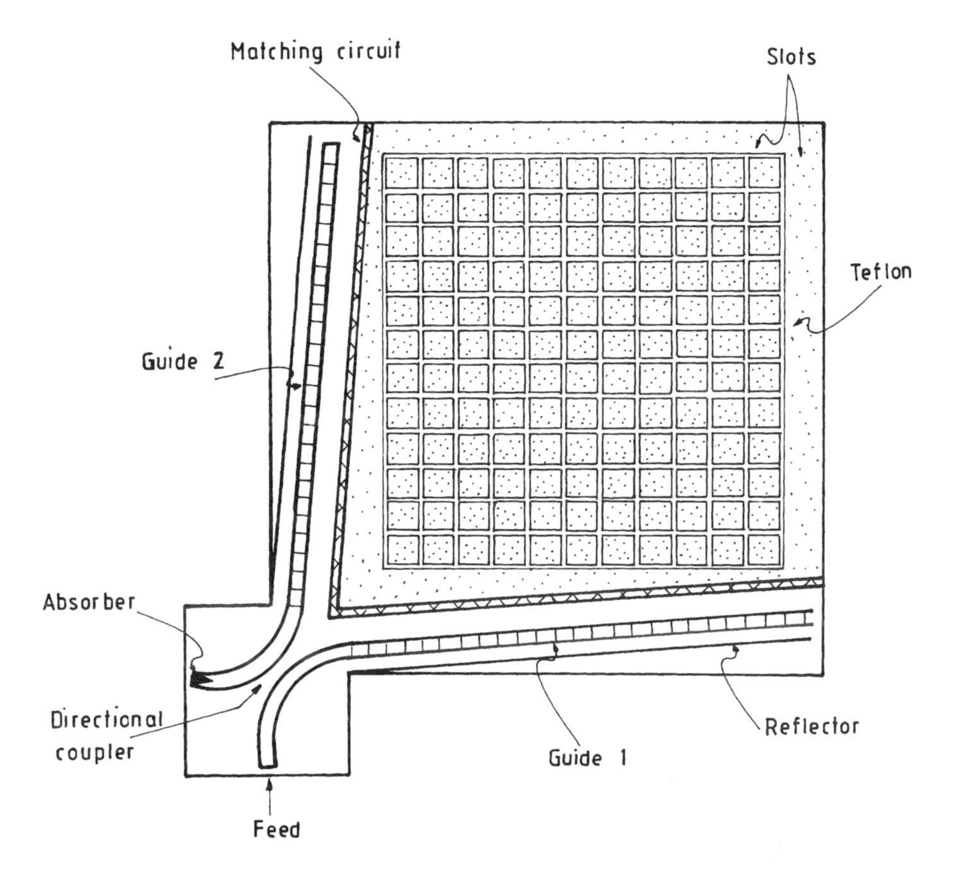

FIGURE 13.30 Structure of leaky NRD guide-fed slot array antenna that can produce circular polarization. (From Maamria et al. [42]. Copyright © 1993 IEEE, reprinted with permission.)

13.6.1 Silicon Guide Antennas

High-resistivity semiconductor materials in the form of low-loss dielectrics can be used for realizing waveguide type antennas. One such antenna is the silicon waveguide grating antenna with metallic strips reported by Klohn et al. [44] and shown in Figure 13.31. The antenna, reported for operation around 60 GHz, is made of a high-resistivity ($\sim$ 30,000 Ω-cm) silicon waveguide having a cross-section of 0.9 mm $\times$ 0.9 mm and length of 7 cm. The metal strip grating on the surface of the guide is comprised of 16 strips each of dimension 0.3 mm $\times$ 1 mm and spaced uniformly with a grating period of 2.5 mm (which is approximately one wavelength in the silicon waveguide). As illustrated in Figure 13.31, the antenna is excited by a metal waveguide and absorbers are positioned at both ends at the dielectric-to-metal waveguide transition for preventing undesired leakage. Klohn et al. [44] have shown that this antenna has a frequency-scanning

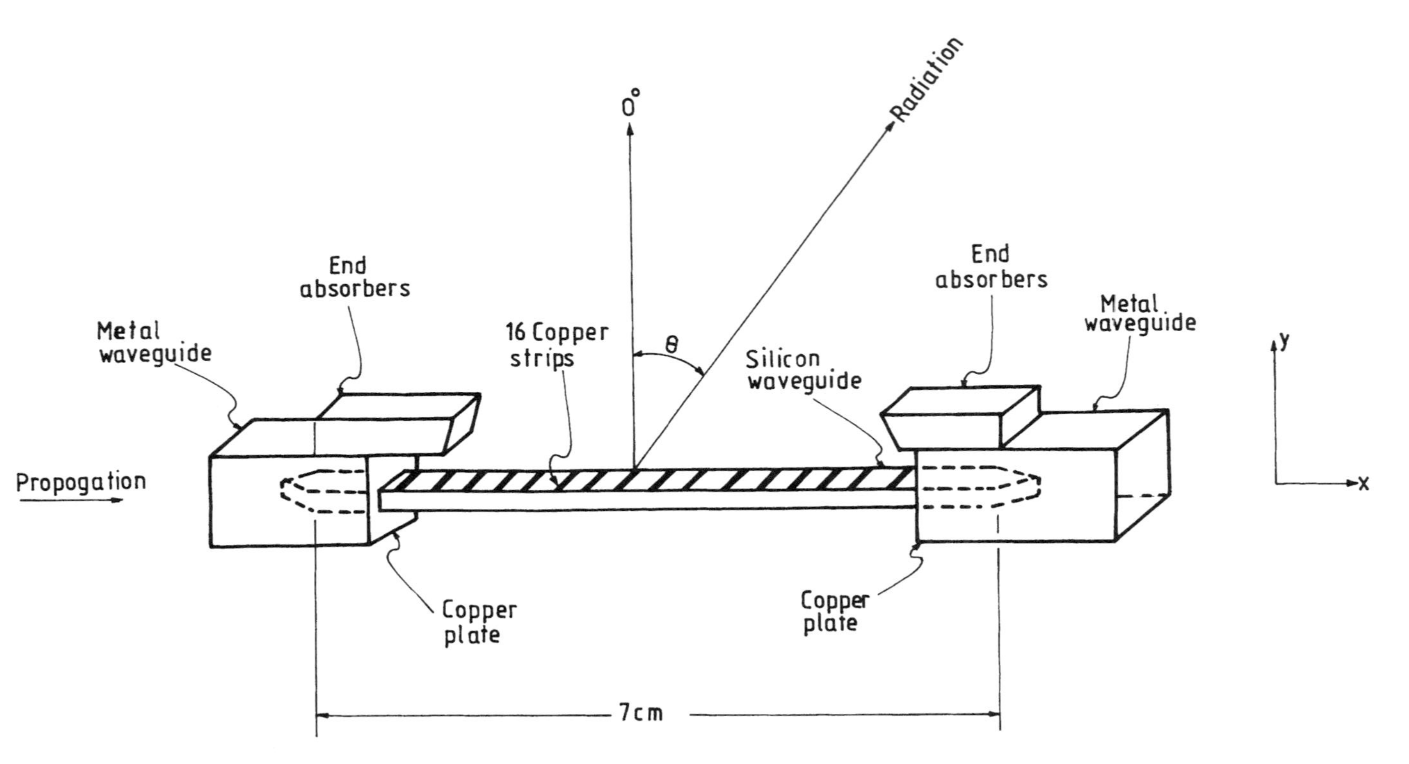

FIGURE 13.31 Silicon waveguide with perturbations for operation around 60 GHz. Si, 0.9 mm × 0.9 mm × 7 cm; strips, 0.3 mm × 1 mm; and grating period, 0.25 cm. (From Klohn et al. [44]. Copyright © 1978 IEEE, reprinted with permission.)

capability of about 6.5°/GHz over a frequency scan of approximately 6 GHz near 60 GHz.

A different type of millimeter wave antenna based on a silicon dielectric guide has been reported by Yao et al. [45]. The geometry is shown in Figure 13.32. It is a tapered trapezoidal dielectric rod etched from a high-resistivity (~ 1000 Ω-cm) Si wafer. The antenna is integrated with a Schottky mixer diode shown located at its vertex along with a V-shaped coupler so that the signal received from the antenna is directly coupled to the diode.

By integrating a p-i-n diode modulator with the silicon dielectric guide, Horn et al. [46, 47] have demonstrated electronic beam scanning at a fixed frequency. The antenna consists of a insular silicon dielectric waveguide with an array of copper strips on its top surface (Fig. 13.33). As has been discussed in earlier sections, the direction of the main beam for a leaky-wave antenna depends on the guide wavelength and the grating period. The radiated beam can therefore be scanned by varying the guide wavelength electronically. This is achieved by means of a distributed p-i-n diode attached to one of the side walls of the guide. With the p-i-n structure in the unbiased state, a portion of the external field of the propagating mode penetrates the diode. When a forward bias is applied to the p-i-n diode, the injected carriers create a conductive wall, thereby displacing the fields of the propagating wave. This new field distribution results in a different guide wavelength, thereby changing the direction of the radiated beam. It has been pointed out that in the unbiased state under which the intrinsic region of the p-i-n diode offers high resistivity, and in the fully biased state under which the diode behaves as a metal wall, the propagating wave experiences very little loss. But with a small bias, under which the intrinsic layer is partially conductive, the propagating wave gets diffracted into the p-i-n diode and incurs certain loss. Separating the p-i-n diode structure from the dielectric guide by a thin insulating layer helps in reducing the coupling to the p-i-n diodes, thereby reducing the loss due to diffraction. For an antenna designed for operation at 64 GHz, Horn et al. [47] have reported an electronic beam scan of about 5° with a maximum insertion loss of 2 dB occurring at intermediate scan angles.

13.6.2 Optically Controlled Silicon Guide Antenna

Alphones and Tsutsumi [48] have investigated leaky-wave radiation from an optically controlled silicon slab guide backed by a ground plane. The antenna and the experimental setup are shown in Figure 13.34. In this arrangement, the surface of the silicon guide is illuminated with certain spatial periodicity by means of an optical source. The photon energy of optical radiation is greater than the bandgap energy of the semiconductor. The electron–hole plasma that is induced results in a permittivity (complex) modulated grating structure. A periodic illumination pattern is achieved by means of a bundle of fibers. Gratings of different periodicity can be induced by suitably spacing the fibers. The silicon guide is tapered at one end and is fed by a rectangular metal horn. The far end of the guide, which is also tapered, is covered with an absorber to prevent unwanted

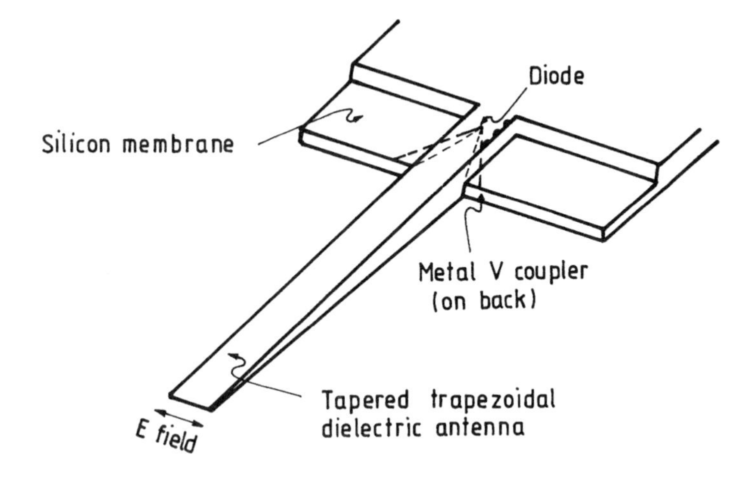

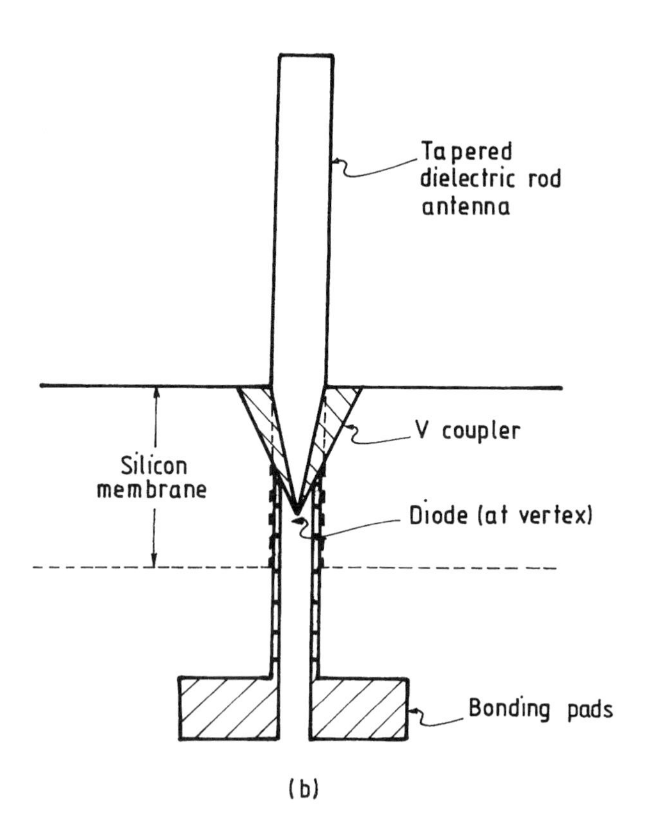

FIGURE 13.32 (a) Monolithic antenna-coupled Schottky mixer. (b) Reverse side of the antenna showing position of V-coupler and low-frequency leads. (From Yao et al. [45]. Copyright © 1982 IEEE, reprinted with permission.)

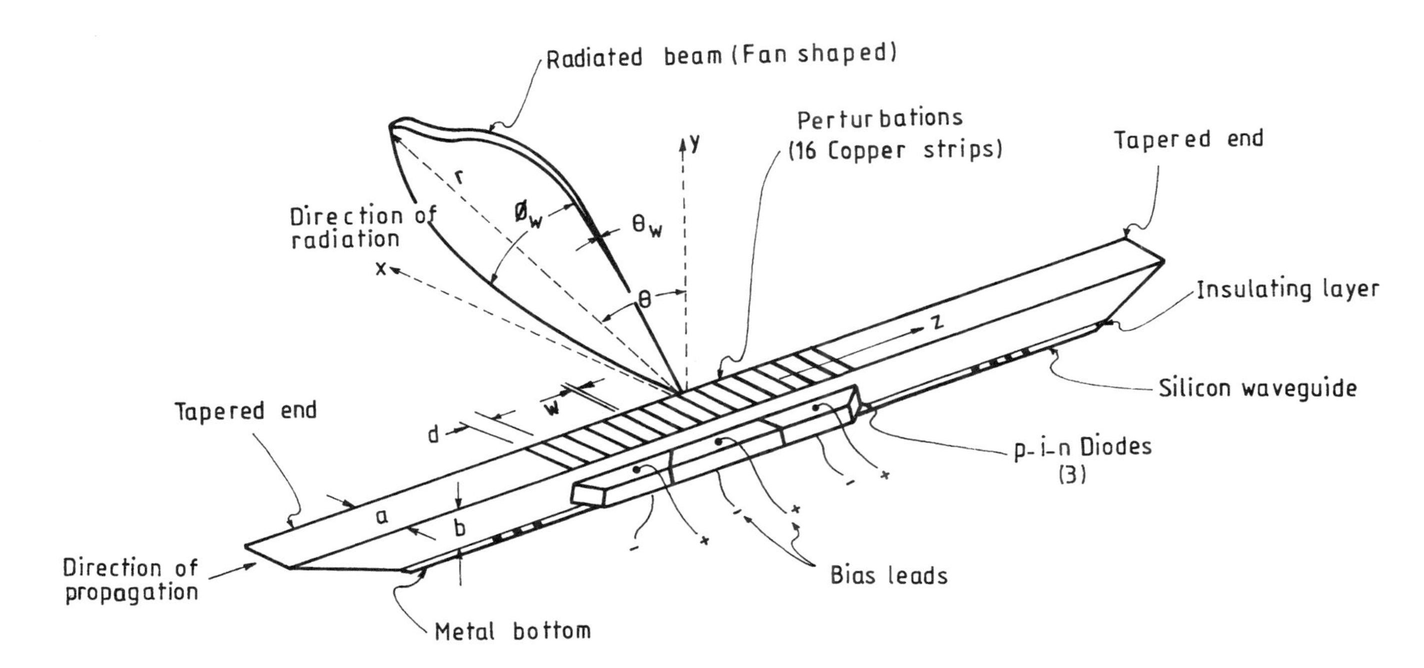

FIGURE 13.33 Electronic beam-steerable grating antenna using distributed p-i-n diode. θ_w, beam width in $r\theta$-plane; ϕ_w, beamwidth in $r\phi$-plane. (From Horn et al. [47]. Copyright © 1982 IEEE, reprinted with permission.)

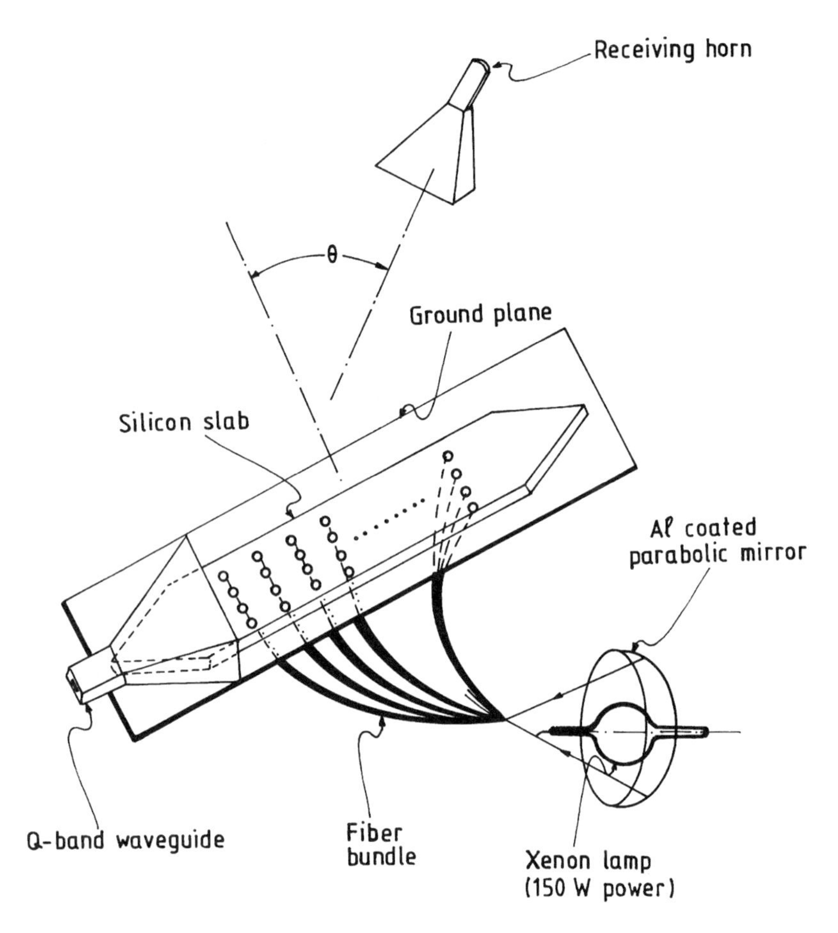

FIGURE 13.34 Experimental setup to observe leaky-wave radiation from optically controlled silicon guide. (From Alphones and Tsutsumi [48], reprinted with permission.)

reflection. The leaky-wave radiation is detected by means of a receiving horn, which is rotatable between the broadside and end-fire directions.

PROBLEMS

13.1 Calculate and plot the radiation pattern of a cylindrical dielectric rod of diameter 1.25 cm and length $10\lambda_0$. Assume $\lambda_0 = 1.5$ cm and $\varepsilon_r = 2.56$. Assume that the antenna is excited in the dominant hybrid (HE_{11}) mode.

13.2 Design a tapered dielectric rod antenna at 30 GHz using polystyrene ($\varepsilon_r = 2.56$) as the dielectric. Determine the diameters at both ends and the gain of the antenna when the length is $5\lambda_0$ and $10\lambda_0$.

13.3 Explain the theory of a leaky-wave grating antenna. Distinguish between surface waves and leaky waves.

13.4 Design a leaky-wave grating antenna in an image guide for operation at 30 GHz. Assume single-beam radiation at $\theta = 45^\circ$ with respect to the broadside direction.

13.5 Explain the differences between image guide leaky-wave antennas realized using (a) periodic gratings and (b) periodic conducting patches. What techniques can be adopted for narrowing the H-plane beamwidth in the two cases?

13.6 A dielectric integrated guide antenna is to be designed for an ESM radar receiver operating over a frequency range of 20–40 GHz. What would be an appropriate antenna for this application? Discuss the design considerations for such an antenna.

13.7 What are the limitations of microstrip patch antennas at millimeter wave frequencies? Enumerate the design and fabrication constraints.

13.8 Compare the relative merits and demerits of a rectangular slot array fed by (a) a rectangular waveguide, (b) an image guide, and (c) a NRD guide.

13.9 Explain the theory of leaky-wave groove guide antennas. Which mode of excitation can cause vertically polarized leaky-wave radiation? Draw typical E-field lines corresponding to this excitation.

13.10 Explain the principle of electronic beam scanning in a semiconductor guide antenna.

REFERENCES

1. D. G. Kiely, *Dielectric Aerials*, Metheun, London, 1952.
2. F. J. Zucker, Surface and leaky-wave antennas. *Antenna Engineering Handbook*, H. Jasik (Ed.), McGraw-Hill, New York, 1961, Chap. 16.
3. R. E. Collin and F. J. Zucker, *Antenna Theory*, Vols. 1 and 2, McGraw-Hill, New York, 1969.
4. R. Chatterjee, *Dielectric and Dielectric-Loaded Antennas*, Research Studies Press, Letchworth, UK, 1985.
5. Y. Shiau, Dielectric rod antennas for millimetre-wave integrated circuits. *IEEE Trans. Microwave Theory Tech.*, **MTT-24**, 869–872, Nov. 1976.
6. S. Kobayashi, R. Mittra, and R. Lampe, Dielectric rod of rectangular cross-section. *IEEE AP-S Int. Symp. Digest*, 27–30, 1980.
7. T. N. Trinh, J. A. G. Malherbe, and R. Mittra, A metal-to-dielectric waveguide transition with application to millimetre-wave integrated circuits. *IEEE Int. Microwave Symp. Digest*, 160–162, Apr., 1980.

8. J. A. G. Malherbe, An integrated antenna for non-radiative dielectric waveguide, Proceedings of the International Symposium on Antennas and Propagation, *ISAP Digest*, 69–72, 1985.
9. K. Solbach, Review of dielectric image line antennas, *Infrared and Millimetre Waves*, Vol. 15, K. J. Button (Ed), Academic Press, Orlando, FL 1986, Chap. 4.
10. F. K. Schwering and S. T. Peng, Design of dielectric grating antennas for millimeter-wave applications. *IEEE Trans. Microwave Theory Tech.*, **MTT-31**, 199–209, Feb. 1983.
11. R. Mittra and R. Kastner, A spectral domain approach for computing the radiation characteristics of a leaky wave antenna for millimeter waves. *IEEE Trans. Antennas Propag.*, **AP-28**, 652–654, July 1981.
12. I. Wolff and K. Solbach, Dielectric image line groove antennas for millimeter waves, Proceedings of the International Symposium on Antennas and Propagation, *ISAP Digest*, 73–76, 1985.
13. K. Solbach, E-band leaky wave antenna using dielectric image line with etched radiating elements. *IEEE MTT-S Int. Microwave Symp. Digest*, 214–216, 1979.
14. S. Kobayashi et al., Dielectric antennas for millimetre-wave applications. *IEEE MTT-S Int. Microwave Symp. Digest*, 566–568, May 1979.
15. S. Kobayashi et al., Dielectric rod leaky wave antennas, Proceedings of the International Symposium on Antennas and Propagation, *ISAP Digest*, 31–34, 1980.
16. M. J. Shiau et al., Mode conversion effects in Bragg reflection from periodic grooves in rectangular dielectric image guide. *IEEE MTT-S Int. Microwave Symp. Digest*, 14–16, June 1981.
17. S. T. Peng, A. A. Oliner, and F. Schwering, Theory of dielectric grating antennas of finite width. *IEEE AP-S Int. Symp. Digest*, 529–532, LA, June 1981.
18. T. Itoh and A. Hebert, Simulation study of electronically scannable antennas and tunable filters integrated in a quasi-planar dielectric waveguide. *IEEE Trans. Microwave Theory Tech.*, **MTT-26**, 987–991, Dec. 1978.
19. T. N. Trinh, R. Mittra, and R. J. Paleta, Horn image guide leaky-wave antenna. *IEEE MTT-S Int. Microwave Symp. Digest*, 20–22, 1981.
20. T. N. Trinh, R. Mittra, and R. J. Paleta, Horn image guide leaky-wave antenna. *IEEE Trans. Microwave Theory Tech.*, **MTT-29**, 1310–1314, Dec. 1981.
21. T. Hori and T. Itanami, Circularly polarized linear array antenna using a dielectric image line. *IEEE Trans. Microwave Theory Tech.*, **MTT-29**, 967–970, Sept. 1981.
22. M. R. Inggs, M. T. Birand, and N. Williams, Experimental 30 GHz printed array with low loss insular guide feeder. *Electron. Lett.*, **17(3)**, 146–147, Feb. 1981.
23. M. T. Birand, N. Williams, and M. R. Inggs, A printed millimetre wave array using a low loss dielectric waveguide feeder. *IEEE Conf. Antennas Propag.*, 321–323, Apr. 1981.
24. M. T. Birand and R. V. Gelsthorpe, Experimental millimeter array using dielectric radiators fed by means of dielectric waveguide. *Electron. Lett.*, **17(18)**, 633–635, Sept. 1981.
25. K. Ito, N. Aizawa, and N. Goto, Circularly polarized printed array antennas composed of strips and slots. *Electron Lett.*, **15**, 811–812, Dec. 1979.

26. N. K. Uzunoglu, N. G. Alexopoulos, and J. G. Fikioris, Radiation properties of microstrip dipoles. *IEEE Trans. Antennas Propag.*, **AP-7**, 853–858, June 1979.
27. T. Itoh, Applications of gratings in a dielectric waveguide for leaky-wave antennas and band reject filters. *IEEE Trans. Microwave Theory Tech.*, **MTT-25**, 1134–1138, Dec. 1977.
28. T. Itoh and A. S. Hebert, Simulation study of electronically scannable antennas and tunable filters integrated in a quasi-planar dielectric waveguide. *IEEE Trans. Microwave Theory Tech.*, **MTT-26**, 987–991, Dec. 1978.
29. T. Itoh and B. Adelseck, Trapped image-guide leaky-wave antenna for millimetre wave applications. *IEEE Trans. Antennas Propagat.*, **AP-30**, 505–509, May 1982.
30. A. A. Oliner, New leaky wave antennas for millimetre waves, Proceedings of the International Symposium on Antennas and Propagation, *ISAP Digest*, 89–92, 1985.
31. A. Sanchez and A. A. Oliner, Accurate theory for a new leaky-wave antenna for millimetre waves using non-radiative dielectric waveguide. *Radio Sci.*, **19(5)**, 1225–1228, Sept.–Oct. 1984.
32. J. A. G. Malherbe, The design of slot array in non-radiating dielectric waveguide, Part I: Theory. *IEEE Trans. Antennas Propag.*, **AP-32**, 1335–1340, Dec. 1984.
33. J. A. G. Malherbe et al., The design of slot array in non-radiating dielectric waveguide, Part II: Experiment. *IEEE Trans. Antennas Propagat.*, **AP-32**, 1341–1344, Dec. 1984.
34. T. Yoneyama, T. Kuwahara, and S. Nishida, Experimental study of non-radiative dielectric waveguide leaky wave antenna, Proceedings of the International Symposium on Antennas and Propagation, *ISAP Digest*, 85–88, 1985.
35. R. S. Elliot, *Antenna Theory and Design*, Prentice-Hall, Englewood Cliffs, NJ, 1981.
36. A. A. Oliner and P. Lampariello, Novel leaky wave antenna for millimetre wave based on groove guide. *Electron. Lett.*, **18**, 1105–1106, 1982.
37. P. Lampariello and A. A. Oliner, Theory and design considerations for a new millimetre wave leaky groove-guide antenna. *Electron. Lett.*, **19(1)**, 18–19, Jan. 1983.
38. T. Yoneyama, Recent developments in NRD-guide technology. *Ann. Telecommun.*, **47(11–12)**, 508–514, 1992.
39. T. Yoneyama, Leaky NRD-guide fed planar antennas, *Proceedings of 1993 International Symposium on Recent Advances in Microwave Technology* (New Delhi, India), pp. 382–385, Dec. 1993.
40. A. Sanchez and A. A. Oliner, A new leaky waveguide for millimeter waves using non-radiative dielectric (NRD) waveguide—Part 1: Accurate theory. *IEEE Trans. Microwave Theory Tech.*, **35**, 737–747, Sept. 1988.
41. A. Sanchez and A. A. Oliner, A new leaky waveguide for millimeter waves using non-radiative dielectric (NRD) waveguide—Part 2: comparison with experiments. *IEEE Trans. Microwave Theory Tech.*, **35**, 748–752, Sept. 1988.
42. K. Maamria, T. Wagatsuma, and T. Yoneyama, Leaky NRD guide as a feeder for microwave planar antennas. *IEEE Trans. Antennas Propag.*, **41(12)**, 1–10, Dec. 1993.
43. F. C. Jain and R. Bansal, Semiconductor antennas for millimetre-wave integrated circuits. *Infrared and Millimetre Waves*, Vol. 15, K. J. Button (Ed.), Academic Press, Orlando, FL, 1986, Chap. 6.
44. K. L. Klohn et al., Silicon waveguide frequency scanning linear array antenna. *IEEE Trans. Microwave Theory Tech.*, **MTT-26**, 764–773, Oct. 1978.

45. C. Yao et al., Monolithic integration of a dielectric millimetre-wave antenna and mixer diode: an embryonic millimetre-wave IC. *IEEE Trans. Microwave Theory Tech.*, **MTT-30**, 1241–1247, Aug. 1982.
46. R. E. Horn et al., Electronic modulated beam-steerable silicon waveguide array antenna. *IEEE Trans. Microwave Theory Tech.*, **MTT-28**, 647–653, June 1980.
47. R. E. Horn et al., Single-frequency electronic-modulated analog line scanning using a dielectric antenna. *IEEE Trans. Microwave Theory Tech.*, **MTT-30**, 816–820, May 1982.
48. A. Alphones and M. Tsutsumi, On the optically controlled leaky wave antenna in silicon slab. **MW91-171**, 45–50, 1991.

Index

Active devices, 29–32
  integration, 29
Amplifier, 500–501
Analysis, dielectric guides, 47–97, 111–143
  approximate methods, 47–97
    effective dielectric constant, 54–66
    electromagnetic fields, 47–51
    network method, 85–92
    scalar potentials, 48
    TM-to-*x* and TE-to-*x* solutions, 50–51
    TM-to-*y* and TE-to-*y* solutions, 49–50
    transverse transmission line technique, 66–85
    vector potential function, 47
  rigorous methods, 111–143
    mode matching technique, 111–136
    potential function, 113–115
    TE-to-*y* modes, 122
    TM-to-*y* modes, 115
Antennas, 509–532
  broadside leaky NRD guide, 542–543
  dielectric rod, 510–513
  groove guide, 538–540
  image guide, 513–527
    grooves in dielectric, 518
    grooves in ground plane, 521
    image/insular guide fed, 527
    leaky wave grating, 513
    periodic conducting patches, 519
  NRD guide-leaky wave, 532–538
  periodic leaky NRD guide, 540–542
  semiconductor guide, 544
    optically controlled, 549
    silicon guide, 547
Attenuation, 132–136, 157–161, 165–166, 197, 200, 202–204
  conductor loss, 132–136
  double strip H-guide, 285–287
  H-guide, 273–274
  image guide, 157–161
  insular image guide, 165–166
  measurement, 398–399
  nonradiative guide, 234–236
  semicircular image guide, 248–252
  semielliptical image guide, 257–258
  silicon image guide, 200, 202
  silicon ridge guide, 203, 204
  silicon waveguide, 197
  Y-dielectric guide, 262

Bandpass filter, 454–455
Bandstop filter, 448
Bandwidth, 153, 156
  image guide, 156
  nonradiative guide, 234
Beam splitter coupler, 421–428
Bends, 384–387
  image guide, 385, 387
  NRD guide, 386
  trapped image guide, 385, 387
Branch guide coupler, 433
Broadside-coupled NRD guide, 239
  propagation constant, 240
Broadside leaky NRD guide antenna, 542–543

Circular dielectric guide, 248–249
  dispersion, 249
  field configuration, 248
Circulator, 474–477
Cladded image guide, 83–85, 183–184
  characteristics, 183–184

Cladded image guide (*Continued*)
  EDC analysis, 83–85
Coaxial cavity coupled oscillator, 488–491
Complex permittivity, 195
  variation with plasma density, 195
Conductor loss, 93–94
  image guide, 93–94
  nonradiative guide, 235
Coupled guide, 127
  characteristics, 170–176
    dispersion, 173–176
    field distribution, 171–172
    wave impedance, 173
  image, 127, 173
  insular image, 127, 175
  trapped image, 127, 173
  trapped insular image, 127
Couplers, 412–435
  beam splitter, 421–428
    analysis, 421
    basic geometry, 422
    design formulas, 426
    with capacitive diaphragm, 426
  branch guide, 433
  image guide-microstrip, 428–433
    coupling coefficients, 432
    design formulas, 432
    schematic, 429
  leaky-wave NRD guide, 433
  parallel guide, 412–421
    broadside-coupled, 416
    edge-guide, 416
    nonsymmetric, 417–421
    π-guide, 416
    symmetric, 413
Coupling coefficient, 238–239
  NRD guide, 238–239
Coupling length, 238
  0-dB, NRD guide, 238–239
Cutoff characteristics, 277
  double groove guide, 301
  H-guide, 277, 279
Cylindrical resonators, 310, 338
  image with top dielectric layer, 337
    characteristic equations, 337–338
  insular image guide, 331
    $EH_{11\delta}$ mode, 335
    $HE_{11\delta}$ mode, 334–335
    $TE_{01\delta}$ mode, 331–333
    $TM_{01\delta}$ mode, 333–334
  isolated, 310
    analysis ($TE_{01\delta}$ mode), 312
    characteristic equation, 313–315, 326–328
    closed form expression, 319
    field configuration ($TE_{01\delta}$), 321
    higher order modes, 323
    radiated power, 318–319
    radiation Q-factor, 315, 321
    resonant frequency, 319, 321, 329
    stored electrical energy, 316–317
  suspended substrate, 335
    characteristic equations, 335–336

Detector circuits, 484–488
  coaxial type, 485
  integrated image guide, 486–488
Dielectric guide, 1–41
  analysis (approximate method), 46–97
  analysis (rigorous method), 111–140
  attenuation (conductor loss), 132, 136
  bandstop grating, 454
  bends, 384–387
  circular, 14–16, 249
  cladded image guide, 183–184
  discontinuities, 369–384
  hollow image, 186
  image, 149
  introduction, 1–41
  inverted strip, 176, 181–182
  junctions, 387–392
  materials, 23–29
  millimeter wave, planar, 16–19
  multistate-reflectometer, 473
  nonplanar, 247–265
  nonradiative, 222–243
  optical, planar, 19–22, 206–215
  π-guide, 416
  rectangular, 11–13
  semiconductor, 194–206
  slab, 6–11
  strip, 176, 178–180
  transitions, 392–397
  wave impedance, 131–132
Dielectric loss, 94–97
  image guide, 94–97
  nonradiative guide, 235
Dielectric resonators, 32–38
  basics, 32
  high temperature superconducting, 357
  image and insular image, 36
  isolated, 34
  optically controlled, 355
Discontinuity, 369–384
  double step, 371
    generalized scattering matrix, 371–373
    mode matching, 374–376
  E-plane dielectric slab, 376
    characteristic equation, 379

field expansion, 378
junction scattering parameters, 380–383
overall scattering matrix, 386
Dispersion, 153
coupled image guide, 173
embedded inverted strip guide, 213
groove NRD guide, 242
H-guide, 277, 280
image guide, 153–156
insular image guide, 161–164
nonradiative dielectric guide, 232–233
optical rib guide, 210–211
semielliptical image line, 255–256
trapped coupled image guide, 173–175
trapped coupled insular image guide, 175–176
trapped image guide, 166–167
tube contacted slab guide, 264
Y-dielectric guide, 261
Distributed bragg reflector (DBR) oscillator, 494–495
Double-groove guide, 298–302
cross-section, 299
cutoff wavelength, 301
field pattern, 300
propagation constant, 302
Double-strip H-guide, 283–287
cross-section, 283
dispersion relation, 284
transmission loss, 285

Edge-loaded isolator, 474
Effective dielectric constant (EDC) method, 51–66
field matching analysis, 54–66
insular image guide (E^x_{mn} mode), 65–66
insular image guide (E^y_{mn} mode), 57–65
principle, 51–54
transverse transmission line technique, 66–85
Electric field probe, 397–398
Equivalent circuit, 490, 492
cavity coupled gunn oscillator, 490
dielectric cavity with gunn diode, 492

Field distribution, 149
circular dielectric rod, 248
coupled guide, 170–173
dielectric rib guide, 208
double groove guide, 300
edge-loaded dielectric guide, 475
groove guide, 292–293
H-guide, 271–272
hollow image guide, 185
image guide, 149–153
insular image guide, 161–164
nonradiative dielectric guide, 230
semielliptical image line, 255
Filters, 435–464
gap-coupled dielectric guide, 454–456
grating, 443–454
parallel coupled, 449–454
single guide, 443–449
periodic branching, 460
transmission characteristics, 462
rectangular resonator coupled, 456
ring resonator, 435–443
basic ring circuit, 436
single-ring, 438–441
two-ring, 442–443
Finite element method, 140–143

Grating antenna, 551
Grating structure, 444–454
inverted strip guide, 444–445
image guide, 445
Groove guide, 287–303
antennas, 538–540
attenuation (TE_{11} mode), 297
characteristic equations, 292, 294
TE_{1n} modes, 292
TM_{1n} modes, 294
cutoff wavelength, 294–295
double, 298–302
field analysis, 287
field configuration, 291
higher order mode, 295–296
Groove NRD guide, 241
circulator, 476
dispersion, 242
Guide wavelength, 398
measurement, 398–399

HEMT, 500–501
amplifier for NRD guide, 501
chip carrier, 502
H-guide, 201–206, 268–287
cut-off characteristics, 279
dielectric attenuation constant, 273
dispersion characteristics, 280
double strip, 283–287
field configuration, 271
hybrid modes, 270
silicon, 201
TE_{0n} antisymmetric modes, 276
TE_{0n} symmetric modes, 275
total attenuation, 274
with grooves, 275

Hollow image guide, 185–186
  dispersion, 186
  field distribution, 185
Hybrid dielectric/HTS resonators, 357. *See also* Resonators

Image guide, 90, 149–161
  attenuation characteristics, 157–161
  attennas, 513–527
    grooves in dielectric, 518
    grooves in ground plane, 521
    leaky wave grating, 513
    periodic conducting patches, 519
  balanced mixer, 497–498
  bandwidth, 156–157
  cladded, 183–184
  conductor loss, 93
  coupled, 127
  coupled insular, 127
  coupler, 427
  detector circuits, 484–488
  dielectric loss, 94–97
  dispersion, 153–156
  edge-loaded isolator, 474
  field and power distribution, 149
  grating filter, 445–449
  hollow, 185–186
  insular, 129, 161
  mechanically variable phase shifter, 478
  microstrip coupler, 428–433
  network analysis method, 90–92
  oscillators, 488–495
  power combiner, 501, 503
  self oscillating mixer, 499
  semicircular dielectric, 247
  semielliptical dielectric, 253
  silicon, 198–201
  trapped coupled, 127
  trapped coupled insular, 127
  trapped image, 129, 166
  trapped insular, 129
  variable attenuator, 467
Image guide-microstrip coupler, 428–433
Insular image guide (IIG), 161–166
  antennas, 527
  attenuation characteristics, 165–166
  cylindrical resonator, 331–335
  field and dispersion, 161–164
Insular layer, 169
Integrated gunn image guide oscillator, 491–493
Inverted strip guide, 181–182
  grating structure, 444–445

Junction, 387–392
  asymmetric Y-, 388
  image guide T-, 390
  NRD guide T-, 391
  scattering parameters, 389
  symmetric Y-, 388
  trapped image guide T-, 390

Leaky-wave NRD guide coupler, 433

Materials (dielectrics), 23–29
  castable, 28
  ceramic, 28
  ferrite, 29
  paste for thick film, 28
  polymer, 28
  semiconductor, 29
Measurement techniques, 397–399
  attenuation constant, 398–399
  field distribution, 397
  guide wavelength, 398–399
  radiation loss, 399
Microstrip-image guide coupler, 430
Millimeter waves, 3
  waveguiding media, 3
Mixers, 497–500
  beam lead diode mount, 499
  conversion loss, 500
  image guide, 498–499
Mode matching method, 111–136
  TE-to-y mode analysis, 122–127
  TM-to-y mode analysis, 115–122
Monolithic antenna, 550
Multilayer thin-film optical rib guide, 212
  modal refractive index, 216
  refractive index distribution, 214

Network analysis method, 85–92
  dielectric step junction, 86–89
  image guide, 90–92
Nonradiative guide, 129, 222–243
  amplifier, 500–501
  bandwidth, 234
  beam lead diode mount, 499
  broadside-coupled, 129, 239
  broadside-coupled insulated, 129
  characteristics, 229–236
  circulator, 476–477
  dispersion, 232–233
  end-coupled, 129
  end-coupled with spacer, 129
  field (LSE_{11} mode), 230–231
  field (LSM_{11} mode), 229–230

gap coupled filter, 455
groove, 241
insular, 129
leaky wave antenna, 532–538
leaky wave coupler, 433
loss characteristics, 234–236
mixer conversion loss, 500
modes, 229
oscillator, 495
parallel-coupled, 237
power divider, 466
transceiver circuit, 503–505
transition, 394–395
Nonplanar dielectric guides, 247–267
semicircular image guide, 247
conductor attenuation constant, 250–253
dielectric attenuation constant, 248–250
semielliptical image guide, 255–258
attenuation characteristics, 258
dispersion, 255
fields, 255
triangular dielectric guide, 259, 262
tube contacted slab guide, 263
Y-dielectric guide, 258–262
Normalized guide wavelength, 256
Numerical methods, 136–143
finite element, 140–143
telegraphist's equation, 137–140

Optically control, 194–206
antennas, 549
dielectric resonators, 355
semiconductor rectangular guide, 195–198
silicon H-guide, 201–206
silicon image guide, 198–201
silicon ridge guide, 201
Optical embedded inverted strip guide, 211–213
dispersion characteristics, 213
Optical rib guide, 209–211
dispersion characteristics, 210–211
Optical strip dielectric guide, 207–209
propagation constant, 209
Oscillator, 488–495
cum power combiner, 501, 503
image guide, 488–494
coaxial cavity coupled, 488–491
distributed bragg reflector, 494–495
integrated gunn, 491–493
NRD guide, 495

Parallel-coupled guides, 405–412
modes, 406
antisymmetric (odd mode), 406
even-like, 409
odd-like, 409
symmetric (even mode), 406
nonsymmetric, 408–412
power transfer, 407
scattering coefficients, 408
symmetric, 405
Parallel plate guide, 222–223
field expressions, 223–229
TE^{y}_{mn} antisymmetric modes, 228–229
TE^{y}_{mn} symmetric modes, 226–228
TM^{y}_{mn} antisymmetric modes, 225–226
TM^{y}_{mn} symmetric modes, 224–225
Passive components, 405–477
circulator, 474–477
directional couplers, 412–435. *See also* Couplers
filters, 435–464
isolator, 474
phase shifter, 466–467
power dividers, 464–466
reflectometer, 468–470
variable attenuator, 466–467
Periodic leaky NRD guide antenna, 540–542
Phase shift, 197
H-guide, 279–282
silicon image guide, 200, 202
silicon ridge guide, 203, 204
silicon waveguide, 197
Phase shifters, 466–468
contact injection, 496
electrically variable, 466–468
electronic type, 495–497
optical injection, 497
Planar optical guides, 206
embedded inverted strip, 211
multilayer thin-film, 212
strip dielectric, 207
rib, 209
Plasma density, 194–195
semiconductor guides, 194
Potential function, 113–115
TE-to-y modes, 114
TM-to-y modes, 113
Power combiner, 501, 503
Power distribution, 149
image guide, 149–153
Y-dielectric guide, 260
Propagation constant, 209
broadside-coupled NRD guide, 240
double groove guide, 302
strip dielectric guide, 209

Q-factor, 315, 321

Rectangular guide, 11–13, 195
  periodic branching filter, 463
  semiconductor, 195–198
  six-port reflectometer, 472
Rectangular resonators, 346–350
  analysis (EDC method), 346
  insular image, 350
  isolated, 348–350
Reflectometer, 468–474
  multistate, 470
  six-port, 469
Refractive index distribution, 214
Resonators, 308–364
  cylindrical image (with top dielectric layer), 337
    characteristic equations, 337–338
  dielectric ring gap, 354
  hybrid dielectric/HTS, 357
  insular image guide, cylindrical, 331
    $EH_{11\delta}$ mode, 335
    $HE_{11\delta}$ mode, 334–335
    $TE_{01\delta}$ mode, 331–333
    $TM_{01\delta}$ mode, 333–334
  isolated, cylindrical, 310
    analysis ($TE_{01\delta}$ mode), 312
    characteristic equation ($EH_{11\delta}$ mode), 328
    characteristic equation ($HE_{11\delta}$ mode), 326–328
    characteristic equation ($TE_{01\delta}$ mode), 313–315
    closed form expression, 319
    field configuration ($TE_{01\delta}$), 321
    higher order modes, 323
    radiated power, 318–319
    radiation Q-factor, 315, 321
    resonant frequency, 319, 321, 329
    resonator height, 319
    stored electrical energy, 316–317
  isolated ring, 339–345
    radiation Q-factor, 342–345
    $TE_{01\delta}$ mode, 339–342
  optically controlled dielectric, 355
  rectangular, 346
    analysis (EDC method), 346
    coupled, 456
    insular image, 350
    isolated, 348–350
  ring in planar dilectric slab environment, 345
  suspended substrate, cylindrical, 335
    characteristic equations, 335–336
Ridge guide, 201
  silicon, 201
Ring resonators, 339–345
  filters, 435–443
  planar dilectric slab environment, 345
  radiation Q-factor, 342–345
  $TE_{01\delta}$ mode, 339–342

Semicircular dielectric image guide, 247
  conductor attenuation constant, 250–253
  cross-section, 249
  dielectric attenuation constant, 248–250
Semiconductor guide antennas, 544–549
Semielliptical dielectric image guide, 253–258
  attenuation characteristics, 258
  cross-section, 253
  dispersion, 255
  fields, 255
Silicon guide, 198–206
  antenna, 547–549
  H-, 201–206
  image, 198–201
  ridge, 201
Six-port reflectometer, 469
Slab guide, 263
  tube contacted, 263
Slot array antennas, 540–547
  broadside leaky NRD guide feed, 544
  circular polarization, 547
Strip dielectric guide, 178–180

Telegraphist's equations method, 137–140
T-guide, 186
Transverse transmission line technique, 66–85
  analysis (E^y_{mn} mode), 66–69
  analysis (E^x_{mn} mode), 69–70
  horizontal slab guide models, 70–78
    single-slab guide with bottom ground plane, 75
    single-slab guide with no ground plane, 78
    symmetric three-slab guide, 78
    three-slab guide with bottom ground plane, 70–75
    three-slab guide with no ground plane, 76
    three-slab guide with top and bottom ground planes, 75
    two-slab guide with bottom ground plane, 75
  vertical layered guide models, 79–83
    asymmetric five-layered guide with metallic side walls, 79

asymmetric three-layered guide with metal side walls, 83
asymmetric three-layered guide with no side walls, 82
symmetric five-layered guide with metallic side walls, 82
symmetric three-layered guide with metal side walls, 83
symmetric three-layered guide with no side walls, 83

Trapped image guide, 166–170
  dispersion, 166–167
  wave impedance, 167–168

Transceiver circuit, 503–505

Transition, 392–397
  image guide, 392–394
    coaxial cavity, 490
    microstrip-slot, 395
    rectangular waveguide, 393
    slot in the ground plane, 394
  metal horn, 485
  NRD guide, 396
    rectangular waveguide, 396
    stripline, 397

Traveling wave resonator, 460

Triangular dielectric guide, 259, 262
  cross-section, 259

Tube contacted slab guide, 263–265
  dispersion, 264
  geometry, 263

U-guide, 186

VSWR technique, 398–399

Wave guidance (dielectric guide), 6–22
  circular, 14–16
  millimeter wave, planar, 16–19
  optical, planar, 19–22
  rectangular, 11–14
  slab, 6–11

Wave impedance, 131–132
  coupled image guide, 173
  trapped coupled image guide, 173–175
  trapped coupled insular image guide, 175–176
  trapped image guide, 167–168

Y-dielectric guide, 258–262
  cross-section, 259
  dispersion, 261
  power distribution, 260
  transmission losses, 262